CORROSION PREVENTION

by

PROTECTIVE COATINGS

Charles G. Munger

Consultant—Coatings and Corrosion

National Association of Corrosion Engineers

Published by
National Association of Corrosion Engineers
1440 South Creek Drive
Houston, Texas 77084

Library of Congress Catalog Card Number: 84-61872
ISBN 0-915567-04-0

Acknowledgments

NACE wishes to thank the many sources of information and graphic materials from which portions of this book were drawn with permission. Sources are acknowledged in chapter references and figure and table captions throughout the book.

To my wife Mary Ann for her love, patience, and understanding during the writing of this book.

C. G. Munger

Contents

Preface

An understanding of corrosion and the development of a corrosion engineering science have come about because of the need to protect materials of all types (*e.g.,* wood, concrete, steel, cast iron, stainless alloys, aluminum, etc.) from disintegration by so-called normal breakdown processes. These processes, which include atmospheric rusting, chemical solution, oxidation, crystallization, and galvanic couple reactions, are the means by which materials return to their original state of oxides, minerals, or elemental carbon.

It is important to note that all structural materials have a strong tendency to revert to their native state. This is because of the tremendous amount of energy used to convert them from their original form to one usable by man. This energy input, whether man-induced or a result of solar radiation, remains latent in the material and is released at every opportunity as the material reverts to a state of equilibrium with nature.

Everyone in modern society is affected in some way by this energy release phenomenon, or *corrosion.* The corporate executive; the marine, chemical, or materials engineer; the petroleum refinery manager; the papermill superintendent; or the amusement park maintenance employee—all are affected by corrosion, and all attempt in their own way to prevent the material under their control from reverting to its original, unusable state. Control of this reversion process is the goal of corrosion engineering.

Some of the most important tools used in corrosion engineering are high performance coatings. Such coatings, as compared with paints, have only been available for a relatively short time (since the late 1930s). The more advanced coatings, however, are presently the most widely used method of corrosion control, and effectively protect more surfaces and substrates from environmental change than any other corrosion prevention system.

In relation to the entire paint field, high performance coatings constitute only a small section. Nevertheless, it is one of the most important sections since it includes products designed for the protection of the most costly and complex structures in the world, *e.g.,* ultra large cargo carriers, LNG ships, chemical transport equipment (ships, barges, and tank cars), offshore drilling and production structures, petroleum refineries, sewage systems, and chemical, nuclear, and paper plants. The importance and social value of such structures and equipment far exceed the material and application costs involved in protecting them. Thus, some of the most highly engineered coating systems are used to prevent their corrosion and disintegration.

High performance coatings provide a true engineering approach to the control of corrosion, and thus form the section of the coating field to which this book is directed. The specific purpose of this book is to supply corrosion engineers and others involved in the selection or application of coatings for corrosion protection basic information that will allow them to understand and use coatings as an engineering approach to the protection of plants and equipment. It is designed primarily to supply the fundamental reasons and philosophy behind coating selection, application, and use so that maximum effectiveness may be obtained from the excellent coating materials available. It is to this most effective and economic use of coatings for corrosion control that this book is dedicated.

1

Introduction to Corrosion

Introduction

Corrosion and *corrosion engineering* constitute a science that has only developed within the lifetime of many people working in the field today. This does not mean that corrosion is strictly a modern phenomenon, for it has, on the contrary, existed since man first discovered ways to make metal from ores. It does, however, mean that prior to this last half-century, the existence of corrosion was passively accepted, along with death and taxes, as inevitable. In fact, many plant managers in this early period insisted that there was no corrosion in their plants; a little rust which eventually called for the replacement of the plant's structures maybe, but never any corrosion. Such evasive attitudes, however, have changed over the years due to a better understanding of the subject and increased awareness of replacement economics. This, in turn, has created a demand for corrosion prevention specialists, now known as *corrosion engineers.*

In the past, there was no formal training for corrosion engineers. They developed into specialists from their various roles as chemists; metallurgists; physicists; and chemical, civil, mechanical, maintenance, industrial, and piping engineers, as well as from their roles as experts in petroleum refining; petrochemicals; paper, water, sewage, and ship operation; and other industrial operations which were adversely affected by corrosion. While today there are some graduates of formalized corrosion engineering programs, the majority were still initially trained in other engineering sciences and have come to specialize in this field because of their experiences with corrosion and their developing interest in its prevention. Despite their interest and expertise in the field of corrosion, and as a consequence of their training in other disciplines, the science and technology of coatings is generally foreign to them. This, however, is understandable given that high-performance coatings represent a relatively recent (*i.e.,* last half-century) and specialized branch of a much broader, established field.

While the general area of coatings (or paint) has only come into its own as an industry during the Modern Age (post WW II era), the discovery and development of materials used by that industry give it roots in more primitive times. A general survey, therefore, of the emergence of such materials may serve as a practical prelude to the development of today's more specialized protective coatings, especially since some of the early materials and even attitudes have carried over into the present industry. The following historical analysis can thus provide a valuable background from which to evaluate many of the materials, procedures, and processes common in today's protective coating field.

Historical Background

Thousands of years before their protective qualities were discovered, coatings were used for decorative and identification purposes. The earliest known paintings, which were found in caves in France and Spain, were made from naturally occurring iron oxides.[1] These were apparently applied without a binder by merely rubbing the iron oxide into the surface of a cave's interior. This, then, was the primitive basis for the subsequent development in paints which took place primarily in two separate areas of the world.

Early Materials

Egypt developed the first synthetic pigment which was known as Egyptian Blue and was made from lime, sand, soda ash, and copper oxide heated together and ground to a relatively fine powder. This was developed in addition to a number of natural pigments which were already in use[2,3,4] (Figures 1.1* and 1.2).

FIGURE 1.2 — An Egyptian painting illustrating that most paints and pigments of this era were used for decorative purposes.

At a somewhat later date, the Egyptians (as well as the Chinese who earlier developed calcined or fired pigments and some organic pigments)[1] began the use of additional paint vehicles. These included gum arabic, eggwhite, gelatin, beeswax, and glue. This led to the Egyptian development of the first actual protective coating which involved the use of various pitches and balsams to coat their ships.[1,4]

The Chinese, as well as the Koreans and Japanese, used a lacquer for the decoration of buildings, instruments, and weapons.[1] Not much is known as to whether this early lacquer was of one type of material or several. It is known, however, that at some point the Japanese developed a coating made from the sap of a "varnish tree."[4] This was used in much of their early decorative painting and in the lacquer ware which we know most about today (*e.g.,* antique trays, vases, boxes, and small bottles and containers). The original sap required cleaning and filtering and was quite poisonous (similar to poison sumac and poison ivy or oak in its effects). When applied in relatively thin layers and kept in a damp, humid atmosphere, it cured or polymerized into a dense, high-gloss coating which had excellent resistance to aging and no poisonous effects.

The Romans, during this same era, learned the techniques of making paint from the Egyptians, many ex-

amples of which can still be seen in the ruins of Pompeii. While the Romans used essentially the same materials as the Egyptians, they also developed several artificial colors. These materials were primarily lead based since the Romans had extensively developed the technology of mining and extracting lead from ores. The Romans also copied the Egyptians in the use of a protective coating for their vessels, adding wax to the Egyptian pitch combination.[4]

The American Indians of the Canadian west coast used a variety of organic materials for their pigments. These included charcoal, lampblack, graphite, and powdered lignite for black; diatomite mined from the bottom of shallow lakes, or calcined deer bones or antlers for white; calcined yellow ocher or roasted hemlock fungus for reds; yellow ocher or ground hemlock fungus for yellow; and copper carbonates and Peziza (the mycelium of a fungus) to prepare blues and greens.[4]

Salmon eggs, both fresh and dried, were used for vehicles, which were prepared by chewing the eggs with the bark of red cedar. Fish oil from eulachon (*Thaleichthys pacificus*) or from perch was also used. Sizes were made by boiling the singed skins of mountain goats or mixing saliva with Peziza fungus, and mountain goat fat was used as the vehicle for their cosmetics.[4]

The early American Indians also made rock paintings, or pictographs, which have lasted for centuries (Figure 1.3*). These were all done in earth colors and made mostly from iron oxide which was probably rubbed into the rock with fingers and twigs.[4]

Burma and Thailand were the original producers of *shellac,* which is possibly the oldest known clear finish. The shellac resin, which is still imported from there today, is made by the lac bug, an insect that during its life cycle attaches to small branches of a variety of fig tree. This insect converts the tree's sap into a brittle reddish-brown resin with which it covers itself. The resin is gathered, crushed, washed, and bleached for use by the present paint industry. In its pure state, it is colorless, odorless, and edible.

Emerging Technology

The late 18th century saw the slow emergence of the paint and varnish industry. Although an actual varnish factory was reportedly established in England at this time,[4] paint was still made by individuals in small volumes for their own use and according to their own method. Much of what was made adhered primarily because of the affinity of the pigment for the surface, and much of it could be rubbed from the surface with a wet finger.

By the early 19th century, some power driven equipment was used to manufacture paint. These manufacturers, however, only used the equipment to manufacture paint ingredients so that the individual painter still combined these ingredients into a usable medium of his own formulation. It was not until the latter part of the century that paint manufacturers put their first prepared paints on

*See color insert.

*See color insert.

the market. This early manufacturing process involved grinding with oils or varnishes on simple stone mills which became the universal machines for dispersing pigments. Some of these mills remained in use until World War II, which indicates that the technology and business of making paint was extremely primitive and still in its infancy well into the 20th century. Table 1.1 provides a more detailed summary of the development of coatings up to this point of an emerging industry.

TABLE 1.1 — Historical Development of Paint

15000 BC	First known painting: caves at Lascaux, France and Altamira, Spain.
8000-6000 BC	Egypt: first synthetic pigment, Egyptian Blue; natural pigments used were red and yellow ochers, cinnabar, Hematite, orpiment (yellow), Malachite, Azurite, charcoal, lampblack, and gypsum; vehicles used were egg white, gelatins, and beeswax; developed crude brushes to apply paint.
6000 BC	Asia: calcined (fired) pigments and organic pigments; vehicles used were egg white, beeswax, and gelatins. Chinese lacquer is prehistoric.
1500 BC	Egypt: imported indigo and madder to make blue and yellow; developed first protective coating, pitches and balsams to coat boats.
1000 BC	Egypt: developed varnish from the gum of the Acacia tree (Gum Arabic).
1122-221 BC	Chinese buildings decorated with lacquer inside and out. Lacquer used on carriages, weapons, and harnesses. Lacquer also used in Japan and Korea.
Roman Era	Same essential materials as Egypt. Also, several artificial colors (white lead, litharge, red lead, yellow oxide of lead, Verdigris, and bone black). Pitch used for tarring ships. Pitch and wax used on ship bottoms.
400 AD	Japan: Japanese lacquer-sap from "Varnish Tree"; sap also used to waterproof drinking vessels.
600 AD	First suggestion for use of vegetable oil in varnish.
1100 AD	First written description for preparing an oil varnish. Dissolved molten rosin in hot oil.
Middle Ages	Considerable use of paint to protect wood.
American Colonists	Made paint using eggs and skimmed milk with earth pigments.
1773	Watin first described the paint and varnish industry. Copals and amber used as resins and turpentine as thinner. One pound of resin used with 1/4 to 1/2 pound of oil.
1790	First varnish factory established in England.
1803	Five classes of varnish described by Tingry.
1833	J. Wilson Neil: first to give details of varnish manufacture.
1840	First reference to use of zinc as a protective coating.
1850	Zinc oxide introduced as white pigment in France; was one of first steps towards a change in the paint industry.
1867	First prepared paint on the market.
1736-1900	Watin book on varnish formulas: varnish makers standard up to 1900; reprinted 14 times.
1800-1900	Red lead-graphite linseed oil paint, first protective coating system.

[Information drawn from the following SOURCES: Encyclopedia Britanica, Macropedia, Vol. 13, Paints, Varnishes and Allied Products, pp. 886-889, 1974; World Book Encyclopedia, Vol. 15, Paint, p. 24, 1978; Mattiello, Joseph J., Protective and Decorative Coatings, Vol. 1, Introduction, Rise of the Industry, John Wiley & Sons, Inc., New York, NY, pp. 1-8, 1941; Neil, J. Wilson, Manufacturers of Varnishes, Vol. 49, Part 2, Trans. Roy. Soc. Arts, pp. 33-87, 1833; Mallet, R., British Assn. for Advancement of Science, Vol. 10, pp. 221-288 (1840).]

A Coating System

The railroad made a significant contribution to the development of protective coatings around the turn of the century. Bridges were a vital part of the rail system and many of the major ones were constructed with riveted

steel; thus, it was necessary they be given maximum corrosion protection. One of the most effective systems consisted of a red lead linseed oil primer applied in one or more coats, followed by a linseed oil-graphite topcoat applied in two or more coats. This marked the first of the truly protective coatings and created the dark grey color which was characteristic of these steel bridges. This first system provided good long-life corrosion protection, except in extremely corrosive marine or industrial conditions. The coating was brush-applied and thoroughly wet the steel surface. The system's heavy red lead primer offered some corrosion inhibition and the graphite topcoat increased water resistance, with its flake structure protecting the vehicle from weathering. This type of coating became a standard of the day for exposed steel structures such as the one shown in Figure 1.4.

FIGURE 1.4 — Complicated bridge structure, all of riveted construction with open-work vertical supports and multiple angle braces, shows clean, rust-free trusses.

As long as the making of paint was considered more of an art than a science, however, progress as an industry was slow. It was not until the turn of the century which brought an increased emphasis on science, and chemistry in particular, that a new era in the coating industry was recognized.

As few as seventy years ago, however, the number of chemists in paint factories was still limited and their sphere of activity restricted. Coating technology consisted largely of empirical data obtained through trial and error. White lead in oil remained standard for exterior house paint, although ready-mixed paint containing white lead and zinc oxide was making headway. Titanium dioxide pigments were still around the corner, and choices in pigments were extremely limited or in some cases nonexistent.

The principal varnish oils used at this time were

linseed and tung. Although many varnish makers resented the intrusion of so-called upstart chemists, new "wonder" varnishes were nevertheless being produced. These varnishes, which dried faster and resisted turning white in water, were developed through a combination of tung oil and ester gum made by heat-reacting rosin and glycerol. The first completely synthetic resin which was introduced was made from phenol-formaldehyde,[5] but even this development was considered only a minor one. It was not until at least a decade later that significant improvements were made following the availability of 100% oil soluble phenolic resins. These vastly upgraded the varnishes of the time by increasing flexibility, decreasing the drying time, and making them much more durable to both weather and water.

World War I produced nitrocellulose which was also a completely synthetic resin. It was made by reacting cellulose with sulfuric and nitric acids, and was produced as an explosive in extremely large quantities during World War I. Thus, as the war came to a close, there were large unwanted volumes of nitrocellulose still available. The material was a hazard to keep in inventory as well as a disposal hazard. DuPont chemists, however, soon determined that if the material were put into solution it would form a clear, continuous film. This discovery was the beginning of the lacquer industry which had a major impact on the automotive industry. Prior to the early 1920s, all automotive finishes were of a varnish type, requiring multiple coats with long drying times. Nitrocellulose, however, dried as soon as the solvent evaporated, leaving a clear film that could be easily pigmented. In addition, nitrocellulose in solution form was no longer an explosion hazard.

Binders

The synthetic resin developments which came out of this field of organic chemistry became the basis for protective coatings used from that time on. Alkyd resins, chlorinated rubber, vinyl copolymers, acrylics, and methylcellulose were all developed during this period. While these materials were extensively tested in coating formulations during this period, progress was still slow. A number of other developments were necessary before the new coating materials could be successful.

One such development was that of proper solvents. The vinyl resins were soluble only in extremely strong solvents and, while acetone was available, its evaporation rate was so rapid that satisfactory films could not be formed from the acetone solution. Many of the new resins were brittle as well, and required the addition of some softening agent in order to provide a durable surface film.

It was actually the development of solvents and plasticizers, occurring at the same time as that of synthetic resins, which made coatings possible. Some early vinyl coatings, as well as some made from chlorinated rubber, were developed in the 1930s. Both had good chemical resistance. The vinyl copolymers, however, would only adhere to a very porous surface and were therefore not satisfactory for application to steel surfaces.

When used alone, chlorinated rubber was likewise unsuitable since it was an extremely brittle, hard, and horny-textured substance. It was soon discovered, however, that its addition to oils and alkyd resins increased drying speed and resistance, and conversely plasticized the chlorinated rubbers. It is interesting to note that the first satisfactory primer for vinyl resin was actually one developed from chlorinated rubber. Not only did it adhere to steel, but the combination of a chlorinated rubber primer and vinyl copolymer body coats and topcoats formed the first of the really chemical-resistant protective coatings. There were, however, some application problems since both materials dried rapidly.

The material nevertheless provided the most resistant protective coating available at that time and was used as a chemical-resistant coating on into the early 1940s. This coating system found application in the chemical, sewer, and marine industries, and as a lining for wine tank cars (Figure 1.5) and for bait and fish storage tanks in the tuna industry. The new chlorinated rubber-vinyl copolymer combination thus represents the first major breakthrough in the development of a corrosion-resistant coating.

FIGURE 1.5 — Riveted tank car coated with chlorinated rubber primer, two vinyl chloride acetate silica-filled intermediate coats, and two or more clear vinyl chloride acetate finish coats.

Surface Preparation

A second breakthrough which also occurred at this time was the increased use of *sandblasting*. Although sandblasting was a well-known technique in the 1930s, it was considered a messy process in which very few plant managers wanted to be involved. It was soon recognized, however, that a sandblasted surface was necessary for proper performance of the new synthetic resin coatings.

This discovery came when the U.S. Navy realized that in order to prolong the life of their ships at sea, it would be necessary to improve ship coating effectiveness. They soon found that the chip-and-scrape procedures were not adequate in achieving the kind of surface preparation essential to improved coating life. Thus, the

Navy's eventual acceptance of sandblasting as a standard method of surface preparation was the beginning of a more general use of proper surface preparation. This new practice of thoroughly cleaning a steel surface prior to coating application, in addition to the improved adhesion of vinyl coatings to steel, provided the foundation for the protective coating industry.

Protective Coatings

This period produced two other breakthroughs which added to the beginning revolution in corrosion control through protective coatings: the Australian development of the first inorganic zinc coating, and, at about the same time, the European development of organic zinc-rich coatings. Both events were significant in that they eventually led to the use of zinc primers for almost all high-performance protective coatings. Their value, however, was not commonly recognized until a practical method of applying the coatings to existing structures was developed more than a decade later.

While World War I marked the beginning of what is now the paint industry, World War II provided the impetus behind the development of the protective coating industry. The need to conserve labor, to lengthen the time a ship could be at sea without returning to drydock, to protect structures against corrosion that developed from rapidly expanding chemical and fertilizer industries, and to conserve existing plants by preventing their deterioration by corrosion, all pushed the development of the protective coating industry away from paint and toward the more resistant polymers and film-forming materials.

Shortly after World War II, the development of another type of material, epoxy resins, had a major impact on the protective coating field. These materials reacted in place to form a protective coating which was easier to apply and had good adhesion and acceptable resistance to corrosion. The epoxies were originally amine cured. Later, however, polyamide epoxy coatings were developed which provided increased adhesion, some flexibility, and increased water and chalk resistance as compared to earlier products. Polyurethane coatings were also developed during this period, but were considered inferior to epoxy because of their original poorer water resistance and tendency to yellow over a period of time.

A major breakthrough in the inorganic zincs field came with the development of self-curing inorganic zinc coatings. These materials eliminated the difficulty of applying the curing agent to the post-cured products, and their rapid cure to insolubility eliminated much of the difficulty with weather during the application stage. While the first self-curing inorganic zinc coatings were water based, the later development of inorganic zinc coatings based on ethyl silicate had a major impact on the protective coating field.

Most of the coating products developed during the World War II period remain in present use. While improvements and developments in both organic and inorganic coatings continue to be made, there has been no major breakthrough in coating technology since the advent of the inorganic zinc and epoxy coatings. Table 1.2 provides an outline of the significant developments in paint and coatings since their emergence as an industry at the turn of the century.

TABLE 1.2 — Significant Developments in Paint and Coatings Since 1900

1909	First completely synthetic resin introduced by Dr. Leo H. Bakeland. Made from phenol-formaldehyde. Was insoluble in oil and not useful until modified in 1920-24. Discovery was the beginning of the Bakelite Corp. which now operates as the Union Carbide Corp.
1918	First sale of titanium pigment.
1920-25	Cellulose nitrate and acetate developed for the paint field. Alkyd resins developed.
1928	100% oil soluble phenolic resin developed.
1929	Urea formaldehyde resin developed.
1930	Chlorinated rubber developed.
1933	Vinyl copolymers developed from vinyl chloride, vinyl acetate, vinyl alcohol, and acrylic.
1935	Ethyl cellulose and cellulose acetobutyrate developed.
1938	First vinyl protective coating developed and used without baking. Five-coat systems: chlorinated rubber primer, two vinyl body coats, two vinyl seal coats.
1939	First inorganic zinc coating developed in Australia.
1940	Organic zinc-rich coatings started in England.
1942-45	Sandblasting specified as surface preparation on Naval ships. Vinyl wash primer developed. Vinyl tripolymer resin developed which provided good adhesion to steel.
	Silicone resins developed.
1945	First self-priming vinyl coating developed and used.
1948	Styronated oils and alkyds developed.
1952-55	First practical inorganic zinc coating developed and used in U.S.A.
	Epoxy resins appeared as coating raw materials, amine cured.
	Polyurethane coatings in U.S.A., were known in Germany during World War II. Coal tar epoxy coatings also developed.
1955-60	Epoxy polyamide coatings developed. First water-base self-curing inorganic zinc coatings.
1960-65	Self-curing, ethyl silicate base, inorganic zinc coatings used in U.S.A.
1965-70	Single-package inorganic zinc coating developed.

NOTE: All of the above dates are approximate. Many materials shown were in the development stage for several years before entering the coating market.

[Information drawn from the following SOURCES: Clark and Hawley, The Encyclopedia of Chemistry, Protective Coatings, Reinhold Publishing Corp., New York, NY, pp. 785-789, 1957; Mattielo, Joseph J., Protective and Decorative Coatings, Vol. 1, Introduction, Rise of Industry, John Wiley & Sons, Inc., New York, NY, pp. 1-8, 1941.]

White Pigments

Without the development of several other ingredients in coatings, the binders themselves would have been of little consequence. The improvement of the white pigments, for instance, which paralleled the development of binders, was a major improvement in the industry. White lead was the primary white pigment available up until the mid 1800s when zinc oxide was developed as a white pigment. The combination of white lead and zinc oxide was soon used almost exclusively for good exterior white paints.

Titanium dioxide was introduced during World War I and also had a significant impact on the paint industry. As shown in Table 1.3, which lists the various white pigments in the order they were developed and used by the paint industry, titanium dioxide has 5 to 10 times the tinting strength and hiding power of the earlier white pigments. (Hiding power is influenced by both pigment con-

TABLE 1.3 — Tinting Strength and Hiding Power of White Pigments

	Tinting Strength	Hiding Power sq. ft./lb.	Hiding Units
Basic Lead Carbonate	100	15	1.00
Basic Lead Sulfate	85	13	0.87
Zinc Oxide	200	20	1.33
Zinc Oxide (35% Leaded)	170	20	1.33
Lithopone	260	27	1.80
Titanium Barium Pigment (a)	380	40	2.67
Titanium Barium Pigment (b)	430	46	3.07
High-Strength Lithopone	400	44	2.93
Titanated Lithopone	400	44	2.93
Titanium Calcium Pigment	450	48	3.20
Titanium Magnesium Pigment	440	47	3.13
Zinc Sulfide	540	58	3.87
Lead Titanate	570	60	4.00
Titanium Dioxide	1150	115	7.67

[SOURCE: Mattielo, Joseph J., Protective and Decorative Coatings, Vol. 1, Introduction, Development of Industry Today, John Wiley & Sons, Inc., New York, NY, p. 13, 1941.]

centration and the type of binder. Various grades of the same pigment type also differ in hiding power.)

Titanium products were originally combined with other inert pigments. The original titanium dioxide was the anatase form of the pigment. This material, which had a number of drawbacks, particularly caused paint to chalk very readily and heavily. The rutile form of titanium dioxide was finally introduced in the early 1940s, and from that time on served as a major white pigment for both the paint and the new protective coating industry. It was used in the protective coatings field primarily because of its very inert chemical characteristics. Titanium dioxide was inert to most acids and alkalis and was therefore incorporated into many chemical-resistant coatings.

Solvent Development

The development of synthetic resins and white pigment was also paralleled by the development of other types of pigment, solvents, and plasticizers.

The large scale of nitrocellulose lacquers is attributed not only to the successful production of lower viscosity nitrocellulose (which was easier to use), but also to a large supply of *n*-butyl alcohol which accumulated as a fermentation process by-product in the production of acetone during World War I. This butyl alcohol was easily converted to butyl acetate, a very desirable nitrocellulose solvent.

As mentioned previously, nitrocellulose served as the fast-drying finish which was needed by a rapidly expanding automobile industry. With nitrocellulose available as a film-former, butyl alcohol and butyl acetate could be produced in quantity as solvents. The nitrocellulose lacquer industry was thus able to increase its volume from 1 to 70 million gallons in 25 years. This provided the impetus for the development of other solvents (ketones, esters, Cellosolves, nitroparafins, chlorinated hydrocarbons, and speciality solvents such as diacetone alcohol) which were eventually used in the protective coatings in-

dustry. They also allowed the use of very high molecular weight resins and combinations of resins which otherwise would have been impossible.

Distinction in Terms

Up to this point, the words "paint" and "protective coatings" have been used rather loosely. Much discussion has revolved around the use of these two terms and there is little doubt that much of the terminology and many of the uses, basic materials, and even manufacturing processes are similar for both paint and protective coatings. On the other hand, there certainly is a case to be made supporting a distinction between the two materials, despite the opinion, particularly of many conventional paint suppliers, that paint and coatings are essentially one and the same. A majority of suppliers of the newer protective coatings agree to the distinction, however, pointing to not only vast differences in composition, but also in use. This perspective is made even clearer by a statement from the *Encyclopedia of Chemistry* which states:

> ...films prepared from oils alone are totally inadequate for protective coatings and combinations of natural resins and derivatives of rosin do not meet the growing needs of industry for protective coatings with adequate resistance to acids, alkalies and weathering and with retention of high elasticity under stress and aging. To meet this need the chemical knowledge of polymerization has been utilized during the past 30 years to develop synthetic resins useful in the surface coating field.[5]

Other references, however, make little distinction between the words "paint" and "coating." In fact, they are often used interchangeably, *i.e.*, all paints are coatings and all coatings are paints. The definition of a coating taken from the "Glossary of Terms" in the *Manual for Coatings of Light Water Nuclear Plants,* for example, is "Coatings (paints) are polymeric materials that applied in fluid stage, cure to a continuous film."[6] This, of course, could serve as a general definition for either a paint or coating because it is a very broad and simple statement. When considering coatings in relation to corrosion, however, it becomes necessary to more specifically define the two terms.

Definitions

Paint may be defined as any liquid material containing drying oils alone or in combination with natural resins and pigments which, when applied to a suitable substrate, will combine with oxygen from the air to form a solid, continuous film over the substrate, thus providing a weather-resistant decorative surface. Paints continue to oxidize over their entire lifetime and gradually become porous to oxygen, water, and ions that may be deposited on the surface. Thus, they provide less permanent protection against corrosion than the more sophisticated protective coating.

A *protective coating* is chemically a substantially different material. It outperforms paint in adhesion, toughness, and resistance to chemicals, weather, humidity, and water. A protective coating is any material composed essentially of synthetic resins or inorganic silicate polymers which, when applied to a suitable substrate, will provide a continuous coating that will resist industrial or marine environments and prevent serious breakdown of the basic structure in spite of abrasion, holidays, or imperfections in the coating.

In order to provide corrosion protection, the protective coating must also: (1) resist the transfer or penetration through the coating of ions from salts which may contact the coating; (2) resist the action of osmosis; (3) expand and contract with the underlying surface; and (4) have and maintain a good appearance, even under extreme weather conditions. In addition, it must be able to fulfill these requirements over a period of time long enough to justify its price and application costs.

In the context of this definition, the protective coating may be formed solely by the evaporation of solvents leaving a resinous film (lacquer), by condensation or internal chemical reaction (epoxy), by reaction with products in the air (alkyd, polyurethane, or inorganic zinc coating), by coalescence (emulsions, plastisols, or dispersions), by heat as a hot-melt (coal tar or asphalt), as a powder which is melted on the surface (epoxy), or by applying a preformed sheet of resinous coating material over the structure (*e.g.,* vinyl sheet).

Naturally, there are exceptions to these definitions, including some alkyds, chlorinated rubbers, and epoxy esters, which are a combination of both oils and synthetic resins. These materials combine some of the properties of both paints and coatings and are very useful for some purposes. On the other hand, since they contain drying oils, their properties are limited because of the chemical nature of oils, and their use is limited to less severe exposures than those of most protective coatings.

For further clarification, the definition of a protective coating can be divided into two additional parts, according to the intended use of the material. The *Encyclopedia of Chemical Technology* makes the distinction as follows:

> A *resistant coating* is a film of material applied to the exterior of structural steel, tank surfaces, conveyor lines, piping, process equipment or other surfaces which is subject to weathering, condensation, fumes, dusts, splash or spray, but is not necessarily subject to immersion in any liquid or chemical. The coating must prevent corrosion or disintegration of the structure by the environment.
>
> A *resistant lining* is a film of material applied to the interior of pipe, tanks, containers or process equipment and is subject to direct contact and immersion in liquids, chemicals, or food products. As such, it must not only prevent disintegration of the structure by the contained product, but must also prevent contamination of the contained product. In the case of a lining, preventing product contamination may be its most important function.[7]

In the context of these definitions, a paint would seldom, if ever, be used as a lining. On the other hand, a protective coating may well be used for decoration purposes only, and thus might qualify under the definition of a paint.

Purposes

The range of reasons for using paint extends beyond the historically prevalent one of decoration, although appearance remains a primary consideration. A chapter in the *Facilities and Plant Engineering Handbook,* entitled "Painting," thoroughly summarizes its purposes as follows:

PROTECTION

The major consideration in painting is preservation of the structure or equipment from the environment. Typical causes of failure are rainfall, water, vapor, sunlight, temperature variations, both overnight and between seasons, mildew and rot. Other less prevalent causes include salt water and vapor, chemicals and chemical fumes, air pollution, and abrasive wear by traffic. Paint acts as a shield protecting the substrate from these elements. Proper paint choice and painting practices will definitely extend the life of the painted object and markedly reduce repair costs.

APPEARANCE

The most important reason for painting (in normal environments) and second most important reason outdoors is the decorative value of paint in producing pleasant and attractive surroundings. The wide choice of colors, even in metallic finishes, the gloss and texture available, as well as its ease of application, make paint the ideal method of producing or changing the appearance of all surfaces on structures including floors as well as operating equipment and even furniture. Paints can be applied to almost all surfaces and substrates and to old painted surfaces, as well, thus enabling the production of any aesthetic effect desired even when structural modifications are made, *e.g.,* installation of new walls, partitions and doorways. Furthermore, paint will readily cover these additions or changes so that they do not stand out.

SANITATION AND CLEANLINESS

Painted surfaces are generally relatively smooth and nonporous so that they are easily kept clean. The coating of rough and porous surfaces seals out dirt and other foreign matter that would otherwise be difficult to remove. Furthermore, light-colored painted surfaces, by contrast, will reveal the presence of dirt, grease, and other undesirable substances and thereby indicate that better housekeeping practices are in order. Therefore, painting for sanitation and cleanliness is especially important in food-processing areas and hospitals.

ILLUMINATION

White and pastel colored paints, when applied to ceilings, are highly efficient in natural and artificial light. Therefore, they can be used to advantage to brighten rooms and work areas and also to reduce lighting costs, when used to cover darker substrates or old paint. On the other hand, darker colors can be used to reduce and soften illumination, where so desired, *e.g.,* in private offices. Furthermore, low-gloss (flat) finishes will soften and diffuse illumination thus reducing glare.

EFFICIENCY

Improved illumination and reduction of glare both aid markedly in improving efficiency of personnel in the area. The ability to use a variety of colors, as well as combinations, also will lead to pleasanter surroundings, another aid to improvement in morale and productivity. Paint can also be used to color-code areas, equipment, piping, and valves, and traffic paints can be used to guide traffic flow, all of which improve efficiency and reduce errors.

VISIBILITY AND SAFETY

Increased illumination improves general visibility in the area, thus also preventing potential accidents. However, colored paints can be used alone or in combination to designate dangerous areas and safety equipment such as fire extinguishers. Certain combinations, such as yellow stripes or letters on a black background or orange combined with white, are visible at much greater distances than single colors. The latter are used on TV, radio, and telephone relay towers to warn aircraft, for example.[8]

The above summary, however, does not cover all of the reasons for using protective coatings, as it does not address the more severe problems for which protective coatings are produced. Protective coatings also act as a barrier, prevent fouling attachment, reduce or increase friction, reduce abrasion, reflect or absorb heat, and, most importantly, prevent disintegration or failure of the substrate by various forms of corrosion (*e.g.,* marine, atmospheric, chemical, or underground corrosion). Thus, the specific purpose of a protective coating is to provide a film which will separate noncompatible materials or conditions.

Protective coatings, or in this case linings, may also be used to prevent contamination, as in the case of nuclear power plants where coatings are used to prevent radioactive contamination of the underlying surface. On the other hand, as a lining for a tank car containing caustic, wine, sugar, syrup, or petroleum products, the coating provides a film separating the steel surface from the contained material. Fire-retardant coatings also apply these principles in that they act as a barrier to heat as well as to oxidation. From a corrosion standpoint, the advertising slogan which was popular a number of years ago, "Save the surface and you save all," aptly applies.

Modern Coatings Industry

Today's coatings industry is very different from what it was at the turn of the century. Its products are now used by almost every individual and certainly by every company in the United States. Approximately one-third of the total paint volume is now used to protect and decorate metal surfaces, and it is probable that as much as three-quarters of the multitude of useful paint formulas are directed at least in part towards the protection of metal from corrosion.

The extrapolated volume of paint shipped in the United States in 1979 amounted to almost 1.4 billion gallons with a value of nearly 6.5 billion dollars. The Department of Commerce breaks these figures into three categories: *architectural products* (stocktype of shelf goods formulated for environmental conditions and general application on new and existing residential, commercial, institutional, and industrial structures); *product finishes, OEM* (products formulated specifically for manufactured items to meet conditions of application and product requirements); and *special purpose coatings* (products formulated for special applications and/or special environmental conditions such as extreme temperature, chemicals, fumes, etc.), which they added in 1979.[9] Unfortunately, the latter classification includes a number of products such as traffic paint, automotive refinish products, and aerosol paint which cannot qualify as high-performance coatings. Table 1.4 gives a volume breakdown for all three classifications.

TABLE 1.4 — Volume of the Three Major Categories of
Paint Shipped in the U.S.A. Extrapolated for 1979
(Thousands of Gallons/Dollars)

1979	Total		Architectural		Product Finishes		Special Purpose Coatings	
	Gal.	$	Gal.	$	Gal.	$	Gal.	$
4 months Actual	340	2,156	174	996	117	784	48.2	376
Year Extrapolated	1,360	6,469	524	2,987	352	2,353	145	1.128
% of Total	100	100	51.2	46.2	34.4	36.4	14.9	17.4

[SOURCE: Current Industrial Reports, Paint, Varnish and Lacquer, U.S. Dept. of Commerce, Bureau of Census, Jan.-April, 1979.]

The paint industry can also be analyzed according to the divisions indicated in Figure 1.6 which compares the National Paint and Coatings Association Sales Survey to the Department of Commerce Bureau of Census Survey for the same period. The Bureau of Census divides the coating industry into two parts: trade sales and chemical coatings. Trade sales includes all sales of stock type items to jobbers, dealers, painters, contractors, builders, and automotive refinish jobbers, and includes sales to company-owned stores as well as those made directly to customers for maintenance of residences, institutions, office buildings, and factories for floors, walls, ceilings, etc. Chemical coatings includes *industrial product finishes* which are defined as any organic product of the paint and coatings industry specifically formulated to meet the conditions of application and end use of the article, and applied to such an article as part of the manufacturing process.

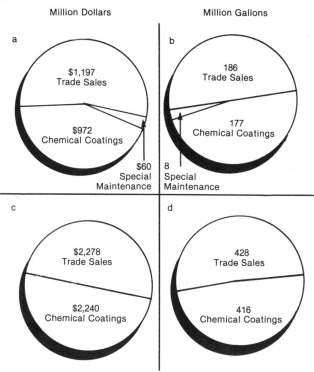

FIGURE 1.6 — Trade sale versus chemical coatings. (SOURCE: National Paint and Coating Association, Sales Survey, 1977.)

The National Paint and Coatings Survey also includes the category of *special maintenance coatings*. They defined these as maintenance paints or liquid coatings applied in the field to nonresidential buildings, equipment, and structures, both stationary and mobile. It is this special maintenance division of the industry which most interests the corrosion engineer. Although the National Paint and Coatings Survey only involved 82 American companies, these figures account for almost 50% of the industry's total dollar sales. When the special maintenance figures in the National Paint and Coatings Survey are extrapolated to the Bureau of Census total figures, special maintenance coatings account for almost

Corrosion Prevention by Protective Coatings

19 million gallons and are of an almost $140 million value for the overall maintenance market.

Another survey specifically directed at the maintenance paint market concluded that:

> ...the maintenance of industrial buildings and equipment (including highways, bridges, marine vessels and equipment, state and municipal structures, transportation facilities, public utility plants, transmission towers, gas holders, pipelines, etc.) is an essential phase of the U.S. economy probably involving an annual expenditure of from $4 to $5 billion.[10]

This is a much larger figure than the extrapolated one above; on the other hand, it does include all parts of the maintenance program (*e.g.*, labor, surface preparation, and material). The term *maintenance paint* as used for the purposes of this survey is defined as:

> Admixtures of vehicles and pigments formulated for the protection of industrial, commercial, marine, institutional, municipal or government structures or facilities whether prepared especially for a particular end application, environment or condition or manufactured as stock or shelf goods and designed for general maintenance application.[11]

Thus defined, maintenance paint is a very broad term, which includes a number of items that are not of interest to the corrosion engineer and do not come under the designation of a high-performance coating.

Table 1.5 gives the breakdown of vehicle type by gallons for the 1975 maintenance paint market. Increasing total gallons by 23% brings the figure to a 1980 total of 112 million gallons (Table 1.6). These tables obviously indicate that at least two-thirds of this total is used for the maintenance of walls, ceilings, houses, roads, and highways, etc. which are not of particular interest to the corrosion engineer. After subtracting these uses, the remainder is much closer to the total arrived at by the National Paint and Coatings survey for special maintenance coatings.

Table 1.7 takes the analysis further by dividing maintenance paints according to types and industry. The table again makes it obvious that a number of industries do not use high-performance coatings, but rather concentrate in latexes, alkyds, and chlorinated rubber (road marking) paint.

Based on information extracted from the Smith Stanley Survey, Table 1.8 uses the same market breakdown, but reduces the overall market to the area of high-performance coatings. This table is a result of extrapolating the various figures provided by the survey so as to provide some indication of the size of the high-performance coating market. The other segments of the market which were included in Table 1.6 are not considered users of high-performance coatings. Thus, Table 1.8 shows that high-performance coatings amount to about 50% of the total maintenance coating market. The market projection for high-performance coatings is incomplete, however, without the addition of the new construction segment. Table 1.8 does not include figures for new construction which are estimated at 50% or more of the volume for high-performance coatings. Taking this into consideration, the total volume of high-performance coatings used per year amounts to approximately 39 million gallons.

In terms of a public image, the paint industry probably has one of the most simplistic, and therefore one of

TABLE 1.5 — 1975 Market for Maintenance Paint by Type of Identified Vehicle

	Gallons	Percent
Total	91,735,400	100.0
Latex	30,550,700	33.3
Alkyd	28,128,700	30.8
Chlorinated Rubber	8,280,300[1]	9.0[1]
Epoxy	5,632,300	6.1
Vinyl	4,693,600	5.1
Chlorinated Paraffin	4,600,000[1]	5.0[1]
Urethane	2,091,800	2.3
Acrylic	704,500	0.8
Inorganic Zinc	658,000[2]	0.7[2]
Bituminous Resin	585,000[2]	0.6[2]
Drying Oils	208,000	0.2
Polyester	133,400	0.1
Other (not identified)	5,469,100	6.0

[1]Primarily used on roads and highways.
[2]Marine application.

[SOURCE: Maintenance Paints, Vol. 3, A Market and Economic Study of the U.S. Markets, Smith Stanley & Co. and American Paint Journal Co., Oct., 1976.]

TABLE 1.6 — Projected 1980 Market for Maintenance Paints

	Gallons	Percent
Total	112,590,700	100.0
Roads and Highways	31,361,000	27.9
Manufacturing	21,270,000	18.9
Federal Government	14,000,000	14.2
Marine	9,000,000	8.0
Office Buildings	8,900,000	7.9
Schools and Universities	8,200,000	7.3
Hospitals and Nursing Homes	4,200,000	3.7
Retail Stores	3,800,000	3.4
Hotels and Motels	3,600,000	3.2
Gas and Electric Utilities	2,723,000	2.4
Railroads	1,800,000	1.6
State and City Buildings	1,800,000	1.6
Eating Establishments	900,000	0.8
Offshore Oil Rigs	598,000	0.5
Water and Sewage	438,700	0.4

[SOURCE: Maintenance Paints, Vol. 3, A Market and Economic Study of the U.S. Markets, Smith Stanley & Co. and American Paint Journal Co., Oct., 1976.]

the most erroneous images of any industry. It is commonly believed that paint consists of a simple mixture of colored pigments and resins. Nothing, however, could really be further from the truth. Today's paint industry is a substantial segment of the broader chemical industry, and

TABLE 1.7 — Maintenance Paints Vehicle Types Analyzed by Industry for 1975 (Number of Gallons)

	Total	Latex	Alkyd	Chlorinated Rubber	Epoxy	Vinyl	Chlorinated Paraffin	Urethane	Acrylic	Inorganic Zinc	Bituminous Resin	Drying Oils	Polyester	Non-Identified
Total	91,735,400	30,550,700	28,178,700	8,632,300	5,632,300	4,693,600	4,600,000	2,091,800	704,500	638,000	585,000	298,000	133,400	5,469,200
Manufacturing	17,424,900	4,914,500	6,674,500	609,900	1,411,600	1,324,400		627,400	592,500					1,272,200
Gas and Electric Utilities	2,082,600	416,000	979,000		167,000	104,900						208,000		208,000
Water and Sewage	410,000		229,600	16,400	73,800	61,500								28,700
Roads and Highways	21,331,000		10,300,000	6,000,000	120,000	180,000	4,600,000							131,000
Railroads	1,770,000	72,600	1,325,700		63,700	152,200								33,600
Marine	7,311,000		1,024,000	930,000	2,486,000	1,316,000		37,200	85,000	658,000	585,000			292,000
Offshore Oil Rigs	540,000	27,000	81,000	54,000	162,000	135,000		27,000	27,000					27,000
Office Buildings	7,850,000	5,652,000	1,491,600		157,000	102,000		109,900					133,400	204,100
Schools and Universities	7,619,000	5,638,000	914,000		199,000	343,000		305,000						229,000
Hospitals and Nursing Homes	3,614,500	2,638,600	578,100		54,300	126,500		72,300						144,600
Retail Stores	3,349,000	2,679,000	419,000		34,000	60,000		67,000						90,000
Hotels and Motels	3,026,000	2,269,000	454,000			70,000		76,000						157,000
State and City Buildings	1,580,000	1,074,000	234,000		55,000	48,000		95,000						24,000
Eating Establishments	826,000	620,000	124,000		8,000	21,000		25,000						28,000
Federal Government	13,000,000	4,550,000	3,250,000	650,000	650,000	650,000		650,000						2,600,000

[SOURCE: Maintenance Paints, Vol. 3, A Market and Economic Study of the U.S. Markets, Smith Stanley & Co. and American Paint Journal Co., Oct., 1976.]

TABLE 1.8 — High-Performance Coatings Projected for Maintenance in 1980 (Thousands of Gallons)[1]

Total All Maintenance Paint Per Smith Stanley Survey	Industry	Alkyd[2]	Chlorinated Rubber	Epoxy	Vinyl	Urethane	Acrylic	Inorganic Zinc	Bituminous	Not Identified (Assumed H.P.)
3775	Roads and Highways (structural only)	377	123	148	221	—	—	—	—	161
21,270	Manufacturing (all industry)	820	749	1736	1629	771	728	—	—	1665
14,000	Federal Gov't.	400	800	800	800	800	—	—	—	3200
9,000	Marine	125	1168	3058	1618	—	—	809	720	359
2,723	Gas and Electric Utilities	120	—	205	128	—	—	—	—	256
1,800	Railroads	164	—	79	187	45	105	—	—	41
598	Offshore	10	66	199	166	33	33	—	—	33
439	Water and Storage	28	20	91	76	—	—	—	—	65
53,605	Total	2044	2926	6316	4825	1649	866	809	720	5780

TOTAL ALL H.P. COATINGS 25,935

[1]Based on information extracted from Smith Stanley Survey.[10]
[2]An estimated 10% of total alkyd could be classified as a high-performance coating. Includes Acrylic, chlorinated rubber, silicone, etc. modifications.

uses more and different raw materials than almost any other single industry.

The paint industry is a technically complex one as well, which has grown in accordance with this century's tremendously increasing production of objects and structures requiring paint or protective coatings. The corresponding increase in paint products has been achieved through intensive efforts of not only paint companies, but of major chemical companies and raw materials suppliers as well. The dye industry has contributed technology and information needed to upgrade and evaluate pigments. The plastics industry, which developed around organic chemical technology, has contributed their development of polymers and gigantic molecules containing hundreds of repeating basic molecules. Thermodynamics has proved a useful tool for understanding the chemical reaction of paint ingredients and the more subtle interactions involved in solution compatibility and pigment wetting. *Rheology,* the science of the deformation and flow of matter, has also had a great impact on the paint industry.

Hundreds of new types of coating materials have evolved as well. These include resins, solvents, plasticizers, pigments, dryers, foam control agents, adhesion promoters, and fire-retardant chemicals and pigments. These materials have helped to achieve corrosion resistance, heat stability, and new methods of application. Paint manufacture has thus progressed from the state of an art to that of a science deeply rooted in the technically complex chemical industry.

The technology of coatings is based on organic and, in more recent years, inorganic chemistry which has expanded tremendously within the last century. A plant manufacturing a broad line of trade sales, maintenance, and industrial paints now requires at least five to six hundred different raw materials. The companies concentrating in high-performance coatings such as the epoxies, vinyls, polyurethanes, inorganics, etc. also require several hundred raw materials (*e.g.,* organic monomers, polymers, resins, pigments, dryers, *extenders* or nonhiding pigments, plasticizers, and solvents). A single formulation may require as many as 15 to 20 individual ingredients which create the chemical forces that are responsible for the finished product's surface adherence.

So many raw materials are needed due to the great diversity of finished products necessary to best serve the innumerable specific purposes for which coatings are applied. Most people think of paint in its relation to houses and buildings since a great deal of so-called ordinary paint is purchased only for appearance purposes. On the other hand, most objects we come in daily contact with are coated with materials that were developed for a specific purpose. Residential aluminum siding, for example, is factory coated through a coil coating process. The finish is applied and hard baked in less than a minute, and is expected to last 15 years once the siding is applied to the house.

Some of the most sophisticated coating systems are applied to automobiles. In the 1920s, as many as 10 or 15 coats of lacquer were applied to automobiles. While the finely rubbed finish was beautiful when new, it lasted only a relatively short time before it started to dull, chalk, and, in many cases, change color. Today, only two or three coats of paint are applied, an outstanding gloss is obtained, color possibilities are almost innumerable, most colors will maintain their gloss and shade for a number of years, and little, if any, hand rubbing is necessary.

Coatings Economics

When the January through April, 1979 Bureau of Census figures are extrapolated for the entire year, the total volume of paint sold amounts to 1.33 billion gallons.[9] Thus, high-performance coatings figures isolated for both maintenance and new construction of approximately 40 million gallons, amounts to approximately 3.5% of the total gallons of paint sold during this period. The significance of this lies in realizing that high-performance coatings comprise a small proportion of the total paint market, yet the economic importance of this volume is many times greater than the 3.5% indicates. To put it in perspective: high-performance coatings are used to protect structures that serve as the production facilities of the world. The value of both the old and new structures is escalating at a rapid rate, making it essential to have them protected against corrosion with high-performance coatings which provide a barrier to keep these essential structures intact and free from failure.

More specifically, Uhlig, in 1979, estimated that annual losses due to corrosion in the United States amounted to $5.5 billion.[12] In the mid 1950s, Vernon calculated the United Kingdom's annual total expenditure, which included metal losses and the cost of corrosion prevention, at 600 million pounds (or approximately $1.25 billion dollars).[13] More recently, it was estimated that the cost of corrosion in the United States is 4% of the Gross National Product. This amounts to somewhere between $48 and 50 billion. As stated by Evans, however,

> ...the true cost of corrosion cannot be reckoned in any money sum representing replacements and maintenance. We have to visualize cases where some plant or machine which had been working perfectly is suddenly brought to a standstill by corrosion breakdown.[14]

A good example of the use of coatings to reduce corrosion cost is in refined oil tankers. In the 1930s and 40s, tankers were built with an expected life of approximately 15 to 20 years. The first part of this life (approximately seven years) was in refined oil service. By the end of this service, the bulkheads in the tanker were corroded to one-half of the original steel's thickness. Any additional corrosion would have reduced the bulkhead below the point of minimum safety. Thus, the ships were usually transferred for the remainder of their existence to black oil service, which for the most part is much less corrosive than the refined products.

Today, however, the life of the tanker is considered 28 to 30 years with no allowance for interior corrosion. This has been accomplished through the use of high-performance inorganic zinc coatings which have proven that with proper application they can maintain bulkheads of a refined oil tanker without loss of metal for 20 years. The entire 30-year life of the refined oil service vessel is thus used in hauling the refined products for which it was designed. The coating, in this case, not only allows the continued transportation of the higher value cargo, but it also reduces the tank's interior maintenance costs to a minimal amount.

In an article in the *Marine Engineering Log* of August, 1974, an example is cited of three 71,000 ton vessels which were treated in 1964 with interior and exterior coatings.

> As of 1974, the coating was almost perfect, 95% to 97% intact, the value of the internal coating has been proved years ago—when coating eliminated the extensive steel renewals previously required.[15]

Today, the bulkhead replacement costs mentioned in the article would amount to approximately $150 to $200 per square foot, compared to approximately $5 per square foot for a protective coating with a 20-year life span.

Coatings thus, in a sense, serve as an insurance policy on the life of the structure. The coating cost amounts to a very small percentage of any structure's total cost, and this small cost increment protects the structure against disintegration for many years.

Coating Manufacture

Coating manufacture started when the first prepared paint was put on the market in 1867. This first product was primarily a mixture of white lead and linseed oil ground together on a burr stone mill. These mills usually consisted of two slabs of granite or similar hard natural stone cut so that the two flat surfaces rubbed together, with the grooves carrying the undispersed paint material in between them. Several runs through the mill were made before the required fine texture or proper pigment dispersion was obtained. Much of the paint manufacturing process involves the dispersion of pigments. The following is a review of the manufacturing process extracted from Unit 1 of the *Federation Series on Coating Technology*.

> In the manufacture of pigmented products, the most critical operation is the incorporation of the pigment in the binder or vehicle, as the case may be. If the ultimate pigment particles are too coarse, it is necessary to reduce them in size by a true grinding process. Most of the pigments used today are sufficiently fine that no grinding is required. However, they contain clusters or agglomerates that must be broken down to the separate particles, which must then be wetted by the binder. This process is correctly designated as dispersion. Some kinds of milling equipment perform both grinding and dispersion, while other types disperse only. In the former category were the stone mills, which were the most common milling equipment 60 years ago. A few plants bore with them until about 30 years ago. They were doomed to the scrap pile by slow production, high maintenance, variable results and need for close attention.
>
> Three roll steel mills appeared about 60 years ago. They operate with a small clearance between rolls and a speed differential between successive rolls. If used for grinding, they are inefficient and the rolls are etched. Their only proper use is for dispersion. Since dispersion depends on the shearing action produced by the differential in roll speed, a high velocity vehicle is essential to good results.
>
> Although largely displaced by newer more efficient equipment, roller mills with three to five rolls continue in limited use. Next came pebble mills and, a little later, steel ball mills, the latter for dark colors only. Collision between the pebbles or balls and with the shell provides sufficient impact to reduce particle size as

well as dispersion. Frequently, they have been used when dispersion only is required. Pebble and ball mills require little attention and are more economical than earlier mills. In general, they do not develop as high a gloss as roller mills.

Although pebble and steel ball mills are still used extensively, there is a strong trend toward newer types of equipment for products that do not require actual grinding.

There was a period during which certain types of formulas required extreme, costly methods of pigment dispersion in order to develop the high gloss demanded. This was true of some nitrocellulose automobile lacquers and, more recently, it applied to some baking appliance enamels. The extra gloss was achieved by mixing in heavy duty mixers a high content of pigment with a viscous liquid at very high consistency, which creates much greater shearing effect than roller mills. The Banbury Mill, designed for rubber compounding, was employed for this purpose. Two roll rubber mills were also used to disperse pigments in thermoplastic resins such as vinyls. Excellent pigment dispersions and a very high gloss were obtained by this method. Later, dough mixers of the Baker-Perkins type were found more suitable for some products. This high cost process holds small and declining interest for the paint manufacturers. However, it is widely used by companies that specialize in difficult-to-make dispersions for sale to paint companies.

Both roller mills and dough mixers depend on a viscous liquid for shearing action at relatively slow speed. Good shearing action can also be obtained with a thin liquid, provided the mixing is done at an extremely high speed. This is the principle of dispersion on which much of today's paint is made. The principle is utilized in numerous designs of equipment, such as the Morehouse Mill, the Kady Mill, the Cowles Dissolver. The economy of high speed mixing makes it first choice whenever it produces satisfactory dispersion without undesirable side effects.

Milling methods that are based on dispersion without grinding owe much of their success to two contributing circumstances. The first is the development of fine particle grades of the common extenders, earth colors and flatting agents. The second is the widespread use of wetting agents, which may be applied on the dry pigment by the supplier or incorporated in the paint formula.

One of the latest types of equipment for pigment dispersion is the first to offer the advantage of a continuous process. A stationary shell holds sand or pea-size porcelain balls. While the mass is being stirred by moving arms, the paint is forced upward through the chamber and out the top. The process is efficient in output and effective in the completeness of dispersion and the degree of gloss development. Its main limitation is unsuitability for short runs because of the problem of cleaning the sand or balls. Since there is no appreciable impact between the grains of sand or the balls, only a shearing and rubbing action, there is dispersion only and no reduction in particle size.[16]

Figures 1.7 and 1.8 are examples of a modern plant capable of manufacturing millions of gallons of paint or protective coatings per year. Of necessity, it is highly automated with all liquid handled in pipelines and mixers which drain the finished product through filters into automatic canning and packaging equipment. This presents a far different picture from the earlier plants of the 19th century with their oil-fired varnish cookers, stone mills, and belt-driven mixers.

While this discussion of manufacturing has primarily referred to paint, high-performance protective coatings are manufactured with the same types of equipment and essentially the same procedures and techniques. Most coatings are made in high-speed mixing tanks using a high-speed impeller for dissolving resins as well as dispersing pigments. Although each manufacturing company has its own procedures that depend on the desired product and available equipment, most protective coatings (as well as most paints) are manufactured according to the batch process. (The average batch size is a thousand gallons or less.) This is particularly true with high-performance

FIGURE 1.7 — Modern paint manufacturing plant showing large scale-supported mixing vessels and pipelines for raw materials and semiprocessed liquids. (Courtesy Ameron, Inc., EFD Facilities, Wichita, KS.)

FIGURE 1.8 — Manufacturing and packaging area of a large paint plant showing overhead storage of liquid materials, pipelines, automatic canning equipment, and end product conveyors. (Courtesy Ameron, Inc., EFD Facilities, Wichita, KS.)

coatings, since numerous different materials are often added to satisfy the product and color specifications of individual purchasers. On the other hand, many companies in the protective coating field have endeavored to standardize coating colors and constituents.

Paint and coating factories vary enormously in size. At the outset of the 1970s, there were approximately 1600 paint factories in the United States. According to the information given in Table 1.9, the majority of paint companies were small and employed fewer than 100 people. On the other hand, the methods used and equipment needed are approximately the same for even the largest manufacturers. The difference exists mostly in scale, since a small manufacturer with the required background and technology can produce an end product equivalent in quality to that produced in much larger plants. The real difference is that most of the smaller companies manufacture paint for distribution in a very restricted area and, for the most part, supply trade sales or small volume specialty products. Many of the large companies, however, make only trade sale products or industrial finishes, although some of them are active in all areas.

Protective coating manufacturers, in particular, have attempted to standardize products for use in various corrosive environments. These products are not limited to local situations, but for the most part are intended for national or international markets. This is due not only to the pervasiveness of corrosion problems, but also to the nature of the industries they serve. Marine installations, chemical plants, refineries, nuclear energy plants, and paper plants, for instance, generally enjoy a worldwide demand for their products.

TABLE 1.9 — Number of Employees and Establishments

Average Number of Employees	Number of Establishments
1-4	468
5-9	242
10-19	311
20-49	350
50-99	171
100-249	113
250-499	36
500-999	8
1000-2499	2

[SOURCE: Maintenance Paints, Vol. 3, A Market and Economic Study of the U.S. Markets, Smith Stanley & Co. and American Paint Journal Co., Oct., 1976.]

Due to this international scope, consistency has become a critical factor in the high-performance coating industry. That is, manufacturers must be sure a given product performs the same regardless of whether it was purchased from a plant in the United States, in Europe, or in Japan. This is often more difficult than it may sound. While most raw materials are standardized in the United States, there can still be differences in pigments or resins that are manufactured in different plants within the same country.

A plant in California, for example, developed a coating using a particular grade of red lead pigment. The coating was well-received by industry, so the formula was eventually transferred to a second plant on the East Coast. The same number and grade of red lead was purchased from the same supplier for both the western plant and the one in the East. There was sufficient difference, however, in the red lead pigment of different origins, that the coating plant on the East Coast could not duplicate the original product, even though it utilized the same formulation. Thus, it was necessary to ship pigment from the West Coast in order to obtain consistent results.

This is only one example of the type of difficulty which often arises when trying to standardize a product manufactured in different places throughout the world. Raw materials which meet a U.S. standard are even more difficult to obtain in foreign countries. Nevertheless, most of the protective coating manufacturers who sell in an international market make every attempt to standardize their product so that when using it for prevention or control of a corrosion problem, the end results will be consistent worldwide. This is something to keep in mind when specifying products both in the United States and overseas where they may come from different manufacturing plants.

Other Coating Terms

Comparisons of various coatings are often made according to *composition.* The composition of a coating is often expressed by dividing the total weight between the pigment and the vehicle as percentages. In this case, the *pigment* includes both the hiding and the reinforcing or extender pigments, and also any material used to regulate the gloss of the coating. The *vehicle* is the complete liquid portion of the coating. Normally, it consists of both nonvolatile matter and volatile materials.

In specifications, the *nonvolatile* portion of the vehicle is indicated as the vehicle nonvolatile, or more commonly as vehicle solids, binders, or film-formers. The *volatile* portion of the vehicle is the *solvent* and is usually designated by that name. The sum of the pigments and the vehicle solids is the *total nonvolatile* or total solids of the coating. This is the part of the coating which remains on the surface after the coating is applied and after the solvent evaporates. It is the part which makes up the thickness of the coating as indicated by the term *mil square feet per gallon, i.e.,* the amount of total solids in one gallon of coating spread one mil thick over a certain number of square feet. A gallon of coating would cover approximately 1600 square feet one mil thick if the coating were 100% solids. If the coating contained 50% solids, it would then cover 800 mil square feet per gallon. An understanding of these terms is important in comparing coatings received from the same manufacturer, from different plants, or from different manufacturers.

Coatings Complexities and Variables

A coating is a complex material made of a whole series of interacting ingredients such as resins, plasticizers, pigments, extenders, catalysts, fungicides, and solvents. All of these materials are then applied as a thin film of only a few microns or thousandths of an inch. The solvents must evaporate and the nonvolatile portion must deposit a continuous film over the surface. In some cases, this film will react with the surface, with internal curing agents, and with oxygen in the air to become insoluble; or

with water in the air to hydrolyze and become insoluble. It must also adhere to the surface and provide an attractive finish that will withstand wind, rain, sun, humidity, cold, heat, oxygen, physical damage, chemicals, biodegradation, and many other physical, chemical, and natural forces.

The variety of materials within a coating and the innumerable conditions under which it must perform thus give rise to hundreds of different types of coatings. Each variation is developed to address differences in material, application, or use. Today's coating process makes the old ''anyone-can-grab-a-bucket-of-paint-and-a-brush'' concept obsolete. Coatings are vital to the protection of all types of structures used by society which are, in themselves, becoming more complex and subject to increasingly more corrosive environments. Thus, coatings are becoming so vital to their protection that they should be considered an actual part of the structure and not simply a last minute detail.

Even more variables are introduced by the drying process. Industrial products finishes have been developed to limit many of the drying variables by controlling the type of application and the speed and temperature at which the coating is cured. Unfortunately, most industrial coatings are applied to structures where the curing of the coating cannot be accurately controlled. This is usually due to variables such as weather, humidity, surface conditions, (rough, smooth, or filled with pinholes), the type of substrate (steel, concrete, wood, plaster, or one of the several nonferrous metals), surface cleanliness, and application techniques, (brushing, rolling, spraying, or the application of a hot-melt). It may also be necessary to deal with a rather wide variety of coatings which often dry in radically different ways, affecting the final dry film.

Types of Coatings

Thermoplastic Coatings

Thermoplastic coatings, or lacquers, dry solely by the evaporation of the solvent (the resin is already in its final form), and there is no chemical or physical change in the nonvolatile portion of the coating that forms the film. In this case, the film-forming process is merely the evaporation of the solvents from the liquid leaving the thermoplastic resins on the surface as a continuous film. This process is not as simple as it sounds, since most coatings are made up of a number of different solvents with different evaporation rates in order to insure that the final film is continuous. If the solvent evaporates too quickly, it may cool the surface of the coating to such an extent that water is condensed on and in the film. This is not an uncommon phenomenon where coatings are applied under high humidity conditions. *Blushing* is the term that refers to water condensation which makes the coating turn white. The film that is blushed is generally porous and does not have the same resistant characteristics as the smooth resin film that has properly formed over a surface. Examples of the thermoplastic-type coatings are vinyls, acrylics, and chlorinated rubbers.

Conversion Coatings

Conversion coatings, on the other hand, dry or react in a whole series of steps. All such coatings undergo a chemical and physical change in the process of film-formation. There are several different types of conversion coatings; the oldest and most familiar are coatings which have a drying oil and a resinous varnish or resin as the binder. These usually dry more slowly than the thermoplastic coatings and the various drying stages are considerably more complex. These stages are solvent evaporation, oxidation, thicking or polymerization, and gelation.

Gelation occurs when the polymers reach a size and concentration that form a continuous network. At this point, although the film is considered dry, it still contains a considerable amount of liquid material and may be somewhat soft. The remaining film continues to cure or dry until the coating becomes hard and ultimately brittle. These latter changes are accelerated by a sunlight and heat mixture. When the films reach their ultimate hardness, they generally tend to increase in porosity and lose resistance to moisture and chemicals.

Epoxy

A much more important conversion reaction, from a corrosion standpoint, is *catalyst conversion* or cross-linking at ambient temperatures. The epoxy coating forms by this process in which the epoxy resin is mixed with an amine just prior to application. The epoxy coating's drying process consists of solvent evaporation followed by a chemical reaction of the amine and the epoxy resin in such a way that *cross-linkage* (the joining of two or more molecules of the epoxy resin through a chemical bond with the amine) takes place. In this case, the amine actually becomes part of the chemical reaction and is an integral part of the new polymer. It is therefore not a true catalyst. This process is temperature sensitive and can take place in the absence of air. Where cross-linkage takes place, the coating is called *thermoset.* Thus, it is no longer soluble in its original solvents, nor is it as sensitive to softening by heat.

Another conversion reaction takes place when an epoxy resin reacts with a second resin, *e.g.,* a polyamide resin. Here, the same mechanism as with the amine takes place, only in this case the two resins, the epoxy and the polyamide, react and cross-link to form a solid resin film. The film is therefore somewhat more resilient and elastic than the films formed using the amine epoxy reaction.

Moisture

A third familiar process of film conversion takes place when water from the atmosphere converts the film from a liquid to a solid. This is one of the processes by which the polyurethane coatings are formed. In this case, water from the air reacts with the polyurethane resins during the initial evaporation stage, cross-linking it and increasing the molecular size of the resin until it becomes solid. The inorganic zinc coatings also require moisture from the air as well as carbon dioxide to change the silicate molecule (*i.e.,* sodium, potassium, or ethyl silicate) into a continuous coating by reaction with the zinc pigment.

Some of the other conversion processes require baking or heating which are not practical where coatings are to be applied to large existing structures or equipment.

The Finished Product

In the coating process, it is important to distinguish between the liquid coating prior to its application, and the finished product. The final step in the coating formation process occurs only after the coating has been applied and is reacting in place. Thus, the completion of a coating is beyond the control of the original manufacturer so that the quality of the coating depends largely on the care taken during this final step of coating formation. This point must not be overlooked by those with responsibilities for proper coating application and curing.

Coatings are different from almost all other purchases, in that the buyer or the user can tell very little about the quality of the product from its appearance when purchased. The purchaser can only see the can, label, color card, and perhaps a set of application instructions. While the label may show the composition of the coating and may outline various safety procedures to be taken during its application, it offers very little information with regard to the ultimate effectiveness of the product. Two products with essentially the same label analysis can differ greatly in price and in performance. The liquid coating, then, is only of temporary concern to the coating user. The user's real interest and long-term concern is in the finished surface film after the drying or curing process has been completed.

The Development of Protective Coatings

Paint, coatings, and eventually high-performance coatings were developed as a need for them arose and as the materials became available which allowed their production. Of prime importance in the development of protective coatings was the petroleum industry, which produced most of the basic ingredients from which all or most of the synthetic resins were developed. The cracking of petroleum produced all types of workable compounds with *unsaturated molecules* (capable of cross-linkage and polymerization) that were important in the building of large resin polymers such as vinyls and acrylics. The solvents that were necessary for the solution of the resins also were derived from petroleum or natural gas. At a somewhat later date, the building blocks for the epoxies and polyurethanes were derived from petroleum refining.

Through natural gas, the petroleum industry was responsible for the rapid growth of the fertilizer industry and of the very broad petrochemical industry. The growth of both of these increased corrosion problems, thus increasing the need to protect buildings, structures, and equipment. In moving to offshore sites, the petroleum industry created additional, massive corrosion problems that could only be solved with high-performance protective coatings.

The basic heavy chemical industries were also operating during this development period. These plants produced basic acids such as hydrochloric, sulfuric, and nitric acid, as well as chlorine and caustic. The steel, paper, and marine industries all had massive corrosion problems, but prior to this time had been dealing with them through replacement rather than protection. Steel was inexpensive, so when a member corroded to the point where it was no longer structurally sound, it was replaced. When pipes and ducts in the paper industry rusted through, they were replaced. When tanker tank bulkheads in the marine industry were reduced to a dangerous level by corrosion, they also were replaced.

There was some thought being given to the protection rather than continual replacement of steel surfaces prior to WW II. The war itself, however, provided the impetus in actually developing a means of corrosion prevention. Not only was steel more costly during this period, but it was also in very short supply, making it necessary to take measures to protect the ships and structures that were already available. It was during this time that the first test application of a coating on the interior of a tanker tank was applied. The test utilized a very small area of only several hundred square feet, yet it did demonstrate that the new synthetic resins had the ability to withstand corrosive attack from salt water and refined petroleum products, and thus provide the protection needed for steel surfaces.

It is interesting to note that the intensified corrosion problems, the increased need for protection, the proper materials (*i.e.*, new synthetic resins), and the tighter economic situation all developed during the same period to create the driving force in the production of new high-performance coatings. Since that time, not only has the economic need for the protection of steel and concrete surfaces increased, but the extremely rapid expansion of the world's various industries have increased corrosion problems as well. Thus, corrosion research and protective coating development has continued at an ever-increasing rate.

The need for these specialized coating products will continue to rise as new industries develop, new chemicals are made, new processes are used to protect the environment, and the cost of basic building materials increases. Their use will also increase as, in providing structural protection, and thus added years of useful life, they become more widely recognized as a major source of energy savings.

The Future of Protective Coatings

While the use of protective coatings is becoming more and more important, there will be changes in the years to come that will drastically affect both the paint and the protective coatings industries. Air pollution and application safety are major areas of concern for several state and federal government agencies, and changes are being demanded that would soon make many presently available products unacceptable for use. Although the industry's corrosion problems are greater now than at any previous time in history, many presently effective solutions may have to give rise to new and different ones because of government requirements.

It will soon be necessary to drastically reduce the emissions of volatile organic compounds or solvents which have been a major ingredient in protective coatings. This is especially true for high molecular weight resins which have the highest and broadest chemical resistance. Air pollution is a major area of concern for high-performance coating manufacturers, with all indications pointing to a major reduction of, if not complete restriction against, the use of solvents in any form. Present research by most high-performance coating production companies is in the area of high solids and 100% solids coatings, and towards water-based or water-dispersed coating materials.

Whatever develops, whether it involves reformulation, higher solids coatings, water dispersed materials, or completely new approaches, the coatings will undoubtedly be even more complicated and demanding in both manufacture and application. More highly technical and highly trained individuals will be required to manufacture, sell, and apply them, and new raw materials and manufacturing techniques will undoubtedly be needed for their production.

This issue of changes within the industry is discussed in an article in *Maintenance and New Construction Coatings* as follows:

> The era of change has arrived for the maintenance and new construction coatings industry. Starting from Government regulations, vibrations of change are impacting the raw materials supplier on one end and the user engineering construction firm on the other. With change there are opportunities.
>
> It's obvious that coatings and raw materials producers are maintaining aggressive research and development efforts on low solvent and/or no solvent non-residential coating projects. It's imperative that all companies involved directly or indirectly with maintenance and new construction coatings be well aware of the changes taking place. Those companies that are taking steps to deal with these changes are the companies that will ultimately benefit.[17]

This, of course, does not mean that the lessons which have been learned in the last half-century are to no avail. The knowledge which has been gained from the study of corrosion and corrosion processes is as valid today as it ever was. The methods of testing coatings under corrosive conditions are equally as valid, the need for proper surface preparation may be even greater with the newer coatings, and while application procedures may change, the principles of application will remain the same.

References

1. Encyclopedia Britanica, Macropedia, Paints, Varnishes and Allied Products, Vol. 13, pp. 886-889, 1974.
2. Encyclopedia Americana, Paint, Vol. 21, p. 107, 1979.
3. World Book Encyclopedia, Paint, Vol. 15, p. 24, 1978.
4. Mattielo, Joseph J., Protective and Decorative Coatings, Introduction, Rise of the Industry, Vol. 1, John Wiley & Sons, New York, NY, pp. 1-8, 1941.
5. Clark and Hawley, The Encyclopedia of Chemistry, Protective Coatings, Reinhold Publishing Corp., New York, NY, pp. 785-789, 1957.
6. ASTM, Manual of Coating Work for Light-Water Nuclear Power Plant Primary Containment and Other Safety-Related Facilities, Appendix A, Glossary of Terms, 1979.

7. Munger, C. G., Kirk-Othmer: Encyclopedia of Chemical Technology, 3rd Ed., Vol. 6, Coatings, Resistant, John Wiley & Sons, New York, NY, p. 455, 1979.

8. Levinson, Sidney B., Facilities and Plant Engineering Handbook, Chapter 6, Painting, McGraw-Hill Book Co., New York, NY.

9. U.S. Dept. of Commerce, Bureau of Census, Current Industrial Reports, Paint, Varnish and Lacquer, Jan.-April (1979).

10. Smith Stanley & Co. and American Paint Journal Co., Maintenance Paints, III, A Market and Economic Study for the U.S. Markets, Oct. (1976).

11. National Paint and Coating Association, Sales Survey, 1977.

12. Uhlig, H. H., Proceedings of the U.N. Scientific Conference on Conservation and Utilization of Resources, Vol. 2, p. 213, 1950.

13. Vernon, W. H., Conservation of Natural Resources, Institute of Civil Engineers, pp. 105, 130, 1956-57.

14. Evans, V. R., Corrosion and Oxidation of Metals, Chapter 1, The Approaches to Corrosion, St. Martins Press, New York, NY, p. 8, 1960.

15. Marine Engineering Log, Cutting Coats for New Ships and Old, August (1974).

16. Federation Series on Coating Technology, Unit 1, Introduction to Coating Technology, Federation of Societies for Paint Technology, Philadelphia, PA, pp. 10, 29-31, 1974.

17. Ozumek, Richard T., Maintenance and New Construction Coatings, Chem. Purchasing, p. 27, April (1979).

2

Corrosion As Related to Coatings

Corrosion, as defined in the NACE's Corrosion Basics book is "...the destruction of a substance (usually a metal) or its properties because of a reaction with its environment."[1] This definition does not make use of the terms *chemical* or *electrochemical* because such terms would define corrosion only as it related to metals and would not allow its application to many other materials which disintegrate due to environmental exposure.

Corrosion of Materials Other Than Metal

As defined, then, corrosion may affect materials other than metals, such as concrete, wood, ceramics, and plastics. It must also be noted that the properties of a material, as well as the material itself, can and do deteriorate. Some forms of corrosion produce no weight change or visible deterioration, yet the material may fail unexpectedly because of certain property changes within the material which may defy ordinary visual examination or weight change determinations. Although these are not necessarily common changes, they are nevertheless important forms of corrosive action which should at least be somewhat familiar to the corrosion engineer.[1]

A good example of such deterioration is the linear breakup of polyethylene sheet after its exposure to weather and sunlight for a period of time. The failure in this case, which is a breakup of the entire sheet structure, is usually sudden. On the other hand, a properly pigmented vinyl sheet may remain essentially unchanged for years with only minor chalking taking place.

Chalking is an example of the way in which coatings themselves may be said to corrode. It occurs when the binder disintegrates from the exposed surface through the depth of the coating until the substrate is exposed. Some coatings, such as the early epoxies, reacted rapidly to weathering; others, such as polyurethanes or vinyls, may last for years.

Early Corrosion Studies

While other materials are affected by corrosion, the area which has received the most intensive study is actually the corrosion of metals. Its attention is warranted by the fact that metals, and steels in particular, constitute the primary structural materials used around the world.

Corrosion studies were first launched at the beginning of the 19th century. In 1824, for instance, Sir Humphrey Davy initiated the use of zinc to control the corrosion of copper-sheathed ships' hulls.[2] Today this same procedure is used to protect steel hulls from corrosion initiated by bronze propellers. Almost a century later, Dr. Willis Rodney Whitney developed a principle that electrochemical reactions form the basis for corrosion. From his experiments involving the corrosion of steel pipe in water, he concluded:

> Practically the only factor which limits the life of iron is oxidation under which name are included all the chemical processes whereby the iron is corroded, eaten away or rusted. In undergoing this change, the iron always passes through or into a state of solution and we have no evidence of iron going into aqueous solution except in the form of ions. We have really to consider the effects: of conditions upon the potential difference between iron and its surroundings. The whole subject of corrosion of iron is therefore an electrochemical one and the rate of corrosion is simply a function of the electromotive force and resistance of the circuit.[3]

Fundamentals of Corrosion

Electrochemical Principles

The application of the electrochemical principles developed by early investigators provides the means for modern methods of corrosion control. The primary reason iron or steel corrodes is that elemental iron, *i.e.,* the condition of iron as it exists after it has been reduced from its ores, is thermodynamically unstable. There does not seem to be any free iron available in nature. Rather, all of the iron that has been found exists in combination with other elements such as oxygen or sulfur. In order to change iron from an oxidized state to that of a metal, it is necessary to force a large amount of energy into the system. This energy is then stored in the metallic iron. The fundamental laws of nature governing the conservation of energy require that, in time, the energy balance must be restored by returning the unstable metal to its oxidized state.

The most common type of iron ore has a composition of Fe_2O_3, and in its oxidized state usually appears as rust, with the red form being the Fe_2O_3. Iron is also found in ore as magnetite or Fe_3O_4, which is also a common corrosion product. The same is true of other metals such as zinc, aluminum, and magnesium. In each case, it takes a massive amount of energy to change the ore into metal. The more energy that is absorbed by the metal, the easier it tends to corrode.

There are some metals, however, which exist in nature in a pure metallic form. These include tin, copper, silver, and gold. They may also exist in the form of ores in combination with other elements. In these cases, however, the extraction from the ore itself requires little input of energy (Table 2.1).

TABLE 2.1 — Metals in Order of Energy Required for Conversion from Their Ores

Common Metals	Energy Required for Conversion
Potassium	Most
Magnesium	
Beryllium	
Aluminum	
Zinc	
Chromium	
Iron	
Nickel	
Tin	
Copper	
Silver	
Platinum	
Gold	Least

[SOURCE: Corrosion Basics—An Introduction, LaQue, F. L., Chapter 2, Basics of Corrosion, NACE, Houston, TX, p. 23, 1984.]

Electromotive Force

The same orientation of the metals exists in respect to electromotive force (Table 2.2). The ones which require the greatest amount of energy input in order to make the metal itself are also the ones with the highest electromotive force. They also have a greater tendency to go into solution and to form ions. Thus, the electromotive force is sometimes referred to as solution potential.

Hydrogen is used as a reference and is rated in the electromotive series as zero. Moving up the list from

TABLE 2.2 — Series of Metals by Decreasing Electromotive Force[1]

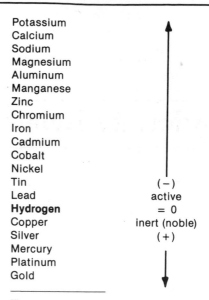

Potassium	
Calcium	
Sodium	
Magnesium	
Aluminum	
Manganese	
Zinc	
Chromium	
Iron	
Cadmium	
Cobalt	
Nickel	
Tin	(−)
Lead	active
Hydrogen	= 0
Copper	inert (noble)
Silver	(+)
Mercury	
Platinum	
Gold	

[1]Table 2.4 gives the complete Electromotive Force Series.

hydrogen, the metals become more active. Moving down the list from hydrogen, the metals become increasingly inert. This means that the metals above hydrogen are increasingly able to corrode, and actually radiate energy as they go into solution. On the other hand, in order for the metals below hydrogen to ionize, or for them to form salts such as copper, mercuric chloride, or silver nitrate, it is generally necessary to add energy. For example, when a piece of metallic potassium is placed in water, the potassium almost explodes in its attempt to ionize. On the other hand, a piece of platinum may be placed in strong nitric acid with no evidence of attack.

The electromotive series given in Table 2.2 also shows which metals can displace another metal in solution and suffer corrosion in the process. Any metal may displace the one below it from solution. Chemically, this is known as a *double displacement reaction*. If iron, for instance, is placed in a solution of copper, the iron will rapidly go into solution and the copper will plate out, thus becoming a metal. This reaction is used extensively in copper mining to obtain the copper which is leached out of copper ores.

As Dr. Whitney first indicated, most corrosion of metals is an electrochemical process which requires the presence of several elements in order to proceed: (1) an anode; (2) a cathode; (3) an electrolyte, *i.e.*, a conductive solution in which the metal finds itself; and (4) an external contact or an external circuit between the anode and cathode, *e.g.*, the metal itself or a wire between two different metals. These four elements constitute what is known as the *corrosion cell* or corrosion battery. Corrosion can only take place when these four elements are present. Oxygen, however, is a fifth element which is also generally required in the corrosion process.

Ionization

The terms ion and ionization are important in an understanding of these corrosion cell elements. An explanation of their roles begins with the concept that all matter is made up of atoms. Each atom, in turn, is made of a nucleus, which contains a given number of particles of unit positive charge (protons), surrounded by particles of unit negative charge (electrons) of like number. Each positive charge balances a negative charge so that the atom itself is electrically neutral.

For example, hydrogen, an element which plays an important part in many corrosion reactions, has only one proton in the nucleus and is associated with a single electron. The hydrogen atom, therefore, represents the simplest form of element construction.

In more complex atoms, the charges are greater than one, but, in each case, the number of negatively charged electrons is equal to the number of positively charged protons in the nucleus. If one or more of these electrons are removed from any atom, the remaining electrons will not be sufficient to neutralize the positive charge in the nucleus, and the residual part of the atom, now called an ion, is positively charged. If one or more electrons are added to the neutral atom, a negatively charged ion results.[4] For example, if a hydrogen atom loses its electron, it becomes a hydrogen ion (H^+). If the hydrogen atom (H) shares electrons with another hydrogen atom, the two atoms form a hydrogen molecule (H_2) (Figure 2.1).

Ions can be defined as atoms or groups of atoms which have either taken up or surrendered one or more electrons from their outer electron rings. These ions bear positive or negative charges: positively charged ions are called *cations* while negatively charged ions are known as *anions*. (Positively charged ions are attracted towards the cathodes, therefore the term cation. Negatively charged ions are attracted towards the anode, therefore the term anion.) The charge carried by ions is largely responsible for their unique properties either in or out of solution, and the properties differ markedly from the neutral atoms or molecules.

Iron is a neutral atom when it is in its metallic form, but it becomes an ion when it loses two electrons and therefore becomes positively charged. Most of our common inorganic chemicals, such as acids (*e.g.*, hydrochloric acid), alkalies (*e.g.*, sodium hydroxide), or salts (*e.g.*, sodium chloride) are strongly ionized when in water solution. Water, for example, contains about 10^{-7} molecules per liter of hydrogen or hydroxyl ions.[5] Although water is not considered highly ionized, if a hydrogen ion is removed from solution, *i.e.*, during corrosion, another immediately takes its place so that the number of ions available from water remains about constant.

Ionization is the state of being ionized. A molecule of sodium chloride for example, is only in molecular form when it is in the form of a solid. As the solid material dissolves in water, the sodium and chlorine separate to become a positive sodium ion and a negative chloride ion. In water solution, sodium chloride is always in a state of ionization. The same is true for ferric or iron chloride.

As previously shown, any atomic element can become an ion by either gaining or losing an electron. Atoms are made up of an equal number of positive and negative charges. If an electron or negative charge is lost, the atom becomes a positive ion. If an electron is gained, it becomes a negative ion. This principle is demonstrated in Figures 2.2 and 2.3.

Ionization of a metal can be expressed in a chemical equation such as the following:

$$\overset{H_2O}{Fe \rightarrow Fe^{++} + 2e^-}. \qquad (2.1)$$

In this case, the symbol "Fe" indicates atomic or metallic iron. The arrow with "H_2O" above it indicates iron in contact with water. The "Fe^{++}" represents an iron ion with a positive charge. The "e" stands for electrons which, as indicated, have a negative charge. Thus, for every iron ion with two positive charges, there are two negative electrons released. The ionization of zinc is indicated in the same way:

$$\overset{H_2O}{Zn \rightarrow Zn^{++} + 2e^-}. \qquad (2.2)$$

The ionization of aluminum, magnesium, or any of the common metals for that matter can be indicated in a similar fashion; although there may be more or less electrons released, depending on the specific metal involved.

Ions in solution form an electrolyte, and the electrical conductivity of the solution depends on the concentration

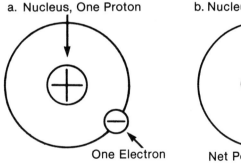
a. Nucleus, One Proton

One Electron

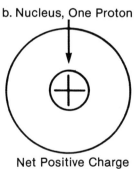
b. Nucleus, One Proton

Net Positive Charge

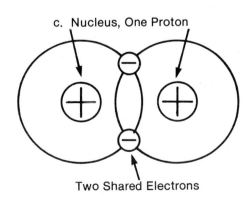
c. Nucleus, One Proton

Two Shared Electrons

FIGURE 2.1 — (a) Hydrogen atom; (b) Hydrogen ion (H^+); and (c) Hydrogen molecule (H_2). [SOURCE: U.S. Dept. of Commerce, Civil Engineering Corrosion Control, Corrosion Control—General, Vol. 1, Dist. by NTIS, AD/A-004082, pp. 16, 33, Jan. (1975).]

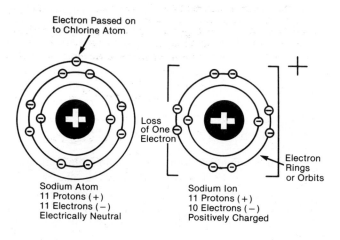

FIGURE 2.2 — Change of sodium atom to a sodium ion.

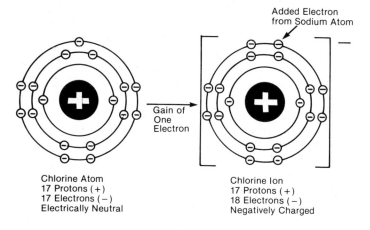

FIGURE 2.3 — Change of chlorine atom to a chlorine ion.

and the mobility of the ions in the solution. The mobility of the ion depends on its size and upon the ion-solvent interaction. The electric current carried by the electrolyte results from the motion of oppositely charged ions in the solution. Each ion type, that is positive or negative, carries a certain proportion of the current. The amount of current conveyed is generally in proportion to the concentration of the ions in the solution. In fact, electrical conduction in all aqueous and other nonmetallic systems is almost entirely carried on by the movement of ions. The energy released in ionic reactions may take the form of electrical energy, as in storage and dry cell batteries, as well as in the corrosion cell.

The Corrosion Cell

The Anode

The first of the four elements necessary to make a corrosion cell is the anode. The *anode* is the area where the metal goes into solution and where the actual metal loss takes place. Figure 2.4 shows the complex chemical reactions which take place at both the anode and the cathode, while Figure 2.5 illustrates the principal anode reactions. The first reaction in the anolyte zone occurs when metallic iron goes into solution as a ferrous ion with the release of two negative electrons. In Zone I, which is the area just beyond the anolyte area where the iron is going into solution, the iron ions react with hydroxyl ions in the area to form ferrous hydroxide. This reaction is important in that it removes the ferrous ion from solution and creates the reasonably insoluble ferrous hydroxide. This changes the equilibrium in the corrosion cell and allows more iron to ionize and go into solution. Ferrous hydroxide is a transitory, whitish precipitate which forms at the interface of the corroding metal.

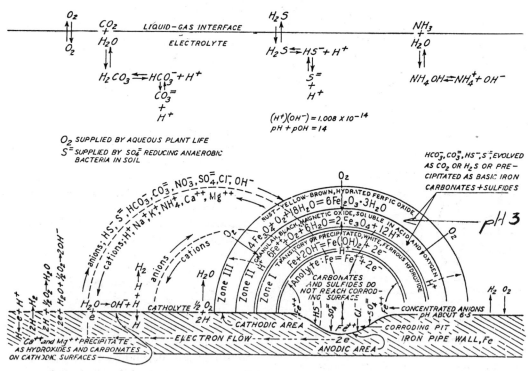

FIGURE 2.4 — Schematic representation of the underground corrosion process. (Courtesy Dr. Gordon N. Scott.)

Corrosion Prevention by Protective Coatings

The reaction which takes place in Zone II to form Fe_3O_4, or magnetic iron oxide, is one of the more complex reactions. As shown in Figure 2.5, iron ions react with oxygen and water to form Fe_3O_4 with the release of hydrogen ions. (Other reactions probably take place as well.) The ferrous hydroxide in Zone I can react with additional oxygen to form Fe_3O_4, or with any carbonate ion that may be in the area to form ferrous carbonate. The Fe_3O_4 is a jet-black material; where it exists in a deep pit, it is a black semisolid material; and where the corrosion product is dry, it is a black scale, often with yellowish cracks running through it.

Zone III is the top of the tubercle over the anode where a more straightforward reaction takes place. The uppermost Fe_3O_4 reacts with oxygen to form hydrated ferric oxide, or Fe_2O_3. This is the yellowish-reddish product commonly known as *rust*.

The negative ions in the electrolyte, such as carbonate, chloride, sulfate, and hydroxyl ions, are attracted toward the anode area because of the positive iron ions available at that point. As the hydroxyl ions react with the ferrous ions to form ferrous hydroxides, an excess of hydrogen ions remain in solution to create a slightly acidic condition. This was recognized by Speller who explained in his book, *Corrosion Causes and Prevention,* that:

> When a tubercle of rust forms with a dense envelope of ferric hydroxide, the inner anodic portion, which is undergoing corrosion, may become decidedly acid. Baylis has observed that the

soluble portion of such rust tubercles has a pH of about 6 with a pH of 8 in the water outside.... For these reasons it can be seen that as the products of corrosion gather over the mouth of a pit, the metal at the bottom of the pit will become more anodic, so that the rate of penetration increases as the pit penetrates deeper into the metal.[6]

It should be noted that the anolyte area in Figure 2.5 is the acidic area with a pH ranging from 6.5 down to 5.0.

The Cathode

The second element in the corrosion cell is the cathode. *Cathode* reactions, although in many ways less complex than anode reactions, are extremely important in controlling the rate that corrosion takes place at the anodes. The anodic reaction cannot occur at a higher rate than the corresponding electrons can be accomodated by the cathodic reaction. The reaction which takes place at the cathode is essentially the neutralization of the electrons which are created as the iron goes into solution. The electrons can be neutralized through one of these three reactions (Figure 2.6):

$$2H^+ + 2e^- \rightarrow H_2 \qquad (2.3)$$

$$H^+ + e^- \rightarrow H \qquad (2.4)$$

$$2H + 1/2\,O_2 \rightarrow H_2O$$

$$2H_2O + O_2 + 4e^- \rightarrow 4OH^- \qquad (2.5)$$

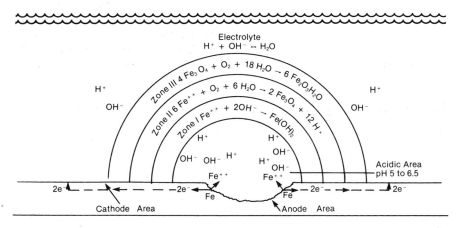

FIGURE 2.5 — Anode reactions.

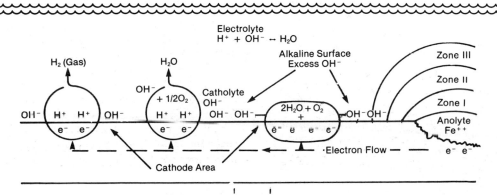

FIGURE 2.6 — Cathode reactions.

The first neutralizing reaction is that of hydrogen ions with electrons to form gaseous hydrogen. Where a massive amount of hydrogen ion is available, as in an acid solution, the gas bubbles form rapidly and hydrogen gas can actually be obtained in volume from the solution of iron by hydrochloric acid. The equilibrium electrode potential (referred to hydrogen) of iron at 25 C in contact with a normal solution of ferrous ions is -0.44 volts. Iron will therefore displace hydrogen from water.[5] This is the first step in the corrosion process and therefore one of the most important chemical reactions involving iron.

The second reaction, also removing hydrogen from the cathodic area on the metal surface, is that of atomic hydrogen with oxygen to form water. In the third chemical reaction, oxygen reacts with water and electrons to form hydroxyl ions. This is an extremely important reaction from a coating standpoint since hydroxyl ions are strongly alkaline. When they are concentrated on the cathode area of a metal, any coating on that metal must be strongly alkali-resistant or it will tend to saponify and disintegrate.

The Electrolyte

The third element in the corrosion cell, the electrolyte, is also very important. The *electrolyte* is the solution which is on, surrounds, or covers the metal. The conductivity of the solution on the metal surface is the key to the speed of the corrosion process. A solution with a low conductivity or high resistance produces a slow corrosion reaction, while a solution with a high conductivity or low resistance makes for rapid corrosion.

Pure water, even though a relatively poor conductor, still contains ions (H^+ and OH^-) so that corrosion will and does take place, although rather slowly, in pure water. Examples of this are often noted in high mountain lakes where tin cans have been thrown in the water. The conductivity of snow water is also relatively low; however, the oxygen content is high and the tin cans form active anodes wherever the iron is exposed to water. In the case of seawater, corrosion cells are formed readily since seawater is almost 100% ionized and is a very good conductor.

The External Circuit

The fourth element, the external circuit, is also important. Where the anode and cathode are on the metal surface, the metal itself acts as the external circuit. If there are two pieces of metal, they must either be in contact or must have an external connection in order for the corrosion process to take place. The conductivity or resistance of the external circuit also helps determine the rate of the corrosion process.

Oxygen as a Factor

Oxygen can be considered the fifth element in a corrosion cell. While corrosion may begin with the presence of only the first four factors, without oxygen, the process soon slows down or stops altogether. Oxygen is extremely important in most all corrosion reactions in order to remove the hydrogen ion from the cathode and to allow additional electrons to be neutralized. When hydrogen ac-

cumulates on the surface as a hydrogen film, the electrons can no longer be easily neutralized and the corrosion cell is said to be *polarized*.

Anode-Cathode Relationship

The formation of an anode-cathode relationship on steel may be easily demonstrated through the use of a gelatin bath with a small amount of dissolved common salt (sodium chloride) to act as the electrolyte. Two indicators must also be added: phenolphthalein, in order to indicate the alkaline area where hydroxyl ions are formed; and potassium ferriferrocyanide, to indicate where the iron is going into the solution or where the iron ions are formed. Prior to allowing the gelatin solution to set, it is shaken in order to add oxygen to the solution. Then, as the gelatin begins to set, a steel plate is placed in the bath. The gelatin is then allowed to gel, and within a very few minutes, the formation of the anode and cathode areas on the steel plate becomes visible (Figure 2.7*).

The steel panel in Figure 2.7* is cold rolled and otherwise untreated. The pink areas where the caustic is forming (a concentration of OH ions) are distributed randomly over the face of the steel. Note that the cathode area is much larger than the total anode, which is generally the case. The blue areas where the steel is corroding (the iron is forming ions) are also formed on the face of the panel, but are also heavily concentrated along the edges of the panel where the panel has been sheared, along any scratches which may appear on the steel surface, or around holes drilled in the panel. The steel readily forms iron ions in these disrupted areas because of the fact that the steel surface in these areas is new and the iron is more active.

It is assumed that if a sandblasted panel were put in the same test, the entire surface would be active. However, the edges, scratches, and area of previous corrosion on this surface will form the blue areas preferentially (Figure 2.8*). It is important to note that corners, edges, welds, and any disrupted areas on the surface of the steel are more corrosion prone than any other area, even though protected with a special coating.

Chemical Concept of a Corrosion Cell

In the chemical concept of a corrosion cell, the iron goes into solution at the anode, reacts rapidly with negative hydroxyl ions, and precipitates. The electrons move through the metal or the exterior circuit to the cathode. At this point, the electrons are neutralized by positive ions, *e.g.*, the positive hydrogen ion on reacting with the electron becomes molecular hydrogen leaving an excess of OH^- ions in the area, or oxygen is reduced by reaction with water and electrons to form hydroxyl ions (Figure 2.9). In either case (the removal of hydrogen or the reduction of oxygen), hydroxyl ions are concentrated on the cathode.

While Figure 2.9 schematically shows the chemical reactions in an iron-copper corrosion cell, Figure 2.10* shows an actual iron-copper galvanic couple with an ex-

---

*See color insert.

ternal connection in a gelatin indicator bath. As would be expected, the iron immediately turns blue, indicating rapid corrosion or solution of iron ions, and the copper panel becomes red indicating the rapid accumulation of hydroxyl ions (OH⁻) on the surface which usually occurs at the cathode.

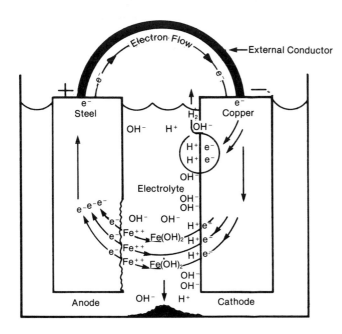

FIGURE 2.9 — Electron flow (chemical concept) in an iron-copper corrosion cell.

Electrical Concept of Corrosion Current Flow

In contrast to the chemical concept, the electrical or conventional concept of current flow is quite different, although nonetheless important since it is used extensively in cathodic protection. According to this concept, the flow of electric current runs from the anode through the solution to the cathode. The anode is often described as the area of the metal surface from which the current leaves the metal and enters the solution. The cathode is used to describe the area of metal surface to which the current flows from the solution and then returns by way of the external circuit to the anode. The electrical current flow, then, is from the positive pole (cathode) through the external circuit to the negative pole (anode) and from the anode or negative pole into and through the solution to the positive pole, the cathode, to complete the circuit. This concept is obviously opposite that of electron flow which runs from the anode to the cathode through the external circuit (Figures 2.11 and 2.12).

Both the chemical and electrical or engineering concept of the corrosion cell are commonly used by corrosion engineers. Table 2.3 summarizes the anode-cathode reactions and indicates the difference between the chemical and engineering concepts.

Polarization

By definition, *polarization* is the shift in potential caused by the passage of a current between anode and

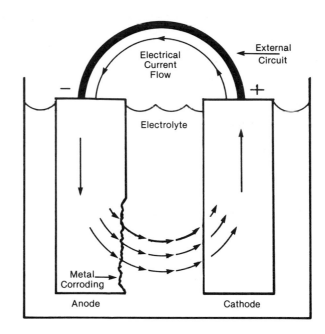

FIGURE 2.11 — Electrical current flow (electrical concept) in a corrosion cell.

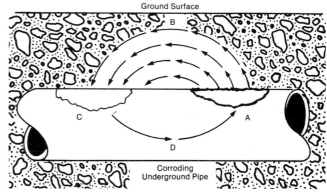

FIGURE 2.12 — The conventional corrosion concept is a more practical illustration of the electrical concept of a corroding underground pipe: (A) Anode area where current leaves the metal to enter the surrounding earth—area where metal is corroded; (B) Current flow through earth from anodic area to cathodic area; (C) Cathodic area where there's no corrosion of pipe surface; (D) Current flows through pipe metal from the cathodic area back to the anodic area to complete the circuit. (SOURCE: Corrosion Basics—An Introduction, Chapter 5, Cathodic Protection, National Association of Corrosion Engineers, Houston, TX, p. 179, 1984.)

cathode half-cells. It has previously been mentioned in this test with regard to the formation of hydrogen on cathodes. The electrode potential is important to the corrosion engineer in that it measures the tendency for an electrode (metal-electrolyte combination) to gain or lose electrons. Two half-cells in combination represent a corrosion cell. The potential difference between the half-cells, *i.e.*, the anode-electrolyte half-cell and the cathode-electrolyte half-cell, represents the driving force for corrosion and therefore the speed at which it takes place. The

TABLE 2.3 — Sign Conventions

Anode	Cathode
Metal loses an electron to become + ion (cation).	An atom that gains an electron becomes a − ion (anion).
The electrode that loses electrons is called the anode.	The electrode that gains electrons is called the cathode.
Oxidation occurs at the anode.	Reduction occurs at the cathode.
Anions (−) move to the anode in the electrolyte.	Cations (+) move to the cathode in the electrolyte.
The anode is positive (+) in the chemical concept.[1]	The cathode is negative (−) in the chemical concept.[1]
The anode is negative (−) in the electrical or engineering concept.[2]	The cathode is positive (+) in the electrical or engineering concept.[2]

[1]Chemically, electrons flow from the anode to the cathode in the metallic (internal) circuit.

[2]Electrically, conventional positive (+) current flows from the cathode to the anode in the metallic (external) circuit.

rate of electrochemical reactions, such as those in the corrosion process, is determined by various environmental factors. Therefore, the potential of a given half-cell is influenced by the condition of the metal, the metal ion concentration and other ions in the electrolyte, and the temperature.

Polarization is thus a very important rate-determining factor in the corrosion process. It is a process which hinders or inhibits the normal corrosion process at the electrode. It can take the form of slow ion movement in the electrolyte, slow combination of atoms to form gas molecules, or slow solvation of ions by the electrolyte. Increasing the reaction area lowers the rate of polarization and allows the corrosion processes to take place more readily by providing more surface on which the reactions can occur. Agitating the electrolytes also lowers the polarization rate by providing a maximum number of ions contacting the electrodes. Electrolyte movement also carries away the products of reaction from the surface, thus decreasing the rate of polarization and increasing corrosion. In the case of hydrogen, agitation would remove the atomic or molecular hydrogen from the cathode surface as soon as it was formed.

Many substances that are included in the electrolytes can greatly affect the polarization of the electrodes. Oxygen and arsenic are examples of substances with opposite effects on the hydrogen electrode reaction. Oxygen effectively depolarizes the electrode or makes the reaction go more rapidly by removing the reaction product, atomic hydrogen. Arsenic, on the other hand, is an effective polarizer in that it makes the combination of hydrogen atoms to gas molecules more difficult. Increasing the temperature increases the rate of most reactions, and therefore would, in general, lower the polarization rate of the cell.

Polarization can be further defined according to its two types: activation and concentration polarization.

These represent the two different methods by which electrochemical corrosion reactions are retarded. *Activation polarization* involves retarding the activity that is inherent in the reaction itself. The rate of the evolution of hydrogen at the cathode is, of necessity, dependent on the speed of the electron transfer from the anode. It is therefore an inherent rate depending on the anode metal. Particular metals vary greatly in their ability to give up electrons; as a result, hydrogen evolution from the cathode likewise varies.

Concentration polarization refers to the electrochemical reaction resulting from concentration changes in the electrolyte adjacent to the metal surface. In this case, if the concentration of hydrogen ions in the solution is relatively low, the neutralization of the hydrogen ion by the electrons will depend on the number of hydrogen ions available and the speed at which they diffuse through the solution. The corrosion reaction would then be controlled by the diffusion rate of the hydrogen ions. With high concentration of hydrogen ions, the electrochemical reaction would go rapidly, as in acid solutions. While activation polarization is usually the controlling factor in corrosion by strong acids, concentration polarization usually predominates when the hydrogen ion concentration is relatively low (*e.g.*, aerated water or salt solutions).

Knowing the kind of polarization taking place allows prediction of the corroding system's characteristics. For example, if the corrosion were controlled by concentration polarization, any change that would increase the diffusion rate of the hydrogen ion in the solution would increase the corrosion rate. Agitating or stirring the liquid would therefore tend to increase the corrosion rate of the metal. On the other hand, if the cathodic reaction is activation controlled (*i.e.*, controlled by the number of electrons available at the cathode), agitation would have no effect on the corrosion rate. There would be little polarization, for instance, on corroding areas of the ship's water line as it passes through the water.

Oxidation and Reduction

Two terms commonly used in chemistry and important to note from the standpoint of corrosion are oxidation and reduction. *Oxidation* may be difficult to pinpoint since according to corrosion terminology, it can mean either the rusting of iron or the development of white oxides on aluminum or zinc. In order to understand their meanings from a chemical standpoint, it is necessary to examine a chemical formula such as that for iron:

$$Fe \xrightarrow{H_2O} Fe^{++} + 2e^-. \qquad (2.6)$$

This formula indicates iron in water in a state of equilibrium where no current flow exists. The "Fe" is iron as a metal, the "Fe^{++}" is the ionized form of iron, and the electrons indicate the negative charges given up when the metal changes to an ion. The movement of the iron from the metal form to the ion form is called *oxidation*. Therefore, in a corrosion cell, the metal is oxidized when it goes into solution as an ion. This occurs at the anode where the term *oxidation* also commonly applies to rust forms.

When proceeding in the opposite direction and adding electrons to the ionized iron, the reaction occurs in the direction of the iron as metal and is referred to as *reduction*. A metal, therefore, which has been changed from its oxidized state to the metal, has been reduced. This is what takes place when iron ore is changed to metal in a blast furnace.

Different metals have different capacities for being reduced and for being oxidized. Gold, for example, exists primarily in the reduced state, *e.g.*, as a metal. Potassium, on the other hand, exists primarily in either the oxidized state as an oxide or in the ionic state as a salt. The symbolic reaction for iron given above and its relative potential for the electrochemical reaction shown is called an *oxidation-reduction potential*. It may also be called a redox potential, half-cell potential, or solution potential.

Galvanic Corrosion

Electromotive Force Series

When the various metals are listed according to their comparative potential, it is referred to as an *electromotive force series* or EMF series. The EMF series in Table 2.4 includes corresponding oxidation reduction potentials. As explained earlier in this chapter, hydrogen is used as an arbitrary reference element. The elements listed above are increasingly more reactive, while the elements listed below hydrogen become increasingly inert with less tendency to ionize or go into solution.

The electromotive force series can be very important in predicting the corrosion of an element in a given environment. A simple rule used for such prediction states that: *In any electrochemical reaction, the more negative half-cell tends to be oxidized and the more positive half-cell tends to be reduced. A half-cell is either the anode plus the electrolyte or the cathode plus the electrolyte.* This means that in comparing two metals, the one with the more negative oxidation reduction potential will tend to go into solution and become an anode, while the more positive metal (or the less negative metal) will tend to be the cathode where hydrogen can be reduced from the ionic form to the atomic form.

The electromotive force series is especially important in galvanic corrosion prediction. This is because the metals which are most negative will corrode in comparison to another metal which is less negative. Thus, in the case of iron and zinc, it could be determined from the EMF series that the zinc would ionize and go into solution or become the anode as compared to the iron which would be the cathode. Similarly, lead would be expected to go into solution if it were in contact with silver in an electrolyte solution. Also, the farther apart the two metals are in the series, the greater the potential difference between the two and the greater the corrosion that will take place on the more negative metal. Aluminum, for example, rapidly corrodes when in contact with copper.

The electromotive force series actually applies only for metals in solution of their own salts. In other electrolytes, *e.g.*, seawater, performances may vary. The general relationships of the metals in the EMF series, however, still hold. Table 2.5, which is a galvanic series of metals and alloys in seawater, serves as an example. While many of the alloys are not listed in the previous series (Table 2.4), and the potentials are different since the metals are in a different electrolyte, the general orientation of metals in the two lists are similar.

TABLE 2.4 — Electromotive Force Series

Electrode Reaction	Standard Electrode Potential E (Volts), 25 C
K = K$^+$ + e−	− 2.922
Ca = Ca^{++} + 2e−	− 2.87
Na = Na$^+$ + e−	− 2.712
Mg = Mg^{++} + 2e−	− 2.34
Be = Be^{++} + 2e−	− 1.70
Al = Al^{+++} + 3e−	− 1.67
Mn = Mn^{++} + 2e−	− 1.05
Zn = Zn^{++} + 2e−	− 0.762
Cr = Cr^{+++} + 3e−	− 0.71
Ga = Ga^{+++} + 3e−	− 0.52
Fe = Fe^{++} + 2e−	− 0.440
Cd = Cd^{++} + 2e−	− 0.402
In = In^{+++} + 3e−	− 0.340
Tl = Tl$^+$ + e−	− 0.336
Co = Co^{++} + 2e−	− 0.277
Ni = Ni^{++} + 2e−	− 0.250
Sn = Sn^{++} + 2e−	− 0.136
Pb = Pb^{++} + 2e−	− 0.126
H$_2$ = 2H$^+$ + 2e−	**0.000**
Cu = Cu^{++} + 2e−	0.345
Cu = Cu$^+$ + e−	0.522
2Hg = Hg$_2$$^{++}$ + 2e−	0.799
Ag = Ag$^+$ + e−	0.800
Pd = Pd^{++} + 2e−	0.83
Hg = Hg^{++} + 2e−	0.854
Pt = Pt^{++} + 2e−	ca1.2
Au = Au^{+++} + 3e−	1.42
Au = Au$^+$ + e−	1.68

[SOURCE: Encyclopedia of Chemistry, Clark and Hawley, Electrochemistry, Reinhold Publishing Co., p. 338, 1957.]

Galvanic Couples

In any galvanic couple, the metal near the top of the galvanic series will be the anode and will ionize and go into the solution or corrode, while the one closer to the bottom of the list will be the cathode and receive galvanic protection. This is illustrated by the metal couples in the gelatin corrosion demonstration (Figure 2.13*). The speed at which galvanic corrosion takes place depends on the difference in electrical potential between the two metals. Metal which is coupled to another relatively close to it in the series will corrode more slowly than when it is coupled with a metal farther down in the series. Zinc coupled to aluminum in a sodium chloride solution will have a poten-

---

*See color insert.

TABLE 2.5 — Galvanic Series in Seawater Flowing at 13 FPS (Temperature, Approx. 25 C)

Material	Steady-State Electrode Potential, volts (Saturated Calomel Half-Cell)
Zinc	− 1.03
Aluminum 3003-(H)	− 0.79
Aluminum 6061-(T)	− 0.76
Cast Iron	− 0.61
Carbon Steel	− 0.61
Stainless Steel, Type 430, active	− 0.57
Stainless Steel, Type 304, active	− 0.53
Stainless Steel, Type 410, active	− 0.52
Naval Rolled Brass	− 0.40
Copper	− 0.36
Red Brass	− 0.33
Bronze, Composition G	− 0.31
Admiralty Brass	− 0.29
90Cu10Ni, 0.82Fe	− 0.28
70Cu30Ni, 0.47Fe	− 0.25
Stainless Steel, Type 430, passive	− 0.22
Bronze, Composition M	− 0.23
Nickel	− 0.20
Stainless Steel, Type 410, passive	− 0.15
Titanium[1]	− 0.15
Silver	− 0.13
Titanium[2]	− 0.10
Hastelloy C	− 0.08
Monel-400	− 0.08
Stainless Steel, Type 304, passive	− 0.08
Stainless Steel, Type 316, passive	− 0.05
Zirconium[3]	− 0.04
Platinum[3]	+ 0.15

[1]Prepared by power-metallurgy techniques, *i.e.,* sheath-compacted powder, hot rolled, sheath removed, cold rolled in air.
[2]Prepared by iodide process.
[3]From other sources.
[SOURCE: Fink, F. W., *et al.,* The Corrosion of Metals in Marine Environment, Battelle Memorial Institute, DMIC Report 254, Distributed by N.T.I.S. AD-712 585-S, pp. 7, 13 (1970).]

TABLE 2.6 — Examples of Galvanic Couples in Seawater

Metal A	Metal B	Comments
Couples That Usually Give Rise to Undesirable Results on One or Both Metals		
Magnesium	Low-Alloy Steel	Accelerated attack on A, danger of hydrogen damage on B.
Aluminum	Copper	Accelerated pitting on A. Ions from B attack A. Reduced corrosion on B may result in biofouling on B.
Bronze	Stainless Steel	Increased pitting on A.
Borderline, May Work, Uncertain		
Copper	Solder	Soldered joint may be attacked, but may have useful life.
Graphite	Titanium or Hastelloy C	
Monel-400	Type 316 SS	Both metals may pit.
Generally Compatible		
Titanium	Inconel 625	
Lead	Cupronickel	

[SOURCE: Fink, F. W., *et al.,* The Corrosion of Metals in Marine Environment, Battelle Memorial Inst., DMIC Report 254, Distributed by N.T.I.S. AD-712 585-S, pp. 7, 13 (1970).]

tial of over 700 millivolts. The greater the potential difference between the two metals, the greater will be the driving force for the more negative metal to corrode. Table 2.6 gives some practical examples of galvanic couples in seawater, showing those which have no chance of working together, as well as some which can be combined with reasonably good results.

The rate of galvanic corrosion is not only determined by the relative position of the metal in the EMF series, but also by the exposed area of the two metals. This is particularly important for metals like carbon steel where the corrosion rate is usually controlled by the total cathodic area available, thus the ratio between the area of the cathode and anode is very important. A small anode of steel coupled to a large cathode of copper both immersed in seawater will result in very rapid attack of the steel. This may have been the case where two copper plates were fastened together with steel rivets, as shown in Figure 2.14.

Galvanic couples are actually an integral part of modern daily activity. Aluminum, magnesium, and other more noble alloys are combined in the construction of aircraft; aluminum deck houses are used on steel ships; and steel fittings are often incorporated into copper pipelines

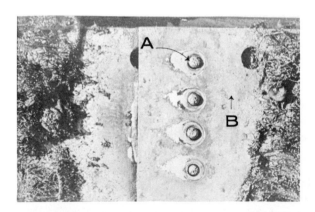

FIGURE 2.14 — A large cathode (B-copper) coupled to a small anode (A-steel) shows intense attack on the steel rivets (A), with little or no corrosion on the copper (B). (SOURCE: Greene, Norbert D., NACE Basic Corrosion Course, Chapter 3, Corrosion Related Chemistry and Electrochemistry, National Association of Corrosion Engineers, Houston, TX, p. 3-14, 1971.)

in the construction of household piping systems. Such couples are both necessary and useful. In order to control or prevent the accelerated corrosion attack which is perpetrated by such galvanic couples, precautions should be taken with at least one of the coupled metals. One possibility is to break the electrical circuit by installing an insulating barrier at the junction of the two metals. This is often done in pipelines through the use of an insulating flange or an insulating coupling. A plastic sheet may also be placed between the two dissimilar metal plates.

If it is not possible to isolate the two metallic surfaces, a break in the electrical conductivity of the electrolyte can be achieved by completely coating both metals. In this case, it is important for the applied coating to thoroughly cover both surfaces, particularly the junction between the two. If it is not feasible to coat both metals, the cathodic member of the couple should be covered with a nonconductive protective coating. By reducing or completely coating the cathodic area, the corrosion of the anode is controlled. Never, under any circumstances, should the anode alone be coated, since any defect or holiday in the coating would then create a small anode and a very large cathode which would result in catastrophic corrosion at the break in the coating.

Corrosion damage resulting from two dissimilar metals in immersion conditions can be considerable even if the two metals are a long distance apart. A ship's hull is a common example of this since the bronze propeller is strongly cathodic to the hull. This aggravates the corrosion at any coating damage, even if it is some distance towards the bow. Such corrosion also occurs on magnesium, zinc, or aluminum anodes hung between the legs of offshore platforms where the actual contact distance between the two metals could be 100 feet or more. On the other hand, where two dissimilar metals are in contact in the atmosphere, galvanic corrosion is confined to a small distance of usually only a fraction of an inch between the junction of the two metals.

Cathodic Protection

Galvanic coupling can also aid in the prevention of corrosion through its role in what is known as *cathodic protection*. In this method, metals with more negative potentials, *i.e.,* higher in the electromotive series, are used to protect metals farther down in the series or the less negative ones. Examples of the use of cathodic protection abound and are generally found wherever metals are buried or immersed. Magnesium, zinc, or aluminum, for example, are commonly coupled to steel structures to provide an excess of electrons on the steel surface, which in turn prevents any of the iron from going into solution as an ion.

Automotive construction necessitates the use of many different metals and alloys for cathodic protection. Leonard C. Rowe, in his paper on automotive engineering design, provides a good list of recommendations for galvanic corrosion prevention. These are applicable to most situations where different metals are combined in a structure:

1. Avoid the use of combinations of metals that have potentials that are widely separated in the galvanic series.
2. Avoid those combinations where the area of the anodic metal is small compared with that of the cathodic metal. Use metals for rivets, bolts and fasteners that are cathodic to the surrounding metal.

3. Insulate joints of dissimilar metals when possible; even paint or plastic coatings will be helpful.
4. Paint or coat all surfaces when possible. Avoid painting the anodic metal only, because corrosion may be accelerated at imperfections or breaks in the coating.
5. Seal faying [close-fit] surfaces.
6. Apply metallic coatings to reduce the potential difference between dissimilar metals.
7. Avoid threaded connections when dissimilar metals are used.[7]

Oxygen Concentration Cells

Other types of corrosion cells may also develop under immersion conditions. One often very destructive type is the oxygen concentration cell. As was previously discussed, oxygen is an important element in the corrosion process, particularly in depolarizing the cathode and thus initiating rapid corrosion at the anode.

An *oxygen concentration cell* commonly develops where two steel plates overlap. Bolted tanks, for example, used particularly in the oil industry to store water and other corrosive solutions, may be made with overlapped plates. Similar conditions conducive to oxygen concentration cell development exist under cocked rivets, washers slightly loosened from the surface, and even on surfaces under a loose paint film like that often found on the bottom of a ship (Figure 2.15). In fact, wherever a crevice exists in immersion conditions with ample oxygen in the solution, the crevice will corrode rapidly. As shown schematically in Figure 2.16, the area outside the crevice forms a large cathode, with the oxygen depolarizing the area and making it very active. A relatively small anode is formed under the crevice, causing the metal to go into solution rapidly.

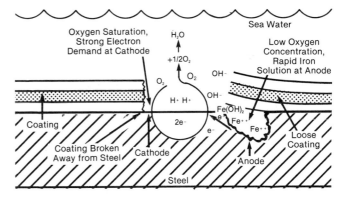

FIGURE 2.15 — Oxygen concentration cell found on a ship bottom.

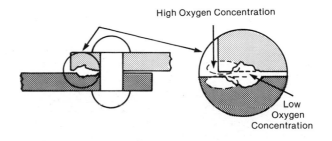

FIGURE 2.16 — Schematic of oxygen concentration cell formed in a crevice.

Similar conditons even exist in tanks or areas where trash accumulates. A pile of sand on a metal surface will likewise create a conditon where the metal ions go into solution under the sand because of the greater oxygen content in the solution surrounding it. The accumulation of mud or sand along with some corrosion on the bottom of a water tank can create an oxygen concentration cell. In this case, the sidewalls act as a large cathode because of their easy access to oxygen in the solution.

Metal Ion Concentration Cell

Corrosion may also be influenced by the concentration of the metal ion in solution. Thus, if the concentration of metal ions corroding from a metal in one place is greater than at another point, the metal at the point of highest metal ion concentration will become the cathode. The area of the metal in contact with the lower concentration will then become the anode. This is a logical development since the area with the greatest metal ion concentration would have less tendency to ionize or go into solution than in areas with less metal ion concentration (Figure 2.17). Evans, in his work, *The Corrosion and Oxidation of Metals,* discusses the concept further:

> The value of the potential of the metal against a solution of one of its salts must depend on the concentration of that solution since if the balance is $M \longleftrightarrow M^{++} + 2e^-$, any increased ion concentration will increase the right to left reaction while leaving the left to right reaction unchanged.

In reference to "Metal Ion Concentration Cells," he continues,

> Local concentration differences can play a part in determining the corrosion patterns of some metals. This is more important on lead than on iron. The intensified attack sometimes met with lead pipes buried in chalky soils may be connected with the removal of the lead ion as basic lead carbonate in places where the lumps of chalk press against the lead. These places would become anodic to the rest going to the concentration cells setup.[8]

Low Metal Ion Concentration

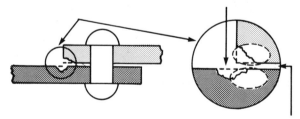

High Metal Ion Concentration

FIGURE 2.17 — Metal ion concentration cell.

A classic experiment in this area was conducted by the Francis L. LaQue Laboratories of the International Nickel Company with a spinning copper disc in seawater. Results indicated that the metal close to the center of the disc moves slower than at the periphery of the disc. This allows the metal ions to accumulate in and under films that develop near the center of the disc, while the ones on the perimeter are rapidly swept away by the rapid movement of the metal. Severe corrosion thus results in the region of the highest velocity, and therefore of the least ion concentration (Figure 2.18). Figure 2.19, which is a photograph of an Admiralty brass disc after rotation in seawater, attests to the original results. Note the heavy metal loss on the outer edge of the disc.

This may also be the case with some pump impellers where the flow of liquid is somewhat less in the center of the impeller compared to its high velocity on the outer surface. While a metal ion concentration cell could exist under these conditions, there are also a number of other factors involved in the operation of impellers.

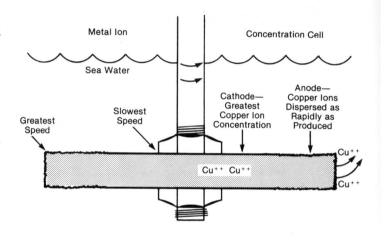

FIGURE 2.18 — Spinning copper disc in seawater.

FIGURE 2.19 — Brass disc after spinning in seawater. (SOURCE: NACE Basic Corrosion Course, F. L. LaQue, Chapter 2, Introduction to Corrosion, National Association of Corrosion Engineers, Houston, TX, p. 2-12, 1971.)

While the spinning disc example does not involve a crevice, most metal ion concentration cells exist in crevices of one type or another (as do most oxygen concentration cells, which are much more prevalent). Some practical suggestions for the prevention of crevice corrosion are listed by Rowe in his previously-mentioned work on automotive engineering design.

Corrosion Prevention by Protective Coatings

1. Use welded joints in preference to bolted or riveted joints.
2. Caulk or seal unavoidable crevices, using durable and noncorrosive materials.
3. Minimize the contact between metal and plastics, fabrics, debris, etc.
4. Avoid contact with materials that are known to contain corrosive elements or that are hygroscopic because they can accelerate the cell effect.
5. Avoid sharp corners, ledges and pockets where debris can accumulate.
6. Where a crevice is inevitable, thoroughly coat both surfaces before they are joined.[7]

Chemical Corrosion

Chemical corrosion is a factor in almost every type of production, ranging from the canning of fruits and vegetables and the bottling of wine, to the manufacture of heavy chemicals such as sodium hydroxide and sulfuric acid. Even many household chemicals can be extremely corrosive when removed from their original containers and allowed to contact only partially protected iron. Certain chemicals, however, may actually protect iron from corrosion. A general rule of thumb in determining the difference is that as chemicals become more acidic, their tendency to corrode iron and other metals increases. Conversely, as chemicals become more alkaline, they are less likely to corrode iron. This tendency is illustrated by the following equation:

$$\text{Acid} = \begin{array}{c} H^+ \\ \leftarrow \\ Cl^- \end{array} \begin{array}{c} \text{Increased } H^+ \\ \text{Corrosion } OH^- \\ \text{of Iron} \end{array} \begin{array}{c} H^+ \\ \text{Water} \end{array} \begin{array}{c} \text{Decreased } Na^+ \\ \rightarrow \\ \text{Corrosion } OH^- \\ \text{of Iron} \end{array} = \text{Alkaline}$$

This does not mean, however, that a 100% ionized strong alkali solution (*e.g.,* sodium hydroxide) cannot be corrosive under certain conditions.

pH

Another method of explaining the corrosive characteristics of chemicals is in reference to the pH scale. Technically, *pH* is the negative logarithm of the hydrogen ion concentration in a solution. In this case, water, which consists equally of both hydrogen and hydroxyl ions, has a pH of 7. The relative amounts of these two ions determine whether any solution has the familiar sour taste, the ability to turn blue litmus paper red, and other acid characteristics; or whether it has a bitter taste, soapy feel, the ability to turn red litmus paper blue, and other alkaline characteristics; or whether it is chemically neutral, neither alkaline nor acid. If hydrogen ions are in excess, the solution reacts as an acid; if hydroxyl ions are in excess, the solution reacts as an alkali; and if both ions are present in equal amounts, the solution is neutral. More specifically, *acids* are identified as substances which, when dissolved in water, increase hydrogen ion concentration; *alkalies* as those substances which, when dissolved in water, increase hydroxyl ion concentration.

The pH varies depending on the chemicals dissolved in the water. Pure water has a pH of 7. Acids have a pH of less than 7 down to zero, while alkalies range in pH from 7 to 14. Figure 2.20 indicates the reaction of various hydrogen ion concentrations on mild steel. As the acid concentration becomes stronger, reactivity on iron be-

comes greater until massive amounts of iron can be dissolved in a short time (*e.g.,* acid pickling of steel). Note that hydrogen evolution starts at an approximate pH of 4 and increases as the pH moves towards 1.

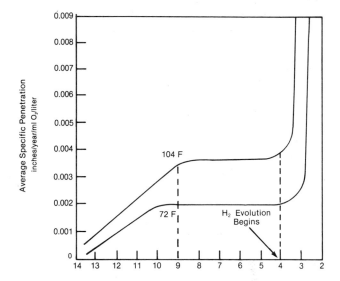

FIGURE 2.20 — Effect of pH on the corrosion of mild steel. [SOURCE: U.S. Dept. of Commerce, Civil Engineering Corrosion Control, Corrosion Control—General, Vol. 1, Dist. by NTIS, AD/A-004082, pp. 16, 33, Jan. (1975).]

pH is not a factor in chemical plant corrosion alone; rather it is an important measure of chemical activity which is an integral part of our everyday lives. pH is a factor in everything from the food we eat and the stomach acid which helps digest it, to the chemical detergents used to wash dishes. Some of the most commonly encountered chemicals and substances are listed according to pH in Table 2.7. Note that most of the familiar substances are on the acid side of water, which means they tend to be more or less corrosive. Some solutions, however, like sodium hydroxide, can be stored in common steel tanks for several years without appreciable corrosion. Some of the stronger organic alkalies, such as amines, can even be used in various processes as corrosion inhibitors.

Acids

Chemicals can be divided into a number of groups, each with its own corrosive characteristics. The first and the most corrosive chemical groups consist of acids. Concentrated and highly ionized acids can corrode metal without the presence of oxygen. Some such acids, when in contact with iron, produce hydrogen as a gas, as shown below:

$$2HCl \, (H^+Cl^-) \text{ hydrochloric acid}$$

$$+ \, Fe \rightarrow FeCl_2 + H_2 \tag{2.8}$$

$$H_2SO_4 \, (2H^+ + SO_4^-) \text{ sulfuric acid}$$

$$+ \, Fe \rightarrow FeSO_4 + H_2 \tag{2.9}$$

$$2HNO_3 (H^+ + NO_3^-) \text{ nitric acid}$$

$$+ Fe \rightarrow Fe(NO_3)_2 + H_2 \qquad (2.10)$$

Some acids, such as sulfuric and phosphoric acid, are nonvolatile, while others, such as nitric and hydrochloric acids are volatile and therefore much more corrosive. Nitric and hydrochloric acids, in fact, are considered two of the most corrosive acids commonly encountered. This is because volatile acids will evaporate, move through the atmosphere as a gas until they contact iron or steel, and then, if there is some moisture present, will react rapidly.

Thus, a plant manufacturing hydrochloric acid, or one which produces it as a process byproduct, is very difficult to protect from corrosion. Only extremely well-applied, strongly acid-resistant materials or coatings can protect the piping and equipment in such plants from rapid failure. Coatings are generally not recommended for immersion in strong acid solutions since any imperfection in the manually applied coating would invite rapid corrosion and failure of the basic structure. Volatile acids (gases in solution) easily penetrate many organic materials, and coatings are no exception. Heavy plastic sheet or rubber linings are the only surface materials which provide satisfactory protection against such strong acid solutions.

In addition to volatile acids, there are also oxidizing acids which can be even more aggressive, particularly towards any material which has some tendency to oxidize. Sulfuric acid in strong concentrations (above 50%) is

TABLE 2.7 — Relative pH of Common Chemicals and Substances

pH 14	Strong alkali (NA$^+$ + O$\overline{H}$) Sodium Hydroxide
13	0.1N NAOH
12	0.01N NAOH
11	0.1N Ammonia
10	Saturated Magnesia
9	0.1N Borax, sodium borate (washing detergent)
8	Sodium Bicarbonate (urine)
pH 7	Water (H$^+$ + O$\overline{H}$) (milk, blood plasma)
6	Saliva (tuna, beans)
5	0.1N Boric acid (turnips, sweet potatoes)
4	Saturated Carbonic Acid (soda water, tomatoes)
3	0.1N Acetic Acid (wine, grapefruit, apples)
2	0.01NHCL (lemons)
1	0.1NHCL (limes)
pH 0	Strong Acid (H$^+$ + CL$^-$) (stomach acid)

highly oxidizing with most organic materials. Furan resin cements, for example, are used to coat pickling tanks for steel since sulfuric acid is used as a pickling agent. While the furan cement lasts four years under hot pickling conditions, it is not satisfactory for concentrations of sulfuric acid above 50%, as it is sensitive to oxidizing conditions. Nitric acid, in addition to being a volatile acid, is also an oxidizing acid in strong concentration. Perchloric acid is so oxidizing that its combination with organic materials may become explosive.

Organic acids fall into another group which has highly individualized characteristics. One of the most common acids in this group is acetic acid. It is volatile, can rapidly attack many metals, and can also act as a solvent for some organic coatings. Naphthenic acid is an organic acid common to the oil industry. Many stove and heating oils contain enough naphthenic acid to make corrosion of its containers a problem. Naphthenic acid, however, consists of rather large molecules so that its corrosive effect can easily be checked through the use of coatings. Highly acidic petroleum products should not be contained in galvanized or inorganic zinc coated tanks, as they slowly react corrosively with zinc. Other examples of organic acids are butyric, stearic, and linoleic acid, which are used in the manufacture of paint and soap.

Salts

Another class of compounds important to a discussion of corrosion is that of salts. Salt is classified as either acid, neutral, or alkaline. A salt is formed by the reaction of an acid and an alkali. An acid salt (*e.g.*, ferrous sulfate) results from the action of a strong acid (*e.g.*, sulfuric acid) with a mild alkali (*e.g.*, ferrous hydroxide). This type of material is relatively soluble and can act as a salt as well as an acid. Neutral salts are a combination of strong (or weak) acids and strong (or weak) alkalies, such as hydrochloric acid and sodium hydroxide. This particular combination yields common table salt, which is a primary corrosive agent encountered in the marine industry as well as in many other industries. The alkaline salts are a combination of weak acids (*e.g.*, acetic acid) and strong alkalies (*e.g.*, sodium hydroxide). These materials are thus on the alkaline side, *i.e.*, a pH of 7 or greater, and are generally less corrosive than either acid or neutral salts. Nevertheless, many are strongly ionized and act as an electrolyte which can therefore create strong corrosion cells.

Many salts, such as sodium or potassium chloride, and sodium nitrate or sulfate are strongly ionized. Others range from being strongly ionized to being completely insoluble (*e.g.*, silver chloride). Generally, as salts become less soluble and less ionized, they also become less corrosive.

Alkalies

Examples of strong alkalies, which are at the upper end of the pH scale and also may be strongly ionized, are sodium and potassium hydroxide. The amount of hydrogen ion present in these alkalies decreases as pH increases so that there is less hydrogen to react with excess electrons. This means that hydrogen is not evolved in the corrosion reaction. Oxygen, however, is normally reduced in accordance with

$$O_2 + 2H_2O + 4e^- \rightarrow 4OH^- \qquad (2.11)$$

and is thus more important to corrosion in alkaline solutions.

Oxidizing Salts

There are a number of alkaline salts which are also oxidizing agents and therefore extremely corrosive. Most common among them is sodium hypochlorite (NaOCl). Chlorine gas is added to sodium hydroxide during manufacturing so that the resulting solution is strongly alkaline. On the other hand, this material is unstable and slowly disintegrates into a nascent oxygen and sodium chloride. This makes it a strong corrosive agent. Any coating which is to come in contact with hypochlorite solution must be resistant to the penetration of nascent oxygen, as well as resistant to strong oxidizing conditions. Any imperfections in a coating will be rapidly corroded, with tubercles of rust forming at the break and deep pits forming underneath. Even dilute sodium hypochlorite solution can quickly attack coating weaknesses. Similar oxidizing salts include calcium hypochlorite, sodium perborate, and sodium perchlorate. It is important to note that even though most oxidizing salts are alkaline, they still can be highly corrosive.

Sulfides

Sulfides constitute another important factor in chemical corrosion. The presence of hydrogen sulfide, for instance, which is a gas and an acid sulfide, can be a major contributing factor to corrosion failure that often goes unrecognized. Other common sulfides include salts of hydrogen sulfide, *i.e.,* ammonium sulfide, calcium sulfide, and sodium sulfide. There are also many complex organic sulfides (*e.g.,* carbon disulfide), particularly in the petroleum industry.

Sulfide reactions are different from the other corrosion reactions in that they can take place even under relatively dry conditions. Hydrogen sulfide, for instance, reacts directly with the metal and does not require an electrolyte for the corrosion reaction to take place. Hydrogen sulfide not only reacts directly with iron, but also with some iron compounds to increase the corrosion rate, as is evident in the following reactions:

$$H_2S + Fe \rightarrow FeS + H_2 \qquad (2.12)$$

$$H_2S + FeCO_3 \rightarrow FeS + CO_2 + H_2O \qquad (2.13)$$

$$H_2S + Fe(OH)_2 \rightarrow FeS + 2H_2O. \qquad (2.14)$$

Figure 2.21 of an oil field production tank is an example of corrosion under high sulfide conditions. This production tank was in service 2 years without being coated. Not only was the roof structure of the tank completely corroded away, but uncoated tank exteriors, roof vents, steel pipes, and other structures in the same area were also severely corroded, even though the sulfide was not as concentrated as it was in the tank's interior.

FIGURE 2.21 — Sulfide attack on a steel petroleum production tank.

Hydrogen sulfide reacts with most other metals in a similar way, especially with copper and silver which turn black when sulfide is in the area (*e.g.,* workers in sewers or other areas where sulfides are present often find that the coins in their pockets have turned black). Iron sulfide is cathodic to iron so that where it forms on an iron surface it can act as a massive cathode, providing that an electrolyte such as seawater is present. This is often a contributing factor in the deep pitting of sour crude tankers. Iron sulfide is also very reactive with oxygen. When oxygen contacts iron sulfide, it reacts according to the following equation to form magnetic iron oxide:

$$6FeS + O_2 + 6H_2O \rightarrow 2Fe_3O_4 + 6H_2S. \qquad (2.15)$$

Hydrogen sulfide released in this reaction can in turn react with any additional iron in the area.

Sulfides are prevalent in many industries, including the rayon industry which uses carbon disulfide and the petroleum industry which encounters hydrogen sulfide in both natural gas and crude oil. Sour crudes, in particular, pose hydrogen sulfide problems that affect everything from the well head, gathering lines, and field production tanks, to the refinery structures and equipment.

The sewage industry also must contend with hydrogen sulfide, which develops as a result of anaerobic bacterial action below the surface of the sewage liquid. The anaerobic bacteria react with sulfates in the water and organic sulfides in the sewage to form hydrogen sulfide gas. The marine industry is also affected by hydrogen sulfide produced by decaying slimes and marine life. While normal marine corrosion may mask the formation of sulfides, they nevertheless significantly contribute to such corrosion and therefore must be considered before proper protection can be achieved.

Displacement Corrosion

Another type of chemical corrosion occurs where metals are attacked in solution that do not contain oxygen or acids. Solutions containing copper, for instance, come into contact with iron. Since copper is more positive in the electromotive series than iron, the copper ion takes up electrons from the iron and precipitates metallic copper, while the iron goes into solution as iron ions. This reaction is called *double displacement reaction* (iron from metal into solution, copper from solution to the metal) and is given in the following diagrams:

$$Cu^{++} + Fe \rightarrow Cu + Fe^{++}$$

or
$$CuSO_4 \text{ (solution)} + Fe \text{ (metal)} \rightarrow$$

$$Cu \text{ (metal)} + FeSO_4 \text{ (solution)}. \qquad (2.16)$$

Copper will react in a similar way with zinc, magnesium, or aluminum. All of these metals must be protected by the proper coatings if there is even a possibility of copper ion being present. The use of uncoated aluminum pipe or tanks in a copper-piped domestic or industrial water system is dangerous. Enough copper ion will be picked up by the water so that the aluminum downstream of the copper pipe will actively corrode. Not only will the copper displace the aluminum as described, but a galvanic cell will then form, accelerating the already precipitated corrosion.

The reduction of ferric salts is another chemical reaction which takes place without the necessity of oxygen. When ferric chloride, for example, comes in contact with iron, the following reaction occurs:

$$Fe + 2FeCl_3 \rightarrow 3FeCl_2. \qquad (2.17)$$

As the iron goes into solution, it relinquishes an electron to the ferric chloride which then becomes reduced to ferrous chloride. Ferric chloride will react similarly with other metals higher in the electromotive series than iron (*e.g.*, zinc) thus necessitating their protection.

Concrete Corrosion

Chemical reactions involving concrete also should be considered since concrete follows metal as a principal building material. Concrete is essentially a hydrated calcium aluminum silicate which is strongly alkaline. Moist concrete can develop a pH of up to 13, which is one reason it is considered a good protective media for steel structures.

All acid materials react readily with concrete, due primarily to the calcium in the complex cement molecule. Cement will even react with an acid as weak as pure water with very low dissolved solids. Concrete will only stand up properly without any solution taking place if the water is somewhat hard. Thus, the acidic materials that must be considered when working with cement range all the way from pure water, fruit juices, and carbonated water, to strong acids such as sulfuric acid. Even extremely dilute sulfuric acid will rapidly react with cement, causing effervescence and precipitation of the insoluble calcium sulfate.

This latter reaction is the one which takes place in sewers where sulfuric acid develops from the bacteria at and above the water line. It has been measured in concentrations as high as 10%, disintegrating concrete at a rate as great as two inches per year under strongly corrosive

sewage conditions. The reactions which create corrosive conditions such as these are shown in Figure 2.22.

Thus, wherever concrete is used as a structural material that may be exposed to acidic compounds, it must be properly coated to assure the preservation of the structure. In many ways, this process can be more complex than that of coating metals (Chapter 11).

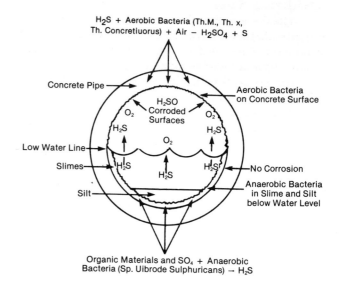

FIGURE 2.22 — The H_2S corrosion mechanism in sewers. [SOURCE: Munger, C. G., Sulfides—Their Effect on Coatings and Substrates, Materials Performance, January (1978).]

Rust

Mill Scale

Mill Scale, or blue scale as it is sometimes called, forms on steel as it is hot rolled, and varies according to the type of operation and rolling temperature. It is not a complex subject, but since the majority of structural steel is hot rolled, the quantity of surface contaminated with mill scale is important.

In general, mill scale is magnetic and contains three layers of iron oxide, although the boundaries between the oxides are not particularly sharp. The thin outer layer of mill scale is essentially ferric oxide (Fe_2O_3), which is relatively stable and does not easily react. The layer closest to the steel surface, and sometimes intermingled with the steel's surface crystalline structure, is ferrous oxide (FeO). This is an unstable substance which is easily oxidized to ferric iron, resulting in a chemical change to ferric oxide. This process, accompanied by an increase in volume, results in loosening the intact mill scale, particularly during weathering or where moisture is present.

The intermediate layer of magnetic oxide is best represented by the chemical formula Fe_3O_4. The actual thickness of mill scale on structural steel, which depends upon rolling conditons, varies from about 0.002 to 0.020 inches, and consists mainly of the magnetic oxide Fe_3O_4 and the FeO layer. Much of the mill scale formed at high initial rolling temperatures is knocked off in subsequent rolling. Thus, it is the oxide that forms on the steel after

Corrosion Prevention by Protective Coatings

rolling but while the steel is still hot that remains intact on the surface.

Mill scale is strongly cathodic to bare steel, which can be demonstrated through the use of gelating indicator solution. Wherever there is a break in the mill scale on the steel panel, the iron goes rapidly into solution. These areas form blue anodes. The remainder of the surface covered by tight mill scale remains a pink color and forms a massive cathode (Figure 2.23*). This same action is schematically shown in Figure 2.24, which indicates that a strong electric current is established between the bare steel (anode) and the mill scale (cathode). When this is coupled with the large areas of mill scale on most hot rolled plates or shapes (area relationship is indicated in Figure 2.23), a massive cathode and small anode relationship is created which causes rapid corrosion to the base steel. The difference in potential between the mill scale and steel can amount to 0.2 to 0.3 volts, which is close to the strength of many galvanic couples such as copper or bronze and steel.

That mill scale can create a condition of very rapid corrosion was demonstrated during World War II when ships were turned out at a very rapid rate. Originally, mill scale was not removed from the bottom hull plates due to the time and expense involved. However, when ships were placed in seawater, deep pitting resulted along welds and in any area where the mill scale was broken. In some cases, penetration of the hull actually took place at the outfitting deck. This problem was overcome by pickling or blasting all of the hull plates before construction.

A problem is also created when mill scale is painted over. The mill scale is then unstable, and when it comes in contact with moisture tends to release adhesion or pop from the metal surface. When painted over, this reaction results in loose areas of the coating with blistering or cracking in these areas. If a steel surface is to be used under corrosive conditions, mill scale must therefore be removed prior to coating in order to obtain a viable and long-lasting coating job.

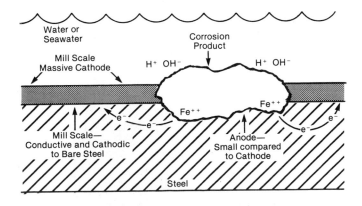

FIGURE 2.24 — Schematic illustration of the strong electric current established between bare steel (anode) and mill scale (cathode).

*See color insert.

Filiform Corrosion

An interesting and unique form of corrosion which can take place on either uncoated or coated metal surfaces is *filiform corrosion*. It can take place when metal surfaces are contaminated with fine solid particles and the surface is exposed to humid atmospheres. It also can take place underneath coatings of various types, either pigmented or clear. It is most dramatic when it occurs under clear lacquer films applied over cold rolled or polished steel surfaces. Figure 2.25 shows filiform corrosion starting at a scribe on a metal panel.

The cause of filiform corrosion is generally thought to be mild surface contamination caused by solid particles deposited from the atmosphere or particles that remain after the metal has been processed. Rozenfeld, in his work, *Atmospheric Corrosion of Metals,*[9] lists a number of sources and surface contaminants which have caused this type of corrosion.

Industrial Atmosphere: Iron oxide, coal dust, silica, ammonium sulfate, organic matter
Industrial and Sea Water: NaCl, $CaCl_2$, $CaSO_4$, $MgSO_4$
Polishing Operations: Carborundum
Prior Processing with Solutions: $NaNO_3$, chromic acid, sodium nitrite, KF
Soldering and Melting Operations: Chloride
Pigments: Iron oxide and zinc chromate
Corrosion Products: Iron sulfate, ferric chloride[9]

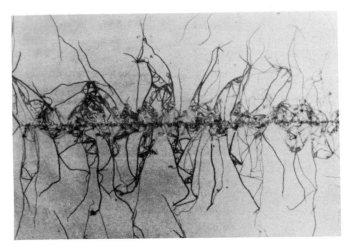

FIGURE 2.25 — Filiform corrosion starting at a scribe on a metal panel. (SOURCE: Pictorial Standard of Coating Defects, Filiform Corrosion, Figure 8-Dense B, Federation of Societies for Paint Technology, Philadelphia, PA, 1979.)

It is interesting to note that some of the solid particles are insoluble in water, but nevertheless cause filiform corrosion. It has been stated that salts which behave normally (rapid corrosion in presence of high humidity) cause only mild filiform corrosion, if any at all, while salts with abnormal behavior (high corrosion rates at low relative humidity) always induce filiform corrosion.[6]

The mechanism of filiform corrosion is not well-known; however, it apparently is similar to crevice corrosion. Sudbury, *et al.,*[6] indicate that the head of the filiment is the anode of an electrochemical cell. The filiment tube is filled with low pH electrolyte, and the resulting growth is due to an oxygen concentration difference between the anode and the cathodes (Figure 2.26).

As shown in Figures 2.25 and 2.26, the corrosion appears to have wormed its way across the surface underneath the coating. As indicated by the type of material which tends to cause filiform corrosion, it most often occurs where coatings are applied under a product finishing operation. Industrial and marine atmospheres can create this condition, as is indicated. Prevention of this type of corrosion, particularly under coated surfaces, involves making certain that the surface has been properly prepared and that the coating is applied over a thoroughly cleaned surface.

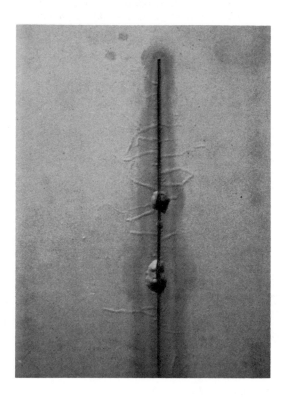

FIGURE 2.26 — Filimentous corrosion under a pigmented coating exposed in a marine environment.

Pitting Corrosion

Pitting is often classified as a separate type of corrosion. Although it may take a variety of forms and is prevalent under many different corrosive conditions, it is primarily an aggravated form of the usual corrosion process rather than a distinctive corrosion type. *Pitting corrosion* is, more specifically, a concentration of corrosion in one particular area so that the metal goes into solution more rapidly at that spot than at any other adjacent area. Oftentimes, it goes directly back to the area relationship of the anodes and cathodes with a relatively small anode and a many-times-larger cathode. This type of corrosion can occur where protective coatings are applied over metal and where there is a break in the coating so that the large coated area acts as a cathode, even though a very weak one.

Pitting corrosion may be caused by a number of

things, including mill scale, which is a very common form of pitting corrosion. When mill scale is placed in an electrolyte, any break in the mill scale surface becomes the anode; the remainder of the mill scale, which is usually many times larger than the break, becomes a very strong cathode. Galvanic action is also a cause of pitting corrosion.

Areas where a brass valve is incorporated into a steel or galvanized pipeline serve as good examples. The junction between the two areas is often badly pitted, and if the pipe is threaded, the thread in close contact with the brass valve rapidly pits, soon causing a leak. This not only occurs frequently in industry, but also in homes and on farms.

Oxygen concentration cells also can cause deep pitting, as might be expected from crevice type corrosion. Pitting also occurs on stainless and similar alloys exposed to marine life. In this case, the marine life dies after a period of time and, as with barnacles, will leave a shell on the surface. Due to the sulfides produced by the dying or dead animal, the oxide film on the metal or passivated surface is destroyed and an active pitting takes place underneath the fouling.

The deep pitting of tankers on the horizontal surfaces of cargo ballast tanks is a particularly aggravated type of pitting (*i.e.*, pits are deep and frequent) (Figure 2.27). In this case, they are caused by frequent changes of cargo and salt water which perpetrate an oxidation reduction corrosion cycle.

Graphitic corrosion, i.e., the preferential dissolution of iron which leaves a soft graphite residue, sometimes takes this form of attack (*i.e.*, pitting). On underground pipelines, corrosion cells may set on the cast iron surface in

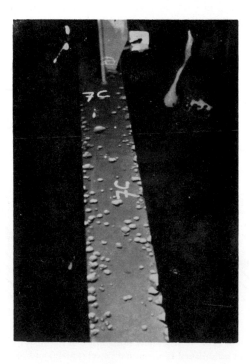

FIGURE 2.27 — Deep pitting in a crude oil tanker cargo ballast tank. Note the depth of the pits compared to the diameter. [SOURCE: Munger, C. G., Deep Pitting in Sour Crude Oil Tankers, Materials Performance, March (1976).]

Corrosion Prevention by Protective Coatings

such a way that the iron goes into solution, leaving the graphite, which is part of the cast iron, intact within the deep pit. The graphite in itself is cathodic to the iron, so that once the cycle starts, pitting occurs rapidly. The graphite in these pits is relatively soft and is easily cut with a knife. It usually proceeds uniformily inward from the surface, leaving a porous matrix. While there may be no outward indication of corrosion damage, in some instances, the corroding area will be covered by a large tubercle of rust. Graphitization occurs in salt waters, acidic waters, dilute acids, and soils (especially those containing sulfates and sulfate reducing bacteria). Coatings and/or cathodic protection can be used successfully to prevent graphitization.

Pitting can also take place under atmospheric conditions. The pitting, in this case, may be caused by breaks in a surface coating over a steel or other metal surface. The corrosion starts at the break and continues to undercut the coating, forming a rather heavy tubercle of hard rust or scale with the pit in the original metal underneath (Figure 2.28). These are common in the marine area as well as in various industries where strong corrosive conditions exist.

FIGURE 2.28 — Pitting in a marine atmosphere. Note the sharp edges on the large pitted area.

Atmospheric Corrosion

Atmospheric corrosion is undoubtedly the most widespread and, from a coating standpoint, the most important type of corrosion. There is more metal area exposed to atmospheric corrosion than to any other type. It is prevalent worldwide and exists not only in marine areas or in industry, but in many rural areas as well where high humidity and damp conditions exist. Table 2.8 gives relative corrosion rates for a number of areas around the world, ranging from those of practically no corrosion to severely corrosive areas.

It has been estimated that half of the total cost of corrosion protection in the United States is spent on protection against atmospheric corrosion. In some countries, the extent of this type of corrosion would be considerably less; on the other hand, there are also countries with primarily marine conditions where this ratio would be much higher than the 50% in the United States. Generally, the more arctic areas are comparatively less corrosive than those in temperate or tropical zones. This does not mean, however, that there are not isolated areas in any climate which differ from the norm.

The nature of atmospheric corrosion is described by Barton in his work, *Protection Against Atmospheric Corrosion,* as follows:

> Atmospheric corrosion is an electrochemical process which occurs in a limited amount of electrolyte. The electrolyte is neutral or slightly acidic (or under exceptional conditions slightly alkaline), and its properties are influenced chiefly by the chemical composition of the atmosphere and the properties of the corrosion products formed. The neutral or slightly acidic nature of the electrolyte and its variable presence on the corroding surface promote the formation of solid corrosion products on all metals which remain unpassivated for thermo-dynamic reasons. This is easily understood, since the solubility product of the reaction product is easily exceeded in the small volume of approximately neutral electrolyte, and so new phases are formed in the system. The properties of this reaction layer are not constant, however, since the temporary presence of the electrolyte layer plays an important role.[10]

TABLE 2.8 — Relative Corrosivity of Open-Hearth Steel Exposed at Various Locations

Location	Relative Corrosivity
Khartoum, Egypt	1
Ablsco, No. Sweden	3
Singapore, Malaya	9
Daytona Beach, FL (inland)	11
State College, PA	25
So. Bend, PA	29
Miraflores, Canal Zone, Panama	31
Kure Beach, NC (800 ft. from ocean)	38
Sandy Hook, NJ	50
Kearny, NJ	52
Vandegrift, PA	56
Pittsburgh, PA	65
Frodingham, UK	100
Daytona Beach, FL (near ocean)	138
Kure Beach, NC (80 ft. from ocean)	475

[SOURCE: Corrosion Basics—An Introduction, Chapter Four, Atmospheric Corrosion, National Association of Corrosion Engineers, Houston, TX, p. 228, 1984.

Components of the Atmosphere

Even though nitrogen and inert gases are the atmosphere's major elements, they do not react with metal surfaces under ordinary circumstances. Nitrogen can be oxidized during electrical storms and is a source of some soluble nitrogen for the soil. The amount involved, however, is so small that it contributes very little to the corrosion of the atmosphere. Oxygen is the second largest component of air and amounts to approximately 20% of the atmosphere. As discussed previously, oxygen is a very reactive material which has a substantial effect on the corrosive character of the atmosphere, as well as on the general corrosion process.

Water is the third most common atmospheric component and may be one of the most important in terms of corrosion. Water exists as ice and snow in its solid form, as rain or condensation in its liquid form, and as humidity in its vapor form. In each of these states it performs a specific role in the corrosion process. In the atmosphere, water is primarily in its vapor form; that is, its individual molecules serve as part of the total atmospheric pressure. The content of moisture vapor in the air varies from almost zero to a point of saturation, and may vary over this entire range on a daily basis.

Water as liquid is an important element in electrochemical corrosion since it is the principal ingredient of the electrolyte that is necessary to carry on the corrosion process. It dissolves various materials from the solid matter and gases that are present in the atmosphere. The pH of water from the atmosphere can fall as low as 3, and it is always saturated with oxygen. When water is condensed or falls on a surface, any material on that surface may be dissolved in rather large amounts to form a relatively concentrated solution. This creates a very strong electrolyte.

The length of time that water is precipitated on the surface, either by rain or condensation, is a major factor in the corrosiveness of the atmosphere. A dry atmosphere, where water is only on the surface of the metal for a short period of time during any one day, will be relatively corrosion-free. On the other hand, an atmosphere where moisture condenses and is held on the surface for many hours at one period can be a very corrosive atmosphere. Such atmospheres are common in marine areas where humidities remain at relatively high stages for most of any daily period. Atmospheric corrosion of metals, then, progresses only in periods where there is a surface electrolyte present. The rate of corrosion during such periods is related to the corrosion activity of the surface electrolyte and the nature of the metal.

Relative Humidity

Moisture in the atmosphere is most commonly measured as its relative humidity. The *relative humidity* is the ratio of the absolute humidity to the saturation value and is expressed as a percentage. Table 2.9 compares the absolute atmospheric humidity as expressed in grams of water per cubic meter at different temperatures and different relative humidities. This compares relative humidity with the actual amount of water vapor in a cubic meter of air. As the table indicates, the actual content of water in the air changes rapidly with temperature.

Temperature and Moisture

Extreme temperatures are of minimal significance from a corrosion standpoint. At temperatures below freezing, water is in its solid form and therefore does not act as a good electrolyte. At the other extreme, high atmospheric temperatures do not allow the moisture to condense and form a film on the surface. Atmospheric corrosion generally does not proceed rapidly, if at all, at temperatures above 25 C (77 F).

TABLE 2.9 — Absolute Atmospheric Humidities at Different Temperatures and Different Relative Humidities (expressed as g. water vapor/M³)

Temperature C	Relative Humidity (%)									
	10	20	30	40	50	60	70	80	90	100
0	0.49	0.98	1.47	1.96	2.45	2.94	3.43	3.92	4.4	4.9
1	0.52	1.04	1.56	2.08	2.60	3.12	3.64	4.16	4.7	5.2
2	0.56	1.12	1.68	2.24	2.80	3.36	2.92	4.48	5.0	5.6
3	0.60	1.20	1.80	2.40	3.00	3.60	4.20	4.80	5.4	6.0
4	0.64	1.28	1.91	2.56	3.20	3.84	4.48	5.12	5.8	6.4
5	0.68	1.36	2.04	2.72	3.40	4.08	4.76	5.44	6.1	6.8
6	0.73	1.46	2.19	2.92	3.63	4.08	5.11	5.84	6.1	7.3
7	0.77	1.54	2.31	3.08	3.85	4.62	5.39	6.16	6.9	7.7
8	0.83	1.66	2.49	3.32	4.15	4.98	5.81	6.64	7.5	8.3
9	0.88	1.76	2.64	3.52	4.40	5.28	6.16	7.04	7.9	8.8
10	0.94	1.87	2.82	3.76	4.70	5.64	6.58	7.52	8.5	9.4
11	0.99	1.99	2.98	3.98	4.97	5.97	6.96	7.96	8.9	9.9
12	1.06	2.12	3.18	4.24	5.30	6.36	7.42	8.48	9.5	10.6
13	1.13	2.26	3.39	4.52	5.65	6.78	7.91	9.04	10.2	11.3
14	1.20	2.40	3.60	4.80	6.00	7.30	8.40	9.52	10.8	12.0
15	1.28	2.56	3.84	5.12	6.40	7.68	8.96	10-20	11.5	12.8
16	1.35	2.72	4.08	5.44	6.80	8.16	9.52	10.90	12.2	13.6
17	1.45	2.89	4.33	5.78	7.22	8.67	10.10	11.60	13.0	14.5
18	1.54	3.07	4.61	6.14	7.68	9.22	10.80	12.30	13.8	15.4
19	1.63	3.25	4.88	6.51	8.13	9.76	11.40	13.00	14.6	16.3
20	1.72	3.44	5.16	6.88	8.60	10.30	12.00	13.80	15.5	17.2
21	1.82	3.65	5.48	7.30	9.13	11.00	12.80	14.60	16.4	18.2
22	1.93	3.87	5.80	7.44	9.67	11.60	13.50	15.50	17.4	19.3
23	2.05	4.10	6.15	8.20	10.25	12.30	14.30	16.40	18.4	20.5
24	2.17	4.34	6.51	8.68	10.85	13.00	15.20	17.40	19.5	21.7
25	2.29	4.58	6.87	9.16	11.45	13.20	16.16	18.30	20.6	22.9
26	2.42	4.84	7.26	9.68	12.10	14.00	16.90	19.40	21.8	24.2
27	2.56	5.12	7.68	10.25	12.80	15.40	17.90	20.50	23.0	25.6
28	2.71	5.42	8.15	10.85	13.50	16.30	19.00	21.70	24.4	27.5
29	2.86	5.72	8.58	11.44	14.30	17.20	20.00	22.90	25.7	28.6
30	3.02	6.04	9.05	12.10	15.10	18.10	21.10	24.10	27.2	30.2

[SOURCE: Barton, Karel, Protection Against Atmospheric Corrosion, Duncan, John, R., Translator, John Wiley & Sons.]

Temperature changes rapidly change the relative humidity. At high humidities, a rapid but small temperature drop in the metal surface can exceed the dewpoint, thus initiating the corrosion process. A rapid increase in surface temperature has the opposite effect in that moisture then evaporates and leaves the surface sufficiently dry so that there is no electrolyte to promote corrosion.

The distance from the ground also makes a considerable difference in humidity values. High humidity values are usually found close to the earth's surface. For example, at sunset, the relative humidity five centimeters above ground is 100%, but it is only 55% at a height of 200 centimeters. In late afternoon, for instance, the hull of a ship may change from being dry to being wet in a matter of only a few minutes. This danger must constantly be guarded against during the application of coatings at times of dropping temperatures. As the temperature drops, the air can no longer contain the moisture and the vapor condenses as water on surfaces which are at a slightly lower temperature than the air. This point at which the air can no longer contain the moisture is called the *dewpoint*.

Humidity, dew, rain, and fog are all forms of water which contribute in a very major way to atmospheric corrosion. Although it is commonly assumed that a continually wet surface is the most corrosive, this is not necessarily the case. Compare, for example, the corrosion conditions along the Oregon coast which is a marine area of high rainfall, and those in San Francisco which also has

Corrosion Prevention by Protective Coatings

a marine atmosphere and a somewhat similar climate. Instead of rain, however, San Francisco has a much higher percentage of fog. Corrosion along the Oregon coast, therefore, is much less than that in San Francisco because the rain washes the surface free of marine salts and therefore creates a more dilute electrolyte. Conversely, the salt from San Francisco's marine atmosphere is precipitated on the surface, and the foggy, very damp atmosphere keeps surfaces almost continually wet, but rarely washes them free of salts.

Corrosion is greatest where rain never washes the surface and where salt air or other contaminants accumulate and are subject to damp, humid conditions. It is true that the sheltered areas also maintain surface moisture for a longer period of time so that the concentration of the electrolyte and the time of exposure are greater. Thus, areas of a bridge which are exposed to rain and washing of the surface are generally less corroded than the areas which are sheltered. Similarly, unwashed sheltered screens have a much shorter life than those which are exposed and washed. Test racks are thus constructed in such a way that panels are exposed on vertical surfaces, upper horizontal surfaces, lower horizontal surfaces, and vertical sheltered areas (Figure 2.29). The underside of these panels has been observed to corrode to a greater extent than the upper side.

FIGURE 2.30 — Test panels showing effect of wind on leading edges. Prevailing wind is from the right where greater corrosion is evident on most panels. This shows the need for special care when coating edges and corners.

FIGURE 2.29 — Typical sheltered exposure rack in a marine atmosphere.

Wind Direction

Wind direction can also have an effect on corrosion. At the International Nickel Company's Kure Beach test site, for example, edges of test panels which face the prevailing wind corrode more rapidly than edges of the panels turned away from the wind. In marine atmospheres, salt is impacted on the leading edge, creating a more concentrated electrolyte at that point. Corrosion is also accelerated on the windward side of a structure which accumulates more windborne contaminates than the lee side. This is only true, of course, if a panel or structure is going to corrode within a reasonable length of time. Inorganic zinc panels, for instance, have shown no corrosion on the leading edge even after 10 years. Figures 2.30 and 2.31 illustrate this edge effect.

FIGURE 2.31 — Close-up view of the leading edge of a panel exposed in a marine atmosphere. Prevailing wind is from the right where heavy corrosion is noted compared to the left edge.

Hygroscopic Salts

One of the mechanisms for creating an electrolyte from atmospheric moisture vapor is the lowering of the dewpoint by hygroscopic water soluble salts which deposit and then form water soluble metal salts on the surface. According to Barton in *Protection Against Atmospheric Corrosion*, this process is much more important than the conversion of water vapor to liquid on the metal surface.

> Water soluble salts coming into contact with humid air after being dried take up water vapor from the atmosphere. If the equilibrium pressure of the water vapor which corresponds to a solid species (solid salt) is exceeded, further water vapor uptake occurs to give a solution whose concentration corresponds to an equilibrium value between the water vapor partial pressure over the solution and that existing in the atmosphere. Particularly in atmospheres which contain large quantities of such species, and in which the metal surface is more or less contaminated with them, this type of electrolyte formation is of special importance. This is especially true in coastal and industrial regions; in the former because of chloride, in the latter because of sulphate.
>
> Aerosol-forming hydroscopic soluble solid species convert into their dissolved forms before deposition. The electrolyte (the concentrated salt solution) falls in droplets to the surface, and these coalesce to form the electrolyte required for corrosion.
>
> Salt particles from the atmosphere are not the sole source of hydroscopic species at the metal surface. During corrosion reactions between metals and gaseous species (normally air pollutants), soluble hydroscopic products form and these lead in turn to electrolyte formation. In this respect sulphate (which arises chiefly as the conversion product of atmospheric sulphur dioxide) and chloride are particularly important. The new phase (the solid corrosion product) formed during the reaction affects the water participation in the reaction system. The fact that accumulations of more or less soluble salts, with anion depending on atmospheric type, are found in corrosion products is an expression of this.[10]

The type of reaction discussed by Barton is one quite common to marine atmospheres. It is also common in areas where there is hydrochloric acid and sodium hypochlorite, as well as in the handling and processing of salt. The chloride ion is, of course, the key to this reaction. If, for example, a steel surface is sandblasted, small droplets of moisture will accumulate in the areas of previous pits. If the humidity is reasonably high, the droplet will grow in a matter of only a few minutes. The original color of the condensation is a light green, indicating a ferrous ion in solution. It soon turns to a brown or yellowish brown, however, which seems to indicate the ferric ion. Following a somewhat longer period of exposure, the spot of moisture will turn black, indicating that it has changed from chloride to oxide, undoubtedly black Fe_3O_4. This reaction partially explains why previously corroded areas on a piece of visually clean steel will be the first to corrode again. This is also responsible for one of the major problems in applying organic coatings without additional surface preparation (Chapter 9).

Moisture retention on a metal surface once corrosion products have already formed is another factor discussed by Barton.

> The anions raise the colloidal component of the corrosion products (especially rust). These colloids bind the water relatively loosely in liquid form below the water vapor saturation pressure so that it is easily freed to act on the corrosion process.
>
> A corrosion process involves the existence of an electrolyte in contact with a metal. Since most corrosion processes yield solid products which contain chiefly hydroxide, hydrated salts and oxides, part of the water is removed from the system in a more or less solidly bound form. These products also have a relatively high sorption capacity for water, which leads to binding of a further fraction of the water. Thus, on the one hand there is partial removal by chemical processes of the very small amount of electrolyte which in effect is atmospheric corrosion (especially at relative humidities below 100%) while on the other hand, there is a layer of corrosion products formed which can be regarded as a

water reservoir, and which under some circumstances (*e.g.*, by exceeding the sorption capacity as the humidity rises) can liberate liquid water. The reverse process is also possible.

As for other cases of electrochemical metal corrosion, atmospheric corrosion should be regarded as a total process composed of simultaneous oxidation (anodic) and reduction (cathodic) reactions. Each reaction type is tied to the other; *i.e.*, the oxidation process (the actual corrosion) could not proceed without there being a simultaneous reduction reaction, though the individual mechanism of the two processes may be completely independent.[10]

Atmospheric Corrosion Products

Atmospheric rust is mainly composed of lepidocroicite and geothite (alpha and y-FeOOH). It also contains magnetite (Fe_3O_4), amorphous and usually hydroxide based components, and either ferrous sulfate ($FeSO_4 \cdot 4H_2O$) or ferrous chloride ($FeCl_2$), depending on atmospheric impurities. Impurities arise from the conversion of atmospheric SO_2, which is found along the coast in primarily industrial and chloride containing atmospheres.

It is often possible to locate three distinct zones of rust while observing corrosion products on a steel substrate, as shown schematically in Figure 2.32. The steel itself is considered the first zone. The second zone is a dark gray and very hard, durable layer which is difficult to remove by either sandblasting or acid pickling because of its tight adherence to the steel. The third zone, however, can easily be broken off of a surface by bending or beating the surface with steel sledges, as is often done on barge decks. This zone is also a hard layer and consists of the usual scale that forms on heavily rusted objects. It has areas of both dark gray oxides and light yellow-brown amorphous material which shows in the cracks and fissures of the dark material. The fourth zone is a relatively soft amorphous, powdery material that can be brushed or scraped from the surface without much difficulty. This material, which is primarily the ferric oxide form of rust (Fe_2O_3), is usually the yellowish-redish-brown color that is typical of most rusted surfaces.

While these zones exist on almost any rusted surface, it is only possible to isolate the distinct corrosion layers on heavily corroded steel (Figure 2.33). Although marine conditions provide some of the obvious examples of layered corrosion, some industrial areas are equally severe.

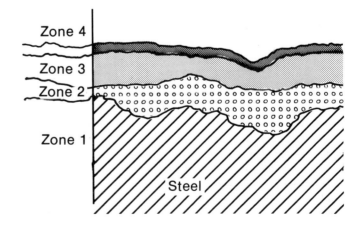

FIGURE 2.32 — Typical rust layers found on heavily corroded steel.

FIGURE 2.33 — Heavy rust layering on steel exposed to marine conditions. Note the heavy rust scale (Zone 3) in both horizontal and vertical surfaces.

Barton has analyzed rust formed in an industrial atmosphere where there was obviously both sulfate and chloride contamination. The heaviest concentrations of both sulfate and chloride anions are found in Zone III since Zone II is extremely dense. Zone III, however, is hard scale that is heavily cracked and fractured so that it is able to hold the greatest amount of soluble materials. Zone IV, on the other hand, is thin and powdery. Table 2.10 gives the composition of the three zones in more detail.

The chemical compositions of the corrosion products of various metals is a complex subject. Not only do they contain miscellaneous oxides and carbonates, but also various crystal forms with their combined water as well as complex sulfates, chlorides, and sulfides. Table 2.11 lists the various chemical compounds commonly found in the atmospheric corrosion products for various metals. The metals listed in this table are the ones normally involved in atmospheric corrosion in both industrial and marine areas.

Atmospheric Dusts

The corrosion process can also be strongly affected by solid particles known as atmospheric dust. This dust accelerates the corrosion of metal surfaces exposed to the atmosphere. These solid particles may consist of soil picked up by wind, smoke, and soot particles, or they may be organic particles of vegetable origin, including microorganisms such as fungi. Many chemical compounds

TABLE 2.10 — Typical Layering of Rust Deposits and the Composition of Individual Layers (Rust Grown for 34 Months in an Individual Atmosphere)

	Zone			
	I	II	III	IV
Amount of Rust (mg cm^{-2})	Steel	29	40	7
Iron Content in Rust, (%) Total	—	66.9	63.9	60.5
Fe^{2+} (%) (ferrous iron)	—	60.5	8.66	0
$(SO_4)^{2-}$ (mg cm^{-2}), Total Sulfate	—	240	1128	281
$(SO_4)^{2-}$ (mg cm^{-2}), Soluble Sulfate	—	40	87	55
Cl^- (mg cm^{-2})	—	130	186	32

Zone I: Steel.
Zone II: removed by chemical methods (*e.g.*, pickling).
Zone III: spring off by bending the specimen.
Zone IV: removed by brushing or scraping.
[SOURCE: Barton, Karel, Protection Against Atmospheric Corrosion, Duncan, John R., Translator, John Wiley & Sons, New York, NY, p. 3, 1976.]

such as ammonium sulfate, coal dust, fly ash, or sodium chlorides, also may be included in the dust, depending on the type of industry in the area. For example, in various industrial districts in England where coal is used as a primary fuel, the release of dust particles approximates 1.2 to 1.4 kilograms of soot and dust per square meter every year. Atmospheric dust measured in Pittsburgh, Pennsylvania had a composition of: organic matter, 35%; mineral (primarily Fe_2O_3), 14.5%; insoluble mineral matter, 3.85%; and soluble matter (mostly sulfates), 9.2%.

The various solid particles that settle as dust can act in several different ways. As discussed previously, the soluble salts, such as ammonium sulfate or sodium chloride, are corrosion active since they readily make strong electrolytes.

There are also dusts which in themselves are not particularly corrosive, but which are capable of absorbing active gases from the atmosphere. These include various

TABLE 2.11 — Compounds Found (Frequently) in Atmospheric Corrosion Products

Metal	Compounds	Notes
Iron	α-FeOOH, γ-FeOOH, B-FeOOH, Fe(OH)$_2$ Fe_3O_4, (Fe_2O_3?)	B-FeOOH only in atmosphere containing chloride
	aFeSO$_4$. bFe(OH)$_2$. cFe(OH)$_3$,	
	FeSO$_4$. 4H$_2$O, FeSO$_4$. 7H$_2$O	In industrial atmosphere
	FeCl$_3$	In marine atmosphere
Copper	CuO, CU$_2$O, CuCl$_3$. 2Cu(OH)$_2$	
	2CUCO$_3$. Cu(OH)$_4$	
	CuSO$_4$. 3Cu(OH)$_2$, CuSO$_4$. 2Cu(OH)$_2$	In industrial atmosphere
	CuSO$_4$. 5H$_2$O	
	Cu$_2$ (OH)$_3$ Cl	In marine atmosphere
	CuS	
Zinc	ZnO, ϵ-Zn(OH)$_2$, β-Zn(OH)$_2$, Zn CO$_3$,	
	ZnCO$_3$. Zn(OH)$_2$	
	ZnCO$_3$. 3Zn(OH)$_2$, ZnCO$_3$. 3Zn(OH)$_2$ H$_2$O	
	ZnCl$_2$. 4Zn(OH)$_2$, ZnCl$_2$. 6Zn(OH)$_2$	In marine atmosphere
	ZnSO$_4$. 4Zn(OH)$_2$, ZnSO$_4$ .x H$_2$O)	In industrial atmosphere
Cadmium	CdO, Cd(OH)$_2$, CdCl$_3$, 2CdCO$_3$. 3Cd(OH)$_2$	
	CdSO$_4$. H$_2$O, Basic sulfate (?)	In industrial atmosphere
	CdCl$_2$. H$_2$O, Cd(OH)Cl	In marine atmosphere
	CdS	
Aluminum	Al(OH)$_3$ gel, γ-Al$_2$O$_3$ γAlQOH, ∂-Al(OH)$_3$	
	Amorphous basic sulfate	In extremely SO$_2$ polluted atmosphere
	Amorphous basic chloride	In marine atmosphere
Lead	PbO, Pb(HCO$_3$)$_2$, PbSO$_4$, Basic sulfate	
Magnesium	MgO, Mg(OH)$_2$, Basic carbonate, MgSO$_4$ MgCl, basic chloride	

[SOURCE: Barton, Karel, Protection Against Atmospheric Corrosion, Duncan, John R., Translator, John Wiley & Sons, p. 3, 1976.]

Corrosion as Related to Coatings

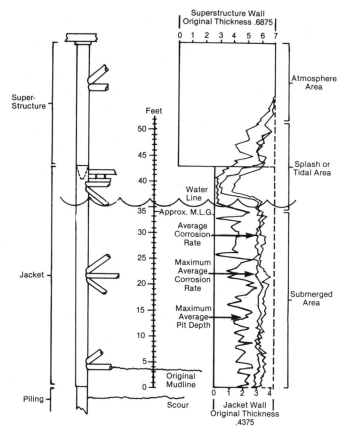

FIGURE 2.34 — Corrosion of an offshore structure after eight years service in the Gulf of Mexico. (SOURCE: Grosz, O. L., Important Methods of Corrosion Control in Offshore Operations, Chevron, USA, New Orleans, LA. Reprinted from Offshore, June, July, 1958.)

forms of carbon such as soot or even coal dust. Coal dusts often promote severe corrosion because of their absorption of SO_2 and because of their sulfur content. They have very small particle sizes and any contained sulfur is oxidized in the atmosphere to SO_2. Soots often absorb SO_2 from stack gases and also can be very corrosive. This is obvious on the decks of tankers which are burning high sulfur crudes in their boilers. When the stacks are blown, soot particles settle out on the deck. These particles are strongly acidic, and wherever they contact bare steel, an active corrosion cell is created. Decks and equipment subject to such fallout require coating with a chemical-resistant coating. The acid in the soot is strong enough to penetrate and pit most inorganic zinc coatings where soot particles have landed.

Rock dust, silica, etc., represent another type of solid material which also settles on surfaces, but is nonreactive in itself. When such materials accumulate, however, it can create corrosion by absorption and retention of moisture on metal surfaces.

While there are a variety of factors that contribute to atmospheric corrosion, the ones which exercise the strongest influence are: (1) water or moisture which form the surface electrolyte; (2) soluble accelerating ions, such as chloride and sulfate ions which are primarily responsible for the conductivity of the electrolyte; and (3) temper-

ature. In general, the higher the temperature, the greater the corrosive activity (as long as the electrolyte is present).

Areas of Atmospheric Corrosion

There are essentially three areas commonly identified with atmospheric corrosion: rural, industrial, and marine. In a rural atmosphere, there are little or no soluble ionic materials to cause a strong electrolyte. The term *rural* indicates an area away from both industrial and marine conditions, so that corrosion in these areas depends primarily on temperature, humidity, and retention of moisture on metal surfaces. In desert areas where humidities are low, water retention on any metal surface is very short, even during periods of rainstorms. Under these conditions, steel will discolor and zinc will change to a dull gray, but very little corrosion will actually take place. In more humid atmospheres, however, where moisture is retained for longer periods, corrosion does occur, but at a relatively slow rate.

In cities or in industrial areas, the corrosion rate also depends on the three factors of temperature, humidity, and moisture retention. In addition, it depends on the amount of soluble ionic materials in the atmosphere which are retained in the moisture layer on the metal surface. In areas where moisture is retained for long periods of time, corrosion can be so severe that heavy rust scale forms where bare steel is exposed.

Zones of Marine Corrosion

Marine atmospheres, on the other hand, are consistently severely corrosive environments. The degree of severity, however, depends on several variables. The humidity in a marine atmosphere is generally high, but the temperature is variable, depending on the climate (tropical, temperate, or arctic) and the amount of sunlight. The chloride content is also variable, depending on the distance from the shoreline.

Three primary areas for the corrosion of marine structures such as offshore platforms can be isolated. The first is the submerged area where corrosion is relatively uniform from the mean low tide area downwards. Even at great depths there is often a large quantity of soluble oxygen in the water which means that corrosion can and does readily take place. A number of factors which influence the corrosion of steel in seawater have been listed by Fink, *et al.,*[11] as shown in Table 2.12.

The second area is the tidal or splash zone. This is an area of maximum corrosion, as it is alternately exposed to seawater and air with maximum oxygen content in the strong electrolyte. The third area is that above the splash zone. It is less corrosive than the splash zone, but is nevertheless a very strong atmospheric corrosion area often characterized by large tubercles of rust and deep pits. Corrosion is most severe in the area close to the tide level and becomes less severe with the height of the structure above the water. This reduction occurs because of two things: (1) the farther into the atmosphere, the less precipitated salt spray there is to form the electrolyte; and (2) the farther above the water level, the higher the temperature is and the less humidity there is.

Figure 2.34 illustrates the corrosion that took place in

two of the three zones on an offshore production platform after eight years in the Gulf of Mexico. The graph of the structure indicates the nature of the corrosion in the two areas. Maximum corrosion occurred at the splash zone where the average corrosion was almost as great as the maximum average pit depth. The totally submerged area exhibited more uniform average corrosion, but with many deep pits. The graph shows little or no corrosion in the upper atmospheric area; however, the structure was coated in this area and therefore somewhat protected.

TABLE 2.12 — Corrosion Factors for Carbon Steel Immersed in Seawater

Factor in Seawater	Effect on Iron and Steel
Chloride Ion	Highly corrosive to ferrous metals. Carbon steel and common ferrous metals cannot be passivated. (Sea salt is about 55% chloride.)
Electrical Conductivity	High conductivity makes it possible for anodes and cathodes to operate over long distances, thus corrosion possibilities are increased and the total attack may be much greater than that for the same structure in fresh water.
Oxygen	Steel corrosion, for the most part, is cathodically controlled. Oxygen, by depolarizing the cathode, facilitates the attack; thus a high oxygen content increases corrosivity.
Velocity	Corrosion rate is increased, especially in turbulent flow. Moving seawater may: (1) destroy rust barrier, and (2) provide more oxygen. Impingement attack tends to promote rapid penetration. Cavitation damage exposes the fresh steel surface to further corrosion.
Temperature	Increasing ambient temperature tends to accelerate attack. Heated seawater may deposit protective scale or lose its oxygen; either or both actions tend to reduce attack.
Biofouling	Hard-shell animal fouling tends to reduce attack by restricting access of oxygen. Bacteria can take part in the corrosion reaction in some cases.
Stress	Cyclic stress sometimes accelerates failure of a corroding steel member. Tensile stresses near yield also promote failure in special situations.
Pollution	Sulfides, which normally are present in polluted seawater, greatly accelerate attack on steel. However, the low oxygen content of polluted waters could favor reduced corrosion.
Silt and Suspended Sediment	Erosion of the steel surface by suspended matter in the flowing seawater greatly increases the tendency to corrode.
Film Formation	A coating of rust, or rust and mineral scale (calcium and magnesium salts), will interefere with the diffusion of oxygen to the cathode surface, thus slowing the attack.

[SOURCE: Fink, F. W., et al., The Corrosion of Metals in Marine Environment, Battelle Memorial Inst., DMIC Report 254, Distributed by NTIS, AD-712 585-S, pp. 7, 13, 1970.]

Figure 2.35 shows the three marine corrosion zones schematically, indicating the type of corrosion and the coatings required for corrosion protection of an offshore drilling or production platform. There is nothing particularly unique about corrosion on offshore equipment, except that severe damage to coatings is the general rule rather than the exception, since they are continually exposed to severely corrosive conditions for years or even decades. Repair of such excessive damage is extremely difficult and costly, therefore maximum corrosion resistance must be designed into the marine service platforms.

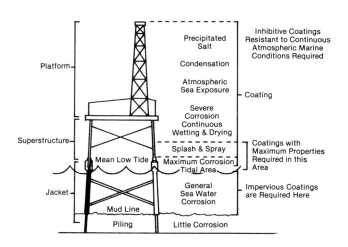

FIGURE 2.35 — Typical corrosion zones on offshore platforms. (SOURCE: LaQue, F. L., Marine Corrosion: Causes and Prevention, John Wiley & Sons, New York, NY, pp. 298-299, 1975.)

Marine corrosion aboard a ship is very similar to that of an offshore platform. The zones and conditions are the same, as well as the coating systems generally used (Figure 2.36). The major difference is that ships are mobile, and thus may be exposed to a variety of climates (e.g., tropical, temperate, or arctic) and of special conditions such as tropical hurricanes or arctic flow ice. On the other hand, their mobility gives them one great advantage; it allows corrosion repairs to be made on a timely basis and under relatively favorable repair conditions. Surfaces can be abrasive blasted and coatings can be applied under other than open sea conditions.

Methods of Corrosion Control

As discussed earlier, the corrosion process takes place because of the natural tendency for materials, particularly the metals commonly used for structures, tanks, ships, etc., to revert from the metallic state to the more stable oxide of the metal. The material is no longer useable in this condition, and if a major metal reversion has taken place, the structure fails. This kind of destruction must therefore be prevented. Several methods of assuring the continued viability and useability of various structures at a minimum cost may be considered:

Corrosion as Related to Coatings

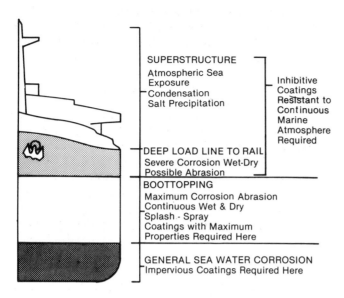

FIGURE 2.36 — Typical corrosion zones aboard ship. (SOURCE: LaQue, F. L., Marine Corrosion: Causes and Prevention, John Wiley & Sons, New York, NY, pp. 298-299, 1975.)

1. Selecting and using specific corrosion-resistant[1] materials of construction.

2. Changing or altering the environment.

3. Using a barrier between the structural material and the environment.

4. Using cathodic protection.

5. Using the principal of corrosion allowance or overdesign.

There are advantages, disadvantages, and areas of the most economical use for each of these methods. An ocean vessel, offshore structure, industrial plant, or underground pipeline, all have numerous, different, and specific areas of corrosion susceptibility. For example, many industrial sites such as paper plants and refineries are a composite of micro atmospheres, each of which may pose a unique corrosion problem. A drip under a pump is certainly a different problem than a tank which contains the liquid being pumped. Similarly, the atmosphere in the area of a blow-down pit in a paper mill is different from that in the area where the pulp is bleached.

No single method therefore, is universal or serves as a cure-all for corrosion problems. Each problem or situation must be individually studied and the specific requirements of that situation taken into consideration, along with such additional factors as the available down time, operating temperatures, appearance, age, and potential life of the area, before a decision can be reached as to which is the best method. If the structure is a new one, the method used or the material selected is most often determined by the design engineer. Since many errors in judgement are made at this design development stage, it is important that the corrosion engineer, when or wherever possible, be consulted by the design engineer to assure that the most practical and cost effective of the above methods will be selected.

Corrosion-Resistant Materials

The use of corrosion-resistant materials is most often indicated in the original design of a structure or piece of equipment. These materials are selected usually for very specific purposes to withstand a particular corrosion condition or to prevent contamination of a contained liquid. In the case of an oxidizing acid solution containing radioactive material, a stainless tank or structure may be the only answer to the problem. High temperatures may also indicate specific materials rather than the ordinary structural grade of steel or concrete. A stack, for instance, may be lined with corrosion- or acid- resistant brick and mortars to prevent the serious breakdown of the structural concrete by fumes and condensation.

Most often, the selection of specific materials is necessitated by a severe corrosion condition. In such cases, the additional material costs are entirely justified by the additional life obtained through their use. Examples of the use of such specific materials in our modern society range from stainless steel pots and pans in the kitchen and in food processing plants, to special alloys used in seawater condensers or materials such as Hastelloy or lead used under strong acid conditions.

The decision to use this type of protection compared to other methods of corrosion control largely depends on the severity of exposure and the ultimate cost of alternate methods. The use of special materials normally can not be economically justified for the majority of structural areas within a plant, even though some corrosion may be present.

There are many other types of materials that may be selected in addition to metals and metal alloys. Plastic materials can be used in many corrosion conditions such as the piping of corrosive solutions. They are also used as ducts for corrosive fumes (*e.g.,* in rayon plants) and as corrosion-resistant gratings. They can be reinforced with synthetic fiber or glass to provide sound structural properties in addition to corrosion resistance, and with such reinforcement are often used as tanks and scrubbers and in large fume processing areas.

Both inorganic and organic mortars are used with acid-resistant brick and tile for specific corrosion-resistant areas. These may be used in such areas as pickling tanks in steel mills or as corrosion-resistant flooring materials in chemical or food plants. The category of corrosion-resistant materials is thus a broad one, with every industrial plant having some use for them.

Changing the Environment

Changing an environment can involve such procedures as the alteration of a piece of equipment in order to prevent splash, spray, or fumes from a corrosive solution coming in contact with adjacent equipment or structural areas. It can also involve a change in humidity by increasing the air flow through the area or by increasing the temperature. It may involve the ducting of corrosive fumes away from their source or the piping of a corrosive liquid rather than allowing free flow to an open floor drain.

The use of chemical inhibitors is also a method of changing the environment. These are often used in steam

Corrosion Prevention by Protective Coatings

and condensate return lines, brine lines, petroleum transportation lines, heat exchangers, and cooling towers. They are most often used in closed systems, although the cooling water which is continually reused in a cooling tower is an important exception. Inhibitors must also be used with some caution, as some may be considered toxic, and improper selection and maintenance of an inhibitor system may accelerate rather than retard corrosion. Nevertheless, if properly used in the areas where they can be effective, inhibitors can provide a simple and relatively low cost answer to corrosion control.

Barriers

The barrier principle represents the most common and widespread use of corrosion prevention. This method utilizes corrosion-resistant materials to isolate concrete, steel, or other structures from a corrosive environment. The variety of barriers available include: acid-proof brick and tile construction on the interiors of tanks or floors; the use of plastic sheeting that is either applied to structures with an adhesive or is fabricated into a unit that may be placed on the interior of a tank to contain corrosive solutions; troweled or sprayed-on plastic cements; and, of course, protective coatings. Since each of these barrier methods is characterized by usefulness, a careful analysis of the problem is required in order to choose the one which would be the most effective and economical.

Cathodic Protection

Of necessity, *cathodic protection* (a reversal of the corrosion process whereby sufficient electrons are maintained on any metal surface to prevent the metal from going into solution) can only be used for immersion or underground conditions. The cathodic current can be supplied either by an externally impressed current or by sacrificial metals used as anodes. Cathodic protection can be an extremely effective method of corrosion control where proper and practical conditions for use exist.

Design Allowance

Overdesign for corrosion conditions is becoming less common as other corrosion prevention methods have become more practical and less costly. Overdesign of a structure through the use of heavier structural members or thicker plates does not eliminate the corrosion problem. It merely adds considerable weight requiring heavier construction. Even though corrosion is anticipated, specific conditions are often difficult to predict. The unprotected structure using the overdesign principle may therefore corrode much faster than anticipated or may corrode much more rapidly in certain areas, thus consuming the corrosion allowance and endangering the life of the structure.

It used to be a common practice in the shipping industry to add a corrosion allowance to hull plates and bulkhead structures. The additional tonnage, however, required additional energy during transportation. These corrosion allowances have thus largely been eliminated by the use of long-lasting protective coatings that are effectively resistant to the type of corrosion found on the interiors of the tanks and the exterior of the hull.

These five corrosion control methods are summarized in Table 2.13. Each of these methods has its own sphere of usefulness, although, under many conditions, a combination of methods may be even more effective. Some stainless steels, for instance, are affected by the chloride ion which results in surface pitting. Thus, the combination of stainless steel and an effective chloride-resistant coating often provides a more effective system than the stainless steel alone. Cathodic protection, more often than not, is also combined with various protective coatings in order to reduce the cost of impressed current or to reduce the number of sacrificial anodes used. In other cases where the surface may be both wet and dry, a combination of cathodic protection and coatings is most effective. A severely corrosive environment may similarly be changed by the combination of additional ventilation and protective coatings.

TABLE 2.13 — Comparison of Corrosion Control Methods

Method	Example	Principal Advantage	Disadvantages
Altering the Environment	Changing process, humidity, or temperature. Use of inhibitors.	General simple changes. Many times low cost. Usually retrofit to existing facility.	May not completely eliminate problem. Inhibitors limited to immersion conditions.
Corrosion-Resistant Materials	Copper, Nickel, Chromium, Molybdenum alloyed with iron or steel. Thermoplastic materials (PVC or Polyethylene).	Long life span. Applicable to only certain situations.	High initial cost. Workability.
Cathodic Protection	Ship hulls, underwater, underground, pipelines.	Simplicity. Effective in presence of good electrolyte.	Limited usefulness in damp or dry areas. Immersion required.
Barriers	Brick linings. Protective coatings, plastic sheetings, Monolithic toppings.	Most effective and versatile. Reasonable cost.	Careful analysis of corrosion problem necessary. Proper surface preparation and application essential.
Overdesign	Heavier structural members or thicker plates than required.		Neither exact length of life nor replacement cost can be predicted. Higher initial cost. Ineffective. Increased weight.

In the words of Frank LaQue,

Knowledge of the reactions involved in the corrosion of steel combined with a knowledge of how a paint system can impede these reactions...can serve as an effective guide to the most effective use of paint to protect steel.[12]

Thus, because no single method is the cure-all or is effective under all conditions, proper selection of a corrosion control method must be preceded by a careful analysis of many factors.

References

1. Corrosion Basics—An Introduction, Chapter 1, The Scope and Language of Corrosion, L. S. VanDelinder, Ed., National Association of Corrosion Engineers, Houston, TX, p. 14, 1984.

2. Denison, I. A., Contributions of Sir Humphrey Davy to Cathodic Protection, Corrosion, Vol. 3, p. 295 (1947).

3. Whitney, W. R., The Corrosion of Iron, J. American Chem. Soc., Vol. 22, p. 394 (1903); reprinted in Corrosion, Vol. 3, p. 331 (1947).

4. U.S. Dept. of Commerce, Civil Engineering Corrosion Control, Corrosion Control - General, Vol. 1, Distributed by NTIS, AD/A-004082, pp. 16, 33, Jan. (1975).

5. Clark and Hawley, Encyclopedia of Chemistry, Ions, Reinhold Publishing Corp., New York, NY, p. 522, 1957.

6. Speller, F. N., Corrosion Causes and Prevention, McGraw-Hill, New York, NY, 1957.

7. Rowe, Leonard, C., The Application of Corrosion Principles to Automotive Engineering Design, Automotive Corrosion by De-icing Salts, p. 243, National Association of Corrosion Engineers, Houston, TX, 1981.

8. Evans, U. R., The Corrosion and Oxidation of Metals, St. Martins Press, pp. 270, 1017, 1970.

9. Rozenfeld, I. L., Atmospheric Corrosion of Metals, English translation, National Association of Corrosion Engineers, Houston, TX, p. 124, 1972.

10. Barton, Karel, Protection Against Atmospheric Corrosion, trans., John R. Duncan, John Wiley & Sons, New York, NY, pp. 3, 1976.

11. Fink, F. W., et al., The Corrosion of Metals in Marine Environment, Battelle Memorial Inst., DMIC Report 254, Distributed by NTIS, AD-712 585-S, pp. 7, 13, 1970.

12. LaQue, F. L., Chapter 1.1, Steel Structures Painting Manual (revised), Corrosion of Steel as Related to Protection by Paint, Vol. 1, Steel Structures Painting Council.

3

Essential Coating Characteristics

Protective coatings are unique speciality products which represent the most widely used method of corrosion control. They are used to give long term protection under a broad range of corrosive conditions, extending from atmospheric exposure to full immersion in strongly corrosive solutions. Protective coatings in themselves provide little or no structural strength, yet they protect other materials so that the strength and integrity of a structure can be maintained. They are the skin, over the skeleton, that both protects and beautifies the bone and muscle of the world's essential structures.

Coating Function

The function of a protective coating or lining is to separate two highly reactive materials; *i.e.,* to prevent strongly corrosive industrial fumes, liquids, solids, or gases from contacting the reactive underlying substrate of the structure. This is another way of saying that the coating or lining acts as a barrier to prevent either chemical compounds or corrosion current from contacting the substrate. This physical separation of two highly reactive materials, the atmosphere and substrate, is extremely important.

That coatings or linings are, in general, a relatively thin film separating the two reactive materials indicates the vital importance of the coating and the concept of a corrosion-free structure. The coating must be, according to this concept, a completely continuous film in order to fulfill its function. Any imperfection becomes a focal point for corrosion and the breakdown of the structure, or a focal point for the contamination of a contained liquid. This relatively thin continuous film concept takes on even greater significance when it is understood that most protective coatings are manually applied to very large areas of structural steel, *e.g.,* tank surfaces, ship hulls, drilling structures, and pipelines. A single coating application may thus involve an area of many thousands of square meters.

Essential Coating Properties

Water Resistance

In order to perform effectively, a corrosion-resistant coating must be characterized by many essential properties. These may vary, depending on the specific use of the coating, but there are several basic characteristics required by all coating materials.

Resistance to water is perhaps the most important coating characteristic since all coatings will come in contact with moisture in one form or another. Water, which affects all organic materials in one way or another, is actually the closest thing to a universal solvent. It is no small wonder, then, that a resistance to it is difficult to achieve. Even rock and concrete gradually dissolve or erode away due to their contact with water, and both iron and steel rapidly oxidize under even normal acidic water conditions. Thus, there is no one coating that can be effective under all water conditions: there are too many different types of structures and forms of water for overall resistance to be that easy.

The protection of iron and steel pipe, for instance, is quite a different problem from the one encountered in above-ground storage tanks leading to the pipe. Similarly, dam gates and trash racks require a different solution from the concrete flume bringing water into the dam. Filters, clarifiers, and floculators present an even different set of conditions that must be addressed. Industry adds its share of difficult situations by requiring deionized water for some processes. This requires the use of special deionization equipment, storage tanks, and piping.

These mechanical problems are only multiplied by the different types of water encountered. Swamp water, which may be pure enough to drink, is ordinarily acidic and will corrode both steel and concrete. Sulfide water, which is prevalent in many areas, reacts readily with most metals (*e.g.*, iron, steel, brass, and copper). High conductivity water or seawater leads to rapid formation of anode-cathode areas on steel which results in severe pitting. Pure water from the snow fields will dissolve the calcium out of concrete at a rapid rate, leaving the aggregate exposed. Water with a high oxygen content will also create anode-cathode type corrosion areas. The problem is thus a very complex one since no single type of material will provide a universal answer.

The water molecule is an extremely small one with the ability to penetrate into and through most all organic compounds. It does this by passing through the intermolecular spaces of the organic material and can either remain there in an absorbed state or can pass through the compound. Moisture generally will come to an equilibrium, with as many water molecules passing into the organic material as evaporate out of the surface. This maintains a relatively constant water content in the organic material, depending on the moisture vapor pressure at any given time. Because of this highly penetrating characteristic, water has more of an effect on organic compounds than any other single material. Since most coatings are organic in nature, they must have the highest possible moisture resistance in order to maintain their properties and be effective over a long period of time.

Water can also affect the permeability of other molecules, depending on their size. Some of the very small molecules (*e.g.*, ammonia, carbon dioxide, and hydrochloric acid) are also extremely penetrating and are aided by water vapor in their penetration into organic compounds. Ammonia is an extremely difficult material to deal with in terms of coatings. Since it is a gas, ammonia becomes alkaline when it is combined with moisture. Its resulting high penetration characteristics cause blistering in many coatings. Hydrochloric acid is also very penetrating since it is both a gas and a small molecule. It not only has a strong affinity for water, but in areas of poor adhesion (even with heavy organic lining materials), it will penetrate and accumulate underneath the lining causing accelerated corrosion.

Carbon dioxide is also very penetrating. In the early days of the automobile, it was considered the ideal gas for inflating tires. Because of the small size of the molecule, however, it was impossible to keep the carbon dioxide from penetrating the innertube and, within a matter of a few hours, causing a flat tire.

Organic materials can, therefore, be permeated by many of the very small molecules such as water. This is particularly true where there is an interface or poor adhesion under a coating where the water vapor can accumulate and possibly condense.

The larger molecules, such as sugar, sodium hydroxide, and even sulfuric acid, do not penetrate the organic molecules. In some cases, they even tend to draw water out of the organic coating by an osmosis reaction. A concentrated sugar solution or sodium hydroxide solution will pull water out of the coating and into the concentrated

solution so that solutions of these materials have little tendency to blister coatings.

For a high-performance corrosion-resistant coating to also have excellent water resistance means that it must not only withstand continuous immersion in water or seawater, but it must do so without blistering, cracking, softening, swelling, or loss of adhesion. It must also withstand repeated cycles of wet and dry conditions, since such coatings are normally exposed to an atmosphere of condensing dew in the evening and night hours and sun drying during daylight hours.

Water Absorption

Water absorption refers to the amount of water which is picked up and retained within the molecular spaces of the coating. Once the coating has formed, the water content comes to equilibrium with the atmosphere, *desorbing* (or evaporating) water under dry conditions and absorbing water when subject to high humidity or immersion (Figure 3.1).

Each coating also has its own level of water absorption. Since the water in the coating is in equilibrium with the moisture in the atmosphere, it is not, in itself, in a critical condition in terms of corrosion. If a coating is strongly adhesive and there is no interface between the coating and the substrate, the moisture absorbed into the coating will remain there in a relatively inert state. At any given moisture vapor pressure, as many molecules leave the coating as enter into it. Thus, the number of absorbed molecules in the intermolecular areas remains constant for each coating type (Figure 3.2).

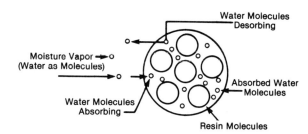

FIGURE 3.1 — Water absorption (SOURCE: LaQue, F. L., Marine Corrosion: Causes and Prevention, Chapter 16, John Wiley & Sons, New York, NY, pp. 285-87, 1975.)

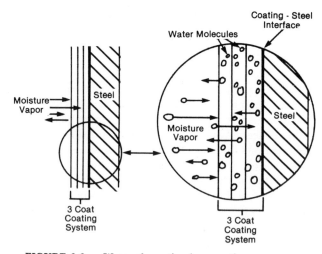

FIGURE 3.2 — Water absorption by a coating.

On the other hand, moisture vapor can contribute to corrosion when combined with other factors, such as the very small molecules of hydrochloric acid, ammonia, or similar materials. Thus, the best corrosion-resistant coating generally has the lowest water absorption.

Moisture Vapor Transfer Rate

The *moisture vapor transfer rate* is the rate at which moisture vapor will transfer through a protective coating when there is a difference in moisture vapor pressure on one side of the coating compared to the other side (Figure 3.3). Each coating and resin has its own characteristic moisture vapor transfer rate. Table 3.1 gives some laboratory measured rates for individual coatings. (Note that these should not be considered standard rates for any generic type of coating.) Depending on the formulation, an epoxy could have a much higher moisture vapor transfer rate than the figures given, while an alkyd could have a lower rate. It is generally held that the lower the moisture vapor transfer rate, the better the protection provided by a corrosion-resistant coating.

In considering the passage of water through a resin film, it should be noted that such passage is not through open pores in the coating. Rather, it is through and into the intermolecular spaces between the resin molecules, with the spacial relationship of the very large resin molecules contributing to moisture absorption and passage. These circumstances are very different from physical imperfections in the coating such as pinholes and voids which allow moisture to penetrate through the opening just as it would in air. In this case, the moisture has easy access to the substrate where it can actively promote corrosion or release the adhesion of the coating.

The transfer of moisture through a coating, as stated previously, depends on the difference in pressure between the two sides of the coating. If a coating has excellent adhesion, then there is no difference in pressure from one side to the other and the coating soon comes to equilibrium with the moisture in the air or the water on the surface of the coating (Figure 3.4). The water molecules merely penetrate into the coating and are absorbed while an equivalent number are evaporated from the coating so that the amount of moisture in the coating (moisture absorption) remains constant.

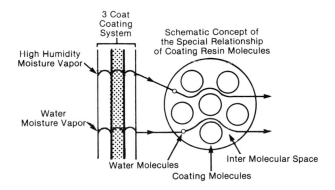

FIGURE 3.3 — Moisture vapor transfer rate. (SOURCE: LaQue, F. L., Marine Corrosion: Causes and Prevention, Chapter 16, John Wiley & Sons, New York, NY, pp. 285-87, 1975).

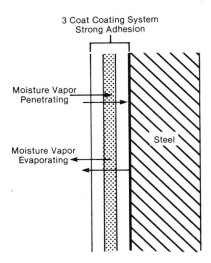

FIGURE 3.4 — Moisture transfer through a coating with excellent adhesion.

TABLE 3.1 — M.V.T. Rates for Specific Coatings

Coating Type	Permeance Perms[1]	Test Thickness Mils	Grams/100 In.²/ 24 Hours
Epoxy Polyamide	0.16	8.0	0.17
Amine Catalyzed Epoxy	0.19	7.5	0.30
Vinyl Chloride-Acetate	0.31	5.5	0.83
Vinyl Acrylic	0.54	5.0	0.83
Alkyd (Short Oil)	2.4	5.0	3.7

[1]Perms = Grains of moisture/1 hour/ft²/P (in. of Hg).

If the coating has poor adhesion, however, either inherently or because it has been applied over a contaminated surface, there is an interface between the coating and the steel, and moisture vapor can transfer into this area. Soon after the coating is applied, there is little moisture vapor pressure in this area so that there is a tendency for moisture to pass in the direction of the poor adhesion. Moisture can condense in this space or, if the temperature of the coating increases, the moisture vapor within the void can develop sufficient pressure to create a blister (Figure 3.5). With poor adhesion, the moisture vapor can penetrate between the steel and the coating, expanding the blister.

Blisters which were six to eight inches in diameter and contained so much water that the coating actually hung down like a bag have been observed on the exterior of tanks. In this case, water, as moisture vapor, penetrated the coating from driving rain and continuing high humidity.

The moisture vapor pressure, within the coating or in a void beneath the coating, is dependent on temperature and would be the same for moisture in vapor, liquid, or condensed liquid form. Moisture vapor and condensed moisture are both pure water forms. The only time there would be a difference in the moisture vapor pressure would be where moisture in contact with the coating contained considerable soluble salts. Pure water, however, has the maximum penetrating power, while water with dissolved salts has a somewhat lesser penetrating force.

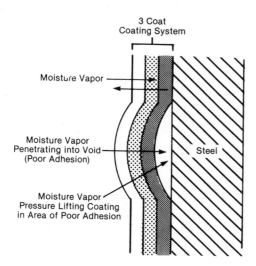

FIGURE 3.5 — Penetration of moisture vapor into an area of poor adhesion under the coating.

Thermal Gradient Across Coating

Another mechanism of moisture as it is in contact with a coating concerns a *thermal gradient* across the coating. This is where the metal or steel substrate is at a lower temperature than the moisture vapor or water on the exterior of the coating. This temperature gradient effect is the basis of what is known as the *Cleveland Coating Tester* which involves forcing moist, warm water to condense on the coating while the steel substrate is exposed to the cooler outside air. The temperature on the inside of the tester is approximately 100 F, while the temperature on the exterior is ambient, thus causing the thermal gradient. This serves as an excellent test for a coating's adhesion characteristics: the warm moisture vapor used in the test will penetrate the coating and tend to condense on the cooler steel substrate underneath the coating, thus creating a water filled blister.

This same principle is utilized in what is known as the *Atlas tester* which is used for testing tank linings. In this case, the liquid that is put in contact with the coating is warmer (or hot), while the exterior substrate is at air temperature. This thermal gradient mechanism should also be considered wherever a tank or pipe containing a warm liquid is lined, or where the exterior of a tank is coated and the liquid on the interior is considerably cooler than the exterior temperature (Figure 3.6).

Osmosis

The mechanism of osmosis also concerns the passage of moisture through a coating. More specifically, *osmosis* is the passage of water through a semipermeable membrane from a solution of less concentration to one of greater concentration. All organic coatings will transmit moisture vapor which makes them semipermeable membranes and therefore subject to this mechanism (Figure 3.7).

Osmosis is an important phenomenon wherever coatings are subject to water immersion, condensation conditions, or even high humidity. Chloride deposition on steel, for example, is not uncommon in marine areas. Contamination can come from a variety of circumstances, including a handprint on the steel surface, a drop of sweat, or poor initial cleaning of a contaminated surface. Once the surface contaminant has dried, it may not be noticed during the application of the coating. Thus, as soon as the coating is put into service under immersion conditions or high humidity, moisture vapor again transfers through the film and, when it comes into contact with chloride or other soluble contaminants, forms a concentrated solution at that point.

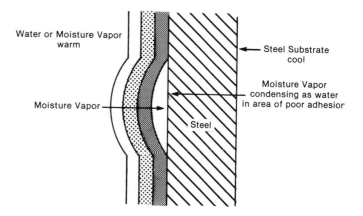

FIGURE 3.6 The thermal gradient effect on a coating with poor adhesion.

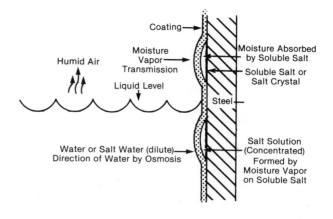

FIGURE 3.7 — The principle of osmosis as it affects coatings. (SOURCE: LaQue, F. L., Marine Corrosion: Causes and Prevention, Chapter 16, John Wiley & Sons, New York, NY, pp. 285-87, 1975.)

Corrosion Prevention by Protective Coatings

Since osmosis transfers moisture from the side of the coating with the least concentration to the area of greatest concentration, moisture is pulled through the coating towards the area of contamination. This principle is a general physical-chemical one which applies in every case where there is a solution concentration differential across a semipermeable film. It is the same principle which applied to reverse osmosis in the purification of water which affects the cell wall during biological activity where water is absorbed by plants from the soil or air. Thus, where a coating is to be immersed or subjected to high humidity, a clean surface is essential to prevent the mechanism of osmosis from occurring.

Electroendosmosis

Moisture vapor is the key reactant in yet another mechanism. *Electroendosmosis* is defined as the *forcing of water through a semipermeable membrane by an electrical potential in the direction of the pole with the same electrical charge as the membrane.* While this may seem to relate only to an isolated series of circumstances, it should be noted that coatings are generally negatively charged and the metal around even a small break in the coating contains an excess of negative electrons, therefore making it a negative surface. Water then tends to be forced through the coating towards the cathode. Thus, wherever a break in a coating occurs, the mechanism of electroendosmosis is possible. The series of circular blisters around a small break in Figure 3.8 shows the effect of electroendosmosis. Figure 3.9 illustrates this mechanism schematically.

The most common example of coating failure by electroendosmosis is where a coated surface is also under cathodic protection. Failures of underground pipe coatings in moist soils have occurred due to excessive cathodic potentials used in the cathodic protection system. The same has been true of ship bottoms. Most of these failures have occurred because of poorly operated and controlled impressed current systems. Zinc and aluminum anodes do not develop potentials that are damaging to most coatings. In fact, coatings with strong adhesion, good dielectric strength, and a low moisture vapor transmission rate, along with controlled cathodic potentials of under -1.0 volts, provide an excellent corrosion protection system.

Water-related reactions cannot be stressed enough as the mechanisms behind the success or failure of any coating. Water is present in one form or another wherever a coating is used to protect a surface. In addition to having a strong effect on coatings due to its highly penetrating characteristic, water also serves as an electrolyte which is required for a corrosion reaction. Moisture can also affect a number of other key coating properties, although less directly than those already discussed.

Dielectric Strength

Dielectric strength is a key coating property since the coating must break the electrical circuit set up during a corrosion reaction in order to be corrosion-resistant. It does so by resisting the passage of any electrons and thus prevents any metal from going into solution at the anode. If the electrons cannot travel to the cathode, the corrosion mechanism is not possible. Dielectric strength is

FIGURE 3.8 — Coal tar coating in a water tank showing typical electroendosmotic blistering. Note the concentric circles of blisters.

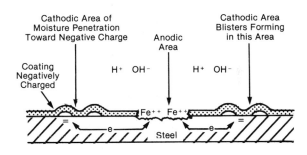

FIGURE 3.9 — Electroendosmosis. (SOURCE: LaQue, F. L., Marine Corrosion: Causes and Prevention, Chapter 16, John Wiley & Sons, New York, NY, pp. 285-87, 1975.)

also a key characteristic wherever coatings are to be used with cathodic protection, since such protection produces a strong excess of electrons on the metal. If, however, a coating has sufficient dielectric strength, it can break the electrical circuit and thus prevent the cathodic current flow. Since the dielectric strength of a coating can be affected by moisture absorption, the lower the moisture absorption, the more favorable the dielectric strength.

Resistance to Ionic Passage

In order for a coating to be effective, it must have a strong resistance to the mechanism of ionic passage. If chlorides, sulfates, sulfides, or similar ions were readily transferred through the coating, it would have little resistance to corrosion, but they would also reduce the dielectric strength in the coating, making it more conductive and, therefore, less corrosion-resistant. Normally, *ionic passage* is the transfer of ions from the exterior of a coating to the substrate. A reverse situation, however, can also take place: the transfer of electrons through the coating from the substrate to the surface. This is most common where a coating is subject to an impressed cathode current and the calcareous or other salt deposits build on the exterior coating surface. The only way the deposits could form on the exterior of an intact coating is if electrons passed through the coating from the substrate, causing the precipitation of calcareous salts on the coating surface.

A coating which has a very high molecular weight and dense molecular structure would have the greatest resistance to ionic transfer through the coating. It appears that the same would be true for electron transfer. An example of this type of material is a high-temperature baked phenolic coating which has a very strong cross-linked structure with excellent water and acid resistance. Such coatings have been used for many years as tank linings and linings for oil well tubing where aggressive solutions are involved.

Coating resins have varying degrees of resistance to ionic and electron passage. Individual coating formulations, even where the same resin binder is used, can also have individual resistance and quite varying properties in this respect. Moisture absorption and moisture vapor transfer rate must also have a bearing on this coating characteristic. Properly formulated vinyl and epoxy coatings both have good resistance to ions. Epoxy coatings have shown some tendency to pass electrons. Alkyds, on the other hand, may have good resistance to ionic passage during the first portion of their life, although as they continue to react in the weather, they become more permeable over a period of time.

Chemical Resistance

Resistance to ionic passage is also a contributing factor to chemical resistance, although the two properties are not necessarily the same. A phenolic coating, for example, which is highly cross-linked and resistant to ionic passage, is not resistant even to reasonably dilute alkali solutions. Although the coating resists the passage of ions, it is rapidly destroyed because of its reactivity with alkalies.

Chemical resistance is the ability of the coating, and particularly the resins from which it is formulated, to resist breakdown by action of chemicals to which it is exposed. Vinyl coatings have a very broad range of chemical resistance, in fact, they probably have the broadest range of any of the coating resins. They are resistant to most acids, alkalies, and salts and to oxidizing conditions, as well as to many solvents such as alcohol. On the other hand, they are readily dissolved by other chemicals. The difference is that while vinyl coatings are generally resistant to inorganic reactions, they are less resistant to the activity of organic materials.

Furan materials, which are used as cements for acid-resistant brick floors and tanks, also have a broad range of chemical resistance to acids, alkalies, and salt, but are rapidly attacked by oxidizing materials. Epoxies have excellent resistance to alkalies and generally to water, but are less resistant than the vinyls to most acid solutions. Epoxies are more resistant than vinyls to many solvents.

Chemical resistance, therefore, depends on both the coating's formulation and on the resins from which the coating is produced. A vinyl coating could be formulated with reactive pigments and would become much less chemical-resistant. On the other hand, when formulated with highly inert pigments, a coating's chemical resistance can be improved. Generally, a coating which is considered chemical-resistant and which would be used for corrosion resistance in a chemical atmosphere should be resistant to salts, acids, and alkalies of a rather wide pH range. It should also be resistant to organic materials such as diesel oil, gasoline, lube oil, and similar materials, since these are compounds found in almost all industrial operations.

Alkali resistance is, of course, extremely important in a primer. Since one of the chemical reactions in the corrosion process is the development of strong alkali at the cathode, any primer which is not highly resistant to alkali will tend to fail in the cathode area, resulting in undercutting of the coating and spreading of corrosion underneath the coating.

Proper Adhesion

A corrosion-resistant coating must also be highly adherent. Since the property of adhesion is essential in preventing the effects of water on the life of the coating and in preventing the problems caused by a temperature gradient across the coating, osmosis and electroendosmosis, adhesion is probably the key requirement in a corrosion-resistant coating. Irrespective of most of its other properties, the coating with very strong adhesion to the surface will retain its integrity much longer than one with less adhesion but other strong characteristics. Adhesion, in fact, has been the sole subject of many books, since each coating is unique in terms of this property.

Adhesion is created by the physical and chemical forces which interact at the interface of the coating and the substrate. The early vinyl coatings which were composed of vinyl chloride-vinyl acetate copolymers had little or no adhesion to the surface and were only practical where the surface was extremely porous. The addition of a third monomer, however, in very small quantities during the vinyl resin polymerization process created a limited number of organic acid radicals on the polymer. These had a strong affinity for metal surfaces and provided a bond between the polymer and the metal surface. Epoxies also have chemical radicals within the molecule which have a strong affinity for metal surfaces. These are often called *polar molecules*. (Strongly polar materials often have the characteristics required for adhesion.)

From the standpoint of the corrosion engineer, adhesion is more a physical problem of applying a coating to a given surface in such a way that the inherent adhesion of the coating will be fully utilized. This involves applying the coating to a clean surface with adequate roughness or tooth, and making sure that the coating adequately wets the surface. This wet film application is necessary so that the liquid coating can come in intimate contact with the substrate. The coating shown in Figure 3.10* demonstrates such adhesion. When cut as shown, there was no evidence of chattering or break-away along the edge of the cut. The cut was smooth, with a feather edge right down to the substrate and no evidence of any improper interface. This coating has proven to have excellent adhesion even under immersion conditions.

Physical roughness can be an important factor in adhesion. Blasted surfaces serve as a good example. As

---

*See color insert.

compared to a smooth, cold-rolled surface, a blasted surface has several times the area on which to adhere in addition to its physical roughness or tooth. This can make a substantial difference in the adhesion of a coating under almost any exposure condition, as shown in Figure 3.11. All four panels in this figure were coated with the same coating. Three were cold-rolled metal and the fourth was sandblasted, cold-rolled steel. Equal thicknesses of coating were applied to each panel. They were all scribed and then immersed in water. The coating with the best adhesion, *i.e.,* applied over a substrate with a greater surface area and some physical roughness, shows no tendency to disbond from the steel. Those surfaces which were not sandblasted, however, disbonded in the area of the scribe.

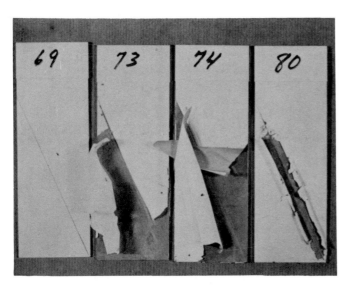

FIGURE 3.11 — Test panels showing the importance of physical roughness to coating adhesion. All panels are cold-rolled steel. Panel 69, however, was blasted and demonstrates excellent bonding, even along the scribe. (SOURCE: LaQue, F. L., Marine Corrosion: Causes and Prevention, Chapter 16, John Wiley & Sons, New York, NY, pp. 285-87, 1975.)

Undercutting

Undercutting is a measure of adhesion. The term applies to the corrosion at a break in the coating, growing back underneath the surface of the coating away from the break (Figure 3.12). To have resistance to undercutting, a coating must be strongly adhesive and must maintain its adhesion even at a raw edge between the coating and the steel. Examples of the undercutting of coatings can be seen on almost any industrial or marine structure. Edges are particularly prone to this type of coating failure since the coating is usually thinner at this point. In fact, extreme cases of undercutting have been observed where the corrosion extended under the coating several feet from the original break.

Organic coatings show the greatest tendency for undercutting by corrosion because of their often variable adhesion to a surface. Also, organic materials produce a definite interface between two very different materials; namely, the metal surface and the organic coating. The adhesion of the coating to the metal is one of a simple

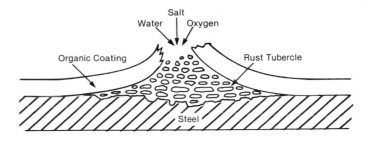

FIGURE 3.12 — Undercutting.

bond between the two materials. Part of this bond is physical, but the remainder has to do with compatibility and the ability of the coating to thoroughly wet the substrate surface. Figure 3.13* shows the results of corrosion starting at a scribe and undercutting an organic coating which otherwise had good resistance to a marine atmosphere.

As compared to organic coatings, inorganic zinc, galvanizing, or even ceramic coatings have a chemical bond in addition to a physical bond between the coating and the steel. This combined bond is much more durable and is not subject to undercutting because of corrosion at breaks in the coatings. The fact that the inorganic zinc coatings have a chemical bond and do not undercut is one of the reasons for their excellent corrosion resistance. Thus, they are often used as a permanent primer over which organic coatings can be applied.

Abrasion Resistance

Coatings which are applied to ships, helicopter decks, barges, offshore platforms, and other similar areas are excellent examples of why many corrosion-resistant coatings also require abrasion resistance. In these areas, coatings are subject to the movement of heavy equipment, foot traffic, possible wheel traffic, and damage by tools and equipment. In order to withstand this type of service and remain effective as a corrosion-resistant film, a coating must be tough, extremely adhesive, hard, and resistant to shock.

Another type of abrasion is the result of sand which is windblown or carried in waves. In this case, the coating is scoured and worn through to the substrate. Sand abrasion occurs most often where waves impact sheet piling close to the sand surface. It has also been observed that when ship bottoms periodically touch river sand bars, their bottom coating is often scoured to the base metal. The scouring, in some cases, is so even and uniform that at the interface between the abraded coating area and the steel, each coat is easily visible.

Organic coatings, which range from soft and rubbery to very hard durable ones, have a wide span of abrasion resistance. Hardness, however, is not a measure of abrasion resistance. The most effective abrasion-resistant organic coatings are the polyurethane coatings. These have exceptional resistance to impact, scouring, and abrasion. Inorganic zinc coatings, however, because of their

---

*See color insert.

good adhesion to the steel surface and because of their silicate and zinc composition, have proven outstanding when applied on the decks of barges and ships and on the boottopping of ships.

Ability to Expand and Contract

Each coating material has a different coefficient of expansion. Any coating which is to withstand corrosive conditions must also have the property of expanding and contracting with the substrate. Thermoplastic coatings, in general, have little difficulty in this area. Inasmuch as they are temperature sensitive, the warmer they get, the more plastic they become and the easier they follow the expansion and contraction of the underlying surface. Since they generally also have good adhesion, the combination of the two properties allows them to withstand most normal expansion and contraction without difficulty.

Thermosetting or cross-linked coatings, such as epoxies or even alkyds, may, after some considerable aging, become brittle and cease to expand and contract with the substrate. This can lead to cracking and spalling from the surface because of the temperature cycling. For instance, where temperature cycling is normal, and as aging progresses, epoxy coatings have been observed to gradually check, crack, and spall from the surface. This reaction is more pronounced when a coating is heavily applied.

Temperature cycling need not be merely from a reaction vessel which heats and cools, but more normally is the result of change in the temperature of surfaces subject to direct sunlight. Dark coatings on a steel surface can rise to a temperature of 65 C (150 F) in daylight and drop to as low as freezing at night under certain conditions. A proper corrosion-resistant coating must withstand such temperature changes without loss of adhesion and without checking or cracking. Inorganic zinc coatings have proven to have exceptional resistance to such changes in temperature.

Weather Resistance

Temperature cycling is also a part of weather resistance. Corrosion-resistant coatings should be thoroughly weather-resistant since many of the surfaces which require corrosion protection are located in unprotected, weather exposed areas. Weather resistance requires the combination of a series of very different properties. A weather-resistant protective coating must withstand the sun's rays, rain, snow, hail, dew, freezing and thawing, expansion and contraction of the substrate, chemical fumes, dusts and particulate fallout, as well as continuing wet and dry cycling—usually on a daily basis. Weather resistance combines into one property almost all of the properties required of a coating for other more specific uses. To be weather-resistant, a coating must resist the above conditions without excessive chalking, checking, cracking, flaking, blistering, loss of adhesion, or substantial color or appearance change for a sufficient number of years to be economical.

A small support structure for an offshore well, which is one of the most weather-affected structures in the world, is shown in Figure 3.14*. It is located in a severe marine atmosphere and therefore subject to all of the above conditions. Coated with an inorganic zinc primer and an epoxy polyamide high visibility topcoat, it remains in good condition after 14 years of exposure.

Resistance to Dirt Pickup

Dirt pickup is a property that concerns appearance more than anything else. Each coating is distinct in this regard. Some coatings will tend to pick up dirt and grime from the atmosphere and hold it very tightly on their surface (Figure 3.15). Others will tend to be unaffected by dirt and grime, with the result that their surfaces stay bright and clean, except for extremely dirty conditions.

Dirt pickup is undesirable from a corrosion standpoint as well. Where dirt accumulates, so do corrosive deposits such as chlorides, sulfates, fly ash, and similar materials that create a good electrolyte and, therefore, a favorable atmosphere for corrosion.

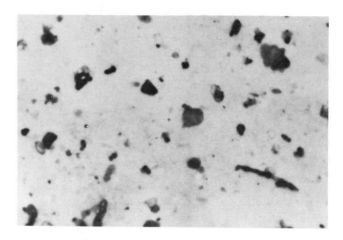

FIGURE 3.15 — Dirt particles imbedded in paint film, 100X. (SOURCE: Pictorial Standard of Coating Defects, Mildew Resistance, Federation of Societies for Paint Technology, Philadelphia, PA, 1979.)

Resistance to Bacteria and Fungus

There are two ways in which bacteria and fungus can affect a coating. First, where they settle on any dirt that has accumulated on the surface of a coating, they tend to live and thrive. This increases dirt buildup and dramatically detracts from the appearance of the coating. They also attack the coating itself and form colonies or areas which not only become unsightly, but which may actually be penetrated by corrosive conditions. Almost everyone is familiar with the dark, blotchy fungus growth on some coatings, particularly on the shady side of a structure (Figures 3.16 and 3.17). These fungus colonies are living on one or more of the coating ingredients and can eventually lead to premature coating breakdown. The susceptibility of coatings to such biological activity can

---

*See color insert.

Corrosion Prevention by Protective Coatings

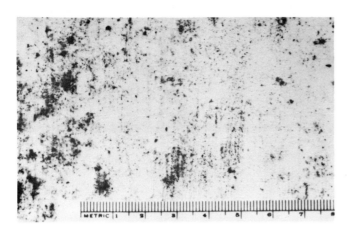

FIGURE 3.16 — Typical appearance of a coating suceptible to fungus attack. The fungus feeds on ingredients in the paint film. (SOURCE: Pictorial Standard of Coating Defects, Mildew Resistance, Federation of Societies for Paint Technology, Philadelphia, PA, 1979.)

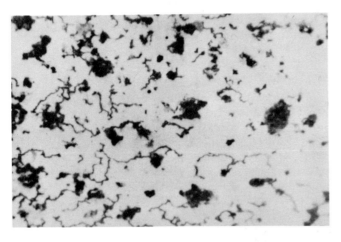

FIGURE 3.17 — Mildew on a coating, 50X. (SOURCE: Pictorial Standard of Coating Defects, Mildew Resistance, Federation of Societies for Coating Technology, Philadelphia, PA, 1979.)

often be offset by the addition of bacteriacides and fungacides to the coating itself during manufacture.

Under some conditions, catastrophic coating failures can occur because of biological activity. One such failure occurred when a polyamide coating was applied to a concrete sewer manhole. Polyamide epoxies have good resistance to water and good adhesion to concrete. In this case, however, the coating became mush within nine months to a year due to bacterial action. An amine cured epoxy, however, was unaffected by the same atmosphere. The difference is that the polyamide part of the molecule is vulnerable to biological attack, therefore making polyamide coatings unsatisfactory for sewer conditions.

Underground conditions can also lead to coating breakdown due to bacteria attack. If a coating contains organic sulfides, they are often subject to breakdown by anaerobic soil bacteria (sulfate reducing bacteria). Sulfur cements were used to join sewer pipe until it was discovered that sulfur-active bacteria used the cements for food with the development of additional quantities of H_2S

gas. This bacteria can also cause metal corrosion by any or all of the following:

1. creating differential electrolyte concentration cells on the metal surface;

2. creating a corrosive environment due to their life cycle and decomposition products;

3. acting as either anode or cathode depolarizers; and

4. generating sulfides which react directly with the metal to form metal sulfides.

Extensive testing was done in this area during World War II to develop coatings that could be used in tropical climates where biological activity created some serious problems. It was found that coatings for use in any area suspected of fungus or bacterial growth should be formulated with resins, pigments, plasticizers, etc., which in themselves cannot be used by biological organisms for food. For the more severe conditions, it is not enough to rely on bacteriacides and fungicides.

Pleasing Appearance

The earlier-mentioned characteristics of weather resistance, dirt retention, and biological action, all contribute to a coating's appearance. Although a coating is primarily used to prevent corrosion and protect the basic structure, it should also be pleasing to the eye and maintain its color. Especially since appearance is, in fact, one of the most obvious properties of a coating, whether it is applied to an automobile, tank, ship, or pipe. If a coating does not serve its purpose of creating a pleasing appearance, it can only be used in seldomly seen, unexposed areas, even though its other properties may justify a much broader use. Thus, if it does not look good, it is considered a poor coating in the eyes of the person who must use it.

Age Resistance

To provide corrosion resistance feasibly, a coating must provide protection for a reasonable period of time. If it does not have a sufficient life span to be a sound economic investment, the coating is not worthwhile. Corrosion-resistant coatings should therefore have the ability to maintain protection effectively over a period of many years under widely differing corrosion conditions. If the coating is defective in even one property, however, such as weather or abrasion resistance, it may not be able to do so. Thus, in order to be age-resistant, all of the properties previously described must be optimally applied.

Easy Application

Application is one of the most important coating characteristics, especially when dealing with intricate structures with many corners, edges, recesses, and similar areas. If a coating is somewhat difficult to apply, these are the areas which suffer and which break down first in a corrosive atmosphere. Most structures, tanks, and offshore platforms are reasonably difficult to coat, even under the best conditions. There are many welds, corners, and edges which are focal points for corrosion, so that if the application is not easily and properly accomplished, the corrosion resistance of the coating suffers.

Essential Coating Characteristics

Additional Coating Properties

Resistance to Extreme Temperatures

For the most part, any corrosion-resistant coating should have the majority of the properties already described in this chapter. Some coatings, however, are used for specific purposes which may necessitate unique requirements. The following are some additional properties which fall into this more specific coating use category.

While all coatings are subject to temperature and some temperature cycling, these conditions are generally moderate. Temperature, however, can be a key factor in coatings used for stacks, pipes, the exterior of process vessels, and for other similar uses. Where coatings are used for excessively cold temperatures (below those which would be found in normal atmospheric conditions), the three general characteristics to be considered are adhesion, shrinkage, and brittleness.

Adhesion is liable to deteriorate under excessively cold conditions, particularly when the cold varies from normal down to very cold temperatures.

Brittleness is closely affected by adhesion. When organic compounds are exposed to excessive cold, they tend to become more brittle and may shatter on impact or because of loss of adhesion.

Shrinkage is also a factor that can affect both adhesion and brittleness. There may be adhesion problems where a coating shrinks to a greater extent than the underlying surface. Thus, with impact or abrasion, the coating can shatter from the surface.

In order for a coating to withstand excessive cold, it should have excellent adhesion. The coating should also be somewhat resilient and retain its plasticity in cold temperatures. Coatings with this characteristic generally do not have shrinkage problems since they follow the expansion and contraction of the underlying surface. Butyl rubber and polyisobutylene polymers aid in the retention of cold weather adhesion.

Cold or heat can also change the corrosion characteristics of various atmospheres. The rates of chemical reaction or rates of corrosion reaction change dramatically as temperatures rise and fall. Epoxy resins which cure well at 21 C (70 F), cure very slowly at 5 C (40 F). If temperatures are 38 C (100 F), some epoxy coatings have been known to react so fast that application was impossible without keeping the mixed (catalyzed) coating in iced containers. Soils also show a marked change in resistance due to temperature changes, thus causing a change in the corrosion rate. Figure 3.18 indicates the rapid increase in soil resistance as the temperature drops below freezing. (As soil resistance increases, the corrosion rate drops.)

High temperatures create additional problems. There are a number of things which happen to coatings at high temperatures or cyclic high temperatures. Chloride containing polymers, such as vinyl chloride and chlorinated rubber, tend to break down over a period of time (depending on the temperature), releasing hydrochloric acid which can severely aggravate a corrosive condition. There are other polymers that do not contain a chloride atom in the molecule which may tend to de-

polymerize and even appear to evaporate from the surface. Some styrene polymers exhibit this characteristic. Other cross-linking polymers, such as epoxies, often overcure at high temperatures and become very brittle, shrink, and lose adhesion.

The inorganic zinc compounds are quite stable to temperatures in the area of 370 C (700 F) and have been used continuously at such temperatures quite effectively. (Figure 3.19). Silicone polymers also have excellent heat resistance in the same temperature range. For relatively short time exposures (e.g., days or weeks, but not continuously), inorganic coatings coated with silicone aluminum topcoats have been exposed to 520 C (968 F) and have maintained their effective coating characteristics. Some nonzinc coatings have also shown good resistance to higher temperatures and alternate high and low temperatures.

Table 3.2 gives some of the common coating types and their comparative temperature resistance. Depending on its formulation, each coating type may vary considerably either up or down in its temperature characteristics. (These figures are essentially for comparison between the coating *types* and not to be considered for the specific coating.)

Radiation Resistance

With the advent of atomic energy and atomic power, coatings have been used extensively for protection against the radioactive contamination of various substrates including steel, concrete, stainless steels, etc. In order to be effectively used on such installations, coatings must be

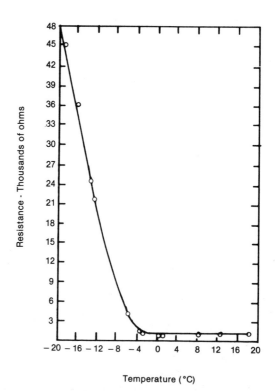

FIGURE 3.18 — Changes in soil resistance due to temperature. (SOURCE: U.S. Dept. of Commerce, Civil Engineering Corrosion Control, Corrosion Control-General, Vol. 1, Distributed by NTIS, AD/A-004082, p. 30, 1975.)

FIGURE 3.19 — Chemical process stack coated with inorganic zinc primer and white inorganic topcoat after several years operation at high temperatures.

TABLE 3.2 — Comparison of Temperature Characteristics for Some Corrosion-Resistant Coating Types

Coating	Immersion	Nonimmersion
Vinyl Copolymer	38 C (100 F)	65 C (150 F)
Chlorinated Rubber	38 C (100 F)	60 C (140 F)
Coal Tar	50 C (122 F)	65 C (150 F)
Coal Tar Epoxy	50 C (122 F)	95 C (203 F)
Epoxy	50 C (122 F)	95 C (203 F)
Urethane	38 C (100 F)	120 C (250 F)
Epoxy Phenolic	82 C (180 F)	120 C (250 F)
Baked Phenolic	82 C (180 F)	120 C (250 F)
Inorganic Zinc	—	370 C (698 F)
Silicone	—	370 C (698 F)

TABLE 3.3 — Radiation Tolerance of Corrosion-Resistant Coatings

Severe Exposure = Greater than 4.5×10^9 Rads
Moderate Exposure = 5×10^8 to 4.5×10^9 Rads
Light Exposure = Less than 5×10^8 Rads

	Maximum Allowable Radiation Dose in Air	
Coating	On Steel	On Concrete
Chlorinated Rubber	1×10^8 Rads[1]	1×10^8 Rads
Epoxy-Amine	1×10^9	1×10^9
Epoxy Coal Tar	5×10^8	5×10^8
Epoxy-Polyamide	1×10^{10}	NA
Inorganic Silicate Finish	1×10^{10}	1×10^{10}
Inorganic Zinc	2.2×10^{10}	NA
Epoxy Phenolic	1×10^{10}	1×10^{10}
Silicone (Baked)	1×10^{10}	NA
Urethane	5×10^8	6×10^9
Vinyl	1×10^8	—

[1]Rad: The unit of absorbed radiation. For most organic material, one rentgen = 1 Rad (ANSI N 5.12, 1973).

[SOURCE: Kirk-Othmer Encyclopedia of Chemical Technology, Munger, C. G., Coatings Resistant, Vol. 6, 3rd Ed., John Wiley & Sons, New York, NY, 1979.]

able to withstand varying amounts of radiation. Some coatings harden, become very brittle, shrink, and crack under heavy doses of radiation; others will tend to depolymerize and become sticky and soft; while still others blister and lose adhesion. Such coatings obviously are not suitable for use under these conditions.

Many coatings have been tested under high-density radiation sources. Table 3.3 shows the comparative radiation tolerance of the principal corrosion-resistant coatings. The inorganic coatings are apparently unaffected by radiation as they show no change after high exposure. Organic coatings, however, appear to be more stable and more resistant if they are strongly cross-linked. The thermoset coatings appear more radiation-resistant than those which are thermoplastic.

Resistance to Soil Stress

Soil stress is particularly important in connection with underground pipeline coatings. The pipe itself expands and contracts due to changes in temperature, as well as to the swelling and contraction of the soil as the water content varies. This is particularly a problem in high clay soils which expand to a great degree when damp and shrink an equal amount when dry. Clay soils move

substantially, which is evident by the large cracks that form as the soil dries out. This cyclic action tends to pull the coating away from the pipe or structure creating cracks, voids, or thin spots.

Coatings are often damaged due to poor backfill procedures which allow rocks or clods to impact the pipe. Nonuniform backfill pressure caused by clods or rocks, whether in the soil or in the backfill, can cause breaks in the coating. Rocks may be round, irregular, or broken with sharp points or edges. The weight of the pipe and the weight of the backfill can cause coating breaks at the points of highest pressure. Roots are also known either to penetrate the coating or to surround the pipe and grow to such an extent that they create sufficient pressure to cause the coating to flow, creating thin spots. Roots surrounding plastic pipe have been known to crush the pipe after a period of time. Coatings used under soil stress conditions must therefore have strong adhesion, high impact resistance, a low tendency to creep or move under pressure, and the ability to resist the abrasion of soil movement.

Resistance to Cathodic Disbonding

Cathodic disbonding is a type of failure characteristic of cathodic protection. In many ways, it is related to electro-endosmosis. The coating must resist the electrical potentials used under cathodic protection. Most coatings will withstand cathodic potentials of approximately -1.0 volts, with an optimum potential of -0.85 volts. Potentials of -1.1 volts and above can create conditions for cathodic disbonding, depending on the coating, its thickness, dielectric strength, water resistance, etc. The coating itself must have a low moisture vapor transfer rate and very high adhesion to help resist cathodic currents. Coal tar epoxies have good resistance to cathodic disbonding, as do many of the heavier pipe coatings which use butyl rubber as a bond coat. Figure 3.20 shows a coating which had disbonded due to excessive cathodic current. Large blisters, at times many inches across, are the usual indication of cathodic disbonding.

Friction Resistance

Some coatings are subject to friction, particularly when they are used as faying surfaces where two sections of metal are riveted or bolted together to form a friction joint. Inorganic coatings have proven to be very satisfactory under such conditions, while most organic coatings are unsatisfactory. Table 3.4 shows the comparative friction coefficient of various surfaces.

Types of Exposure

There are three essential types of exposure which corrosion-resistant coatings and linings are subjected to: atmospheric exposure, immersion, and underground exposure. Each of these has a different priority from the standpoint of the coating characteristics described previously.

Atmospheric

The obvious major difference between atmospheric exposure, immersion, and underground use is that of weather resistance. Coatings which are immersed or which are designed for underground use are not usually exposed to weathering conditons. On the other hand, weather resistance is the major consideration from the standpoint of atmospheric exposure. A coating under atmospheric exposure must withstand a multitude of conditions which include actinic radiation, heating and cooling, maximum exposure to oxidation, fallout from airborn chemicals, and alternate wetting and drying, in addition to the more basic requirements of strong adhesion, low moisture vapor transfer rate, the need for some inhibitive properties to reduce corrosion and undercutting at damaged areas, general chemical resistance, and the requirement to maintain a good appearance.

Atmospheric coatings are usually relatively thin films, which makes the retention of the above properties and the general aging of the coating extremely important. The conditions of exposure for an atmospheric coating are

FIGURE 3.20 — Cathodic disbonding of a coating under excessive cathodic protection potentials (-1.2 volts in seawater). Note white deposits on unperforated blisters.

TABLE 3.4 — Coefficient of Friction of Various Surfaces When Used as a Faying Surface

Surface Condition	Coefficient of Friction
Solvent-Based Inorganic Zinc	0.52
Rusted and Wirebrushed	0.51
Water-Based Inorganic Zinc	0.48
Rusted	0.48
Sandblasted	0.47
Mill Scale	0.30
Galvanize	0.25
Rust Prevention Paint	0.11
Red Lead Paint	0.06

NOTE: A coefficient of friction less than that of sandblasted steel is not recommended for riveted or bolted joints.

extremely broad, ranging from coatings which are used in very hot, dry atmospheres and those used under tropical conditions to those which are used essentially in cool or cold climates.

Because of their use in such a very broad range of conditions, atmospheric coatings require careful development and formulation, as well as application, in order to obtain the corrosion resistance needed. Good examples of this are the coatings for automobiles which undoubtedly are the most highly engineered coatings in present use. They are required to maintain their appearance and prevent corrosion to the underlying surface for many years when exposed to all types of atmospheric conditions, temperatures, humidity, and chemicals. Their application is controlled to a much higher degree than that of almost any other coating. The surface is very thoroughly prepared, and is pretreated so that the highest degree of adhesion is obtained with a minimum of undercutting due to road damage. They are undoubtedly the finest coatings made for use under the broadest conceivable atmospheric conditions.

There are other coatings which are more effective under more specific atmospheric conditions, such as in marine atmospheres (*e.g.*, ship hulls and barge decks) or in industry where the coatings are exposed to chemical fumes, splash, or spillage.

Immersion

Immersion coatings, as compared with atmospheric coatings, are primarily subject to water solutions ranging from very pure water to ones containing high concentrations of various chemicals. Examples range from snow water or deionized water used in various industries, to seawater and on to higher concentrations of various materials such as acids, alkalies, and salts, or organic solutions such as sugars or glycols. There are, of course, specific immersion situations, such as the lining of petroleum and solvent tanks, where water solutions are not encountered. Water, however, is the key factor in most immersion conditions, since the effect of water on most coating materials can be quite severe.

The primary coating requirements for immersion, then, are adherence and resistance to moisture vapor transfer, ionic penetration, osmosis, chemicals, cathodic disbondment, and varying temperatures. Immersion coatings, for the most part, do not require any specific weather resistance since, once under the surface of a liquid or water, solar radiation, air oxidation, and similar conditions do not apply. They are, however, subject to continuous contact with water and/or various chemicals, either alone or dissolved in the water. This means that water absorption, the absorption of various ions, and the passage of moisture through the coating, are all at their maximum driving force and the coating must be designed to withstand these immersion forces.

Snow water, distilled water, or deionized water are close to, if not the most penetrating of all of the chemicals in which a coating is immersed. On the other hand, as the salt content of water is increased, it becomes more aggressive from other standpoints. It is more conductive and therefore corrosion can take place at a rapid rate. It is generally conceded that seawater is more aggressive than pure synthetic seawater. Polluted fresh water is more aggressive than pure fresh water and can be more destructive to some coatings than seawater.

Corrosion engineers must be prepared to determine which coatings are best under the particular conditions that exist. As an example, coal tar epoxy coatings are very effective under many water conditions, including contaminated water. On the other hand, they are not satisfactory for potable water because of possible taste and odor contamination. Inorganic zinc coatings can be used for continuous exposure to refined oil and various solvents, while they are not particularly satisfactory for immersion in water or seawater without topcoating. Vinyl coatings have been used as coatings for water storage tanks for many years. They are noncontaminating and can be used for potable water. On the other hand, they are not satisfactory for use where there may be organic solvent contamination of the water or for immersion in many solvent type materials. Certain epoxy coatings have good resistance to many organic chemicals, and yet are not satisfactory for many dilute acid solutions.

Thus, wherever a coating is to be used under immersion conditions, the conditions should be precisely determined prior to the selection of any coating. Solutions with minor contaminants which were scarce enough to be deemed unimportant, have caused many coating failures under immersion conditions. A good example is minor quantities of naphthenic acid in diesel oil (acid value of more than 0.5). Inorganic zinc coatings exposed to such conditions react rapidly with the naphthenic acid, forming zinc naphthenate which contaminates the diesel fuel.

Underground

The coating requirements for underground conditions are quite similar to those for immersion: adhesion, moisture vapor transfer, resistance to ionic passage, and resistance to osmosis. Underground conditions are in many ways similar to immersion, or actually can be immersion conditions where a pipe is subjected to submarine conditions or to high ground water tables. Ground water, then, is the key element to be protected against in underground conditions. Coatings must have an extremely high water resistance and resistance to moisture vapor transfer in order to be effective. On the other hand, other factors such as soil stresses come into play and can be extremely damaging. Biological damage also can exist here, due to the activity of sulfate reducing bacteria or various fungus conditions.

Coating thicknesses, however, are probably more important underground than in either of the other two coating exposures. Pipes or structures underground are subject to varying backfill conditions, varying soil movement, and expansion and contraction due to more or less moisture in the soil. Soil forces can be strong enough to actually pull the coating from the surface of the metal which then cracks it and allows other external forces to react on the metal itself. Cathodic disbonding and electro-endosmosis both are factors contributing to underground coating failure. High dielectric strength is also a requirement. Thickness is important since rock points and damage during backfill are quite common. Thickness also helps contribute to moisture impermeability as well as to the impermeability of various soil chemicals.

In general, many coatings applied to pipe and underground structures are thicker than atmospheric coatings or linings. Many pipes have an extruded plastic coating which may vary from 50 to 250 mils in thickness. Pipe wraps of various types using hot-melt coatings are commonly used to build a reinforced laminated coating over pipe surfaces. Thin coatings are generally less satisfactory and less durable underground than the heavier built-up coatings or those which are extruded over the exterior of the pipe.

Again, it should be stressed that the corrosion engineer must understand the soil conditions under which the coating must operate in order to be able to select the best coating for the job. Many factors have to be taken into consideration. For example, a coal tar epoxy was applied to an underground pipe which was alternately subjected to steam and cooling water. The soil was soft, sandy loam containing some but not excessive moisture and no particular chemicals. The coal tar epoxy coating was considered a good selection inasmuch as the temperature involved was too high for coal tar or asphalt coatings. After a short period of cycling between steam temperatures and cooling water, the pipe was perforated from the exterior because of the disintegration of the coal tar epoxy due to the cycling temperature conditions.

All conditions must be taken into consideration, even the minor ones, when selecting a coating for either atmospheric, immersion, or underground exposure since no two coating jobs are ever the same, even though the conditions may seem to be the same. Table 3.5 lists the various coating characteristics that have been described and rates from 1 to 10 (the most to the least important) the characteristics which are most important for atmospheric, immersion, and underground conditions. There are a number of coating characteristics which are not numerically identified for any of the conditions, but which may be important where specific conditions are encountered.

When the corrosion engineer intends to build or bury a structure for long service life, one that will be relatively stable and free of excessive maintenance, there are certain basic fundamentals to consider. A selection of the coating type best suited for the task to be performed must be made to also meet the design requirements of operation and maintenance. In a paper on "Coating Fundamentals," Sal Bellassai listed these fundamental considerations.

   1. The type of environment in which the structure will be buried or submerged.
   2. The characteristics of the environment as to the texture, uniformity, varying degree of moisture or soil shrink factor.
   3. Operating temperature ranges.
   4. The nature of the product carried in the pipeline.
   5. Can quality of backfill be controlled?
   6. Use of cathodic protection and possible range of current potentials.
   7. Method of application, whether in the plant or in the field.
   8. The degree of handling to which the coated structure will be subjected prior to final resting location, such as loading, unloading, bending, and so forth.
   9. The experience with similar materials in this environment or on paralleling structures.[1]

All of the above factors must be considered since each represents a force of deterioration that, by itself or in conjunction with any of the others, will result in the degradation of the coating system. While the above list is designated for pipe, the principles are relevant for the selection of any coating application.

TABLE 3.5 — Comparison of Coating Characteristics Required for Atmospheric, Immersion and Underground Exposure

Coating Characteristic	Atmospheric Exposure	Immersion	Underground
Weather Resistance	1	—	—
Water Resistance			
M.V.T. Rate	5	2	2
Adsorption	—	—	—
Osmosis	—	3	—
Ionic Passage	6	4	3
Electroendosmosis	—	3	4
Adherence	2	1	1
Undercutting	4	—	—
Inhibition	3	—	—
Temperature Resistance	—	8	—
Temperature Cycling	7	—	—
Thermal Gradient	—	—	—
Chemical Resistance	9	6	9
Dielectric Strength	—	5	5
Cathodic Disbondment	—	7	6
Biological Damage	—	—	8
Thickness	—	9	10
Resistance to Soil Stress	—	—	7
Radiation Resistance	—	—	—
Abrasion	—	—	—
Appearance	8	—	—
Dirt Pickup	—	—	—
Age Resistance	10	10	—
Easy Application	—	—	—
Easy Repair	—	—	—

NOTE: "1" indicates best resistance required; "10" indicates comparatively lesser resistance is needed. Characteristics without numbers may be critical under special conditions.

Measurable Coating Characteristics

For every coating material there are individual, measurable coating characteristics which are important in identifying the coating. They are: (1) specific gravity (weight per gallon), (2) total nonvolatile measured by weight, (3) total nonvolatile measured by volume, (4) viscosity, (5) fineness of grind, (6) color, and (7) solvent tolerance. These seven characteristics serve to identify the liquid coating material and indicate certain quantitative data in connection with its use. Such characteristics are determined directly on the liquid.

Specific gravity, or the weight per gallon, generally indicates something of the material ingredients used in the coating. A high solid coating will, for example, generally weigh more than one which is very high in solvent, and a coating which contains red lead will weigh more than one which contains carbon black.

The *nonvolatile by volume* figure indicates how much of the coating must be used to develop the required film thickness over a given area. Nonvolatile by volume data is useful in determining the wet film thickness required to develop a specified dry film thickness.

A total *nonvolatile by weight* provides a measure of the solids in the coating compared to the amount of solvent in the coating. The coating with a total nonvolatile by weight of 50%, means that it would have 50% resin and pigment by weight and 50% volatile solvents by weight.

The *fineness of grind* indicates the degree of dispersion

of the pigment in the vehicle and provides some determination as to the appearance of the coating. A coating with a very fine grind may be a high gloss or a very smooth semigloss finish.

The *viscosity* indicates the type of applications which are possible. Viscosity is generally expressed in two ways: (1) in seconds, which is the time required for a measured amount of the coating to flow through a measured orifice (This equipment is generally used for the lower viscosity coating materials such as the ones suitable for air spray.); and (2) in Krebs Unit where the viscosity is determined by a one point rotational viscosity meter (Krebs Stormer: The viscosity is ordinarily expressed as KU value, which means Krebs Units. This equipment is used for higher viscosity materials, generally those used for brushing or for airless spray. KU values in the range of 75 to 95 KU indicate brushing or spraying viscosity, and KU values in the range of 95 to 100 KU indicate a coating that may be intended for high build use.)

There are also some other terms which relate to viscosity. The first of these describes a material which retains the same viscosity no matter how rapidly it may be stirred. This is referred to as a *Newtonian solution*, and coatings of this type are generally the low-build, easy flowing type. The second term describes a material which appears to be very heavy when stationary, but when stirred or agitated decreases in viscosity and is quite liquid, although it becomes heavy in consistency when the stirring or mixing is stopped. This is known as a *thixotropic material*. Thixotropic materials are low in viscosity as they pass through the head of the spray gun. However, as soon as the shear action of the spray equipment is finished, they again increase in apparent viscosity after being deposited on the surface. This allows the coating to remain on the surface in relatively high volume without drips or sags.

Solvent tolerance indicates a type of solvent which may be added to the coating as a thinner or type of solvent used for cleanup. Using improper solvent for thinning can be extremely damaging. As an example, vinyl coatings will tolerate some mineral spirits in the liquid form; however, once they are applied, the active solvents evaporate from the vinyl coating and leave the mineral spirits behind. These mineral spirits are gradually squeezed out and accumulate underneath the coating, leaving the coating without any adhesion whatsoever. There are many instances where this type of failure resulted when proper thinning instructions were not followed.

In terms of color, it is essential when coating large structures to make sure that all containers of liquid coating are from the same batch. If they are not, spot checks should be made to assure that the color is consistent.

There are also inherent coating characteristics which are not apparent in the liquid coating, but are measurable only after the coating is applied. These characteristics include gloss, drying time, time to tack-free, time to touch, time to recoat, and hardness. These are generally different for each coating material being used, and each may be an important consideration in the selection of a coating.

Whenever a protective coating is being considered as a means of dealing with a corrosion problem, the first step to be taken involves determining the desired coating characteristics. Without first analyzing the specific, essential coating requirements, improper materials may be selected that could result in costly failures.

References

1. Bellassai, S. J., Coating Fundamentals, Materials Performance, No. 12 (1972).

4

Coating Fundamentals

The fundamentals of a coating could refer to several things, depending on the purpose or the use of the coating. For example, an antifouling paint would have the fundamental property of inhibiting the growth of animal or vegetable organisms on the coating. A fire-resistant coating must fundamentally resist burning, or at least retard the burning of the substrate. A coating to be applied over concrete must have a fundamental property of resistance to strong alkali.

All corrosion-resistant coatings, however, must fundamentally resist the corrosive atmosphere and prevent it from reaching the basic structure. Thus, there are as many variations in the types of coatings as there are in the forms of corrosion. The design of an effective anticorrosive coating is a complex task which requires an extensive knowledge of not only corrosion principles, but of the science and chemistry of coating formation as well. Without such inclusive information, the development of effective corrosion-resistant coatings would be impossible.

A coating is not a self-supporting structure. It is part of an overall system which includes the basic structure that supports the coating. Although it is always on a substrate of one kind or another, a coating can be thought of in the same light as a building. In order to be strong, a building must have a heavy, carefully constructed foundation; in order to be durable, a coating must also have a carefully designed (formulated) and constructed (applied) foundation (substrate and primer). A building also consists of a number of interlocking parts; the foundation, the superstructure, and the roof, and each one has a different function. The corresponding parts of a coating are the primer, intermediate coats, and topcoat.

In the case of a small building which has a relatively short, useful life, the foundation and superstructure may be minimal. The same is true of a coating applied only for decorative purposes where surface preparation, applica-

tion, and long life may be easily overlooked. In the instance of a substantial industrial structure, however, durability, reliability, and long life are required. Again, the same holds true for an industrial corrosion-resistant coating which likewise must be engineered with a properly prepared substrate, a sound foundation coat or primer, a strongly reinforced intermediate coat, and long-lasting weather- and corrosion- resistant topcoats. In constructing a building, the substrate (the soil or ground) indicates the type and extent of the foundation since sand, clay, or rock all have different foundation requirements. The same, of course, is true of coatings. The primer must be designed specifically for the substrate, whether it is steel, aluminum, concrete, or wood. In fact, the surface over which a coating is applied may be more important from the standpoint of long life and durability than the design of the coating itself.

The fundamental concepts involved in corrosion-resistant coatings, then, include those of coating protection, component design, component function, and coating formulation. Many coatings contain as many as 15 to 20 ingredients, each of which has its own function in the overall performance of the coating.

A coating system may employ one or more of the basic coating concepts of impermeability, inhibition, and cathodic pigments. While many coating systems employ only one of these concepts, some of the most successful anticorrosive systems combine two of the concepts into one coating system.

Basic Coating Concepts

Impermeability

Impermeability is a concept basic to most available anticorrosive coatings. While no coating is totally impermeable to moisture vapor, an *impermeable* coating contains no materials which will react with moisture vapor. Each ingredient is designed to be unaffected by the moisture vapor and to only allow the vapor to accumulate within the coating to the point of normal moisture absorption content.

An impervious coating is most often used as an immersion coating and must therefore be inert to surrounding chemicals. It must also be impervious to air, oxygen, carbon dioxide, and the passage of ions and electrons. It must be dielectric and have very high adhesion to the underlying surface, and also it must wet the surface well enough to prevent any voids at the coating substrate interface. All in all, an impervious coating forms an inert barrier over the surface (Figure 4.1). This concept has been responsible for some of the most effective anticorrosive coatings available.

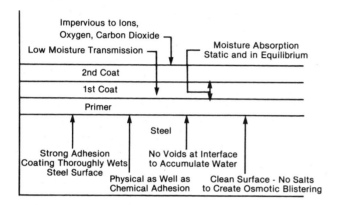

FIGURE 4.1 — An impervious coating serves as an inert barrier to protect the surface. (SOURCE: Munger, C. G., Kirk-Othmer: Encyclopedia of Chemical Technology, Coatings, Resistant, 3rd Ed., Vol. 6, John Wiley & Sons, New York, NY, pp. 456-7, 1979.)

This type of coating prevents corrosion of steel by interrupting or providing a block to the normal processes necessary for corrosion. Coatings used for water immersion have long proven the effectiveness of this type of coating. Many hundreds of miles of steel pipe have been lined with coal tar enamel or a hot coal tar pitch applied to the interior of the pipe, usually by centrifugal means. It consists of a primer and a thick hot-melt coat of the coal tar enamel which forms an entirely impervious coating. Many inert vinyl coatings have also been used for water immersion. Figure 4.2 shows the interior of a large scroll case which is part of a water turbine at the outlet of a dam for the generation of electricity. The scroll case is the outer portion of the hydraulic turbine which swirls the water into the turbine wheel. This coating, which has been in service five years, is also fully impervious and is without any inhibitive or cathodic primer.

Inhibition

The second concept involves an *inhibitor* which usually is only in the primer and consists of pigments that react with the absorbed moisture vapor within the coating. These then react with the steel surface in order to passivate it and decrease its corrosive characteristics. In-

hibitive pigments are sometimes characterized as anodically active, which means that the pigments within the coating sufficiently ionize in the water vapor to react with the steel or metal substrate. This maintains that area in a passive or inactive condition. Instead of a completely inert coating film, as with impervious coatings, the inhibitor coating uses the absorbed water in the film to aid in the passivation of the substrate (Figure 4.3).

FIGURE 4.2 — A vinyl coating on the interior of an electric turbine scroll case subject to continuous immersion in fast flowing water. After five years use, note the smooth coating without any evidence of blistering, even around the rivets.

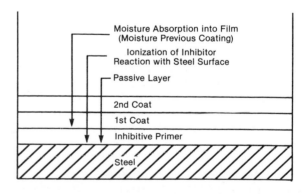

FIGURE 4.3 — In inhibitive coatings, moisture penetrates to the inhibitive primer where the reactive pigments are activated to passivate the metal substrate at the coating-metal interface. (SOURCE: Munger, C. G., Kirk-Othmer: Encyclopedia of Chemical Technology, Coatings, Resistant, 3rd Ed., Vol. 6, John Wiley & Sons, New York, NY, pp. 456-57, 1979.)

In contrast to the coatings developed on the basis of impermeability, the inhibitive coatings are used, for the most part, in atmospheric exposures; that is, as coatings for steel or other metals which are subject to weathering but not to immersion. There are some inhibitive coatings used for immersion; however, these are few and far between compared to those used for strictly nonimmersion, atmospheric purposes. This is due primarily to the inhibitive pigment's reactivity with water. Many pigments are so water-sensitive that upon immersion, osmosis takes place and the inhibitive pigment solution draws so much water into the coating that blisters develop. This was one of the difficulties encountered with many of the zinc chromate inhibited marine primers used during World War II. While corrosion was generally inhibited, blistering was sufficient to limit the useability of the coating.

One of the most unique and more effective coatings was a vinyl coating developed during World War II which used the vinyl wash primer as a metal treatment. This material contained zinc chromate as well as phosphoric acid so that there was a definite reaction with the metal substrate. This primer was followed by intermediate coats of a special vinyl resin solution pigmented with red lead. The red lead in the intermediate coats added to the inhibitive character of the metal treatment primer, while the final coats were primarily inert vinyl resin pigmented coatings used to provide weather resistance and color.

Inhibitive coatings have been used in marine atmosphere applications practically as long as steel vessels have been used. These coatings were originally oil based and heavily loaded with red lead. In fact, some such coatings used today are still oil modified. Many of the more recent coatings, however, such as the vinyl, epoxy, and urethane coatings, use an inhibitive pigment primer as a base when subject to atmospheric, marine, or industrial conditions. Figure 4.4 shows the application of a vinyl coating system to the downstream face of a large drum gate, an area subject not only to water, splash, and spray, but to weathering as well.

Cathodically Protective Pigments

The concept of cathodically protective pigments is, in many ways, an extension of the inhibitive primer principle. The reactions which take place, however, are entirely different. In the case of an inorganic zinc primer or an organic zinc-rich primer, the zinc acts as a cathode to the steel, and whenever there is a break, the cathodic action tends to protect the basic steel substrate from corrosion. Many times, where scratches or damage to an inorganic zinc coating have occurred, the zinc reaction products have proceeded to fill in the scratch or damaged area and seal it against further atmospheric action.

The inorganic zinc coatings (and this concept should include galvanizing) may be used alone or as a permanent primer over which topcoats may be applied. In order to satisfactorily topcoat over a reactive base coat containing zinc, the topcoats must be highly alkali-resistant. Such topcoats would include vinyls, chlorinated rubbers, epoxies, and coal tar epoxies. When zinc primers are overcoated with alkali-resistant coatings and in proper coat-

FIGURE 4.4 — Application of a vinyl coating system to the downstream face of a drum gate. Note the many corners, edges, and rivets which must be fully protected by the inhibitive coating.

ing thickness, the zinc primer remains inactive until a break occurs in the coating. At this point, the cathodically protective active primer reacts to protect the steel substrate (Figure 4.5).

Inorganic zinc primers are also highly adherent, reacting with the substrate to form a chemical bond in addition to the physical bond with the steel surface. The high adhesion of the zinc primer prevents undercutting of the organic topcoats so that the breaks in the coating which occur because of abrasion or other causes do not expand and enlarge, as is the case with many organic inhibitive primer systems (Figure 4.6). Although organic zinc-rich primers protect in a similar manner, provided the zinc is in particle-to-particle contact within the primer, the organic binder is not chemically reacted to the substrate. Thus, the coating may be undercut if corrosion occurs.

Figure 4.7 shows a large offshore structure under construction which was coated with a cathodically protective coating prior to being topcoated. Coating systems such as this (inorganic zinc primer plus vinyl or epoxy intermediate and topcoats) have provided long term protection for millions of square feet of steel surface under severe marine conditions.

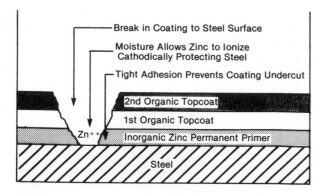

FIGURE 4.5 — An inorganic zinc primer reacts to protect the steel substrate at breaks in the alkali-resistant topcoat. (SOURCE: Munger, C. G., Kirk-Othmer: Encyclopedia of Chemical Technology, Coatings, Resistant, 3rd Ed., Vol. 6, John Wiley & Sons, New York, NY, pp. 456-57, 1979.)

FIGURE 4.7 — Offshore platform coated with inorganic zinc primer.

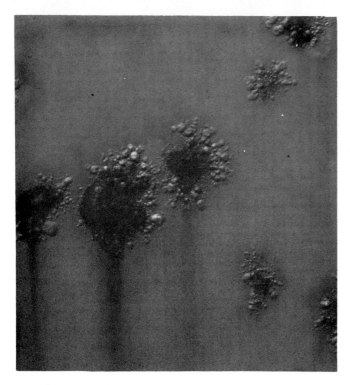

FIGURE 4.6 — Blistering of organic inhibitive coating with undercutting. Even though inhibitive, this is typical of breaks in many organic coatings subject to marine or otherwise corrosive atmospheres.

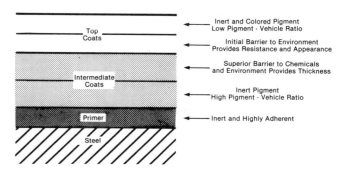

FIGURE 4.8 — Five-coat impervious coating system.

The concept of the cathodically protective pigment can be used for coatings which are to be subject to either atmospheric or immersion conditions. Under immersion conditions, particularly marine, the zinc coatings must be overcoated, preferably with inert, impervious coating systems. This prevents the gradual buildup of zinc salts on and within the zinc coating which inhibit its cathodic action. Proper cathodically protective pigmented systems, including the inert topcoats, have proven very effective under many immersion conditions.

The Coating System

For serious corrosion situations, the system approach (primer, intermediate coat, and topcoat) provides an excellent answer to many specific coating requirements. Figure 4.8 indicates a five-coat impervious coating system and the purpose for each of the three different kinds of coats.

Components

Primers

The primer is universal for all anticorrosive coatings and is considered one of the most important components of the coating system. The primary purposes of a primer are listed as follows:

Corrosion Prevention by Protective Coatings

1. Adhesion (strong bond to substrate).
2. Cohesion (high order of internal strength).
3. Inertness (strong resistance to corrosion and chemicals).
4. Intercoat bond (high bond to intermediate coat).
5. Distension (appropriate flexibility).

The *primer* is the base on which the rest of the coating system is applied. As a base, it must have strong adhesion to the substrate surface. If the coating system is an inhibitive one, it must contain the inhibitive pigments and be capable of using these pigments in a way which will passivate the metal surface and reduce its tendency to corrode. In a cathodically protective active primer, the coating preferably reacts to a certain extent with the steel surface in order to obtain an even greater and stronger adhesion. In addition, this primer must react with moisture and electrolytes from an outside source in order to cathodically protect the steel substrate. Primers are actually the key to the adhesion of the total coating system.

The primer must also provide a proper and compatible base for the topcoats. It must be thoroughly wetted by them, and, by its generally flat, nonglossy surface, provide some physical adhesion to the topcoats. Primers, then, have dual requirements: adhesion to the substrate and provision of a surface which will allow proper adhesion of the following coats.

Primers are often applied and allowed to stand for many days or months prior to the application coats.

Therefore, they must also have sufficient resistance to the atmosphere to protect the steel substrate from any corrosion during the period between the time the primer is applied and the time that the topcoats are applied. If it allows corrosion to take place during this period, it is not performing the whole purpose for which it was designed.

A primer, generally, must have the ability to stifle or retard the spread of corrosion discontinuities such as pinholes, holidays, or breaks in the film. The primer for the impervious coating system must be, in itself, highly adherent and very inert, so it tends to stifle corrosion due to these factors. The primers using the inhibitive system contain anticorrosive pigments which aid the coating in preventing corrosion, while the primers which use the cathodic protection principle inhibit the corrosion due to actually providing cathodic protection to the underlying steel surface.

Primers also, under certain conditions (particularly where they are used for immersion conditions or tank linings), must have chemical resistance equivalent to the remainder of the coating system in order to satisfactorily protect against the chemical solution in which it is immersed. This property is usually associated with the impervious coating system, and unless the primer is as fully resistant as the remainder of the system, the underfilm breakdown would soon cause coating difficulties with rapid corrosion. Table 4.1 is a comparison of the various primer types.

TABLE 4.1 — Primer Comparison Table

Requirement	Alkyd or Oil Primer	Primer Type Inhibitive Primer (May be Mixed Resin System)	Impervious Primer (Resin May be Identical to Topcoats)	Cathodic (Zinc) Primer
Bonding to Surface	Usually wets and bonds to most surfaces. Somewhat tolerant of substandard surface preparation. May be used on metal or wood surfaces, but not concrete. Not recommended for immersion.	Adhesive properties are major consideration. Not as tolerant of substandard surface preparation as oil primers. Used primarily on metal. Inhibition not necessary on wood or concrete. Usually not for immersion.	Surface must be properly prepared. Primers require maximum adhesion. Used on metal or concrete. Used for immersion.	Inorganic zinc: outstanding adhesion to properly cleaned steel or iron surfaces. Chemical as well as physical adhesion. Organic zinc: adhesion depends on base resin.
Adhesion of Topcoats	Satisfactory for oil types. Usually unsatisfactory for vinyls, epoxies, and other synthetic polymers. Soften and lose integrity by attack from solvent systems of synthetic topcoats.	Formulated for adhesion of topcoats. Specific coating systems may require specific primer.	Usually part of specific generic system. Primer designed for specific intermediate or topcoats.	Fits into wide range of systems. "Tie Coat" may be required. Specific recommendation should be obtained for immersion systems.
Corrosion Suppression	Limited. Alkali produced at cathode attacks film (saponification). Spread of underfilm corrosion results.	Usually formulated with good resistance to alkali undercut. Contain inhibitive pigment for a degree of corrosion resistance.	Relies on inert characteristics. Very strong adhesion.	Inorganic zinc: outstanding ability to resist disbonding and underfilm corrosion. Anodic property of metallic zinc protects minor film discontinuities.
Protection as Single Coat	Limited by severity of exposure.	Limited by severity of exposure.	Limited by severity of exposure. Usually suppresses corrosion alone for some period of time.	Will protect without topcoat with very few exceptions.
Chemical Resistance	Typical of alkyds. Not recommended for alkali exposure.	May be of lower order of resistance than that of topcoat due to inhibitor.	Typical of coating system.	Not resistant to strong acids and alkalies. Inorganic: outstanding solvent resistance.

Intermediate or Body Coats

Intermediate or body coats are usually used in coating systems which are designed for specific purposes. One coating system which uses this principle is the wash primer, vinyl red lead, vinyl topcoat system. In this case, the wash primer is considered a metal preparation, the red lead vinyl is considered the primer, and the topcoats provide resistance to the atmosphere. Irrespective of the names applied to these coats, the principle remains the same since the intermediate vinyl red lead coat is filled with red lead and inert pigments, providing a heavy body coat which adds thickness and resistance to the coating system. The primary purposes of an intermediate coat are to provide:

1. Thickness for total coating.
2. Strong chemical resistance.
3. Resistance to moisture vapor transfer.
4. Increased coating electrical resistance.
5. Strong cohesion.
6. Strong bond to primer and topcoat.

The formulation of the intermediate coats is important, primarily as it increases thickness. Physical thickness improves many other essential properties of a coating, such as increased chemical resistance, reduced moisture vapor transfer rate, increased electrical resistance, increased abrasion, and impact resistance. The body coat or intermediate coat must also provide strong adhesion to the primer, as well as a good base for the topcoats. The intermediate coat usually has a rather high pigment to vehicle ratio, so that it is a flat coat with physical adhesion. Without the ability of this material to properly adhere to the primer and to provide proper adherence to the topcoats, the problem of intercoat adhesion would cause early coating breakdown.

Another important role of the intermediate coat is in providing a superior barrier with respect to aggressive chemicals in the environment or when immersed. The intermediate coats are usually deficient with respect to appearance properties so that they are generally not used as finish coats. They may also be used to add physical resistance. Most intermediate coats are used with the impervious type coating system.

Topcoats

Topcoats also perform several important functions in that they:

1. Provide a resistant seal for the coating system.
2. Form the initial barrier to the environment.
3. Provide resistance to chemicals, water, and weather.
4. Provide a tough and wear-resistant surface.
5. Provide a pleasing appearance.

In the primer, intermediate coat, and topcoat system, the topcoats provide a resinous seal over the intermediate coats and the primer. The first topcoat may actually penetrate into the intermediate coat, thus providing the coating system with an impervious top surface. The topcoat is the first line of defense against aggressive chemicals, water, or the environment. It is the initial barrier in the coating system. It is more dense than the inter-

mediate coats because the topcoats are formulated with a lower pigment to vehicle ratio. Although they are a much thinner coat than the intermediate coats, as would be expected from the lower pigment volume, with the higher resin ratio they provide a tough upper layer. It also provides the coating system with the characteristic of appearance through its color, texture, and gloss.

There are a number of situations, however, where the intermediate coats provide the primary barrier to the environment, while the finish coat is applied for entirely different purposes. The topcoat, for instance, can be used to provide a nonskid surface, while the intermediate coat and the primer provide the barrier to the environment, as in a marine environment. The finish coat or topcoat may also provide resistance to marine fouling, such as shell growth and algae. In other cases, a topcoat may be applied for appearance alone. A summary of coating component functions is given in Table 4.2.

TABLE 4.2 — Functional Summary of the Component of a Coating System

Coat	Main Function	Specific Requirement	General Requirements
Primer	Adhesion	Adhesion to substrate	Adhesion Cohesion Resistance
		Bond to intermediate	Flexibility Internal bond
Intermediate	Thickness and structure	Bond to primer	Cohesion Intercoat bond
		Bond to topcoat	Thickness Strength Resistance - Chemical - M.V.T. - Electrical
Topcoat	Resistance to atmosphere	Atmosphere and/or environment resistance	Seal surface Strength Resistance Flexibility
		Bond to intermediate	Appearance Toughness

Variations

While a coating system is obviously only for severely corrosive conditions, modification of the numbers of coats can effectively be used for a variety of corrosive situations. Several combinations of this system have proven very effective and have provided many years of corrosion-free service on offshore platforms.

A coating system, however, need not be composed of the three different parts; even a single coat can provide a coating system, depending upon the requirements of the coating. Inorganic zinc coatings, for example, provide an excellent one-coat system for the storage of refined oil products and many solvents. A single coating formulation applied in two or more coats may provide the best answer to a specific problem. A self-adherent vinyl system, for example, has been applied to the interior of water tanks in three or four coats for many years with outstanding performance.

A two-material system may provide the best answer under other circumstances. An epoxy-based primer, for example, topcoated with one or more coats of coal tar epoxy provides a system with excellent resistance to water and seawater, particularly on the bottom of ships.

The three-material system (*i.e.,* primer, intermediate, and topcoats) is often used where chemical resistance is needed. Another example where the three-material system has proven itself is on the face of dam gates, where the coating is subject to abrasion, immersion, and continuous weather. A vinyl system (which included a vinyl primer, vinyl intermediate coats using a silicate pigment, and vinyl topcoats pigmented with aluminum), provided a long-lasting system for the upstream faces of rotary and floating drum gates. A similar system has been used on the interior of scroll cases and on impellers for the generation of hydroelectric power.

Coating systems, then, may consist of any number of coats and combinations of materials. In most cases, a coating system is based on the same or similar resin combinations. More recent technology has allowed some combinations of entirely different coating materials, such as an inorganic zinc permanent primer, an epoxy intermediate coat, and coal tar topcoats. In this case, each one of the different binders in the coating system layers provides its own series of properties, characteristics, and benefits to the total coating system. Such systems are generally for specific purposes such as immersion in seawater, for use in cargo ballast tanks, or in chemical storage.

Mixing Coating Systems

There are also dangers in mixing different coating systems. This situation usually occurs when individuals take it upon themselves to apply a mixed coating system. The use of a mixed coating system simply for expediency or an individual's whim is not a recommended practice, as it frequently causes problems. An attempted application of vinyl or epoxy coatings over an alkyd or shop primer is an example of this kind of mix. It generally does not work since the solvents from the topcoat systems penetrate and break up the alkyd so that it no longer maintains its integrity. The reverse could also occur where a vinyl primer and intermediate coat is applied, with an alkyd topcoat applied over the entire system. The alkyd, in this case, may work, but more often than not, it will check and crack away from the underlying vinyl after a certain period of time. Mixed coating systems which have been researched by the manufacturer are generally more sound and should perform as recommended. Materials or systems should not, however, be used simply because it seems like a good idea at the time.

Basic Coating Formation

The basic formation of a coating is a highly technical reaction, and the type of reaction is extremely important to the effectiveness of the coating in its particular use. In order to produce a film which will perform practically and satisfactorily in a given environment, the coating, after its application, must convert to a very dense, solid membrane which is resistant to that particular environment. This conversion from the liquid resin to the solid resin film is the most important reaction which takes place in the formation of a coating.

Molecular size, weight, and complexity of the coating resin often determine the type of coating film which forms. Generally, for corrosion-resistant applications, a very dense, tight, chemical-resistant film is desired. Resins which form this type of film by evaporation are of very high molecular weight, and are reacted into their finished form prior to being formulated into a coating. For the high molecular weight film to form on the surface, it is only necessary for the solvents to evaporate from the resin solution. Because of the high molecular weight of these resins, they are often difficult to put into solution, requiring strong solvents with the total coating formulation having relatively low solids.

On the other hand, resins which are of relatively low molecular weight may be liquids in themselves. These require reaction in place, either by catalytic action, reacting with other resins, or by reaction with oxygen or moisture from the air to form films. These materials, which are of lower molecular weight to begin with, have the advantage of building a higher solid combination into the coating so that there can be less volatile material or, in some cases, no volatile material in the coating. The polymerization, or condensation, creates the high molecular weight coating resin in place. In this case, the conditions during application are critical for the film-forming reaction to take place.

There are several different types of binders, or film-formers, which are used to formulate protective coatings for corrosion-resistant applications. Each of these have their own characteristics and requirements for film formation. The types of binders are those formed by solvent evaporation, oxidation, polymerization (co-reaction), condensation, reaction with moisture from the air, coalescence, and the formation of inorganic coatings.

Materials which form a film may generally be classified under two categories: thermoplastic and convertible. A *thermoplastic* material is one that becomes soft or fluid when heated. On cooling, it regains its original physical and chemical properties. *Convertible* film-formers undergo a chemical change or conversion. This results in a definite alteration of their physical properties. The word convertible is used broadly to include all of the methods of conversion mentioned above with the exception of evaporation of the solvent and inorganic materials. Since most of the film formers, regardless of class, are synthetic, a brief description of synthetic resins should serve as a background for their use in forming coating films. Unit II of the Paint Federation's Series on Coating Technology, "Formation and Structure of Paint Films," provides a general description of resins and their formation, as follows.

General Nature of Resins

Resins are polymers. They are made by combining single units (monomers) of chemical compounds, such as styrene, vinyl chloride, vinyl acetate, ethyl acrylate, phenol, formaldehyde, and urea. Activated by heat and catalyst, the monomers unite to form molecules that are many times larger. When only one kind of monomer is used, the resin is a homopolymer. If two or more kinds are used, the resin is a copolymer. Each type and grade of resin has characteristic properties. This difference in properties determines which resin or combination of resins is most suitable for a specific purpose, and it permits a resourceful paint chemist to satisfy a wide diversity of requirements. To be capable of polymer formation a monomer must contain chemical groups that have potential chemical reactivity. Frequently resins take their names from these chemical groups. Among such groups are: hydroxyl ($-OH$); carboxyl ($-COOH$); amine ($-NH_2$); vinyl ($CH_2=CH-$); epoxy ($-\underset{O}{C}-CH_2$); and isocyanate ($-N\overset{2}{C}=O$). The functions of heat and catalyst are to energize and direct the reactions of polymerization. A double bond in a formula indicates unsaturation. Unless the double bond is in a benzene ring structure, it carries a strong tendency to react in a manner that replaces it with a single bond.

Next to the kind of monomers, the most important feature of a resin is the degree of polymerization. With a given type of resin increasing the size of the polymer molecules, or molecular weight, results in a general improvement in film properties; greater hardness; greater film strength; better resistance to water, chemicals and solvents; and better exterior durability. On the negative side, higher molecular weight means reduced solubility and higher viscosity or lower solids of solutions. This relationship between viscosity and film properties has great importance. It may be the deciding factor in the choice of resin for a particular purpose.[1]

Film Formation by Solvent Evaporation

The resins which dry by solvent evaporation are all thermoplastic film-formers. Shellac is an old example of this type of material. The original gum shellac was dissolved in alcohol and applied to a wooden surface; the alcohol evaporated leaving the dry, continuous film of shellac on the surface. A more modern example would be a solvent vinyl solution where the vinyl resin is dissolved in ketone solvents and, when these evaporate, a smooth, clear, continuous film of vinyl resin is obtained. Film formation by solvent evaporation, which appears to be the simplest type of film formation, is actually far from simple. The film formation does not commence until the evaporation of the solvent has reached an advanced stage which brings the molecules of the resin into such close contact that their mutual chemical attraction draws them together. Film properties are influenced by the molecular arrangement or structure within the film. A homogeneous, dense structure is promoted by a solvent that maintains maximum dispersion and mobility of the polymers during film formation. The opposite is true when the resin is precipitated out of the solvent. The attraction between the polymer molecules is not just limited to films formed from solution, but it is the underlying basis of all films, and is the force that holds the molecules together.

In order to obtain a smooth, continuous resin film, it is usually necessary to use a combination of solvents. In order to form a good vinyl resin film, for example, it is necessary to use a combination of solvents which are classed as active solvents, latent solvents, and diluents.

The *active solvents* are those which easily dissolve the resin and are the primary ones for putting it into solution. The *latent solvents* are less active, but still act as solvents, while the *diluents* are materials which will tend to soften the resin, but will not actively dissolve it. The combination of these three types working together often provides a better and stronger film than a single solvent alone. They

evaporate at different rates, with the active solvent usually being the last solvent to leave the film, creating the conditions whereby the resin molecules orient themselves properly to form a smooth, clear, continuous film. The skill in mixing the three types of solvents for any given polymer is what makes a solvent evaporating coating easy to apply as a smooth film with a good gloss. The aim in mixing the solvents is to provide a uniform evaporation. If the solvents evaporate too fast when the coating is applied, the resin will tend to dry before it hits the surface and overspray will result. Fast-evaporating solvents also eliminate any possibility of brushing. Solvent combinations which evaporate too slowly make for a very slow film formation. The film generally remains tacky and the solvent is retained for a long period of time, making the resin film less water- and chemical-resistant.

There is a considerable difference in the way that various resins will release the solvents from the film. In the case of vinyl resins, there is a *mutual solubility;* that is, the resin dissolves in the solvent, but there is also some of the solvent which dissolves in the resin. The solvent which dissolves in the resin is the solvent which is retained the longest and the one which is most difficult to remove from the resin. In many ways, it acts as a permanent plasticizer. Resins which hold the solvent in this way are said to have poor solvent release. On the other hand, there are other materials, such as chlorinated rubber, which have a rather rapid release of their solvents and quickly form a very hard film. The solvent formulation of the coating oftentimes has a considerable effect on the solvent release. If two solvents have the same volatility, the one with the higher solvency volatilizes more slowly from the resin itself.

Thermoplastic Film Formers

There are many thermoplastic film forming materials. Nitrocellulose, which had a great deal to do with the automobile industry, is one example. There are also the acrylic ester resins, such as methyl methacrylate, styrene butadyrene resins, vinyl acetate resins, vinyl chloride acetate resins, celulose acetate butyrate resins, various petroleum resins, including asphalt and coal tar, chlorinated rubber, and some rubber solutions. All types of solvent combinations are required in order to properly develop films from each of these materials, and no two would require exactly the same solvent combination.

There is also a wide variation in molecular weight between these materials. As the molecular weight increases, it is more difficult to dissolve the resin in solvents, resulting in a generally higher viscosity. This limits the amount of resin which may be placed in solution and practically applied to form a coating film. Thus, the dry film thickness of a number of the solvent evaporating coatings such as nitrocellulose, vinyls, and the acrylic esters, is rather low.

Films Formed by Change-of-Phase

Thermoplastic resins may also be formed into coatings by another method called a *change-of-phase*. This generally means that the resin is changed, usually by heat, from a solid to a liquid and then back to a solid again. The process is more commonly referred to as *hot-melt*.

Use of this hot-melt technique to form a coating dates back to the Egyptians and their method of applying rosin and tars to the bottom of their ships. Today, it is still an extensively used method, particularly in the piping field. The principal materials used in this process are asphalts and coal tars. Both of these are relatively easily converted to a liquid through the use of heat. The liquid resin can then be applied to the metal surface by *daubing,* a process of brushing the material on the surface while it remains in a liquid state. This method, which is often used on the interior of large water storage tanks and other similar structures, has a number of disadvantages. Because the liquid resin cools rapidly, it is particularly difficult to obtain a smooth, even film by this brushing technique.

For the interior of pipe, the hot-melt materials are often applied by centrifugal force; that is, the liquid materials are poured into the pipe while the pipe revolves. The centrifugal force of the turning pipe spreads the resin uniformly over the surface and a very smooth, even coating can be obtained.

The application of the hot resin to the exterior of pipe can also be done by pouring the resin evenly over the pipe surface as it turns. The common method of external application, however, involves inserting a number of pipe wraps into the liquid resin as the pipe revolves. This reinforces the hot-melt resin and aids in holding it in place, while at the same time making a more uniform, even coating. Pipe wrap materials can be organic fabric, fiberglass matte or fabric, asbestos matte, or other similar high-strength materials.

The hot-melt technique is effective wherever a basic resin can be melted to a reasonably liquid form. Since there are no solvent or volatile materials involved, 100% of the resin material is applied to the structure. Thick coatings are easily built-up from the hot-melt materials and, when done properly, the hot-melt materials wet the surface of the metal very well.

The formation of a good coating depends on the control of the temperature of the resin, the condition of the metal, the temperature of the metal, and, many times, the weather and humidity. Under most circumstances, the majority of the pipe applications are done under plant conditions where pipe is easily handled, so that a very good coating can be obtained at a relatively low cost. These materials can also be handled by over-the-ditch hot-melt wrapping machines, but only at some added expense and with less application control.

This change-of-phase or hot-melt process is similarly used for the application by extrusion of polyethylene coatings as on the exterior of pipe. In this case, the polyethylene is heated in the extruder and smoothed on the surface of the pipe by the extruder die in an even film. The film is formed and complete as soon as the polyethylene cools.

This change-of-phase mechanism is also used in the application of powdered coatings where the solid film-former is made into a very fine powder. The powder is sprayed on a preheated metal surface, which is above the melting point of the resin in order to hold it in place. The metal is then reheated to a temperature high enough to fuse the resin onto the surface. A fluidized bed process also uses the change-of-phase principle. In this case, the fine resin is fluidized by the use of compressed air, and parts are heated and then dipped into the fluidized resin until the proper thickness is obtained. Again, the part is reheated and the resin fused on the surface. In all of these change-of-phase coatings, the resin is applied at 100% solids so that no solvent evaporation is necessary to form the film.

Plastisols and Organosols

Another example of the use of this change-of-phase process for corrosion control is in the application of plastisols or organosols. In this case, the resin is dispersed in a plasticizer which remains in the coating permanently, or in a latent solvent which only becomes active when heat is applied. In the case of the organosols, it is possible to obtain a hard finished film which contains little or no plasticizer. In the case of the plastisols, the resin is dispersed in a plasticizer which acts as a latent solvent and does not tend to solvate the resin particles until a certain temperature is reached. At that temperature, the resin dissolves in the plasticizer, forming a semiliquid gel, and then, upon cooling, becomes a plasticized resin coating. This film-forming process is often used on steel pipe, particularly in the chemical and mining industries.

Film Formation by Oxidation

The formation of films in this category is primarily from drying oils which are natural materials of vegetable or fish origin. They are chemically classified as *triglycerides;* that is, compounds of one molecule of glycerine and three molecules of long-chain fatty acids.

The use of drying oils is one of the oldest methods of forming paints. The oils are applied in relatively thin films and allowed to stay in place until they have reacted with oxygen in the atmosphere long enough to become hard and dry. Originally, and for a considerable time after their original drying period, the oils are quite resistant to atmospheric conditions. They continue hardening, however, until they eventually check, crack, and chip away from the surface.

Oxygen reactions with unsaturated oils are varied and complex. The long-chain unsaturated oil molecule reacts with oxygen irrespective of whether it is attached to an alkyd, epoxy, or urethane to form the base coating film. Oxidation of an oil can isomerize, polymerize, and cleave the carbon-carbon chain, as well as form oxidation products. Blown oils of varying viscosity are manufactured through the reaction of oxygen in sufficient amounts to give what appear to be polyethers. The steps involved in film-forming from drying oils may be summarized as follows:

1. An induction period in which little visible change in physical or chemical properties of the oil occur but antioxidants present in the film are being destroyed.
2. Oxygen uptake becomes measurable and hydroperoxides and conjugation form.
3. Decomposition of the hydroperoxides occurs to form free radicals and their concentration increases and the reaction becomes autocatalytic.

4. Polymerization and cleavage reactions begin and high molecular weight cross-linked polymers as well as low-molecular weight scission products including carbon dioxide and water are formed. Oxygen absorption reaches a maximum at about the time the film forms and oxygen continues to be absorbed but at a much slower rate.[2]

Film Formation by Polymerization

There are several reactions which create large molecules out of small ones. Most often these are processes which must be carefully controlled under strict manufacturing conditions and are therefore not applicable to on-site coating formation. Vinyl resins, however, are an example of a resin which is completely polymerized before it is cast into a film from a solvent. These films cannot be polymerized in place. When coating films are reacted after application, the primary consideration is the coating reaction in place. Liquid resins are converted to a solid continuous film after application to the surface. Thus, the process actually being referred to is polymerization by cross-linkage.

Polymerization by Cross-Linkage

Polymerization by cross-linkage broadly includes many types of baked coatings, such as those used on appliances of all sorts. Its meaning in this discussion, however, is limited to the cross-linkage which occurs at normal or ambient temperatures and under the conditions in which protective coatings would be applied. The polymerization takes place between a monomer and one or more polymers of different types to produce the resin film which is cross-linked, as compared to the linear polymer described under vinyl resins. In this case, a rigid, three-dimensional molecular structure is created on-site to form a coating film which is thermoset, *i.e.,* the coating becomes insoluble in its own solvents and is not softened appreciably by heat.

The two most important examples of this type of reaction are both epoxies. One is the basic epoxy resin which is reacted with a monomeric amine, such as a diethylene triamine or an amine adduct, to form a cross-linked film. This reaction takes place at ambient temperature to form a hard, somewhat brittle film. The second example is the reaction of the epoxy resin with a polyamide resin. This cross-links the epoxy through the amine groups on the polyamide resin. In this case, by varying the proportion of the viscous polyamide resin, films can be formed with a very wide range of physical properties.

Polyurethane resins which use an amine curing agent are also of the cross-linked type. The basic resin used in this case is an isocyanate prepolymer, to which is added an amine curing agent just prior to application. The film is formed at ambient temperature and is a truly cross-linked, three-dimensional polymer. The amine in both the epoxy and the urethane is often called a catalyst. This is a misnomer, since the amine becomes a part of the much larger polymer molecule. A true catalyst initiates a reaction, but does not become part of the end product.

Cross-linked coatings are much more resistant from a corrosion standpoint than the oxygen convertible coatings. This is especially true in terms of hardness and chemical, water, and solvent resistance.

Catalyst Polymerization

Unsaturated polyester resins are the only good examples of catalyst film formation. Polyester resins are not generally used as coatings, but rather are often combined with fiberglass or other reinforcing materials to form highly corrosion-resistant linings. These heavy linings are then most commonly used in chemical tanks, plating tanks, and tanks for the electrolytic refining of metals.

The polyester resins are made by reacting dibasic acids, such as maleic anhydride or phthalic anhydride with a polyhydric alcohol, usually propylene glycol. The resulting polyester is dissolved in styrene in order to develop a liquid solution for easy application. Styrene also reacts with the unsaturated polyester to provide a copolymer resin. In order to bring this about, a cobalt compound is added to the liquid resin, and, shortly before the application, a solution of a material such as methyl ethyl ketone (MEK) peroxide is blended with the resin solution. These two materials catalyze the reaction between the polyester and the styrene to form a copolymer resin through the unsaturation of both components. The MEK peroxide is a true catalyst and is not a part of the resulting polymer. Since the liquid styrene becomes a part of the compound, the polyester resin solution is essentially 100% solids. Except for the small amount of styrene that evaporates initially, the entire solution converts to a solid resin film.

Most often, the polyester linings are applied as rather thick films. A thin coating is difficult to obtain because of the rapid loss of exothermic heat which occurs during the polymerization and because of the loss of styrene. The reaction of the polyester resin from a liquid form to a solid film is rapid, so that ordinary methods of application are not practical once the catalyst is mixed. The pot life is very short as the reaction increases rapidly when the material is held in volume. A two-component spray gun is therefore usually used in the lining's application, with the catalyst mixed at the head of the gun. The resulting film is three-dimensional with good chemical and water resistance. There is also extensive film shrinkage because of the continuing polymerization reaction even after the film or lining becomes solid.

Inorganic Zinc Film Formation

The formation of a coating from inorganic or organic silicates and zinc represents quite a different series of reactions from that which takes place in organic films. While the molecules of organic films are primarily made up of carbon atoms combined into long-chain linear polymers or cross-linked polymers; the basic building blocks of inorganic zinc coatings are silica, oxygen, and zinc. In liquid form, they are relatively small molecules of metallic silicates such as sodium silicate, or organic silicates such as ethyl silicate. These essentially monomeric materials are cross-linked into a silica-oxygen-zinc structure which is the basic film-former or binder for all of the inorganic zinc coatings. This occurs through a chain of rather complex chemical reactions, some of which take place rather rapidly while others come about slowly.

There are essentially three stages in the formation of the inorganic coating. The first reaction is a very simple one: the concentration of the silicates in the coating by evaporation after the coating has been applied to the surface. As the solvent evaporates, the silicate molecules and the zinc come in close contact and are in a position to react with each other. This initial solvent evaporation provides for the primary deposition of the film on the surface.

The second reaction is the ionization of the zinc metal which initiates the reaction of the zinc ion with the silicate molecule to form a zinc silicate polymer.

The third reaction is the completion of the film reaction over a long period of time by continuing formation of zinc ions which react to increase the size of the zinc silicate polymer and cross-link it into a very insoluble, resistant, three-dimensional structure. In the field of inorganic chemistry, this is a unique reaction since inorganic materials generally do not form a coherent thin film. The only other inorganic film is one formed by fusing the inorganic material to a basic metal in order to create a ceramic enamel.

Coating Component Functions

Individual coating components may be combined into certain categories, and any one coating may be made of more than one component type, all of which serve different functions. In order to obtain a gray coating, for example, it may be necessary to use several colored pigments. Figure 4.9 shows the general components of a coating and the manner in which they are ultimately combined into the complete and finished product.

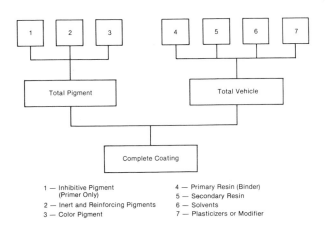

1 — Inhibitive Pigment (Primer Only)
2 — Inert and Reinforcing Pigments
3 — Color Pigment
4 — Primary Resin (Binder)
5 — Secondary Resin
6 — Solvents
7 — Plasticizers or Modifier

FIGURE 4.9 — General components of a coating.

Inhibitive Pigment (Primer Only)

While the primary function of the inhibitive pigment is to react with the substrate to provide a passive surface and therefore cathodically protect it, they can also contribute to the overall prime color or lend opacity of the coating.

Inert and Reinforcing Pigments

The inert pigments are often added to improve the density and corrosion resistance of the coating, as well as to increase the thickness. The reinforcing pigments, on the other hand, are to do what the word indicates, i.e., reinforce the paint film so that it becomes tougher with less tendency to check and crack after extended periods of weathering. Both of these pigments tend to increase the hardness and tensile strength of the film. They can also increase the chemical and atmospheric resistance. They very definitely aid in the adhesion of the primer, although how they do so is not completely understood. It is known, however, that properly pigmented films can be substantially more adherent than a clear film of the same resin. These pigments improve the bonding surface for topcoats. They are used to build film thickness and contribute substantially to the viscosity, thixotropy, and overall workability of the coating in its liquid stage.

Color Pigment

The primary function of color pigment is to provide a pleasing and decorative color, to impart opacity, and to hide the underlying surface. Another very important function, however, is to protect the resinous binder from the penetration of the sun's ultraviolet rays into the coating itself. This is important inasmuch as many binders are rather rapidly affected by the sun's rays when used as a clear coating.

Primary Resin (Binder)

The primary resin has a number of functions. It binds the various pigments in the coating together into a homogeneous film. There must also be sufficient resins present so that the binder wets the individual pigment particles and is thus able to bind them together. In addition, the binder resin must provide the adhesion of the overall coating to the substrate. Again, there must be sufficient resin left after wetting the pigment for the resin to also be able to wet the surface of the substrate sufficiently for adhesion to be obtained. The resin is the primary barrier to all of the various materials that may come in contact with the coating, either when subject to atmospheric or to immersion conditions. It must also maintain its integrity in a corrosive environment.

Secondary Resin

While many coatings have only one primary resin as their binder, many others incorporate more than one resin in order to develop specific properties. Secondary resin is also part of the overall binder. Its function is to extend the primary resin functions, to increase the amount of resin available for both wetting the surface and the pigments, to aid in the adhesion of the coating to the substrate, to increase the overall resistance of the coating, and to build the coating thickness. One primary requirement is that it be compatible with the primary coating resin so that the two maintain a cohesive resin structure.

Solvents

A function of the solvents is to dissolve the binder into a compatible and workable liquid. Many of the resin binders are solids and would be unworkable without proper solution. Solvents aid in the control of the viscosity of the overall vehicle. They not only aid in making the binder-pigment combination workable, but they transport the combination to the substrate. By providing a more liquid, viscosity-controlled coating, they also aid in the wetting of the substrate surface.

Plasticizers

Not all coatings contain plasticizers. However, many resins require the addition of a plasticizing material in order to provide satisfactory coating properties. Plasticizers in many ways are permanent solvents in that they are both dissolved by and dissolve into the resins. Plasticizers provide many coatings with flexibility, extensibility, and toughness.

Basic Coating Components

Binders

In order to perform in a practical way in any given environment, a coating, after its application, must convert to a dense, solid, adherent membrane with all or most of the properties that have been previously discussed. The *binder* is the material which makes this possible. It provides uniformity and coherence to any coating system. Not all binders are particularly corrosion-

resistant so that only a few serve as the basis of all protective coating systems. Those most commonly used are listed in Table 4.3.

The binder's ability to form a dense, tight film is directly related to its molecular size and complexity. Binders which generally have the highest molecular weight are those which form films by the evaporation of solvent only. The resin or the binder molecule is in its completed form prior to application. Additional binders must chemically react in place, and generally the molecular weight of the finished binder resin is considerably less than those which form a film by solvent evaporation or heat conversion.

Binders can generally be classified, according to their essential chemical reactions, as one of the following types:

1. Oxygen reactive
2. Lacquer (thermoplastic)
3. Heat conversion
4. Co-reactive (thermoset)
5. Condensation (thermoset)
6. Coalescent (nonimmersion)
7. Inorganic

TABLE 4.3 — Binders Commonly Found in Corrosion-Resistant Coatings and Linings

Rating: E = Excellent, G = Good, F = Fair, P = Poor

Binder Type	Generic Type	Alkali	Acid	Resistant Properties Water	Weather	Temperature	Primary Use
Lacquer	Copolymer-Vinyl Chloride-Vinyl Acetate	E	E	E	E	to 65 C/150 F	Resistant intermediate and topcoats
	Polyacrylates	F	F	F	E	to 65 C/150 F	Resistant topcoats
	Chlorinated Rubber	E	E	E	G	to 60 C/140 F	Resistant intermediate
Co-reacting	Epoxy-Amine Cure	E	G	G	F	to 93 C/200 F	Resistant coatings and linings
	Epoxy-Polyamide	E	F	G	G	to 93 C/200 F	Resistant coatings and linings
	Urethane (2 package)	G	F	G	F	to 120 C/250 F	Abrasion-resistant coatings
	Urethane (moisture cure)	G	F	G	G	to 120 C/250 F	Abrasion-resistant coatings
	Urethane-Aliphatic Isocyanate	G	F	G	E	to 120 C/250 F	Weather- and abrasion-resistant topcoats
Condensation (requires added heat to cure)	Phenolic	P	E	E	F	to 120 C/250 F	Chemical- and food-resistant lining
	Epoxy Phenolic	F	E	E	F	to 120 C/250 F	Chemical- and food-resistant lining
	Epoxy-Powder Coating (requires high heat to fuse and cure)	G	G	G	F	to 93 C/200 F	Pipe coating and lining
Inorganic	Zinc Silicate	P	P	G	E	to 315 C/600 F	Permanent primer or single coat weather-resistant coating
	Glass (fused to metallic substrate)	F	E	E	E	to 260 C/500 F	Chemical- and food-resistant lining

(SOURCE: Kirk-Othmer Encyclopedia of Chemical Technology, Munger, C. G., Coatings Resistant, Vol. 6, 3rd Ed., John Wiley & Sons, pp. 456-578, 1979.)

Oxygen Reactive Binders

Oxygen reactive binders are generally low molecular weight resins which are only capable of producing coatings through an intermolecular reaction with oxygen. This reaction is often catalyzed by metallic salts of such metals as cobalt and lead. (The chemical reactions involved are discussed in Chapter 5.) There are several types of coatings in this category that are important to the corrosion engineer.

Alkyds. To produce alkyds, natural drying oils are chemically reacted into a synthetic resin in such a way that film curability, chemical resitance, weather resistance, etc. is improved over the original drying oil. Nevertheless, the drying oil part of the molecule is responsible for the conversion of the liquid coating into the solid state, and the properties of the drying oil usually dominate.

Epoxy Esters. Epoxy resins are combined chemically with drying oils to form epoxy esters. The drying oil part of the molecule determines the basic properties of the epoxy ester coating. The coating dries by oxidation in the same manner as an alkyd. The chemical and solvent resistance is less than an unmodified epoxy, but greater than an alkyd or olioresinous paint. The coatings generally are hard and have good adhesion and flexibility. Many epoxy resin oil combinations can be made, with the properties being dependent on the type and amount of oil used in the formulation of the resin. Epoxy ester coatings have made excellent machinery enamels and interior coatings. They are not, however, entirely suitable for exterior finishes as they chalk readily and heavily.

Urethane Alkyds. Epoxy resins are also chemically combined with drying oils as part of the molecule which is further reacted with isocyanates to produce urethane alkyds. Upon application as a liquid coating, the resin oil combination converts by oxidation to a solid. The oil properties are again dominant, with the isocyanate contributing to the characteristic abrasion resistance and toughness of this coating.

Silicone Alkyds. Alkyd resins are combined with silicone molecules to form an excellent weather-resistant combination known as silicone alkyds. Heat resistance is also increased in spite of the drying oil part of the overall resin molecule.

Lacquers

Lacquers are coatings that are converted from a liquid material to a solid film by the evaporation of solvents alone. Lacquers, in general, have a relatively low volume of solids as compared to materials formed from lower molecular weight resins and then converted into a solid film.

Polyvinyl Chloride Copolymers. The principal corrosion-resistant lacquer is made from polyvinyl chloride copolymers. The vinyl molecule is a large one and will only effectively dissolve in solvent in the 20% range. The film build is therefore low (in the neighborhood of 1 to 2 mils of thickness per coat). The overall chemical resistance is broader than almost any other binder.

Chlorinated Rubbers. In order to be effective, chlorinated rubbers must be modified by other resistant resins, not only to obtain higher solids, but to decrease brittleness and increase adhesion. The chemical and water resistance of chlorinated rubbers depend on the type of modifier used. Again, the solids are relatively low, providing coatings of from 1 to 3 mils thick per coat. Low molecular weight resin modifiers have proven to be the best chemical and corrosion-resistant coatings. Alkyd or oil-base modifiers do an excellent job of plasticizing the chlorinated rubber, but these combinations suffer from reduced corrosion resistance because of the oil.

Acrylics. Acrylics are also of high molecular weight and may be combined with vinyls to improve exterior weatherability and color retention. When used alone, they have excellent color, gloss, and weatherability. However, their chemical and water resistance is not as high as that of vinyl copolymers or chlorinated rubbers.

Bituminous Materials. Asphalts and coal tars are often combined with solvents in order to form lacquer type films. Hard asphalt, gilsonite, or coal tar is dissolved in solvents in what is called an asphalt or coal tar cut back. These provide good chemical- and corrosion-resistant films, but can only be applied where appearance is not a factor.

Heat Conversion Binders

Hot-melts. As discussed previously, *hot-melts* usually involve asphalt or coal tar which are melted and applied as 100% solids in the hot liquid condition. The resin binder is generally not combined with any other resins, but is used essentially as a basic coating material.

Organisols and Plastisols. As also discussed previously, *organosols* are high molecular weight resins which are dispersed in solvents. The solvent does not actually solvate the resin until heat is applied. It is applied as the solvent resin mixture and is heated to a point where the resin is solvated and fuses not only to itself, but to the surface as well. *Plastisols* are primarily high molecular weight vinyl materials dispersed in a plasticizer which also does not solvate the resin until heated. The film is formed only after the plastisol has been applied to the substrate and then heated to convert the liquid to a solid.

Powder Coatings. Powder coatings can be high molecular weight thermoplastic resins or semithermoset resins, such as certain epoxies, which are usually converted to a very fine powder and applied to a heated substrate which first slightly melts the resin. Then, the entire object is heated above the resin's fusion temperature to form the completed coating.

Co-reactive Binders

Co-reactive binders are formed from two low molecular weight resins which are combined, usually just before application, and which, after having been applied to the surface, co-react with each other to form a solid film. film.

Epoxies. Epoxy binders are made up of relatively low molecular weight resins in which the epoxy group is at the end of each molecule. These low molecular weight resins are then reacted with ammonia type compounds called amines. The amines may be liquid, low molecular weight materials or they may be higher molecular weight resinous semiliquids with the amine groups scattered over longer chain molecules, such as the polyamides. In either case, the amine group on the molecule reacts with the epoxy resin to form the solid binder. These materials not only react quickly, but also over a longer period of time to form higher molecular weight binders with good solvent and chemical resistance. As previously indicated, these materials are cross-linked and form a thermoset structure. In some cases, the epoxy materials are combined with other lower molecular weight materials, such as asphalt or coal tar. The coal tar epoxy, particularly, results in a binder which combines the good properties of both of the base materials. Water resistance is improved over the epoxy binder alone and the solvent resistance of the coal tar is improved by the reaction of the epoxy.

Polyurethanes. Polyurethanes are co-reactive binders in which relatively low molecular weight resins containing alcohol or amine groups are reacted with diisocyanates into an intermediate resin prepolymer. This urethane prepolymer is then capable of reacting with resins or chemical groups containing amines or alcohols to form the finished coating. As previously indicated, another possibility exists where the isocyanate group comes in contact with water and the moisture from the air converts the binder from a liquid to a solid.

Condensation Binders

Condensation binders are based primarily on resins which interact to form cross-linked polymers when subject to relatively high temperatures. These are the so-called high baked materials which are used as tank and pipe linings. Some powder coatings may also come into this category. *Condensation* is essentially the release of water during the polymerization process. The oldest of this type of coating is the pure phenolic. Thin films of a phenolic resin are applied as a coating and baked at approximately 375 F to form an extremely hard, adherent, chemical-resistant film. Coatings of this type are modified from the pure phenolic by the use of epoxies and other reactive resins. These condensed materials are strongly cross-linked and are very chemical-resistant. The exception is the pure phenolic which is attacked by caustic solutions. It is, however, strongly resistant to waters and acids.

Coalescent Binders

Coalescent binders include the coatings where the binders of various resin types, such as vinyls, acrylics, or epoxies, are emulsified to form a liquid binder. They are primarily emulsified with water, although some solvent dispersions have been made from various resins. In this case, the binder is in a dispersed form in the emulsion. When applied to the surface, the water or other medium must evaporate, leaving the coating in such a way that the binder resin gradually flows into itself, or coalesces, to form a continuous film. Water, chemical, and corrosion resistance are generally decreased because of the fact that the dispersed phase never forms as tight and effective a film as does a solvent solution.

Inorganic Binders

Inorganic binders are primarily inorganic silicates which are dissolved in either water or solvent and which, when once applied to the surface, react with moisture in the air in order to form an inorganic film. The type of inorganic binder depends on the form of the silicate during its curing period.

Post-Cured Inorganic Silicates. Water solutions of alkali silicates, such as sodium or potassium silicate, combined with zinc dust form very hard, rock-like films referred to as post-cured inorganic silicates. In this state, however, they cure to water insolubility at an extremely low rate. The coating must therefore be reacted with an acidic curing agent to achieve the conversion of the silicate film from the water-susceptible stage to the completely insoluble zinc silicate.

Self-Curing Water-Based Silicates. Self-curing water-based silicates are also mixtures of alkali silicates often combined with coloidal silica to improve the speed of cure. Once the material has been applied to the surface, they develop water insolubility from the absorption of carbon dioxide and moisture from the atmosphere.

Self-Curing Solvent-Based Silicates. Organic esters of silica which are liquids and are converted into solid binders by reaction with moisture from the air, form self-curing solvent-based silicates. In this final form, they are very similar to those binders formed from the water-based silicates. A major advantage of these materials is their conversion to rain or moisture resistance shortly after their application. All of these materials contain zinc which is part of the reactive mechanism that forms the silicate binder. The zinc in the liquid coating acts as a pigment; however, once applied to the surface, the zinc also reacts with the silicate in such a way that a zinc silicate matrix is formed which surrounds all of the zinc particles. This makes a very hard and extremely corrosion-resistant binder.

Pigments

While binders are responsible for many of a coating's primary properties, pigments also contribute several properties important to their effective use. In fact, proper or improper pigmentation can either make or break a coating in terms of corrosion resistance. Several different pigments may be used within the same coating, all of which contribute to the coating's general characteristics and perform several functions.

Primary Pigment Functions

Color. Pigment produces an aesthetic effect (decoration) and hides substrates.

Protection of Resin Binder. Pigment absorbs and reflects solar radiation which can cause breakdown of binder.

Corrosion Inhibition. Chromate salts and red lead in primers act as passivators. Metallic zinc, when in high enough concentration, gives cathodic (sacrificial) protection.

Corrosion Resistance. Proper pigmentation can increase both the chemical and corrosion resistance of a coating. Conversely, improper pigment use can seriously reduce resistance, *e.g.*, calcium carbonate pigments in an acid-resistant coating.

Film Reinforcement. Finely divided fibrous and plate-like particles of pigments increase hardness, toughness, and/or tensile strength of a film, as well as increases cohesive strength.

Nonskid Properties. Particles of silica or pumice roughen a film's surface and increase abrasion resistance.

Sag Control. So-called thixotropizing pigments prevent sagging of the wet film and also reduce the tendency of other pigments to settle in the container during storage.

Increased Coverage. Properly selected inert pigments can increase the volume of solids (or coverage) of a coating without reducing its chemical or corrosion resistance. There is a limitation on how much inert pigment can be used with a given resin composition. This constraint is termed the *critical pigment volume concentration,* and indicates the volume of pigment which can be bound by the resin without leaving voids in the film.

Hide and Gloss Control. Increasing the color pigment concentration improves hiding, while an increase in either color or other pigmentation generally decreases gloss, depending on the critical pigment volume concentration.

Adhesion. Certain pigments, particularly plate-like or flake pigments, can increase coating adhesion over that of the binder alone.

In order for pigments to properly function in a coating, they must be thoroughly wet with the binder. This means that each individual particle of pigment must be surrounded by a layer of the binder resin. In other words, the pigment must be dispersed in a matrix of the binder resin. If the pigment is added to the binder in an amount greater than the critical pigment volume concentration, the pigment will not be completely covered by the binder, and a porous, flat film with little strength will be produced. Pigments vary in their wetting characteristics. They also vary in particle size, and certain binders will have greater wetting characteristics for one pigment than another. Thus, each pigment/binder combination has its own critical pigment volume concentration.

A large amount of energy is needed in order for the binder to completely wet each particle of pigment. This is put into the system by milling the binder and the pigment together. *Milling* does two things: it breaks up the agglomerates of pigment into the individual particles, and by physical action the resin or binder is rubbed over the surface of the particle so that it is completely surrounded. While this was at one time done with two revolving granite stones, present milling operations properly disperse and wet the pigments with intensive mixers or high-speed impellers. Using a much more liquid mix, the high-speed mixer simply transfers enough energy into the system that all particles are properly wetted.

Manufacturers of pigments have also increased the wettability of the pigment so that it disperses much more readily. The dispersion of the pigment makes a difference in the viscosity or thixotrophy of the coating, the hiding qualities, the ability of the pigment to remain in suspension, as well as the gloss or lack of gloss of the finished coating. Unless a pigment is properly dispersed in the vehicle, the quality of the coating will be seriously impaired.

Pigment Classes

Pigments can be separated into classes as either coloring pigments, reinforcing pigments, inhibiting pigments, or metallic pigments. Its designated class depends on the pigment's purpose within the coating.

Color Pigments. Color pigments, of course, provide the pleasing color and decorative characteristics expected of a coating. While this is more true in the case of ordinary paint, corrosion-resistant coatings must also provide pleasing as well as utilitarian surfaces. A knowledge of the nature of color and its cause and effect is often helpful to the corrosion engineer, and is thus discussed in Unit 8 of the Federation Series on Coating Technology.

The Nature of Color. The usual first response to the word color is to think of it as a human reaction or color sensation. This discussion will be concerned mainly with the stimulus that evokes the sensation or the physical basis of color.

Color has its origin in light. Sunlight or white light from any source consists of the relatively narrow band of radiant energy or electromagnetic waves that comprise the visible spectrum. When white is passed through a quartz prism, the various wave lengths composing it are bent (refracted) at different angles, producing a spectrum according to Figure [4.10].

A surface that reflects all wave lengths of the visible spectrum appears white. If it absorbs all wave lengths, it appears black. If, on the other hand, a surface absorbs some wave lengths and reflects others, it has the color of the reflected wave lengths. For example, if the only wave lengths reflected are those above 610 millimicrons, the color will be red. Most simply stated, color is the result of selective reflection of the wave lengths of the incident light. Implicit in this statement is the fact that the kind of illumination influences the color. The light reflected by a particular color pigment may not conform sharply to a primary color. In the case of a green pigment, it may be a yellow green, a medium green, or a blue green.

The terms absorption and reflection indicate the results of a series of complex phenomena. In a pigmented film, incident illumination is partially reflected from the surface of the film and partially transmitted into it. The transmitted portion suffers refraction, diffraction, and absorption. Any part of the light that reaches the substrate is partially reflected from it and traces another complex path back through the film. Finally, part of it is refracted into the air.

The relationship between the primary colors is made clearer by the color circle in Figure [4.11].

Colors that are opposite in the circle neutralize each other when mixed and produce neutral gray. They are said to be complementary. When any two hues that are not adjacent on the circle are mixed, there is a tendency toward grayness or lack of purity. It is a principle of decoration that complementary colors harmonize well.

Color Attributes. The preceding discussion of color has been limited to hue, which is one of the three attributes of color perception. Hue is the property that differentiates the primary colors: green, yellow, orange, red, purple, and blue. There are also intermediate gradations of hue. Another attribute of color is lightness or value, the percentage reflectance of "white" light. Certain hues, notably yellow, have higher lightness than other hues such as blue, at equal chroma or purity. The lightness of any color is raised by admixture with white, lowered by admixture with black and may be varied at will by admixture with a range of grays. The third attribute of color is chroma, also known as purity or saturation. High chroma means that the color is saturated or intense, in contrast with being diluted with gray. More definitely, chroma expresses the degree of departure from the gray of the same lightness.

The color sphere shown in Figure [4.12] is a device for assisting in an understanding of the attributes of color.[3]

In the use of corrosion-resistant coatings, every care must be taken in the selection of pigments to make certain that corrosion resistance is not reduced by them. There are many pigments used in paint which should not be

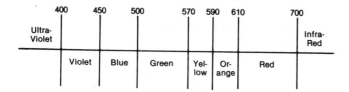

FIGURE 4.10 — Wavelengths in millimicrons (mμ).

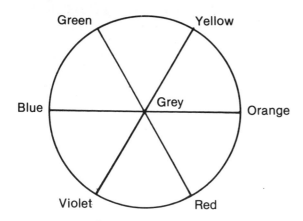

FIGURE 4.11 — Color circle.

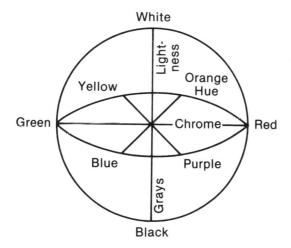

FIGURE 4.12 — Color sphere.

used in corrosion-resistant coatings because their incorporation would drastically reduce the resistance of the basic binder. The number of color pigments suitable for corrosion-resistant coatings is therefore limited. Fortunately, the best of the white pigments, titanium dioxide, is also inert and chemical-resistant. Titanium dioxide is the whitest, brightest white available with high opacity. Not only can good white coatings be made from it, but many pastel shades as well.

Black pigments (lamp black and carbon black) are also inert and chemical-resistant. Iron oxide pigments, which are inorganic, have some very desirable corrosion-resistant properties. Prior to the 20th century, iron oxide

pigments were produced from various deposits of iron oxide found in nature. Colors then covered a range of reds, purples, browns, yellows, and black. As the coating business became more sophisticated around the turn of the century, chemical processes were developed for the manufacture of iron oxides of the same chemical types found in nature. These products, however, were purer in composition, richer in color, finer in particle size, and more uniform in their properties. Thus, they are the ones primarily used in corrosion-resistant coatings as we know them today.

With a few exceptions, the desirable coating properties of the iron oxide pigments are:

1. High hiding power (up to 1,000 sq. ft./lb.).
2. High tinting strength.
3. Color fastness.
4. Heat resistance (excepting yellow and black oxides).
5. Nonbleeding (insoluble in solvents).
6. Chemical resistance (unaffected by alkalies and weak acids).
7. Ease of dispersion in all vehicles (oils, resins, and water).
8. Many grades have very fine particle size, permitting dispersion in impeller mills and sand mills.
9. High infrared reflectance (for camouflage paints).
10. High ultraviolet absorption.
11. Low price.[3]

Due to their favorable chemical and physical properties, iron oxide pigments are desirable for corrosion-resistant coatings when the proper color can be obtained from them. They are used not only as mass colors or individual colors, but are also used in connection with titanium dioxide to form the pastel colors.

The color of iron oxide pigments generally is not as clean as some of the organic reds, yellows, greens, and blues. However, their resistant characteristics for corrosion-resistant coatings outweigh this disadvantage. Table 4.4 shows a number of color pigments which are used in corrosion-resistant coatings. Depending on the coating's intended use, the properties shown for the various pigments indicate the influence that the pigment will have on the effectiveness of the coating. Only those which are indicated as good or excellent should be used where severe corrosion conditions exist. Table 4.5 indicates the chemical composition of pigments which are common to corrosion-resistant coatings.

Reinforcing Pigments. *Reinforcing pigments,* often called extender pigments, are usually considered of minor importance as fillers or bulking materials which are used primarily to lower cost. Reinforcing pigments, however, actually have a profound effect on the performance of most pigmented coatings and, in most cases, are far more important from a corrosion-resistant standpoint than the color pigments. Most of these pigments are inorganic and, for the most part, are relatively inert materials which can therefore be used in corrosion-resistant coatings. The carbonate pigments, which are generally not used in corrosion-resistant coatings, constitute the one exception.

Several types of reinforcing pigments are given in Table 4.6. Although there are many others which are not listed, these are considered representative. Most of them, except for the carbonate pigments, are commonly used in corrosion-resistant coatings.

Corrosion Prevention by Protective Coatings

TABLE 4.4 — Color Pigments Useful in Corrosion-Resistant Coatings

F = Fair; G = Good; P = Poor; E = Excellent; B = Borderline

| | Type of Pigment | | Resistant Properties | | | |
Color	Organic	Inorganic	Alkali	Acid	Exterior Durability	Heat Resistance
Red	Toluidine Red		P	P	G	P
	Monastral Red		G	G	G	G
		Iron Oxide Red	G	G	G	G
		Molybdenum	G	B	G	B
		Cadmium Red	B	B	G	E
Orange		Red Iron Oxide	G	G	G	G
		Molybdenum Orange	G	B	G	B
Yellow		Yellow Iron Oxide	G	G	G	G
		Chrome Yellow	F	F	G	P
		Zinc Chromate (primer only)				
		Nickel TiO_4 Titanate	G	G	G	G
Green		Chrome Oxide Green	G	G	G	G
		Chrome Green	F	F	G	P
	Phthalocyanine Green		E	E	E	E
Blue		Prussian Blue	P	G	G	P
	Ultramarine Blue		P	G	P	P
	Phthalocyanine Blue		E	E	E	E
Black	Lamp Black		G	G	G	G
	Carbon Black		G	G	G	G
White		Titanium Dioxide	G	G	G	G
		Zinc Oxide	F	F	G	G
Metallic		Aluminum Flake	P	P	E	E
		Stainless Steel Powder	E	G	E	E
		Lead Flake	G	E	E	G

Reinforcing pigments are available in many different particle sizes and in many shapes, including spheroids, needles, fibers, and plates. The particle shape is most useful in developing film density, flexibility, and film strength. Particle size and shape also influence such characteristics as opacity, viscosity, film porosity, and gloss.

The film is strengthened by various means, such as:

1. interlocking of pigment particles;

2. pigment relieving stresses developed in the binder;

3. lowering of the thermal coefficient of expansion;

4. conducting heat away from the localized heat sites; and

5. forming a barrier to ultraviolet radiation.

The ratio of reinforcing pigment to binder is very important. It is important for the reinforcing pigment to be sufficiently covered and wetted by the binder. As indicated previously, the wetting of the pigment by the binder is a major step in the production of maintenance- or corrosion-resistant coatings. The completeness of the dispersion, and therefore the wetting, has a very important bearing on the durability, gloss, and application characteristics of the coating. Table 4.7 lists some of the properties of the principal reinforcing pigments used in corrosion-resistant coatings.

Inhibitive Pigments. Inhibitive pigments are principally used in primers or first coats and in coatings which use the concept of inhibition rather than impermeability. These are pigments which react with the moisture absorbed by the coating to form sufficient ions which react

TABLE 4.5 — Chemical Composition of Common Pigments

Pigment	Class[1]	Color	Approximate Composition
WHITE HIDING PIGMENTS			
Antimony Oxide	Inert	White	99% Sb_2O_3
Basic Carbonate White Lead	Slight	White	62-75% $PbCO_2$ 38-25% Lead Hydroxide
Basic Sulfate White Lead	Slight	White	15-28% PbO
Zinc Oxide	Slight	White	98.5% ZnO (min)
Titanium Dioxide, Rutile	Inert	White	98% TiO_2
Titanium Dioxide, Anatase	Inert	White	98% TiO_2
Titanium Calcium, Rutile	Inert	White	30% TiO_2 70% $CaSO_4$
Titanium Calcium, Anatase	Inert	White	30% TiO_2 70% $CaSO_4$
Lithopone	Inert	White	30% ZnS 70% $CaSO_2$
Zinc Sulfide, Barium Sulfate	Inert	White	50% ZnS 50% $BaSO_4$
Zinc Sulfide	Inert	White	98% ZnS
EXTENDERS			
Barium Sulfate (Barytes)	Inert	—	99% $BaSO_4$
Barium Sulfate (Blanc Fixe)	Inert	—	97% $BaSO_4$
Calcium Carbonate, Natural	Reactive	—	98% CaCO
Calcium Carbonate, Precipitated	Reactive	—	98.7% CaCO
Magnesium Silicate (Talc)	Inert	—	
China Clay	Inert	—	
Mica	Inert	—	
COLORED HIDING AND/OR INHIBITIVE PIGMENTS			
Iron Oxide, Synthetic	Inert	Red	98% Fe_2O_3
Iron Oxide, Natural	Inert	Red	95% Fe_2O_3
Iron Oxide, Natural	Inert	Red	70% Fe_2O_3
Yellow Iron Oxide	Inert	Yellow	99% $Fe_2O_2:H_2O$
Chrome Orange	Inhibit.	Orange	65% $PbCrO_4$, 35% PbO
Chrome Yellow	Inhibit.	Yellow	70% $PbCrO_4$, 30% $PbSO_4$
Zinc Yellow	Inhibit.	Yellow	38% ZnO, 7% H_2O, 43.5% CrO_2
Chrome Green C.P.	Reactive	Green	85% Cr Yellow 15% Fe Blue
Chrome Green Reduced	Reactive	Green	55% BaSO, 25% China Clay 20% Color
Chrome Oxide	Inert	Green	98.5% Cr_2O_2
Iron Blue	Reactive	Blue	
Lamp Black	Stimul.	Black	96% Carbon 3% Moisture
Carbon Black	Stimul.	Black	92% Fixed Carbon
Black Iron Oxide	Inert	Black	97.75% Iron Oxides
Red Lead	Inhibit.	Orange-Red	96% Pb_2O_4, 4% PbO
Litharge	Inhibit.		99% PbO
Blue Lead	Inhibit.	Gray-Blue	
Aluminum Paste	Reactive	Aluminum	65% Aluminum 35% Solvent
Aluminum Powder	Reactive	Aluminum	
Zinc-dust	Inhibit.	Gray	96% Za
Metallic Lead Paste	Inert	Gray	90% Metallic Pb 1% Stearic Acid 9% Mineral Spts.

[1]Class - classified as rust inhibitive, slightly inhibitive, inert, rust stimulative, or reactive. (SOURCE: Steel Structures Painting Council, Steel Structures Painting Manual, Vol. 1, p. 82, 1966.)

TABLE 4.6 — Types and Sources of Extenders or Reinforcing Pigments

| | | Source | |
Chemical Type	Natural Ores	Synthetic (Mfg.)	Common Name
CARBONATE			
Calcium	x	x	Whiting
Magnesium	x	x	—
OXIDE			
Silicon	x (Amorphous)	x (Hydrogel)	Silica
Silicon	x (Crystalline)	x (Aerogel)	Silica
Silicon	x (Diatomaceous)	x (Pyrogenic)	Silica
		x (Arc)	Silica
		x (Precipitated)	Silica
SILICATE			
Aluminum Hydrate	x (Many types)	—	Clay
Calcium	x (Wollastonite)	x	—
Magnesium	x (Fibrous)	—	Talc
Magnesium	x (Platy)	—	—
Magnesium	x (Granular)	—	—
	x (Acicular)	—	—
K, Na, Al, Fe, Li	x (Many minerals)	—	Mica
SULFATE			
Barium	x	x	Barytes, blanc fixe
Calcium	x	x	Anhydride
Calcium Hydrate	x	—	Gypsum

(SOURCE: Madison, W. R., Federation Series on Coating Technology, Unit 7, White Hiding and Extender Pigments, Federation of Societies for Coating Technology, Philadelphia, PA, 1967.)

TABLE 4.7 — Reinforcing Pigments Used in Corrosion-Resistant Coatings

F = Fair; G = Good; P = Poor; E = Excellent; B = Borderline

| Generic Type | Common Name | Resistant Characteristics | | | | Physical Characteristics |
		Alkali	Acid	Water	Weather	
Magnesium Silicate	Talc; Asbestine Asbestos	F	G	E	E	Fibrous-platelike fibrous
Barium Sulfate	Barytes	G	G	G	G	Cubical, heavy
Silica	Diatomite; Silica Flour	P	E	E	E	Porpos, hard, sharp crystals
Aluminum Silicate	Clay	F	G	F	G	Platelike
Potassium-Aluminum Silicate	Mica	G	G	G	G	Platelike, used to reduce moisture vapor transfer

with the underlying metal surface to passivate it and make it more corrosion-resistant. These pigments are used primarily as atmospheric coatings and not for immersion or constantly wet conditions, since some are so soluble that osmosis could draw water into the coating, creating blisters. This has been one of the difficulties with the use of inhibitive pigments. They are, however, effective in reducing underfilm corrosion in many atmospheric conditions when used properly. Table 4.8 lists a few of the inhibitive pigments found in primers for corrosion-resistant coatings. This table also gives the solubility of the various materials, the ones with the lower solubility being the ones preferred for corrosion resistance.

TABLE 4.8 — Inhibitive Pigments Used in Some Primers for Corrosion-Resistant Coatings

Pigment	Solubility in 1 L H_2O at Equilibrium, g of CrO_3
Zinc Chromate	1.1
Strontium Chromate	0.6
Basic Zinc Chromate	0.02
Barium Chromate	0.001
Lead Chromate	0.00005
Lead Silica Chromate	0.00005
Red Lead	essentially insoluble
Zinc Powder	provides cathodic protection to substrate

(SOURCE: Munger, C. G., Coatings Resistant Kirk-Othmer Encyclopedia of Chemical Technology, Vol. 6, 3rd Ed., John Wiley & Sons, New York, NY, 1979.)

Metallic Pigments. Metallic pigments are categorized separately because of their unique properties. They are metals which, except for zinc dust, are generally in the form of flakes or flat platelets. This is true not only for aluminum and stainless steel pigment, but also for the various colors of bronze which are primarily metallic copper flake. The flat plate-like structure is important in that it tends to reinforce the binder. In leafing pigments, it creates a shingle effect which prevents actinic (ultraviolet) rays of the sun and light from penetrating into the binder. The flat, plate-like structure often improves the adhesion of the coating as well.

The most common of the metallic pigments is produced through a wet ball milling process which uses steel balls as the grinding media. The aluminum metal is in the form of an atomized powder and is charged into the ball mill in addition to stearic acid and mineral spirits. It is then milled long enough to produce flakes of the desired fineness. The stearic acid aids in the production of the flat plates, and it also imparts the ability to float on the surface of the coating vehicle. This same general procedure is used for the other plate pigments such as stainless steel, lead, and copper.

Aluminum pigment is divided into two classes: leafing pigments, which float to the surface of the coating film and impart the metallic color; and nonleafing aluminum pigment, which remains uniformly distributed through the film and gives the natural gray color of aluminum without the metallic luster. From a corrosion resistant standpoint, the leafing aluminum type pigment is the most important. The force that causes the leafing pigments to concentrate on the surface is the convection currents caused by evaporation of the solvent in the coating. Once the flakes reach the surface of the coating prior to drying, they are held in place by interfacial tension, and as the solvent continues to dry, the viscosity of the coating increases and prevents further movement of the aluminum pigment.

Aluminum coatings have characteristic qualities almost irrespective of the binder. The hiding power of the coatings is excellent as one coat will completely hide the underlying surface. This does not mean that in corrosive situations only one coat should be used. As with any coating, the thickness is important both to the durability and the effectiveness of the coating. The flake of the aluminum or other metal is completely opaque to light and the shingle effect of the flakes assures a continuous film. The binder is thus completely shaded from the effect of actinic rays and other forces which can damage the binder. As mentioned previously, the shingle effect of the leafed flakes materially reduces the moisture vapor transfer rate through the coating. Any moisture or other gas must travel a long distance around each individual flake in order to reach the substrate. Since sunlight and moisture are the principal factors in film deterioration, aluminum paints have outstanding durability as exterior topcoats.

The use of aluminum pigment is not limited to thin vehicles alone, but has been incorporated into heavy mastic coatings and even very heavy roof coatings which still provide a bright, metallic surface to reflect the sun's energy. The principal way in which aluminum coatings protect iron from corrosion is in eliminating access of air and moisture to the substrate. The aluminum metal itself does not have any cathodic effect, as do zinc-rich coatings. So for best results, they must be applied over a good anticorrosive primer of some of the best zinc-rich materials. Some excellent corrosion-resistant coatings have been developed through the use of inorganic zinc-rich primers followed by aluminum pigmented topcoats using vinyl, asphalt, or other binders.

Leafing aluminum pigments are outstanding in heat resistance and have been used effectively for coating stacks at 800 to 1000 F. Silicone alkyds have been used as the binder; however, the most effective for high temperature is a pure silicone binder with aluminum leafing pigments.

The other leafing metallic pigments, such as stainless steel, lead, and copper, act in much the same way as the aluminum. Lead flake is preferred over the other materials from the standpoint of chemical resistance. Aluminum flake is affected by both acid and alkali, while the lead flake has excellent resistance to both. The same is true of the stainless steel powder.

Zinc is a metallic pigment which is usually used as a duct, although some zinc has been formed into flakes. It is principally used in anticorrosive coatings which use the cathodic protection effect of the zinc metal. (A more detailed discussion of zinc-rich coatings is contained in Chapter 6.) Table 4.9 summarizes the general properties of the metallic pigments.

Solvents

Most painters do not realize the importance of *solvents,* or thinners, in the formation of the coating and in the development of the most effective coating film. Although solvents do not remain in the coating, they can affect the coating in many different ways, *i.e.,* by creating porosity, discoloration, poor gloss, floating of pigment, fisheyeing, poor coating strength, and lack of adhesion. All of these things can happen if the proper solvent or solvent combination is not used in a protective coating. The proper use of solvents will create a smooth, clear resin film with a good gloss, and the coating film will have the inherent strength and other properties of the basic resin.

Most coatings are made with multiple solvents; in fact, there are very few that use a single solvent alone. As

---

TABLE 4.9 — Metallic Pigments Used for Corrosion Resistance

E = Excellent; G = Good; F = Fair; NR = Not Recommended.

Generic Type	Common Name	Alkali	Resistance Acid	Water	Weather	Physical Characteristics
Aluminum	Aluminum Flake	NR	NR	E	E	Creates shingle effect, protects binder, increases moisture upon transfer resistance.
Stainless	Stainless flake	E	E	E	G	Does not leaf as well as aluminum flakes. Reinforces binder without reducing chemical resistance.
Lead	Lead flake	E	E	E	E	Does not leaf as well as aluminum. Excellent chemical and water resistance.
Copper	Copper flake	NR	NR	G	F-G	Leafs well, good copper color, chemical resistance only fair. Has good antiflaking properties.
Zinc	Zinc powder	NR	NR	E	E	Provides cathodic protection to steel. Reacts with inorganic vehicles to form hard adherent coating.
	Zinc flake	NR	NR	E	E	Provides some cathodic protection to steel. Reinforces some organic binders. May be used with zinc powder for reinforcing purposes.

in most cases, particularly with the synthetic resin binders, the combination of the solvents will provide a better film than where one solvent is used alone.

The choice of solvents influences viscosity, flow properties, drying speed, spraying or brushing characteristics, and gloss. Each type of binder will have its own specific combination of solvents which will provide the best film. There is no universal solvent for protective coatings; a best solvent for one may not be practical with another. Asphalt, for instance, is readily dissolved by hydrocarbon solvents such as mineral spirits or toluene; on the other hand, it would not be dissolved by alcohol. Shellac and some epoxy resins are readily dissolved by alcohol, but would not be readily dissolved by aliphatic hydrocarbons. Vinyl resins are readily soluble in ketone solvents, but are precipitated out of solution by alcohols or aliphatic hydrocarbons.

It is important to emphasize solvents and the fact that they are specific for individual binders because, not realizing this, many painters add any solvent that happens to be available. This, more often than not, causes coating problems after a period of time, if not immediately. A painter, for instance, may add mineral spirits to a vinyl coating as a thinner. Vinyl coatings will tolerate a small amount of mineral spirits without precipitation; however, because it is a slow drying material and the active solvents in the vinyl solution are more rapid, it remains in the coating until the last of the active solvents has evaporated. At this time, it is squeezed out both underneath and on top of the film. The resulting coating has no adhesion whatsoever.

Solvent Types

Active solvents. The primary solvents for a particular binder are known as active solvents. In the case of vinyl resins, the active solvent would be the ketones. In the case of chlorinated rubber, it would be aromatic hydrocarbons. Active solvents are the ones which completely dissolve the resin and form a true solution of it.

Latent solvents. Latent solvents are not good solvents for the binder and may only swell the binder at room temperature. At somewhat elevated temperatures, they may become sufficiently active so as to form a solution. On cooling, however, it is likely that the solution would turn to a gel. Latent solvents are used in coatings with active solvents to regulate the solvent evaporation and in some cases improve film properties.

Diluents. Diluents are materials which are not true solvents for the resin, but which, when combined with active solvents, can be used to dilute the solution. Diluents are used for a number of purposes. One is to improve the film properties; just why they do so, though, is not entirely known. A combination of an active solvent and a diluent, however, may provide a smoother, stronger film than when the film is cast from the active solvent alone. In many cases, they are also used to reduce cost. Care must be used to make certain that the diluents leave the coating film prior to the majority of the active solvents or a poor film will result.

Solvent Combination Characteristics

Compatibility. More often than not, two or more resins are combined in the formulation of one coating. The solvent combination is often used to improve the compatibility of the various resins, one with another. This is very important. If all of the resins are not fully compatible, a poor film oftentimes results. On the other hand, with a proper solvent combination, the compatibility of the resins may be maintained and a good film cast from the solvent combination. All three of the above solvent types (active, latent, and filuent) can have an effect on the compatibility of the solution.

Sovlent Retention. Another important characteristic of the solvent combination is solvent retention. We know that the resin dissolves in the solvent and that the solvent is used to put the resin in a state which provides proper application. On the other hand, the solvents can also dissolve into the resin, causing high solvent retention in the coating. In many cases, solvent retention is not a positive characteristic in that it can reduce adhesion and the water and chemical resistance of the coating. The retention of a slow evaporating solvent in a film can reduce the water resistance and result in blistering of the film and loss of adhesion over a period of time.

Common Solvent Groups

Aliphatic Hydrocarbons. Aliphatic hydrocarbons may also be called paraffins. Chemically, they are open-chain hydrocarbons which are sometimes called straight-chain hydrocarbons, although this can be somewhat of a misnomer. Their schematic chemical diagram is:

Hexane

The most common of these are mineral spirits or V M & P naptha. Mineral spirits is often called *painter's naptha.* It is a relatively high boiling petroleum product used for dissolving asphalts, oils, and alkyds. Table 4.10 lists the common aliphatic hydrocarbons and their properties.

Aromatic Hydrocarbons. Aromatic hydrocarbons are chemicals which have a closed-chain six-carbon group as a principal part of the molecule. The simplest chemical of the family is benzene. This chemical family includes toluene, xylene, and some of the higher boiling homologs. They are major solvents for chlorinated rubber, coal tar, and certain alkyds, and are used as diluents in combination with solvents for vinyl, epoxies, and polyurethane materials. A chemical diagram for the aromatic hydrocarbons is:

Benzene   Toluene

Table 4.11 gives the properties of the principal aromatic solvents used in protective coatings.

Ketones. Oxygenated hydrocarbons of the acetone family are ketones, which include methyl ethyl ketone and methyl isobutyl ketone. They are the most effective solvents for vinyls and are often used in epoxies and other resin formulations. The chemical formula for acetone is:

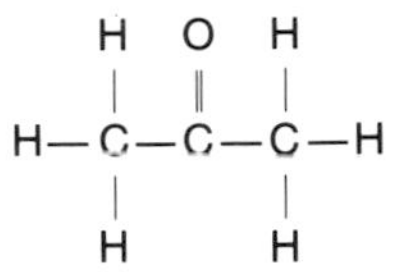

Acetone

Table 4.12 gives some of the properties of the principal ketones used for coatings.

Esters. Esters are also oxygenated hydrocarbons. They usually have a very distinctive and, for the most part, pleasant odor. (It is usually a fruity one since banana oil is one of the esters.) They can be used as latent solvents for vinyls and are commonly used in epoxy and polyurethane formulations. The chemical formula for one of the ester family is:

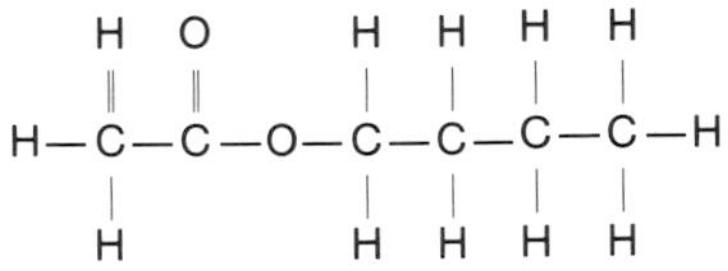

Butyl Acetate

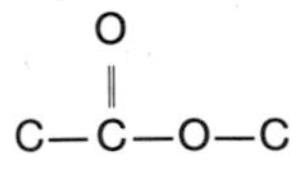

Ester Group of the Molecule

Table 4.13 lists properties of the principal esters used in protective coatings.

Alcohols. Alcohols are oxygenated hydrocarbons and are good solvents for highly polar binders such as phenolics. Some alcohols are also used in connection with epoxies. Methanol is the lowest homolog in the alcohol series. The chemical formula for ethyl alcohol, the most common of the alcohol family, is:

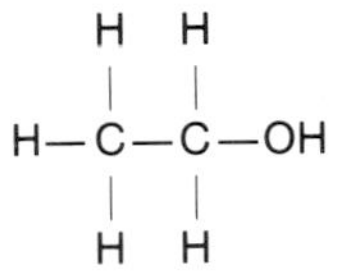

Ethyl Alcohol

C—OH

Alcohol Group of the Molecule

The properties of a series of alcohols used in protective coatings are given in Table 4.14.

Ethers. Ethers, such as ethyl ether (shown below), are not usually used as solvents for synthetic resins since they are very flammable. They are, however, excellent solvents for some of the natural resins, oils, and fats. The usual form of the ether which is used in protective coat-

---

TABLE 4.10 — Aliphatic Hydrocarbon Solvents

Solvent	Evaporation Rate N. Butyl Acetate = 1	Distillation Range °C	Flash Point Closed Cup °C	Pounds/Gal.
Laquer Diluent	4.0	96 - 105	6	6.1-6.3
V M & P Naptha	1.5	114 - 126	13	6.2-6.4
Mineral Spirits (odorless)	0.10	180 - 185	55	6.3-6.5

(SOURCE: Federation Series on Coating Technology, Unit 6, Solvents, Federation of Societies for Coating Technology, Philadelphia, PA, p. 46, 1967.)

TABLE 4.11 — Aromatic Hydrocarbon Solvents

Solvent	Evaporation Rate N. Butyl Acetate = 1	Distillation Range °C	Flash Point Closed Cup °C	Pounds/Gal.	Sp. Gr.
Benzol	5	79 - 80	- 12	7.34	0.885
Toluol	2	110 - 111	5	7.25	0.870
Xylol	0.6	139 - 141	28	7.25	0.869
High Flash Naptha	—	170°	38	7.25	0.870

(SOURCE: Federation Series on Coating Technology, Unit 6, Solvents, Federation of Societies for Coating Technology, Philadelphia, PA, p. 46, 1967.)

TABLE 4.12 — Ketones

Solvent	Evaporation Rate N. Butyl Acetate = 1	Distillation Range °C	Flash Point Closed Cup °C	Pounds/Gal.	Sp. Gr.
Acetone	9	57 - 58	– 10	6.59	0.792
MEK (Methyl Ethyl Ketone)	4	79 - 81	– 4	6.71	0.806
MIBK (Methyl Iso Butyl Ketone)	1.6	114 - 117	22	6.68	0.802
MIAK (Methyl Iso Amyl Ketone)	0.5	140 - 148	40	6.77	0.814
Cyclo Hexanone	0.2	154 - 161	54	7.88	0.945
Diacetone Alcohol (Tech.)	0.2	60 - 170	15	7.64	0.929
Diisobutyl Ketone	0.2	163 - 173	49	6.72	0.808
Isophorone	0.03	205 - 220	96	7.68	0.923

(SOURCE: Federation Series on Coating Technology, Unit 6, Solvents, Federation of Societies for Coating Technology, Philadelphia, PA, p. 46, 1967.)

TABLE 4.13 — Esters

Solvent	Evaporation Rate N. Butyl Acetate = 1	Distillation Range °C	Flash Point Closed Cup °C	Pounds/Gal.	Sp. Gr.
Ethyl Acetate (95%)	4.1	75 - 80	13	7.47	0.897
N. Propyl Acetate	2.3	88 - 103	18	7.35	0.885
N. Butyl Acetate	1.0	118 - 128	38	7.34	0.883
Amyl Acetate (95%)	0.4	135 - 150	41	7.29	0.876

(SOURCE: Federation Series on Coating Technology, Unit 6, Solvents, Federation of Societies for Coating Technology, Philadelphia, PA, p. 46, 1967.)

TABLE 4.14 — Alcohols

Solvent	Evaporation Rate N. Butyl Acetate = 1	Distillation Range °C	Flash Point Closed Cup °C	Pounds/Gal.	Sp. Gr.
Methyl Alcohol	6.0	65	16	6.6	0.793
Ethyl Alcohol	2.3	78 - 80	24	6.76	0.812
Propyl Alcohol	1.0	96 - 99	31	6.70	0.806
Iso Propyl Alcohol (91%)		79 - 84			
Butyl Alcohol	0.5	116 - 119	46	6.75	0.811
Cyclo-Hexanol	0.1	160 - 162	68	7.91	0.948

(SOURCE: Federation Series on Coating Technology, Unit 6, Solvents, Federation of Societies for Coating Technology, Philadelphia, PA, p. 46, 1967.)

ings is alcohol ether, such as ethylene glycol mono methyl ether (as shown below). This is commonly known in the trade as Cellosolve,[1] and there are a number of the glycol ethers sold under this designation.

Cellosolve is a good solvent for many oils, gums, natural resins, and synthetic resins such as alkyds, ethyl cellulose, nitro cellulose, polyvinyl acetate, polyvinyl butyral, and phenolics. It is a slow solvent and is used in many lacquers to improve the flow-out and gloss. These materials are often combined with other solvents to help achieve solvent formulations with higher flask points. Table 4.15 gives properties of some of the common glycol ethers.

Water. Water cannot be ruled out as a volatile sol-

[1]Trademark—Union Carbide Co., Danbury, CT.

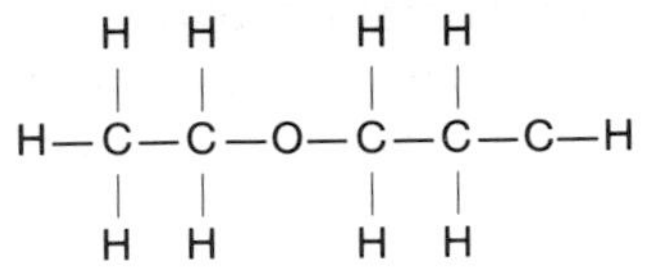

```
      H   H       H   H
      |   |       |   |
  H — C — C — O — C — C — C — H
      |   |       |   |
      H   H       H   H
```

Ethyl Ether
C — O — C
Ether Group of Molecule

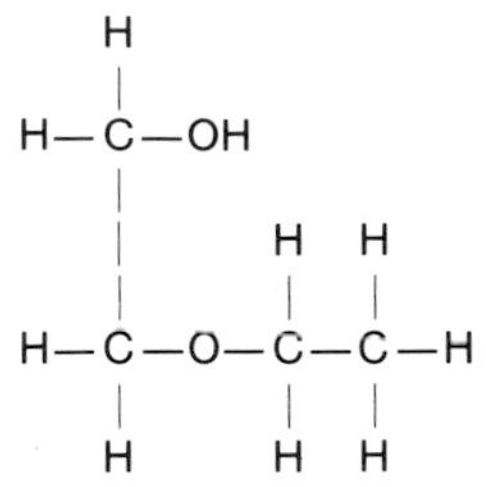

```
      H
      |
  H — C — OH
      |
      |           H   H
      |           |   |
  H — C — O — C — C — H
      |           |   |
      H           H   H
```

Ethylene Glycol Mono-Ethyl Ether
(Cellosolve)

vent since it is the most general solvent of any existing material. It is used as a thinner for latex and emulsion coatings since, in this case, the resins are dispersed rather than dissolved in the water. On the other hand, water is used for several of the inorganic zinc silicate materials, and in that case, acts as a true solvent. There are also some resins currently becoming available which likewise are truly dissolved in water. While these have not been used to any great extent for corrosion-resistant coatings, the possibility of water-dissolved corrosion-resistant resins does exist.

Miscellaneous Solvents. Thetrahydrofuran is a cyclic ether with very strong solvent characteristics for a wide variety of resinous materials such as acrylates, styrene, PVC, rubbers, and epoxies. It is even a solvent for high molecular weight homopolymers. The nitroparaffins are also good solvents for many synthetic resins. The principal member of the group is 2-nitropropane which has relatively low toxicity and evaporates at about the same rate as butyl acetate. It is a solvent for nitrocelullose, acrylics, epoxies, and, when mixed with toluol, is a very good solvent for vinyl chloride acetate resins. Table 4.16 indicates the properties of these solvents.

Plasticizers

Plasticizers are important in many coatings since many of the coating resins are too hard or too brittle in their original state to form a good permanent coating. A good example of this is chlorinated rubber which, in its original state, is a very hard, horny-textured substance that does not form a good coating because it is too brittle and gradually cracks up. On the other hand, chlorinated rubber is a very chemical-resistant material and does have film-forming properties. In order to become useable as a coating, it is necessary to add other materials to chlorinated rubber which will soften it, provide some extensibility, and decrease its brittle characteristics. To

TABLE 4.15 — Ethers - Alcohols

Solvent	Evaporation Rate N. Butyl Acetate = 1	Distillation Range °C	Flash Point Closed Cup °C	Pounds/Gal.	Sp. Gr.
Ethylene Glycol Mono Methyl Ether (Methyl Cellosolve)	0.5	124 - 126	46	8.03	0.966
Ethylene Glycol Mono Butyl Ether (Butyl Cellosolve)	0.3	132 137	54	7.74	0.931
Ethylene Glycol	0.06	166 - 173	74	7.51	0.902

(SOURCE: Federation Series on Coating Technology, Unit 6, Solvents, Federation of Societies for Coating Technology, Philadelphia, PA, p. 46, 1967.)

TABLE 4.16 — Miscellaneous Solvents

Solvent	Evaporation Rate N. Butyl Acetate = 1	Distillation Range °C	Flash Point Closed Cup °C	Pounds/Gal.	Sp. Gr.
Tetra Hydro Furan	6.0	65 - 68	− 15	7.39	0.888
2 Nitro Propane	1.1	118 - 122	38	8.24	0.992
Trichlor Ethylene	4.5	86 - 88	—	12.23	1.466

(SOURCE: Federation Series on Coating Technology, Unit 6, Solvents, Federation of Societies for Coating Technology, Philadelphia, PA, p. 46, 1967.)

Coating Fundamentals

form highly corrosion-resistant coatings, softer resinous materials, such as chlorinated paraffin, are added to it to act as a plasticizer. Where some oil characteristics are practical, alkyd resins can be used as a plasticizer for the coating. Thus, materials which are not practical coatings in themselves, *e.g.,* chlorinated rubber, can be formed into good useable coating materials by the use of modifying resins.

There are a number of theories concerning how plasticizers work. The easiest and most understandable is that the plasticizers are permanent solvents for the resin. As with any solvent, they tend to separate the large molecules so that they then become more plastic, less brittle, more extensible, and pliable. Plasticizers, as such, can be almost any material, depending again on the material which is to be plasticized (*e.g.,* water acts as a plasticizer for clay, resins can act as plasticizers for chlorinated rubber, and high molecular weight or high boiling point oxygenated solvents act as plasticizers for polyvinyl chloride materials). Almost all polyvinyl chloride materials are plasticized in one way or another, and all of the vinyl sheets are made with a substantial amount of plasticizer for handling and workability.

It is essential that coatings utilize the plasticizer principle in order to obtain materials with practical properties from an application, corrosion resistance, and film standpoint. Proper plasticizers add flexibility to the coating, help prevent cracking and checking upon aging, and add compatibility to the resin system. This is particularly true where two or more resins are mixed in order to obtain desired coating characteristics. A plasticizer as a permanent solvent for both resins aids in maintaining a mixed resin in a compatible state.

Plastisols are a good example of the use of plasticizers in a coating. In this case, there are no other solvents involved. The resins and the plasticizer are mixed, and, as long as they are cold, there is little activity of the plasticizer on the resin. In other words, the resin is not soluble in the plasticizer in a cold state. On the other hand, when the mixture is heated, the plasticizer becomes an active solvent and the resin and plasticizer become liquid. On cooling, they then form a continuous coating.

There are both internal and external plasticizers for coatings. *Internal plasticizers* are those where a large molecule is copolymerized into the resin molecule in such a way that the polymerized resin becomes permanently flexible and extendable to a greater degree than the original material without copolymerization. The use of vinyl acetate as a copolymer for vinyl chloride is an example of this. The combination of the two is a softer, more flexible and extendable material. The advantage of an internal plasticizer is that it is permanent and a part of the molecule which does not leech out, nor is it extracted by water or other solvents. Other characteristics of the original resin, however, may be lost in this process so that internal plasticization is not without its drawbacks.

External plasticization is the use of a second material as a permanent solvent in order to accomplish the plasticizing action. It can generally be said that as the external plasticizer content of a coating is increased, the water permeability also increases (Table 4.17). Also, with the

same molar concentration of plasticizer, the more efficient plasticizers give higher permeability. Tricresyl phosphate, for example, being a poorer plasticizer than some of the others, shows lower permeability to moisture. All factors being equal, the lower permeability would indicate increased corrosion resistance.

There are exceptions, however. In tests on plasticized PVC film exposed to 12 environments ranging from distilled water to various acids, bases, and salts, the phthalates were the least attacked, the next in order were the phosphates, and then the adipates and polymerics. Polyvinyl chloride maintenance paints, plasticized with a low concentration of tricresyl phosphate and chlorinated biphenols, have withstood weather, seawater, and chemical plant atmospheres for up to 20 years with little degradation of the paint film and no corrosion of steel or concrete substrates.

TABLE 4.17 — Water Permeability of Plasticized PVC Films

$P \times 10^8$ [(g/(hr)(cm^2)(mmHg)/(cm))]

Plasticizer	Mole % of Plasticizer					
	0	4	6	8	10	12
Tricresyl Phosphate	0.50	0.55	0.65	0.92	1.74	2.06
Dibutyl Phthalate	0.50	0.60	0.94	1.33	2.46	44.02
Dioctyl Phthalate	0.50	0.64	1.16	1.98	3.05	4.97
Dibutyl Adipate	0.50	0.96	1.72	2.67	4.08	5.89
Dioctyl Adipate	0.50	1.45	3.00	4.02	6.81	10.95
Dibutyl Sebacate	0.50	1.09	2.13	3.30	5.00	8.64
Dioctyl Sebacate	0.50	1.64	3.20	5.32	8.03	12.05

NOTE: For comparison of mole %, DOP concentration in PHR (Parts/Hundred of Resin) are: 0   26   40   54   69   86.

(SOURCE: Sears, Ken. Federation Series on Coating Technology, Unit 22, Plasticizers. Federation of Societies for Coating Technology, Philadelphia, PA, 1974.)

Some plasticizers can improve the exterior weather and light resistance of polymers, particularly PVC. Plasticizers containing epoxy groups, such as low molecular weight epoxy resin, can help the stability of plasticized PVC formulations where temperatures are rather high, as in southern desert areas. Weather resistance is enhanced since heat stability is improved by the epoxy compounds.

Organic phosphates as a sole plasticizer in PVC are not good. 2 ethyl hexyl diphenyl phosphate is perhaps the best. In blends with dioctyl phthalate (DOP) and similar plasticizers, however, they can add significantly to outdoor durability. This is true of both pigmented films and thin, clear films. Unit 22 of the Federation Series on Coatings states that a 4 mil film with no UV screener, plasticized with ethyl hexyl dephenyl phosphate will last one year in Miami, Florida. Plasticized with DOP, it will last one-and-a-half years.[4] If 10% of the DOP is replaced with 2 ethyl hexyl diphenyl phosphate, however, the film will last about two-and-a-half years.[5]

Other coating resins can also use plasticizers. Acrylics

are primarily internally plasticized for use as surface coatings. This is done by copolymerization with other compatible monomers. Actually, a wide variety of external plasticizers are compatible with acrylics. These include the phosphates, phthalates, adipates, and some epoxidized oils. One of the materials which has been most widely used is butyl benzol phthalate, which served as the standard for acrylic lacquers for comparison to plasticizers. Lacquers of this type are used for such things as automobiles and aircraft.

Chlorinated rubbers have been previously discussed. For highly corrosion-resistant coatings, polymeric plasticizers, such as chlorinated biphenyls and terphenyls, have been used. Since these are no longer available, chlorinated paraffins and aromatic phthalates are used. Very corrosion- and chemical-resistant coatings can be obtained from these combinations. Without the plasticizers, the excellent resistant properties of chlorinated rubber could not be utilized.

Epoxy resins may be modified with common plasticizers which will reduce the original viscosity of the epoxy and lead to easier handling and application. On curing, however, the final resin or coating is softer, has poorer impact resistance and less resistance to extraction by water and solvents. Epoxy resins used for coatings are, for the most part, internally plasticized. The materials most used are the various polyamide resins which react with the epoxy resin in almost any concentration. The coatings derived can vary from a very hard to a very soft coating with considerable elongation. Polyamide acts both as a reactive internal plasticizer and a cross-linking and curing agent for the epoxy. Good corrosion-, chemical-, and water-resistant coatings are obtained from the epoxy polyamide combination.

Plasticizers thus have a significant use in the formation of corrosion-resistant protective coatings. They not only add flexibility, increase compatibility, and prevent cracking and checking of the film, but the actual resistance of the films to the environment can be increased as well.

Miscellaneous Components

There are many coating components which are included in various formulations for many different reasons. They are usually in small quantities, with the level of use seldom exceeding 1 to 2% of the entire formulation. Nevertheless, they contribute to ease in manufacture, package stability, ease of application, appearance, and quality of resistance. Only one of these components, listed below, actually has an effect on corrosion resistance.

Biological Inhibitors

The growth of fungus on a coating surface gives it a very dirty, objectionable appearance. No surface is free of mildew growth since even a very thin film of dirt and moisture can prove to be a nutrient for the organisms. The effect of mildew will be more pronounced, however, if the coating contains ingredients which are also nutrients for the specific fungii. If these nutrients are present, the entire coating can be destroyed (Figure 4.13). A poly-

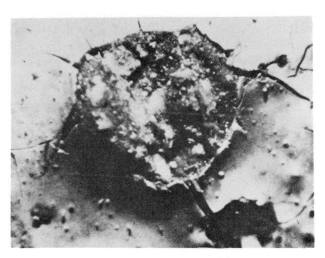

FIGURE 4.13 — A photomicrograph of mildew erupting through the coating surface. In this case, the mildew is living on the coating ingredients and so is not just on the surface. (SOURCE: Stewart, Wm. J., Federation Series on Coating Technology, Unit 11, Driers and Additives, Federation of Societies for Paint Technology, Philadelphia, PA, 1969.)

amide epoxy, for example, was rapidly destroyed in a sewer because of the nutrient value of the polyamide. An amine cured epoxy, however, showed no sign of decomposition under the same conditions.

Vehicles vary greatly in their resistance to biological growth. Common oils, such as linseed oil, are most susceptible. Reaction products of oils and other resins, such as alkyds, epoxies, and urethane esters, increase in resistance, while a high polymer, such as a vinyl chloride-acetate resin, is highly resistant. Most other polymers, such as epoxies, chlorinated rubber, urethane, etc., are also highly resistant unless they are modified with a nutrient resin.

The most fool-proof way of combating fungus or bacterial attack is to formulate with nonnutritive materials, starting with the basic resin and continuing through the pigments, plasticizers, etc. Even if biological growth inhibitors are necessary, a highly resistant formulation will allow maximum advantage to be contributed by the biocides. Mercury and copper compounds have proven most successful in preventing objectionable fungus growths. Phenyl mercury compounds provide the best protection. Their use, however, is objectionable and even prohibited under many conditions. Two organic phtalamide fungicides have been found effective at low concentrations while at the same time having a low toxicity.

Algae are one of the serious fouling problems aboard ships, particularly the very large cargo carriers, because of the depth of the boottopping (area between the light and heavy load lines). Algae seriously reduce the cruising speed. Copper, copper oxide, and tributyl tin compounds have been used to prevent this problem. Coal tar, vinyl, epoxy, and inorganic zinc coatings containing either copper or tin compounds, or both, have been effective. Shell fouling is also prevented by the same coatings, with some new inorganic zinc-tin coatings having promise of several years fouling protection.

The finished coating is the sum of all its individual parts. The resins, pigments, plasticizers, solvents, and possible special additives all contribute to the effectiveness of the coating and its life under corrosive conditions. While all of these ingredients can be combined and sealed in their container, the liquid coating is still of no value until it is applied to the substrate, formed, and cured in place. It is not a coating, nor is it at all effective until this last step is taken. Time, or its life after formation, will determine the true effectiveness of the material.

References

1. Munger, C. G., Kirk-Othmer: Encyclopedia of Chemical Technology, Coatings, Resistant, 3rd Ed., Vol. 6, John Wiley & Sons, New York, NY, pp. 456-57, 1979.
2. Craver, Kenneth J. and Tess, Roy W., Applied Polymer Science, Organic Coatings and Plastics Div., Amer. Chem. Soc., p. 518, 1975.
3. Fuller, W. R. and Love, C. H., Federation Series on Coating Technology, Unit 8, Inorganic Color Pigments, Federation of Societies for Paint Technology, Philadelphia, PA, 1968.
4. Sears, Kern, Federation Series on Coating Technology, Unit 22, Plasticizers, Federation of Societies for Paint Technology, Philadelphia, PA, 1974.
5. Good Painting Practice, Steel Structures Painting Manual, Vol. 1, Steel Structures Painting Council, Pittsburgh, PA, p. 82, 1966.

5

Corrosion-Resistant Organic Coatings

Corrosion-resistant organic coatings are the high-performance materials which give long-term protection to industrial, marine, chemical, and petroleum structures from serious corrosion conditions caused by marine atmospheres, chemicals, and industrial processing. They are, for the most part, specialty products designed to resist many different corrosive conditions as linings or coatings. They are not only resistant to corrosion, but are used to prevent contamination of liquids as well (*e.g.,* refined oil, iron-sensitive chemicals, and food products). Whichever the problem (protection against corrosion or contamination), the corrosion-resistant coating is the key material which separates two reactive materials.

Corrosion-resistant coatings are also known as maintenance coatings, as compared with industrial product finishes. The former are, for the most part, applied in the field to either existing or new structures and are subject to all of the difficulties of field application and field curing. The latter, however, are applied under very carefully controlled conditions of both application and cure. The actual coating materials could generically be quite similar and yet, because of the differing methods of application, the two types of products may be quite different in their basic formulation.

There is an important distinction to be made between these two approaches to protective coatings because there is an overlap in the materials used and because many pieces of equipment (*e.g.,* motors, pumps, etc.) are coated at the time of manufacture (in-plant). Because of such controlled application conditions, it could be assumed that these plant-applied coatings are superior to the field-applied ones. This may be a poor assumption, since an epoxy product finish can be a very different coating from a corrosion-resistant epoxy applied in the field.

The product finish, for instance, consists of two coats of an epoxy ester machinery enamel compared to a 10 to 12 mil high-build, field-applied epoxy coating. The life expectancy of these two epoxy materials would be quite different under the same actual operating conditions. This often becomes obvious in the field when equipment is supplied by the manufacturer and the new structure is coated by the general contractor or an on-the-job painting contractor. Once the unit is in operation, the manufactured item, even though it is supposedly coated with a similar generic coating, often fails a considerable amount of time before the field-applied material. This chapter will consider only the corrosion-resistant, field-applied organic coatings.

Protective coatings can be divided into several classes according to the basic chemical reactions involved in film formation. These classifications are given in Table 5.1, together with the principal generic coating materials in each class. Some of these generic materials can be further classified according to the individual materials which also may be significant from a protective coating standpoint. Examples of this are epoxy coatings, which can be broken down into epoxy amine coatings and epoxy polyamide coatings, or urethane materials, which can be broken down into aromatic isocyanate-cured urethanes and aliphatic isocyanate-cured urethane. Each of the subdivisions has its own properties.

Natural Air-Oxidizing Coatings

The classification of natural air-oxidizing coatings is included here not because of its overall importance in the corrosion-resistant coating field, but mainly because the corrosion engineer may encounter areas where such materials can do an effective job. Today, oil-type coatings are primarily used for wood house paints or for other wood structures, since it is difficult to improve on the penetrating characteristics of an oil as compared with the much higher molecular weight synthetic resin materials that

TABLE 5.1 — Protective Coating Classification

Basic Coating Formation	Generic Coating Material
Natural Air-Oxidizing Coatings	Drying Oils Tung Oil Phenolic Varnish
Synthetic Air-Oxidizing Coatings	Alkyds Vinyl Alkyds Epoxy Esters Silicone Alkyds Uralkyds
Solvent Dry Lacquers	Nitrocellulose Polyvinylchloride-acetate Copolymers Acrylic Polymers Chlorinated Rubber Coal Tar Cutback Asphalt Cutback
Coreactive Coatings	Epoxy Coal Tar Epoxy Polyurethane Polyesters Silicone
Emulsion-Type (Coalescent) Coatings	Vinyl Acetate Vinyl Acrylic Acrylic Epoxy
Heat-Condensing Coatings	Pure Phenolic Epoxy Phenolic
100% Solid Coatings	Coal Tar Enamel Asphalt Polyesters Epoxy Powder Coatings Vinyl Powder Coatings Plastisols Furan Materials

tend to remain on the surface of wood. Because of their highly penetrating characteristics, there may be areas and conditions where a drying oil coating could prove to be more effective than some of the more sophisticated generic materials.

Oils are also combined with many other resins to form drying oil varnishes. One of these materials which appears to be worthy of comment is tung oil phenolic varnish. Tung oil is a fast drying material, since it has three conjugated double bonds, and, when combined with phenolic resins, provides an air-drying film of maximum film hardness, excellent water resistance, and good flexibility and toughness.

Prior to the development of resins such as vinyls or epoxies, the tung oil phenolic varnish, when properly formulated, made a very corrosion-resistant coating. For example, two coats of a red lead tung oil phenolic primer and two coats of black tung oil phenolic topcoat were applied to floats carrying a dredging pipe from a large sea-going dredge. The coating, which was applied over a sandblasted surface, was a good, even coating. Even after several years in this severe service, the coating maintained the pontoons in essentially a corrosion-free condition.

Chemically, natural oils are triglycerides, i.e., a combination of one molecule of glycerine (glycerol) and three molecules of long-chain fatty acids. These are combined through an ester linkage $-C{\overset{O}{\parallel}}-O-C-$ to form the oil. Oils are generally liquids, while fats, which are also triglycerides, are solid at room temperature. Lard is the most common example of a fat. The difference between fats and oils is that fats are triglycerides which are formed with saturated fatty acids (no double bonds) and oils are formed with unsaturated fatty acids (one or more double bonds). Glycerine is a trihydric alcohol, chemically designated with the formula:

$$
\begin{array}{c}
H \\
| \\
H\text{-}C\text{-}O\text{-}H \\
| \\
H\text{-}C\text{-}O\text{-}H \\
| \\
H\text{-}C\text{-}O\text{-}H \\
| \\
H
\end{array} \qquad (5.1)
$$

The formula for linolenic acid, which has a triple double bond, is shown in Figure 5.1. The numbers in parenthesis indicate the carbon atom which is unsaturated in the 18-carbon chain molecule.

When the glycerine and the fatty acid are combined into an oil, the chemical formula shown in Figure 5.2 results.

A description of the drying oil reaction is provided in the Federation Series on Coatings Technology, Unit II, as follows.

> The shortest possible description of the drying of an oil is that it is convertible from a liquid to a solid. When a drying oil is exposed to the air in a thin film, there is a variable induction period during which oxygen absorption is negligible and there is no polymerization, which would be evidenced by thickening. The explanation is that the oil contains natural antioxidants and they must be destroyed by oxygen before normal oxidation can commence. Following the induction period the oil absorbs oxygen from the air. In the case of thin linseed oil the total amount of oxygen taken up is 10 to 12% by weight. Concurrently, there is a build up in the oil of unstable peroxide and hydroperoxide compounds. They increase

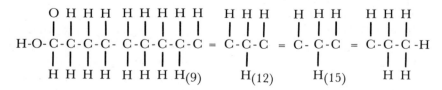

FIGURE 5.1 — Chemical formula for linolenic acid.

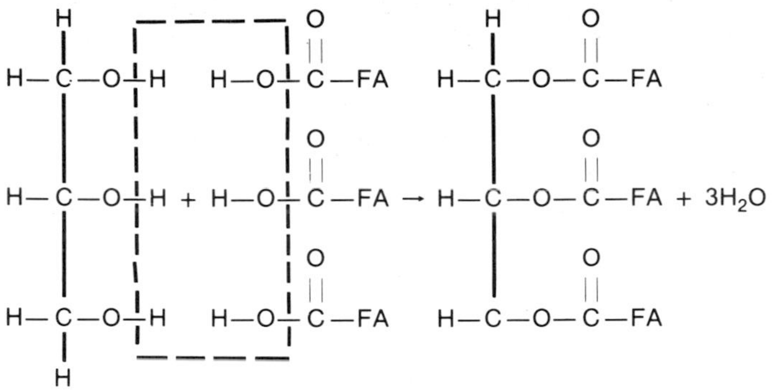

1 Glycerol + 3 Fatty Acid → Triglyceride (Oil) + 3 Water

FIGURE 5.2 — The chemical combination of glycerine and fatty acid results in an oil: FA = fatty acid molecule.

to a maximum and then decrease, even while oxygen absorption is continuing. Polymerization starts with the development of the peroxides and is most rapid at the time when the peroxides are disappearing the fastest, indicating that decomposition of the oil-peroxide compounds introduces cross-linkage of the oil. Since the fatty acid chains that are crosslinked are in different planes, the resulting polymers are three dimensional. The dried films contain ether linkages (-O-). The polymerization is accompanied by evolution of water, hydrogen peroxide and carbon dioxides.

There is wide overlapping of the various stages. When the oil polymers reach a certain stage in size and number, they join in a structure or framework throughout the film. It is now "set-to-touch." More technically, it is a gel. The film is very soft because it consists of liquid material, within a framework of solid material. Naturally, the reactions are more rapid at the surface of the film where there is contact with the most oxygen. The continued conversion from liquid to solid results in a stage when the surface of the film is all solid material—it is "surface dry." The process extends slowly downward through the films until finally it becomes "through dry." Even then the film contains a small amount of liquid material, which plasticizes or flexibilizes the film. Complete conversion to solid material is greatly retarded by poor accessibility to oxygen and, perhaps, by the non-drying stearic and oleic components. Fully recognizing that the stages of drying overlap, it may clarify the process to think of it as oxidation, polymerization, gelation and consolidation.[1]

An understanding of this oil reaction is important since it is peculiar to wherever the unsaturated oil molecules are found. The air oxidation of linseed oil, alkyds, epoxy esters, urethane oils, etc. all follow this process through the unsaturated oil attached to the resin molecule.

Natural Drying Oils

Driers

Metallic driers are common in the use of all oil-type paints. Oils dry slowly without the use of metallic driers; thus, it is conventional to use them to achieve practical drying times. Driers, which catalyze the action of oxygen with the drying oil, are the metallic salts of metals such as lead, manganese, cobalt, etc. The salts are usually the lead, manganese, or cobalt napthanates, and most formulations are both cobalt napthanates for the rapid surface-dry, and lead napthanate in order to create a through-dry. Cobalt alone creates a rapid surface reaction which often can cause wrinkling.

Oil Coating Advantages

Drying oils and oil-modified coatings have a number of advantages. Their number one advantage probably is ease of application. These materials are generally brush-applied: they brush easily and smoothly over both metal and wood surfaces, with the oil acting as a lubricant. They may also be applied by air spray; however, this method is not as successful. Oil-modified materials have excellent characteristics when applied to wood. They tend to penetrate the wood and eventually cure in place. The oil-modified film allows controlled moisture penetration and helps to control the expansion and contraction of the wood. Oil-based materials are sufficiently flexible, except after very long aging, to expand and contract with the summer and winter grain of the wood without cracking and chipping. Eventually, however, cracking and chipping does take place, although well-formulated oil coatings will withstand the summer wood expansion much better than most more sophisticated coatings such as vinyls and epoxies.

Oil-modified materials tend to wet metal surfaces very well. They have an affinity for metal surfaces and are able to penetrate wire-brushed, rusted surfaces to a much greater degree than many other coating materials. Oil coatings generally have good weather durability and will provide protection for a number of years, until the resinified oil becomes sufficiently brittle to check and crack.

Oil Coating Disadvantages

Oil paints or coatings have a number of disadvantages. They are not highly corrosion-resistant. This is due to the relatively high moisture vapor transfer rate, the transfer of ions through the coating, and the fact that they are not sufficiently alkali-resistant to withstand the alkali buildup at the cathode. The lack of resistance to alkali is one of the major drying oil deficiencies. Even mild alkalies tend to react with the oil part of the molecule, causing disintegration of the coating. Since they are not alkali-resistant, they are not satisfactory for application to concrete or other alkaline surfaces. Most oil-based films become brittle with aging, at which time they have a greater tendency to allow moisture vapor transfer as well as ionic transfer. Eventually, they tend to crack, craze, and then chip from the surface. Table 5.2 lists their general properties.

Synthetic Oxidizing Coatings

While drying oils may be used without other resin additives for oil-base house paints, there are a number of other uses of oils, such as incorporation into varnishes, alkyd resins, epoxy ester resins, oil-modified urethane resins, and other similar materials. This diversity of utility shows that the oils still maintain a prominent and essential position as coating components. They are more widely used as a portion of a binder in organic coatings than any other product of natural origin. For industrial use, they are gradually being replaced by acrylic esters, vinyl copolymers, epoxies, and synthetic materials.

From a chemical standpoint, all of the oil-modified

TABLE 5.2 — Natural Air-Oxidizing Materials

Property	Linseed Oil Paint	Tung Oil Phenolic Pigmented Coating
Physical Properties	Soft-Flexible, excellent for wood	Hard-Tough
Water Resistance	Fair High M.V.T. Rate	Good
Acid Resistance	Poor	Fair
Alkali Resistance	Poor	Poor
Salt Resistance	Poor-Fair	Fair
Solvent Resistance	Poor	Fair
Weather Resistance	Good	Good
Age Resistance	Gradually hardens and brittles over several years	Good, better than oils alone
Temperature Resistance	—	—
Recoatability	Good	Fair-Good
Best Characteristic	Penetration of wood	Moisture and Weather Resistance
Poorest Characteristic	Brittle on aging	Alkali Resistance
Primary Coating Use	Paint for wood	Mild Corrosion Resistance

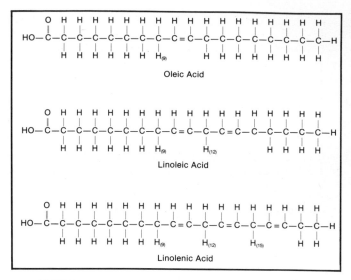

FIGURE 5.3 — Comparison of some unsaturated C^{18} fatty acids. The C = C bonds are chemical groups which react with oxygen and change the liquid oil to a solid.

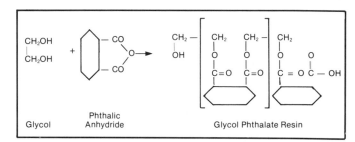

FIGURE 5.4 — Typical polyester reaction.

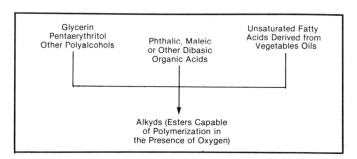

FIGURE 5.5 — Formation of an air drying alkyd resin.

resins are based on unsaturated fatty acids. Figure 5.3 shows the structural formulas for three of the acids which are incorporated into alkyds, epoxy esters, oil-modified urethanes, and varnishes.

Alkyd Resin Formation

There are other oils used in the paint industry in addition to the preceding three. Some of these are safflower oil, soya, tall oil acids extracted from waste materials in paper processing, tung oil, oiticica oil, castor oil, and various fish oils.

Figure 5.4 shows the simplest reaction of two materials to form a polyester resin. This is an example of the basic reaction that takes place in the formation of alkyd-type resins, although other materials may be used. The reaction of glycol and phthalic anhydride actually produces a hard, brittle resin which is heat-convertible or thermoplastic.

To form an air drying alkyd resin, a drying oil fatty acid, such as one of those shown in Figure 5.3, is added to materials as phthalic anhydride and glycerine. In this case, the phthalic anhydride and the glycerine react to form a resin similar to the one shown in Figure 5.4. However, since the glycerine has an additional hydroxyl group, the drying oil fatty acid reacts with it and is incorporated into the resulting resin. A simplified indication of this reaction is shown in Figure 5.5.

This same reaction can be shown using letters as symbols of the chemical molecules: G = glycerol, PA = phthalic anhydride, and FA = fatty acid, similar to those

shown in Figure 5.3.

$$-G-PA-G\ -PA-G\ -PA-G-PA-$$

with

$$\begin{array}{cccc}
| & | & | & | \\
O & O & O & O \\
| & | & | & | \\
H & FA & FA & H
\end{array}$$
(5.2)

Glycerol-Phthalic Anhydride-Fatty Acid Alkyd Resin

In this case, an essentially linear resin is formed with the long-chain fatty acid, providing solubility and flexibility. If the glycerine were changed to a four-hydroxyl molecule (pentaerythritol), as is often done, an even more complex resin structure would result. The fatty acid molecules attached to the alkyd resin can then react with other similar FA groups (in the same way that an oil reacts with oxygen, in that it polymerizes, gels, and hardens). All of the various oxygen-convertible resins, including the alkyds, epoxy esters, and urethane oils, react to form a film through the drying oil fatty acid group and follow the same chemical reaction pattern.

Alkyds

Any resin that is a polymer of ester-type monomers is a polyester resin. Alkyd resins, in general, can be considered polyesters. However, the term *alkyd* generally applies to polyesters that are modified with a triglyceride oil or fatty acids. Alkyd resins, then, generally mean the reaction product of a polybasic acid, polyhydric alcohol, and a monobasic fatty acid or oil, as has been previously shown. The name alkyd comes from the "al" of alcohol and the "cid" ($\approx$ kyd) of acid.

Alkyds are undoubtedly the most common of all of the oil-modified materials, as they have the broadest usage. Alkyds generally surpass all other present types of coatings in versatility and volume, providing a broad spectrum of performance with general economy. Alkyds are the "work horses" of the coating industry. They derive their versatility from the numerous variations in raw materials from which they can be made. Alkyds are made from three basic materials (*i.e.*, a polybasic acid, a polyhydric alcohol, and a fatty acid), each of which has an almost infinite number of variables, making the com-

binations available to form alkyds almost endless.

Alkyds can be classified in a number of different ways, the most common of which is according to the oil length. This, in general, means the percentage of oil in the alkyd, based on the total nonvolatile components. They are designated as short, medium, long, and very long oil alkyds. Table 5.3 shows the properties and the classification of alkyds by oil length.

The alkyds that have been described so far are *pure alkyds,* as differentiated from modified (*e.g.,* vinyl) alkyds. In the pure alkyds, the oil length influences all properties and is the primary factor in determining solubility, viscosity, flexibility, and hardness.

The alkyds which are most important from the standpoint of corrosion protection are the medium oil alkyds. They have a wide range of properties and are common for most of the applications familiar to corrosion engineers. As far as alkyds go, they are not as easy to apply as are the longer oil materials. On the other hand, because of the reduced oil content and higher amount of the polyester resins, they are more corrosion-resistant. Figure 5.6 shows a typical use for an industrial-type alkyd, showing both the color and the gloss characteristics. Exterior equipment enamels of this type are common throughout the world.

Figure 5.7 shows the installation of an alkyd red lead primer to the deck and deck structures of a small coastal tanker. This is a typical marine use of alkyd coatings.

Short oil alkyds are primarily used for baking

TABLE 5.3 — Classification and Properties of Alkyds by Oil Length

	Short	Short Medium	Medium Medium	Long Medium	Long	Very Long
Oil (% TNV)	35-43[1]	43-48	48-53	53-59	59-74	74-85
Fatty Acids	39-39	40-45	45-50	50-55	55-70	70-80
Phthalic Anhydride	50-38	38-36	36-33	33-30	30-20	20-10
Solvent Type	Aromatic Hydrocarbon	Aromatic	Aliphatic Semi-Aromatic	Aliphatic	Aliphatic	Aliphatic
Usual Solvents	Xylol, Toluol	Xylol	Mineral Spirits VM&P Naptha Aromatic Naptha	Mineral Spirits	Mineral Spirits	Mineral Spirits
Normal Solids of Solution	45-50	50	50-60	60	60-70	70-100
Normal Cure	Bake	Bake, force air dry	Force or air dry	Air dry	Air dry	Air dry
Normal Application	Spray or Dip	Spray or Dip	Spray Dip Brush	Dip Brush	Brush or Roll	Brush or Roll

[1]Formulas modified with non-fatty monobasic acid may have an oil length as low as 25%.
(SOURCE: Blegen, James R., Federation Series on Coating Technology, Unit 5, Alkyd Resins, Federation of Societies for Paint Technology, Philadelphia, PA, p. 9, 1969.)

FIGURE 5.6 — Alkyd-coated air cooler. (SOURCE: LaQue, F. L., Marine Corrosion: Causes and Prevention, Chapter 16, John Wiley & Sons, New York, NY, pp. 189-190, 1975.)

FIGURE 5.7 — Alkyd red lead primer applied to hand-cleaned steel in a marine atmosphere.

finishes and are therefore particularly well suited as product finishes. They also have fair corrosion resistance. The long oil alkyds, however, are the most popular materials used in trade sales, both interior and exterior. While they have some use from an industrial standpoint, they are primarily used as coatings for exterior wood structures and as exterior trim enamels. These materials are much closer to oil paints than they are to the alkyds used for corrosion-resistant purposes.

Modified Alkyds

Alkyds are modified with a large number of other resinous materials, and this is one of their distinct advantages. Each of the other resins provides some of its properties to the alkyds and, in doing so, makes for a versatile series of coating materials. The primary materials that can be combined with alkyds are listed in Table 5.4. Some, but not all of these combinations are of interest to the corrosion engineer. Most of these materials are actually reacted into the basic alkyd resins and are considered copolymers. They usually have increased corrosion-resistance when compared with a pure alkyd, depending on the modifying resin. The following alkyd modifications will be taken up individually.

Vinyl Alkyd

Alkyd resins can be combined with hydroxyl-modified vinyl chloride-vinyl acetate resins to form a combination that adds some of the vinyl properties to the alkyds. The vinyl alkyd, as compared with an unmodified alkyd, has decreased drying time, improved adhesion, very good water resistance, and very good exterior weather durability. Acrylics can be used to modify alkyds. The method of preparing the vinyl and acrylic alkyds is generally quite similar and consists of three steps:

1. The blending of vinyl or acrylic solutions with that of the alkyd.

2. A copolymerization method which involves the unsaturation of the fatty acids.

3. The chemical combination of the polymers by other functional groups such as the hydroxyls, usually forming an ester linkage between the two resins.

TABLE 5.4 — Modifiers for Alkyd Resins

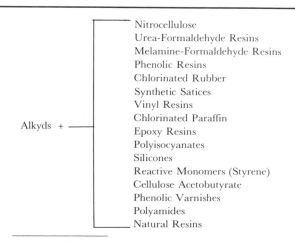

Alkyds +
- Nitrocellulose
- Urea-Formaldehyde Resins
- Melamine-Formaldehyde Resins
- Phenolic Resins
- Chlorinated Rubber
- Synthetic Satices
- Vinyl Resins
- Chlorinated Paraffin
- Epoxy Resins
- Polyisocyanates
- Silicones
- Reactive Monomers (Styrene)
- Cellulose Acetobutyrate
- Phenolic Varnishes
- Polyamides
- Natural Resins

(SOURCE: H. J. Lanson, "Chemistry and Technology of Alkyd Resins," Applied Polymer Science, Chapter 37, J. K. Craver and Roy W. Tess, Eds., Organic Coatings and Plastics Chemistry Division of the American Chemical Society, Washington, DC, pp. 543, 545, 1975.)

Corrosion Prevention by Protective Coatings

The vinyl alkyds, with their good water resistance and durable weather properties, have been used extensively by the U.S. Navy for application to their ships. These materials are listed under a government standard specification for this purpose.

Silicone Alkyds

Like the vinyl alkyds, the silicone alkyds are a definite step above the pure alkyds in their resistance to general weathering conditions and particularly to resistance to high temperatures. In this case, the reactive hydroxyl group on the long oil air dry alkyd resins reacts with hydroxyl groups on the silicone intermediates to give a copolymer structure that chemically combines the alkyds and the silicone. This creates quite a different molecule, even though it still retains a high percentage of unsaturated fatty acids. It therefore combines the workability of the alkyds with the durability, gloss retention, general weather resistance, and heat resistance of the silicones. The silicone alkyds are increasingly used as maintenance and marine finishes because of these improved properties. Silicone alkyds also are used rather extensively for stack coatings and for similar areas where moderately high temperatures are involved. The improved weather resistance and durability of the silicone alkyds are shown in Figure 5.8, comparing the silicone-modified product with standard air drying alkyds and epoxy.

Silicone intermediates are also reacted with hydroxylated polyesters. The simplest of these is the glycol phthalate resin, as shown in Figure 5.4. This resin has two terminal hydroxyl groups which readily react with the silicone intermediates to give silicone polyester copolymers. These have outstanding gloss retention and weather resistance. Silicone-modified polyesters such as these are extensively used in strip coating for steel and aluminum building panels where maximum exterior durability is a requirement.

Uralkyds

Uralkyds or urethane oils are materials which again utilize the reactive hydroxyl groups on the basic resin's molecule. In this case, they are reacted with toluene diisocyanate or other isocyanates to upgrade the properties of the alkyd or oil-base coating. The alkyd molecules still retain the oxidizing oil fatty acid group, so that the curing of these materials is primarily through the oil oxidation reaction. A reaction of the isocyanate with alkyds greatly improves the abrasion resistance of the final product, and these materials are often used for wooden floor coatings to withstand excessive abrasion. Abrasion resistance appears to be their outstanding property. Hydroxylated polyesters are also used in connection with isocyanate materials to provide high-quality maintenance finishes. These, however, are more appropriately classed under the urethane label and will be discussed later.

Epoxy Esters

Epoxy esters are a type of alkyd in that a high molecular weight epoxy resin is reacted with one of the alkyd resins. This reaction is an important one because of the rather extensive use of epoxy ester coatings. The reaction that takes place is usually between solid grade epoxy resins and drying oil fatty acids. This reaction requires elevated temperatures to form the epoxy ester; therefore, this reaction does not take place except under manufacturing plant conditions. The end product, the epoxy ester, can then be used to make the air drying epoxy ester coatings. The reactions to form the epoxy ester are given in Figure 5.9.

While the epoxy resin becomes a chemical part of the epoxy ester molecule, the drying or curing of the coating results from the oxidizing reactions of the unsaturated fatty acid part of the resin. It therefore dries like an oil-base enamel. Such a combination improves the chemical resistance of the basic polyester or alkyd resin and makes it a step above the ordinary alkyd resin for chemical resistance, particularly under alkaline conditions. Epoxy esters provide a very hard coating and make a good machinery finish or similar type product, primarily for interior use. They are not satisfactory for exterior weathering conditions, however, as they tend to chalk rapidly and excessively.

In a consideration of all of the various alkyds and

FIGURE 5.8 — Performance comparison of silicone alkyds and standard air drying alkyds and epoxy. (SOURCE: H. J. Lanson, "Chemistry and Technology of Alkyd Resins," Applied Polymer Science, Chapter 37, J. K. Craver and Roy W. Tess, Eds., Organic Coatings and Plastics Chemistry Division of the American Chemical Society, Washington, DC, pp. 543, 545, 1975.)

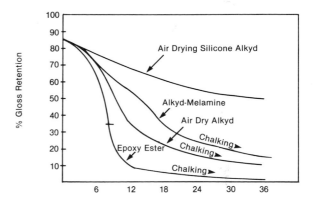

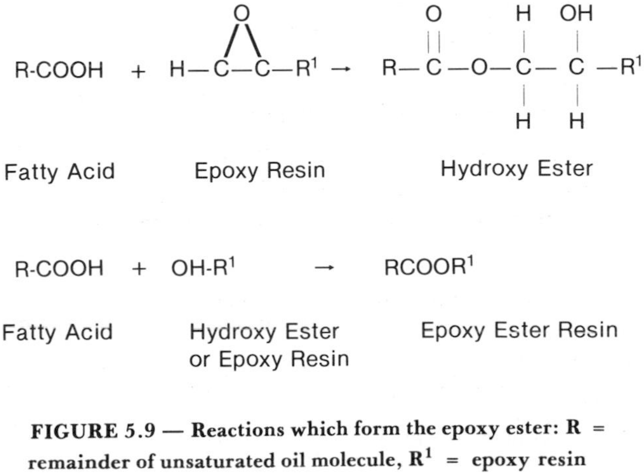

FIGURE 5.9 — Reactions which form the epoxy ester: R = remainder of unsaturated oil molecule, R^1 = epoxy resin molecule.

alkyd modifications for corrosion-resistant coatings, it is necessary to keep in mind that all of them are based on unsaturated fatty acids, and that this is the reactive part of the molecule which provides the curing mechanism for the coating. The alkyds are a step above linseed oil paint from the standpoint of corrosion resistance, and the various alkyd modifications are again a step above the plain oil-base alkyds. Nevertheless, the words *vinyl alkyds, acrylic alkyds, silicone alkyds,* etc., do not include the durability that is indicated by the generic names of vinyls, epoxies, urethanes, etc. These are strictly modifications of the basic alkyd coating, improving its general resistance while maintaining its essentially easy application. The oil-base part of the alkyds is the critical part from the standpoint of corrosion resistance, and these materials should not be compared to other coating materials that are used for severe corrosion resistance.

The alkyds are generally used for atmospheric coatings in areas which are not considered severely corrosive. They provide interior coatings for houses, shops, factories, and offices, as well as exterior coatings for tanks, silos, buildings, structural steel, railroad equipment, bridges, and topside marine surfaces. The general properties of the various alkyd resins are shown in Table 5.5.

Lacquers

Lacquers are materials which form a coating by evaporation of solvents. The resins which have been discussed are permanently soluble and, for the most part, thermoplastic. They are higher polymers that are inert and do not react to cure in the environment, and they are permanently soluble in their own solvents.

Nitrocellulose

As was described in Chapter 1, nitrocellulose is the oldest of the synthetic lacquer-type materials, and its growth paralleled that of the automobile industry inasmuch as it provided a high production finish for automobiles. Nitrocellulose is developed from natural cellulose, primarily from wood fibers and cotton linters. In its normal state, cellulose is an insoluble material. However, the molecule does contain a number of hydroxyl groups and can therefore be esterified. Through esterification, it can be made soluble and thus into a useable product.

The manufacture of nitrocellulose consists of reacting refined cellulose from either wood or cotton with a combination of nitric and sulfuric acid, so that all or most of the hydroxyl groups on the cellulose molecule are nitrated. With this reaction, the insoluble cellulose molecule becomes soluble in oxygenated organic solvents and is capable of making a good, solid, continuous film when the solvents evaporate from the nitrocellulose solution. Nitrocellulose is still used today in a variety of prod-

TABLE 5.5 — Alkyds: General Properties

Property	Medium Oil Alkyd	Vinyl Alkyd	Silicone Alkyd	Uralkyd	Epoxy Ester
Physical Properties	Flexible	Tough	Tough	Hard Abrasion Resistant	Hard
Water Resistance	Fair	Good	Good	Fair	Good
Acid Resistance	Fair	Best of group	Fair	Fair	Fair-Good
Alkali Resistance	Poor	Poor	Poor	Poor	Fair
Salt Resistance	Fair	Good	Good	Fair	Good
Solvent Resistance	Poor-Fair	Fair	Fair	Fair-Good	Fair-Good
Weather Resistance	Good	Very Good	Very Good Excellent gloss retention	Fair	Poor
Temperature Resistance	Good	Fair-Good	Excellent	Fair-Good	Good
Age Resistance	Good	Very Good	Very Good	Good	Good
Best Characteristic	Application	Weather Resistance	Weather & Heat Resistance	Abrasion Resistance	Alkali Resistance
Poorest Characteristic	Chemical Resistance	Alkali Resistance	Alkali Resistance	Chemical Resistance	Weathering
Recoatability	Excellent	Difficult	Fair	Difficult	Fair
Primary Coating Use	Weather-Resistant Coating	Corrosion-Resistant Coating	Corrosion-Resistant Coating	Abrasion-Resistant Coating	Machinery Enamel

uct finishes, particularly for furniture. It is not, however, used as a corrosion-resistant coating.

The chemical reaction which takes place to form nitrocellulose is shown in Figure 5.10. Cellulose also may be reacted in quite a different way to form ethycellulose, which is a cellulose ether. Ethycellulose and hydroxyl ethycellulose are specialty film-forming resins. They are not particularly used in the corrosion-resistant coating field, but they are of considerable importance to the ink, paper coating, pharmaceutical, and other miscellaneous industries.

Vinyl Coatings

Vinyl resins and vinyl resin coatings are considered one of the standard specialty coatings which will perform for years where other materials have failed. They have one of the broadest and most useful ranges of properties of any coating yet developed. Polyvinylchloride is nearly 150 years old. It was not until the 1930s, however, that vinyl materials were developed to the point where coatings could be made from them. The copolymerization of vinyl chloride and vinyl acetate in the United States began in about 1928. The actual development of practical coatings from the vinyl resins, however, did not take place for several more years.

In 1936 the Union Carbide Corporation[1] began its first commercial production of their vinyl chloride copolymer resins, which was the beginning of serious coatings development work in vinyls. From a corrosion-resistant coating standpoint, the vinyl copolymer containing 86% vinyl chloride and 14% vinyl acetate, or this ratio combined with a small amount of maleic acid, is the basis for almost all of the corrosion-resistant vinyl coatings as they are known today. Polyvinyl acetate is widely used for decorative wall coatings, but is of no particular interest to corrosion engineers.

Theoretically, a vinyl resin is any material composed of a basic vinyl molecule which is a carbon-double-bond-carbon group. The most well-known basic material in this class is a polyvinylchloride resin. It is not, however, used as a maintenance coating, as the PVC is relatively insoluble.

Soluble resins have been made by copolymerizing various vinyl molecules to form excellent coating materials. Examples are vinyl chloride copolymerized with vinyl acetate, vinylidene chloride, vinyl cyanide (acrylonitrile), or acrylates. These vinyl copolymers have properties that lend themselves particularly to surface coating applications.

Double Bond or Addition Polymerization

Some of the most important polymers used in the coating industry are those formed by double bond or addition polymerization. Materials which use this mechanism primarily contain the vinyl group ($CH_2 = CH_2$). Chemically speaking, vinyl resins include a number of other materials, such as polyethylene, polystyrene, and the polyacrylates and methacrylates, in addition to polyvinyl chloride, polyvinyl acetate, and their copolymers. All of the above contain the basic vinyl group, and all of the monomers containing this group, when combined

[1]Union Carbide Corp., Danbury, CT.

Corrosion-Resistant Organic Coatings

with the proper catalyst, react together to form long-chain polymers.

This reaction is schematically shown in Figure 5.11. In this case, the double bond in one molecule breaks, reacting with similar molecules to form a saturated, very high molecular weight polymer. The formation of a polymer by the double bond process does not sacrifice or give off any part of the molecule, as is the case during polymerization by condensation. The molecular weight of the polymer is a multiple of that of the monomer.

Vinyl chloride, one of the simplest of the unsaturated monomers, can be made to react with other vinyl chloride monomers to form a linear straight-chain polymer, as represented in Figure 5.12.

When the vinyl acetate monomer and the vinyl chloride monomer are reacted together, they also form a long-chain linear polymer. As mentioned previously, commercial polyvinyl chloride-acetate resin contains about 86% vinyl chloride and 14% vinyl acetate. The two materials are mixed in this proportion prior to the polymerization reaction, and the finished polymer results with the same composition. The copolymer is formed as shown in Figure 5.13. These polymers can be made up from hundreds or thousands of the single monomer units, and molecular weights from 25,000 to 100,000 are not uncommon. This type of polymerization is interesting since no byproducts (*e.g.*, water) are formed during the reaction. None of the atoms which make up the original monomer molecule are lost during the process. The chemical bonding between the two monomers after polymerization is a carbon-carbon bond which makes a very stable and resistant polymer. This is the primary reason for the vinyl polymer's excellent chemical and age resistance.

As mentioned previously, all of the materials which

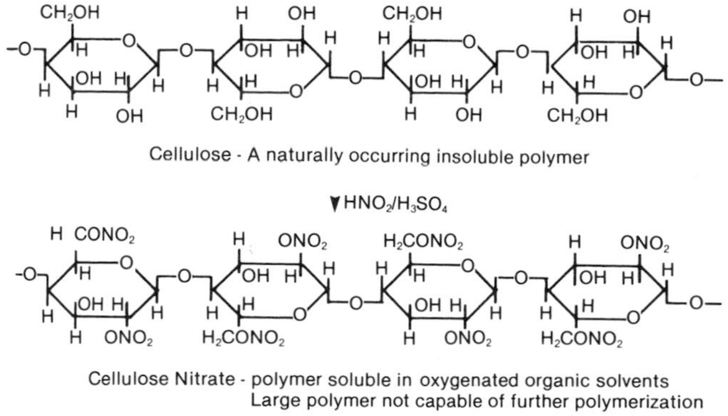

Cellulose - A naturally occurring insoluble polymer

▼HNO₃/H₃SO₄

Cellulose Nitrate - polymer soluble in oxygenated organic solvents
Large polymer not capable of further polymerization

FIGURE 5.10 — Chemical formation of nitrocellulose.

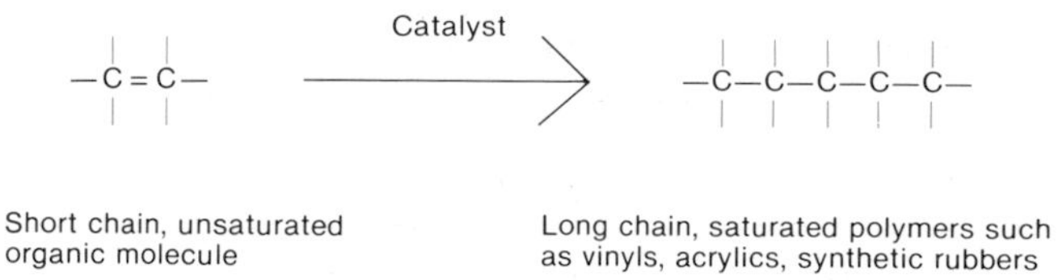

Short chain, unsaturated organic molecule

Long chain, saturated polymers such as vinyls, acrylics, synthetic rubbers

FIGURE 5.11 — Short-chain molecules, reacted with the proper catalyst, form long-chain polymers in the double bond polymerization process.

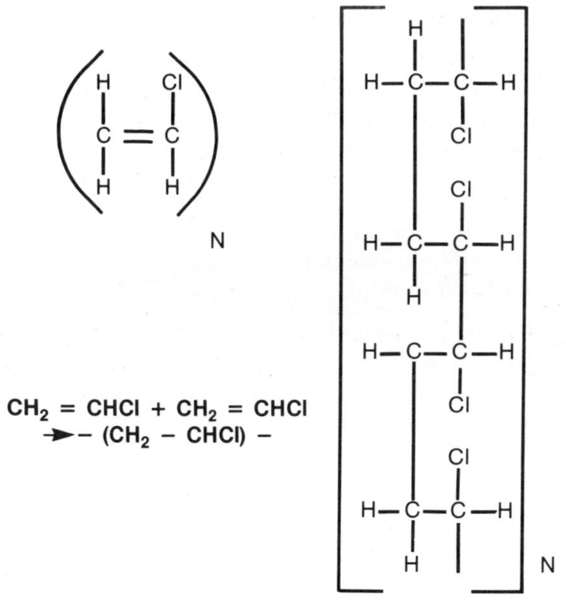

FIGURE 5.12 — Reaction of vinyl chloride monomers can form a linear short-chain polymer.

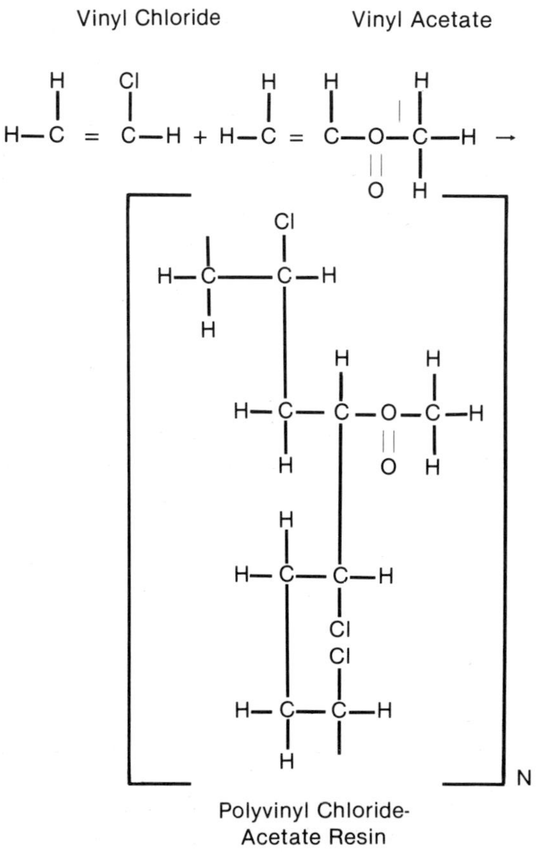

FIGURE 5.13 — Formation of a long-chain linear polymer from the vinyl acetate and vinyl chloride monomers.

contain the vinyl double bond grouping could technically be called vinyl materials. On the other hand, it has become customary in industry to limit the term vinyl resin to vinyl acetate, vinyl chloride-vinyl acetate copolymers, and other similar copolymers. The acrylate

and methacrylates are generally called acrylics, and polyethylene and polystyrene are referred to by these respective names.

The long-chain carbon molecules of vinyl resins make them thermoplastic, which therefore makes it possible to produce solution-type coatings from them. These long, saturated carbon chains also make them physically strong and resistant to a wide range of chemical reactions, such as those involving acids, alkalies, and many salts. These vinyl resins have a broad chemical resistance when compared with the more common coating polymers.

Coatings which are made from the vinyl resin polymers have most of the positive characteristics of the basic resins, such as hardness, strength, toughness, and resistance to water and to a broad range of chemicals. As indicated earlier, these materials follow the solution type of coating formation. The polymers are completely polymerized to their final state prior to being put in solution, and once the solvents have dried from the resin solution and the coating formed, all of the original properties of the vinyl resin are incorporated into the film.

The outstanding properties of vinyl coatings (polyvinyl-chloride acetate-type) are as follows.

Film Formation

The ability to form a homogeneous, tight film over a surface is a basic property. The film formed by the vinyl resins is different from many other coating materials in common use because it is prereacted or polymerized and because it goes through no oxidation, chemical reaction, or other change in the formation of the film or in aging. This is important, since these coatings do not rely on temperature, humidity, solar radiation, or moisture vapor in order to cure in place. As long as the solvents will evaporate from the film, the finished coating will be formed, having all of the properties inherent in the basic resin. This dense, homogeneous, continuous film accounts for much of the long life of vinyl coatings.

Chemical Resistance

A vinyl coating is basically inert to almost all inorganic substances (*e.g.,* acids, alkalies, and salts), as well as to water, oil, grease, alcohols, and similar materials. Because of its polymeric structure and its independence of other materials such as catalysts to develop its basic resistance, vinyl coatings are inherently chemical-resistant. There are only a few materials, its own solvents or similar products, which will destroy its film integrity. There are few other coatings that will resist both strong acids and alkalies, and strong oxidizing chemicals as well. Vinyl coatings, for example, have been used as a coating for the belly band on sulfuric and nitric acid tank cars where concentrated acid spills are common. They also have been used as a lining for caustic storage tanks and tank cars holding 50% or greater concentration of sodium hydroxide. Vinyl coatings also have been used to line perchloric acid fume ducts, where any other organic material would have caused an instant fire or explosion. These extremes indicate the versatile chemical resistance of these coatings.

Water Resistance

Vinyl coatings are generally unaffected by continuous water exposure. Some of the oldest vinyl coatings in existence are on the interior of water storage tanks or penstocks. Fifteen or more years without any maintenance or coating breakdown has been the common experience. Two coatings, Bureau of Reclamation specifications VR-3 and VR-6, were standard products for use on penstocks, dam gates, and similar structures for many years, with actual use experience ranging over 15 to 20 years with little or no maintenance. Some of this experience involved a water flow of 18 to 20 feet per second.

Age Resistance

Continuous use in various water installations is an indication of the age resistance of vinyl coatings. There is no other coating, with the exception of hot-applied coal tar enamel, that has the long-term exposure of vinyl coatings in water. Another similar exposure is the vinyl lining in sewer pipe and structures. The oldest installation was in 1948 and it is still in service, unchanged in well over thirty years. It is anticipated that the vinyl lining will last the life of the sewer, which is 100 years or more.

A related report was made by A. K. Doolittle at the annual NACE meeting in March, 1963, where he reported on the oldest fully-documented coating installation in the United States that could be inspected at that time. Table 5.6 shows the exposures of 9 years and older which he inspected. It is important to note that out of all of the oldest exposures, only one was not a vinyl coating. These long-time exposures were not mild ones either, but were considered extremely severe conditions for any coating to withstand. These included equipment coated for protection against 15 to 18% caustic soda, viscose storage tanks, and filter presses exposed to caustic, carbon disulfide, hydrogen sulfide, etc. There were also potable water tanks, dam gates, pipe racks in chemical plants, atomic energy installations, and a phosphoric acid plant included in the inspection.

The uses of vinyl coatings are quite broad, particularly in areas where severe corrosion exists. These uses include marine exposures (both exterior atmospheric and immersion), chemical and paper plants, petroleum refineries, food processing units, water and sewer plants, offshore structures, and many others.

TABLE 5.6 — Nine-Year-Plus Exposures

Number of Exposures	Number of Years	Substrate	Surface Preparation	System
1	19	Steel	White Metal Blast	Vinyl
1	18	Concrete	Acid Etch	Vinyl
1	15	Steel	White Metal Blast	Vinyl
1	15	Steel	Commercial Blast	Vinyl
1	14	Concrete	None	Vinyl
1	12	Steel	White Metal Blast	Vinyl
1	11	Steel	White Metal Blast	Vinyl
2	10	Steel	White Metal Blast	Vinyl
1	9	Concrete	None	Vinyl
1	9	Steel	Commercial Blast	Inorganic Zinc
3	9	Steel	White Metal Blast	Vinyl
14				

Chlorinated Rubber Coatings

Chlorinated rubber coatings consist of natural rubber reacted with chlorine. This forms a very hard, horny-textured resin lacking in the elastic and resilient characteristics of rubber products. The coating has a specific gravity of 1.64, which is almost twice that of pure rubber. It is odorless, tasteless, and nontoxic, and will not support combustion or burn.

The chlorinated rubber coating also is stable to heat for considerable periods up to 125 C. At higher temperatures, there is a tendency toward chemical decomposition with the evolution of hydrochloric acid. At about 135 C, the chemical breakdown becomes significant. Boiling water or steam is destructive to chlorinated rubber and is not recommended for coating applications which require durability under these conditions. In spite of this, chlorinated rubber has a low water absorption.

In the absence of sunlight, aging has very little effect on chlorinated rubber. Pigmented compositions are durable in the sunlight; however, sunlight causes both discoloration and embrittlement in clear, unstabilized films.

The electrical properties of chlorinated rubber are excellent, as might be expected from a chlorinated polymer. Dielectric constant at 25 C is 2,200 volts per mil. It has excellent abrasion and shock resistance if properly compounded with other materials.

Coatings made with chlorinated rubber, particularly those which are modified with chlorinated resins, have very good chemical resistance, particularly to inorganic chemicals. Chlorinated rubber by itself, however, is not practical and must be modified in order to be useable. It is essentially modified in two ways: One way is to use chlorinated resins as the modifiers and the second is to use alkyd resins.

There are actually about four categories in which chlorinated rubbers can be placed. The first category consists of formulations containing high proportions of chlorinated rubber modified with nonreactive resins and plasticizers. These nonreactive materials are most often highly chlorinated materials (*e.g.,* chlorinated paraffin), and this category is often called an *all-chlorinated system.* The materials are used in situations requiring maximum chemical or flame resistance. A considerable number of the chlorinated rubber coatings fall into this category.

The second category differs from the first in that part of the nonreactive resin or plasticizer is replaced with an alkyd or an oleoresinous varnish. The chlorinated rubber contents still remain high; however, some of the excellent chemical resistance which is available in the first category is sacrificed. Corrosion resistance is still good, however, and the coating gains in adhesion, weatherability, and ease of application.

In category three, the alkyd resin becomes a major ingredient, and chlorinated rubber is, in essence, fortifying the alkyd. Coatings in this category are often called *chlorinated rubber-fortified alkyds.* They have good weatherability, adhesion, gloss, and brushability, plus some improved chemical and water resistance over the alkyd alone.

Category four is intended for alkyd or oleoresinous

materials which require some upgrading, particularly in their speed of drying. Table 5.7 shows the categories of chlorinated rubber-based coatings and some of their properties.

The properties of chlorinated rubber are better understood in light of its chemical makeup. In the chlorination process, rubber reacts with chlorine in an amount sufficient to yield a product of approximately 64 to 65% chlorine. Rubber is made up of isoprene units. These undergo a complex chlorination reaction with both addition and substitution of chlorine. The chlorination process is brought to an end when the double bonds have disappeared and the chlorine content is high enough to assure optimum stability, compatibility, and resistance to fire. The chlorine atoms are probably scattered along the available carbon atoms with little tendency for more than one chlorine atom to be on a single carbon atom. Generally, there is less than the theoretical chlorine content for the chlorine substituted isoprene unit, $C_5H_6Cl_4$. The isoprene-chlorine reaction is illustrated in Figure 5.14.

Chlorine-containing organic chemicals generally have very good chemical-resistant properties. As can be seen from the formula for chlorinated rubber or chlorinated isoprene, the amount of chlorine in the molecule is substantial. It is this, as well as the saturated carbon bonds, that give it good chemical resistance.

One of the positive properties of chlorinated rubber is that it has the ability to fast through-dry so that rapid application is possible. It increases the drying speed of alkyds substantially and, when used as a chlorinated rubber lacquer, dries rapidly and hard. These coatings, *i.e.,* the fully chlorinated coatings, are primarily used in the chemical and marine industries, although one of its major uses is in coating concrete swimming pools. As chlorinated rubber topcoats, they provide a very good maintenance coating. They are hard, tough, and chemical-resistant; they have good resistance to extended weathering; and they can be readily repaired and recoated.

Acrylic Polymers

Acrylic resins are primarily polymeric derivatives of acrylic and methacrylic acid. More important resins are polymers of methyl and ethyl esters of these acids or copolymers of mixtures of these monomers. Propyl, butyl, and isobutyl esters are also used, as well as acrylamides, acrylonitriles, and other similar materials. These monomers may be blended in any number of different proportions and then polymerized into finished resins. Because of the variation in esterification and the copolymerization of the various esterified acrylic monomers, there is almost an untold number of combinations that can be used. The polymers can vary from very hard, brittle materials to very soft, flexible plastics.

The solution and film properties of the various acrylic polymers, when they are made into coatings, are regulated by the molecular weight, the nature of the polymer solution, and the composition of the polymer or copolymer chemical structure. The effect of molecular weight is rather obvious. Film formation of any solution coating depends either on the formation of the primary chemical bonds or upon the entanglement of the polymer chains by secondary chemical interaction. In the case of thermoplastic coatings, the longer the polymer chain, the more thoroughly the chains will be mixed and the tougher and more coherent the film will be. This is not completely beneficial, even though the high polymers have better

---

TABLE 5.7 — Categories of Chlorinated Rubber-Based Coatings

	Category			
	1	2	3	4
Chlorinated Rubber, %	50-60	45-60	20-35	5-50
Nonreactable Resin and Plasticizer, %	40-50	0-25	—	—
Alkyd or Varnish (selected for job in hand), %	—	20-30	65-80	0-25
Other Polymers, Resins, and Plasticizers, %	—	—	—	0-95
Chemical Resistance	Excellent	Good	Fair (much higher than conventional enamels)	—
Outdoor Weathering	Good	Very Good	Excellent	—
Adhesion to Metal	Fair	Good	Excellent	—
Typical Applications	Paints and mastics for indoor, extremely corrosive environments; unusually severe outdoor environments; extreme nonflammability.	Paint for typically tough corrosive environments outdoors; water- and chemical-resistant enamels; for masonry, railroad, marine, swimming pool, food processing, buildings; excellent non-flammability.	Product finishes for fast dry with excellent water and chemical resistance; traffic paints; Navy flame-retardant paint.	Embraces a host of specialty finishes from chlorinated rubber to acrylics with excellent chemical resistance and durability, to marginal varnishes fortified with a little Parlon to make them salable.

(SOURCE: Properties and Uses of Chlorinated Rubber, Hercules, Inc., Wilmington, DE.)

Corrosion Prevention by Protective Coatings

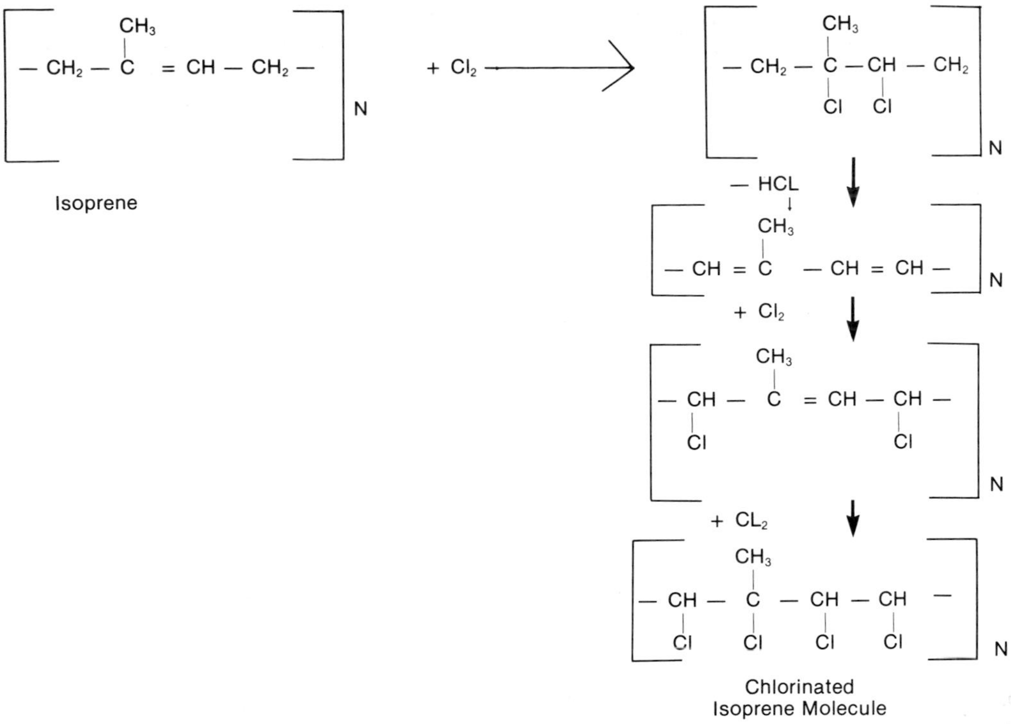

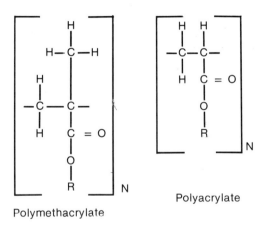

FIGURE 5.14 — Formation of chlorinated rubber.

properties, since the viscosity of the coating solutions increases exponentially with the molecular weight. The molecular weight of the acrylics, therefore, must be kept down to a reasonable level at which a workable viscosity can be obtained with the polymer in solution.

The acrylic polymers used in coatings are primarily those of the polymethacrylates and polyacrylates. The chemical structure is shown in Figure 5.15.

The properties of the polymers are dependent on three different items: (1) the presence of the CH_3 or hydrogen on the alpha carbon, (2) the length of the ester side chain indicated by "R," and (3) the functionality in the ester side chain.

Commercial acrylic polymers are almost always copolymers of several monomers so that a wide range of hardness, strength, and flexibility can be achieved. Table 5.8 shows the effect of various monomers on coating film properties.

Acrylic resins are characterized primarily by their water-white color, resistance to change in color over time, and their perfect transparency. Acrylics generally have excellent durability properties due to the chemical nature of the polymer itself. The main polymer chain is comprised entirely of carbon-to-carbon single bonds (which, as previously discussed, are relatively inert), and are not as susceptible to chemical change as are the ester, ether, and amide linkages. The ester side chain can be hydrolyzed; however, such change does not necessarily result in a breakdown of the polymer carbon chain. The general durability characteristics of some of the esters of methacrylates and acrylates are shown in Table 5.9.

The acrylates increase in flexibility, durability, and water resistance as the ester chain link increases. Most of

FIGURE 5.15 — Chemical formulation of the acrylic polymers, polymethacrylate and polyacrylate.

the acrylic materials with which corrosion engineers will be involved are those acrylics which are coreacted with other resins (e.g., epoxies, vinyls, and isocyanate modifications). The use of the acrylic in these combinations is to increase the exterior durability and weather resistance and to retain the appearance of the coatings over long periods of time.

Many of the acrylics also are applied as emulsions and water dispersions. For the most part, however, these are applied as decorative coatings rather than for corrosion resistance. Except for the modifications of the other corrosion-resistant coating resins, the acrylics are primarily used as product finishes on automobiles, refrigerators, and similar products which require excellent durability in the long-lasting, factory-applied coating.

TABLE 5.8 — Effect of Various Monomers on Film Properties

Film Property	Contributing Monomers
Exterior Durability	Methacrylates and Acrylates
Hardness	Methyl Methacrylate Styrene Methacrylic and Acrylic Acid
Flexibility	Ethyl Acrylate Butyl Acrylate 2-Ethylhexyl Acrylate
Stain Resistance	Short-Chain Methacrylates
Water Resistance	Methyl Methacrylate Styrene Long-Chain Methacrylates and Acrylates
Mar Resistance	Methacrylamide Acrylonitrile
Solvent and Grease Resistance	Acrylonitrile Methacrylamide Methacrylic Acid
Adhesion to Metals	Methacrylic/Acrylic Acid

(SOURCE: W. H. Brendley, G. V. Calder, and L. A. Wetzel, "Chemistry and Technology of Acrylic Resins for Coatings," Applied Polymer Science, Chapter 55, J. K. Craver and Roy W. Tess, Eds., Organic Coatings and Plastics Chemistry Division of the American Chemical Society, Washington, DC, pp. 862, 865, 1975.)

TABLE 5.9 — General Durability Characteristics of Acrylic Homopolymers

	Methacrylate	Acrylate
Methyl	Very Good	Poor
Ethyl	Excellent	Fair
Iso-Butyl	Excellent	Good
N-Butyl	Excellent	Excellent

(SOURCE: W. H. Brendley, G. V. Calder, and L. A. Wetzel, "Chemistry and Technology of Acrylic Resins for Coatings," Applied Polymer Science, Chapter 55, K. J. Craver and Roy W. Tess, Eds., Organic Coatings and Plastics Chemistry Division of the American Chemical Society, Washington, DC, pp. 862, 865, 1975.)

Coal Tar Coatings

There are a number of coal tar coatings made by dissolving processed coal tar pitch, or a blend of these pitches, in suitable solvents. They dry entirely by evaporation of the solvents and their properties depend to a great extent on the type of coal tar raw materials and the blending of these materials. Generally, they are all quite similar. The outstanding property of coal tar coatings is their extremely low permeability to moisture and their high dielectric resistance, both of which contribute to their corrosion resistance.

Coal tar coatings are made in different consistencies; those without any inert fillers, and those which contain inert materials in order to build film thickness. Coal tar coatings, in general, are not affected by mineral oil, but can be dissolved by a vegetable or animal oil and grease and detergents. They have good resistance to weak acids, alkalies, salts, seawater, and other aggressive atmospheres. They derive their corrosion-resistant properties from an impermeable film concept; as such, they provide protection by exclusion of moisture and air from the underlying surface. They are alkali-resistant and can be applied to concrete as well as to steel.

One of the problems with coal tar coatings is their tendency to alligator when exposed to direct sunlight. This is brought about by the hardening of the upper layer of the coal tar film because of the exposure to the sun's ultraviolet rays. The upper layer of the film contracts and slips over the softer under layer causing alligator marks. Coal tar cutback coatings should be protected from the sun in order to retain their corrosion-resistant properties. The coal tar cutback coatings can be considered true solvent dry lacquers.

Asphalt

Asphalt cutbacks are also extensively used as corrosion-resistant coatings. They not only are made with distilled asphalt from petroleum, but also are made from natural asphalt, or gilsonite. It is a very hard, asphaltic material and is mined in the same way that coal is mined. It has excellent chemical resistance and good weather resistance. Most of the better grades of asphaltic coatings will contain a proportion of gilsonite in order to upgrade the petroleum-derived asphalt. In general, the higher the proportion of gilsonite, the more resistant the coating will be.

Asphalt coatings are more weather-resistant than the corresponding coal tar coatings since they do not tend to alligator. On the other hand, they are inferior in water resistance to the coal tar products.

Solvents used in asphalt coatings are very mild. Since they do not impart taste to water, many asphaltic coatings are used for painting steel water tanks and concrete reservoirs for storing drinking water.

Asphaltic coatings also are combined with higher polymer petroleum resins which increase the strength and provide more flexibility to the asphalt gilsonite coatings. Asphaltic coatings are also made with inert fillers which add to the thickness of the coating and, in many cases, to its impervious characteristics. Some of these asphaltic-type coatings have excellent resistance to industrial fumes, water, moisture, condensation, and exterior weather exposure. They do tend to lose gloss and to chalk when exposed to the weather; on the other hand, this apparently does not detract from their other good characteristics or reduce their overall weather resistance.

Both asphalt and coal tar coatings have the disadvantage of being black, and while some very dark iron oxide colors can be obtained from the bituminous-type coatings, they must generally be used where a black coating with generally poor appearance is adequate. Bitmunious coatings can be pigmented with leafing aluminum and provide a bright aluminum color in this manner. This, however, is the limit to which bituminous coatings can be pigmented for decorative purposes.

Advantages of Solvent-Evaporating Films

There are a number of very distinct advantages in the use of solvent-evaporating films. The advantages are general for the whole group of resins which are in this thermoplastic category. One of the first important characteristics is fast film formation. As has been described, the thermoplastic resins only require the evaporation of the solvent to form their final film so that film formation is in a simple stage, as compared with conversion coatings where two or more stages are required in order to properly form the film. This fast film formation, and the fact that the resin is completely reacted prior to its going into solution, enables one coat to follow another in rather rapid succession. This is important in many chemical areas where chemical fallout may be a problem. It is also true in marine areas where chloride-containing mist or spray may precipitate on the coating. In this case, one coat of the coating may follow another as soon as the major amount of solvent has evaporated, laying down the completed coating of one or more coats in a relatively short period of time and thus eliminating the intercoat contamination.

Where necessary, most lacquer-type coatings may be handled quickly without the necessity of going through additional stages of cure before they become hard enough to handle. There is no critical reaction time or critical reaction conditions which are necessary for most solvent evaporating coatings. They may be applied in subfreezing conditions as long as the solvents will evaporate from the film. They also may be applied under rather warm or hot conditions as long as the solvents are formulated to evaporate somewhat more slowly. Again, they do not have to go through the other stages of cure which can be very critical, depending on the temperature, humidity, and air circulation. Being permanently soluble in their own solvents, repair and maintenance of lacquer-type coatings is generally easier than with thermosetting or conversion coatings. From the corrosion engineer's standpoint, these favorable factors may more than outweigh the advantages of some of the other coating types.

Disadvantages of Solvent-Evaporating Films

Solvent dry coatings also have a number of disadvantages. As mentioned previously, if the solvent formulation is too fast, it will tend to overspray, causing a rough film which may not be continuous and which may allow rapid corrosion. Improper solvent formulation may also cause the film to blush. Under high humidity conditions, a solvent combination may be such that the liquid coating dries sufficiently fast so that the surface cools and moisture precipitates on the surface before the coating is dry. This is called *blushing*. When this happens, the surface of the film becomes milky and white, many times causing the resin to precipitate on the surface, leaving a noncontinuous film and one which is not satisfactory for corrosion resistance. Solvent retention also has been mentioned. Where there is a solvent in the solvent formulation which does not release readily, a measurable percentage remains in the film. This can cause poor water or chemical resistance.

Oftentimes with solvent-evaporating coatings or lacquers, there is somewhat more difficulty in the application of the coating than there is in conversion coatings. This, again, is primarily due to the solvent formulation. During the application, however, it must be realized that a considerable amount of solvent is lost between the atomization point at the gun and the impact point on the surface. Gun distance must be properly controlled and a continuous wet film applied. If this is not done, then a noncontinuous film will result.

The application by brush of many lacquer-type coatings also is difficult. This is particularly true when they are formulated with fast-drying solvents. They pull on the brush, leave brush marks, and may not provide a smooth, even film. Vinyls and chlorinated rubber coatings are more susceptible to this than are the bituminous cutbacks.

Probably the most obvious property of solvent-applied coatings which can cause difficulty is that of being permanently soluble in their own solvents. Each one of the thermoplastic resins is, of course, susceptible to any of the solvents in which it is readily dissolved. Resins which require oxygenated solvents to dissolve (*e.g.*, alcohols, esters, ketones, or ethers) usually have good resistance to aliphatic hydrocarbon solvents. On the other hand, resins which are easily dissolved in hydrocarbon solvents generally have poor resistance to the oxygenated solvents as well. As an example, vinyl resins are excellent materials for use in gasoline, diesel oil, and many of the other aliphatic hydrocarbon solvents. On the other hand, they are not practical for use with aromatic solvents, such as toluene or xylene. These are the diluent materials used in making the solvent formulations.

Coal tar and asphalt coatings also create difficulty where they are to be overcoated with other coatings, even oil-base paints. In both cases, the bituminous resins dissolve sufficiently to bleed and cause discoloration of the overcoating material. Harder coatings applied over the bituminous materials may also tend to shrink and alligator on drying. The general properties of solvent dry lacquers are shown in Table 5.10.

Co-reactive Coatings

As previously noted in Chapter 4, co-reactive coatings are considerably different from the coatings which dry by solvent evaporation. The co-reactive coatings which are discussed in this chapter are those which react at room temperature to form corrosion-resistant coatings on large, new, or existing structures. Some of the advantages of these materials are that they are usually relatively low in molecular weight during application and then, due to cross-linking with added materials or from moisture in

TABLE 5.10 — Solvent Dry Lacquer Coatings: General Properties

Properties	Vinyl Chloride Acetate Copolymer	Vinyl Acrylic Copolymer	Chlorinated Rubber Resin Modified	Alkyd Modified	Acrylic Lacquers	Coal Tar	Asphalt
Physical Property	Tough Strong	Tough	Hard	Tough	Hard-Flexible	Soft Adherent	Soft Adherent
Water Resistance	Excellent	Good	Very Good	Good	Good	Very Good	Good
Acid Resistance	Excellent	Very Good	Very Good	Fair	Good	Very Good	Very Good
Alkali Resistance	Excellent	Fair-Good	Very Good	Poor-Fair	Fair	Good	Good
Salt Resistance	Excellent	Very Good	Very Good	Good	Good	Very Good	Very Good
Solvent (Hydrocarbon) Aromatic Aliphatic Oxygenated	Poor Good Poor	Poor Good Poor	Poor Okay Poor	Poor Okay Poor	Poor Okay Poor	Poor Fair Poor	Poor Poor Poor
Temperature Resistance	Fair 65 C (150 F)	Fair 65 C (150 F)	Fair	Fair 60 C (140 F)	Fair	Depends on softening point	Depends on softening point
Weather Resistance	Very Good	Excellent	Good	Very Good	Excellent	Poor	Good
Age Resistance	Excellent	Excellent	Very Good	Good	Very Good	Good	Good
Best Characteristic	Broad Chemical Resistance	Weather Resistance	Water Resistance	Drying Speed	Clear Color Retention, Gloss Retention	Easy Application	Easy Application
Poorest Characteristic	Critical Application	Critical Application	Spray Application	Chemical Resistance	Solvent Resistance	Black Color	Black Color
Recoatability	Easy	Easy	Easy	Easy	Easy	Easy	Easy
Primary Coating Use	Chemical-Resistant Coatings	Exterior Chemical-Resistant	Maintenance Coatings	Weather-Resistant Coatings	Weather-Resistant Coatings	Water-Resistant Coatings	Chemical-Resistant Coatings

the air, cross-link in place to form high molecular weight thermoset coatings on the surface.

The process of cross-linkage is the key to the formation of co-reactive coatings and it is also the key to their general resistance. The cross-linkage increases the size of the molecule, which generally can increase the corrosion resistance and definitely does increase the resistance to various solvents. Many of these co-reactive materials, then, have a relatively broad resistance not only to chemicals in general, but also to solvents which, in the case of solvent-deposited coatings, would destroy the coating in short order.

Of the number of co-reactive coatings available, the epoxies are undoubtedly the most important from a corrosion engineer's standpoint. They are not only important because of their properties, but also because of the numerous reactions which take place with the various curing agents for the epoxy resins. These many different curing agents create a multitude of epoxy coatings, all of which have some good and different properties from a corrosion resistance standpoint. Some of the most impor-

tant properties will be considered in the discussion of epoxy resins.

Epoxy Resins

The principal types of epoxy resins that are used for corrosion-resistant coatings are those based on the condensation of bisphenol A and epichlorohydrin. In this case, the polyhydroxy bisphenol A is reacted in the presence of alkali with epichlorohydrin to form the basic epoxy resin. This is shown in Figure 5.16, which depicts the two epoxy groups at the terminal ends of the polymer.

The success of the epoxy coatings has been the result of the chemical reaction of curing agents through the secondary hydroxyl groups on the body of the epoxy resins and the terminal epoxide groups, both of which are shown in Figure 5.16. These reactive groups on the basic epoxy resin react separately, for the most part, so that the hydroxyl groups may be reacted with one type of curing agent while the epoxy terminal groups can react with other, quite different curing materials. The most common co-reactants are the aliphatic amines, the epoxy adducts, the low molecular weight polyamides, the aromatic amines, acid and anhydrides, latent curing agents, and

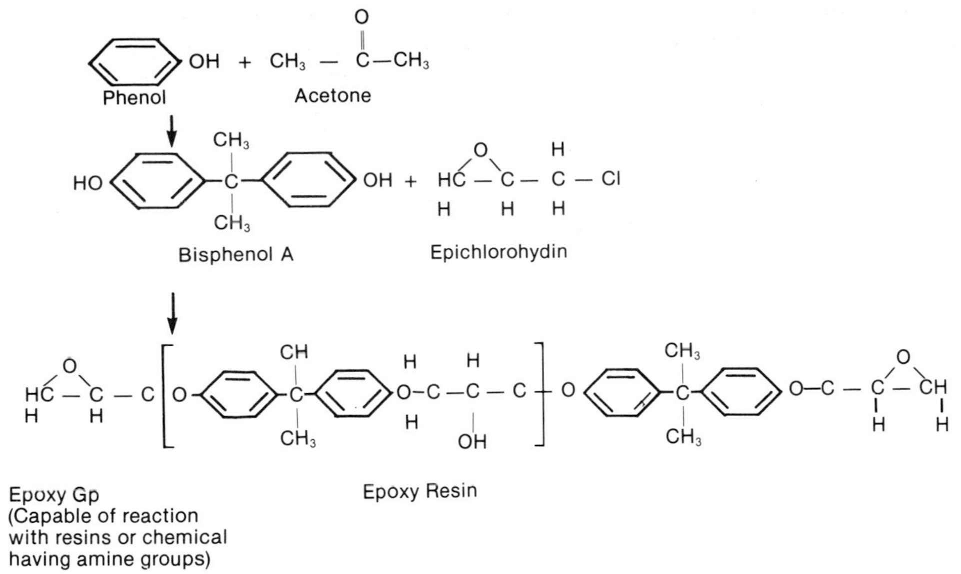

FIGURE 5.16 — Epoxy groups form at terminal ends of the polymer.

phenolformaldehyde resins. The curing mechanism of these materials can involve either the epoxide or the hydroxyl groups on the resin molecule or a combination of both.

The basic epoxy resins may be in the form of relatively low viscosity liquid resins or they may be solid resins of increasing hardness, depending on the size of the molecule. Both the liquid and the solid resins may be used to form corrosion-resistant coatings, or a combination of both the liquid and solid epoxy resins can be mixed into a coating in order to develop the reactivity necessary for the particular material desired. With the number of curing agents outlined earlier and the various combinations of epoxy resins that are available or can be combined, the epoxies in general are a very versatile type of material from which to develop corrosion-resistant coatings.

Aliphatic Amine-Cured Epoxies

The principal corrosion-resistant epoxy coatings are formed by the use of the amine-epoxide reaction. In this case, the terminal epoxy groups on the basic resin react with the active hydrogen groups on the primary and secondary amines, as shown in Figure 5.17. This reaction takes place readily at room temperature. Depending on the amine, the reaction can be almost explosive or it can be quite slow. These reactions are temperature-sensitive so that at low ambient temperatures the reaction takes place rather slowly or, in some cases, not at all. At higher ambient temperatures, the reaction takes place much faster.

Diethylene triamine is a common trifunctional amine used to cure epoxy resins and to form the cross-linked coating. This is shown in Figure 5.18, which indicates a trifunctional amine reacting with the epoxide resin to tie the resin molecule together in a typical cross-link pattern. Note the number of hydroxyl groups that form due to the amine-epoxide reaction.

As might be expected from the reaction shown, it is necessary for such an epoxy coating to be marketed in a

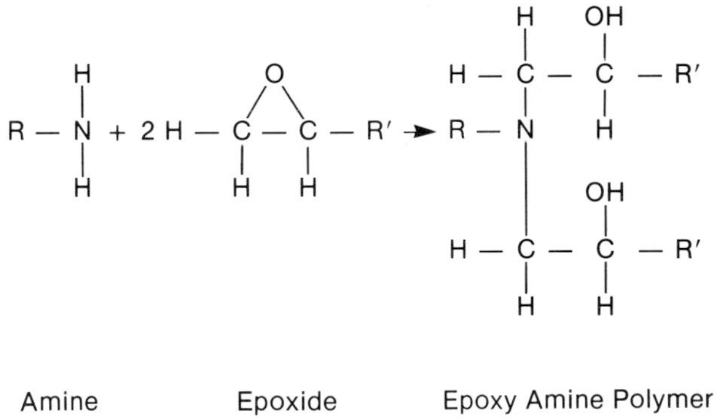

FIGURE 5.17 — Amine epoxy reaction: R = aliphatic part of amine molecule, R' = epoxy resin molecule.

two-package system, the ingredients of the two packages being mixed just prior to application. Various amines react at different rates so that the pot life of the mixed system can be regulated to react within just a few minutes or after several hours or even days. The usual pot life of such a system for corrosion-resistant coatings is in the neighborhood of four to eight hours. The reaction actually continues over a number of days, with the coating becoming harder and more resistant.

The aliphatic amine-cured epoxies present certain handling hazards. The amines are moderately toxic and are skin irritants which may cause allergic reactions. Care should thus be taken during the application of an amine-cured epoxy to prevent contact with the skin and to avoid inhalation of the vapors containing the amine.

The aliphatic amine-cured epoxy systems form very hard, adherent coating systems which have very good chemical and corrosion-resistant properties. They are tightly cross-linked and therefore have good solvent resistance. They have excellent alkali and salt resistance and good water resistance. This being the case, they are used as protective coatings under many highly corrosive conditions. Their weather resistance, however, is not the best since they tend to chalk rather readily.

Corrosion-Resistant Organic Coatings

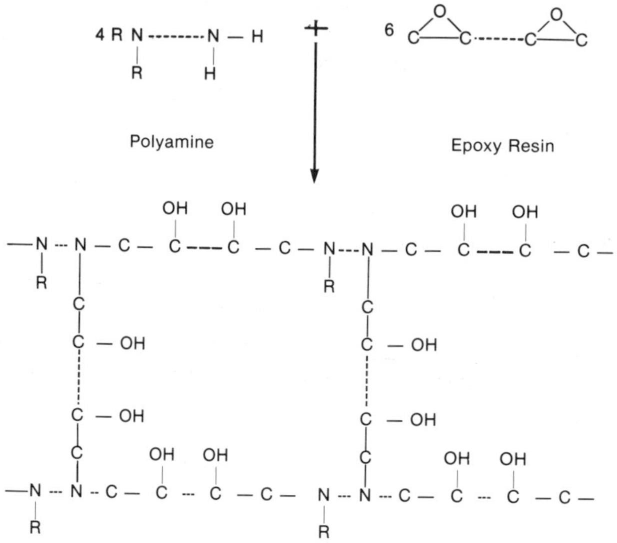

FIGURE 5.18 — The reaction of a diethylene triamine and an epoxy resin forms a typical cross-link pattern.

Polyamide Epoxy

The reaction of the basic epoxy resin with the polyamide curing resins forms the basis for probably the most widely used corrosion-resistant epoxy coatings at the present time. While the polyamides might seem to be a type of curing compound, the polyamides are actually resinous materials which have amine groups attached, and it is these amine groups which react with the epoxide group on the epoxy resins. The reaction is the same as described with the aliphatic amine, with the much larger polyamide resin acting as the cross-linking mechanism.

There are a number of different viscosities and molecular weights of polyamide resins, and they can be used in different quantities with different epoxy basic resins. Again, there are a number of coating characteristics that can be changed in this manner. However, the polyamide epoxy coatings are generally considerably softer and more resilient and flexible than the amine-cured epoxy coatings.

They also differ in other properties. Polyamides have excellent alkali resistance, but their acid resistance is not as satisfactory as the amine-cured epoxy coatings. The weather resistance of the polyamide epoxy coatings is considerably better than that of the amine-cured type. Thus, the former materials are primarily used for exterior atmospheric corrosion resistance, while the latter are used to a greater extent as tank linings and interior coatings for acidic corrosion conditions. The polyamides have less solvent resistance than the amines. The polyamide epoxy coatings cure readily at room temperature and, as usually formulated, have good pot lives for easy application. The polyamide resins, being rather large molecules, form a more bulky curing agent than the amines, so that, for the most part, they are packaged in two containers on an almost equivalent volume basis for the epoxy resin solution and the polyamide resin curing agent. This also makes for easy handling in the field during application.

Amine Adducts

Polyamine adducts are made by reacting epoxy resins of relatively low molecular weight with an excess of typical polyamines, such as diethylene triamine. Amine terminated epoxy resin is the resulting product which can then react with the basic epoxy resin in the usual amine reaction. The resulting product is similar to the amine-cured epoxy; however, it has the advantage of lowering the volatility and the safety hazards of the low molecular weight amine. The amine adduct, which has very little free monomeric amine included in it, has a much lesser tendency to blush or for the monomeric amine to migrate to the surface of the coating. This is a distinct advantage. Also, the greatly increased combining weight of the adduct, as compared to the monomeric amine, makes for a much larger volume and therefore a less critical curing agent: basic resin mixture. Again, in this case, the curing agent can be in an almost equivalent volume to the base resin to facilitate mixing in the field.

Ketimine Epoxy Coatings

Another approach which is used to form corrosion-resistant coatings is the ketimine curing agent. In this case, the amine curing agent is a so-called "blocked" amine curing agent. This is prepared by reacting a primary amine with a ketone, as shown in Figure 5.19. The active hydrogens on the primary diamine have been removed in forming the ketimine and, in this form, the curing agent is not active and is therefore referred to as blocked. The ketimine does not become active until water from the air acts with it, regenerating the original amine and the ketone. The amine then proceeds to react in the same manner as any primary amine, as shown in Figure 5.18. The ketone merely evaporates so that it no longer enters into any of the coating reactions.

This type of curing agent has its advantage in high-solid epoxy coatings where amine curing agents would react so rapidly that the formation of the coating could not be controlled. It is also used in high-build, solventless coatings. The ketimine approach prevents the highly exothermic amine reaction from occurring until the amine is gradually released by moisture from the air, thus making for a well-controlled coating formation and one which is reasonable from an application standpoint.

High-Build Solventless Epoxies

The high-build solventless epoxies are becoming more and more popular due to the problems involved with the use of solvents and air pollution. There are a number of materials presently on the market which are

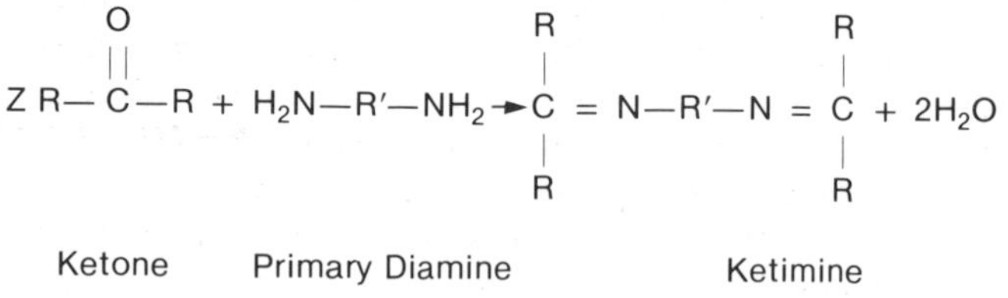

FIGURE 5.19 — Ketimine curing agent prepared by reacting a primary amine with a ketone.

Corrosion Prevention by Protective Coatings

referred to as high-solid epoxy coatings. These are made with a sufficient amount of reactive diluent to reduce the viscosity so that they can be properly handled by normal spray units in the field. Actually, they are not 100% solids epoxy coatings, since they normally contain a very small amount of solvent in addition to the reactive diluent. Most of these materials react to form a coating according to the amine-type reaction, and many use the ketimine-blocked amine-type of approach.

There are a number of advantages to the high solids or 100% solid epoxy coatings. They can be applied by ordinary paint spray equipment, particularly the coatings with somewhat less than 100% solids. They build a heavy film and react in place to form a very good corrosion-resistant coating. Truly 100% solids, epoxy coatings, for the most part, use a two-part gun in which the amine is metered into the system at the gun tip. These materials also apply a very heavy coating onto the surface and one which reacts in place to create the coating.

In both of these cases, although to a greater extent with 100% solids coatings, there are some difficulties in building the coating film to too great a thickness. In this case, the coating may crack, check, and, in some cases, pull itself away from the substrate. This actually has happened in a number of instances, resulting in dramatic failures. The high-solids materials oftentimes also are applied as a hot mix coating. If such a hot resin mixture is applied to a cold substrate, poor wetting can result with very poor adhesion of the coating, particularly if it is applied as a very heavy film. These problems should be taken into consideration during application where high-solids epoxies are involved.

Aromatic Amine Curing Agents

Here again, we are dealing with the amine:basic epoxy resin reaction. In the case of the aromatic amines, the curing is much slower, and, under most conditions, must be stimulated by an increased temperature of around 300 F in order to properly cure. Aromatic amines are primarily used for structural plastics; on the other hand, there are a few corrosion-resistant coatings which use aromatic amines and which are considered to be air curing. In these cases, the aromatic amine is dissolved in a strong solvent for both the basic epoxy resin and the amine, which allows intimate contact of the amine group on the aromatic molecule with the epoxide groups on the epoxy molecule. A reaction does take place and at ambient temperatures. It is, however, generally slower than the reaction with most aliphatic amines. This process is included here because the aromatic amine cures of the epoxy provide the maximum chemical and solvent resistance available with an air-cured epoxy. These materials have excellent corrosion resistance and, where they can be used as linings or coatings subject to strong chemical atmospheres, perform well.

High-Build Epoxy Coatings

High-build epoxy coatings should be differentiated from high-solids or 100% solids epoxy coatings in that they are usually coatings of the amine or polyamide type to which are added inert pigments, such as silica, mica, talc, titanium dioxide, and similar pigments. These are added to the maximum pigment volume which provides complete wetting of the pigment and the surface over which it is applied. These high-build materials are generally very inert and are designed on the impervious coating concept; *i.e.*, there are no inhibitive pigments included and the resistance of the coating is due to its completely inert characteristics. These coatings are usually quite flat, have a semigloss or no gloss, and are generally used under severe corrosion conditions. Many have been used as linings. They may be used alone, in one or more coats over a substrate, applied over inorganic zinc primers, or as either a base coat or an intermediate coat over which lower solids epoxies with a higher gloss are applied. These are practical materials and are used where the corrosive conditions indicate the need for added coating thickness.

Silicone-Modified Epoxy Resin

A silicone modification of the epoxy is accomplished by reacting the methoxy groups of a methoxy polysilicone intermediate with the secondary hydroxyl groups on the solid basic epoxy resin. In this case, the OH group on the epoxy resin is being used rather than the epoxide group. The silicone modification is a plant reaction and cannot take place in the field inasmuch as the two reacting materials must be at elevated temperatures in order to provide the new molecule. About 10 to 20% of the silicone intermediate is incorporated with the basic epoxy resin. The epoxide groups on the molecule are not affected, and the resulting silicone-modified epoxy resin is cured in the field by the use of an amine curing agent. These materials are rather new and have improved water resistance, acid and chemical resistance, and weather resistance. The reaction with the silicone intermediate and the basic epoxy resin is shown in Figure 5.20.

Epoxy Phenolic Coatings

The epoxy phenolic coating is another instance where the hydroxyl group on the epoxy resin is a primary area of activity. The phenolic resin may be any number of reactive phenolic resins. The specific chemistry of this is given later in this chapter under "Heat-Condensing Coatings"; however, the essential reaction is shown as follows:

$$R\text{-}OH + HO\text{-}CH_2\text{-}R^1 \rightarrow R\text{-}O\text{-}CH_2\text{-}R^1 + H_2O \quad (5.3)$$

where R = epoxy resin, and R^1 = phenolformaldehyde resin

As the molecular weight of the epoxy resin is increased, chemical and solvent resistance is increased and the flexibility and impact resistance is improved. Increasing the level of the phenolic resin improves the general chemical and solvent resistance of the epoxy phenolic coating at the expense of some of the flexibility and resistance to alkalies. Many of the epoxy phenolic coatings are baking coatings requiring several hundred degrees of temperature to properly cure. These are discussed later in this chapter ("Heat-Condensing Coatings").

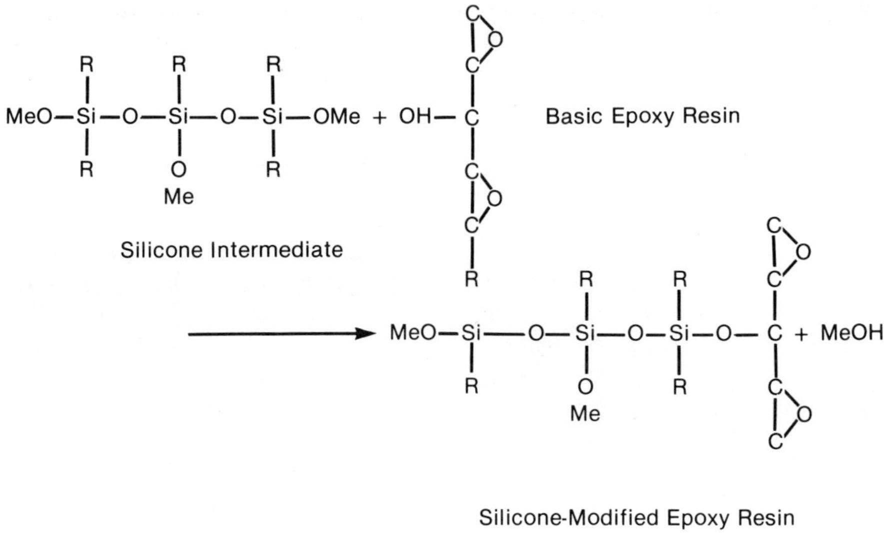

FIGURE 5.20 — Formation of a silicone-modified epoxy resin.

There are also epoxy phenolic coatings available which will cure at substantially lower temperatures. They are very slow to cure at ambient temperatures; however, at approximately 60 C (140 F), they do cure to a chemical-resistant coating with very nearly the equivalent resistance to those of the high bake. Many of these coatings are used for solvents, salt water, and particularly for the lining of tank cars or barges transporting concentrated sodium hydroxide solution. These epoxy phenolic coatings or linings provide excellent chemical and corrosion resistance.

Epoxy Urethane Coatings

Basic epoxy resins also can be cross-linked at room temperature with polyisocyanate compounds. The isocyanate reacts primarily with the hydroxyl groups from the higher molecular weight solid epoxy resins (Figure 5.21).

The reaction of the isocyanate with the hydroxyl does not interfere with the epoxide groups on the epoxy resin molecule; thus, some excellent corrosion-resistant coatings have been developed using this reaction. In this case, the isocyanate was used to react with as many of the hydroxyl groups on the epoxy resins as possible, increasing the size of the epoxy resin molecule and eliminating the hydroxyl group which reduces water and chemical resistance. The isocyanate-reacted epoxy resin was then cured in place as a coating through use of the amine reaction between the terminal epoxide groups on the epoxy resins. The resulting material, cured at room temperature, had a very high degree of solvent, acid, and corrosion resistance.

Coal Tar Epoxies

Coal tar epoxies are a combination of the basic epoxy resin with coal tar. The coal tar is in the form of a semiliquid pitch and is blended with the basic epoxy resin and solvent. The curing mechanism for the coal tar epoxy is the amine reaction with the terminal epoxide groups on the epoxy resin. It is not known whether a reaction, if

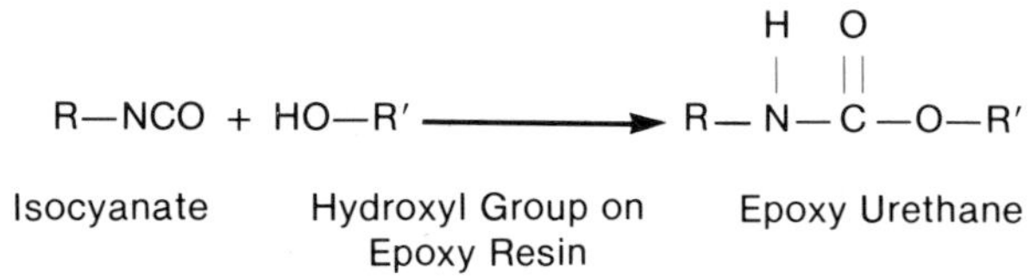

FIGURE 5.21 — Formation of an epoxy urethane.

any, takes place between the coal tar and the epoxy resin. It appears possible that some of the hydroxyl or methylol groups on the phenolic compounds within coal tar may react with some of the hydroxyl groups on the epoxy, causing some cross-linking between the two materials. This, however, is merely a supposition. On the other hand, there is some evidence that this might take place because of a reduction in the solubility of the coal tar when combined with the epoxy. Some reaction also may be indicated because of the fact that some coal tars, even those in the same softening range, will produce better and more effective coal tar epoxy coatings than others. There is a distinct difference in effectiveness between many of the coal tar epoxies produced around the world because of the source of the coal tar.

A combination of the two materials appears to combine the good properties of both the epoxy and the coal tar to form a superior water and saltwater-resistant coating. It is not often that a combination of two widely different materials such as epoxy and coal tar result in a finished material which is superior to either when used alone. This does, however, appear to be the case in the coal tar epoxy combination. The combination is resistant to a wide variety of aqueous conditions as well as to materials such as hydrochloric acid, sodium hydroxide, and sour crude oil. The combination of the two materials resists sagging up to 200 C and is less brittle and more resistant to impact than unmodified coal tar alone. As previously stated, the curing mechanism of the coal tar epoxy is by an amine curing agent. The aliphatic amine provides a very hard, tough, adherent coating. The coal tar epoxy also may be mixed with the polyamide curing agents to form a much more resilient, somewhat softer coating, also with good adhesion. The polyamide coal tar epoxy has excellent water resistance.

The polyamide coal tar epoxy also has some serious drawbacks. It tends to delaminate between coats for any number of different reasons. Water or moisture on the surface, too long a period between coats, and even exposure to strong sunlight for even a few hours will cause delamination of the following coat. The amine-cured coal tar epoxy also will show some of this deficiency, but not to the extent that the polyamide-cured material does. This is something for a corrosion engineer to keep in mind when using coal tar epoxy coatings. The problem has been serious enough so that in certain instances polyamide coal tar epoxies have not been used for exterior coating work. In order to overcome this situation, several suppliers of polyamide coal tar coatings have produced a high-solids material which can be applied in one coat at from 15 to 20 mils in thickness. The material applies well and produces a smooth, even coat. There is no problem with delamination since no additional coats are necessary.

Two areas where coal tar epoxy coatings are most effective include the sewage and marine industries. In the sewage industry they are used for both steel and concrete surfaces. In this case, the amine-cured coal tar epoxy is essential because of the bacterial attack on the polyamide curing agent. Coal tar epoxy is not only resistant to the continued water immersion, but it also provides good protection against oxidized hydrogen sulfide which is the corrosive agent found in sewage conditions. In the marine industry they are used as coatings for underwater ship hulls, ballast tanks, and combined cargo and ballast tanks, and for resistance to both salt water and crude oil. Here again, there is a hydrogen sulfide problem and the coal tar epoxy has excellent resistance to the sour crude.

Another property of the coal tar epoxy coating is its excellent resistance to cathodic protection currents. In the marine industry, it is used as a shield around impressed current anodes to spread the current away from the anodes and to reduce the current density in that particular area. It is one of the few materials that will withstand the strong current densities found in this area.

Water-Based Epoxy Coatings

Much research has been conducted on water-based epoxy resin systems as a result of the various regulations limiting the use of organic solvents in coatings. These coatings are usually two-package epoxy polyamide systems in which the epoxy component is an emulsion based on a proprietary mixture of a liquid epoxy resin and an aliphatic epoxy monomer. The low viscosity imparted to the epoxy resin by the aliphatic diluent is desirable to ensure proper coalescence of the resin co-reactants during film formation as the water evaporates. The polyamide curing agent component is supplied in a mixture of a high-boiling aromatic solvent and a hydroxyl-free water miscible solvent such as ethylene glycol monoethylether acetate. As in the case of a conventional two-package epoxy system, the curing agent component is packaged

separately. Where weather conditions permit (average humidity and temperatures near 15 C) coatings of the water-based type have given good service, even under some reasonably severe corrosion conditions. Table 5.11 gives the characteristic properties of the various epoxy coating combinations.

Polyurethanes

Polyurethane coatings are another group which, like the epoxies, can have a number of coating combinations which create different properties. This, also like the epoxies, is due to the reactivity of the isocyanate with many basic materials of various properties. The polyurethanes are capable of being made into foams or soft, rubbery materials, as well as into very hard, tough, abrasion-resistant products. From the standpoint of coatings, however, the groups discussed will be limited to those polyurethane reactions which have some definitely good coating characteristics.

Polyurethane coatings contain resins made by the reaction of isocyanates with hydroxyl-containing compounds, e.g., water, mono- and diglycerides made by the alcoholysis of drying oils, polyesters, polyethers, epoxy resins, and numerous others. As a matter of fact, wherever there is an active hydroxyl group, the isocyanate will react with it.

Several chemical reactions enter into both the formation and curing of urethane coatings. Foremost in resin manufacture and also useful in film forming is the reaction of an isocyanate group with hydroxyl groups present in polyethers, castor oils, polyesters, or polyhydric alcohol derivatives of drying oils.[2]

$$\underset{\text{Isocyanate}}{\text{R-N} = \text{C} = \text{O}} + \underset{\text{Hydroxyl}}{\text{R'OH}} \rightarrow \text{R-}\overset{\text{H}}{\underset{}{\text{N}}}\text{-}\overset{\text{O}}{\underset{}{\text{C}}}\text{-O-R'} \quad (5.4)$$

Polymer formation is made possible by using di- or poly-functional isocyanates and hydroxyl-terminated compounds. Typical is the reaction between two, 4-toluene diisocyanate (TDI) and a polyether such as polypropylene glycol[2] to form an isocyanate-terminated polyurethane or prepolymer. Commercially available polyethers have molecular weights ranging from a few hundred to several thousand, so that the polymer can vary widely in ultimate molecular weight (Figure 5.22).

Such an isocyanate-terminated polyurethane can be further reacted with additional hydroxyl-containing compounds to form even higher polymers. One of the other important reactions of the isocyanate groups is with water.[2]

$$2\text{R-N} = \text{C} = \text{O} + \text{H}_2\text{O} \rightarrow \text{R-}\overset{\text{H}}{\underset{}{\text{N}}}\text{-}\overset{\text{O}}{\underset{}{\text{C}}}\text{-}\overset{\text{H}}{\underset{}{\text{N}}}\text{-R} + \text{CO}_2$$
$$(5.5)$$

This takes place with the isocyanate-terminated polyurethane shown here, which is the general mechanism of all moisture curing urethanes. Linear polyurethanes are

TABLE 5.11 — Coreactive Coatings: Epoxy

Properties	Aliphatic Amine Cure	Polyamide Cure	Aromatic Amine Cure	Phenolic Epoxy	Silicone Epoxy	Coal Tar Epoxy Amine Cure	Coal Tar Epoxy Polyamide Cure	Water Based Epoxy
Physical Property	Hard	Tough	Hard	Hard	Medium-Hard	Hard (brittle)	Tough	Tough
Water Resistance	Good	Very Good	Very Good	Excellent	Good-Excellent	Excellent	Excellent	Fair-Good
Acid Resistance	Good	Fair	Very Good	Excellent	Good	Good	Good	Fair
Alkali Resistance	Good	Very Good	Very Good	Excellent	Good	Good	Very Good	Fair
Salt Resistance	Very Good	Very Good	Very Good	Excellent	Very Good	Very Good	Very Good	Fair-Good
Solvent Resistance (Hydrocarbons)								
Aromatic	Very Good	Fair	Very Good	Very Good	Good	Poor	Poor	Poor-Fair
Aliphatic	Very Good	Good	Very Good	Very Good	Very Good	Good	Good	Good
Oxygenated	Fair	Poor	Good	Very Good	Fair	Poor	Poor	Poor
Temperature Resistance	95 C	95 C	120 C	120 C	120 C	95 C	95 C	95 C
Weather Resistance	Fair, Chalks	Good, Chalks	Good	Fair	Very Good, Chalk Resistant	Fair	Fair	Good
Age Resistance	Very Good	Very Good	Very Good	Very Good	Very Good	Very Good	Very Good	Good
Best Characteristics	Strong Corrosion Resistance	Water and Alkali Resistance	Chemical Resistance	Chemical Resistance	Water and Weather Resistance	Water Resistance	Water Resistance	Ease of Application
Poorest Characteristics	Recoatability	Recoatability	Slow Cure	Very Slow Air Cure	Recoatability	Black Color Recoatability	Poor Recoatability Black Color	Proper Coalescence
Recoatability	Difficult	Difficult	Difficult	Difficult	Difficult	Difficult	Difficult	Difficult
Primary Coating Use	Chemical Resistance	Water Immersion	Chemical Coating	Chemical Lining	Weather Resistance	Water Immersion	Water Immersion	Atmospheric Corrosion

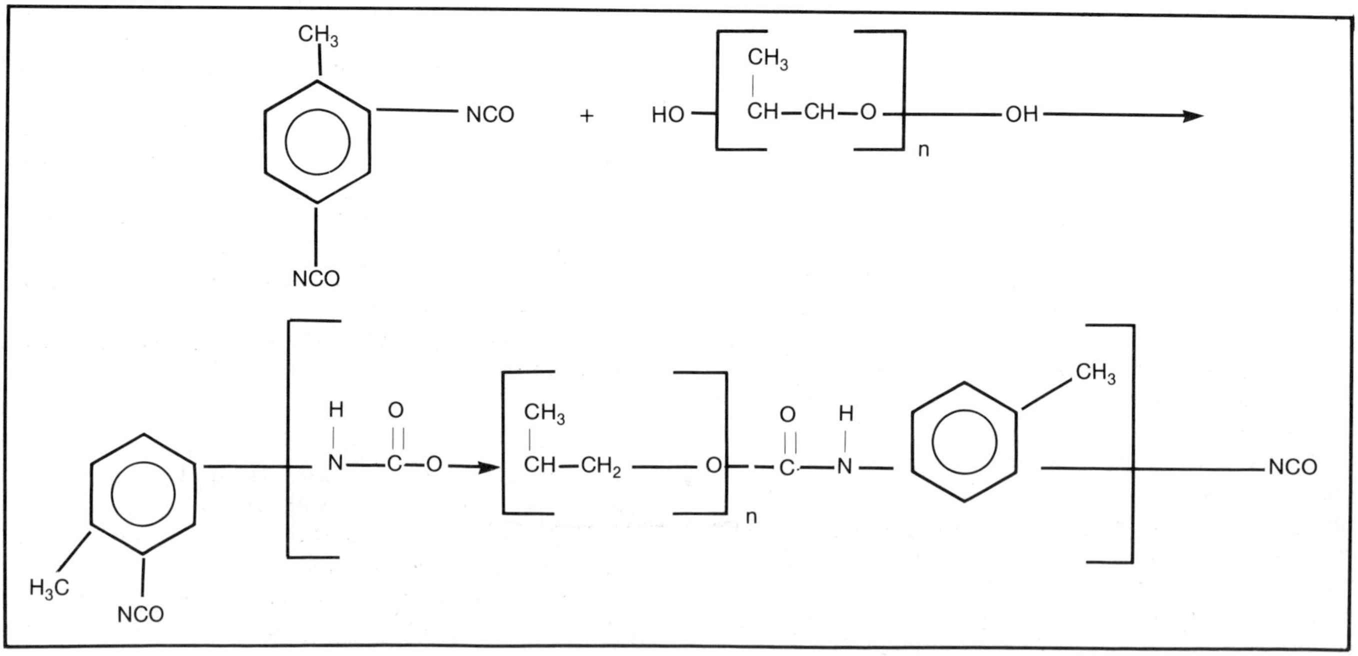

FIGURE 5.22 — Formation of an isocyanate-terminated polyurethane.

produced by the reactions shown. For highly insoluble cross-linked coating, however, it is important to use higher functional polyols, such as triols, so as to create the degree of cross-linking desired in the molecule. The greater the degree of the cross-linking, the harder and less flexible will be the resulting coatings.

The reaction of isocyanate with hydroxyls is the primary basis for all of the various isocyanate or urethane coatings. As outlined by ASTM, there are five different urethane coating types.

Oil-Modified Polyurethane

In the case of the oil-modified type, alcoholysis products of drying oils are reacted with isocyanate. This forms a polymer, with the unsaturated drying oils as a part of it. The alcoholysis of an oil to a monoglyceride is shown in Figure 5.23, as well as the reaction of the diisocyanate with the mono- or di-glyceride. The drying oil portion of the polymer is then oxidized to a coating film, as was previously indicated.

These coatings contain no active isocyanate at the time of application, since it is completely reacted during manufacture. They look and handle like high-quality marine spar varnishes. They are finding wide use as clear wood finishes, floor varnishes, and spar varnish replacements. They are very abrasion-resistant and have good gloss retention and weatherability.

Moisture Cure Polyurethane

Moisture-reactive polyurethanes are formed with resins

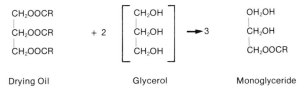

Drying Oil   Glycerol   Monoglyceride

R = fatty acid radical, unsaturated

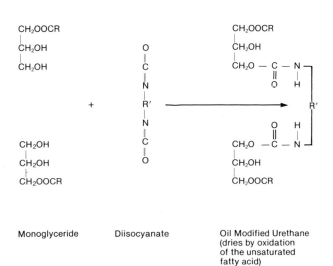

Monoglyceride   Diisocyanate   Oil Modified Urethane (dries by oxidation of the unsaturated fatty acid)

FIGURE 5.23 — Conversion of a triglyceride to an oil-modified urethane.

having a terminal isocyanate group in the molecule, as previously shown for the formation of a prepolymer. The prepolymer, after application, reacts with moisture in the atmosphere to form the final cross-linked coating.

In general, higher molecular weight dialcohols increase flexibility and abrasion resistance at the expense of hardness and some chemical resistance. Increasing triols imparts additional toughness through the cross-linking. Many of these single-package polyurethanes are actually mixtures of diols and triols in order to arrive at the proper combination of hardness, flexibility, and toughness. The urethane prepolymer coatings are useful because of the combination of properties that are possible in a single coating. In particular, the abrasion resistance is outstanding, combined with flexibility, hardness, and tensile strength. The chemical and solvent resistance of the moisture cure polyurethanes is also good.

The fact that these urethane coatings are cured from moisture in the air is sometimes also a disadvantage. The curing time is reduced rapidly at high humidities, while it is lengthened to the point of no cure if the humidity is very low. This must be taken into consideration whenever coatings of this type are being used. Also, because of the cross-linking and increase in solvent resistance of these materials, recoating should be accomplished before the polymer reaches its complete cure. This is usually within 24 hours after application. Otherwise, the coating will require abrasion of the surface in order to secure intercoat adhesion. One of the most popular uses for the moisture cure polyurethanes is for wood floor finishes. The coating gives maximum wear and mar resistance and excellent appearance. It also may be used on concrete where the one-package finish provides a very tough abrasion-resistant coating.

Blocked Urethane

With the blocked urethane coatings, the prepolymer is used as was previously indicated. In this case, however, it is formed into an adduct by reacting the isocyanate groups with a material such as a phenol, and in this manner makes it unreactive at room temperature. This material can be packaged in one can with other polyols and pigments with good package stability. The curing of the blocked resins, however, requires heating, and a threshold temperature must be reached before any curing can take place. In the case of phenol-blocked resins, this is approximately 140 C. Cure may be greatly speeded up by use of an appropriate catalyst. Because of the requirement of heating, these materials usually are not used for anticorrosive coatings in the field and are primarily used as product finishes.

Prepolymer Plus Catalyst

In the prepolymer plus catalyst coating, the isocyanate prepolymer is used with a reaction essentially the same as outlined under moisture cure, with the exception that separate catalysts are mixed with it to increase and accelerate the cure. The catalysts used are metal driers of the same type used for drying oils or some of the amines such as diethanol amine.

Two-Component Polyurethanes

In the case of the two-package coatings, the prepolymers differ from those used in the moisture cure product by being of relatively low molecular weight. These prepolymers are reacted with relatively low molecular weight polyols, such as alcohols, to form adducts. These adducts then form one part of the two-can system. The curing is obtained from the second component, which can be any of the polyols which have been considered in the other polyurethane coating types. The more hydroxyl groups that there are on the polyol, which is in the second package, the greater the cross-linking, which produces somewhat less flexible films, but with higher chemical and solvent resistance.

Urethane Lacquers

Urethane lacquers are a rather recent development, particularly the nonyellowing type. They dry by simple solvent evaporation. The urethane lacquers are fully polymerized, thermoplastic coatings which are relatively high in molecular weight and dissolved in suitable solvents. The polymers are prepared by the reaction of the polyisocyanates with the polyols until no pre-isocyanates remain. The resulting polymer is then dissolved in solvents, which may be dimethylformamide, tetrahydrofuran, or solvent blends such as methyl cellosolve acetate and xylene. These lacquers have relatively low solids; however, they produce films of a number of different consistencies ranging from very hard to soft and rubbery. Even rubber itself can be coated and protected with these materials. In fact, most of the exterior flexible or semirigid articles for automotive use, such as rubber bumpers, are being coated with this type of material.

Uses for these lacquers are not presently in the anticorrosive area. However, these materials are some of the newer products developed and could have worthwhile corrosion-resistant characteristics.

Nonyellowing Urethanes

One of the most important developments in urethane coatings has been the use of aliphatic isocyanates in the development of prepolymers. Hydrogenated derivatives which do not break down into highly colored color bodies also are used. These polyurethane coatings have changed from ones which yellow badly with aging to an excellent nonyellowing, glossy, long-lasting coating with an outstanding depth of color. One of the best of the aliphatic isocyanates, hexamethylene diisocyanate, is extremely toxic. Many other aliphatic isocyanates are available; however, these materials tend, for the most part, to be somewhat softer and slower in curing than is desirable in resistant coatings. It is often necessary to alter the basic polymer in order to obtain the desired results with the aliphatic isocyanate. The chemistry of the aliphatic isocyanates is the same as previously has been described, and the general properties are much the same, except for the nonyellowing characteristics.

One common use of these materials is for aircraft where the properties of the urethanes are used to their best advantage (*i.e.,* any aircraft coating must be abrasion-resistant to ice, dust particles, rain, etc.). They must: (1) withstand continual exposure to weathering conditions; (2) have a good, long-lasting gloss; (3) have top quality color retention; and (4) be resistant to hydraulic fluids, gasoline, jet fuels, and similar solvents. The general properties of the various urethane coatings are given in Table 5.12.

As can be seen from Table 5.12, the general corrosion and chemical resistance of the urethane is not as great as some of the other corrosion-resistant coatings. The outstanding urethane property is that of abrasion resistance. There are probably no other coatings available which have the resistance to abrasion that these materials do. Thus, they make excellent floor and deck coatings and withstand much more traffic and hard use than most other coatings. Their chemical and corrosion resistance is also sufficient for this type of use when combined with the excellent abrasion and impact resistance. The exceptions would be those areas where there is continuous acid or chemical spillage requiring acidproof brick and tile floors. For example, the Tabor Abrasor values for a polyurethane will have approximately one-tenth the loss of coating per 1000 revolutions, as will an amine-catalyzed epoxy. Urethane coatings have excellent impact properties and can withstand both direct and reverse impact, which is a property that many other resistant coatings do not have. The same urethane coating can be designed to have flexibility, impact resistance, and high abrasion resistance, all combined into one coating.

New nonyellowing urethanes, in addition to the abrasion and impact characteristics, also provide a coating with an excellent depth of color, good color retention, and good gloss retention over long periods of time in weather exposure. In fact, they are equal, if not superior, to some of the better acrylic coatings.

One of the areas where urethanes may have a problem is in the retention of adhesion when exposed to water, either in immersion or as moisture. It is a characteristic that must be kept in mind for corrosion-resistant applications. Generally, urethane coatings are used as topcoats and applied over good epoxy primers for best results for corrosive conditions. (The NACE Technical Practices Committee report of Task Group T-6B-25 on *Urethane Protective Coatings for Atmospheric Exposures* is recommended for more specific information on urethane coatings.)

Polyester Coatings

The application of polyester coatings in the area of alkyd resins was discussed earlier. From a coating standpoint, the alkyds are undoubtedly the most important of the polyester coatings. Their use is extremely widespread, and they are used for many different mildly corrosive conditions. Polyesters are included in the discussion of coreactive coatings because they have been used as linings for tanks, for portions of tanks, or for the bottoms of tanks for many different corrosive conditions. The polyester, as used for these purposes, has excellent corrosion resistance. Polyester resins are used mostly in the reinforced plastics industry, where they are used as binders for syn-

TABLE 5.12 — Co-reactive Coatings-Urethanes: General Properties[1]

Properties	Type 1 Oil Modified	Type 2 Moisture Cure	Type 3 Blocked	Type 4 Prepolymer Catalyst	Type 5 Two Component	Aliphatic Isocyanate Cure (Non-Yellowing)
Physical Property	Very Tough	Very Tough, Abrasion Resistant	Tough, Abrasion Resistant	Tough, Abrasion Resistant	Tough-Hard, Rubbery	Tough-Rubbery
Water Resistance[2]	Fair	Good	Good	Fair	Good	Good
Acid Resistance[2]	Poor	Fair	Fair	Poor-Fair	Fair	Fair
Alkali Resistance[2]	Poor	Fair	Fair	Poor	Fair	Fair
Salt Resistance[2]	Fair	Fair	Fair	Fair	Fair	Fair
Solvent Resistance (Hydrocarbon)						
Aromatic	Fair	Good	Good	Poor	Good	Good
Aliphatic	Fair	Good	Good	Fair	Good	Good
Oxygenated	Poor	Fair	Fair	Fair	Good	Fair
Temperature Resistance	Good 100 C	Good 120 C	Good 120 C	Good 100 C	Good 120 C	Good 120 C
Weather Resistance	Good, Yellows	Good, Yellows	Good, Yellows	Good, Yellows	Good, some yellowing, chalk	Excellent, good color and gloss retention
Age Resistance	Good	Good	Good	Good	Good	Good
Best Characteristic	Exterior, Wood Coating	Abrasion, Impact	Abrasion, Impact	Speed of cure	Abrasion, Impact	Weather Resistance, color and gloss retention
Poorest Characteristic	Oil Base Chemical Resistance	Dependent on humidity for cure	Heat required for cure	Chemical Resistance	Two package	—
Recoatability	Fair	Difficult	Difficult	Difficult	Difficult	Difficult
Primary Coating Use	Clear Wood Coating	Abrasion Resistance, Floors	Product Finish	Abrasion Resistance	Abrasion Resistance, Impact	Exterior Coatings

[1]The properties of urethanes vary over a wide range due to the many and varied basic polyols and isocyanates. The above listings are only indicative. Manufacturers must be contacted for specific properties of specific materials. Harder coatings are more resistant than softer, more rubbery types.

[2]Resistances are for nonimmersion conditions.

thetic, glass, or graphite fibers, to form a myriad of high-technology products.

There are two areas where polyester resins are used particularly for coatings. The first is the so-called *gel coat*, which is used primarily in connection with the formation of reinforced plastics. It generally is used as the first material on the mold in order to provide a dense, glossy, weather- and corrosion-resistant surface which also has a good apearance. Another area where polyesters may be effective in the future is that of 100% solids coatings which have good corrosion resistance properties. Considerable research is going into this area. The use of these materials for coating applications undoubtedly will become more and more important.

The polyesters with which we are involved as co-reac-tive coatings are the unsaturated polyesters, which are linear polyester resins based on dihydric alcohols, and dibasic acids which contain sufficient unsaturation so that they have the capability of cross-linking with vinyl-type monomers to form a thermoset prepolymer. The unsaturation in the polyester usually comes in the unsaturated acids or anhydrides which are used to form the polyester. Cross-linking of the coating is then accomplished by use of such monomers as styrene, vinyl toluene, methyl methacrylate, and others. The formation of the coating is in place using the double bond reaction between the polyester resin and the styrene activated by a catalyst such as one of the organic peroxides. The chemistry involved in this process is shown in Figure 5.24.

The addition of the styrene or a similar vinyl-type

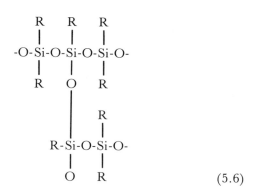

FIGURE 5.24 — Chemical reaction between the polyester resin and the styrene when activated by a catalyst such as one of the organic peroxides.

monomer is necessary to further polymerize the polyester resin, cross-link it, and render it insoluble. The styrene acts in two ways: (1) it acts as a solvent for the basic polyester resin, and (2) when a catalyst is added (*e.g.,* methyl ethyl ketone peroxide and a cobalt accelerator), a rapid reaction takes place at room temperature, cross-linking the polyester and forming a solid resin coating or lining. One of the disadvantages of the polyester-styrene copolymers has always been their tendency towards shrinkage. This is due to the internal shrinkage of the molecule when polymerization takes place. This has been improved by the incorporation of glass fibers or flakes and siliceous fillers. Considerable shrinkage, however, still takes place when a large area of polyester resin and glass is applied to a tank bottom or similar structure. Thus, shrinkage must be taken into consideration during application.

A polyester glass structure provides a good corrosion-resistant coating or lining. Its thickness may range from 100 to 250 mils, and its resistance to acidic water or acids is very good. It is not, however, satisfactory for alkaline solutions since both the glass and polyester tend to break down in a caustic solution. A polyester glass structure on tank bottoms, sidewalls, and similar areas is usually applied using a double-tip spray gun in which the catalyst is added to the resin at the spray gun tip. Such spray guns ordinarily include a glass chopper which chops fiberglass roving into sections from one-half inch to four inches in length so that the glass, resin, and catalyst are all blown on the surface at the same time. The surface is then rolled in order to smooth it out prior to the final set of the resin. Where shrinkage has been properly accounted for, a very good corrosion-resistant lining material can be obtained.

Silicones

Silicones are formed by chemical modification of quartz, sand, or silicon, and they may be thought of as hybrids of glass and organic resins. They have much the same inertness as glass, but at the same time can be incorporated into coatings in the same manner as organic polymers. They are a silicon-based polymer which has primarily silicon-oxygen-silicon linkages. The silicon atom is tetrafunctional in the same way that the carbon atom is, and one or more of the attached functional groups may be based on organic carbon.

The organic groups which have the most desirable properties combined with the silicon are the methyl and phenyl groups. High phenyl-containing resins tend to have better heat and oxidation resistance than the methyl substitutes. Most of the silicone resins used in the coating industry are a combination of the two, forming methyl phenyl silicone polymers. Properties can be varied widely, depending on the ratio of the methyl and phenyl groups. Polymers used by the coating industry, for the most part, have the following configuration.

$$
\begin{array}{ccc}
R & R & R \\
| & | & | \\
-O-Si-O-Si-O-Si-O- \\
| & | & | \\
R & O & R \\
& | & \\
& R & \\
& | & \\
R-Si-O-Si-O- \\
| & | & \\
O & R &
\end{array}
$$

(5.6)

Silicon Polymer

where R = organic groups

A silicone polymer may have a number of different organic groups attached to the silicon atoms. As the number of organic groups on the silicone polymer increases, the resin becomes softer, more flexible, somewhat slower curing, and more thermoplastic. Most of the silicone resins are dissolved in hydrocarbon solvents. The solvents are usually included in the manufacture of the resins to improve control of the hydrolysis of the basic silicon molecule to prevent gelation. Modified silicone resins, such as those represented above, generally require curing at high temperatures to obtain their optimum characteristics. They do, however, respond to catalysts. Zinc catalysts, such as zinc napthenate, are used most often and yield good results. The cure temperatures can be reduced to 400 F (205 C) using zinc.

Some of the properties of silicone coatings are as follows:

Hardness

Silicones are generally somewhat softer than many organic coatings, and hardness properties can be improved if silicone resins are modified with organic coating materials.

114

Adhesion

Silicone adhesion is generally good to most substrates, depending, as in the case with organic materials, on the cleanliness of the surface.

Abrasion Resistance

Silicone polymers have generally poor abrasion resistance. Again, modification with organic materials can considerably improve resistance.

Chemical Resistance

In general, chemical resistance of silicones is good, including water, mild acids, alkalies, and other corrosion-resistant materials. Modified resins generally have poor solvent resistance.

High Heat Resistance

Heat resistance is one of the outstanding properties of silicone coatings. They can withstand temperatures of 540 to 640 C (1004 to 1184 F) when pigmented with aluminum or black pigments. When pigmented with ceramic frits, silicone coatings have gone as high as 760 C (1400 F). A modified silicone resin based on the methyl phenyl polymers is often used to topcoat inorganic zinc coatings for continuous use at temperatures of 360 to 540 C (680 to 1004 F).

Weather Resistance

Silicone resins have the property of being transparent to ultraviolet light. Because of this, they are not generally subject to the kind of degradation and coating failure associated with organic coatings. They have excellent water resistance and resistance to thermal change which undoubtedly contribute to their weather resistance. Unmodified silicone coatings have been exposed to weather conditions for over 12 years with no loss of gloss, color change, chalking, or other similar types of failure.

Corrosion Resistance

Silicone resins have excellent electrical properties. They are good electrical insulators. These properties, combined with their natural water repellancy, make them good candidates for corrosion-resistant coatings. One of the long-time uses of the silicone resins has been on the exterior of smoke stacks where they are not only subject to varying temperatures, but to sulfur dioxide and other combustion products as well. They have no problem in passing U. S. government specifications requiring exposure at temperatures up to 480 C (96 F) and then passing a 20% salt spray test.

It has been mentioned that silicones are copolymerized with other organic polymers in order to improve the organic material's properties of heat resistance and weathering. In some cases, silicones can be coblended with other organic materials. Generally, however, copolymers have more desirable characteristics. Copolymers are made from several silicone intermediates which have hydroxyl or methoxy groups attached to the silicone molecule. These are the groups that can react with the organic polymers which have active hydroxyl groups. These silicone intermediates are shown in Figure 5.25.

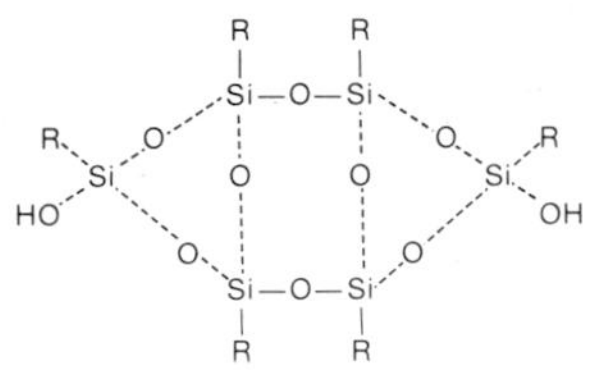

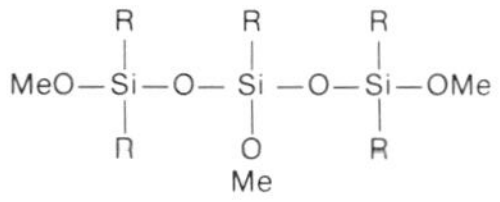

FIGURE 5.25 — Silicone intermediates used to react with other organic polymers containing active hydroxyl groups. (SOURCE: Clope and Glazer, Federation Series on Coating Technology, Unit 14, Silicone Resins for Organic Coatings, Federation of Societies for Paint Technology, Philadelphia, PA, pp. 16-17, 1970.)

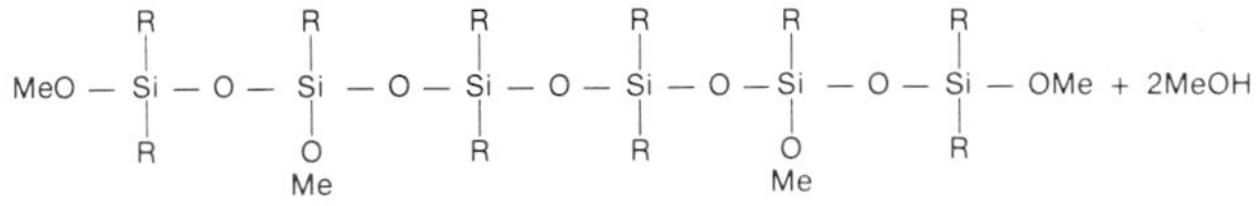

The organic materials which form satisfactory copolymers with the silicone intermediates are alkyds, polyesters, polyols, epoxies, epoxy esters, uralkyds, acrylics, and others. Generally, the copolymers have increased heat and weather resistance over and above the organic coating material alone. The silicone organic copolymers cure into a final coating using the same mechanism as the organic part of the copolymer. This has been shown previously under alkyds, where the silicone alkyd is cured by the oxidation reaction of the oil-modified alkyds. Silicone copolymer maintenance coatings generally have good color and gloss retention, and tend to chalk at a much slower rate than the organic finish alone. Improved water and corrosion resistance is also common.

One of the rapidly developing uses of the silicone copolymers is for coil coating. This is where strips of metal are coated in a continuous process, rolling from a coil of metal passing through the coating and baking process, and then recoiled within a matter of only a minute to a minute and a half. The temperature in this case is a high bake (600 F). The materials used are silicone polyester copolymers and are generally made with the linear silicone intermediates for flexibility. Such materials have been exposed in South Florida for over seven years and have been unaffected by the weather exposure with no color change or appreciable chalking.

Many of the silicone copolymers are important to the corrosion engineer and may become even more important because of their water-, weather-, and corrosion-resistant characteristics. Silicone alkyds, epoxy silicones, and similar materials have upgraded temperature, weather, and chemical resistance over the organic materials alone.

Emulsion-Type Coatings

The formation of coatings from emulsions is not new. It is, however, becoming increasingly important because of the newer and more stringent air pollution regulations. There is increasing effort and research on the development of emulsion-type coatings of the high-performance type. Water dispersed coatings would be ideal from a solvent standpoint since there would be no solvent pollution problem if all coatings could be made in this way. Unfortunately, the water-base, high-performance coatings that are of interest to the corrosion engineer are not yet entirely practical. The basic reason is the process of coalescence. The resin particles are discrete in the emulsion and must completely join together to form a continuous film. So far, this has not been possible. Even unpigmented latex films are more porous to water and ions than are solvent-applied films of the same resin. This is aggravated by the many additives which are necessary to make the discrete resin particles disperse in the water and then to flow together and form a continuous film. Heat of fusion of the resin particles makes for good latex coatings. This, however, is not practical for in-place applied coatings.

Most of the present latex paints are made from thermoplastic resin types, such as styrene-butadiene, vinyl acetate, vinyl-acrylic copolymers, acrylic esters, vinyl acetate-acrylic, styrene acrylic, etc. In this case, the resin is suspended in water in the form of very minute spherical particles. In order to keep these particles separated, each one is coated with an extremely thin layer of emulsifier, which is necessary to keep the particles apart and to keep them from flocculating.

The films of latex coatings are a complex mixture and consist of a number of materials, *i.e.,* hiding pigment, extender pigment, pigment wetting and dispersement aids, emulsifiers, thickener and protective colloids, fusion or coalescing aids, freeze-thaw stabilizer, wet edge promoter, defoamer, fungicide, water, and the dispersed resin. Table 5.13 indicates some of the necessary emulsion coating formulation ingredients and their functions. In addition, Table 5.14 is a series of helpful materials for use in emulsion-type coatings. Most of these materials are required in the coating in order to help it to coalesce into a usable coating film. The film formation is the result of the fusion or coalescence of the resin particles, including pigment and the other materials included in the tables.

In the liquid coating, the dispersion factors are the emulsifier, thickener, and protective colloid. These predominate; otherwise, the resin pigment dispersion tends to semicoalesce and precipitate to the bottom of the container. After the coating has been applied to the surface and the water starts to evaporate, the resin particles come closer together and the coalescence must begin in order to fuse the resin particles to form a continuous film. At this point, the dispersion factors must cease to predominate and the coalescent or film-forming factors take over. Figure 5.26 is a schematic representation of the resin particles as they gradually coalesce into a complete film.

The process of coalescence occurs as water evaporates from an emulsion coating. Note that a com-

TABLE 5.13 — Necessary Emulsion Coating Ingredients

Ingredient	Function
Prime Pigment	Hiding, Color
Extender Pigment	Hardness, H_2S Resistance
Reactive Pigment	Rust Inhibition
Pigment Dispersant	Disperse Pigment
Defoamer	Prevent foaming of paint
Thickener	Provide Viscosity
Can Preservative and Mildewcide	Inhibit bacteria and mildew growth in the can

[SOURCE: Mercurio, Andrew and Flynn, Roy, Latex Based All Surface Primers, Journal of Coatings Technology, Vol. 51, No. 654, July (1979).]

TABLE 5.14 — Helpful Emulsion Coating Ingredients

Ingredient	Function	Example
Surfactant	Pigment Wetting	Polyglycol
Coalescent	Film-Forming Aid	TBP
Glycol	Freeze-Thaw Stability	Ethylene Glycol
Oil Modifier	Chalk, Metal Adhesion	Alkyd + Cobalt
2nd Reactive Pigment	Additional Rust, Stain Resistance	Barium Metaborate, CA-Zn, Molybdate, Zn Phoso Oxide

plete film is not formed until Stage 4 is reached. Stage 1 shows the dispersed resin particles as they would appear in the coating liquid. Stage 2 is after partial water evaporation. At this stage, there would be poor film characteristics and lack of toughness. In Stage 3, there is some coalescence, but it is not complete. In this form there would be a high permeability to water and atmospheric salts. Stage 4 shows the complete coalescence of the resin particles. While they never form a really smooth film, such as is possible with solvent evaporation, particles have flowed together in order to provide a continuous coating. Figure 5.27 shows a completely coalesced, non-pigmented film of latex, indicating the packing orientation of the resin particles as they coalesce into a film.

Ideally, when a resin emulsion is pigmented, each pigment particle in the finished coating is completely surrounded by the matrix of fused resin particles. This, however, is never entirely the case, although the results

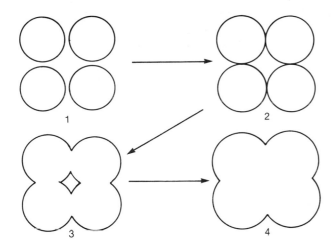

FIGURE 5.26 — Degrees of coalescence.

FIGURE 5.27 — Coalesced unpigmented latex, 28, 580X. (SOURCE: Fuller, W. B., Federation Series on Coating Technology, Unit 2, Formation and Structure of Paint Films, Federation of Societies for Paint Technology, p. 15, 1965.)

are approachable when using a coating with a low pigment volume concentration. For use as a corrosion inhibiting coating, this is the type of formulation which should be considered. With high pigmentation, such as the coatings used for flat wall paints, there would be little chance of reaching a state where the resinous phase is continuous. Even in solvent films, which are highly pigmented, this is difficult.

The most successful products up to the present time are of the flat wall type or semigloss materials, primarily for interior application. Emulsion technology is leading to anticorrosive primers and medium gloss topcoats which have some good atmospheric and reasonable corrosion-resistant properties. Epoxies, coal tar epoxies, and similar coatings have been developed using the water emulsion film-forming processes. The newer corrosion-resistant emulsion-type coatings are approaching the continuous resin phase, with some, such as the epoxies, providing good films.

Emulsion-type epoxy coatings are for the most part two-package epoxy polyamide systems. The epoxy portion is a liquid epoxy resin with some aliphatic epoxy monomer as a diluent to lower the viscosity of the epoxy resin and to help in the coalescence of the two film-forming resins during the drying periods. The polyamide part of the system, which is used as a curing agent, also has some added materials to reduce the viscosity and to aid in the coalescence of the polyamide and the epoxy resins to form the finished coating. These are usually high-boiling aromatic solvents and a water miscible solvent. The two parts of the system are packaged separately and, when mixed, have a reasonable pot life. Such epoxy emulsions have been used in seamless floor base coats, as block fillers, and both as maintenance primers and maintenance finish coats. Some of the water-based epoxy systems have provided good results in the field, although primarily for atmospheric corrosion-type conditions.

At this time, the water-based epoxy coatings are not as effective as the solvent-type epoxy coatings of the same type. This is particularly true under damp, moist, or immersion conditions. Some have been used as topcoats for inorganic zinc base coats with good results. Under certain conditions, vinyl acrylics and acrylic water-base materials have been used as topcoats for corrosion-resistant primers, both organic and inorganic, where conditions are right and the corrosion-resistant base coat will withstand the corrosion conditions of the area. These coatings have been effective as decorative topcoats. Generally, however, they do not contribute substantially to the anticorrosive characteristics of the total coating. There is some danger in the use of these materials as topcoats over inorganic zinc coatings when humid conditions exist. White salt reaction products from the zinc in the base coat may form underneath and penetrate through the porous water-based topcoat. (This problem will be discussed in more detail in Chapter 6.)

None of the water-base emulsion-type materials will coalesce properly under cold, humid atmospheres. This factor should be taken into consideration whenever these materials are considered for coating use. Table 5.15 gives the general properties of the presently available water emulsion-type coatings which may be considered for maintenance purposes.

Heat-Condensing Coatings

This group is essentially composed of two types of coatings: a so-called "pure" phenolic and the epoxy phenolic. These are two of the most corrosion- and chemical-resistant coatings available and are used under some very highly corrosive conditions.

Phenolic Coatings

The type of resins in which we are primarily interested are the unsubstituted heat reactive resins. This is only one of a number of phenolic resins which can be used for coatings. On the other hand, these coatings are the type used for highly corrosion-resistant purposes. An unsubstituted phenol is one which does not have any methyl groups attached to the phenol. Phenol itself is the most

TABLE 5.15 — Coalescent-Emulsion Type Coatings
(Atmospheric Use Only)

Properties	Vinyl Acetate	Vinyl Acrylic	Acrylic	Epoxy
Physical Property	Scour Resistant	Scour Resistant	Scour Resistant	Tough
Water Resistance	Fair	Good	Good	Good
Acid Resistance	NR[1]	NR	NR	Fair
Alkali Resistance	NR	NR	NR	Good
Salt Resistance	Fair	Fair	Fair	Good
Solvent Resistance				
Aliphatic hydrocarbon	Fair	Good	Fair	Good
Aromatic hydrocarbon	NR	NR	NR	Good
Oxygenated hydrocarbon	NR	NR	NR	NR
Temperature Resistance	60 C (140 F)	60 C (140 F)	60 C (140 F)	70 C (158 F)
Weather Resistance	Good	Very Good	Very Good	Fair
Age Resistance	Good	Good	Good	Good
Best Characteristic	Weather Resistant	Weather Resistant	Weather Resistant	Reasonable Corrosion Resistance
Poorest Characteristic	Porosity	Porosity	Porosity	More porous than solvent base
Recoatability	Good	Good	Good	Fair-Good
Principal Use	Decorative Topcoat	Decorative Topcoat	Decorative Topcoat	Topcoat

[1]NR = Not Recommended.

important of all of the unsubstituted phenols. This material is reacted with formaldehyde which then forms a series of hydroxybenzoalcohols; these then continue to react, forming polymers of higher and higher molecular weight until it is possible to obtain a jelled cross-linked structure. After this, further heating increases the cross-link density until a completely insoluble material is obtained.

The basic chemistry of the phenolformaldehyde is given in Figure 5.28. The depicted five methylol phenols are formed during the initial phenolformaldehyde reaction (Figure 5.29). The second step is for two or more of polymer. This process continues with heat until a completely insoluble, highly cross-linked polymer is formed.[3]

The phenolic resin that is used in coatings of the high-performance type may be very viscous or may be carried to the solid stage while it is still soluble in polar solvents such as alcohol. Unsubstituted heat-reactive resins are multifunctional and therefore can form cross-linked film. They are soluble in alcohols, ketones, esters, and glycol ether. However, they are insoluble in aromatic and aliphatic hydrocarbons. Such resins are compatible with amino resins, epoxies, polyamides, and polyvinyl butyral. These phenolic resins in solution are very heat reactive and it is often necessary to store them under refrigeration. Typical alcohol solutions of the most reactive of the resins will gel in three to six months, even at ambient temperatures, and the solid resins when stored at ambient temperatures will continue to polymerize to the point of being insoluble and infusible, and therefore of no value from a coating standpoint. This is important when using these heat-reactive phenolic coatings since average liquid coatings may be sufficiently polymerized so that the end result is poor adhesion or other application problems.

These coatings are usually applied from an alcohol solution by spray, dip, or roller. Some of the solid resins even find application as powder coatings. The application of these coatings is critical. During their condensation, they release water, and it is necessary to remove this water during the curing process. The general procedure of the application is to apply the phenolic resin solution to a very clean, steel surface at a maximum thickness of about one

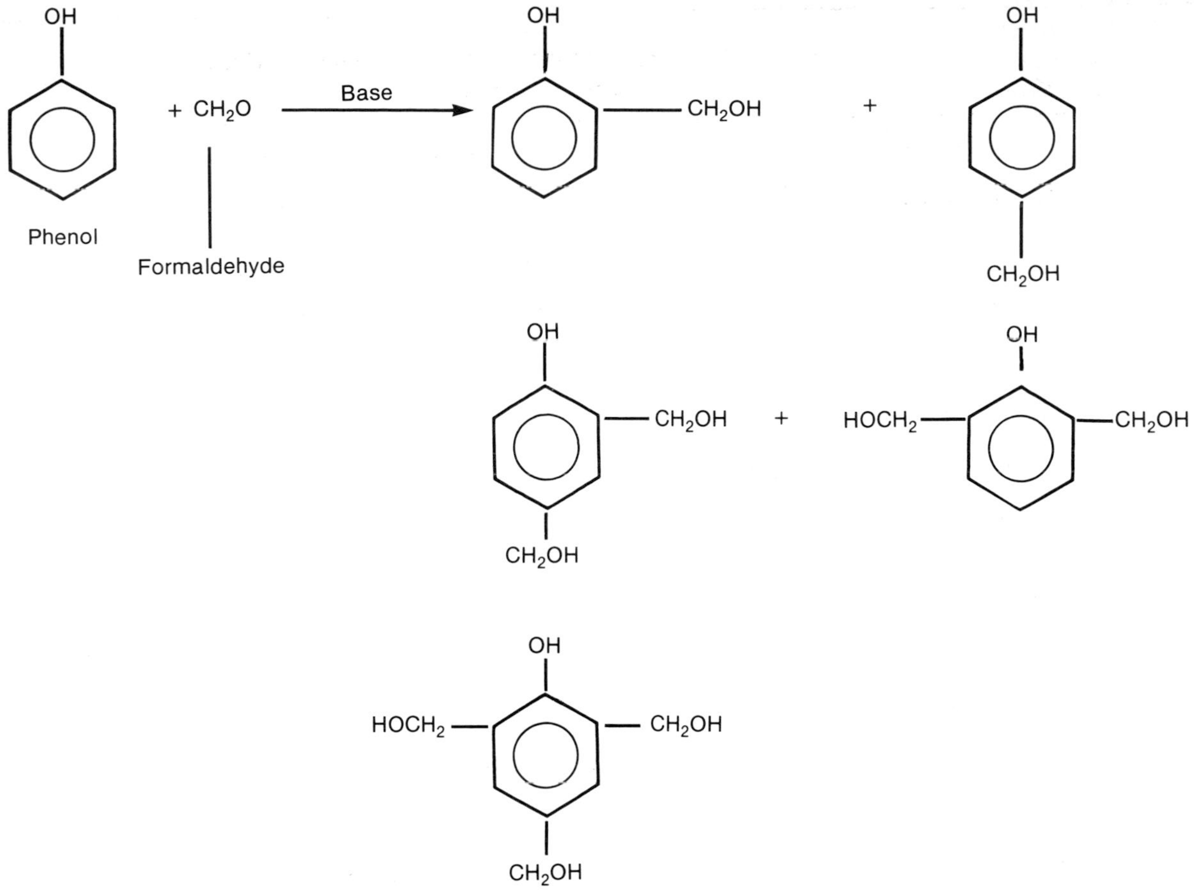

FIGURE 5.28 — Chemical formulation of phenolformaldehyde.

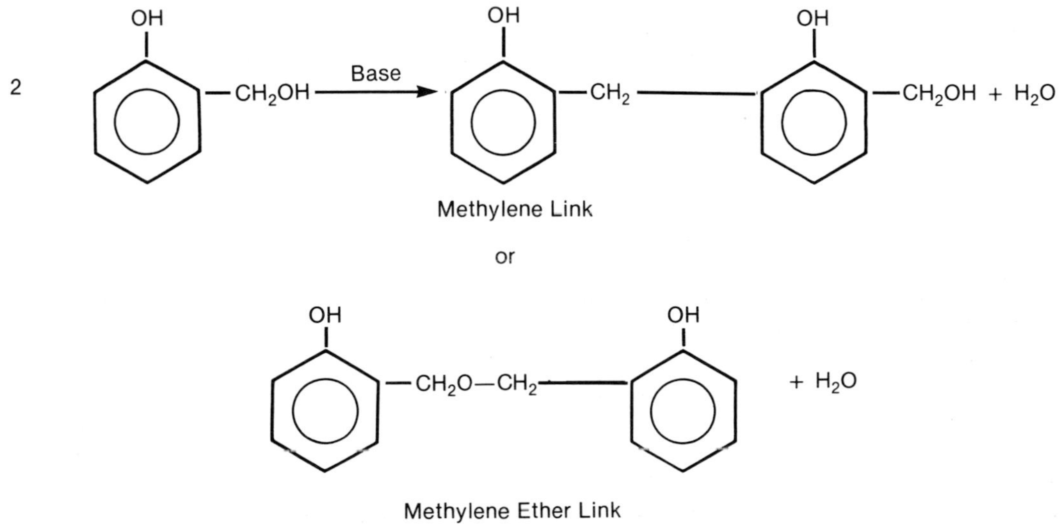

FIGURE 5.29 — Five methylol phenols are formed during the initial phenolformaldehyde reaction.

mil. The coating is then heated to temperatures of 135 to 300 C for a few minutes. This results in a partially cured state. Additional coats are applied using the same procedure for approximately four to six coats.

Following the last coat and the short bake as previously outlined, the coating is heated for several more hours at a temperature of approximately 230 C. Increased cure time and higher temperatures result in increased cor-

rosion and solvent resistance and decreased flexibility. Short, high-temperature bakes produce a more resistant film than long, low-temperature bakes. The color of the coating darkens upon baking, going from a rather golden color upon application to a very dark reddish-brown. Most of the phenolic tank linings are cured to a medium reddish-brown color during the curing period.

These cured coatings are affected less by solvents

than any other type of organic coating, and have remained unaffected by exposure to alcohols, ketones, esters, aromatic and aliphatic hydrocarbons, and even chlorinated solvents for a number of years. They also are odorless, tasteless, and nontoxic in the fully cured form and are therefore noncontaminating to contained material. They often are used for storage and processing of food products such as wine, beverage alcohols, various sugar syrups, beer, alcohol fermentation, and many similar applications. They also have excellent resistance to boiling water, aqueous solutions of mild acids, and acidic and neutral salts so that their chemical resistance is quite broad. Their primary use is in tank linings, although they are extensively used as a lining for downhole petroleum tubing which is subject to high temperatures, high-temperature water, salt solutions, hydrogen sulfide, and various petroleum products.

The one "Achilles heel" of these coatings is their lack of alkali resistance. They are not satisfactory for use in alkali of any concentration. While these phenolic resins constitute an older type of high-performance coating, they still continue to be used as previously mentioned because of their inert characteristics and their excellent resistance to hot water at near boiling temperatures.

Epoxy Phenolic Coatings

Epoxy phenolic coatings, which are also cured at relatively high temperatures, have exceptional chemical and solvent resistance and are used primarily as tank linings to contain chemicals of various types. The coatings rely primarily on the reaction between the methylol groups of the phenolic resin and the secondary hydroxyl groups on the epoxy resin. Epoxy resins used are those where the hydroxyl groups predominate on the molecule. These are usually the higher molecular weight epoxy resins.

The reaction of the epoxy resin with the methylol groups on the phenolic resins is an important one from the standpoint of coating technology. It occurs between the secondary hydroxyl groups of the higher molecular weight epoxy resins and the methylol group of the phenolformaldehyde resins, as shown in Figure 5.30. These coatings are usually heat cured at 180 to 204 C (356 to 400 F) and often have an acidic catalyst to improve the reaction between the phenolic and the epoxy resin.

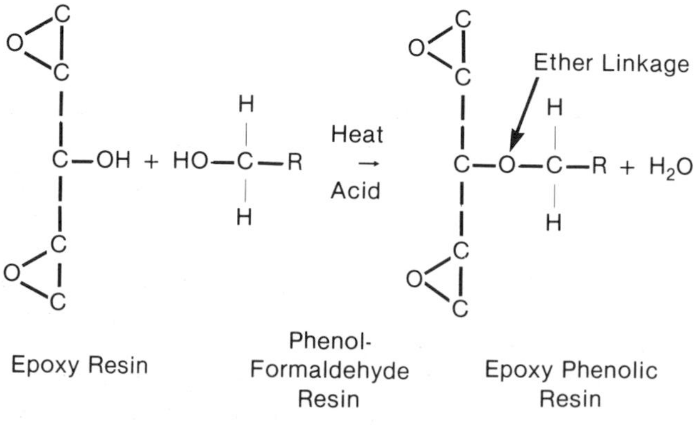

FIGURE 5.30 — Formation of an epoxy phenolic resin: R = phenolformaldehyde polymer.

The ether linkages, as shown in Figure 5.30, are highly resistant to chemical attack. The large number of alcohol groups on the epoxy and methylol groups on the phenolic make for the highly cross-linked structure that is obtained from the fully cured epoxy phenolic resin coatings. When the coating is cured, it is very hard, tough, and solvent- and chemical-resistant.

The application of these coatings is somewhat less critical than the phenolformaldehyde coatings. These coatings are also a little less sensitive to curing conditions and may be applied in thicker coats while still obtaining a thorough cure. The procedure is nevertheless primarily the same: a coat of the resin is applied to a very clean steel and baked for a few minutes at a relatively high temperature, after which another coat is applied and the same treatment follows. Once the coating is completed, the entire coating is then cured for a longer time at the same or higher temperatures for the completed coating.

These coatings, when completely cured, are also nontoxic and may be used for food containers. Heat-cured epoxy phenolic coatings are widely used as interior drum coatings and also for industrial tank linings. This is due to their high level of chemical resistance and their resistance to water and extreme temperatures. They are also used as internal linings for pipe and for coating downhole tubing.

Table 5.16 shows the broad range of chemical resistance of these coatings. One of their best properties is their alkali resistance. This type of coating has been used extensively for the lining of tank cars containing 50 and 73% caustic soda which is transported at 110 to 120 C (230 to 248 F). The temperature-resistant characteristics of this coating are good, giving it excellent hot water resistance. In many ways, it is very comparable to the phenolformaldehyde resin coating, having nearly the same general chemical resistance of the phenolics plus the advantage of being alkali-resistant. Table 5.17 lists the properties of heat-condensing coatings.

100% Solids Coatings

In recent years, there has been a continuing effort to develop 100% solids coatings of many different types. There are many advantages to such coatings. There are no waste materials from the coating process such as solvents which must evaporate. The contamination to the atmosphere is generally low and, on a cost-per-mil of thickness, they are very cost effective. While application procedures may be a little complex, the film which forms is either ready for service at that point or will be as soon as the curing reactions take place. Most 100% solids coatings have the advantage of additional thickness which helps to increase the physical properties and water and chemical resistance. Because of these very obvious advantages, there will undoubtedly be many more 100% solids protective coatings in the future.

There are also some disadvantages and deficiencies as far as these coatings are concerned. One of these is the ability to obtain uniformly good adhesion. Adhesion can be variable, depending on the heat of the object to be coated at the time of the application of the resin coating.

TABLE 5.16 — Chemical and Solvent Resistance of Epoxy Phenolic Baking Coatings

A. Films unaffected by a three-month immersion in the following reagents at room temperature (all solutions are aqueous):

Solvents	Chemicals
Ethanol	Sodium Hydroxide Conc
Isopropanol	Ammonium Hydroxide (10%)
Sec-Butanol	Acetic Acid (1%)
N-Butanol	Linseed Fatty Acids
Methyl Isobutanol	Sulfuric Acid (up to 75%)
Diacetone Alcohol	Hydrochloric Acid (up to 20%)
Hexylene Glycol	Nitric Acid (up to 10%)
Glycerine	Phosphoric (up to 85%)
Carbon Tetrachloride	Liquid Detergent (100%)
Allyl Chloride	Liquid Detergent (50%)
Methyl Isobutyl Ketone	Solid Detergent (1%)
Toluene	Sodium Methoxide (40% in methanol)
Xylene	Sodium Chlorite (25%)
Diethyl Ether	Sodium Hypochlorite (5%)
Bis (B-Chloroethyl) Ether	Calcium Hypochlorite (5%)
	Ferric Chloride (5%)
	Water
	Salt Spray at 38 C for 500 hrs

B. Films unaffected by the following materials, all exposed for three weeks at 66 C, except as noted:

Isopropanol	Methyl Isobutyl Ketone
Sec-Butanol	Allyl Chloride
Methyl Isbutyl Carbinol	20% Sodium Hydroxide (boiling 24 hrs)
Ethanol	73% Sodium Hydroxide (138 C, two weeks)
Diacetone Alcohol	Glycerine
Hexylene Glycol	Glycerine (77 C, six weeks)
	Water

C. Films soften slightly after one month at room temperature in acetone, methyl ethyl ketone, ethylene dichloride, hydrochloric acid (36%), sulfuric acid (78%), and hydrogen peroxide (15%).

(SOURCE: Allen, R. A., Federation Series on Coating Technology, Unit 20, Epoxy Resins in Coatings, Federation of Societies for Paint Technology, Philadelphia, PA, pp. 12, 41, 1972.)

Some of this can be alleviated by the use of a proper primer, as is done with coal tar enamel coatings. On the other hand, if the hot enamel is applied to too cold a surface, the coating will be chilled too rapidly with possible adhesion problems. This has definitely been the case where 100% solids epoxy-type coatings have been applied as a hot spray. On contacting cold metal, the coating chilled rapidly, shrinking, while at the same time reducing the cure reaction—all of which affects adhesion.

There have, in fact, been some dramatic failures of this type of coating; some of which were caused by this problem. In the case of coal tar enamel, if moisture contacts the coal tar primer prior to the application of the enamel, the adhesion is often poor. In fact, where a knife cut was made longitudinally down the length of a pipe, coal tar was seen peeled away and laid back completely from the surface of the pipe. Conditions of application are therefore critical for these types of coatings. Powder coatings must be properly fused to the surface in order for adhesion to be adequate and for a proper impervious film to be formed.

Coal Tar Enamel

Coal tar is a material which is derived from the coking of coal. When the coal is heated without air at about 1100 C (2000 F), it is decomposed into gases, liquids, and coal tar. The coke, or almost pure carbon, is the residue. The coal tar is further distilled and dehydrated, producing coal tar pitches of varying properties, depending on the source of the coal. The coal tar pitch is then reinforced with inert fillers to form the 100% solids coal tar enamel which is applied to various surfaces by heating the coal tar to a liquid condition and applying it to the surface. This type of coal tar enamel is probably the oldest of all of the high-performance protective coatings since coal tar pitches have been available ever since the advent of the steel industry. It is certainly the oldest of the 100% solids coatings and has been used for the exterior of pipelines and various steel and concrete structures for many years. It also is used as an interior lining for water pipe and has provided service without appreciable failure for 50 or more years.

The coating performance is based on the inert impervious film concept, and the outstanding properties of coal tar are very low moisture permeability, high resistance to electrical currents, and permanence when continuously immersed in various water solutions. Water resistance is its outstanding property and there are very few if any materials, either old or new, which have the same water resistant characteristics. The coal tars are not affected by aliphatic oils and greases. They may, however, be softened by vegetable-type oil and certainly will be softened or completely dissolved by aromatic hydrocarbons. The resistance to all types of dilute water solutions, including acids, salts, and dilute alkalies, is very good, and their resistance to soil chemicals and soil conditions is excellent. One of the advantages of coal tar enamels is that they are applied as a thick film (100 to 150 mils). The coating as such is hard, tough, and sufficiently heavy to resist abrasion due to installation and soil stresses after the pipe or other structure is in the ground.

Coal tar enamels are generally supplied in three different types. Type 1, the regular enamel, is a very hard product and is designed for service from 0 to 50 C (32 to 122 F). This type has the highest resistance to moisture, petroleum oils, and soil stresses of any of the three grades, but it also has the least flexibility in the narrow service temperature range. Type 2 is a semiplasticized coal tar pitch. This product has a wider service temperature range of −18 to 60 C (0 to 140 F). It is somewhat more flexible than Type 1. Type 3 is a fully plasticized coal tar pitch and has a service temperature range from −29 to 70 C (−20 to 158 F). This type is used for pipelines which are subject to low storage and handling temperatures and where good flexibility is required.

While there are some other uses for coal tar enamel

TABLE 5.17 — Heat-Condensing Coatings

Properties	Phenolic	Epoxy Phenolics
Physical Property	Very Hard	Hard-Tough
Water Resistance	Excellent 100 C	Excellent 100 C
Acid Resistance	Excellent	Good
Alkali Resistance	Poor	Excellent
Salt Resistance	Excellent	Excellent
Solvent Resistance (Hydrocarbon)		
Aliphatic	Excellent	Excellent
Aromatic	Excellent	Excellent
Oxygenated	Very Good	Good
Temperature Resistance	120 C (250 F)	120 C (250 F)
Weather Resistance	Good (darkens)	Good
Age Resistance	Excellent	Excellent
Best Characteristic	Acid and Temperature Resistance	Alkali and Temperature Resistance
Worst Characteristic	Brittle, poor recoatability	Poor Recoatability
Recoatability	Poor	Poor
Principle Use	Chemical and Food Lining	Chemical Lining

besides application to pipe, the primary use is as a pipe coating and lining for soil and water service. As previously stated, the coal tar is applied as a hot-melt, and when used on the exterior of steel pipelines, may be used with or without a reinforcing wrapper. The coal tar is usually applied to the exterior of pipe by a machine, the pipe being revolved and moved forward under a pouring head where liquid coal tar is evenly distributed on the surface, usually followed by a glass or asbestos pipe wrap pulled into the liquid coal tar coating in order to force any air out of the wrap and to make a dense reinforced coal tar surface. The process is a rapid one, with the coal tar cooling quickly so that the pipe can be handled within a matter of minutes.

The over-the-ditch application of coal tar is similar to that of the plant application, except that the equipment used to distribute the molten coal tar and to apply the pipe wrap is operated on the pipe itself, making a continuous application of the coating over joints and welds as well as the pipe surface.

In some cases the hot-applied coal tar is applied by the process of *daubing*. This is the use of long handled brushes dipped into the molten coal tar and spread over the surface to be protected. The interior of many large water tanks has been protected by this method. Brush marks, however, are a problem during this type of application, and two or more coats are usually required in order to build up a proper film thickness and eliminate holidays.

This type of application also is used for other types of steel structures where automatic applications cannot be used. The coal tar enamel is applied to the interior of steel pipe by revolving the pipe, inserting a lance in the interior, and distributing the liquid coal tar lengthwise

through the pipe as the pipe revolves. A smooth, even thickness is obtained by this method because of the centrifugal force applied during the revolution of the pipe.

The application of the hot-applied coal tar to interior joints in the field is usually by daubing. Exterior joints may be daubed, but are more often poured, using a so-called diaper which allows the hot-applied coal tar to flow around the pipe and over the joints. Some of the properties of such a coal tar coating are as follows.

Water Resistance

As previously stated, the water resistance of coal tar is excellent in terms of both moisture vapor transfer rate and water absorption (Figure 5.31).

Electrical Resistance

The electrical resistance of coal tar enamel is also high, which makes it particularly suitable for both water and underground service (Figure 5.32). This is particularly true where the pipeline is under cathodic protection.

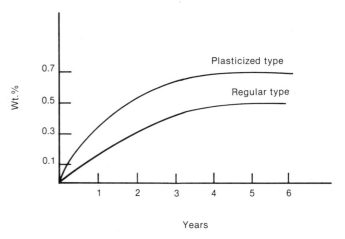

FIGURE 5.31 — Water absorption rates for coal tar enamel. [SOURCE: Kemp, W. E., Coal Tar Enamels Coatings for Underground Pipeline, Materials Protection, June (1970).]

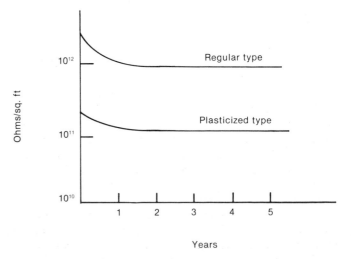

FIGURE 5.32 — Electrical resistance rates for coal tar enamel. [SOURCE: Kemp, W. E., Coal Tar Enamels Coatings for Underground Pipeline, Materials Protection, June (1970).]

Adhesion

Coal tar enamels, when properly applied, also have good adhesion. Even though the enamel is quite thick, it is still difficult to remove from the surface, and, unless there is some surface contamination, neither water nor cathodic protection currents tend to release the adhesion. *Cathodic disbonding* is the process of a loss of adhesion due to moisture and cathodic currents. Properly applied coal tar enamel has excellent resistance to cathodic disbonding. The coal tar enamel is also resistant to soil bacteria and to hydrogen sulfide which may be developed by the anaerobic bacteria in the soil.

Chemical Resistance

The general chemical resistance of coal tar enamel is improved by the thickness of enamel and by its excellent water resistance. It is generally resistant to most chemicals with which it might come in contact. All in all, while coal tar enamel is an old answer to pipe and underground corrosion problems, it is also a very effective one with excellent corrosion-resistant characteristics.

Asphalt

Asphalt is a naturally occurring material which is derived either by mining (which is the method of obtaining gilsonite, a naturally occurring asphalt), or it is a residue from the distillation of asphaltic petroleum. Gilsonite is a solid material mined in much the same way as coal. It is soluble in aliphatic and aromatic solvents and has excellent chemical resistance as well as good weather resistance. Asphalts vary in their chemical and physical characteristics, depending on the distillation process and temperatures to which they are subjected. They also may be subject to steam distillation and to air blowing during the distillation process.

Unlike coal tar, asphalt has comparatively better weather resistance and does not tend to alligator in the same manner as coal tar when subjected to weather conditions. This is one of the reasons why hot-applied asphalt is applied for roof coatings where it is subject to continuing temperature changes and weather conditions. Depending on the grade of asphalt, it is usually a less brittle material than coal tar and is therefore used to a greater extent on steel structures subject to weathering and temperature changes. Asphalt properly applied to steel or concrete surfaces has good adhesion and can be built up to the same type of thicknesses as coal tar, *i.e.*, from 100 to 250 mils. Chemically, asphalt is a stable aliphatic hydrocarbon which has good resistance to water and most chemicals and salts. It is solvent sensitive, even to aliphatic solvents, and is softened and dissolved by vegetable oils. The water resistance of asphalt is good. It does not, however, compare with coal tar in this respect.

The application of asphalt as a 100% solids hot-melt coating is very similar to the methods used with coal tar and glass. Asbestos wrap also are used to reinforce the asphalt when applied over the exterior of pipe. Its use as a pipe coating is not nearly as extensive as coal tar. On the other hand, many steel pipes, particularly in relatively small diameters, have been dipped in asphalt and have been in service for many years with good results. This type of pipe, for example, was used throughout California as an irrigation pipe. Some of these have been recently dug up, as the ground usage has changed from agricultural to residential, with the asphalt still providing corrosion protection.

The major differences between coal tar and asphalt are the excellent water resistance of coal tar and the good weather resistance of asphalt. The asphaltic materials are used more for above ground uses, while coal tar coatings are used underground. Many water tanks are coated with asphalt because of its nontoxic and relatively tasteless nature. Asphalt generally has good corrosion-resistant properties and it may be used wherever a decorative coating is not required. While some asphalt coatings are heavily pigmented, it is primarily a black coating, and this characteristic is difficult to change. It may effectively be applied over other coatings which have sound adhesion to a surface, although it is a very difficult material to overcoat because of its sensitivity to oils and solvents of all types.

The chemical makeup of asphalt is not complex. The various types of asphalt are primarily based on different melting point or softening point materials. They are all polymeric aliphatic hydrocarbons and their chemical properties are those that you would expect from such a material.

Polyesters

100% solids polyesters are used as tank linings or for the coating of tank bottoms. They have some very good properties, as was previously noted. The general resistance of polyester linings is good. They have good acid resistance and good resistance to oxidizing materials. They are, however, subject to hydrolysis when in contact with alkalies. They are satisfactory for acid salts and neutral salts, but strong alkaline salts are not recommended. The problem involved is the ester group, which is the part of the molecule subject to the alkaline hydrolysis, breaking the molecule at that point and causing the coating to disintegrate. Alkali resistance also is affected by the glass cloth or mat reinforcing often used in connection with the use of polyesters as tank linings.

While the weather resistance of polyester resins is satisfactory, these materials are seldom used as coatings and therefore are not often subject to weathering conditions. Some polyester materials are incorporated into cements for acidproof brick and tile, and used particularly where oxidizing products are encountered. Because of the inherent shrinkage of polyester linings (as was covered previously), they are seldom used without some type of filler or reinforcer, either of the pigment type or more often glass or synthetic fabrics or glass mats. In the case of polyester cements, they are reinforced with siliceous aggregate in order to hold down shrinkage. The polyesters have some very good properties; on the other hand, their shrinkage characteristic limits present widespread use.

Epoxy Powder Coatings

Powder coatings are materials with which the corrosion engineer should be familiar. Although they are not practical for maintenance and field use, the corrosion engineer should be familiar with them because they are rapidly becoming one of the standard product finishes. They can be found on many different types of equipment that can be heated and dipped into a fluidized bed to form the coating. They may also be applied by electrostatic spray or by a flocking gun. However, the surface over which they are applied must be at fusion temperature and generally kept at that temperature for a sufficient period for the coating to fuse, flow, and cure in order to form a continuous film.

The properties of the powder coatings depend to a considerable extent on the curing agent used. Some of the curing agents, such as the aromatic amines, provide coatings with excellent chemical and corrosion resistance. The curing agents are usually friable solids.

Powder coatings can be made by two different processes: the fusion process followed by pulverization, or dry grinding of the various ingredients in a pebble mill. In the fusion technique, the powdered epoxy resin, pigment additives, and powdered curing agent are first dry blended and then passed through a specially designed extruder which accomplishes both the fusion and mixing at temperatures that are low enough to minimize any curing that might start. The extruded material is then pulverized to the desired size. This process intimately mixes the diverse ingredients of the formulation and provides a good degree of pigment wetting and dispersion. High-gloss coatings can therefore be obtained. Such coatings are usually relatively thin (one or two mils in thickness).

Powdered coatings which need to be applied at thicknesses of 10 to 20 mils are usually made by the dry grinding process. In this case, the ingredients are all charged into a cooled pebble mill and the mill is allowed to grind for up to 24 hours to yield the completed coating formulation. The fact that the pebble mill is cooled allows a curing agent with highly reactive characteristics to be used. Most of the curing agents used in connection with these powdered coatings are ones which are heat activated so that little curing takes place at ambient temperatures.

Once the powder is ready to use by either of these processes, the powder is applied to the heated metal surface either by the fluidized bed technique which lowers the heated object into the powder, or it is electrostatically sprayed on the heated metal. The object, which is above the fusion temperature, rapidly accumulates the powder and fuses it onto the surface. Some of these systems will form a coating and cure it within a matter of 30 seconds at 230 C (446 F). Other curing systems applied in heavier coating thicknesses can require as much as 15 to 30 minutes at 200 C (392 F).

One common use of epoxy powder is as a coating for underground pipe. In this case, one of the heavier coatings is applied, although it is still considered a thin film coating (10 mils) in contrast to the heavy coal tar enamel-type coating. Epoxy powder coatings have excellent adhesion, impact resistance, and chemical resistance consistent with the curing agent; good thermal stability; excellent insulation properties; and resistance to corrosive soil and cathodic disbonding. These characteristics have led to their use as exterior pipe coatings.

There also have been difficulties where powdered coatings are applied, particularly where fusion temperatures have gotten out of control and the surface preparation of the metal or other surface was not as satisfactory as it should have been. In these cases, the adhesion has been less than satisfactory, and blistering and loss of adhesion has resulted on pipelines. Some other uses for the epoxy powdered coating includes coatings for electrical equipment and housing finishes for office and other types of equipment, including light features, handrails, automobile accessories, tubular steel furniture, cabinets, and other similar areas. Most of these are product finish uses which are not necessarily of direct interest to the corrosion engineer.

Vinyl Powder Coatings

The vinyl powder coatings are in many ways much simpler than the epoxy coatings because the vinyl is a thermoplastic material which does not require curing in order to fuse and obtain adhesion. The vinyl powders are mixed with pigments, milled to a stable consistency, and, in the case of a fusion milling technique, reground. As they do not contain any reactive curing agents, the entire process of mixing, milling, and grinding is simplified. They may also be extruded to obtain a uniform mix prior to being ground to the proper powder size.

The application techniques for vinyl powder coatings are the same as those used for the epoxy powder coatings. They may be applied by either the fluidized bed technique or by electrostatic spray. Where properly applied over a thoroughly clean surface, the vinyl powder coatings can have all of the good properties found in lacquer-type coatings. Most of the powdered vinyl coatings, however, are somewhat thinner than many of the vinyl lacquers. Again, these coatings are important to the corrosion engineer only because equipment is sometimes received that has been coated in this manner.

100% Liquid Epoxy Coatings

100% solids liquid epoxy coatings have been a goal for some time. Ideally, they would be epoxy coatings which were sufficiently thin in viscosity so that they could be applied to either structures or equipment by normal spray equipment in the field. There are a number of high-solids epoxy coatings which have been developed. Most of these, however, do contain some volatile material in order to reduce the viscosity of the epoxy resins to a point where they can be handled. There are also some 100% solids epoxy coatings applied from two-headed spray equipment where the catalyst and the epoxy are mixed at the point of spray. Many of the epoxy compounds used in this case are heated during the application process, thus the reaction with the curing agent is usually rather rapid. Heavy film thicknesses can be achieved this way, although in some cases this has been a hindrance rather than an advantage. Unless every care is taken during the application of such materials, adhesion and disbonding problems can result.

While it is true that 100% solids epoxy coatings are now available, a readily and easily useable, effective, truly 100% solids epoxy has not yet been developed. The properties of the 100% solids epoxy coatings which are available are similar to those of the solvent-applied epoxy coatings of the more chemical-resistant type.

Plastisols

Plastisols are 100% solids in that they are essentially vinyl resin milled into a plasticizer which does not solvate the vinyl resin until heat is applied. The amount and type of plasticizer determines the physical as well as chemical characteristics of the plastisol. One common use of plastisols is in the product finish field where plating racks, trays on the interior of automatic dishwashers, and other similar objects are coated. The coatings in these cases are subject to continuous and difficult operating conditions. Many plastisols are used for lining objects that can be heated, as well as for lining pipe. Much plastisol-lined pipe has been used by the mining industry for tailing pipe and for pipe handling strongly corrosive solutions such as copper sulfate.

The chemistry of this coating is relatively simple. The powdered vinyl resin is mixed into the plasticizer or combination of plasticizers where it remains suspended in a very viscous state. It is then applied to its intended object, and both are heated to the point where the plasticizer solvates the resin and is absorbed into it. The vinyl plasticizer mixture becomes homogeneous with the vinyl resin, and the plasticizer completely dispersed, one within the other.

Chemical characteristics depend on the plasticizer used since it is a major part of the coating film. The vinyl resin itself is generally a homopolymer of vinyl chloride which is extremely resistant. However, when a large amount of plasticizer is included, chemical characteristics can be reduced considerably, depending on the plasticizer involved. Plastisols are ordinarily rather soft, rubber-like coatings and are applied in rather thick films (from 30 to 125 mils), so that the end result is a flexible, resilient coating having the benefits of an appreciable thickness. These properties have led to its successful use as lining for pipe of some substantial size in mining operations. The resilience of the coating, as well as its chemical resistance, provide a protection against the abrasive and corrosive mine liquids.

Furan Materials

A corrosion engineer may conceivably face a problem where furan products could serve as a valuable solution. Furan resins have one of the greatest potentials of any of the synthetic organic compounds available for the corrosion protection field. They have a broad span of chemical resistance, ranging from strong acids to strong alkalies with good resistance to solvents as well. Their temperature resistance is well above other organic resins. On the other hand, furan resins cure explosively, are brittle and brash, have poor adhesion characteristics, and are black. Recent research, however, has led to a reduction of the materially negative characteristics without changing the valuable properties of these compounds. There presently are patents on over a thousand of these materials, although from a practical, industrial standpoint, there are only a few which are of value. These primarily are based on furfural, furfuryl alcohol, or combinations of the two.

The source of furfural, the base of furan resins, is vegetable matter which contains cellulose and hemicellulose. The cellulose compounds are built up from long-chain sugar units, mainly pentosans (xylan and araban) and hexosans (mannan and galactan). Of these components, the pentosans serve as the major source of furfural.

A pentosan is a five-carbon sugar with vegetable sources of corn cobs, oat hulls, cottonseed hulls, bagasse rice hulls, etc.

The chemical formation of furfural and furfuryl alcohol from pentosans consists of: (1) acid hydrolysis of the raw cellulose which contains the pentosans producing pentoses; (2) continued acid hydrolysis of the pentose to furfural; and (3) catalytic hydrogenation of the furfural to furfuryl alcohol (Figure 5.33).

Furfuryl alcohol is the primary basic material for furan resins, and it is a mobile, amber-colored liquid. It is extremely sensitive to strong acids and polymerizes readily in their presence. The effect of hydrogen ion concentration on the polymerization process of furfuryl alcohol was studied many years ago when it was found that the rate of polymerization is a simple function of pH, at least during the early stages of the reaction. The first step in the reaction is one of intermolecular dehydration which depends upon condensation between the hydroxyl group of one molecule of the alcohol with the labile nuclear alpha-hydrogen atom of another molecule, as shown in Figure 5.34.

There are two stages of a furan resin: a liquid stage and a hard, cross-linked solid stage which constitutes the ultimate product. The first liquid stage is the one used in most industrial products which is later reacted to the solid state as desired. The furfuryl alcohol or furfuryl alcohol-furfural resins are fluid or viscous liquids, dark in color with a reddish-black appearance, and generally stable in storage under normal conditions for extended time periods. Further polymerization takes place with the addition of acid catalysts to the furfuryl alcohol resins. When catalyzed to the final thermoset polymer, the resin becomes black.

Chemical resistance is one of the furan resin's outstanding characteristics since it is not limited to any one type of chemical reaction. Furan resins are resistant to strong acid solutions at high temperatures, strong alkali solutions, most salt solutions, and many solvent-type materials as well. Some of the more specific properties of the furan resins are as follows.

Water Resistance

The water resistance of furan resins is excellent. Furan resins have been made into acid-resistant cements for use in the steel industry to line acid pickling and rinse tanks in galvanizing plants and continuous steel strip lines. They have remained unchanged for many years, even though temperatures range from 69 to 85 C.

Inorganic Acids

Furan resins have shown excellent resistance to most inorganic acids at temperatures as high as 90 C for

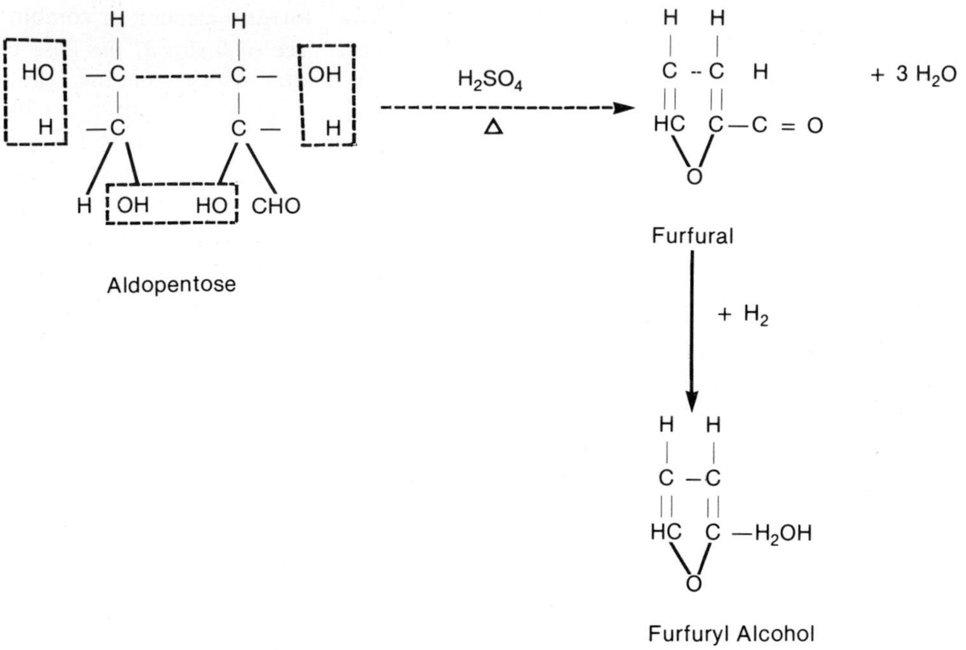

FIGURE 5.33 — Chemical formation of furfural and furfuryl alcohol.

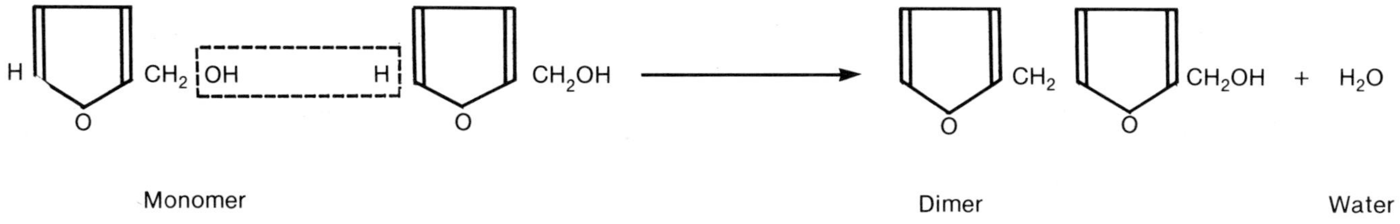

FIGURE 5.34 — Effect of hydrogen ion concentration on the polymerization process of furfuryl alcohol.

periods of many years. Concentrated sulfuric acid is the exception. Concentrations above 60% of sulfuric are not recommended.

Oxidizing Conditions

Oxidizing conditions represent the one area where furan resins are deficient. Furan resins generally are not satisfactory for exposure to strong oxidizing chemicals such as hypochlorites, hydrogen peroxide, chlorine dioxide, and chromic acid.

Organic Acids

The furan resins have very good resistance to organic acids, even at high temperatures. These include lactic, oxalic, and many fatty acids.

Alkalies

Furan resins demonstrate excellent resistance to alkalies in general. Many chemical tanks containing strong sodium hydroxide have been constructed using furan resin mortars and carbon brick.

Salt Solutions

Very good resistance to continuous exposure to most

salt solutions at 90 C is another characteristic of furan resins. Those salts which hydrolyze and release oxidizing acids are the only exception.

Solvents

Furan resins generally are resistant to most solvents, particularly where the resins are cured at above ambient or room temperature. Resins should not be cured at room temperature for extended periods. Also, materials such as aniline and methylene chloride tend to soften the resins over a period of time.

Animal, Mineral, and Vegetable Oils

Furan resins are unaffected by mineral, animal, and vegetable oils.

Temperature Resistance

Temperature resistance is one of the outstanding properties of furan resins, since little heat distortion occurs with these resins, even at temperatures as high as 190 C. This is shown in Table 5.18 where the compressive strength of furan is compared with polyester and epoxy polyamide resins exposed to the same temperature conditions over the same period of time. The furan resin main-

TABLE 5.18 — Compression Strength in psi at Various Temperatures						
Temperature, C	22	65	95	120	150	175
Crosslinked Furan	10,572	13,863	14,691	13,331	12,908	10,466
Polyester	16,172	18,577	16,476	13,588	9,554	7,366
Epoxy Polyamide	7,706	10,059	2,059	1,116	806	136

(SOURCE: Munger, C. G. and Ignatius Metil, Industrial Applications of Furan Resins. Presented at Third Interamerican Congress on Chemical Engineering. Mexico City. Mexico, October, 1966.)

tains its original compressive strength from ambient temperature through 175 C, while a substantial change is evident in both of the other resinous materials.

The broadest use of furan resins for corrosion applications, is in the steel and chemical industries where chemical-resistant cements are made. Furan has been used for a number of years primarily as an acidproof cement used for laying up brick in acid pickling tanks in steel plants. These are large installations operated on a continuous basis with steel passing through the hot sulfuric acid solution at a rapid rate. In these cases, the furan cement not only is utilized for its acid resistance, but also for its ability to retain its properties under strong, hot acid conditions, remaining sufficiently tough and adherent to the brick so that abrasion and impact from the steel has limited effect on the structure.

In addition to the acid pickling tanks and continuous acid strip mills in steel plants, there are many other similar tanks throughout industry using equivalent acid-proof construction. In steel galvanizing, the steel must be pickled in a way similar to that in the steel mills. Rinse water tanks are also needed and lined in a similar way, since the rinse water is hot and gradually picks up sufficient acid to create a very corrosive condition. Chemical companies, particularly heavy chemical plants, use several tanks of this type for storage of a variety of severely corrosive liquids. Stacks in chemical processing and paper mills are lined with acid brick and furan mortars, as are many floors that are subject to continuous exposure to corrosive solutions. Some furan components have been formulated into troweling compounds and used for application on pump bases and similar areas where active corrosion exists.

The furan materials are versatile corrosion-resistant compounds. While there can be significant problems with their use, they may, in some instances, provide the only practical answer to a severe corrosion problem.

As evident from the foregoing discussion, 100% solids coatings, with the possible exception of 100% solids liquid epoxy coating, are primarily specialty products which are only for limited use in specific situations. They cannot be sold for general plant maintenance or other similarly broad purposes. Nevertheless, they can perform where other materials would fail or have short useful lives. Table 5.19 gives a summary of the general properties of the 100% solids coatings.

Corrosion-resistant organic coatings have numerous and varied properties which are necessary to match the variety of corrosion conditions which exist. For this reason, organic coatings are the most widely used method of corrosion protection available. Without such versatility, adequate corrosion protection for modern industrial structures and equipment would be very difficult.

					TABLE 5.19 — 100% Solids Coatings: General Properties			
Properties	Coal Tar Enamel	Asphalt Enamel	Polyester	Epoxy Powder	100% Solids Liquid Epoxy	Vinyl Powder	Vinyl Plastisol	Furfural Alcohol Resins
Physical Properties	Hard	Somewhat Resistant	Hard	Hard	Hard-Tough	Hard-Tough	Soft-Rubbery	Hard-Brittle
Water Resistance	Excellent	Very Good	Good	Good	Good	Good	Good	Excellent
Acid Resistance	Very Good	Very Good	Excellent	Good	Good	Good	Very Good	Excellent
Alkali Resistance	Good	Fair-Good	Poor	Very Good	Very Good	Good	Good	Excellent
Salt Resistance	Very Good	Very Good	Very Good	Very Good	Very Good	Very Good	Very Good	Excellent
Solvent Resistance (Hydrocarbons)								
Aliphatic	Good	Poor	Very Good	Very Good	Very Good	Very Good	Good	Excellent
Aromatic	Poor	Poor	Fair	Very Good	Very Good	Fair	Poor	Excellent
Oxygenated	Poor	Poor	Poor	Good	Good	Poor	Poor	Good
Temperature Resistance	60 C (140 F)	50 C (122 F)	65 C (150 F)	93 C (200 F)	93 C (200 F)	60 C (140 F)	60 C (140 F)	120 C (248 F)
Weather Resistance	Poor	Good	Good	Good (Chalks)	Good (Chalks)	Very Good	Very Good	Good
Age Resistance	Excellent	Very Good	Good	Good	Good	Very Good	Very Good	Good
Best Characteristic	Water Resistance	Water and Weather Resistance	Acid and Oxidizing Chemical Resistance	General Water and Alkali Resistance	General Chemical and Alkali Resistance	General Chemical Resistance	Water and Chemical Resistance	Temperature and Acid Resistance
Poorest Characteristic	Weather Resistance, Black	Solvent Resistance, Black	Alkali Resistance	Critical Application	Critical Application	Critical Application	Adhesion	Adhesion, Brittleness
Recoatability	Poor	Good	Fair	Poor	Difficult	Good	Good	Good
Principle Use	Lining and Coating Pipe	Waterproofing Structures	Tank Lining	Exterior Pipe Coating	Chemical Lining	Chemical-Resistant Product Finish	Pipe Lining	Cement for Acid-proof Brick

Corrosion-Resistant Organic Coatings

References

1. Fox, Fred L., Federation Series on Coating Technology, Unit 3, Oils for Organic Coatings, Federation of Societies for Paint Technology, Philadelphia, PA, p. 10, 1969.

2. Lasovick, Daniel, Federation Series on Coating Technology, Unit 15, Urethane Coatings, Federation of Societies for Paint Technology, Philadelphia, PA, pp. 8-9, 1970.

3. Applied Polymer Science, Chapter 48, Chemistry and Technology of Phenolic Resins and Coatings, K. J. Craver and Roy W. Tess, Eds., Organic Coatings and Plastics Chemical Div., Amer. Chem. Soc., pp. 725-726, 1975.

6

Corrosion-Resistant Zinc Coatings

Zinc is a unique and very useful metal, particularly when in the form of a thin coating. It can be used as a pure metal coating or it can be combined with other materials and still provide full corrosion protection when applied over steel surfaces. Metallic zinc is resistant to most atmospheric conditions, yet it remains sufficiently reactive to cathodically protect steel where zinc and iron are in contact. It does not develop a continuous and inert oxide film, as does aluminum. The aluminum oxide film makes an aluminum surface nonreactive and thus prevents the degree of cathodic protection developed by zinc.

This characteristic of being relatively inert to atmospheric conditions, yet sufficiently reactive to protect steel is unique. Another important advantage of zinc is that it is readily available. In fact, among the nonferrous metals, zinc is one of the least expensive and one of the most readily available.

While there are other metals which also can be used to coat steel (e.g., magnesium, aluminum, and cadmium), none of them have proven to be as useful or as effective as zinc. Over a million tons of zinc are used in the United States annually, and approximately half of that tonnage is used as a coating to protect steel. The majority of it is used for galvanizing or electrodeposited zinc coatings. Approximately 10% of the tonnage used for coatings is applied in the form of zinc-rich coatings. Approximately ten million tons of steel are coated in the United States each year with some type of zinc coating.

Protection by Zinc Coatings

Zinc coatings are protective in two different ways: they serve as a barrier and also as a galvanic protector of steel surfaces, regardless of the type of zinc coating involved. Metallic zinc protects steel from corrosive attack by most atmospheres by acting as a continuous and long-lasting barrier between the steel and the atmosphere. Zinc has a much lower corrosion rate than steel, so that in all

except very polluted (acid or alkaline) atmospheres, the coating of zinc will provide protection against rust for long periods of time. Historical data to prove this point has been released in reports by ASTM and by the Zinc Institute as a result of actual zinc coatings service and field tests.

Table 6.1 gives the estimated life of zinc-coated products in the atmosphere for various thicknesses of zinc. These are only estimates, but they do indicate that, when applied as a continuous film, zinc does provide corrosion protection in proportion to the thickness of the coating. The greater the thickness (up to an optimum point), the longer the time before the underlying steel will begin to corrode.

There are several different methods of applying zinc to steel surfaces (Table 6.2). Each of these processes has its own unique characteristics, in spite of the fact that the protection is derived from metallic zinc. The processes are generally complimentary rather than competitive in nature so that each one fits a specific need.

Application of Zinc Coatings

Galvanizing

Galvanizing is the primary process by which zinc is applied to steel. It was first suggested by two different French engineers at approximately the same time (1840).[1] Since then, its use has steadily increased.

Galvanizing is the process of cleaning steel free of all mill scale or other impurities and then dipping the steel into molten zinc. The molten zinc wets clean steel very readily and alloys with the steel, making a strong bond between the zinc and the steel surface. One of the great advantages of galvanizing is that once the object is dipped, removed, and cooled, the process is finished and the galvanized object can be handled without fear of damage.

TABLE 6.1 — Estimated Life of Zinc-Coated Products in the Atmosphere

Thickness, in.	Weight in oz./sq. ft. of Surface[1]	Life in Years under Atmospheric Conditions					
		Rural	Tropical Marine	Temperate Marine	Suburban	Urban	Highly Industrial
0.0036	2.00	50	40	35	30	25	15
0.0023	1.25	35	30	25	20	17	9
0.0018	1.00	25	20	15	12	10	7
0.0011	0.60	10	8	7	5	4	3
0.00066	0.37	7	6	5	4	3	2
0.00044	0.25	5	4	3	3	2	1

[1]In the case of galvanized steel sheets, the weight of zinc is specified in terms of total zinc on both sides of the sheet; *i.e.,* a 2-oz. sheet has 1 oz. of zinc per sq. ft. of surface.

[SOURCE: Zinc Controls Corrosion, Z1-51-10M/71, Zinc Institute, Inc., New York, NY, p. 11 (1971).]

TABLE 6.2 — Principal Methods of Corrosion Protection with Zinc

1. Galvanizing
    a. Hot Dip
    b. Continuous Line Galvanizing
2. Electrogalvanizing
3. Zinc Plating
4. Sherardizing
5. Zinc Spray
6. Zinc Coating
    a. Organic Zinc Rich
    b. Inorganic Zinc

The hot dip galvanized coatings are rugged and provide an impervious and long-lasting barrier against most atmospheric corrosion processes.

Hot Dip Galvanizing After Fabrication

Hot dip galvanizing is the process with which the corrosion engineer will probably have the most contact. It is the earliest of the methods to be used for zinc coatings and, like many simple processes, is quite effective, continuing to be used year after year with only minor improvements. It is a flexible process in that it can apply zinc to steel parts ranging from extremely small ones such as nuts and bolts, to large fabricated pieces such as small tanks, containers, transmission towers, pole line hardware, guard rails, etc.

A great deal of structural steel is galvanized for corrosion protection. As mentioned previously, the hot dip galvanizing forms an alloy with the steel which provides for maximum adhesion between the two metals. Figure 6.1 is a photomicrograph of a hot dip galvanized coating on a steel surface. Starting with the basic steel at the bottom, which is 100% iron, there is an adjacent, thin layer which is approximately 75% iron and 25% zinc. The next layer is approximately 90% zinc, the third layer approximately 94 to 95% zinc, and the final top layer, which is the thickest, is pure zinc. There is not any real line of demarcation between the iron and the zinc, but rather a gradual progression of iron and zinc alloy from the pure iron to the pure zinc. This provides a powerful bond between the two materials.

The structure of the zinc coating and its thickness depend on the composition and physical condition of the steel being treated as well as the temperature, time in the bath, and other factors, some of which are under the control of the galvanizer. Heavier coatings tend to be deposited on rough surfaces or coarse grain steel. The total thickness of the alloy layer tends to be slightly greater in corners and similar areas than it is in hollows. The thickness of the coating can be controlled by the length of time in the bath and the speed by which it is removed. Many times when a thin coating is required, the zinc is mechanically wiped from the surface as the object is being removed. Small parts and threaded parts are often centrifuged after being hot dipped in order to remove the excess zinc.

Some other elements also may be added to the galvanizing bath. Tin and antimony give a spangled effect, and some lead added to the bath also is considered desirable. Aluminum aids in the ductility and adds to the

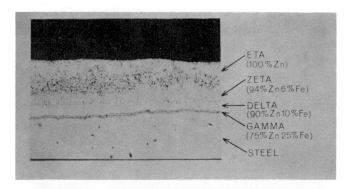

FIGURE 6.1 — Photomicrograph of a section through a typical hot dip galvanizing coating showing alloy layers. [SOURCE: User's Guide to Hot Dip Galvanizing for Corrosion Protection in Atmospheric Service (TPC-9), National Association of Corrosion Engineers, Houston, Texas, p. 1, 1983.]

corrosion resistance of the coating. The thickness of zinc applied by the hot dip process usually varies from a maximum of 2.75 ounces per square foot down to an ounce or less per square foot. If the steel is thoroughly cleaned, a continuous coating is formed over the entire surface. Even the rivets, welds, and edges of complicated fabricated steel structures are well covered.

One of the disadvantages of hot dip galvanizing is the possibility of warping the steel structure due to the heat of the galvanizing bath. There also is some possibility of the embrittlement of malleable cast iron. Such difficulties, however, usually can be overcome through proper galvanizing techniques. All in all, it is a very useful process, particularly for complicated, relatively lightweight objects or steel fabrications.

Continuous Line Galvanizing. Continuous line galvanizing is a process of hot dip galvanizing developed in the 1930s, whereby coils of sheet steel could be continuously hot dipped. A small amount of aluminum is added to the zinc bath which provides a coating with essentially no iron zinc alloy, yet one of good adhesion and sufficient ductility to allow for deep drawing and folding without appreciable damage to the coating. Nearly all of the hot dip galvanized sheet steel that is used in metal building fabrication is produced by the continuous strip method. Approximately 6.5 million tons of steel a year are coated by this process, which accounts for the greatest use of zinc for corrosion protection.

The weights of coatings produced by this process vary from 2.75 ounces of zinc per square foot down to as low as 0.5 ounces per foot. The Zinc Institute standard for galvanized sheet is 2 ounces of zinc per square foot. In the case of galvanized sheet, 2 ounces per square foot refers to both sides of the sheet. This equals 1 ounce of zinc per square foot on each surface.

The zinc coating applied by this process is considerably different from the hot dip coating in that there is little alloying of the steel with the zinc. Figure 6.2 is a photomicrograph of continuous strip line galvanizing which shows a distinct demarcation between the two metals.

FIGURE 6.2 — Photomicrograph of a continuous galvanized coating cross section. Aluminum addition minimized alloy layers provides uniform coating. (Photo courtesy of Zinc Institute, Inc., New York, NY.)

Electrogalvanizing

Electrogalvanizing is an essentially cold process compared to the heat involved in hot dip galvanizing. Most electrogalvanizing is continuous and is applied to sheet, wire, and electrical conduit or similar objects. It produces a thin, pure zinc coating which has excellent adherence. The coating is smooth, free of spangled characteristics, and can be readily prepared for painting by phosphatizing. Electrogalvanized steel is generally painted since it is applied as a relatively thin film. The coating produced on strip coils or sheets generally has a coating weight of from approximately 0.06 to 0.2 ounce per square foot. This is a thickness of 0.0005 to 0.00017 in. on each side of the sheet. Electrodeposition steel offers a process for applying zinc coatings to parts that cannot be hot dipped. It is especially useful where high processing temperatures may injure the part.

Electrogalvanized steel can easily be prepared to receive organic coatings. Many organic coatings have good adhesion to the zinc surface, and the zinc base tends to increase the life of the organic coating over the surface, as compared with a similar coating applied to steel alone. This is similar to the effect that inorganic zinc coatings have on the life of organic coatings.

For outdoor surfaces, electrogalvanized products are commonly painted to increase corrosion resistance and protection of the thin zinc coating. Such protection is adequate for many mild services.

Zinc Plating

Zinc plating is similar to electrogalvanizing, although it is not continuous and is applied as a batch process. In zinc plating, the thickness of the zinc also may be controlled by the plating process and time in the plating bath. This is an effective method of applying zinc to small objects. Barrel plating of many small objects at the same time is a common process where the parts are tumbled in a barrel which is in the plating bath.

The normal plated zinc coating is dull gray in color and has a matte finish. The coating is pure zinc and is of uniform composition. It adheres by means of a metal-to-metal bond and is not alloyed.

Sherardizing

Sherardizing also is a process for relatively small parts. It applys zinc coatings to clean steel by rotating the parts in a sealed drum in the presence of zinc dust and at a temperature in the range of 700 to 800 F. Tubing, conduit, nuts, bolts, and small castings are handled in this manner. This is the process which was most commonly used earlier in the century. At the present time, however, it is rarely used in the United States.

Zinc Spray

Zinc spraying is a process whereby zinc is melted in the spray gun and is atomized and projected onto a steel surface. The steel is usually sand or grit blasted. The sprayed zinc should be applied as soon as possible after the surface has been prepared in order to reduce oxidation on the steel and to make sure of an effective metal-to-metal bond. The bond can be affected both by the oxidation and many times by the temperature of the steel at the time the zinc spray is applied.

The spraying of zinc is accomplished by two methods. One is the wire process in which zinc wire is fed into the center of a very hot flame. A stream of compressed air disperses the molten metal and sprays it out of the nozzle in a fashion similar to the spraying of coatings. The zinc wire is fed continuously into the gun as long as the gun is operating. The second process involves the use of zinc dust or powder. The finely divided zinc is transported into the gun by gas and heated by a flame surrounding the nozzle, with compressed air again providing the driving force for the steam of molten zinc to impact the base metal. Sprayed zinc of these types can be applied to structures of almost any size or shape and can be done either on-site or in a plant. Coating of the steel is dependent on the gun operator, and strict care is required in order to obtain a smooth and even film over the surface. Zinc spraying is one of the only satisfactory methods of depositing a heavy zinc coating of 0.01 inches or more on the steel surface (Figure 6.3).

Zinc spraying is difficult, if not impossible, in cavities, depressions, corners, and similar areas. Due to the porosity of the molten metal application, zinc spray is usually sealed or used as a base for an organic topcoat.

One of the older yet continuing uses of zinc coatings and galvanized steel has been by electrical utilities. Hundreds of thousands of major transmission towers have been protected by hot dipped zinc galvanizing. Thus, a good record of the life of hot dip galvanized coatings has been kept by this industry (Figure 6.4).

The steel frame in an electric substation in Figure 6.5 is typical of the complex structures on which zinc coatings or galvanizing does an excellent job of corrosion protection. Such structures could not be satisfactorily protected by organic coatings alone because of the hundreds or thousands of linear feet of edges, corners, welds, rivets, etc. Only a coating with the properties of zinc can provide the corrosion protection required. These properties are summarized in Table 6.3.

Zinc Dust Coatings

When the galvanizing process was first used in the mid 1800s, it had one substantial drawback. This was the inability of zinc to be applied to large, existing structures or new structures that were too complex to fit into a bath of molten zinc. The size and weight of structures, both substantial factors in the galvanizing process, increased several times during this period. Structures also were becoming more costly and processes were becoming more corrosive, so that a substantial need was developing for zinc coatings that could be applied to large or existing structures. The use of zinc dust as a basis for an anticorrosive coating was conceived in two areas of the world at approximately the same time (*i.e.,* 1930s): zinc dust was incorporated into organic coatings in Europe and it also was being tried as an anticorrosive in inorganic coatings in Australia.

Organic Zinc-Rich Coatings

The development of organic zinc-rich coatings in Europe came slightly before its counterpart in Australia

FIGURE 6.3 — Metallizing is one way of applying a zinc coating. Molten zinc is sprayed at high pressure onto a clean steel surface. (Photo courtesy of: Metalweld, Inc., Conshohocken, PA.)

and, in many ways, proceeded more rapidly. Drying oils were the first type of materials used which, for the most part, were not entirely successful. Today, however, one of the standard zinc dust products is a Federal Specification, TT-P-641, in which Type 1 is a zinc dust, zinc oxide, linseed oil paint for outdoor exposures. It is used for air drying only and is recommended as a primer or a finish coat for galvanized steel, particularly where the galvanizing has started to show rust penetration and pinpoint rust on the steel surface. It is specified that this product have a minimum weight of 23 pounds per gallon.

Type 2 included in this specification is a zinc dust, zinc oxide, alkyd resin coating which may be either air dried or baked at temperatures up to 300 F. This is a so-called heat-resistant paint and is oftentimes used as a coating for low-temperature stacks. It also can be used as either a primer or a finish coat under outdoor conditions where corrosion is not too severe. A minimum weight per gallon of this product is specified at 16 pounds.

Type 3 is a zinc dust, zinc oxide, phenolic resin paint which may be air dried or baked, again at temperatures up to 300 F. Because of the phenolic resin, this coating may be used for water immersion and other similar conditions. It has a minimum weight of 16.4 pounds per gallon. (These are all current Federal specifications.)

During the 1930s, many other resin and oil types were used with zinc dust. These included such materials as polystyrene, chlorinated rubber, vinyl resins, and similar products. It was not until considerably later that zinc dust was used with epoxy resins.

The degree of variation in the performance of coatings depended primarily on the vehicle. In general, the performance was in direct proportion to the resistance of the vehicle to corrosive conditions. Some early tests of the organic zinc-rich coatings illustrate the type of variable results which were obtained. The following test directly compared zinc-rich coatings made from drying oil, chlorinated rubber, styrene, and epoxy. The equivalent formulations were exposed in seawater immersion and in a seacoast atmosphere. The results were as follows:

1. Immersion in Florida Seawater, Four Years: (a) Drying Oil-Type Zinc-Rich Coating—Coating was completely dissipated, panel was solid rust; (b) Chlorinated Rubber Zinc-Rich Coating—Rust starting at scribe and edges, some active anode-type corrosion along edges; (c) Styrene-Type Zinc-Rich Coating—Severe edge pitting, scribe badly corroded with pits 3/4 of an inch wide, one or more penetrating the panel; and (d) Epoxy Zinc-Rich Coating—Edges corroding with active pitting, scribe has several small pits.

2. Seacoast Atmosphere, Eight Years and Six Months: (a) Drying Oil-Type Zinc-Rich Coating—General pinpoint rusting, 80% failure; (b) Chlorinated Rubber Zinc-Rich Coating—Light carbonate deposit, otherwise okay, no score corrosion; (c) Styrene-Type Zinc-Rich Coating—Light score rusting, pinpoint rusting on edge of panel; and (d) Epoxy Zinc-Rich Coating—Light score rusting, pinpoint rusting near bottom edge.

The effectiveness of the various coatings was rated as follows:

1. Chlorinated rubber zinc-rich coating
2. Epoxy zinc-rich coating
3. Styrene-type zinc-rich coating
4. Drying oil-type zinc-rich coating

The drying oil coating is obviously a failure because of the breakdown of the vehicle due to the alkaline zinc reaction products. However, the more resistant vehicles under other environmental conditions might have performed equally with the chlorinated rubber types.

The early commercial zinc-rich products which were used for corrosion resistance were based on chlorinated rubber, which today is one of the primary resins used in zinc-rich organic coatings. It is an ideal resin inasmuch as the resin itself is rather heavy, containing a sizable amount of chlorine, and thus aids in the suspension of the heavy zinc dust. It is primarily a lacquer-type material and therefore may be dried rapidly, providing a base coat which can then be quickly followed by other topcoats.

Later (1940s), epoxy resins became available and were soon found to provide a good vehicle for zinc-rich coatings. Today, there are two epoxy-type vehicles which, along with chlorinated rubber, make up the majority of the organic zinc-rich products presently in use. These are based on the epoxy polyamide resins and the phenoxy resins. The phenoxy resin is a long-chain thermoplastic resin which has epoxy groupings at the ends of the molecule. Each of these is a very resistant material, particularly to alkalies, so that any alkaline reaction of the zinc would not have any effect on the vehicles.

These organic vehicles also are generally dielectric. In fact, all three have good dielectric properties, which makes it necessary for the zinc to be in sufficient quantity so that when the coating is dry and ready for use, the zinc is in particle-to-particle contact throughout the film. It is generally conceded that the dry film in an organic zinc-rich coating must contain 90 to 95% zinc by weight in order to have the zinc in particle-to-particle contact. The

TABLE 6.3 — Summary of Zinc Coating Properties

Process	Typical Products & Specifications	Zinc Coating Weight Range
GALVANIZING	Sheet & Strip ASTM A525-65T	1.25-2.75 oz./sq. ft. of sheet
Application	Roofing Sheets ASTM A361-65T	1.25-2.75 oz./sq. ft. of sheet
Hot dip galvanizing is the most commonly used process for coating steel with zinc. It is employed on a wide range of items from heavy structurals to house hardware, individual fasteners, etc.	Structural Sheets ASTM A446-65T	1.25-2.75 oz./sq. ft. of sheet
Technique	Wrought Iron ASTM A163-63	1.25-2.75 oz./sq. ft. of sheet
Two hot dip galvanizing techniques are available:		
1. Continuous Galvanizing of Sheet, Strip, or Wire. Modern continuous galvanizing allows uninterrupted passage of carefully prepared steel sheet, strip, or wire through a bath of molten zinc. This results in an exceptionally good zinc-to-steel bond.	Tube & Pipe ASTM A120-65	2.00 oz./sq. ft.
2. Hot Dip Galvanizing (after fabrication). Items to be coated are dipped into a bath of molten zinc. This technique is particularly suitable to irregularly shaped articles.	Wire Strand ASTM A475-62T	0.15-3.00 oz./sq. ft.
Properties	Farm & Right-of-Way Fencing ASTM A116-65	0.20-0.80 oz./sq. ft.
1. Galvanized coatings are metallurgically bonded to the iron or steel base. Length of the coating's protective life is directly proportional to the coating thickness applied.		
2. Properly treated galvanized surfaces form an ideal, permanently bonded base for paint.	Chain Link Fence Fabric ASTM A392-63T	3.2-2.0 oz./sq. ft.
3. Galvanized coatings withstand rough usage on applications such as highway guard rails.	Prods. fabricated from rolled, pressed, forged steel shapes, plates brass & strip ASTM A123-65	2.0-2.3 oz./sq. ft
4. Continuous galvanized sheet can be severely formed without damage to the coating.		
5. Prepainted galvanized steel sheet can be fabricated without damage to the paint film. A range of colors is available.	Coatings, nuts, bolts, washers, pole line hardware ASTM A153-65	1.0-2.00 oz./sq. ft.
6. Prepainted galvanized siding is available for residential or industrial applications.	Transmission Tower Nuts & Bolts ASTM A394-65	1.25 oz./sq. ft.
	Assembled Steel Prods. ASTM A386-65	1.00-2.00 oz./sq. ft.
ELECTROGALVANIZING	Sheet & Industrial fasteners ASTM A164-55	0.00015-0.001 in.
Application	Chain link fence fabric woven after plating ASTM A392-63T (hot dip also)	1.2-2.0 oz./sq. ft.
Generally used for wire, conduit, hardware, and fasteners.		
Technique		
Zinc coating is electrodeposited. Usually thin coatings, but heavier coats can be built up.		
Properties		
Bright, ductile adherent coating, when phosphate treated provides an excellent base for paints and finishes.		
METALLIZING		
Application	Large structures, bridges, fabricated assemblies, conduit U.S. MIL. SPEC. M6874-1950	0.003-0.016 in.
Generally used where:		
1. Hot dip galvanizing unavailable		
2. No distortion of welded sections is permissible		
3. High-alloy steels are used	American Welding Soc. C2.2.52	0.001 in. thickness equals 0.5 oz./sq. ft. of surface
4. Coating thicknesses must vary		
5. Repairs to galvanized surfaces needed		
Technique		
After shot blast cleaning, molten zinc is sprayed on surface. Zinc may be powder, wire, or molten.		
Properties		
Coating may be in shop or at site. Excellent as paint base.		

NOTE: (1) Coating weight requirements on sheet and strip are in ounces per sq. ft. of sheet (both sides inclusive, i.e., 2 sq. ft. of surface); (2) "Seal of Quality" galvanized steel sheet carries 2 ounces of zinc coating to meet the specification of Zinc Institute; and (3) One ounce of zinc per sq. ft. of surface is equivalent to a coating thickness of 0.0017 inches.

[SOURCE: Zinc Controls Corrosion, 21-51-10M/71, Zinc Institute, Inc., New York, NY, p. 11 (1971)]

zinc particles must be in contact from the base metal clear through to the surface of the coating in order for the coating to be conductive and to prevent corrosion by cathodic protection (Figure 6.6).

There is little chemistry involved in the formation or

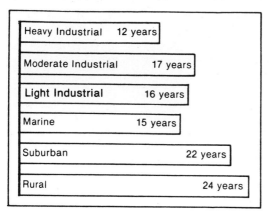

Transmission Towers

Heavy Industrial	12 years
Moderate Industrial	17 years
Light Industrial	16 years
Marine	15 years
Suburban	22 years
Rural	24 years

Heavy Industrial	9 years
Moderate Industrial	16 years
Light Industrial	18 years
Marine	19 years
Suburban	21 years
Rural	24 years

Substations

FIGURE 6.4 — Average service life—Years before first painting of galvanized transmission towers (75 companies) and of galvanized substations (87 companies). [SOURCE: Zinc Controls Corrosion, Z1-51-10M/71, Zinc Institute, Inc., New York, NY, p. 11 (1971).]

FIGURE 6.5 — Thousands of edges, corners, and bolt heads make this electric substation extremely difficult to protect. Inorganic zinc or galvanizing, applied before erection, is a practical answer to long term corrosion protection.

formulation of organic zinc-rich coatings other than to provide a resistant vehicle which does not react with the zinc. The chemistry consists primarily of the curing of the vehicle. The zinc exists within the film as a heavy loading of pigment. The formation of the film is therefore very similar to any organic coating, with the exception that the volume of pigment within the vehicle is, of necessity, high. Once the coating is formed, only the zinc on the surface reacts to form ions to provide cathodic protection. Once ionized, it may react with other atmospheric ions to form such compounds as zinc oxide, zinc hydroxide, zinc carbonates, etc., on the surface of the coating.

Characteristics of Organic Zinc-Rich Coatings

Organic zinc-rich primers have some important characteristics, particularly as related to and compared with the inorganic zinc coatings.

Compatibility

The most outstanding characteristic of organic zinc-rich primers is their compatibility with both organic and steel surfaces. This is extremely important in coating repair and may be important during original construction where many different types of surfaces are involved which all require excellent corrosion protection. Organic zinc-rich coatings may extend from bare steel out over various primers and topcoats (organic or inorganic) and provide adhesion to each surface. They should be compatible with oleoresinous topcoats as well as synthetic resin types, which is not possible with inorganic zinc coatings.

Cathodic Protection

Organic zinc-rich coatings do provide cathodic protection, as long as the formulation is such that particle-to-particle zinc contact is maintained.

Application

With an organic binder, the application of organic zinc-rich coatings covers a very wide range of conditions. Organic binders may be fast or slow drying, and curing conditions can vary widely, depending on the requirements of the application.

Binder Resistance

A binder in an organic zinc-rich primer may be more or less chemical-resistant, depending on the binder and its use requirements. Chlorinated rubber, epoxies, or vinyls

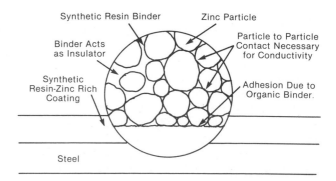

FIGURE 6.6 — Particle-to-particle contact of organic zinc-rich coating.

Corrosion Prevention by Protective Coatings

are those which would provide chemical resistance; silicones provide weather and temperature resistance; while some alkyds provide intermediate temperature resistance and good weather durability. Numerous organic binders can be used, depending on the intended use of the zinc-rich coating.

Surface Preparation

It often is claimed that the organic zinc-rich primers are less subject to critical surface preparation than are the inorganic zinc materials. This may be true for the initial application since they would be less subject to problems from organic contamination. On the other hand, eliminating the organic contamination factor, light rust coloration on the steel surface to be coated may be more easily tolerated by an inorganic zinc coating than by an organic-based material. This is due to the ability of the inorganic zinc vehicle to thoroughly wet the oxide and to react with it.

Inorganic Zinc Coatings

Zinc has proven its effectiveness as a corrosion-resistant coating, not only through galvanizing, but to a lesser degree through organic zinc-rich coatings. Nevertheless, prior to the advent of inorganic zinc coatings, maintaining industrial and marine structures was a difficult and continuous chore. There was no effective long-time protection for large or existing new structures. Coatings were applied and reapplied at short intervals in order to maintain structures in a safe condition for satisfactory operation.

Inorganic zinc coatings, as a single coat applied to a clean steel surface, have completely changed the old concept of paint and repaint on a continued basis in order to maintain critical industrial and marine structures. Inorganic zinc has proven its resistance to corrosion in the Arctic, in the Tropics, and in all of the intermediate climates. One of its first uses was in the marine area, on offshore structures and ships, as well as on shore structures along seacoasts throughout the world. It continues to be the standard of the industry in these areas for long-term corrosion protection.

Inorganic zinc coatings have been one of the true technological developments of our time which has made a real and positive impact on modern society. While this impact is not as dramatic as that of television or space travel, inorganic zinc coatings nevertheless have made a solid contribution towards the preservation of scarce materials, thus eliminating the need for the replacement of existing structures, reducing the cost of steel structures, saving manpower, reducing the energy required for metal replacement, and providing new structures with a substantial increase in life expectancy. A good example is the refined oil ship tankers mentioned in an earlier chapter, which not only have an increased lifetime when constructed of the same quality materials and protected inside and out with inorganic zinc coatings, but also are able to spend their longer life in the transportation of maximum value cargo.

Development of Inorganic Zinc Coatings

Credit for this contribution of inorganic zinc coatings is given to Victor Nightingall of Australia who first mixed zinc dust with sodium silicate to make a coating that would replace galvanizing. He spent several of the last years of his life studying ways in which a chemical compound could be made to duplicate the durability of zinc ores and at the same time provide long-time corrosion protection. Nightingall's goal is well stated in one of the opening paragraphs in his United States patent on inorganic coatings.

> Protective coatings for metals surfaces must meet several requirements if they are to be considered effective. Such coatings must be continuous and impervious to corrosive elements, must be sufficiently hard to withstand the normal abrasion and mechanical shock to which metal articles are frequently subjected, and must have a high degrees of adherence to the surface of the metal article. If a coating does not adequately meet these requirements, it fails to qualify as a protective coating. This is particularly true with respect to protective coatings for ferrous metal articles where a single flaw in the coating will permit widespread corrosion and even disintegration of the ferrous article beneath the coating.[2]

A reference was made in a previous paragraph of the patent to zinc ores. It was Nightingall's idea that if he could form a coating which would closely simulate the chemical characteristics of willemite or zinc silicate, then he would be able to accomplish his goal. He was a man of purpose who set out to make his approach to the problem work. The following is a quotation from his paper on dimetalization in 1940.

> When the right ingredients had been determined by research in one of the zinc-iron silicate ores, it was found that a coating painted or spread on a pickled iron surface would dry in quite a short period of time, usually about half an hour; and that this coating, when dry, would have placed all the chemicals in their right proportions and molecularly closely associated together, but this was nothing more than a mechanical method of placing the requirements of the chemical reaction in position on the iron surface, and if in this stage, the article was again wetted with water, the coat would dissolve and the zinc washed away. To complete the chemical reaction, it was necessary to place the completed iron goods in an oven and heat to a temperature over and above 180° F, the reason for this being that to complete the molecular combination of a silica, zinc and iron, it is necessary that this should be done in the presence of a body that will bring about a release of silicic acid when a rapid silication of silica-zinc iron takes place; such a body was found to be carbon dioxide.[3]

He did this by mixing a rather alkaline sodium silicate with zinc dust and a small amount of red lead and sodium bicarbonate. This was the start of the inorganic zinc coatings as we known them today. The billion or more square feet of surface which is presently protected by these zinc coatings was the result of the effectiveness of these early and very crude products.

The first well-documented field test of this product was installed in 1942 and was a section of steel pipe in the Woronora pipeline which was part of the water system for the city of Sydney, Australia. The line ran above ground, close to the bay, and within a few feet of some large oyster beds. The pipe section was inspected in 1950 and was found in perfect condition (Figures 6.7 and 6.8). Many similar trials were made on sections of pipe, bridges, gasoline storage tanks, and even part of a water seal on a gas holder. To justify Nightingall's own conviction and to prove to the world that the product was reliable, its first application on 250 miles of above-ground 30-inch exterior coated steel pipe was guaranteed to the Australian government for 20 years.

The Morgan Wyalla Pipeline

The coating of the Morgan Wyalla pipeline, which is the 250-mile pipeline now famous in inorganic silicate history, was negotiated in 1941-42 and completed in 1944.[1] The negotiations included the 20-year guarantee on the coatings. This was completed with only the experience with the Woronora pipe section to rely on. The Woronora and the Morgan Wyalla pipelines were thus the beginning of the present era of inorganic zinc coatings.

Today, after almost 40 years, the Morgan Wyalla pipe is still in excellent condition, with its guarantee never having been challenged. In fact, a parallel line was constructed and an inorganic zinc coating of the same type was used for its protective coating. The pipeline runs through the Australian bush country and grasslands with its large population of emus, kangaroos, and sheep (Figure 6.9). The corrosion there is not severe, but since the line is above ground it is subject to severe brush fires. (Both the tall grass and the eucalyptus brush are explosive in the dry season.) Organic paints do not stand a chance under these conditions. The pipe also runs through a humid zone along the Spencer Gulf where it is subject to salt marshes and salt beds. The crystallization of the salt, in fact, has severely spalled the concrete pipe supports. The exterior of the all-welded pipe is subjected to practically all forms of corrosive atmosphere except direct immersion in water or salt water (Figure 6.10).

The coating used was a very simple formula made at the time the pipe was coated. As for surface preparation, the pipe was pickled, freed of mill scale (Figure 6.11), scrubbed with fiber brushes to remove the black pickling deposit, and rinsed with dilute phosphoric acid (Figure 6.12).

As soon as this process was completed, the coating was mixed by weighting about 10 pounds of sodium silicate and a small amount of sodium bicarbonate into a bucket. Twenty pounds of zinc dust and about two

FIGURE 6.7 — First application of inorganic zinc coating on above ground pipeline near Sydney, Australia.

FIGURE 6.9 — Morgan Wyalla pipeline in Australian bush country. Inorganic zinc is used as protection against grass and brush fires.

FIGURE 6.8 — Inspection of original field test of an inorganic zinc coating after 8 years of exposure.

[1]The Morgan Wyalla waterline runs from a small pumping station at a point called Morgan on the Murray River in South Australia across the Australian bush to the Spencer Gulf and a town named Port Perie. Here, it runs around the Gulf to the site of a steel mill at Wyalla. The steel mill was installed as a World War II measure, and the Morgan Wyalla pipeline supplied it with water.

Corrosion Prevention by Protective Coatings

FIGURE 6.10 — Morgan Wyalla pipeline crossing salt marshes near the Spencer Gulf in Australia.

FIGURE 6.12 — Rinsing and scrubbing pipe before coating application.

FIGURE 6.11 — Pickling steel pipe for inorganic zinc application.

FIGURE 6.13 — Brushing a zinc-sodium silicate mixture onto a pipe with 6-inch brushes.

pounds of red lead followed, and the whole thing was stirred vigorously with a stick. When mixed, it was then applied to the exterior pipe surface with 6-inch brushes (Figure 6.13). The coating was well-worked onto the surface to eliminate holidays. The pipe was then moved in front of a large burner which blew flame, hot air, and combustion products into the pipe at one end and out the other (Figure 6.14). This procedure raised the temperature of the steel pipe to be between 300 and 500 F. The time involved amounted to approximately half an hour per pipe.

Since the Morgan Wyalla line was designed to transport water, its interior also required coating. This was done by spinning a concrete lining on the interior of the pipe. The pipe was supported on rubber belts, which tended to polish the zinc coating where it spun on the belt (Figure 6.15). There was apparently no loss of thickness at this point, and, after 28 years of exposure, there is no evidence of coating inperfection or breakdown in these areas (Figure 6.16).

The Nightengall-McKenzic conviction that the zinc coating was a permanent one was entirely borne out by the guarantee period passing in 1965. In 1970 the South Australian Government duplicated the original Morgan Wyalla line, using the same type of, although somewhat more refined, exterior coating. A section of the first line

near Wyalla was inspected firsthand in 1972 after 28 years of service. The pipe was in perfect condition and showed no evidence of rusting, chalking, or any change due to long exposure to the atmosphere (Figures 6.16 and 6.17). This particular area was adjacent to the Spencer Gulf as well as to the steel plant. It therefore was subject to both a mild marine atmosphere and an industrial atmosphere. Even the field welds which were touched up after installation and simply allowed to dry, showed no corrosion.

Early U.S. Research Work

Following the inspection of the Morgan Wyalla line in 1949, it was recommended that the dimetalization process be brought to the United States. As mentioned previously, the original Australian inorganic zinc silicate required stoving (heating to 350° for half an hour or more). This simple Australian concept was brought to the United States in 1950. While stoving proved effective on several different installations, it was soon noted to have a severe limitation. While it was a satisfactory method for steel plates and even small structures which could be placed in an oven to increase the temperature to the proper point, it could not be applied to large structures, such as offshore drilling platforms, ships, large storage tanks, refinery

FIGURE 6.14 — Stoving a coated pipe by flame and combustion products from large open burner. The coating was cured in approximately a half-hour.

FIGURE 6.15 — The burnished areas on a zinc silicate coating were caused by pipe rotation on rubber belts at about 400 RPM during the concrete lining process.

FIGURE 6.16 — Morgan Wyalla pipeline in 1972 after 28 years of service. A steel mill can be seen in the background.

FIGURE 6.17 — Close-up of Morgan Wyalla pipeline after 28 years of service. A longitudinal weld displays the excellent protection offered by the original, rather crude coating.

structures, or even to relatively small equipment such as workboats, fishing boats, etc.

The Australians had completed several large storage tanks with zinc silicate, but these were done plate by plate, with the individual pieces being placed in an oven for stoving. This same process was still being used in Western Australia in 1972. It became obvious, however, that in order for a coating with the advantages of inorganic zinc to be widely used, it had to be more easily and effectively applied, and essentially by existing coating procedures.

Several years of intensive research was carried out in the United States to develop the first practical inorganic zinc coating which could be applied without stoving and to any existing structure by ordinary painting methods. There were a number of early attempts to apply the coat-

ing by spray and then to cure it through the use of heat. One application was on approximately one mile of bypass water piping which was coated on the exterior and heated in the same manner as the pipe had been heated in Australia. This pipe and the coating lasted for several years, but because it was handled by bulldozers and other similar equipment, both the coating and the pipe wore out after a few years of service.

Several installations also were made where portable weed burners were used to bring about a cure of the coating. This process, however, was clumsy and dangerous. While the steel temperature never did rise a great deal, the water vapor and the carbon dioxide in the combustion products undoubtedly provided some limited cure. Some of the applications actually lasted several years before pinpoint rust started.

The original laboratory work in the United States was an intensive study of the chemical processes involved in silicate chemistry, with the object of obtaining a cure for the zinc silicate which would take place at ambient temperatures and under the widely varying weather conditions that exist throughout the industrial and marine world. This not only included sodium silicate, but also studies of potassium silicate, lithium silicate, ethyl silicate butyl titanate, borates, phosphates, and other similar chemicals.

There were many attempts to use various salt solutions as a cure. These were primarily fairly concentrated solutions in water or alcohol of magnesium chloride, zinc chloride, aluminum chloride, and some soluble phosphates. One manufacturer during this period actually recommended that the coating be washed with seawater in order to bring about a cure. These trials, including the seawater, had some basis in fact, as the following equation suggests.

$$2Zn + NaCl + 3H_2O \rightarrow$$

$$ZnOZnCl_2 + NaOH + 2H_2 \qquad (6.1)$$

The zinc oxychloride or a basic zinc chloride is insoluble, and this material, together with zinc carbonate which would undoubtedly be part of the reaction products, could provide a sufficiently insoluble product so that the coating would hold together until further zinc silicate reactions could take place.

Actually, none of these curing procedures worked very well and only indicated that a post-cure was possible. Finally, it was determined that a solution of phosphoric acid neutralized with an amine would hydrolyze when in contact with the zinc silicate so that a slow and thorough cure was obtained. The final post-cure material was dibutylamine phosphate, and the resulting product using this phosphate cure had all of the good characteristics of the stoved inorganic zinc, as originally conceived by Victor Nightingall. This product marked the beginning of the inorganic zinc revolution, and it is still recognized today as the standard of the industry with an exceptionally long life under very difficult corrosion conditions. That the post-cured product was equivalent to the heat-cured Australian formulation is demonstrated in Figures 6.18 and 6.19.

Research on inorganic zinc coatings has continued by many companies throughout the world. Several of the materials tried include different sodium oxide-silica ratios of sodium silicate, potassium silicate, lithium silicate, ammonium silicate, various phosphates, titanates, borates, zinc oxychlorides, magnesium oxychlorides, coloidal silica, various silica colloids in solvents, ethyl silicates, cellosolve silicates, and combinations of these.

Much of the research effort up to the present time has been directed toward simplifying the product and its application and developing products which will self-cure. This has been accomplished effectively by many different companies and through the use of a number of different basic raw materials. Not only has the research resulted in self-curing products, but also in single-container or single-package products where the silicate and the zinc are combined in a single container ready for application in a

FIGURE 6.18 — The original Australian formulation (stoved) shows pinpoint failure just beginning after 25 years of exposure at the 80-foot lot at Kure Beach, NC.

FIGURE 6.19 — Post-cured inorganic zinc after 25 years of exposure at the 80-foot lot at Kure Beach, NC. The damaged area on the right was the result of a hurricane 10 years after installation. Only limited rusting has occurred after several years of marine exposure.

similar manner to standard paints. One of the early materials of this sort was a dry mixture of dry sodium silicate and zinc powder to which water was added just prior to application. This was an interesting concept which was overshadowed by a number of other products developed at approximately the same time.

Another interesting product was the development of the zinc phosphate coating. This coating actually went beyond the research stage to the point where it was applied to some substantial structures. It also was self-curing, and tests of the product show it to be unaffected even after 15 years in a marine atmosphere. It again was overshadowed by some of the other developments, plus the application of the zinc phosphate was not as easy or straightforward as some of the other products.

The use of ethyl silicate was one of the developments that took place not long after the development of the original post-cured inorganic zinc and was patented in the late 1950s or early 1960s. It also served as the basis of the first self-curing zinc coating. Actually, ethyl silicate had been used as a binder for other pigments during the early 1950s; however, these products were never widely marketed. The ethyl silicate product has been extensively used throughout the world and has been modified by many different companies to provide products which they believe to be superior to others. Basically, however, the products are all formulated with partially hydrolyzed organic silicates and modified by various small amounts of soluble metal salts.

The effective products which are presently on the market are based on sodium silicate, ethyl silicate, potassium silicate, lithium silicate, and colloidal silica. The single-package systems are primarily based on partially hydrolyzed ethyl silicate, as are a number of the two-package products. That many of these products have proven effective is indicated in Figures 6.20 through 6.23. These test panels, exposed for eight years in a marine atmosphere, compare four different inorganic zinc vehicles. The protection offered by each of these applied in an equivalent thickness appears to be almost equivalent after the eight-year exposure period at Kure Beach, North Carolina. Only the colloidal silicate-based product shows beginning pinpoint corrosion.

Table 6.4 shows a series of interesting combinations of silicate coatings. Many of these have proven most practical and successful. Progress has therefore been made in the last forty years, from the original product which was made by mixing the individual ingredients just prior to application, to the present finished inorganic zinc products which are in a single package and may be used essentially like paint. These zinc coatings have spurred a coating revolution which has developed to the point where coating maintenance is minimized even under severe corrosive conditions. Painting crews are no longer retained aboard ship since any required repair or maintenance is done in drydock. Critical industrial structures have been in service for 20 years or more without any maintenance. Some of these products in a single coat have provided protection in severe atmospheres better and considerably longer than galvanized surfaces under the same conditions. Other inorganic zincs, when overcoated, have in-

creased the life of organic topcoats several times over that of the organic material alone.

Composition of Inorganic Zinc Coatings

Inorganic zinc coatings, whether composed of sodium silicate, potassium silicate, lithium silicate, colloidal silica, or hydrolyzed organic silicates, are reactive materials from almost the moment that they are applied. Inorganic zinc coatings are in a state of constant change. While the degree of change depends to a great extent on the atmosphere to which they are exposed, there is nevertheless a slow continuing reaction which takes place up until the time when the zinc in the coating has practically been consumed in protecting the steel over which it is applied. In certain cases, the surface of inorganic zinc coatings can become inactivated by the accumulation of zinc salts on the coating surface to the point where the coating becomes an inert film rather than one which protects by cathodic protection.

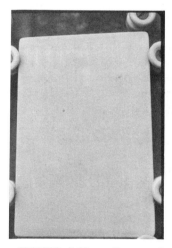

FIGURE 6.20 — Post-cured sodium silicate zinc coating after 8 years of exposure in a marine atmosphere.

FIGURE 6.21 — Inorganic zinc coating based on colloidal silica after 8 years of exposure in a marine atmosphere.

FIGURE 6.22 — Self-cured water-base inorganic zinc silicate coating after 8 years of exposure in a marine atmosphere.

FIGURE 6.23 — Ethyl silicate-based inorganic zinc coating after 8 years of exposure in a marine atmosphere.

TABLE 6.4 — Typical Examples of Various Zinc-Rich Coating Formulations

Type	Vehicle	Pigment	Weight Ratio Pigment/Vehicle
Water-Based Inorganic Zinc-Rich Coatings			
Post-Cured			
A	3.2 ratio sodium silicate; 22% SiO_2; sodium dichromate	Zinc dust + red lead	2.8
B	3.2 ratio sodium silicate; 24% SiO_2; potassium dichromate	Zinc dust	3.2
Self-Cure Potassium Silicate			
A	2.9 ratio potassium silicate; 14% SiO_2; manganese dioxide; sodium dichromate	Zinc dust	2.9
B	2.4 ratio potassium silicate; 9.25% SiO_2; acrylic emulsion	Zinc dust	2.0
C	2.8 ratio potassium silicate; 18% SiO_2; quaternary ammonium hydroxide; soluble amine; carbon black	Zinc dust + red lead	2.8
D	3.2 ratio potassium silicate; 15% SiO_2; quaternary ammonium hydroxide; soluble amine; carbon black	Zinc dust	2.5
Self-Cure Lithium Silicate	Lithium-sodium silicate; 19% SiO_2; sodium dichromate	Zinc dust + iron oxide	3.3
Self-Cure Silica Sol	Silica sol; 32% SiO_2; soluble amine potassium dichromate; carbon black	Zinc dust + red lead	4.1
Self-Cure Quaternary Ammonium Silicate			
A	Quaternary ammonium silicate; 32% SiO_2	Zinc dust	2.5
B	Quaternary ammonium silicate; sodium silicate; 20% SiO_2	Zinc dust	2.5
Solvent-Based Inorganic Zinc-Rich Coatings			
Self-Cure			
A	Partly hydrolyzed ethyl silicate; 10% SiO_2; clay fillers	Zinc dust	2.2
B	Partly hydrolyzed ethyl silicate; 22% SiO_2	Zinc dust	3.4
C	Basic hydrolyzed ethyl silicate; 15% SiO_2; clay fillers	Zinc dust + iron oxide	2.4
D	Polyol-Alkyl Silicate; 20% SiO_2	Zinc dust	2.2

(SOURCE: Steel Structures Painting Manual, Vol. 1, Chapter 5, Section 2, Zinc-Rich Primers, Steel Structures Painting Council, 1966.)

Some of the typical zinc reactions are:

$$Zn\ (metal) + H_2O \rightarrow Zn^{++} + 2e^- \qquad (6.2)$$

(the normal corrosion reaction for zinc)

$$Zn + 2H_2O \rightarrow Zn(OH_2) + H_2 \qquad (6.3)$$

$$Zn + H_2O + CO_2 \rightarrow ZnCO_3 + H_2 \qquad (6.4)$$

$$2Zn + 2NaCl + 3H_2O \rightarrow$$

$$ZnOZnCl_2 + 2NaOH + H_2 \qquad (6.5)$$

Galvanized surfaces or pure zinc react with the carbon dioxide and oxygen in the air to form zinc carbonate or zinc oxide on the surface almost as soon as they come out of the galvanizing bath. The original bright zinc surface, after a few days in the weather, becomes a dull gray, and will even, at times, accumulate a rather substantial quantity of white salts on the surface. While the initial reaction takes place rather quickly and visibly, it is entirely a surface phenomenon and only progresses at an extremely slow rate over many years, even in fairly corrosive atmospheres.

The inorganic zinc coatings are considerably more complex than metallic zinc in their reactions to the atmosphere. They are composed of powdered metallic zinc which is mixed into a complex silicate solution so that silicate chemistry is actually the key to the reactions and to the cure which takes place within the inorganic zinc coating. The chemistry of silica which applies directly to the inorganic zinc coatings is explained by E. G. Rochow in a paragraph on polysilicic acids as follows.

Corrosion-Resistant Zinc Coatings

Since silicon forms no classical double bonds to oxygen, it is characteristic of all non-ionic compounds of silicon with oxygen to form siloxane chains and networks in which each oxygen atom is bound to two different silicon atoms. It follows that metasilicic acid and disilicic acid must be polymeric, and that it is but a short step from their structures (with a few terminal OH groups) to those of the polysilicic acids (with still fewer terminal OH groups). Thus the polymeric H_2SiO_3 may be represented as:

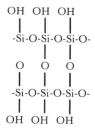

$$\begin{array}{ccccc} OH & OH & OH & OH & OH \\ | & | & | & | & | \\ -Si-O-Si-O-Si-O-Si-O-Si-O- \\ | & | & | & | & | \\ OH & OH & OH & OH & OH \end{array}$$

and that cross-linking with neighboring chains will correspond to condensed or polysilicic acids with lower – OH content:

$$\begin{array}{ccc} OH & OH & OH \\ | & | & | \\ -Si-O-Si-O-Si-O- \\ | & | & | \\ O & O & O \\ | & | & | \\ -Si-O-Si-O-Si-O- \\ | & | & | \\ OH & OH & OH \end{array}$$

The cross-linking can be occasional (in the lower polysilicic acids), frequent (in the more highly condensed acids) or even complete (in SiO_2) itself. Since Si-OH groups are inherently acidic, as is shown so clearly in the organosilanols (all of which are more acidic than their carbinol counterparts), all condensed HO-bearing siloxane structures may be expected to be weakly acidic in relation to proportion of Si-OH groups they contain.

In the more condensed polysilicic acids it is quite possible that the interior of each small particle is nearly completely cross-linked with interlocking Si-O-Si chains, but still the surface oxygen atoms necessarily are unsaturated and will accept hydrogen, so that the outside of the particle must still be covered with weakly acidic Si-OH groups. In this sense there is no distinction between highly condensed polysilicic acids and colloidal silica.[4]

With this chemistry in mind, it becomes obvious that in spite of the different starting point, *e.g.,* sodium silicate, potassium silicate, lithium silicate, ethyl silicate, or colloidal silica, the ultimate reaction product as it exists on the steel surface is quite similar for each one of the coatings. Each vehicle, water soluble or organic silicate or colloidal silica, reacts or hydrolyzes to form a polymer of silicic acid, and when zinc is added to the system, a silica oxygen zinc polymer is created. This combination is very insoluble and forms the strong matrix surrounding the zinc powder to form the coating. Since the silicic acid is a mild acid, and since it is known that ions will be formed when iron is exposed to water or moisture, it is reasonable to assume that in addition to the zinc reacting with the silicic acid, iron ions from the sandblasted steel surface also will react with the silicic acid to form an insoluble reaction product at the interface of the metal and the silicate coating. This provides the basis for the adhesion of the coatings as well as for their excellent resistance to undercutting by corrosion.

Reactions of Different Silicate Types

The actual reaction, starting from the time the coating is applied to the surface, is somewhat different for each type of silicate, even though the end product appears to be similar for each of the various zinc silicates. The exception is the post-cured coating. Coatings based on sodium or potassium silicate, or ethyl silicate or colloidal silica (since the latter is usually stabilized with sodium or potassium hydroxide) react as follows. The initial reaction is that of concentration of the ingredients as the water evaporates from the coating. This brings the zinc and the silica compounds into close relationship and provides for the initial deposition of the coating on the surface. This initial reaction is an important one since this is the time at which the coating thoroughly wets the metal surface and is in intimate contact with the steel. Should there be organic contamination on the surface, this wetting does not take place and the coating will have no adherence, beginning at the initial deposition of the coating.

The second reaction to occur takes place shortly after the coating deposition. This is the initial insolubilization of the coating caused by the reaction of zinc ions with the silicic acid to form the initial zinc silicate. Only a few zinc ions are necessary for the silicic acid to become insoluble, even though there are a number of other acid groups on the polymer which can react at a later time. The coating is now a solid insoluble coating on the metal surface. Even so, it is not completely reacted.

The third stage of the reaction is one which occurs over a long period of time, usually taking many days, weeks, or months to come to completion. The third reaction is the continuing activity of carbonic acid, formed by carbon dioxide and moisture in the air, acting on and within the coating to complete the formation of the zinc silicate matrix. The coating, as originally formed on the surface, is shown in Figure 6.24.

The action which forms the zinc silicate matrix within the coating and surrounding the zinc particles does not react all of the zinc within the coating. Zinc ions are formed from the surface of the zinc particles. These diffuse into the silicate gel forming the insoluble matrix; however, there is also a layer of zinc silicate surrounding each of the zinc particles. In many ways this is the way cement reacts in concrete. The water hydrates the cement

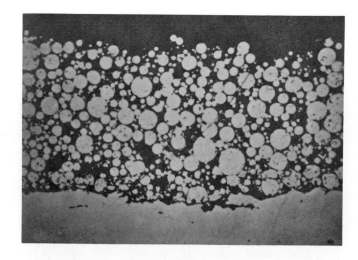

FIGURE 6.24 — Cross section of an inorganic zinc coating soon after coating application, showing the character of the sandblasted surface (bottom).

Corrosion Prevention by Protective Coatings

particle primarily on the surface and forms a matrix between and around the sand and rock particles in concrete. The center of the cement particle remains unhydrated until the concrete is cured for a month or more. Likewise, most of the zinc is left as metallic zinc within the coating (Figure 6.25).

This does not mean that the original coating does not contain pores or is not porous. It certainly is porous, as evidenced by the problems involved in applying organic coatings over a newly formed inorganic zinc. Air reactions, *e.g.*, the reaction of carbon dioxide, water, and carbonic acid, continue, forming zinc carbonate and zinc hydroxide within the coating and within the porous areas within the coating. This is the reaction which decreases the porosity over a period of time and makes the final coating into a continuous film (Figure 6.26).

Reactions in the Formulation of Various Inorganic Zinc Coatings

There are essentially four different general chemical reactions which take place in the formation of various zinc coatings. The first, the reactions of the post-cured inorganic zincs, are as follows.

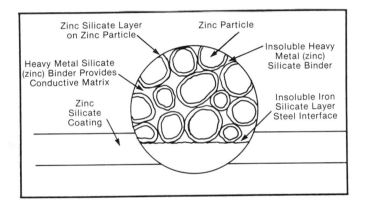

FIGURE 6.25 — The zinc silicate matrix within the inorganic zinc coating and surrounding the zinc particles does not use all of the zinc within the coating.

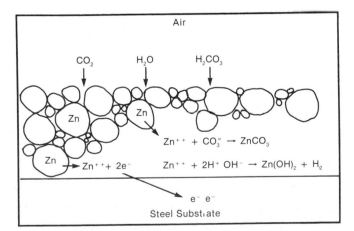

FIGURE 6.26 — Continuing reaction of CO_2 and H_2O within the inorganic zinc coating. These reactions provide the mechanism for continued corrosion protection by cathodic protection, as well as filling the coating voids with the zinc reaction products.

1. The first step is primarily the dehydration process in which the water, which is the basic solvent for the silicate, evaporates from the clean steel surface, leaving the zinc and the alkaline silicate on the surface as a coating. At this point, when the coating is free of water, it is very hard and metallic. Scraping with a coin merely polishes the zinc in the coating. It is not, however, cured to water insolubility.

2. Once the initial drying process has taken place, a solution of an acid amine phosphate salt is sprayed or brushed over the surface of the coating in order to thoroughly wet out the surface with the curing compound.

3. The next reaction consists of the acid salt gradually neutralizing the sodium in the sodium silicate solution and creating a mildly acidic condition on and within the coating.

4. This neutralization of the sodium opens up the reactive silica groups to which the sodium was attached.

5. The addition of the acidic phosphate solution on the surface of the zinc coating ionizes the zinc in the coating, as well as the lead oxide which is mixed with it. As this takes place, there occurs a rapid reaction between the active silica groups and the lead and zinc, which insolubilizes the silica matrix and forms some zinc and lead silicate polymers from the basic molecular structure of the silicate.

6. The acid phosphate also reacts with the ionized zinc and lead to form very insoluble zinc and lead phosphates. This reaction forms the zinc and lead phosphates within the silicate matrix. At this point, the coating has become insoluble to water and is unaffected by exposure to weather.

7. Following the application and reaction of the curing agent, all of the remaining soluble salts on the surface are removed with clean water. This eliminates the sodium phosphate and the amine salts so that the majority of all soluble materials are removed from the coating. The coating at this point is dense and nonporous, as well as being insoluble in water and unaffected by marine atmospheres. At this stage, it can be readily overcoated with organic coatings.

8. While insolubility has been achieved, along with resistance to weathering and marine activity, the curing of the coating continues. The complete cure comes gradually over a period of many months with the reaction of carbon dioxide and water on the coating further reacting the silica matrix with the zinc ions which are formed by the carbonic acid. This reaction is well known and increases the toughness, hardness, and adhesion of the coating over a period of many months or years.

Figure 6.27 traces the initial solution through the initial cure and on to the final carbon dioxide cure.

The second type of coating, the water-based sodium or potassium silicate self-curing coating, generally reacts as follows. First, the alkali silicates become concentrated through the evaporation of the water. This provides the initial drying and primary deposition of the coating. In this case, the alkali silicate is much less alkaline than might be the case in the post-cure coating, so that more of the silicate is in the form of a polysilicate acid. This allows for faster reaction of the zinc into the coating and forms

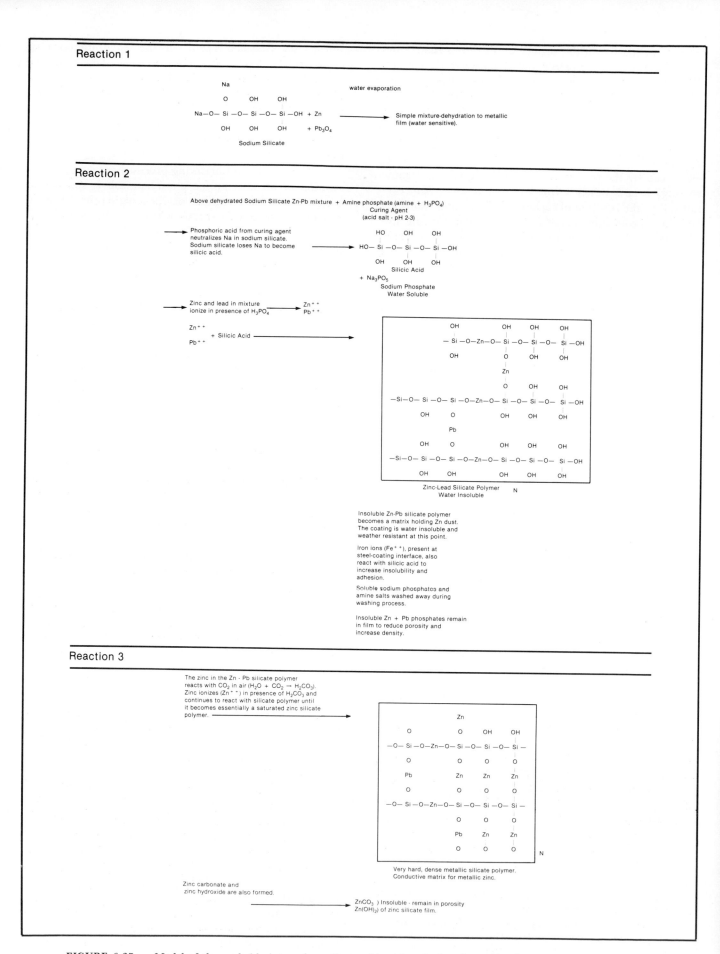

Reaction 1

Na
O   OH   OH   water evaporation
Na—O— Si —O— Si —O— Si —OH + Zn   Simple mixture-dehydration to metallic
                                                ⟶   film (water sensitive).
OH   OH   OH   + Pb₃O₄

Sodium Silicate

Reaction 2

Above dehydrated Sodium Silicate Zn-Pb mixture + Amine phosphate (amine + H₃PO₄)
Curing Agent
(acid salt · pH 2-3)

⟶ Phosphoric acid from curing agent   HO   OH   OH
   neutralizes Na in sodium silicate.
   Sodium silicate loses Na to become   ⟶   HO— Si —O— Si —O— Si —OH
   silicic acid.
                                                    OH   OH   OH
                                                       Silicic Acid
                                                  + Na₃PO₅
                                                    Sodium Phosphate
                                                    Water Soluble

⟶ Zinc and lead in mixture   Zn⁺⁺
   ionize in presence of H₃PO₄ ⟶   Pb⁺⁺

Zn⁺⁺
Pb⁺⁺   + Silicic Acid ⟶

              OH   OH   OH   OH
            — Si —O—Zn—O— Si —O— Si —O— Si —OH
              OH   O   OH   OH
                          Zn
                          O   OH   OH
      —Si—O— Si —O— Si —O—Zn—O— Si —O— Si —O— Si —OH
         OH   O   OH   OH   OH
               Pb
         OH   O   OH   OH   OH
      —Si—O— Si —O— Si —O—Zn—O— Si —O— Si —O— Si —OH
         OH   OH   OH   OH   OH   OH

   Zinc-Lead Silicate Polymer   ₙ
   Water Insoluble

Insoluble Zn-Pb silicate polymer
becomes a matrix holding Zn dust.
The coating is water insoluble and
weather resistant at this point.

Iron ions (Fe⁺⁺), present at
steel-coating interface, also
react with silicic acid to
increase insolubility and
adhesion.

Soluble sodium phosphates and
amine salts washed away during
washing process.

Insoluble Zn + Pb phosphates remain
in film to reduce porosity and
increase density.

Reaction 3

The zinc in the Zn · Pb silicate polymer
reacts with CO₂ in air (H₂O + CO₂ → H₂CO₃).
Zinc ionizes (Zn⁺⁺) in presence of H₂CO₃ and
continues to react with silicate polymer until
it becomes essentially a saturated zinc silicate
polymer.

                    Zn
          O   O   OH   OH
       —O— Si —O—Zn—O— Si —O— Si —O— Si —
          O   O   O   O
          Pb   Zn   Zn   Zn
          O   O   O   O
       —O— Si —O—Zn—O— Si —O— Si —O— Si —
                    O   O   O
          Pb   Zn   Zn   Zn
          O   O   O   O   ₙ

   Very hard, dense metallic silicate polymer.
   Conductive matrix for metallic zinc.

Zinc carbonate and
zinc hydroxide are also formed.

                              ⟶   ZnCO₃) Insoluble · remain in porosity
                                  Zn(OH)₂) of zinc silicate film.

FIGURE 6.27 — Model of the probable internal reactions taking place during the curing of a post-cured inorganic zinc coating.

the basis for the self-cure. Other metal ions also may be included to speed up this self-curing process.

The second reaction is the insolubilization of the silicate matrix through a reaction with zinc ions from the surface of zinc particles, and probably some reaction with iron ions from the sandblasted steel surface. As the zinc reacts into the silicate or silicic acid polymer, the coating becomes insoluble, forming a silica-oxygen-zinc polymer. This is the basis of the matrix which holds the zinc powder in place (Figure 6.24).

The third reaction which takes place in the coating is the one which occurs over a long period of time, making it a matter of days, months, or years before the final saturation of the silicate polymer with zinc occurs. This continuing reaction is the result of humidity from the air, condensation of moisture on the surface, or rain, which creates a very mild acidic condition and continues to reduce the alkali and aids in the ionization of the zinc so that it continues to react with the silicate acid polymer. This reaction gradually proceeds through the coating to the interface of the steel, increasing the adhesion of the coating to the steel surface and making the coating extremely dense and metal-like. Over a period of time, zinc carbonate and zinc hydroxide are formed on and within the coating (as was shown in Figure 6.26) to decrease any porosity which might exist and to form the film into a continuous coating.

The third type of inorganic coatings is the organic silicate type. The general reactions for formation of the coating are similar to the water-base self-cure products. There is, however, a sufficient difference to make it worthy of description. The first step consists of applying the hydrolyzed ethyl silicate and zinc mixture to a sandblasted steel surface. The initial reaction involves the evaporation of the solvent in the organic silicate vehicle, leaving the organic silicate zinc mixture on the metal surface.

In the second reaction, which is the one that takes place during application or shortly thereafter, the moisture from the air continues to hydrolyze the ethyl silicate to silicic acid. This then reacts with the zinc, and possibly other metal ions that have been added to the mixture, to give the coating its initial insolubility. The organic silicate coatings are noted for their rapid resistance to water, which is due to both the precipitation of the ethyl silicate zinc mixture on the surface, and the rather rapid reaction that takes place in a humid atmosphere, as outlined in Reaction 2.

The third reaction (the long-time curing mechanism) is similar to the reactions which take place with the other inorganic zinc coatings. The polysilicic acid further polymerizes and further reacts with zinc and the other metal ions that may be incorporated in the formula. These other metal ions are incorporated as very small quantities primarily for initial cure, although some (e.g., lead) tend to improve the silicate film over a period of time (Figure 6.28).

The fourth type of inorganic zinc coating, i.e., the ones made with colloidal silica, react similarly to the second type of inorganic zinc coating already described. In the case of colloidal silica, it is primarily the silicic acid

polymer which is stabilized with an alkali, such as sodium or potassium hydroxide. The primary difference lies with the greater concentration of polysilicic acid. As Rochow stated, "There is no distinction between highly condensed polysilicic acids and colloidal silica."[4] The reactions are therefore essentially the same for both the sodium or potassium silicate self-cure and the colloidal silica self-cure (Figure 6.29).

One additional type of zinc silicate has been used extensively and reacts in a somewhat different fashion due to the alkali silicate from which it is made. This is the lithium silicate-based coating. Reaction 1 is similar to the other alkali silicate coatings. Reaction 2 is different in that the lithium hydroxides in the film react with carbon dioxide, readily forming lithium carbonate. This is a very insoluble product and becomes part of the inorganic film. As the lithium carbonate is formed, acid groups are created on the silica polymer which react with the zinc and any other heavy metal ions present. The lithium zinc silicate coating is somewhat less porous than those coatings based on sodium, potassium, or ammonium silicate where the alkali reaction products are soluble. Lithium carbonate, which is very insoluble, is then rapidly formed. This aids in reducing porosity. Reaction 3 is similar to those of the silicate-type coatings previously described.

Additives Used in Inorganic Zinc Coatings

There are a number of vehicle additives which are incorporated by various manufacturers to improve their inorganic zinc coatings in one way or another. Dean Berger, in his *Current Technology Review—Zinc Rich Coatings*, describes a number of these additives.

In U.S. Patent 3,392,036, McLeod offers several vehicle modifications using trimethyl borate with a siloxane. It can be shown that these materials will react with partially hydrolyzed ethyl silicate in the presence of moisture to form a complex borosilicate. A boron content of five to 20 percent of the silica is useful. The ethyl dimethylsiloxane is used to help prevent mudcracking of the applied zinc-rich primer. Many paint companies incorporate this technology, including Mobil, Carboline International and Standard Paint.

The first polyvinyl butyral-modified ethyl silicate zinc-rich paint is described by Robert A. Rucker of Zinclock in U.S. Patent 3,392,130. This product which is now produced by Porter Paint, Louisville, KY offers extremely good flexibility of the final film. In addition, Zinclock is easy to apply and to top coat. It will not, however, withstand high temperatures. D. P. Boaz (Standard Paint) also uses this technology plus borates and silanes, in U.S. Patent 3,730,746.

Combinations of cellosolve silicates with borates, glycols and silanes are mentioned as vehicles for zinc paints in Patent 2,147,299 to Anderson and 2,147,804 and 2,147,865 to Gordon McLeod and in South Africa, Patent 71/3993 to Nils Trulsson.

Aaron Oken of DuPont, in U.S. Patent 2,649,307 describes a borosilicate vehicle for zinc-rich paint. Dean Jarboe of Plaskem describes a tri-methyl borate and aluminum oxide modified zinc-rich paint in U.S. Patent 3,412,063.

The use of two ethylhexoic acid and monoethanolamine is described by Blake F. Mago in U.S. Patent 3,634,109 to Union Carbide. The amine salt is used to partially hydrolyze the ethyl silicate. W. R. Keithler of Plaskem describes the use of di-2-ethylhexylamine in U.S. Patent 3,202,517 and describes the first zinc-rich based on an amine hydrolysis.

Various colored zinc-rich paints are described by Robinson

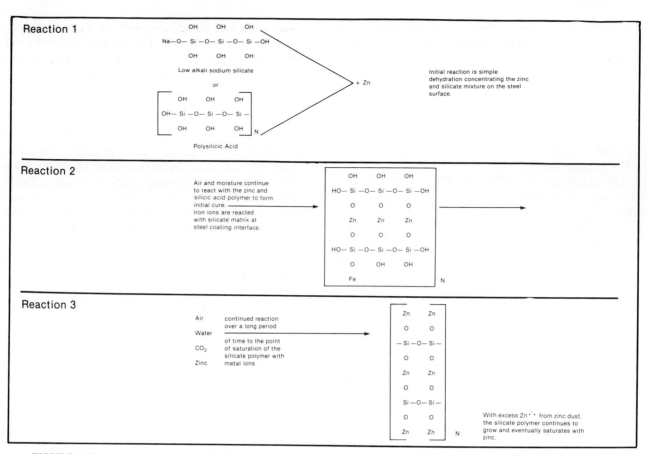

FIGURE 6.28 — Model of the probable internal reactions of a water-base self-cure or a colloidal silica self-cure inorganic zinc coating.

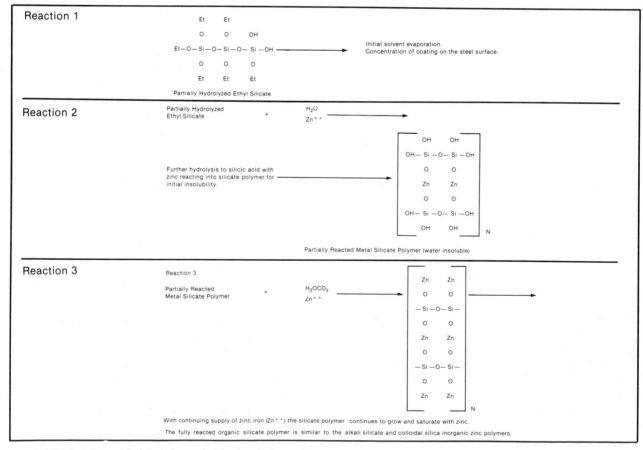

FIGURE 6.29 — Model of the probable chemical reactions within an organic silicate zinc coating.

Corrosion Prevention by Protective Coatings

in British Patent 1,205,394. These coatings have become very popular. Recently Napko Chemical and Carboline have also introduced zinc-rich paints of various colors. One-coat protection is offered and a top coating is not required.[5]

Pigmentation

Zinc Nitride Formation

There is also a rather obscure chemical reaction which takes place during the formation of zinc dust that also may be a factor in the insolubilization and the curing of inorganic zinc dust coatings. This reaction begins during the formation of the zinc dust. Zinc dust is formed by the distillation of liquid zinc and the condensation of zinc vapor into a large chamber. The temperature of the zinc vapor and the temperature of the chamber regulate the size of the zinc particles.

There are three principal materials from which zinc dust is manufactured. These are zinc dross [which is the waste (surface skimmings) from the galvanizing process], scrap die castings (primarily from the automotive industry), or prime zinc ingots. The first two represent the most common sources of zinc dust. Since the dust is formed by the distillation of the zinc, very few contaminants carry over into the dust itself. The reaction product under discussion, however, may come from two sources: (1) the ammonium chloride used as a fluxing agent on the top of galvanizing baths, and (2) nitrogen from the air. At the temperature of the distillation of the zinc, there is some zinc nitride formed on the surface of the condensing particles. When subject to moisture, this little-known compound can hydrolyze to zinc hydroxide and ammonia.

That this compound exists and breaks down into zinc hydroxide and ammonia under ambient temperatures and humid conditions has been varified. For instance, zinc dust was stored for several weeks or months at temperatures from 25 to 35 C (75 to 95 F) in a high-humidity area. During this storage process, ammonium gas developed in the zinc containers, creating enough pressure to actually bulge the tops of the metal containers. When they were opened, the emerging strong ammonia odor was unmistakable. Also, the reaction of zinc nitride with moisture is the reason for the use of the desiccating packages of silica gel which are found in almost all containers of zinc dust used for zinc coatings. Zinc nitride is very reactive and has been responsible for the instability of liquid zinc silicate coatings shortly after they are mixed and prior to application. When ammonia is found in a zinc container and the zinc is mixed with the silicate liquid, the solution becomes grainy and therefore unusable. This reaction is prevented by the dessicant in the can, which absorbs any moisture that may enter the can due to the high humidity on the outside of the can versus the zero humidity in the can's interior. Moisture is actually pulled through the mechanical joint on the lid of the can.

$$Zn_2N_3 + 6H_2O \rightarrow 3Zn(OH)_2 + 2NH_3 \qquad (6.6)$$

This nitride water hydrolysis also may play a role in the original insolubilizing reactions of the coatings, and possibly in the continuing reactions that take place with time. The zinc nitride hydrolyzes and zinc ions are formed which can react with the silicate molecule, helping to polymerize it and causing it to become insoluble.

Grades of Zinc Dust

Zinc thus enters into a number of different reactions in the inorganic zinc coatings. Some of these reactions, such as the absorption of moisture and the ionization of zinc in the various vehicles, is due to the fine particle size of the zinc that is used in the coatings. Actually, there are several grades of zinc dust which are used for pigments. For so-called *regular* coating grade, which has the largest average particle size, the zinc particles range from approximately 6 to 10 μ, with a median size of 7.6 μ; an intermediate grade which is used in some formulations has a median particle size of 6.3 μ; and a superfine grade has a median particle size of 4.5 μ and a range of 2 to 5 μ. There is yet another grade used primarily in the automotive industry for the zinc coating of critical automotive parts. This has a median particle size of 5.5 and a range from 4.5 to 6.5 μ. This grade, however, is not particularly important in the coating of large structures, pipes, tank exteriors, etc., for corrosion protection.

The regular coating grade is used in most of the water-based inorganic coatings and many of the ethyl silicate-based two-package systems. The superfine grade is primarily used in the single-packaged systems and for improved suspension of the zinc dust in the vehicle. Since these materials are prepackaged and must remain in the container for considerable periods of time, the settling of the zinc dust is extremely important. The finer grades improve this property.

Two grades of zinc dust, the regular and the superfine grade, also often are mixed in order to obtain a more uniform gradation of particle size throughout the coating and to minimize porosity by the packing of the zinc particles. This process is probably best demonstrated by relating the zinc dust coating to concrete. In concrete, the aggregate is graded in such a way that the fine sand particles pack in between the larger rock particles to form a concrete with a maximum density. The same packing principle is applied to the use of zinc dust in zinc dust coatings. The size of the zinc particles also makes a difference in the surface area of metallic zinc subject to chemical reactions, with, as might be expected, the finer particle size having some increased reactivity over the larger particle size.

The packing characteristics of zinc dust are illustrated in Figure 6.30, which is a photomicrograph of a cross section of an inorganic zinc film and demonstrates the typical distribution of the particles in the coating. The various sizes of zinc dust are well distributed throughout the film, forming a dense structure of the zinc aggregate. The area between the zinc particles is filled with the silicate matrix and, as it reacts with the zinc ions, it becomes a very strong binder, both to the metal surface and in and around the zinc particles. This is shown in Figure 6.31 where the area between the zinc particles and surrounding zinc particles is filled with the reactive zinc silicate.

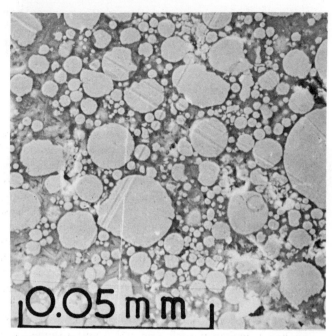

FIGURE 6.30 — Cross section of an inorganic zinc coating showing typical packing characteristics of the zinc particles.

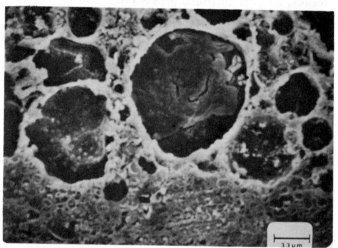

FIGURE 6.31 — SEM photograph of a cross section of a zinc silicate coating showing silicate matrix and the zinc reaction products surrounding the zinc particles.

Manufacture of Zinc Dust

The manufacturer of zinc dust determines the sizes of the zinc particles and thus the various grades. As was previously stated, zinc dust is manufactured by the distillation process. This consists of melting the metallic zinc in a retort. The zinc is heated to the distillation point where the zinc metal evaporates and forms a zinc gas. The metallic gas is then transferred from the retort into a relatively large metal condenser where the zinc gas cools and is precipitated as a fine zinc dust.

The control of this process is critical in the determination of various particle sizes. The controlable variables consist of the temperature of the gas, the temperature within the condenser, and the speed by which the gas is propelled from the retort into the condenser. This,

of course, is determined by the zinc gas pressure. As might be expected, oxidation to zinc oxide is not desirable. Therefore, the retort and the condenser must be kept as free of oxygen as possible in order to precipitate the zinc metal without a large content of zinc oxide on the surface. The control of the process determines the particle size, the particle size distribution, and the content of zinc oxide on the surface of the particles.

Pigments Other Than Zinc

There are a number of other pigments in addition to zinc dust which may be incorporated into any given formulation for an inorganic zinc coating. Lead oxide, or red lead, was one of the first materials used in the formulation for the original Nightingall work in Australia. The red lead not only reacted into the final silicate matrix as lead silicate, but also apparently helped to regulate the speed of the reaction once the silicate and the zinc dust were mixed. The addition of the lead oxide appears to increase the hardness and toughness of the inorganic film, as well as improve its resistance. The addition of the red lead also slightly changes the color of the zinc coating to a light reddish-gray, which aids in visibility when it is applied over a sandblasted surface. For these reasons, red lead is still used in a number of formulations.

Iron oxide is incorporated in some inorganic zinc coatings. In this case, it is primarily used as a secondary, nonreactive pigment in order to change the color of the coating to one which is slightly reddish, which again differentiates it from the color of the sandblasted steel during application.

Some chromates are added to the inorganic zinc formulations primarily to control the speed of reaction of the zinc dust with the vehicle. Chromates are also corrosion inhibitors and, as such, may in some small way aid the zinc in providing a corrosion-resistant coating.

A number of various color pigments have been added to inorganic zinc coatings in order to change not only the color from the light gray of the zinc, but to attempt to add some decorative character to the coating itself. Because of the large quantity of zinc dust used in an inorganic zinc coating, changing the color with a small amount of pigment is difficult, since the zinc itself masks the other color to a great degree. Secondarily, when sufficient amounts of these color pigments are added so that a substantial color change can be made, the amount of zinc in the coating is diluted and the corrosion-resistant characteristics of the coating itself tend to be reduced.

Some of the other chemical reactions which take place in the coating also reduce the effectiveness of color pigments. Formation of zinc oxide and zinc hydroxide on the surface of the coating over a period of time tends to reduce the effectiveness of the color pigments. The primary color pigments which are effective are based on inorganic oxides, such as iron and chrome oxide. Some black pigments have been added in order to change the color of the basic zinc gray.

There are quite a number of inert pigments which are added to the zinc dust in order to provide reinforcing for the coating. This tends to increase the coating toughness and to reduce mud cracking where the coatings are applied at too great a thickness. The materials usually used are ascicular-type pigments, such as asbestine or very fine asbestos fiber. Short glass filaments also have been tried. Many of the commercial formulations contain such reinforcing materials (Figure 6.32).

Another inert pigment which has been widely acclaimed for use with inorganic zinc coatings is di-iron phosphide. It is claimed that 30% or more of the di-iron phosphide can be used as a substitute for zinc in zinc coatings, and a number of commercial products contain some of this material. There have been some good test results in the area of 30% substitution. However, there is still considerable controversy at this time as to the full effectiveness of the substituted material as compared with nonsubstituted zinc coatings.

There is one area which appears to be well documented, and this is the property of weldability. The di-iron phosphide is a conductor, and many of the zinc coatings, both inorganic and organic, where weld-through characteristics are a requirement, contain the di-iron phosphide pigment. This has been particularly true in the zinc-rich coatings used by the automotive industry where spot welding is of vital importance. Norbert Intorpe has described some of the automotive uses of zinc-containing di-iron phosphide in a paper entitled ''Enhanced Zinc-Rich Primers,'' as follows:

> Major U.S. automotive companies have been using zinc-rich primers containing di-iron phosphide pigment since 1974. These primers are applied to automobiles in areas where extra corrosion protection is required. Because di-iron phosphide pigment is highly conductive, both electrically and thermally, it also improves spot welding. This is an important benefit in the automotive industry, where spot welding is so widely used.
>
> Spot weld tests were made on steel sheet coated with zinc-rich primers containing various levels of di-iron phosphide. The welding rate was 15 welds per minute until 2,000 consecutive welds were produced. Weld nugget size was measured at intervals of 250 welds. Results show that the formulation containing 40% di-iron phosphide with 60% zinc passed the test with a nugget size of 0.61 to 0.69 cm (0.24 to 0.27 in.) at 2,000 welds. One hundred percent zinc-rich primers typically result in nugget sizes of 0.46 to 0.56 cm (0.18 to 0.22 in.) for the same number of welds. Since weld strength is directly proportional to nugget size, the advantages provided by including di-iron phosphide in the primer formulation is quite apparent.
>
> The increase in diameter of the electrode face, measured after 2,000 welds was 0.05 to 0.15 cm (.02 to .06 in.) when a primer containing 40% di-iron phosphide was used. This can be compared with a 0.05 cm (.02 in.) increase in the electrode face diameter after only 1,250 welds were made when a 100% zinc-rich primer was used.
>
> Since electrode wear is attributed to the combination of copper with zinc vapor, which results from the low melting point of zinc, decreasing the amount of zinc in a primer with the use of di-iron phosphide pigment can also reduce the amount of electrode wear. This is important because electrode wear decreases electrical current density and weld strength.
>
> Furthermore, by including di-iron phosphide in a zinc-rich coating system:
>
> 1. Welding can be performed on coated steel in applications where it was previously not possible.
>
> 2. Coatings with higher film thicknesses can be welded (unusually heavy coatings occasionally occur in production operations, causing the welding process to stop due to the loss of electrical conductivity).
>
> 3. Welding speed can be dramatically improved without sacrifice to weld strength.[6]

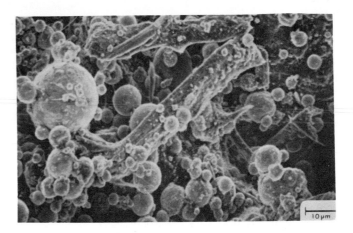

FIGURE 6.32 — SEM photograph of reinforcing pigments in a solvent-base inorganic zinc film, showing the size relationship of the reinforcing pigments to the zinc dust.

Types of Inorganic Zinc Coatings

Preconstruction Primers

One of the important anticorrosive uses for inorganic zinc coatings has been their use as preconstruction primers. These materials are a little different in formulation from the full thickness inorganic zinc coatings; on the other hand, they are formulated with the same basic materials as are used with the standard zinc coating. The primary difference is that they are formulated to be applied at approximately 3/4 to 1 mil in thickness and are usually applied by automatic spray equipment. The purpose of the inorganic zinc preconstruction primer is to provide a corrosion-free surface for steel during its prefabrication and fabrication stage. The majority of these types of coatings have been used in the marine industry where the steel plate or steel shapes, as they enter the shipyards, are automatically blasted free of mill scale and rust. Within a matter of minutes, the steel is sprayed with the preconstruction primer, and again within minutes, the plate is transferred to storage or to the fabrication area. The coating must dry quickly so as not to be damaged by handling with large magnetic lifting devices.

The preconstruction use of the inorganic zincs requires that they be tough and abrasion-resistant, inasmuch as the steel plate and shapes are handled continuously through the fabrication stage with heavy equipment. The coating must be easily welded and cut both by hand and by automatic welding and cutting equipment. There should be little or no burnback at the welds or cut areas, and the life of the coating should be such that it fully protects the steel from any corrosion for the full period of construction up to the point where additional coatings are applied. The preconstruction primer can act as a base for topcoats, or it can be lightly blasted and followed by additional inorganic zinc coatings for maximum corrosion resistance. It is in use by many shipyards and other fabrication facilities where it not only performs as required, but provides a light gray surface with good visibility during fabrication as well.

Figure 6.33 shows the application of a preconstruction primer to deck plates with the automatic weld between the plates. This photograph was taken almost two years after the application of the preconstruction primer and, even with all of the abrasion which takes place on the deck, there was little or no wear-through of the thin primer and no corrosion to the steel itself. It provided an excellent base for an additional inorganic zinc coating following a light blast of the surface to remove surface contamination and to prepare the weld areas. Surface preparation for the final coats was reduced to a minimum by the use of this material.

Figures 6.34 through 6.39 demonstrate the application and handling of the preconstruction primer in a steel fabrication shop.

The rolling of steel also is common during the fabrication process, and it demonstrates the excellent adhesion and abrasion resistance of the preconstruction primer. Such treatment is not uncommon in tank fabrication or in ship building, and the coating must withstand such treatment without appreciable abrasion damage or corrosion after fabrication.

FIGURE 6.35 — Automatic application of inorganic zinc preconstruction primer.

FIGURE 6.33 — Inorganic zinc preconstruction primer applied to deck plates showing an automatic weld. Note the minor burnback at the weld edge and the lack of any pinpoint corrosion after 2 years of service. The reddish color was used to contrast the color of the blasted steel.

FIGURE 6.36 — Magnetic pickup of inorganic zinc-coated plate directly after application. Coating is solid but uncured with no damage due to pickup.

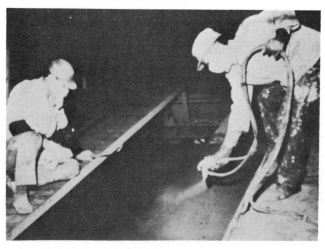

FIGURE 6.34 — Manual application of inorganic zinc preconstruction primer.

FIGURE 6.37 — Magnetic handling of inorganic zinc-coated plate during fabrication.

Corrosion Prevention by Protective Coatings

FIGURE 6.38 — Automatic cutting of inorganic zinc-coated plate. Primer is 18 to 25 microns (0.75 to 1 mil) in thickness. Note the absence of white zinc fumes.

FIGURE 6.39 — Forming plate coated with inorganic zinc preconstruction primer. Formed plates show excellent corrosion resistance.

Advantages of Inorganic Zinc Preconstruction Primer. There are a number of advantages to using an inorganic zinc (IOZ) preconstruction primer, particularly where large volumes of steel are to be coated. The two principal reasons for using a preconstruction primer have to do with surface preparation.

1. Preparing steel by blasting is most economically done by automatic blasting processes. Large volumes of steel can be handled and blasted at a very low cost before it is cut and fabricated into larger or more complex sections. The difficulty is the retention of this surface preparation over a long period of time. This is where the IOZ preconstruction primer performs a service.

2. The IOZ primer is applied as the steel plate or section comes out of the automatic blast machine. Often, the steel is preheated so that it is slightly warm—a few degrees above ambient temperature. The time lapse between blasting and coating is less than one minute, so that the blasted surface is immediately protected. With the coating thickness approximately 25 μ (one mil), it dries to handle very rapidly, particularly if the steel is warm. The whole blasting and coating process is very low cost as compared to hand blasting after fabrication. Even with the IOZ material added to the automatic blasting cost, the total is less than hand blasting after fabrication.

3. Where the IOZ preconstruction primer is applied evenly, the steel can be stored, formed, cut, welded, and fabricated, and still be corrosion-free after many months or even years. This retains the original low cost surface preparation so that additional full-scale blasting is not required.

4. The original IOZ primer can be recoated with a full thickness coat of inorganic zinc by brushblasting to remove any surface contamination and to blast the welds, cut surfaces, etc., down to bare metal. Many organic topcoats can be applied directly to the preconstruction primer by merely washing or steam cleaning the surface and spot blasting the welds. The original intact IOZ primer provides a good base for any of the above coatings.

5. An added advantage of the IOZ preconstruction primer is its light color. Template marks, punch lines, and other construction marks are easily visible, making the cutting and fabrication work much easier.

There are essentially three types of zinc-base preconstruction primers. The first type, water-base inorganic preconstruction primers, have exceptional corrosion resistance. They harden quickly, are easily handled, and have high abrasion resistance. They are excellent for application to warm steel; on the other hand, they are not as satisfactory for cold steel under highly humid conditions.

The second type is based on the solvent-base inorganics, and is the most common type in current use. It gives excellent results under cold and under high-humidity conditions, and may be applied readily by automatic airless equipment. It does not harden as rapidly as the water-base primers, but nevertheless may be handled within minutes after application.

Presently, the most satisfactory of the inorganic preconstruction primers are based on the single-package technology of the solvent-base inorganic zinc coatings. These materials are handled and used in much the same way as paint through automatic spray equipment, and yet they provide the majority of benefits of the two-package primer in terms of handling and durability.

The third type is the organic-base zinc-rich preconstruction primer. This is a common product in Europe and Japan, primarily based on epoxy vehicles. It is an easily applied material. It is much slower to harden than the inorganic zinc preconstruction primers and never reaches the hardness of the inorganics. Applied in the same thickness range, it is much more subject to abrasion and is somewhat less durable than the inorganic-base coatings.

There are also a number of nonzinc preconstruction primers based on several different vehicles. Generally, these can be counted on for proper protection only for a matter of a few weeks, compared with many months or years of full protection where the inorganic zinc preconstruction primers are used.

Single-Package Inorganic Zinc Coatings

The development of the inorganic zinc single-package products has been a breakthrough in the application and handling of inorganic zinc coatings. The original single-package products were recommended primarily for use where the inorganic zinc was to be coated, as their durability was not comparable to the two-package inorganic zinc products, either water-base or solvent-base. The single-package products are based on the solvent-base or organic silicate inorganic zinc materials. As might be expected, the manufacturer of these products is more critical, with particular attention having to be given to the storage life of the mixed material prior to use. Continued research has improved these materials until they are comparable in durability to the solvent-base inorganics with the advantage of single-package handling.

The chemistry involved in these single-package products is similar to that of the organic silicates. The formulations have evolved in such a way that any reaction is delayed until after the application of the product and its exposure to humidity in the air. Once the initial drying stage has been complete, *i.e.,* the removal of the solvent and the initial deposition of the coating, the humidity of the air is then able to react with the organic silicate and the zinc in the same manner as previously described for the ethyl silicate-base products. Some additional information as to the reaction which takes place is provided by Dean Berger in a review paper, as follows.

U.S. Patents 3,615,730 and 3,653,930, and Dutch Patent 6,900,749 describe amine initiated hydrolysis and colloidal suspension of silica in various solvents. This technology led to the first commercially produced, single-package zinc rich paint.

Some interesting aspects of U.S. Patent 3,653,930 include the use of amines, such as cyclohexylamine or triethanolamine. These amines provide a hydroxyl source that is non-reactive with the organic polysilicate vehicle.

When the coating is applied, the amine reacts with atmospheric moisture to yield OH ions. The alkyl polysilicate then undergoes basic hydrolysis to form the polysilica matrix and the resultant alcohol by-product is lost by vaporization.

$$RHN_2' + H_2O \rightleftarrows RNH_3 + OH^-$$

$$C_2H_5O \left[\begin{array}{c} O\text{-}C_2H_5 \\ | \\ \text{-Si-O-} \\ | \\ O\,C_2H_5 \end{array} \right] - C_2H_5 + OH^- + H_2O \rightarrow$$

$$\rightarrow N(SiO_2) + N(C_2H_5OH)^{[(2)]}$$

[(2)]Author's Note: The $N(SiO_2)$ represents the silica matrix or polymer, but without the zinc in the molecule. With zinc, as would be the case in an inorganic zinc coating, the silica polymer would be as represented in the model of probable chemical reactions within an organic silicate zinc coating previously shown. The $N(C_2H_5OH)$ represents the alcohol produced by the hydrolysis reaction and which evaporates out of the coating.

It has been found that gas evolution may be controlled by adding a few percent of 1-nitropropane. Thus, it is suggested:

$$Zn + H_2O \rightarrow Zn(OH)_2 + H_2$$

$$CH_3 - CH_2 - CH_2 - NO_2 + 3H_2 \rightarrow$$

$$CH_3CH_2CH_2NH_2 + 2H_2O$$

1-Nitropropane.

Close examination, however, indicates the following combined reaction likely:

$$3\,Zn + 4H_2O + RNO_2 \rightarrow 3\,Zn(OH)_2 + RNH_2$$

Red lead has also been found to reduce the gas evolution of such inorganic single-package systems. The patent further claims the use of non-polar solvents, which contribute to non-settling characteristics of the paint. In addition, a cyclic ketone is also claimed as a hydrogen scavenger.[5]

There is a significant number of manufacturers in the United States manufacturing inorganic zinc coatings, each of which has at least a slightly different product from the others. Table 6.5 gives the range of the type of products available and shows some of the variations between the different commercial products. The products on this list do not necessarily mean that they are equivalent in effectiveness. However, they all could be considered to be supplied by reputable manufacturers.

The chemistry involved in the various inorganic zinc coatings can be quite complex. The effectiveness of the end product is due to the research undertaken and to the skill of the formulator in making changes and minor additions of ingredients which react into the zinc silicate structure to make it more or less conductive, harder, tougher, and more insoluble than would be the case with a simple mixture of the basic ingredients. Since these minor changes are all proprietary, they are not common knowledge within the industry. Nevertheless, it is believed that all of the truly inorganic zinc coatings ultimately have matrices composed of heavy metal silicates, with the primary heavy metal being zinc. This is true of either the water-based or the solvent-based inorganic zinc silicate coatings. The proprietary minor quantities of other heavy metal salts which may be reacted into the silicate matrix include lead, magnesium, aluminum, calcium, barium, iron, etc.

There have also been many attempts to use metals other than zinc to provide an inorganic coating. None of the other common metals react in the same way that zinc does. Metallic aluminum and metallic magnesium are both cathodic to steel and therefore should, if formed into a coating, provide an effective corrosion-resistant product. This, however, has not been the case. Even when used as a pigment in an organic-base coating, the results have not been comparable to a zinc-base coating. Zinc is a uniquely reactive metal, and the other metals just do not react in the same way, either in combination with silicic acid to form an inorganic film or in being able to develop a cathodic surface over a steel substrate. None of the efforts to use these other metals have been even reasonably successful in the formation of the basic coating or in obtaining corrosion resistance.

TABLE 6.5 — Typical Inorganic Zinc Coatings and Compositions

Coating Type	Powder #/Gal. (Two Component)	Zinc Dust #/Gal.	Volume Solids %	Mil Sq. Ft. Coverage	DFT Mils	Ounces Zinc Dust Per Sq. Ft. at DFT	Pigment Other Than Zinc
Reconstruction Primer—Single Package	N.A.	6.91	35.0	561	3/4	0.15	—
Preconstruction Primer—Water Base	14.15	14.00	35.0	561	3/4	0.30	—
Water Base—Post-Cure	23.0	19.89	66.2	1052	3	0.90	Red Lead
Water Base—Self-Cure	25.0	21.62	75.4	1209	3	0.86	Red Lead
Water Base—Self-Cure	19.4	16.78	67.8	1088	3	0.74	Red Lead
Organic Base—Self-Cure	14.94	14.79	66.1	1060	2.5	0.56	Iron Oxide
Organic Base—Single Package	N.A.	10.0	50.0	800	2.5	0.50	—
Organic Base—Self-Cure	14.6	14.6	62.3	1000	3.0	0.70	—
Organic Base—Self-Cure	18.0	16.82	65.0	1042	2.5	0.65	Celite
Organic Base—Self-Cure	15.0	12.0	63	1010	2.5	0.475	Celite
Organic Base—Self-Cure	8.0	7.42	31.0	497	1.0	0.24	Celite

Inorganic Zinc vs Galvanizing

Although inorganic zinc coatings are made with metallic zinc, they should not be considered a metallic coating, *e.g.,* galvanizing. There has been considerable discussion and controversy with regard to inorganic zinc coatings and galvanizing, with most of the proponents of either material taking a rather strong stand in favor of their particular product. Actually, inorganic zinc coatings and galvanizing should not be considered competitive. Rather, they should be considered complementary, since both of them provide an excellent corrosion-resistant application under the conditions where each one operates best. They are two entirely different concepts of coating, even though they both rely on metallic zinc for the basis of their corrosion resistance. Both are chemically bonded to the metal surface, the galvanizing by an amalgam of zinc and iron, while the inorganic coating is bonded by a chemical compound of iron and silica. Actually, galvanizing can be considered an inorganic zinc coating and, in many ways, it will do the same things that an inorganic zinc-rich coating will do.

There are also some basic differences. The zinc in an inorganic zinc coating is not continuous as it is with galvanizing. It is made up of many individual zinc particles which are surrounded by and reactive with an inert zinc silicate matrix. This matrix is very chemically inert and, except for very strong acids or alkalies, is unreactive with most environmental conditions where coatings would be used. This does not mean that in an acid atmosphere the zinc in the inorganic zinc coating might not be dissolved. However, because it is in a chemical-resistant matrix as discrete particles completely surrounded by the matrix, the solution of the zinc is slowed down in a major way. On the other hand, zinc in galvanizing is pure zinc, and any acid in the atmosphere reacts directly with it with no inhibition of the reaction, as is the case in the inorganic zinc coating. This is an important difference between the two materials and is the reason why, under many difficult corrosion conditions, the inorganic zinc coating will have a much longer life than the galvanizing under the same conditions. This has proven to be the case not only in laboratory testing over a number of years, but also in both industrial and marine atmospheres.

Figure 6.40 shows the direct comparison of an inorganic zinc coating and a three-ounce-per-square-foot hot dip galvanized coating under tidal conditions. In this case, the top panels were continuously just above the high tide level. The lower panels were continuously immersed. These panels were exposed for two years with no appreciable corrosion on the three mils inorganic zinc coatings compared with an almost complete breakdown of the metallic zinc coating (galvanizing) by pinpoint rusting. Many similar tests have been run with similar results.

Inasmuch as the zinc in a zinc coating is surrounded and interlocked into an inert matrix, the coating has controlled reactivity and controlled conductivity. This was shown in a test in which inorganic zinc panels and galvanized panels were placed in a dilute acid solution and coupled to a plain iron panel. Both the voltage that developed and the current flow were measured. In the case of the voltage developed, both the inorganic zinc and the galvanizing developed a potential of approximately 0.5 volts, showing that the zinc in each one provided the same potential under these conditions. On the other hand, when the current flow was measured between the coated panels and the bare steel, the inorganic zinc coating provided 52 milliamps of current, while the plain zinc surface provided 92 milliamps of current, thus showing that the metallic zinc was considerably more reactive than the zinc which was protected by the inorganic zinc matrix (Figures 6.41 through 6.44).

While galvanized surfaces provided a malleable zinc surface, the inorganic zinc coating, because of the hard, rock-like character of the zinc silicate matrix, results in a much harder and more abrasion-resistant coating than the metallic zinc. All of the above differences generally indicate, on an exposure-for-exposure basis, that the inorganic zinc will tend to have a longer life span under most conditions than will the normally galvanized steel surface.

Characteristics of Inorganic Zinc Coatings

A number of the characteristics of inorganic zinc coatings have been discussed. However, it appears useful

FIGURE 6.40 — The left panels are 3-oz-per-sq-ft galvanized panels. Those on the right are 75 micron (3 mil) water-base inorganic zinc. Exposure time was 2 years in tidal seawater conditions. Note that the galvanized panels are completely covered with pinpoint rust. However, there is no corrosion evident on the inorganic zinc panels, even in the scribes.

FIGURE 6.41 — Post-cured inorganic zinc silicate panel coupled to a plain steel panel in a dilute acid solution developed a potential of approximately 0.5 volts.

FIGURE 6.42 — Galvanized steel panel coupled to a plain steel panel in a dilute acid solution developed a potential of approximately 0.5 volts.

FIGURE 6.43 — Post-cured inorganic zinc silicate panel coupled to a plain steel panel in a dilute acid solution developed a current flow of 52 milliamps.

FIGURE 6.44 — Galvanized steel panel coupled to a plain steel panel in a dilute acid solution developed a current flow of 92 milliamps.

to list the major general characteristics of inorganic zincs as a whole so that they may be compared with the many organic coatings described in previous chapters.

Corrosion Resistance

As mentioned in a review of the chemical characteristics of inorganic zincs, their general and chemical resistance is extremely good compared with metallic zinc by itself. This characteristic simply cannot be overemphasized since galvanizing has been a standard of corrosion resistance since the beginning of this century.

Cathodic Protection

One of the most significant properties of inorganic zinc coatings is the cathodic protection they provide. The simple test of a zinc-rich coated panel connected to a steel panel in an indicator gelatine bath is ample indication of this characteristic, as discussed in earlier chapters. Other tests where bare areas are left on a coated panel and the panel immersed in a conductive solution show the same protection to the bare area. The inorganic matrix is conductive and allows the zinc to go into solution in a controlled manner, making it anodic to steel and thus able to cathodically protect any breaks that may occur in the

coating. Eventually, any minor holidays, pinholes, scratches, or scars heal by the formation of zinc reaction products, such as zinc hydroxide and zinc carbonate. This action is important since it provides an added increment of protection to damaged areas of the coating.

Weather Resistance

An inorganic zinc coating, being completely inorganic, is unaffected by weathering, sunlight, ultraviolet radiation, rain, dew, bacteria, fungus, or temperature. Since it is essentially inert to these weather-oriented factors, the coating does not chalk or change with time. The inorganic zinc film remains intact and with essentially the same thickness, even after many years of exposure. This has been demonstrated by making pencil marks on the coating shortly after application. After 10 years of marine exposure, the pencil marks were still intact and readable. Any surface change in the coating would have eliminted such marks in a short period of time. Thickness measurements made over a several-year period also show no change, even under severe weathering conditions.

Undercutting

The prevention of undercutting of a coating on steel is also an important anticorrosive property. Organic coatings generally do not have it. Inorganic zinc-base coatings do prevent rust from undercutting. Both galvanizing and zinc silicate coatings have this property. As explained previously, the galvanized coating is amalgamated with the iron. Zinc silicate coatings provide this adhesion by the inorganic binder chemically reacting with the underlying steel surface in a similar way to its reaction with the surface of the zinc particles. This reaction occurs at the interface between the steel and the coating, forming a permanent chemical bond between the two. This is the property responsible for the effectiveness and life of the coating in addition to the prevention of undercutting of the coating by corrosion. This adhesion cannot be overemphasized since the majority of organic coating failures under severe corrosion conditions is by underfilm corrosion starting at small breaks in the coating. This property of the inorganic zinc base coat multiplies the effective life of an organic topcoat several times.

Figure 6.45 shows two Ken Tator panels coated with a two-coat epoxy which were exposed for more than a year in a very humid, slightly acid industrial area. The panels were adjacent to each other and subject to equivalent conditions. Many similar tests as well as actual uses have proven the benefit of the inorganic zinc base coat in protecting organic topcoats from undercutting and corrosion.

Shrinkage

Another important characteristic of inorganic zinc coatings is that they do not shrink upon drying or curing as do organic coatings. Once applied, the inorganic material follows the configuration of the surface over which it is applied. This is due to the method by which the film is formed and is a major advantage in overcoating rough, pitted, and corroded surfaces or rough welds. The liquid coating wets the metal surface well, and when applied, flows into the small imperfections in the surface. As the film forms, the water or solvent evaporates, leaving the coating in place on the surface. At this point, the coating has no strength or toughness, as do organic films; it merely lies on the surface. Any reduction in volume by solvent evaporation is in depth rather than parallel to the surface. The only exception is where the film is applied too thick and then mud cracking occurs. Generally, no bridging of pits, cavities, or inside corners occurs (Figure 6.46). Zinc coatings have been observed which were applied over steel surfaces that were completely covered with corrosion pits, yet showed no evidence of bridging in the concave pit areas. This is an important characteristic in coating rough areas, welds, or previously corroded surfaces.

FIGURE 6.45 — **Condensation on two-coat epoxy panels.** The right-hand panel was also coated with an inorganic zinc primer. Note the heavy undercutting and blistering on the left-hand (epoxy only) panel.

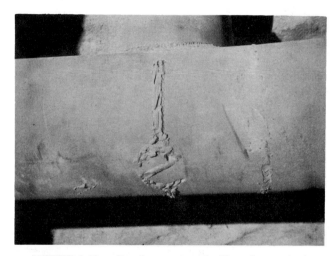

FIGURE 6.46 — **Rough areas coated with an inorganic zinc coating.** Note the complete coverage of the deep pits and undercuts.

Temperature Resistance

Inorganic zinc materials are relatively unaffected by temperatures at or somewhat above the melting point of zinc. Used as a primer and topcoated with silicone-base topcoats, the combination has provided protection even at temperatures of 1000 F for long periods of time. Steel stacks, hot processing equipment, and similar structures have been fully protected for many years without coating breakdown.

One example of this concerns a steam exhaust muffler and water separator. It was heated with steam to 160 C (320 F) and then cooled with water to ambient temperature on a 20-minute cycle. During the steam exhaust, it was completely wet with water. This rather large steel installation was coated with a single coat of inorganic zinc which fully protected the surface for more than 10 years of this heating and cooling cycling.

Steel stacks from a large steam generator station, shown in Figure 6.47, serve as another example. They were adjacent to the seacoast and had already been in use several years. Only one coat of inorganic zinc provided the protection.

Solvent Resistance

Inorganic zinc coatings, being completely inorganic, are unaffected by organic solvents, even the high-strength ones such as ketones, chlorinated hydrocarbons, aromatic hydrocarbons, etc. They also are unaffected by gasoline, diesel oil, lube oil, jet fuel, and many similar refined products. Therefore, they may be used alone or in connection with topcoats for continuous exposure to such chemicals. One coating manufacturer, for example, has all of his solvent storage tanks coated with inorganic zinc to protect against rust contamination.

Generally speaking, water-based inorganics are preferred for continuous immersion in solvents. Some solvent base inorganics may show slight softening when continuously exposed to ketone, ester, or similar solvents. The classic solvent use of water-base inorganics has been as a lining for refined oil ship tankers (Figure 6.48). These transport alcohols, toluol, and xylol, as well as all types of gasoline, jet fuel, and lube oil. They are at the same time subject to seawater washing.

Radiation Resistance

Inorganic zinc coatings are unaffected by gamma ray or neutron bombardment. These coatings have been exposed to atomic radiation up to and beyond 2×10^{10} R without any change in properties. Table 6.6 indicates the comparison of inorganic zinc coatings with several organic coatings in this regard. Most of the organic materials showed some failure at the indicated radiation dosage. The test on the inorganic zinc and inorganic topcoat was discontinued without any evidence of failure. Figure 6.49 is typical of many of the reactor containment shells that are coated on both the interior and exterior with inorganic zinc coatings.

TABLE 6.6 — Examples of Typical Coating Resistance to Radiation Based on Comparative Tests

Type	Substrate	Radiation Resistance[1] Air	Water
Inorganic Zinc	Steel	2.2×10^{10}	Discontinued before failure
Inorganic Topcoat	Steel	1.0×10^{10}	
Epoxy Amine	Concrete	$>3 \times 10^9$	$>3 \times 10^9$
Epoxy Amine	Steel	6×10^9	—
Epoxy Polyamide	Steel	1×10^{10}	—
Epoxy Polyamide	Concrete	1×10^{10}	—
Modified Epoxy Phenolic	Steel	1×10^{10}	7.0×10^9
Modified Epoxy Phenolic	Concrete	$>3 \times 10^9$	$>3 \times 10^9$
Epoxy Surfacer	Concrete	1×10^{10}	—
Vinyl	Concrete	4.42×10^9	$5 \times 9 \times 10^8$
Vinyl Acrylic	Steel	5.6×10^9	—
Chlorinated Rubber	Steel	1×10^8	—
Chlorinated Rubber	Concrete	1×10^{10}	—
Urethane	Steel	5×10^8	—

[1]Rads = The unit of absorbed dose, which is 100 ergs/gram in any medium. For most organic materials: 1 RAD = 1 Roentgen; Roentgen = exposure dose; and RAD = absorbed dose.

(SOURCE: Munger, C. G., Coatings for Nuclear Plant, NACE Western Regional Conference, Seattle, WA, October, 1974.)

FIGURE 6.47 — Steel stacks from large steam generator plant after 5 years service coated with one coat of inorganic zinc coating. The coating was applied to stack sections on the ground and then erected.

FIGURE 6.48 — One coat of water-base inorganic zinc coating on the interior of a refined oil tanker. Coating is free of rust after 2 years of exposure.

FIGURE 6.49 — Typical atomic reactor shell prior to coating the interior and exterior with an inorganic zinc coating.

Coefficient of Friction

The strong permanent bond to steel surfaces and the rock-like character of the inorganic film form a base which has proven to have excellent friction characteristics. Therefore, the inorganic zinc coatings may be used as a coating for faying surfaces (friction interfaces between bolted or riveted structural steel sections) on structural steel buildings, bridges, towers, etc. These joints are subject to severe weathering conditions, including both industrial and marine atmospheres, and corrosion often starts at the joints (Figure 6.50). Such areas are hard to prepare and hard to coat properly. On the other hand, inorganic zinc coatings, because of their good friction resistance, can be applied before the joint is made, providing full protection both within and at the junction of the two steel sections. Organic coatings, and even galvanizing, act as lubricants and do not allow a proper coefficient of friction to be developed in the joint.

Table 6.7 provides some comparative figures based on actual tests of various surfaces. Note that the inorganic zinc coatings provide a coefficient of friction equal to or above a sandblasted surface. Any coefficient of friction less than a sandblasted surface (0.47) is usually not acceptable for steel construction and is the reason for masking the joints on bolted or riveted structures which are shop primed with ordinary paint prior to erection.

Abrasion

This same property of a hard, metallic, abrasion-resistant surface is also important in areas of severe abrasion. The boottopping and upper hull of a ship is a good example. Abrasion due to docking and mooring can be extremely severe in these areas. Where an inorganic zinc coating is used as a permanent primer, even though the topcoats are abraded away, the zinc coating remains to provide full corrosion protection.

Fire Resistance

One of the important characteristics of inorganic zinc coatings is their resistance to burning and fire. They will not support combustion because of their inorganic nature and will not burn even though exposed to severe flames. The welding and cutting of inorganic coated steel demonstrates this characteristic. This point also is dramatically demonstrated by their use on the missile launching structures and launching pads at Cape Canaveral and other launching sites. Even the extreme heat caused by the rocket launch does not cause the coating to oxidize or burn. Even petroleum storage tanks coated with inorganic zinc which have accidentally burned still had the protective coating intact after the fire was extinguished.

Welding

Joints and connections on steel structures usually are subject to welding; thus, there has been a great deal of pro and con discussion with regard to welding inorganic zinc-coated steel. The structural and physiological problems of welding galvanized steel are well known, since some reduction of strength in the welds due to zinc inclusions results and the welding fumes cause various health problems for welders.

Battelle Memorial Institute, along with several foreign laboratories, has made tests to determine the ef-

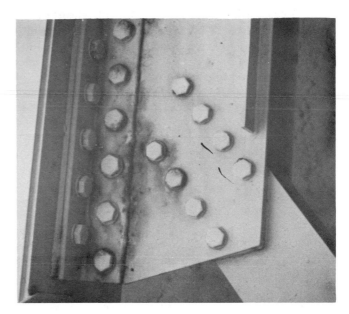

FIGURE 6.50 — Typical bolted bridge member with corrosion starting at the overlapping joint. This is a characteristic joint failure with the corrosion undercutting the organic coating.

TABLE 6.7 — Actual Results of Coefficient-of-Friction Tests for Various Surface Conditions

Surface Conditions	Coefficient of Friction
Solvent-Based Inorganic Zinc Coating	0.52
Rusted and Wirebrushed Surfaces	0.51
Post-Cured Inorganic Zinc	0.48
Rusted Surfaces	0.48
Water-Based Inorganic Zinc Coatings	0.47
Sandblasted Surfaces	0.47
Mill Scale Surfaces	0.30
Galvanized Surfaces	0.25
Rust Preventative Paint	0.11
Red Lead Paint	0.06

[SOURCE: Munse, W. H., Static and Fatigue Tests of Bolted Connections Coated with Dimecoat, Univ. of Illinois, Urbana, IL, private report to Ameron Corp., March (1961).]

fects of inorganic zinc coatings on both the welds and the welders. These tests have shown that the inorganic zinc coatings can be welded without any reduction of weld strength either for preconstruction steel primers or full thickness [75 microns (3 mil)] inorganic zinc coatings. High-speed production welding cannot be accomplished with full thickness coats without some porous welds. Inorganic preconstruction primers or weld-through primers 18 to 25 microns (3/4 to 1 mil) do not cause any problem with the speed of cutting or welding. The welds obtained by either automatic or hand welding techniques are equivalent to those on bare plate.

The following test results are indicative of the weld effectiveness using a submerged arc method. Test plates were welded using square butt joints with all surfaces, including the edges, coated with an inorganic steel primer. All tensile strengths of the welded specimens exceed the requirements of the American Welding Society Specification AWS D2.0-66. All welding materials, currents, voltages, etc., are shown in Table 6.8. Speed changes within the range of 35 to 45 inches per minute did not affect the results. Similar results also were obtained using a shielded metal arc and a gas metal arc.

Many millions of square feet of preconstruction primed steel have been cut and welded without creating hazardous working conditions. Nevertheless, carefully controlled tests were made to determine if any hazards were present. Another project was conducted in which the safety conditions when welding inorganic zincs were thoroughly studied. Measurements of the zinc oxide, lead, and iron oxide content in a welding atmosphere were made with a Gelman GM4 membrane filter. Measurements were made during the welding of vertical tee joints made in A-36 steel that was either uncoated or coated with standard inorganic zinc and inorganic preconstruction steel primers. The filter was positioned about 12 inches from the tee specimen and slightly above it, *i.e.*, in approximately the same location as the welder's face. The results are summarized in Table 6.9.

In general, the conclusions from the tests were as follows.

1. The concentration of zinc, lead, and iron in the air was well below the recommended threshold limits of 5, 0.2, and 15 mg/m^3, respectively.

2. The concentration of zinc in the fumes generated by welding uncoated A-36 and A-36 coated with post-cured inorganic zinc increased from a background value of 0.09 mg/m^3 to 2.1 mg/m^3, respectively. The same results were noted when welding was performed on plate coated with self-cured inorganic zinc or preconstruction inorganic zinc primer.

3. The concentration of zinc in the fumes produced when welding galvanized A-36 plate was about three times that produced when welding inorganic zinc, but was within the allowable limits.

Manual welding of an inorganic zinc structure is not only possible, but practical. Even with the full thickness coatings of 75 microns (3 mil), proper welding procedures will provide welds with equivalent strength to uncoated steel. This makes structural repairs of existing inorganic zinc-coated surfaces equivalent to the new surface.

The above characteristics indicate that inorganic zinc coatings are unique and different from any other corrosion-resistant coating on the market. It is no wonder then that these coatings have revolutionized the corrosion protection of steel structures of all types throughout the world. Not only do they provide longer and better protection, but their use as a base coat multiplies the life of organic topcoats many times over.

Types of Zinc-Rich Coatings

SSPC Designations

The general properties of the various zinc-rich type coatings have been discussed so that the principal coating types can be more easily differentiated. In order to bring the whole area of zinc coatings into some kind of an organized picture, the Steel Structures Painting Council, in their *Specification SSPC Paint 20X,* describes the basic types of zinc-rich coatings that are in general use at the present time. There are two obvious types: Type 1, the inorganic zinc-rich coatings; and Type 2, the organic zinc-rich coatings. These are outlined in the specifications as follows.

Type 1-A: Inorganic post-curing vehicles—water soluble and include alkali metal silicates, phosphates, and modifications thereof. These coatings must be subsequently cured by application of heat or a solution of a curing compound.

Type 1-B: Inorganic self-curing vehicles—water reducible and include water soluble alkali metal silicates, quarternary ammonium silicates, phosphates, and modifications thereof. These coatings cure by crystallization after evaporation of water from the coating.

Type 1-C: Inorganic self-curing vehicles—solvent reducible and include titanates, organic silicates, and polymeric modifications of these silicates. These systems are primarily dependent upon moisture in the atmosphere to complete hydrolysis, forming the polysilicate.

Type 2: Organic vehicles—include phenoxies, catalyzed epoxies, urethanes, chlorinated rubbers, styrenes, silicones, vinyls, and other suitable resinous binders. The organic vehicles covered by this specification may be chemically cured or may dry by solvent evaporation. Under certain conditions, heat may be used to facilitate or accelerate hardening.[7]

Such a variety of zinc-rich coatings is available for

TABLE 6.8 — Submerged Arc Method

	Amps	Volts	Speed I.P.M.	Tensile Strength PSI	Free Bend Test Gage Reading		
					Orig.	Final	% Elong
Uncoated Steel	—	—	—	69,300	—	—	36.5
Uncoated Steel	—	—	—	70,020	—	—	34.0
Coated Steel	650/700	30/32	35	73,520	0.433	0.591	36.5
Oxweld 36	650/700	30/32	35	75,560	0.361	0.512	41.8
Unionmelt 50	650/700	30/32	45	73,580	0.290	0.413	42.4
	650/700	30/32	45	73,521	0.263	0.374	42.2
Coated Steel	650/700	30/32	35	72,971	0.414	0.611	47.6
Lincoln L-61	650/700	30/32	35	73,797	0.409	0.571	39.6
781 Flux	650/700	30/32	45	74,260	0.249	0.335	34.5
	650/700	30/32	45	73,830	0.283	0.349	23.3

NOTE: Tensile Strength Base Metal = 74,700 psi. Minimum acceptable elongation is 23%. All specimens passed.

(SOURCE: Munger, C. G., Inorganic Zinc Coatings, Proceedings of II Symposia Sul-Americano de Corosao Metalica, Rio de Janeiro, Brazil, 1971.)

TABLE 6.9 — Analyses of Welding Fumes Collected with Gelman GM4 Filter
(Within 12 in. of Welder's Face)

Elemental Metal	Threshold Limit mg/m^3	Not Coated		Post-Cured Inorganic		Self-Cured Inorganic	
		mg	mg/m^3	mg	mg/m^3	mg	mg/m^3
Fe	5.0	0.20	3.7	0.05	1.1	0.05	1.1
Pb	0.2	0.005	0.09	0.005	0.10	0.002	0.04
Zn	15.0	0.005	0.09	0.10	2.1	0.10	2.1

[SOURCE: Pattee. H. E. and Monroe. R. E., "Effect of Dimecote Coatings on Weldability of Selected Steels." Welding Journal. Vol. 48. No. 6. pp. 222s-230s. 1969.]

many reasons. One of the primary reasons is the wide variability of atmospheres around the world, which is a particularly important consideration during the application of the various coatings. In Japan and many other more northern areas, for instance, it is rather cool and highly humid during most of the year. Thus, it is difficult under many conditions to apply water-base materials. On the other hand, in the relatively dry, hot areas of the Southwestern United States and Mexico, there is difficultly applying solvent-base materials because of rapid drying and the formation of overspray. When the humidity is very low, they do not cure properly.

Basic use of the materials, then, as well as the conditions under which they are to be applied, dictate which type of formulation should be used.

Type 1-A

The Type 1-A post-cured coating has a broad range of applications. Not only is it a heavy-duty inorganic zinc coating, but it has been used effectively on an almost worldwide basis, from the Arctics to the Tropics. It is a water-base material, so that wherever water will effectively evaporate from the initial coating, this product can be used. This initial evaporation is the key to the proper formation of the coating and to the development of the hard, metallic film characteristic of this coating type. It may be applied in cool areas, such as the northern part of the United States and Canada or Northern Europe, and it has been readily applied in areas close to the equator, as long as proper drying conditions exist. It forms a good coating under either warm or cool conditions due to the use of the post-curing agent which gives it a fast initial cure and insolubility as soon as the curing agent has been applied.

These coatings generally are not effective under freezing conditions or under conditions of very high humidity when the water will not effectively evaporate from the coating system within a short period of time. Generally, the evaporation time should be between a few minutes and one to two hours. A longer drying period than this may cause the zinc to separate in the vehicle, which makes for a poor coating with little or no corrosion resistance.

It is also difficult to use and apply this coating where rain showers are frequent, since the coating must dry to a solid state and then be cured with the curing solution before additional water can come in contact with the coating. This is because the alkali silicates form a gel during the original drying stage of the coating. If this gel becomes rehydrated, it breaks and no longer forms a continuous or adherent film. A rain shower on a Type 1-A coating prior to the application of the curing agent will break up the silicate gel into a powder and coating becomes useless. Most of the Type 1-A coatings are most effective where they are used alone, without topcoats, since the removal of the curing agent residue is essential wherever topcoats are to be applied. On the other hand, after washing the post-cured solution from the coating, it is relatively non-porous, and topcoats are relatively easily applied without the difficulty of solvent penetration into the voids of the coating, causing solvent blistering.

Type 1-B

The Type 1-B self-curing inorganic zinc coatings have a water base. They can be based on a number of different alkali silicates, or on phosphates or silicasols. Ammonium silicate coatings, for example, come under this type. As with Type 1-A, it is essential that the water be able to evaporate from the coating readily in order to form the initial film. Once this takes place, leaving a hard, metallic coating on the surface, the film becomes water insoluble in a short period of time and continues to cure to full hardness and adhesion by reaction with carbon dioxide and humidity from the air. Some of these formulations require more humidity for a complete cure than others. Many of these types of coatings are effective under warm, dry conditions. Sufficient humidity is required to continue the chemical reaction and to insolubilize the coating.

These types of coatings have been applied in many areas of the world. One of their most useful applications has been on the interior of tanks and tankers under warm or hot climatic conditions. Since there are no solvents in the film, they can be applied on the interior of closed areas without difficulty. The water which evaporates from the coating itself creates sufficient humidity to continue the cure of the coating to its insoluble and useable state. These materials are not effective under cold, highly humid conditions where water will not properly evaporate from the surface within a reasonable period of time.

Type 1-C

The Type 1-C solvent-base, self-curing inorganic zinc coatings include many different formulations. The primary type, and the one in most general use, is the organic silicates such as partially hydrolized ethyl silicate. There are a number of other organic silicates also used, either alone or in combination with the ethyl silicate, in order to broaden the range of application of these coatings. This coating type is widely used because it can be applied effectively under cold conditions where the humidity is also high.

This does not mean that it should be applied to areas where there is condensation on the surface, since this type of coating reacts readily with water, and water on the surface prior to its wetting the surface prevents the formation of a proper film or one with proper adhesion. In many areas of the world, however, the temperature and humidity are such that water does not readily evaporate during certain times of the day. This coating type will form a good film in any except condensation conditions or under very warm, dry conditions. Thus, it has performed well in plants and shipyards throughout many areas of the world. These coatings do not generally form a film unless there is adequate moisture in the air to allow proper hydrolysis of the silicate.

Since these materials are also solvent base, they are less effective than Type 1-A and Type 1-B under rapid drying, windy conditions. A powdery, dry film is the usual result when these materials are applied in a warm area with considerable air movement. A number of manufacturers have made formulations with slower evaporation than the usual ethyl silicate zinc coating. They also have made faster evaporating formulations for use under less favorable conditions.

There are two specific types of solvent-base inorganics which are additional variations of the hydrolyzed ethyl silicate product. These are both assumed to be included in the Type 1-C category. The first is a single-package inorganic. As previously noted, this product combines all of the ingredients needed to form the coating, including the zinc, into a single package which can be opened, stirred, and used like paint. The product characteristics are generally similar to the two-package system, and the general curing mechanism is the same. As previously discussed, modifications of the single-package inorganic zinc coating also are used for a preconstruction primer for steel prior to its being processed into a structure. The decided advantage of this type of product is its handling characteristics and the fact that two materials do not have to be mixed in the field. Also, only one package has to be handled, stored, and transported to the job. Most of the single-package products are used as primers for organic topcoats in areas where topcoats are necessary, either for additional protection or cosmetic purposes.

The second type is the modified inorganic zinc primer. Here, the solvent-base (Type 1-C) inorganic is modified by the addition of a compatible organic resin, usually a vinyl butyral which is soluble in alcohol solvents. The product characteristics are a compromise between the completely inorganic zinc coating and the organic zinc-rich primers, with some of the good properties of each appearing in the modified product. Any deficiency which might come about would be due to the life of the organic resin incorporated into the system. The advantages claimed for this type of material are improved application properties, a smooth film, easy and rapid overcoating, adhesion to most clean steel surfaces, and good repair properties for previously zinc-primed and overcoated surfaces. This type of product is usually used where topcoats are to be applied.

Type 2

SSPC Type 2 is the organic zinc-rich classification. These materials are made with many different organic vehicles. As previously discussed, the most important are those made from phenoxies, epoxies, and chlorinated rubber. They can be applied under almost any condition where an organic vehicle applies effectively. However, they also are subject to all of the basic problems inherent in organic vehicles, such as weathering, undercutting, release of adhesion from water absorption, blistering, etc. One of the useful applications of the organic-based zinc-rich coatings is as a repair primer for inorganic zinc primers and galvanized surfaces which have been topcoated and then damaged during use. Through the use of the organic zinc-rich primer, the zinc-base coating is maintained over the bare steel area, while the organic vehicle is compatible with the organic topcoats, allowing it to be feathered out over the edge of the existing organic material.

With the many different formulations of both inorganic and organic zinc-rich primers, ones with high zinc loadings and those with a minimum loading, some with additives and others without, some precautions should be taken in selecting a product for use.

It is suggested that information be obtained on the total solids content, the theoretical and practical coverage, the percent of zinc in the dry film, the type of binder, and the scope and duration of actual filed applications or field tests of the several materials which may be considered for the job. Where there is a requirement for a high-performance coating, the best is none too good, and since materials cost is only a small part of the completed coating job, only the best materials for the purpose, not the cheapest, should be selected. The data suggested above should be readily available from the manufacturer and can be a good basis for comparing the various zinc-rich coatings offered for a project.

Comparison of Zinc-Rich Coatings

There may be times when some test comparison of the various zinc-rich types of material is desired. There are always the relatively long-term tests, such as salt spray, the Cleveland humidity test, salt water, and immersion. However, these tests generally take a considerable period of time. One quick test is the so-called "V" type test. In this case, a panel approximately 4 in. by 12 in. is sandblasted and prepared for coating. On the middle of one side is placed a "V" made from masking tape or scotch tape which goes from a point at the bottom to a 1/2 in. width at the top. The panel is then coated and allowed to dry or cure, at which point the "V" of masking tape is removed. The panel or panels are then subjected to weather conditions for periods of up to a week and are observed daily. The measure of the test is the area of the "V" which is fully protected from rust or corrosion. This provides a quick comparison between different inorganic zinc or organic zinc coatings. While it is not infallible, it does provide for a comparison which generally has correlated well with field results.

Figure 6.51 shows a series of different inorganic and organic zinc products, including galvanized steel, which were compared in this manner. These panels were exposed to 100% humidity for three days. The various coatings used on the panels which were marked with identical "V" holidays, and their performances are summarized in Table 6.10.

This test can be considered a test of the activity of the zinc in the individual zinc-rich formulations and its throwing power or its ability to provide cathodic protection to bare steel areas. A numerical rating can be developed from the test by measuring the amount of the "V" from the bottom of the panel up to the point where there is a first indication of rust. The examples given are commercial formulations, and it must be stressed that every formulation will perform differently since this is simply a quick comparison test. It is used exclusively for the direct comparison of two or more materials, therefore it does not yield conclusions as to whether one generic type of zinc-rich paint is better than another.

FIGURE 6.51 — Various zinc-rich coatings compared by using the "V" method to determine the degree of initial protection to the bare steel. Note the difference in protection by the various coating types after 3 days of weather exposure. Note also the zinc reaction products which have covered some of the "V," providing complete protection.

Topcoating

Inorganic Topcoats

In many ways, the technology involving inorganic zinc coatings is similar to that of inorganic zinc topcoats, except that the inorganic topcoats do not contain metallic zinc. Some of these topcoat products have been in service for a good many years in spite of the fact that their materials have never really "caught on" in the corrosion field. They have been used successfully, however, in chemical plants, refineries, storage tanks, and the interiors of steel stacks.

One of the larger geothermal plants, for instance, located south of Mexicali in Mexico at Cerro Prieto has been coated with an inorganic zinc base coat and an inorganic topcoat for over ten years. The first section of the plant was coated with a post-cured inorganic zinc coating followed by the inorganic topcoat in white, and is the section which has been in service for over ten years. Within the last two years, a second section has been added to the plant and, because of the excellent service in resisting the hot, sulfur-containing fumes and liquid, it has been

TABLE 6.10 — Coating Type and Performance for Test Panels

Type I	Zinc Lead Silicate Post-Cured	(SSPC Type 1-A)
Type II	Zinc Silicate	(SSPC Type 1-B)
Type III	Zinc Silicate	(SSPC Type 1-B)
Type IV	Zinc Phosphate	(SSPC Type 1-B)
Type V	Zinc Lead Silicate	(SSPC Type 1-B)
Type VI	Hydrolyzed Ethyl Silicate and Zinc	(SSPC Type 1-C)
	Epoxy and Zinc	(SSPC Type 2)
	Chlorinated Rubber and Zinc	(SSPC Type 2)
	Galvanized Steel	

After three days of exposure to 100% humidity:

Order of Performance

Type I	Zinc Lead Silicate	100% protection
Type V	Zinc Lead Silicate	100% protection
Type IV	Zinc Phosphate	100% protection
Type II	Zinc Silicate	100% protection
	Galvanized Steel	95% protection
Type III	Zinc Silicate	85% protection
Type VI	Hydrolyzed Ethyl Silicate and Zinc	70% protection
	Chlorinated Rubber and Zinc	60% protection
	Epoxy and Zinc	40% protection

coated in the same manner. Figure 6.52 shows one of the condensers in the plant which has been in service for a ten-year period. Figure 6.53 shows a condensate tank in a chemical plant which operates continuously at 250 F. The white inorganic topcoat has maintained a clean white appearance for several years.

These inorganic topcoats may be made from either water-base or organic-base silicates, and considerable development work is taking place at the present time in the improvement of these products to provide top grade corrosion protection. This, in fact, represents an area where a major coating breakthrough may take place with some widespread consequences. Inorganic base and topcoats could provide the principal anticorrosive coating systems of the future. At the present time there are two developments in progress which indicate the versatility that may be obtained with inorganic systems. The first is an inorganic antifouling and the second an inorganic tank lining which has very good resistance to SO_2 fumes, particularly fumes from the burning of high-sulfur crudes and coal. Both of these are based on some new approaches to silicon chemistry. Although they are proprietary at the present time, research in this field is nevertheless progressing and new inorganic film-forming reactions may open up a broad, new coating field.

The inorganic antifouling is one of these new approaches. Actually, the heavy zinc content of almost any inorganic coating will provide some antifouling protection due to the zinc ion which is present. In the inorganic antifouling, organic tin compounds have been reacted into the vehicle system and the combination of the heavy metal zinc and the organic tin compounds have made for a unique antifouling product. One of the advantages of this material is that it can be applied directly over an inorganic zinc coating on the base steel. It adheres well to the base zinc coating and reacts between the coats to form an excellent bond.

One of the problems which always has been present in using an inorganic base coat for underwater marine use

FIGURE 6.52 — Barometric condensor located in a geothermal power plant coated with one coat of inorganic zinc and one coat of inorganic white after approximately 10 years of exposure.

FIGURE 6.53 — Condensate tank coated with inorganic zinc base coat and white inorganic topcoat exposed to a chemical plant atmosphere.

is the organic antifouling topcoat. Often, these had to be separated by a substantial nonconductive impervious barrier in order to prevent undercoat reactivity with eventual breakdown of the antifouling and the base coat. With an all-inorganic system, the zinc-base coat can be applied, the inorganic antifouling applied, and the system placed

in service. Later it can be touched up using the original zinc-base coat, even allowing it to be applied over the inorganic antifouling. A coat of inorganic antifouling topcoat can then be applied overall with the total system being entirely compatible. This inorganic antifouling has been under observation for several years. Although there have been no large applications, extended field tests have performed well for over a two year period.

Atmospheric Changes

As has been outlined previously, inorganic zinc coatings change in the atmosphere. They change from their initial reaction, merely the evaporation of the solvent which laid down a rather porous film, to, upon aging, a much denser film. This can be seen in the scanning electron microscope photographs of a newly applied solvent-base zinc coating compared to a weathered coating of the same type and shown at the same magnification. Note in Figure 6.54 how the particles are layed down in a relatively discrete way, and also the porosity which exists in the initial coating. Each individual zinc particle is visible, along with some of the ascicular reinforcing pigments. Some reaction has been started, as is indicated by the circle of light-colored reaction product around each of the particles of zinc.

On the other hand, Figure 6.55 shows that after several years of exposure to the weather, the zinc particles are much less discrete; they look more as though they were encased in a matrix, and even some of the ascicular particles appear the same way. It appears to be a much denser film than that where the coating is initially applied.

A cross section of a similar coating, when newly applied, shows the porosity that exits in the coating all the way down to the metal surface. Figure 6.56 shows the discrete zinc particles as they are formed in the body of the coating, and the excellent wetting of the coating as it is applied over the sandblasted steel surface (the solid area at the bottom of the photograph). Note the manner in which the coating conforms to the sandblasted surface.

Figure 6.57 shows a solvent-base coating after it has been applied to the surface for several years. The cross section displays the individual zinc particles and thus indicates the buildup of the matrix around the zinc particles and, in general, the reduction of the porosity within the coating. Also note the continued excellent wetting of the steel surface and the appearance of a light reaction product at the coating-steel interface.

Immersion

Marine atmospheres have proven to be an excellent exposure for inorganic zinc coatings. Marine immersion, however, is quite a different condition. Inorganic coatings are seldom used alone when in direct contact with seawater. However, it has been shown that excellent galvanic protection of the underlying steel can be obtained for periods of time up to two years or more. Not only will the coating provide full corrosion protection, but it will be antifouling as well. The two properties go hand in hand; as long as the zinc provides corrosion protection, it also will provide protection from marine fouling. The zinc ions are responsible for both reactions.

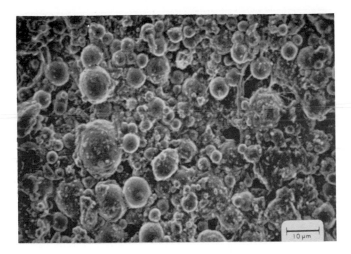

FIGURE 6.54 — SEM photograph (surface view) of a newly applied solvent-base inorganic zinc coating. Note the initial porosity.

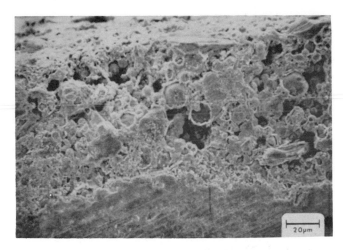

FIGURE 6.56 — SEM photograph of a cross section of newly applied solvent-based inorganic zinc coating. Note the porosity, the coating surface, and the intimate contact with the sandblasted steel.

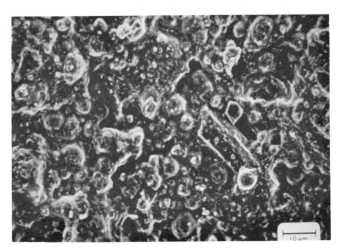

FIGURE 6.55 — SEM photograph (surface view) of a weathered solvent-base inorganic zinc coating. Note the dense structure of the coating.

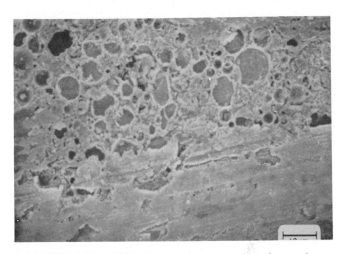

FIGURE 6.57 — SEM photograph cross section of a weathered solvent-base inorganic zinc coating. Note the coating density, the zinc reaction products surrounding the zinc particles, and the excellent contact with the sandblasted steel.

This condition, however, tends to change with time, and at some period from 18 months to beyond two years, the coating ceases to provide cathodic protection and becomes inert. At this time, the coating also will start to allow growth of marine organisms. This inerting reaction of the coating reverses its polarity so that instead of being anodic, it becomes cathodic. This phenomenon also has been noted in the literature by M. O. P. Velsboe, who states,

A current will start to flow if a steel panel coated with zinc paint... is connected to a bare steel panel....The current can be measured on the meter....The current will drop as time goes by and so, when measured against a reference electrode, the current will become zero after some time telling us that the galvanically active period of the paint is over....But no rust will be visible on the coated panel even if it is left for months in water. This means that the film protects the steel from rusting even if it is no longer galvanically active. The pores have been blocked completely. If the paint on the panel is then scraped down to the steel, the current will start to flow again because there is still excess metallic zinc in the paint film.[8]

Figure 6.58 serves as a good example of this. These panels were fully immersed in seawater approximately one foot below the surface for eleven years. Note the corrosion along the edges of the panels where fouling made a break in the coating. Even where the oysters were tightly attached to the center portion of the coating, it is completely intact and completely protecting the steel.

Figure 6.59 shows the coating surface which existed under one of the tightly adherent oyster shells. Some of the coating was broken away when the oyster was removed, but the zinc particles continue to exist along with the dense inorganic structure surrounding the zinc particles. The surface of the coating had become completely inert and the coating was acting as an inert coating over the surface of the steel, fully protecting the metal by its inert film rather than by cathodic protection. Dr. Uhlig has recognized the same reaction.

At room temperature, in water or dilute NaCl, the current output of zinc as an anode decreases gradually because of insulating corrosion products which form on its surface. In one series of tests, the current between a couple of Zn and Fe decreased to zero after 60 to 80 days and a slight reversal of polarity was reported.[9]

Corrosion-Resistant Zinc Coatings

FIGURE 6.58 — Water-base inorganic zinc-coated panels immersed in seawater for 11 years. Note the excellent corrosion protection (except at edges), even at the scribes.

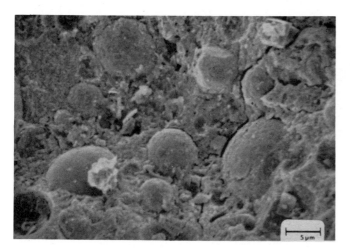

FIGURE 6.59 — SEM photograph of the surface of the coating in Figure 6.55 after removal of a large oyster. Note the dense coating and encapsulated zinc particles.

In this case, the zinc referred to is metallic zinc or galvanizing. It also is interesting to note that the same reversal of polarity does not take place in the tidal area where the coating will continue to protect for much longer periods than where the coating is in full immersion.

The initial reactions which occur on zinc coatings in a marine environment are all similar to those previously shown for a rural atmosphere. The one substantial difference is the presence of the seawater salts which create a highly conductive solution on the coating (Table 6.11).

It has been indicated by Pourbaix that zinc is thermodynamically unstable in the presence of water and aqueous solutions and tends to dissolve with the evolution of hydrogen in acid, neutral, or very alkaline solutions. He also indicates that zinc in the presence of moderately alkaline solutions of pH between 8.5 and 10.5 can cover itself with a film of zinc hydroxide.[10] When CO_2 is present, this pH range is extended from approximately 7 to 10.5. This is the pH range where the corrosion rate of zinc is at a minimum due to passivation by the hydroxide film.

TABLE 6.11 — Composition of Seawater and Ionic Constituents

Constituent	G/Kg of Water of Salinity, 35 o/oo	Cations %		Anions %	
Chloride	19.353	Na $^+$	1.056	Cl $^-$	1.898
Sodium	10.76	Mg $^{++}$	0.127	SO$_4$ $^-$	0.265
Sulfate	2.712	Ca $^{++}$	0.040	HCO$_3$ $^-$	0.014
Magnesium	1.294	K $^+$	0.038	Br $^-$	0.0065
Calcium	0.413	Sr $^{++}$	0.001	F $^-$	0.0001
Potassium	0.387				
Bicarbonate	0.142	Total	1.263	Total	2.184
Fluoride	0.001				

[SOURCE: Lyman, J. and Abel, R. B., Chemical Aspects of Physical Oceanography, J. Chem. Education, Vol. 35, No. 3, pp. 113-115 (1958).]

The presence of the insoluble zinc hydroxide on and within the coating is believed to make inert or passivate the zinc coatings when immersed in seawater. Seawater has a pH of approximately 8. It also contains a substantial amount of bicarbonate so that conditions appear proper for the formation of the zinc hydroxide as the principal corrosion product.

The reaction of metallic zinc, ionizing and forming an insoluble hydroxide film, is a slow process in the presence of other sea salts such as chlorides and sulfates. This accounts for the period of 18 months to 2 years that zinc coatings protect the steel over which they are applied. Enough time is required for the zinc hydroxide to form in all of the pores of the coating and on the surface to prevent any additional zinc ions from forming. When this occurs, the coating surface becomes inert and thus cathodic to the iron substrate. If there are no breaks in the coating, it fully protects the surface; however, if a break does occur, it creates a condition of a very large cathode and a relatively small anode. Such a condition makes for rapid pitting at the break, which is what occurs. This phenomenon takes place irrespective of the type of zinc coating, *i.e.,* galvanizing, inorganic zinc coatings of all types, and even organic zinc-rich coatings (Figure 6.60).

FIGURE 6.60 — Corrosion anodes formed along scribe and on edges of panel coated with inorganic zinc coating immersed in seawater. Note perforation of panel in lower anode area.

Corrosion Prevention by Protective Coatings

The passivation of the zinc coating by the hydroxide deposit is also the reason for fouling organisms to deposit on the surface. As long as the heavy metal zinc ions are available, it makes an unsatisfactory surface for fouling attachment. As soon as the surface ceases to provide the soluble metal ions, the marine organisms no longer find it objectionable.

Most zinc coatings, because of the above type of reaction, generally are not recommended for use alone for seawater immersion conditions. They have been used effectively as a basic primer on steel surfaces for marine uses, ship bottoms, offshore platforms, and similar structures where the zinc coating has been overcoated with coatings such as vinyls, epoxies, and epoxy coal tar coatings.

Another observation has been made for zinc coatings under marine conditions. While under immersion, they protect for approximately two years. In the tidal zone, the zinc does not become passive and will protect for a much longer period, *i.e.*, until most of the zinc has dissolved.

The pH of a solution has a substantial influence on the solution rate of zinc. It is believed that the same essential reaction takes place as was indicated for immersed zinc, except that during the rise and fall of the tide, carbon dioxide from the air tends to lower the pH slightly and the reaction product may be zinc carbonate in place of zinc hydroxide.

As indicated in Table 6.12, the solubility of zinc carbonate is substantially higher than for zinc hydroxide. As the zinc carbonate is slowly dissolved, the tidal surface is kept active, with the zinc coating continuing to provide cathodic protection.

TABLE 6.12 — Solubility of Zinc Topcoats

Porous Topcoating	Solubility in Cold Water
Zinc Carbonate	0.001 gr/100 m.l.
Zinc Oxide	0.00016 gr/100 m.l.
Zinc Hydroxide	0.00000026 gr/100 m.l.

(SOURCE: Handbook of Chemistry and Physics, 47th Ed., Chemical Rubber Co., Cleveland, OH.)

There is also an interesting environmental reaction on zinc coatings in marine atmospheres. This is a reaction which takes place when zinc coatings, galvanizing or inorganic zinc, are coated with a relatively thin porous coating. The coating in this case must be sufficiently porous to allow sodium and chloride ions to contact the zinc. In this case, the metallic zinc and sodium chloride react together in the presence of moisture to form a zinc oxychloride compound and sodium hydroxide. The reaction is as follows.

$$2Zn + 2NaCl + 3H_2O \rightarrow ZnOZnCl_2 +$$

$$2NaOH + H_2 \qquad (6.7)$$

It is interesting that above the pH of approximately 10.5, zinc reacts rapidly to form soluble zincates. As the sodium and chlorine ions penetrate the thin organic topcoat, they are held within and under the coating, causing a buildup of pH from the NaOH formed from Equation (6.7). The zinc oxychloride is insoluble, and with continuing contact with moisture, salt, and carbon dioxide, voluminous white salts are formed which penetrate the porosity of the topcoating, forming a heavy, white precipitate on the exterior, as well as building up underneath and disrupting the coating. Such a reaction is not uncommon aboard ship or on offshore structures and has created conditions whereby both the zinc coating and the topcoat have been severely damaged. This reaction consumes substantial amounts of zinc, with the underlying zinc coating being rapidly depleted so that the entire coating system must be removed and replaced (Figure 6.61*).

Fortunately, such a condition can be overcome at the time that either the galvanizing or the inorganic zinc coating is originally topcoated. The topcoat, under these conditions, should be adherent, inert, impervious, and above all, thick enough to provide a completely nonporous film through which ions, such as sodium chloride, cannot penetrate. Thin vinyl, epoxy, and other coatings have shown the above reaction; on the other hand, when these materials have been properly applied and applied in proper thickness, no evidence of such a reaction takes place.

When only a cosmetic coat is to be applied, under conditions where sodium chloride ions are available, it may be better to leave the zinc coating bare rather than to put a coating of improper thickness or porosity over it.

As indicated by the various exposures for zinc and inorganic coatings, the environment has an important effect on these coatings. It is responsible for conditions whereby the coating is much more resistant and long-lasting, as well as conditions which can lead to its breakdown. Fortunately, the reactions to the environment are such that the majority of zinc-coated surfaces, either galvanized or inorganic zinc, have provided protection to steel surfaces which is not possible by any other form of protective coating.

The use of inorganic zinc coatings for actual immersion service has long been an area of debate. Many coating and corrosion engineers claim they should be used for atmospheric purposes only, while others recommend immersion when properly topcoated. This "properly topcoated" criteria seems to be the key to immersion resistance, particularly in water and salt water, and the organic primer applied directly over the inorganic zinc is the key to its success. One very effective immersion coating system consists of:

1. One coat water- or solvent-base inorganic zinc
2. One coat epoxy polyamide primer
3. Two coats epoxy coal tar
4. A total thickness (complete system) of 500 microns (20 mils)

This has been an effective system for ship bottoms

---

and ballast and ballast crude oil cargo tanks. Another system extensively used for ship bottoms and boottopping areas has been:

1. One coat inorganic zinc
2. One coat vinyl primer
3. Two coats vinyl topcoat
4. A total thickness (complete system) of 200 to 250 microns (8 to 10 mils)

Figures 6.62 and 6.63* show the boottopping on a very large crude carrier (VLCC) after five years of continuous service. Even with the extreme abrasion caused by the dock shoes there is no corrosion or other coating failure in evidence. The ship utilizes the same coating system on the bottom, and it has been equally effective.

Topcoating of Inorganic Zinc Coatings

Millions of square feet of surface have proven that inorganic zinc coatings are very effective when used alone and not topcoated. However, topcoating to improve chemical resistance, general corrosion resistance, or appearance is the more common practice. Better appearance and conformance to a color scheme is by far the primary reason for topcoating. Most inorganic coatings are a nondescript gray and may become mottled by weather exposure. This in no way changes their corrosion resistance; in fact, under some conditions the uncoated inorganic may perform better than one which is poorly topcoated *e.g.,* voluminous white deposit buildup. Topcoating has become the rule with IOZ coatings rather than the exception, and most of the high-performance coatings not only adhere well to the inorganic zinc, but the lives of the topcoats are multiplied several times due to lack of undercutting. The inorganic zinc, organic topcoat system provides maximum corrosion resistance and life for coatings in difficult corrosion areas.

While topcoating is a preferred coating method, it also can create a serious problem under many conditions. If an organic topcoat is applied over a newly formed inorganic zinc-base coat, immediate blistering of the topcoat occurs while the coating is still wet. These blisters may break, leaving a pinhole within the film, or the organic coating may be sufficiently strong so that the blister forms and stays in the film. Oftentimes, as the coating dries, the blister falls back and flattens out in the surface. The areas under the blister will have no adhesion. This problem is due to the volatile materials or solvents evaporating into the porous zinc coating as the organic coating dries. If the surface of the organic dries rapidly and forms a skin, the vapor pressure from the volatile solvents within the zinc coating exerts sufficient pressure to push up underneath the topcoating and form blisters (Figure 6.64).

If the surface is warm, the solvent evaporation is more rapid, the vapor pressure is greater, and the blistering is exaggerated. The opposite is, of course, true for cooler surfaces. There are several approaches to overcoming the problem.

1. A nonsolvent topcoating may be used. In this case

FIGURE 6.62 — Close-up view of boottopping (right) on a VLCC after 5 years exposure. Note the smooth, corrosion-free surface. The boottopping is immersed, except when the ship is light (without cargo or ballast). Note severe abrasion from dock bumpers (left).

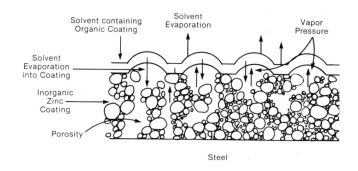

FIGURE 6.64 — An inorganic zinc topcoat blistered from the vapor pressure created by a rapidly drying surface.

(there are a number of nonsolvent or very low solvent topcoatings available), there is little or no solvent to evaporate. As the coating is applied, it flows into the pores and dries in place, sealing the surface of the inorganic zinc coating. This is the best approach to the problem of blistering.

2. A coating containing a high boiling or slow evaporating solvent may be used. This acts in many ways like

*See color insert.

the nonsolvent coating. With the high boiling, low volatility solvents, the coating can penetrate and flow on the surface without any buildup of vapor pressure. This eliminates any blister problem. Care should be exercised, however, on warm or hot surfaces such as tank tops.

3. Some applicators and coating manufacturers recommend the application of a flash or mist coat followed within minutes of a full coat of the organic topcoat. The objective is to apply a very thin coat of the organic over the inorganic zinc. The solvent flashes off of the thin membranes and semiseals the inorganic zinc surfaces. This process is less than foolproof since the zinc coating is still porous, solvents can penetrate, and blisters still form. It does help on cool and on vertical surfaces that are not warm, but it is not completely effective on surfaces warm from the sun.

4. Water-base topcoats also are used. In this case, the vehicle is a water emulsion, and the topcoat acts in many ways like the coatings with a slow evaporating solvent. Water evaporates slowly and allows the coating to flow without building up much vapor pressure. Also, and this may be more important, the vehicle is a dispersion (emulsion) and forms a film in a manner different from that of a solvent-base coating. The emulsion is a series of discrete particles until almost all of the water has evaporated. At that time, the particles coalesce into a film, while prior to that time the coating is porous to water and water vapor. Thus, if a thin water-base emulsion is applied for cosmetic purposes and the structure is in a humid area where sodium chloride or other chemical salts are present, the porous topcoat could cause the zinc coating to self-destruct by forming voluminous white salts on the surface, as was previously described.

This does not mean, however, that all inorganic zinc coatings will cause topcoats to blister, or even that this problem occurs in most inorganic topcoat systems. In fact, coatings which have cured from a few days to several months, depending on the area and weather conditions, usually do not cause blistering. Post-cured inorganics are normally free of blistering problems, since the post-cure seals the surface with insoluble salts during the curing process. There is little chance that the thin preconstruction primers will cause blistering either. This type of coating is sufficiently thin so that little porosity exists. The organic-modified inorganic zinc coatings also have been used to help eliminate this problem. In this case, the organic modifier in the film helps in terms of volume to reduce the coating porosity.

All in all, the blistering of topcoats over an inorganic zinc can be a frustrating problem. On the other hand, advance planning and timing of the topcoat application during cool conditions and the use of one or more of the previously described approaches can prevent this application difficulty.

Rapid Topcoating of Inorganic Zinc Coatings. Under most application conditions, inorganic zinc coatings undergo considerable exposure times so that the initial cure is complete before they are overcoated. This has always been considered necessary. Recent work, however, indicates that rapid overcoating of some inorganic

zinc formulations may be possible. Special formulations of solvent-based inorganics have been utilized for this purpose. By increasing the degree of prehydrolysis of the silicate polymer, the amount of moisture needed from the air is reduced to a minimum for a complete cure. Since organic films have a definite moisture transfer rate, the cure can be completed after the topcoat has been applied. Daniel H. Gelfer, in his paper, *Rapid Topcoating of Inorganic Zinc Rich Primers,* explains this mechanism.

Mechanism of Curing Under Topcoats

In an attempt to explain these results, the following hypothesis was evolved:

1. The silicate binder in the primer film at the moment of topcoating is at the same degree of prehydolysis and inorganic polymer formation as formulated in the wet package condition.

(a) The brief time interval between primer application and topcoat application is not sufficient for any moisture uptake from the atmosphere.

2. During the saturated humidity exposure period, moisture vapor permeates through the topcoat, in an amount determined by the moisture vapor permeability of the topcoat.

3. Depending upon the moisture vapor permeability of the topcoat, sufficient water will permeate to and be made available to the silicate vehicle of the primer and will react with and complete the hydrolysis of the vehicle.

4. The volatile byproduct of the hydrolysis reaction, ethyl alcohol or other alcohol, permeates the organic topcoat in the opposite direction, without disrupting the topcoat, and escapes into the external atmosphere.[11]

Differences in the moisture vapor permeability characteristic of the organic topcoat were reviewed to see how they would affect the 72-hour test results obtained above. The permeance of these topcoats, plus other candidate topcoats, is recorded in Table 6.13.

Since Table 6.14 consists only of laboratory tests, additional tests and field trials are necessary before the rapid overcoating of certain inorganic zinc primers can become a fully proven process. The current research, however, indicates a trend which can become important to corrosion engineers.

TABLE 6.13 — Permeance of Organic Topcoats

Topcoat Type	Permeance-Metric Perms (ASTM E-96) (Gm/24 Hour. Sqm. mm Mercury)
High-Build Polyamide Epoxy	0.35
High-Build Fast-Drying Epoxy	0.30
Epoxy/Urethane	0.105
Coal Tar Epoxy	0.042
High-Build Vinyl	0.092
Vinyl-Acrylic	0.115
Chlorinated Rubber	0.089

NOTE: Comparing these permeance values with onset of primer hardness reveals a correlation between rapid development of primer hardness and the higher permeance values of the topcoat.

[SOURCE: Organic and Inorganic Zinc-Filled Coatings for Atmospheric Service, 6B173, NACE, Houston, TX (1973).]

Comparison Summary

As an indication of the effectiveness of various zinc-rich coatings, Table 6.15 provides some results of actual full-scale applications after several years of service. Both the coatings and the service conditions are compared. Zinc-rich coatings, both alone and overcoated, are represented, with the results indicating the excellent corrosion protection provided by most zinc-rich products. Generally, the inorganic zinc coatings provide longer continuous service than do the organic zinc-rich products.

Tables 6.16 and 6.17 give a direct comparison of the general resistance and physical properties of both organic and inorganic zinc-rich coatings. Such tables summarize the difference between the two systems and provide general information about the two coating types. The information is not specific for any individual coating.

Inorganic zinc coatings have come a long way since they were first conceived by Victor Nightingall. Their use is expressed in acres rather than in square feet, and they have proven effective in hundreds of areas of severe corrosion. They have provided revolutionary corrosion control irrespective of the size, complexity, or location of the structure, and in one coat they have reduced both initial capital cost and the continuing cost of maintenance. Even with such extensive use and service, however, many believe that their development is still in the beginning stages. These inorganic-based materials may one day provide the means by which all metal surfaces are both protected and decorated. Their outstanding qualities have been proven; it requires only imagination and research to match them with the work to be done.

TABLE 6.14 — Immersion Performance: Single-Package Inorganic Zinc Topcoat Systems
5500 Hour Synthetic Seawater Immersion After
1200 Cleveland Humidity Chamber Exposure

Topcoat Type	Primer Dry Time (Hours)	Adhesion to Steel	Intercoat Adhesion	Blistering (ASTM D-714)	Scribe Corrosion (ASTM D-1654)
High-Build Polyamide Epoxy	1	Good	Good	None	None
	4	Good	Good	None	None
	7	Good	Good	None	None
	16	Good	Good	None	None
Fast-Drying Epoxy	1	Fair	Good	None	None
	4	Good	Good	None	None
	7	Good	Good	None	None
	16	Good	Good	None	None
Coal Tar Epoxy (16 mils)	1	Poor-Fair	Poor-Fair	None	None
	4	Poor-Fair	Poor-Fair	None	None
	7	Poor-Fair	Poor-Fair	None	None
	16	Poor-Fair	Poor-Fair	None	None
Epoxy Tiecoat/Urethane Topcoat	1	Satisfactory	Satisfactory	Few No. 8	None
	4	Satisfactory	Satisfactory	Few No. 8	None
	7	Good	Good	Few No. 8	None
	16	Good	Good	None	None
Chlorinated Rubber	1	Good	Good	None	None
	4	Good	Good	None	None
	7	Good	Good	None	None
	16	Good	Good	None	None
High-Build Vinyl	1	Satisfactory	Satisfactory	Dense No. 8	None
	4	Good	Good	Dense No. 8	None
	7	Good	Good	Dense No. 8	None
	16	Good	Good	Dense No. 8	None
Vinyl Tiecoat/Vinyl Topcoat	1	Satisfactory	Satisfactory	Few No. 6	None
	4	Good	Good	None	None
	7	Good	Good	None	None
	16	Good	Good	Few No. 8	None
Vinyl Tiecoat/Vinyl-Acrylic Topcoat	1	Fair	Fair	Dense No. 8	None
	4	Good	Good	Dense No. 8	None
	7	Good	Good	Few No. 8	None
	16	Good	Good	Dense No. 8	None

[SOURCE: Gelfer, Daniel H., Rapid Topcoating of Inorganic Zinc-Rich Primers—A Case for Improved Productivity, CORROSION/80, NACE, Houston, TX (1980).]

TABLE 6.15 — Experience Records on Various Zinc-Filled Coatings

Binder Type	Type of Surface	Geographic Location	Exposure	Surface Preparation	Coating System	Performance
Inorganic Silicate, Post-Cure	Structural steel in a chemical plant	Los Angeles, CA	High humidity, salt air, industrial fumes	Sandblast Near White NACE No. 2	1 coat zinc silicate 1 coat tiecoat 2 coats vinyl; total thickness 14 mils	10 yr maintenance free service and still offering excellent protection
Inorganic Silicate, Post-Cure	Steel piping on cooling tower	Lake Charles, LA	High humidity	Sandblast Near White NACE No. 2	1 coat 3 mils	12 yr continuous service
Inorganic Silicate, Self-Cure	Structural steel	Cape Kennedy, FL	High humidity, salt air	Sandblast Near White NACE No. 2	1 coat zinc silicate, 1 coat vinyl tiecoat 2 coats vinyl	Applied in 1959, touch-up of topcoat, primer still intact and in service (1967)
Inorganic Silicate, Self-Cure	Structural steel	Texas Gulf Coast	High humidity, industrial fumes	Sandblast Near White NACE No. 2	1 coat zinc silicate 2-3 mils	5 yr service (1967)
Inorganic, Phosphate	Tank exterior	Texas Gulf Coast	Coastal industrial atmosphere	Sandblast Near White NACE No. 2	1 coat zinc primer 3 mils 1 coat vinyl-5 mils	In excellent condition after 8 yr service.
Organic Epoxy Ester	Unitized automobile body, blind fenders, box sections	Detroit, MI	High humidity, road salts	Pickled and/or phosphated steel	1 coat 2-3 mils	Prevented corroding of car body for 8 yr.
Organic Chlorinated Rubber	Structural steel	Chicago, IL	Industrial fumes	Near White Sandblast NACE No. 2	1 c chl. rub. zinc 1 c chl. rub. fin. 5 mils	6 years condition perfect
Organic Phenoxy	Structural steel, paper mill	Charleston, SC	Coastal area, acid fumes	Shop blasted	1 coat zinc primer 2 coats epoxy 2-component 8-10 mils total dry film	6 yr service, no touch-up or repaint to date.
Organic Epoxy, two-component	Structural steel	New Jersey	Industrial fumes, coastal atmosphere	Sandblast Commercial Blast NACE No. 3	1 coat epoxy zinc primer 1 coat epoxy topcoat	After 5 yr, topcoat has weathered to expose primer. Complete topcoat recommended.
Organic Polystyrene	Galvanized structural steel	Philadelphia, PA	Rural and mild industrial	Repair of damaged galvanized steel, field repair.	1 coat	Requires touch-up in 4-5 yr. Pinhole rusting observed

[SOURCE: Organic and Inorganic Zinc-Filled Coatings for Atmospheric Service, 6B173, NACE, Houston, TX (1973).]

TABLE 6.16 — Resistance of Zinc-Filled Coatings in Various Environments

Organic	Inorganic
Water and Moisture	
Excellent resistance to high humidity, splash, and spray conditions in both fresh and saltwater atmospheres.	Excellent resistance to high humidity, splash, and spray conditions in both fresh and saltwater atmospheres. The inorganics in single-coat applications have a long history of excellent service in humid environments.
Inorganic Acids, Oxidizing Agents, Organic Acids	
Not recommended for direct splash, spillage, or fume from inorganic acids, oxidizing agents, and organic acids.	
Alkalies	
Will tolerate mild alkaline atmospheres. Alkyd vehicles are not recommended for alkaline service.	Not recommended for prolonged exposure to alkaline atmospheres. Silicate binders are attacked by dilute alkalies.
Salt Solutions	
Excellent resistance to splash, spillage, and fog of neutral salt solutions. Resistance to acid and basic salts will depend on pH of the media.	Excellent resistance to splash, spillage, and fog of neutral salt solutions.

(continued)

Organic	Inorganic

Solvents

Very limited resistance to solvents. Epoxy, two components, alkyds, and epoxy esters can be used in aliphatic hydrocarbon service.

Excellent resistance to aliphatic, aromatic, and dry chlorinated hydrocarbons, petroleum products, esters, and ketones.

Oils and Fats

No information available.

Limited information indicates satisfactory performance. Not satisfactory in fatty acids or oil with high acid number

Gases

Coatings are permeable and will be attacked by wet acidic gases such as SO_2, SO_3, Cl_2, and also by alkaline gases such as NH_3.

[SOURCE: Organic and Inorganic Zinc-Filled Coatings for Atmospheric Service, 6B173, NACE, Houston, TX (1973).]

TABLE 6.17 — Physical Properties of Applied Zinc-Filled Coatings

Organic	Inorganic

Temperature Limitations

Maximum dry service temperature at which a coating can be used is determined by the type of binder. The normal maximum temperature limits of the various binders are listed. These same binders loaded with zinc dust may yield films with different service temperature.

Provide protection at a maximum service temperature of 315 C (600 F). Some have been reported to withstand intermittent service to 593 C (1100 F) when topcoated with silicone aluminum. The latter temperature is above the melting point of zinc (420 C, 788 F).

Chlorinated Rubber	71 C (160 F)
Styrene-Butadiene	71 C (160 F)
Polystyrene	71 C (160 F)
Epoxy Ester	107 C (225 F)
Phenoxy	120 C (250 F)
Epoxy, 2-Component	120 C (250 F)
Silicone Alkyd	232 C (450 F)
Silicone	400 C (750 F)

Abrasion and Impact Resistance

Epoxy esters, thermoset epoxies, and thermoplastic epoxies have good abrasive resistance. Chlorinated rubber and polystyrene binders lack this resistance initially. As organic films age, their abrasive resistance increases but never reaches that of the inorganics.

Possess outstanding impact and abrasion resistance when thoroughly cured.

Weathering

Excellent resistance to atmospheric weathering when applied in 2- to 3-mil dry film thickness. Typical matte gray appearance will lighten on exposure. White rust common to galvanized steel can form on zinc-filled coatings. Optimum performance is obtained from multiple coats of zinc paint or when topcoated with appropriate finish coat.

Excellent resistance to atmospheric weathering in one coat. Typical matte gray appearance will lighten on exposure. White rust common to galvanized steel can form on zinc-filled coating.

Toxicity

In general, the cured films are considered nontoxic. The use of extender pigments, plasticizers, and resin modifiers may change the nontoxic status significantly.

Cured films using zinc dust as the only pigment are considered nontoxic and have been used as a container lining for dry food products. The use of certain extender pigments, particularly lead compounds, will influence the toxicity rating.

Weight of Applied Coatings

Dry film specific gravity normally ranges from 4.5 to 5.5, which will yield films varying in weight from 0.02 to 0.03 pounds per square foot per mil.

Electrical Properties

Should be considered electrically conductive coatings. The degree of conductivity will depend upon zinc content, binder type, as well as the amount and type of extender pigment.

Adhesion

Excellent adhesion to sandblasted and acid-pickled steel. Polystyrene, phenoxy, and chlorinated rubber coatings are applied also over galvanized steel, aluminum, and copper surfaces with excellent results.

Excellent adhesion to properly sandblasted steel surfaces. Adhesion to sandblasted aluminum is good.

Appearance

Organic and inorganic zinc-filled coatings are flat. Color may not be uniform.

[SOURCE: Organic and Inorganic Zinc-Filled Coatings for Atmospheric Service, 6B173, NACE, Houston, TX (1973)]

References

1. Mallet, R., Brit. Assoc. Advancement of Science, Vol. 1, pp. 221-388 (1840).
2. Nightingall, Victor, U.S. Patent 2.440.969, May (1948).
3. Nightingall, Victor, Dimetalization for the Prevention of the Corrosion of Iron, Steel and Concrete, Melbourne, Australia (1940).
4. Rochow, E. G., Comprehensive Inorganic Chemistry, Chapter XV, Chemistry of Silica, Pergamon Press, New York, NY.
5. Berger, Dean M., Current Technology Review—Zinc Rich Coatings, Modern Paint and Coatings, June (1975).
6. Intorp, Norbert B., Enhanced Zinc Rich Primers, CORROSION/80, Preprint no. 114, National Association of Corrosion Engineers, Houston, TX, 1980.
7. Steel Structures Painting Council, Spec. SSPC Paint 20X.
8. Velsboe, O. P., Organic Zinc Coatings, presented at the International Ship Painting and Corrosion Conference and Exhibition, May, 1974.
9. Uhlig, Herbert H., Corrosion and Corrosion Control, John Wiley & Sons, Inc., New York, NY, pp. 202-204, 1965.
10. Pourbaix, Marcel, Atlas of Electrochemical Equilibria in Aqueous Solutions, Chapter IV, Sec. 51.1, National Association of Corrosion Engineers, Houston, TX, 1974.
11. Gelfer, Daniel H., Paul VanDorsten, Rapid Topcoating of Inorganic Zinc Rich Primers—A Case for Improved Productivity, CORROSION/80, Preprint No. 113, National Association of Corrosion Engineers, Houston, TX, 1980.

7

Structural Design for Coating Use

Design is a key word in our culture. It not only dictates fashion styles, but influences the wide world of industrial products and plants, offshore structures, ships, aircraft, household equipment, and consumer products as well. Design and engineering are as important to our complex, mechanical society as raw materials, for without the ability to visualize an end product and the skills to realize that vision, most basic materials would be useless.

Even the formulation of protective coatings involves designing. *Formulation*, in fact, is the design of a coating for a specific job, for the results of improper selection and combination of materials to form a coating could be much worse than having no coating at all. Coatings in themselves, however, cannot be designed as end products. Since a coating must be applied and formed in place before it can perform the role for which it was designed, it is part of a much larger design, *e.g.*, that of a structure, a piece of equipment, or an entire plant. The design and construction of a coated structure under corrosive conditions are the keys to its life span and ultimate effectiveness. Thus, a large part of expensive service failures due to corrosion would not occur if proper precautions were taken during the design stage.

One factor in the proper design of a coating is its acceptability for application to the plant or structure for which it was intended. On the other hand, the structure or object to which the coating is to be applied must be designed in such a way that it can easily accept the coating and provide a proper base for it. This is an important concept for the architect or engineer who is responsible for the design of a building or plant. Many of them do not understand that if a structure is not properly designed to accept a coating and allow it to form a continuous film of even thickness, it is difficult to keep the structure free of corrosion.

Bridges are often a good example of this, since many of them are built with open box beams, hidden recesses,

rivets, bolts, overlapping joints, and many other untenable problems from a coating and a corrosion standpoint (Figure 7.1). Figure 7.2 is a closer view of the junction of several key bridge members showing the open box section, the overlapping bolted plates, and hidden and inaccessible areas which are all exposed to a marine atmosphere. Such design may be adequate from a strictly engineering standpoint, but can soon lead to costly maintenance and even steel replacement because of poor design for proper coating application.

Plants of all types, from those for food processing to refineries, may be designed in the same way, *i.e.*, with little regard for corrosion resistance. This lesson of design for coating application and corrosion resistance was soon learned by the offshore engineers who quickly changed from structural shapes to pipe or cylinders in order to reduce the surface area and eliminate as many corners, edges, and discontinuous areas as possible, thus forming a smooth structure with as many plain or rounded surfaces as was practical. Figure 7.3* is a good example of the original design used in offshore construction with its angles, channels, and H-beams. This section of the offshore platform, however, was removed as being unsafe due to corrosion.

Figure 7.4* shows an offshore platform using the more modern design with rounded smooth surfaces. There is no question about the severity of the exposure, yet this photograph was taken after 14 years of service with only one coat of inorganic zinc coating. The corroded areas in the lower region are open work gratings which were galvanized. Of all structural forms, open work grating is the most difficult to coat and properly protect.

Figure 7.5 is a closer view of the pipe design showing the smooth, uninterrupted surface, even where one sec-

---

*See color insert.

FIGURE 7.1 — Typical bridge construction where design for coating and corrosion protection has been overlooked. Note the angles, corners, edges, and open box sections.

FIGURE 7.2 — Structural framing on a bridge in a marine atmosphere. Coating application is approximately 2 years old. Note the heavy scale (left center) which has dropped from the back of the channel above where coating access was impossible. Note also the rusting of edges, boltheads, and overlapping steel sections.

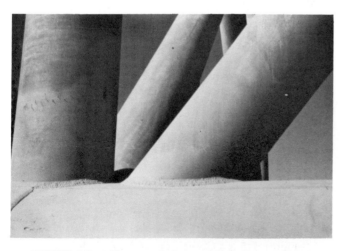

FIGURE 7.5 — Close-up of pipe section design showing the smooth flow of one section into another area where they are joined.

tion joins another. Such design provides the best possible surface for coating and obtaining maximum corrosion protection. The entire surface, including the welds, is freely available for proper surface preparation and coating application.

This offshore design was quickly adopted for a number of reasons: (1) because of the severe continuous corrosion encountered; (2) because the use of steel as the basis of a structure was the only practical material, from both an ease of workability and a cost standpoint; (3) coatings are the only practical corrosion control method for large complex structures subject to continuous severe corrosion; (4) the most effective coating is one applied to a plain, smooth surface with a minimum of discontinuities; and (5) the combination of a steel structure and high-performance coatings provides the most economical method of construction for large, complicated corrosion-resistant structures.

Principle of Design for Coating Use

The basic principle underlying design for the use of coatings on corrosion-resistant structures is to keep the structure as simple as possible and to reduce the surface area to be coated to the smallest area practical. This includes the deliberate reduction of complicated areas, *i.e.,* those inaccessible to proper coating application; the elimination of all overlaps and riveted or bolted joints wherever possible; and the reduction of sharp edges, corners, and rough areas.

These considerations must then be balanced with the necessary engineering requirements for a safe and effective structure able to perform the service for which it was intended. While this may not be an easy task, it is much more effective than adhering to conventional design methods which ignore the need for a good coating base. Examples of this more effective design approach are given by Rudolf in an early paper on design.

A 4 × 5 inch T can be used to replace two 3 × 3 inch angles with the following design figures. Two 3 × 3 × 1/4 inch angles weigh 9.8 pounds per foot of run, a 4 × 5 inch T weighs 8.5 lbs. Two 3 × 3 inch angles have a radius of gyration in x direction of 0.93 and the 4 × 5 inch T is about equivalent. The total exposed area of the two 3 × 3 inch angles is about two square feet per foot of run, of the 4 × 5 inch T, about one and one half. Of much greater importance is the fact that the area between the backs of the angles, 25 percent of the total, is a foul area which cannot be cleaned or coated adequately and which is a source of maintenance trouble all its life. The T on the other hand has no foul area, is readily cleaned and coated, does not harbor corrosion and presents even less area to clean and coat than the shape it replaces.

A comparative newcomer to this field of substitution of simple shapes for a combination of shapes is the welded hollow tube, sealed at both ends. The most familiar and striking example of this is the water tower leg made from rolled plate and welded. With this substitution surface is not only decreased, but half the total surface is removed from corrosive influence altogether. Consider the perimeter of a standard combination of shapes, a tower leg made up of two channels, plate riveted to the channels for a third side and the fourth side a lattice of bars riveted to the channels. If 12-inch channels had been used, the total exposure to corrosion per foot of run of such a column could scarcely be less than ten square feet. Whereas a 12-inch diameter pipe sealed at the ends has a total corrosion exposed area of three square feet per foot of run [Figure 7.6].[1]

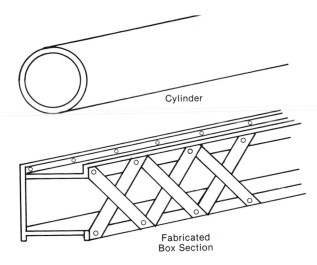

Cylinder

Fabricated
Box Section

FIGURE 7.6 — The comparative simplicity of essentially equivalent structural designs. Note the greatly reduced surface area of the cylinder.

It should be noted that in addition to reducing the exposed area (pipe vs made-up beam), the use of the pipe for construction also eliminates numerous angles, corners, overlaps, and rivets, all of which are difficult to coat properly as well as being focal points for corrosion.

The above examples not only influence the effectiveness of the coating and the corrosion resistance of the structure, but they also influence effective coating application. Pipe is by far the easiest structural section to coat, while the made-up box section cannot be properly coated by any practical means. If the structure cannot be properly coated, and there are literally hundreds of areas in presently designed structures where this is the case, the structure cannot be made even reasonably corrosion-resistant. The basic design, then, controls the effectiveness of the coating application and thus definitely influences the corrosion resistance of the structure.

The selection of materials and structural forms solely for their functional use may not, then, be the best answer to all design problems. Many such designs have proven unreliable, unsafe, expensive, and very corrosion prone. The designer, therefore, must look beyond the functional aspects of his design and also consider the life of the structure under the environmental conditions in which it is to be placed. In the practice of responsible engineering, corrosion control cannot remain subordinate to any other portion of the design work any more than weight, wind loads, or earthquake stresses can be overlooked.

Interior and Exterior Design

There are two distinct areas where design for the use of coatings is quite different: (1) the interior of containers for use in processing materials, and (2) the exterior of structures subject primarily to atmospheric conditions. When dealing with the interior of tanks or similar structures, design for coatings and linings has generally followed the basic principle outlined above, i.e., reduction of the area exposed to any corrosive substance.

The interior of a tank using dished heads is a good example. The entire interior surface is a perfectly smooth, continuous surface. The only breaks in the surface are at welds, at the manhole, and at inlets or outlets to the tank. Thus, the welds are generally ground smooth and the outlets in the tank, including the manhole, are ground around the perimeter of the opening in order to provide a smooth radius rather than a sharp corner or edge. In other words, the interior is designed to accept the coating or lining in order to gain its full effectiveness. All welded tanks, no matter what size, generally follow this design principle.

Large storage tanks may have a roof problem where structural supports are required to support the roof. Many of them use a central column or series of columns throughout the tank to support the roof. If the material to be contained is particularly corrosive, all supports, trusses, and beams other than the columns can be placed on the exterior rather than the interior, and all columns can be made from pipe. This type of construction maintains the smooth interior surface throughout, thus making for the best possible coating application.

Bolted tanks, which are a common design option, generally are used for temporary storage where tanks can be moved from one place to another. Unfortunately, the temporary storage often becomes permanent storage and the bolted design then presents a problem if a coating is required to protect the interior from serious corrosion.

The storage of sour crude is a good example of the type of product which can cause considerable corrosion damage. Figure 7.7 indicates the extensive corrosion which may occur with improperly designed and coated tanks used for the storage of sour crude. The complete destruction of this tank required less than two years. An all-welded tank, properly coated, would have eliminated this problem. Much larger tanks in the same area which were properly designed and coated are performing satisfactorily.

Generally, interior surfaces subject to corrosion are more likely to be designed with coatings in mind than are exterior structural surfaces. Many modern plants are not fully enclosed, so that the exterior of all the surfaces is subject to whatever atmosphere may be in contact with the structure. This type of design is particularly true of chemical plants and refineries where much, if not all, of the processing equipment and pipe is subject to the plant atmosphere.

Figure 7.8 shows a pipe rack in a typical chemical plant fabricated from a standard structural shape. Joints are overlapped and bolted, increasing the surface area and providing additional edges, corners, cracks, and crevices where corrosion can initiate and continue. The exterior design of such structures could vastly improve the corrosion resistance of the unit. Structures that have been designed for the acceptance of a coating, i.e., those with flat, cylindrical, and smooth surfaces, as well as surfaces joined by continuous welding with a minimum of overlapping joints, are much more easily maintained and coated.

The cost of maintaining a structure also is directly related to its design. A structure with a minimum of edges, corners, crevices, etc. can be much more easily maintained and will have a much longer useful life under

FIGURE 7.7 — A bolted steel production tank used for the storage of sour crude. The tank was completely corroded from both the interior and exterior in 2 years.

FIGURE 7.8 — A design showing hundreds of linear feet of edges, as well as overlapped bolted joints. Such a design is difficult to both properly coat and maintain.

corrosive conditions than one which is designed without regard to the structural requirements of coatings.

Coating Problems Related to Design

Structural Steel Shapes

Structural steel shapes almost always pose coating problems, particularly those in smaller areas. Unfortunately, they represent a basic building material and therefore, in many cases, must be used regardless of the coating difficulties they create. These basic structural shapes consist of the angle, channel, H-beam, and I-beam (Figure 7.9).

The angle contains the most problem areas. The outside of an angle forms a continuous right-angled edge and is a danger area where coatings are generally thin. Thus, the area is subject to considerable corrosion damage because of its location. The same holds true for the arm ends of an angle since the edges and exterior corners are more vulnerable to coating failure than the flat portion of the angle. Most high-performance coatings, with the exception of the inorganics, exhibit considerable surface tension upon drying which causes the coating to pull away from edges and exterior corners, making it much thinner in these areas than on the flat section. Where a coating is applied to an area making a sharp change in direction (Figure 7.10), the internal forces in the coating draw it away from the edge, leaving it either exposed or only minimally covered.

The interior corner is also an area of difficulty and a danger point. During surface preparation, considerable dust or other contaminants can build up in this area where removal of these materials is much more difficult than on a flat surface. From a maintenance standpoint, it is an area where dirt accumulates, particularly if the angle is in a horizontal upright position. It also accumulates moisture which can quickly create areas of coating failure. Interior corners are also difficult to reach by either spray or brush and therefore they make it difficult to maintain a constant coating thickness. Air spray tends to bounce out of a corner, while airless spray may apply much too great a volume of coating in that area.

In many cases, the high-strength organic coatings tend to shrink somewhat upon drying, due to both surface tension and decrease in volume because of solvent evaporation. If there is dust on the surface prior to coating which then prevents proper wetting of the interior corner area, the high-tensile film may tend to bridge over the inside angle, causing an air space to form between the coating and the metal (Figure 7.11). This coating reaction becomes a focal point for corrosion in any corrosive environment.

Channels cause problems similar to those of angles, with the exception that the problem areas on channels are double that of the angle. I-beams and H-beams also have similar danger points to angles and channels except for the sharp right angle exterior corner. They are perhaps somewhat less of a problem than the other two shapes since their dimensions are often greater, thus allowing a greater proportion of flat area which can be properly coated as compared with the area of coated corners and edges. Thus, wherever coatings are involved, structural shapes should be regarded with additional care in surface preparation and coating application so that a continuous, smooth, even coating thickness is created over the entire surface.

In order to reduce surface preparation costs, many coating systems are applied either in part (e.g., primer or base coat only) or in full to sections of the structure prior to erection. Applying the coating "on the ground," where there is ready access and easy movement around the section, makes for better surface preparation and better ap-

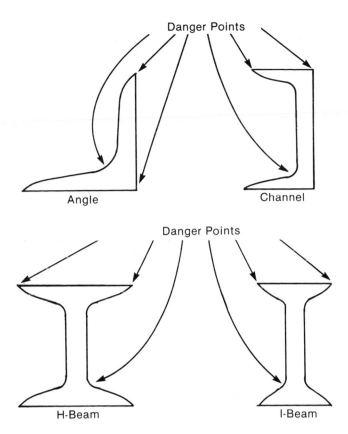

FIGURE 7.9 — Basic types of structural shapes.

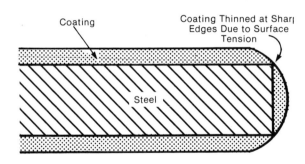

FIGURE 7.10 — Coating applied to a square-cut steel section.

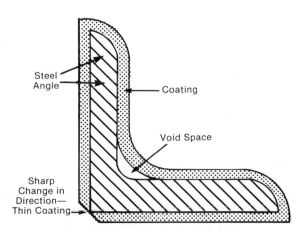

FIGURE 7.11 — A coating danger point is created where the coating bridges over an inside angle, causing an air space to form between the coating and the metal.

plication as compared with the same structure coated upon completion hundreds of feet off the ground. Similarly, many block sections of ships are precoated before erection because of greater access to the surface, better visibility, better ventilation, and better overall coating application conditions. Where structural shapes are used, such prefabrication and design, with coating before erection, increases the probability of a good corrosion-resistant structure. This is particularly the case where inorganic zinc coatings are applied as a base coat since they are relatively unaffected by handling after coating because of their excellent abrasion resistance.

One problem with precoating a prefabricated section before erection lies with the construction joints. Where structural steel is erected using high-strength bolts or rivets, the overlapping surface can create a corrosion problem, particularly where it is left bare of coating prior to joining. Since inorganic zinc coatings have a good coefficient of friction, they can be applied to faying surfaces prior to erection. Coating the faying surfaces in this way reduces the possibility of corrosion at the steel joints.

In large fabricated sections, such as the block sections in the construction of a ship, the joining welds are always a problem. Again, a base coat of inorganic zinc can be used on the block sections and junction welds made with very little burnback of the coating. After installation and prior to coating application, the welds must be smooth and thoroughly prepared. There are many instances of block section designs where the body of the coating on the block section is in excellent condition, yet all of the construction welds have failed within a short period of time due to poor surface preparation and lack of care during coating application.

Sharp Edges

Sharp edges of steel on overlapping plate, or edges left by shearing or cutting can present serious coating problems. Coatings will fail first in areas with sharp edges almost without exception. This is demonstrated in Figure 7.12 where an exposure panel has been placed in a marine atmosphere. The panel was a cold-rolled panel sheared to shape, and then sandblasted and coated. Extra care was even taken in coating the edges to prevent edge corrosion, but nonetheless, when the coating started to fail, it first failed at the square cut edges rather than in the smooth area of the panel.

While square cut edges are the most vulnerable to failure, the same principle holds true to the smoother edges of most structural steel which come to a small radius and are still much more difficult to coat than plane surfaces. Not only does the surface tension tend to pull the coating away from these areas, but more often than not, the application of the coating is tangent to the edges, rather than direct, thus creating a thin area of coating along the edge. This is further aggravated wherever there are corrosive dusts or the precipitation of corrosive fumes or mists. These accumulate on the flanges of structural steel, since corrosive products tend to concentrate at the edges of any horizontal shape (Figure 7.13).

In order to provide coating protection for edges, it is necessary to apply coating directly to the edge before each

Structural Design for Coating Use

FIGURE 7.12 — Corrosion beginning at the edges of a panel.

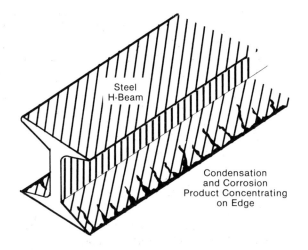

FIGURE 7.13 — Buildup of corrosive products at the edges of a horizontal shape.

complete application and then apply the normal coating layers out over the edge in order to develop an equivalent coating thickness in those areas. In many cases, the best protection is achieved by actually increasing the thickness on edge areas, if at all possible.

Rivets and Bolts

Rivets and bolts, used for connecting steel sections in a structure, represent another area where the coating will preferentially fail. This is due to the increased surface area as well as the increased edge area where this type of construction is used. Each of these methods increases the chance of crevices and similar surface discontinuities around the edges of the bolts or rivets.

There are three common types of rivets: round, pointed (conical), and countersunk (Figure 7.14). The countersunk rivet is the easiest to coat since the finished rivet is almost level with the bare steel, leaving only a small line around the edge of the rivet which requires attention from a coating standpoint. Round rivets are the most common. They have an almost hemispherical head and fit tightly to the surface when properly installed. Conical rivets are becoming less and less common. They used to be a standard material in riveted tanks and riveted tank cars. They create more of a coating problem than the other two types, however, because of the conical shape of the rivet which comes to a relative, though not sharp point at the top. It thus creates an area which can easily be damaged and one from which the coating tends to pull away and form a thin, vulnerable area.

One of the problems in using the last two types of rivets, round and conical, is that of a *cocked* rivet. This occurs where one side of the rivet is driven tight while the other side is cocked away from the surface, causing a crevice between the rivet and the surface (Figure 7.15).

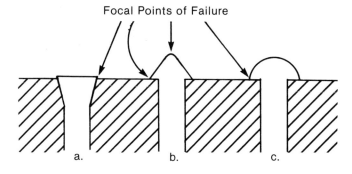

FIGURE 7.14 — Basic types of rivets.

This is an extremely difficult area to coat properly and thus is an area where corrosion can easily take place. All rivets used in corrosive areas should be caulked tight, which involves going around the edge of the rivet with an impact gun and caulking tool, driving the edge of the rivet against the surface to make tight contact so that no crevice areas exist.

Riveted joints between steel sheets represent another area to watch. The overlapping plate, unless thoroughly caulked tight, will provide a crevice area containing air and one which moisture and chemical solutions can easily penetrate. It is also an area which is almost impossible to cover with a coating and thus serves as a focal point for corrosion.

All hot driven rivets have an oxide scale on the exterior. Therefore, they must be thoroughly blasted in

Corrosion Prevention by Protective Coatings

order to remove the black iron oxide which is strongly cathodic to steel and thus an aggravation to corrosion in any crevices which may exist.

In order to obtain proper coating protection on a riveted structure, it is necessary to double coat all riveted areas in the same manner as described for a sharp edge. It is preferable to brush the first coat around each rivet in order to assure full adhesion and coverage. Since rivets are always a focal point for corrosion, it is preferable to use welded construction rather than rivets or bolts to join steel sections in any area where corrosion is a factor.

Bolted Joints

Bolted joints are more difficult to coat than riveted joints, and many modern steel structures are installed with high-strength bolts. Not only do they consist of many sharp edges, including the threads on the bolt and the sharp edges on the hex nut, but washers are usually used which create an additional area for crevices (Figure 7.16). In order to coat bolted areas properly, essentially the same procedure should be used as that for rivets. The first coat (primer) should be thoroughly brushed around each bolthead, which is difficult because of the many edges and crevices that are possible. When spraying subsequent coats, it is necessary to spray completely around each bolthead to cover each surface properly along with the many edges.

In Figure 7.17 the overlapping steel bolted joint shows active corrosion beginning at the joint as well as around several of the individual bolts. While situations such as this often are blamed either on poor coating application or a poor coating, the actual difficulty is the conventional design of the structure which was not suited for a corrosive atmosphere.

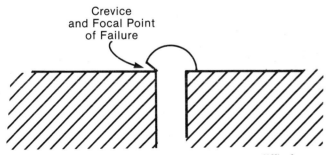

FIGURE 7.15 — A cocked rivet showing the many difficult-to-coat areas.

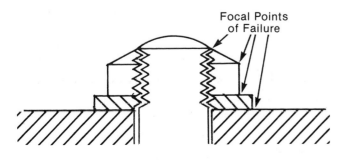

FIGURE 7.16 — Points of failure on a steel bolted joint.

Welds

The majority of modern steel construction uses welding to join steel sections. The process has become extremely sophisticated with automatic machines making most of the welds. Figure 7.18 shows the exterior of the hull of a ship and demonstrates two types of welds. The upper weld on the smooth hull is an automatic weld joining two hull plates. It is smooth and continuous with no undercutting along the edges. The three welds made below the machine weld involve the stabilizer fin which is attached to the hull. These are considered very good hand welds. Notice, however, that they are somewhat rough and are thus more vulnerable to coating application problems and possible corrosion failure than are the machine welds. Thus, machine welds are preferable and made on a structure wherever possible.

FIGURE 7.17 — Bolted steel sections exposed to a marine atmosphere. Note the corrosion at the overlapping joints on and around the bolthead.

Even in ship building, however, where much of the welding is done by machine, a large part of welding footage is still done by hand. Figure 7.19 shows a hand weld between two plates which were sandblasted to emphasize the details. Notice the even roughness of the weld which is created by the manual application. Note also that along the edges there are some minor undercuts and shallow crevices which pose difficulties in coating. This is true particularly when using organic coatings which often bridge such areas. Inorganic zinc coatings, however, which do not shrink on curing, would wet and remain in these irregular areas without difficulty. Weld areas also should be thoroughly blasted to remove scale and weld slag. It is preferable to brush organic primers over a weld to work the coating into all rough and uneven spots.

From a design standpoint, a welded joint is highly preferable to a bolted or riveted joint, both from a corrosion protection and a coating application standpoint. Unfortunately, hand welds are only as poor or as good as the welder who makes them.

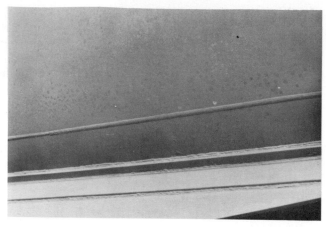

FIGURE 7.18 — Typical machine (top) and hand (bottom) welds.

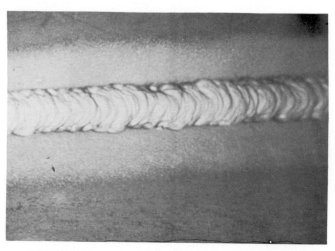

FIGURE 7.19 — Typical hand weld with undercut areas along the edges.

Although rough welding is not designed into a structure, specifications should be written in such a way that rough welds are eliminated from any installation where corrosion is a factor. If they do exist in critical areas, they should be ground smooth and special care taken in coating the areas to ensure full coating thickness and continuity. The best protection for rough welds is an inorganic zinc base coat (Figure 7.20).

Figure 7.21 shows the results of a rough weld on an otherwise well-coated bulkhead. The rough weld and its location close to a reinforcing member caused a porous coating over the weld and therefore early corrosion. Care in the treatment of the weld and in the coating application over the weld could have prevented this costly corrosion problem.

One of the major difficulties along welds occurs because of the weld *spatter*. These are small adherent balls of metal which fly away from the hand weld and stick to the adjacent metal surface. These small, adherent balls of metal provide points from which the coating flows away and becomes thin. Small crevices also develop in and around the base of the metal ball which creates an area where coatings do not tend to penetrate well (Figure 7.22). These rough areas of weld spatter should therefore be removed before the surface is prepared for coating.

Figure 7.23 shows a weld with some weld spatter made in a design which can only cause difficulty. Corrosion is already starting around the weld and in the crevice between the two pieces of metal. Note also the coating break in the middle of the weld caused by the weld slag that was allowed to remain on the surface and was then coated over.

FIGURE 7.20 — Rough hand welds showing areas where metal brackets have been cut from the surface. Inorganic zinc protects such poor workmanship.

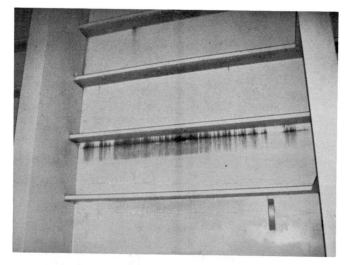

FIGURE 7.21 — Rough welded seam which creates a corrosion problem.

Seal Welds

Smooth seal welds should be used along seams between two metal pieces to seal them against penetration of moisture and other corrosive elements. *Seal welding* differs from the standard forms of welding. Its primary purpose is to provide a metallic seal between two surfaces to prevent penetration of gases or fluids into the crevice and to provide a continuous metal surface over which to apply coating. The secondary purpose of the seal weld is to provide a fastening between the two surfaces.

Weld Flux

Weld flux is necessary for the proper welding of the metal. It prevents oxidation of the molten metal and

Corrosion Prevention by Protective Coatings

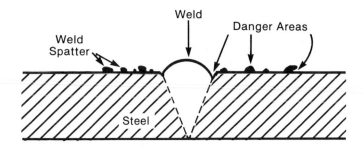

FIGURE 7.22 — Schematic showing a weld with some weld spatter.

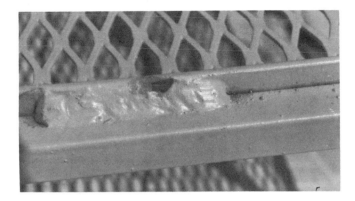

FIGURE 7.23 — A poor design and a poor weld for corrosion protection. Note the weld splatter and the crevice between the two metal sections.

FIGURE 7.24 — Rough steel remaining after a construction bracket was cut from the surface. Note the rough metal and sharp undercuts, all of which create coating problems.

FIGURE 7.25 — Scaffold bracket remaining in a coated tank showing corrosion underneath the bracket. Such permanent brackets should be seal welded to allow for proper coating.

allows it to flow together to make a completely fused joint. On the other hand, especially where there are hand welds, it can be retained in rough areas, particularly along the edges of the weld and even where the weld is blasted. Care in both blasting and coating is necessary, as slag is difficult to remove and its strongly alkaline hydroscopic nature creates a spot where early coating failure will occur. See spot of failure in Figure 7.23.

Although welding is one of the best methods for joining two steel surfaces, there are problems involved in its use which must be taken into consideration wherever coatings are to be applied.

Brackets and Holddowns

Numerous types of brackets, holddowns, permanent scaffolding, and other fabricating aids are used in the construction of steel structures and are welded to the steel surface. These may or may not be designed into the structure, but are oftentimes necessary for the construction of the unit. As long as they do not pose any mechanical problems in terms of the finished structure, they are often left in place. Otherwise, they are either cut from the surface with acetylene torches or merely broken away with a hammer (Figure 7.24).

The rough metal left from such an installation, however, causes a focal point for coating breakdown and therefore serious corrosion. This should be considered in the design specifications which should call for the complete removal of the units and grinding down of the steel to create a smooth surface. Proper specifications also

should call for the seal welding of any brackets which are to be left permanently attached to the structure since corrosion can occur rapidly underneath the overlapping steel (Figure 7.25).

Skip Welds

It is a common design practice to call for the use of a skip welding technique in the construction of many structures. *Skip welding* is a process of welding two or three inches, skipping several inches, and then applying additional welds of a similar length, until the designated area is covered. It is used mainly for reinforcing purposes where it is not considered necessary to provide a continuous weld (Figure 7.26). Skip welding is commonly used for the reinforcing ring around the top of an open-top tank. It also may be used on roof structures; some areas of trusses, machinery, or tank bases; and many other similar areas.

Skip welding, however, creates several problems since corrosion easily penetrates the area between the welds and is thus a constant source of difficulty. It also makes coating application difficult since shortly after

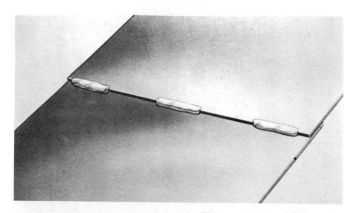

FIGURE 7.26 — Skip welding.

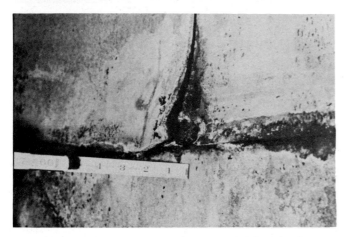

FIGURE 7.27 — A lap-welded seam on a tank interior showing corrosion between the overlapped plates.

welding, air and moisture accumulate in the openings between the welds (Figure 7.26). Buildings or structures which are exposed to moist, corrosive atmospheres should therefore have continuous seal welds in between the heavier skip welding. The seam should be sealed to prevent corrosive materials from accumulating and to provide a continuous surface over which to apply a coating. Without the continuous seam weld, skip welding not only creates all of the welding difficulties that we have previously covered, but it leaves a deep crevice which is impossible to fill with a coating. In any design for corrosion service, seal welds should be used to augment skip welds.

Lap Welding

In some tank construction and often in tank roofs, a *lap welding* technique is used. This is the welding of steel continuously on the exterior of the tank or roof while leaving the plate merely lapped on the interior. This provides good weather resistance and makes a tight tank. On the other hand, it also creates a crevice on the interior into which moisture, corrosive gases, or corrosive liquids can penetrate and accumulate (Figure 7.27).

If such an overlapping technique is called for in a design, the only possible answer to overcoming the corrosion problems which it creates is to seal weld the interior of the lapped steel continuously in order to close off the

area and to provide a smooth area on which to apply the coating. A much preferable design, however, is to butt weld the tank plates and roof sheets. This immediately eliminates the problem with little or no additional cost.

Back-to-Back Angles

A common design in the construction of steel truss buildings is to use the back-to-back angle technique to form the trusses. Such angles are ordinarily separated by washers and then riveted or bolted together so that the two back-to-back angles form a T. Part of the design is to join other members to the truss by using the area between the back-to-back angles, bolting the third member between the two angles. Usually, there is a space of from 1/8 to 3/8 inch wide left in between the two angles. This space is then left completely unprotected from any corrosive dust, fumes, or moisture (Figure 7.28). This type of truss construction is almost impossible to protect properly (Figure 7.29).

The space between two angles is impossible to clean properly and there is no way of satisfactorily apply-

FIGURE 7.28 — Typical back-to-back angle design with a deep crevice between the angles.

FIGURE 7.29 — Back-to-back angle trusses before installation.

Corrosion Prevention by Protective Coatings

ing a coating into such a deep crevice. There are, however, several design alternatives. One is to use a T shape and thus eliminate the back-to-back angle construction. A cylindrical pipe shape also can be used in place of the back-to-back angles. If the structure already exists and coatings are required, such areas can be sealed by the use of heavy mastics applied in and along the seam between the two angles.

Box Beams

Box beams are a relatively common type of construction in areas where the box cross section provides the strength, and where some of the steel can be eliminated to reduce weight. There are a number of box designs. Some are completely sealed and are not responsible for any particular corrosion problems (Figure 7.6).

One box beam design involves the periodic removal of steel to reduce weight (Figure 7.30). These open box designs should never be considered where corrosion problems are involved. With periodic openings, the interior of such a box beam is impossible to protect properly against corrosion. Heavy rust, corrosion, and scale build up on the interior and most often cause failure of the structure. In the bridge shown in Figure 7.30, there were several thousand linear feet of such beams, and heavy scale corrosion had built up on the interior of the boxes within a matter of only two to three years.

Some box sections are designed to be almost entirely closed. Even though the openings are small, however, the atmospheric heating and cooling of the box will make it breathe, causing condensation and even salt buildup. Corrosion then takes place on the interior where it goes unnoticed until the structure fails. Such hidden areas are a real safety hazard to any critical section of a structure such as the beams shown in Figure 7.30. To prevent such corrosion, box sections or other enclosed spaces where proper maintenance is impossible should be seal welded with installed fittings for pressure testing. A low pressure air test will reveal openings or pinholes which would allow breathing with consequent internal corrosion.

If a void area can retain 5 psi air pressure, it is an indication that the void space or box section is tight enough to prevent breathing of moist air into the void. Pipe sections which are used in a similar manner for support or columns, should be seal welded closed and pressure tested in a similar manner in order to ensure a corrosion-resistant structure. The exterior of such fully closed box beams or pipe columns can be easily maintained and protected by proper coating application.

Tank Construction

Many tanks are constructed with a cone roof and an umbrella-type roof construction. This design usually calls for a center pole with I-beam rafters extending from the center to the outer edge of the tank. The rafters may be simple I-beams or they may be reinforced by internal truss members to provide the proper strength. The steel roof plates are then laid directly on top of the rafters, completing the roof construction. In any such tank, regardless of the liquid it contains, the cone roof area is subjected to

FIGURE 7.30 — Typical box beam design with an open interior where both proper surface preparation and coating are impossible.

constant high humidity and fumes, with condensation taking place on the underside of the roof, particularly during the cooler night hours. The entire area of the tank above the liquid level is subject to constant oxidation and continual wet, corrosive conditions. The underside of the roof, the rafters, and the trusses must withstand such conditions (Figure 7.31).

There are two extremely difficult areas to protect with coatings in this construction. The first is the area between the I-beam rafters and the steel plate roof. The steel plate roofs are not ordinarily welded to the rafters because of the continuous expansion and contraction of the roof plate. Thus, there is an area between the roof plate and the rafter which provides a crevice that is extremely difficult to protect. If a coating is the only corrosion protection to be used, the rafters and the underside of the roof should be coated prior to installation, thus allowing proper application. If the coating is applied after construction, it is necessary for the roof plate to be raised by wedges or other means and coated by whatever means is possible. The coating must be applied to both the underside of the roof plate and to the top of the I-beam rafter to provide good protection. Coating of such areas is difficult at best and generally requires continuous maintenance in order to prevent internal corrosion of the roof to the point where leaks occur.

FIGURE 7.31 — Typical umbrella-type storage tank with complex rafter design.

FIGURE 7.32 — A typical design for an offshore drilling structure in the Gulf of Mexico. Excellent protection was provided by the coating even after several years of exposure.

The second area is caused by the overlapping of the steel roof plates, creating a shingle effect from one plate to another. This provides a lapped joint and the crevice between the lapped roof plates provides a reservoir for corrosive solutions, moisture, and condensation, and corrosion takes place very rapidly. Most often, the roof plates are welded on the exterior to provide a continuous roof structure, while the interior of the lap is left open. Where this type of construction is necessary, a seal weld should be used on the inside of the lapped joint in order to provide a continuous surface for proper coating. Preferable construction is the use of butt welded roof plates. Also, since the roof plates have a tendency to expand and contract, a rubber or plastic cushion should be applied, where possible, to the top of the rafter in order to prevent abrasion between the roof and the rafter caused by the metal movement. Such a rubber or plastic strip, with the proper adhesive, will adhere to the rafter and protect it against corrosion.

A completely different type of tank design is used for process vessels. Since most vessels are designed for a lining, the interior is smooth, the tank heads are dished, and all welds are ground level with the tank steel. This provides a continuous surface for the coating or lining. If such tanks are to receive baked coatings, however, such as a high-baked phenolic, a coating problem can develop which may not have been anticipated in the original design. Tank supports, reinforcing, tank outlets, and similar areas on the tank exterior act as heat sinks and may remain at temperatures below the curing temperature of the baked lining during the heating process. This creates areas of coating in these spots which are uncured and are thus focal points for coating breakdown. Such exterior heat sinks require heavy insulation during the baking process, and the heating of the tank should be continued for a sufficient time for the heat sinks to reach the same temperature as the body of the tank in order to prevent serious coating failure.

Pipe or Cylindrical Construction

As discussed previously, pipe or cylinders provide the smallest amount of surface area subject to corrosion compared to ordinary structural shapes. They also have no angles, corners, edges or other surface configurations which can cause coating problems. The offshore industry, because of the corrosion problems involved with other structural shapes, have gone to cylindrical members for most all of their critical strength areas in order to reduce the possibility of coating failure and corrosion (Figure 7.32).

This structural use of pipe, however, is not without some problems. Extra care is required wherever pipe must be coated because of its cylindrical shape. The larger the cylinder, the better the application conditions. Since the application of a coating to the pipe is usually made longitudinal with the pipe, many holidays may be formed in the coating due to insufficient overlapping during the coating process. Spray application to pipe is made with a fan from a gun; thus, some of the spray on each side of the fan is applied to the pipe at a tangent to the pipe surface rather than directly impacting the surface as it does in the center of the fan. While these tangent sections may appear to be coated, the coating in these areas is actually both thin and porous. This is the cause of longitudinal holidays on pipe sections. Actually, it is necessary to coat pipe from at least four positions, and it is preferable to spray from at least six in order to make certain that the coating is fully overlapped. The number of passes naturally depends on the size of the pipe.

In addition to the cylindrical structure, pipe flanges, threaded joints, pipe hangers, and similar changes in surface configuration are common to piping. All of these areas are focal points for corrosion because they are difficult to coat. Crevices are formed in threaded couplings which allow penetration of moisture and subsequent corrosion. While such areas occur frequently on pipe racks and similar installations, they generally do not exist where pipe is used for primary construction sections such as shown in Figure 7.32. Despite the possibility of insufficient overlapping or additional surface configurations, a cylinder is still preferable to almost any other shape from a coating and a corrosion standpoint (Figure 7.33).

FIGURE 7.33 — Pipe bridge with a simple but effective design for corrosion resistance.

Water Pockets and Recesses

Water pockets, recesses, low spots, horizontal flat areas, or catch basins where channels or other structural shapes are used in a horizontal position are all areas which should be eliminated during the initial design of the unit or structure. If they are not noted at the design stage, they oftentimes can be altered without any change in structural strength during actual construction. One example is cited by Rudolf, as follows.

> One further point should be noted on the part of the designer, this is the matter of drainage of moisture from every part of the structure. Angular coupling of shapes usually provides some sort of haven for moisture which cannot drain away but has to evaporate in place. Angle, channel or I-beam should never be placed to leave a sort of vessel to retain moisture. Either the position of the shape should be changed or drain holes cut to provide rapid movement of moisture downward away from the piece. With good drainage of all members of a structure, one of the strongest measures of corrosion mitigation has been taken.[1]

One area where water tends to accumulate and evaporate on a continuing basis is on the upper surface of a floating roof tank. These are large, flat areas and it is extremely difficult to fabricate such a roof without the plate buckling in numerous areas. When rain or condensation occurs, the low areas in the roof accumulate water. If there are corrosive fumes in the area, these also are concentrated in the water as the water evaporates. A strong corrosive solution often develops with the consequent danger of severe corrosion taking place in these spots. Such depressions should be eliminated during initial design, if possible. During construction, low areas should always be watched as extra care is required during the coating process in order to make sure that a full coating thickness is applied. Even if the entire floating roof is well coated, low spots are still the areas where the coating will initially break down, first by blistering, followed by pitting.

This is also true of the horizontal areas on the interior of tankers and storage tanks. These areas occur on the large horizontal stiffeners and other structural members. While weep holes may be designed into the stiffeners, many times low spots occur because of warped steel during formation for tank fabrication (Figure 7.34). Such areas should be eliminated either by drilling holes in the stiffeners to drain the water away, or the area should be filled with an epoxy troweling cement.

Flanges

Flanges of all types are difficult areas to protect with coatings. Figure 7.35, which shows a christmas tree on an offshore well, is a good example of the complexity that can arise in piping systems due to the use of flanges. In this case, the protection was particularly good since the entire wellhead was coated with inorganic zinc. It does, however, show the various focal points for corrosion, including the myriad of boltheads, threads, valve handles, and the crevice between flanges. Only a very thorough coating application can keep such a unit corrosion-free.

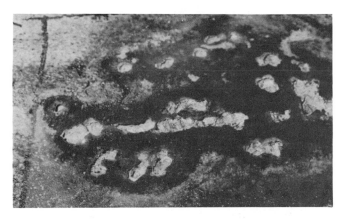

FIGURE 7.34 — Typical pitting in low areas of horizontal steel in a tanker. The dark area around the pits is discoloration caused by water standing in the low spot. Such pits were 1/4 to 3/8 in. in depth. Note the blistering in the coating around the edge of the low spot (upper left). [SOURCE: Munger, C. G., Deep Pitting Corrosion in Sour Crude Tankers, Materials Performance, No. 3, p. 22 (1976).]

FIGURE 7.35 — Valve manifold (christmas tree) on an offshore production platform. Note the boltheads, nuts, threads, and crevices between flanges which serve as focal points for corrosion and coating breakdown.

These corrosion focal points are almost impossible to eliminate by design. Flange spaces should therefore be coated wherever possible, and galvanized nuts and bolts should be used as a base for whatever organic coating may be applied later. Where piping is already installed and flanges exist, it is often practical to use a glass-reinforced vinyl or polyethylene tape with a butyl rubber adhesive to seal the opening between the two flanges. This can then be overcoated with vinyl coatings or other similar materials. Taping, in fact, has successfully eliminated corrosion between the flanges in many areas and prevented rust staining which inevitably flows from such areas.

Flange connections are not an isolated problem, but rather one that exists in almost all industries. Figure 7.36 is a typical cooling tower piping manifold which shows the many necessary flange connections.

Stairways

Designs for stairways, ladders, and similar access structures are, for the most part, still done conventionally with the possible corrosion factors generally overlooked. Although areas like stair treads and handrail risers are critical safety areas, they are often seriously corroded due to their original design and subsequent coating difficulties. Figure 7.37, which is a typical stairway on the exterior of a tank, demonstrates the numerous focal points for corrosion and areas where it is most difficult to obtain a proper coating. In this case, the rail is an angle, the riser and braces for the handrail are angles, and the side rails on the stair proper are channels with the stair risers fitted in the channel itself, creating many isolated areas difficult both to clean and to coat. Not only are these isolated areas the first to corrode, but corrosion is less obvious than on the exterior of the structure.

In the example shown in Figure 7.37, if the stair tread had been seal welded to the flat area of the channel with the arms of the channel extending out rather than in, a much more corrosion-resistant structure would have resulted with better access to clean and coat the steel properly. Also, if the handrail risers, braces, and handrail itself had been made from pipe, both cleaning and coating would have been more effective. Open grating should not be used for stair treads, unless it is absolutely necessary, because of the complexity of the grating structure and the difficulty of providing proper corrosion protection to such areas. Heavy galvanizing is suggested wherever such stair treads are used in order to provide an effective anticorrosive base for organic coating.

Open Grating

Open grating is a common structural material used for elevated walkway design on ships, refineries, offshore structures, chemical plants, and many similar structures. At best, it is an extremely difficult shape to protect, but more often than not, it is an area where corrosion rapidly decreases the safety of such construction. Figure 7.38* is an example of a galvanized grating used as a platform in

---

*See color insert.

FIGURE 7.36 — Cooling tower intake piping manifold coated with inorganic zinc. The piping was subject to elevated temperatures and continual water spray.

FIGURE 7.37 — Typical stairway on the exterior of a tank with its many sharp edges, corners, and hidden recesses.

the loading area of a large tanker. The galvanizing has rapidly disappeared from the grating which displays general corrosion. In contrast, the deck in the foreground is coated with one coat of inorganic zinc and has been fully protected for the same time span as the galvanized platform treads.

Where it is necessary to use open grating from a weight or safety standpoint, it is recommended that the grating be galvanized and coated as well as possible with organic topcoats to provide a maximum life span. When corrosion starts, such tread should be regalvanized or coated with inorganic zinc before corrosion has progressed to the point of dangerously reducing the steel cross section. It is possible to coat grating effectively with inorganic zinc, even though it is difficult, and the grating must be blasted and coated from several different angles on both sides (Figure 7.39).

Corrosion Prevention by Protective Coatings

FIGURE 7.39 — Drawbridge using open work grating as a roadway, with a walkway of deformed steel plate. The entire structure was recoated with inorganic zinc which has provided excellent protection after 10 years of continuous use.

Another option in dealing with the corrosion of open work grating is the use of plastic grating made from glass-reinforced corrosion-resistant resin. Such grating has been used rather extensively in serious areas of chemical corrosion. Its design is not completely without problems, since the plastic can weather and become quite brittle over a period of time, thereby reducing the safety of the structure. Wherever its use is practical, however, it is preferable to use a deformed plate (diamond tread) as stair treads and walkways, since it can be effectively protected and maintained by the use of coatings.

Pipe Supports

Pipe supports are a constant problem in corrosive areas. The pipe hanger or fastener is always closed with a bolt which serves as a focal point for corrosion. Since this design is not easily eliminated, pipe supports and hangers should be galvanized to provide a solid base for an organic anticorrosive coating. The innumerable edges, corners, and inaccessible places presented by pipe supports design must all be treated with care in order to provide a proper continuous coating.

Figure 7.40* shows a typical pipe support or pipe hanger. Condensation is obvious on the overhead, and corrosion has started on the pipe clamps. While the pipe in this example is small, similar installations are used for many types of pipe, all with the same problems involved. Where pipe clamps such as these are used, it is recommended that a plastic or rubber gasket be used around the pipe to insulate the pipe from the clamp and to protect the pipe from corrosion.

Where pipe is installed on pipe racks, and particularly where the pipe moves, it is recommended that pipe slides be installed on the pipe and on the rack in order to prevent abrasion at these points. These can be made of heavy plastic sheet laid on the pipe rack itself. Pipe slides have been designed and made from anticorrosive plastic or graphite which fit both on the pipe and on the pipe rack. When these are used, the pipe movement is taken up by the plastic slides and the pipe does not fret, thus preventing an aggravated corrosion problem. Coatings can be applied under the pipe slides so that both the pipe and the structure are protected. Since pipe supports can create serious corrosion problems, every precaution must be taken in the design, steel preparation, and coating stages in order to keep them corrosion-free.

Blind Openings, Remote Areas, and Small Areas

In nearly all steel fabrications, there seem to be at least a few areas that are remote or form blind openings which cannot be either cleaned or properly coated. These openings accumulate blasting materials, dust, dirt, moisture, and similar contaminants so that it is impossible, or at least very difficult, to provide an effective coating in such areas. Steel grit, sand, or similar abrasive media, when lodged in such places, form focal points for corrosion over which a coating cannot be effectively applied. These areas should therefore be noted and eliminated in every steel fabrication, preferably at the design stage, but at least during construction.

Welding of Precoated Structures

Even with the best designs, there are inevitably areas of changes and additions to the structure where welding is required. If the structure has been precoated, there may be serious damage to the coating at those points. Even more of a problem are tank areas which have been coated on the interior followed by repairs or changes on the exterior of the tank. The heat from the welding on the exterior damages the coating on the interior and the coating must be thoroughly repaired before the interior lining is effective (Figure 7.41*). While this problem is not one involved in basic design, it certainly is associated with design changes during construction. The problem demonstrates that changes in a design after a structure has been essentially completed may cause considerable coating damage and therefore must be taken into consideration before the unit can be placed in service.

Tanker Interiors

Tanker interiors are included in this chapter because of the complex design they require. Not only are these areas subject to many stresses, but they also are subject to some of the most severe corrosion problems possible. The transportation of refined products or sour crudes makes these areas subject to such extreme corrosion that any design which can be used to eliminate the complex nature of these tanks should be used (Figures 7.42 and 7.43).

One design which has vastly improved the corrosion resistance of sections of tanker interiors is the use of corrugated bulkheads. In this case, the reinforcing members on the bulkhead are not necessary since the corrugated design develops the strength without such reinforcing. The interior of tanker tanks can be effectively coated in spite of their complexity. Proper coating materials are

FIGURE 7.42 — An internal bulkhead block section of complex design with many coating danger points.

FIGURE 7.43 — A main transverse framing section in a tanker with many angles, corners, and blind areas.

available, and the key to their effectiveness is the care with which they are applied. Inorganic zinc coatings alone or as a base for organic coatings have been extremely effective in reducing the corrosion in such areas.

Galvanic Cells

The use of two or more different metals in corrosive areas should be carefully considered and, wherever possible, eliminated in the initial design stage. A variety of metals are often used due to the particular requirements of a process or design. One area which has caused difficult corrosion problems is the use of aluminum deck houses on ships. These ships, ranging from tuna clippers to destroyers, are often designed improperly so that the aluminum has rapidly deteriorated. This is due to the marine atmosphere and the fact that the aluminum is anodic to steel.

Massive amounts of both aluminum and steel are used in these installations. Coating is therefore not the entire answer for such extreme conditions. The coating can protect both the steel and the aluminum and insulate the surface of the two from each other. There are, however, inevitable crevices, cracks, and similar openings between the steel and the aluminum which collect water and quickly create difficulties. Where aluminum deck houses or aluminum structures are joined with steel, the aluminum should be thoroughly insulated from the steel through use of inert plastic or rubber gaskets which are sufficiently thick and sufficiently extensive to make a complete break between the two metals.

Bronze valves which act as a small cathode in relation to the deck (large anode) also create corrosion problems at the junction between the two different metals. Wherever bimetallic couples are formed, the basic design should provide for insulating the two metals from each other, and for the proper coating of both metals in order to increase the insulation.

Precoated Equipment

On any new project there is usually at least some equipment or structural parts which have been fabricated away from the job site and precoated. In the case of structural steel, precoated, fabricated units are often coated with a shop primer which was applied at the fabricating plant and failure has already initiated before erection takes place. This is because such primers are usually low cost products, made primarily for dressing up the steel as it goes out of the fabricating plant. These products make an extremely poor primer for the more sophisticated coatings. They have little corrosion resistance and result in poor adhesion of the topcoat to the steel. Also, in many cases, they are applied over mill scale which in itself is an invitation to failure for the better grade coatings.

The corrosion engineer should be on the watch for such shop-primed units, particularly if the units are to be used in a corrosive area. If such units are received on the job site, they should be completely blasted free of the shop primer prior to the application of any of the more corrosion-resistant materials. The best procedure, however, is to specify in the original design that such units be completely blasted and that a compatible and satisfactory preconstruction primer be applied so that it can be used as a base for the corrosion-resistant topcoats. Preferably, the preconstruction primer should be an inorganic zinc coating which would provide a proper base for the organic topcoats.

Precoated equipment is also common. In this case, the coating on the equipment may be adequate for the corrosive atmosphere involved. Usually, however, it is a standard coating used by the equipment company and applied in one or two coats. In most cases, it is an enamel-type coating cured with heat or infrared to a hard, glossy, thermoset state which may be entirely adequate for limited equipment usage. It probably is not, however, satisfactory for severely corrosive atmospheres. Glossy thermoset enamels, for example, might consist of a low or medium oil-base alkyd. These generally make a poor base coat for the adhesion of the more corrosion-resistant topcoats.

In order to protect the equipment properly, it is therefore necessary to determine the compatibility of this equipment enamel with the corrosion-resistant topcoats. If it is compatible, then it is necessary to at least break the gloss of the coating in order for the topcoat to adhere. The best procedure, however, is to remove the original coat and then apply a complete system of the anticorrosive product required. Again, the most acceptable procedure is to specify in the original design and specifications that all of the equipment to be used in the corrosive area must be either primed with a proper primer or coated with the full anticorrosive coating system at the equipment manufacturer's shop.

Such problems involving shop coating are frequently overlooked in the design stage of a project and can thus cause serious problems once the structure or the equipment is put into use. The corrosion engineer should be aware of this problem and should be on the lookout for equipment which may be coated with less than the required coating for the atmosphere in which it is to operate.

Mill Scale

There is yet another serious problem encountered wherever structural steel shapes are used: mill scale. Although this is not a problem in design, it should be covered in the design specifications for any job in an area where corrosion is a factor. *Mill scale,* or iron oxide scale, invariably appears on the exterior of steel plate, steel shapes, or pipe which have been hot rolled. It has been proven many times, and under many different circumstances, that mill scale is cathodic to the steel itself, and because of this, will cause the steel to pit badly and rapidly.

During World War II, it was found that if ships were not descaled prior to being placed in service, such severe pitting took place in a six-month period, that in many cases there was actual penetration of the bottom plates. While overcoating with coatings helps to lengthen the life of the plate under such conditions, it is only a temporary stopgap, since moisture will pass through the coating and into black iron oxide, causing it to swell slightly and pop off or loosen from the steel surface. When this happens, coating failure proceeds rapidly and the steel at these spots goes into solution at an exaggerated rate.

Steel need not be under continuous damp or immersion conditions for this reaction to take place. There are instances on almost every steel structure that has been coated prior to removal of mill scale where the coating has been broken because a piece of the scale has lifted from the steel surface. Acidic fumes or chlorides greatly

exaggerate such a condition, and as soon as the breaks are formed, rapid corrosion takes place. In designing a structure to resist corrosion, the specifications should call for the complete removal of mill scale prior to the application of any coating.

Insulation

Another danger spot which requires attention and which is often overlooked in the design stage, is the coating of insulation. In areas of severe corrosion in a plant, the insulation itself may disintegrate if there is any way for the fumes to reach it. Also, the steel underneath the insulation may corrode rapidly in an area where it is impossible to notice. It is essential, therefore, to coat the underlying pipe or structure properly and to seal the insulation against water and fume penetration. If fabric-covered insulation is used, it is usually difficult for a thin coating to seal all of the openings in the fabric. If these are not sealed, the fabric disintegrates rapidly in many corrosive atmospheres and the insulation itself absorbs moisture and disintegrates. The pipe or structure being insulated can then corrode underneath.

One remedy is to apply a heavy mastic-like coating over the surface of the insulation in order to make certain that all of the holes are filled. This can then be followed with a thinner, chemical-resistant coating, if required. Wherever pipe hangers enter into the insulation, such areas should be thoroughly sealed by a heavy mastic that is compatible with the topcoat to be applied. Metal insulation cover also can be used to protect the insulation if all joints are properly sealed against moisture and fumes.

Summary

The following is a summary list of design details which should be taken into consideration by the corrosion engineer on any structure subject to corrosive atmospheres.

1. Structure Steel Shapes: Particular attention should be paid to the coating of all angles, channels, and H- or I-beams on both edges and in corners.

2. Sharp Edges: Sharp edges and corners should receive additional attention. They should be ground smooth or to at least 1/8-inch radius before surface preparation and coating.

3. Rivets and Bolts: Particular attention should be paid to riveted and bolted areas. These are focal points for corrosion. Use galvanized or other corrosion-resistant bolts wherever possible. Unnecessary corrosion problems often result from using ungalvanized, electrogalvanized, or cadmium-plated bolts and nuts. Hot dip galvanized bolts and nuts are preferable. Where the bolt size is small, stainless steel bolts or fasteners can provide a much more corrosion-resistant system.

4. Welds: Welds should be checked for rough areas, undercut areas, and areas which retain weld slag. Rough welds should be ground to a smooth contour, and surface imperfections such as weld spatter, metal slivers, and sharp protrusions should be eliminated by chipping or grinding prior to surface preparation.

5. Seal Welds: Use seal welds wherever there are joint crevices, skip-welded areas, or where brackets, plates, or other attachments are left on the structure.

6. Skip Welds: Skip-welded areas should be seal welded in order to seal all crevice areas and to prevent corrosive liquids or fumes from penetrating into the joint.

7. Void Areas: Seal weld plates over openings of difficult-to-reach, inaccessible void areas, pipe columns, or closed box girders. An additional precaution is to pressure check the seal-welded void areas by installing an air connection in the plate. If the void area can maintain air pressure, it prevents the breathing of moist air and corrosive fumes into the void and thus prevents internal corrosion.

8. Steel Trusses: If steel trusses are formed with back-to-back angles or similar construction, particular attention should be paid to the area between angles or structural shapes. This is a dangerous area for corrosion to start and must therefore be sealed before coating the structure.

9. Adjacent Structural Members: Avoid placing two structural members close together so that the adjoining surfaces cannot be properly prepared or coated.

10. Lap Welds: Where lapped steel plates are used and welded on only one side, serious corrosion can occur. Such lapped plates should be seal welded on both sides if they are to be coated.

11. Pipe Construction: Where pipe design or construction is used, make certain that all openings are seal welded to prevent internal corrosion. When coating the exterior, each spray pass should be overlapped 50%.

12. Flanges: Rust and staining from the corrosion of flange faces and the use of ungalvanized stud bolts can be reduced by using hot dip galvanized flanges and bolts, or by precoating the flange face up to the raised base before the flange is installed. If coating must be done after the flange is in place, seal the opening between the flanges through use of reinforced adhesive plastic tape which is compatible with the proposed topcoat.

13. Pipe Contact Surfaces: When long runs of pipe are on a pipe rack, the pipe may move back and forth on the rack causing wear-related corrosion. Plastic or steel wearplates should be specified.

14. Water Pockets: Eliminate recessed areas, low spots, or water pockets by draining through a one-inch drilled hole or by filling such recesses with a trowelable epoxy or other grout.

15. Scaffold or Other Permanent Brackets: Scaffold or permanent brackets are usually skip welded. They should be permanently seal welded to prevent corrosion and provide a continuous surface for proper coating.

16. Precoated Equipment: Precoated equipment or structural steel receiving the shop primer should be given particular attention to make certain that the existing coating provides a proper base for anticorrosive topcoats. If it does not, it should be removed.

17. Electrical Boxes and Connection: Electrical connection boxes and conduits consistently give corrosion problems. Cast aluminum or die cast boxes must be given particular attention during coating in order to provide sufficient corrosion resistance. These corrode readily from the interior as well as the exterior if not properly sealed. Electrical conduit must have particular attention during the coating process in order for a continuous coating of adequate thickness to be applied.

18. Bimetallic Couples: Avoid bimetallic couples wherever possible. If it is impossible, make sure that the two metal surfaces are thoroughly insulated from each other in order to prevent galvanic corrosion.

19. Structural Modifications: Structural modifications are a continual problem during construction, particularly where changes are made on a portion of the structure already coated. Such modifications made by cutting and welding destroy the coating on both sides of the plate or structure.

20. Open Gratings: Deformed plates should be used in place of open gratings, whenever possible. The plate can be properly protected by coatings, while the open-work grating is a constant coating problem. Any open-work grating used should be galvanized prior to topcoating.

21. Insulation: Insulation can both disintegrate and cause corrosion to the underlying pipe. Both pipe and insulation should be properly coated.

All of the above types of construction and construction problems are common throughout today's industry. Coatings are usually the last item to be considered by a design engineer and they are the last item to be accomplished on a construction project. It is little wonder then that less attention is paid to coatings than to any other part of the structure. Coatings are, nevertheless, critical to the life of any structure where corrosion is a problem. The design and construction of a structure should be such that as many of the difficult-to-coat areas as possible are eliminated during the design stage or, when necessary, during construction. Also, every care should be taken on existing buildings and equipment to make sure that such problem areas are properly protected.

The possibility of corrosion should be taken into consideration in the planning of any structure, and it must be considered as much of a design problem as any other engineering factor. A structure which is designed with corrosion in mind, and where all of the possible problems are eliminated (e.g., sharp edges, corners, crevices, rough welds, etc.) will operate much longer, more effectively, and more economically than one which is conventionally constructed and then coated and recoated at short intervals because of areas which cannot be properly protected.

References
1. Rudolf, Henry, T., Design Against Atmospheric Corrosion, Corrosion, Vol. 2, No. 8, pp. 35-38 (1955).

Bibliography
Corrosion Control—Principles and Methods, Ameron Protective Coating Division, Brea, CA.

Munger, C. G. Coatings—Good Structural Design Aids Battle Against Corrosion. Iron Age, July (1955).

Munger, C. G. Good Design Can Prevent Corrosion Problems. Plant Engineering, July (1964).

Munger, C. G. Industrial Coating Performance Improved by Good Design. Corrosion, Vol. 15, No. 6, p. 102 (1959).

Sparling, Rebecca H. Corrosion Prevention Should Begin at the Drawing Board. Materials Protection, Vol. 2, No. 12, p. 8 (1963).

Thompson, Lewis J. Design Features That Reduce Corrosion Problems on Offshore Structures, CORROSION/79, Preprint No. 34, National Association of Corrosion Engineers, Houston, TX, 1979.

8

The Substrate - Importance to Coating Life

Usually, we think of the coating as the most significant factor in the protection of a surface, and in one sense this is quite true. On the other hand, consider the substrate (surface) and its effect on the permanence, durability, and effectiveness of the coating. The *substrate,* or surface over which the coating is applied, is the groundwork or foundation of the coating, thus its characteristics have a direct bearing on the life of the coating.

The construction of a building on sand, clay, or rock demonstrates a similar relationship. The same house could be built on each base; however, the one built on clay would have a shorter life and therefore be less satisfactory than the one built on rock. Also, the foundation necessary for a house built on sand would differ from that needed to build one on clay.

In the same way, coating systems vary according to the substrate. For example, consider a coating exposed to sodium hypochlorite. Sodium hypochlorite is a reactive chemical which gradually breaks down into nascent oxygen and sodium chloride. While vinyl coatings are resistant to this reagent, if the coating is applied over steel, the nascent oxygen will penetrate any coating defect and thus initiate rapid metal corrosion. However, the same coating applied over a smooth concrete surface will last for a much longer period of time. This is because the concrete surface does not react with either the nascent oxygen or the sodium chloride, so that full coating effectiveness can be realized.

Another example is that of a coating applied over both steel and aluminum in an alkaline atmosphere. The coating can be thoroughly resistant to the alkali and yet fail over the aluminum because of that metal's reactivity with the alkali. Any break or imperfection in the coating will allow the aluminum to corrode, leading to early coating failure. The same coating applied over steel, however, would have a long, effective life because of the low reactivity of the steel and alkali.

Thus, the surface to be protected is a key factor in the life and effectiveness of a coating. This is an important point to be considered when selecting structural materials for any corrosive atmosphere.

There are many types of substrates over which coatings are applied: *e.g.,* steel, concrete, wood, aluminum, copper, lead, zinc, stainless steel, cast iron, concrete block, plastic, plasterboard, masonite, reinforced plastics, and even rock. Each of these has a different effect on the coating applied over it, and different coatings may be required to provide effective protection. Some of these substrates and the characteristics which are important in terms of the life of a coating are described in the following sections.

Types of Substrates

Steel

Steel is the most common surface over which high-performance coatings are applied. Steel makes up the mild hot-rolled plate, sheet, and structural shapes that are commonly used for construction throughout the world. There are many other forms of steel available; however, none is as widely used as the carbon hot-rolled variety. Fortunately, that variety provides one of the best surfaces for coating application. Steel, with the exception of areas that might have been previously corroded, presents a relatively uniform, even, smooth surface without any changes in the basic metal surface that might cause variable coating adhesion. This is particularly true of new steel, either in the form of cold-rolled plates or hot-rolled steel which has been pickled or blasted free of all original mill scale. This is important, since two of the other most widely encountered surfaces that require coating are concrete and wood, neither of which has a uniform surface.

The steel surface is also dense and nonporous, so that

a coating which has good adhesion characteristics and is properly applied should have relatively uniform adhesion over the entire surface. Steel surfaces are reactive with acids, chlorides, sulfides, and numerous other chemicals. Where these materials have access to the surface, rapid corrosion results. On the other hand, with the proper adhesion of coatings over the steel surface and with a smooth, dense, uniform surface, coatings can prevent these reactive materials from coming in contact with the steel and therefore provide full protection. Most coating materials are compatible with steel surfaces, which with proper surface preparation such as abrasive blasting or acid pickling, provide a good surface over which to apply high-performance coatings.

Figure 8.1 illustrates the normal surface of new steel after sandblasting. It is relatively uniform in texture, with few variations in surface structure to cause uneven coating reaction or adhesion.

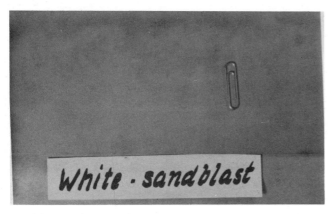

FIGURE 8.1 — White sandblasted steel panel providing an excellent coating substrate with a smooth, uniform texture and even color.

Low-Alloy Steels

Low-alloy steels have the advantages of good strength and mechanical properties, which make them useful for particularly critical areas of a structure. Thus, many questions have been asked about the effectiveness of coatings over these products. This issue is addressed in a paper by Copson and Larrabe, as follows.

It is quite well known that most of the low-alloy constructional steels also have improved resistance to atmospheric corrosion. The extent of the improvement depends upon the composition, with the copper, nickel, chromium, and phosphorous contents being particularly advantageous. Some of these grades have roughly four to six times the atmospheric corrosion resistance of carbon steels. This pertains to bare uncoated steels which are free to rust.

It is not so well known that this improved corrosion resistance makes paint coatings more durable. Rust will form, of course, at any breaks or discontinuities in the paint coating. Rust also forms to some extent underneath many paint coatings. The accumulating rust tends to break down the paint film. Owing to their better corrosion resistance, much less rust forms on the low-alloy steels than on carbon steels. The smaller amount of rust from the low-alloy steels causes much less damage to paint coatings than does the more voluminous rust from carbon steels. Any rupturing of the paint film permits larger amounts of moisture to contact the steel surfaces, and this accelerates the deterioration.

By prolonging the life of paint, the use of low-alloy steels makes it possible to go much longer before repainting is necessary. The corrosion damage at any rusting areas and the spread of rust from a scratch or other holiday is lessened greatly. The labor in preparation for repainting is reduced. The net result is that the use of low-alloy steels can effect a considerable savings in the cost of labor and materials for maintenance painting.[1]

This work, along with similar work reported by Hudson[2] and LaQue[3] is further indication of the importance of low-alloy steel to effective coating use.

The comparative resistance of several commonly used steels in a marine atmosphere are shown in Figure 8.2. These tests were conducted at the International Nickel Testing Station at Kure Beach, North Carolina on bare steel panels. The composition of the steels is shown in Table 8.1. Even relatively small changes in steel composition can change the life and effectiveness of any coating applied over it.

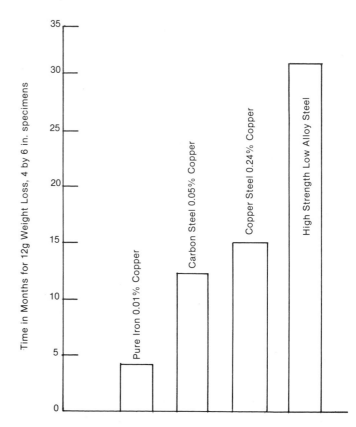

FIGURE 8.2 — Comparative resistance of commonly used steels exposed bare in a marine atmosphere. [SOURCE: Copson, H. R. and Larrabee, C. P., Extra Durability of Paint on Low Alloy Steels, ASTM Bulletin, December (1959).]

Stainless Steel

The term "stainless steel" is a generic term which represents the many different types of stainless available. Stainless is a good substrate over which to apply an organic coating because of its inert characteristics, both to the atmosphere and to chemicals. It does provide a smooth and dense surface which can, in itself, create difficulties in terms of coating adhesion. Where coated stainless steel is exposed to high humidity or immersion conditions, it may provide a sensitive surface over which to ap-

TABLE 8.1 — Composition of Steels Shown in Figure 8.2

Material	Carbon	Manganese	Silicon	Composition, percent		Nickel	Copper	Chromium
				Sulfur	Phosphorus			
Open-hearth iron	0.02	0.02	0.002	0.03	0.005	<0.05	0.014	Nil
Carbon steel	0.04	0.39	0.007	0.02	0.007	0.01	0.05	0.07
Copper steel	0.06	0.32	<0.01	0.02	0.010	<0.05	0.24	Nil
Low-alloy steel	0.08	0.37	0.29	0.03	0.089	0.47	0.39	0.75

[SOURCE: Copson, H. R. and Larrabee, C. P., Extra Durability of Paint on Low-Alloy Steels, ASTM Bulletin, December, 1959].

ply organic materials because the dense smooth surface does not allow maximum adhesion. Coating adhesion is improved over stainless where the surface is lightly etched, blasted, or treated to a dust blast prior to the application of the coating.

While a stainless surface is inert to many common chemicals, chlorides can often reduce the protective oxide film on the stainless surface causing it to pit at breaks in the coating. Some coatings may tend to break down and release hydrochloric acid when heated or exposed to strong sunlight. These have most often been clear coatings, and, when used over stainless, have caused underfilm corrosion and cracking.

Concrete

Concrete could be considered the antithesis of steel in terms of the characteristics which may affect coating life. While concrete is a porous material, this property in itself may vary widely from one portion of the surface to another. Cast concrete surfaces tend to contain innumerable water and air pockets; some of which are small and others which are quite large. Some may extend into the concrete surface as much as one inch.

Figure 8.3 is a view of a typical poured concrete surface showing the air pockets and pinholes that are always present. This is an average concrete surface shown after the forms were removed.

The difference in cast surfaces depends on the way the concrete was poured into the forms, the surface of the form itself, the water:cement ratio, and the mix of the concrete. A concrete which has the optimum water content and is well vibrated or puddled will provide the best cast surface, but even this will be somewhat porous and will contain some water and air pockets. Concrete which is improperly cast with little regard for proper placement may have sizable rock pockets and other areas which cannot be coated. This makes for a nonuniform surface which varies from one pour to another, and may change in character within a matter of a few inches. It often varies from a smooth, hard, trowelled surface that is almost glasslike, to one that is rough, and porous, and requires considerable surface preparation before the application of any coating.

Concrete surfaces may also contain a surface residue called *laitance*. This material is derived from some of the nonreactive materials in the cement or in the aggregate that are worked to the surface during placement. The in-

FIGURE 8.3 — Typical poured concrete surface soon after form removal showing the form tie hole, air pockets, pinholes, and the texture of the wood form.

terior surface of centrifugally spun concrete pipe, for example, contains a heavy layer of concrete laitance. Trowelled surfaces that appear to have a glaze often have a thin layer of the laitance which gives the glazed appearance. The laitance, either heavy or thin, is a weak, powdery material that has little adhesive strength. It will ordinarily act as a parting agent between the concrete surface and the coating.

In order for coatings to adhere satisfactorily to the concrete, laitance must be removed by acid etching or by a light sandblast. However, the best concrete surface for coating application is a hard-trowelled surface, since it produces the most uniform surface for concrete. It is also quite dense so that a smooth, continuous coating film can be easily applied. Figure 8.4 illustrates the uniform, dense, slightly granular surface provided by trowelled concrete.

One of the major problems in coating all substrates, is the development of a continuous film. Poured concrete is filled with pinholes so that it is almost impossible to obtain a continuous coating. As a matter of fact, in any area where concrete corrodes readily, such as in a septic sewer, the surface porosity is such that no sprayed-on coating will provide a completely continuous film and prevent corrosion. Under these conditions, concrete must be specially

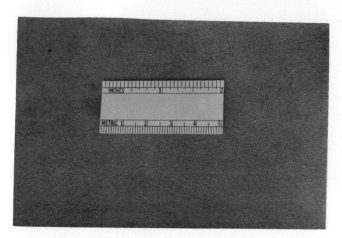

FIGURE 8.4 — A trowelled concrete surface.

treated by trowelling a coat of cement plaster over the surface, or by using some of the newer resinous surfacers to fill the pores and still provide a base over which the coating may be applied.

Concrete surfaces are chemically reactive with any material of an acidic character. This includes pure water. Many examples have been seen of concrete surfaces in mountain areas where pure snow water has flowed over the surface and left it completely etched, with the aggregate heavily exposed after only a few years in service.

Some of the most advantageous coatings for concrete surfaces are those which use the porosity of the concrete to the advantage of the coating. These are usually epoxy-type products which penetrate the pores of the concrete and react in place. This provides for maximum adhesion of the coating.

The penetration of coatings into the concrete pores is important, and even vinyl coatings, which are applied as a dilute solution, will penetrate into a concrete surface and obtain good adhesion. Coatings that remain mostly on the surface, such as water-based emulsion-type coatings, are not recommended for concrete where severe corrosion is a problem. Generally, coatings for the corrosion protection of concrete require greater thickness than coatings applied over less porous and more uniform surfaces. Thicker coatings are able to provide a more continuous coating than a thin material.

Aluminum

Aluminum provides a generally smooth, dense surface covered by a thin film of aluminum oxide. Coating adhesion over an aluminum surface is often improved by the use of chemical surface treatments. However, this may not always be possible in the field. If a coating is desired on the exterior of an aluminum tank, process vessel, or deckhouse on a ship, the best surface treatment is to lightly dust blast the exterior of the aluminum. This increases the surface area and vastly improves the adhesion of most high-performance coatings.

Aluminum reacts with a large number of chemicals, particularly those which are alkaline. Thus, coated aluminum surfaces in an alkaline atmosphere tend to corrode wherever there are thin areas, holidays, or breaks in

the coating. Where sulfides are the corrosive agents, aluminum may do quite well since it does not react as readily with sulfides, as do many of the other metals such as iron, copper, silver, lead, etc. For example, coatings applied over aluminum in a sulfide atmosphere will last much longer than those applied over a steel surface because of the sulfide resistance of the aluminum.

Generally, aluminum makes a good surface over which to apply high-performance coatings. This is especially true of lightly blasted surfaces. For instance, a small aluminum boat was prepared in this manner, followed by three coats of a vinyl coating, and the coating has maintained perfect adhesion over a period of 40 years.

Zinc

Zinc as a substrate, either in the form of hot-dipped galvanizing, electrogalvanizing, or metallic zinc, provides a smooth, uniform surface which is quite reactive. The surface of zinc reacts with carbon dioxide over a period of time to form a dull, gray surface. Zinc also reacts readily with most acids, with galvanizing only lasting a short period of time when exposed to an acidic atmosphere.

Zinc also has some beneficial reactions, notably its being anodic to steel and thus providing cathodic protection to steel surfaces. The effectiveness of coatings that are compatible with zinc and thus applied over it is increased in a substantial way because of this cathodic protection reaction. Holidays or damaged areas in the coating are protected from undercutting by the tightly adherent zinc on the steel surface, as was mentioned previously. Compatible organic coatings applied over zinc surfaces such as galvanizing or inorganic zinc coatings will have a much longer life than the same coatings applied over bare steel in most corrosive atmospheres. Even in acidic atmospheres, it has been shown that a zinc-base coat underneath an organic coating will prevent undercutting and will reduce blistering and corrosion at any break in the coating because of the strong adhesion of the zinc to the steel.

Zinc, then, as a substrate for organic coatings, generally adds to the life and effectiveness of coatings which are compatible with it, in spite of the chemical reactivity of metallic zinc. Not all coatings, however, are entirely satisfactory over a zinc surface. There are many (such as oil-base and alkyd coatings) that will react with the zinc, forming zinc soaps which do not allow long-time continued adhesion of the coating over the zinc surface. The coatings which benefit most from a zinc surface are the high-performance, nonreactive coatings used throughout the chemical, marine, petroleum, and other similar industries.

As has been previously described, zinc may react rapidly underneath a rather thin and porous topcoat. This is particularly true in a chemical or chloride atmosphere when water and chlorides are able to penetrate through the organic topcoats to the zinc. Alkali is developed and zincates may be formed. Voluminous white powdery deposits are built up under a porous organic film, which destroy the effectiveness of the topcoat as well as the zinc surface.

Copper

Copper has a dense surface covered with a copper oxide film. Because of the smooth oxide film, some coatings may fail to obtain proper adhesion to this surface. In many ways, copper is similar to aluminum as a substrate, with its dense, tightly adherent oxide film. Copper does not react readily in many chemical exposures and has a long history of permanence when used as flashing or roofing in an urban atmosphere. However, a copper surface is reactive in a sulfide atmosphere, where the copper surface turns black in a short period of time.

Where sulfides are a problem and copper is involved, a good protective coating is required. Breaks in the coating will allow the sulfides to react with the copper substrate, causing failure of the copper at the break. Breaks in the insulation on copper wire have been noted to cause the copper to be reacted away (dissolved) in a matter of a few weeks in a sulfide atmosphere.

Most acids react readily with copper. This is true of both the inorganic acids (sulfuric, hydrochloric, nitric, phosphoric) and the organic acids (acetic, butyric, stearic, etc.). Ammonia reacts readily with copper, as demonstrated by a copper water pipe which was penetrated after being exposed to soil where a large amount of ammonia-type fertilizer was present.

These factors indicate that any coating applied over copper should be well applied, since holidays or breaks are focal points for substrate failure. Copper generally presents a good surface over which vinyls, epoxies, asphalt, chlorinated rubber, and similar inert coatings can be applied.

In order for most coatings to obtain good adhesion to a copper surface, it is suggested that the surface be dust blasted prior to application. Dust blasting merely breaks up the copper surface and reduces the amount of copper oxides that are present. Dust blasting is recommended over even a fine sandblast since the impact of the sand grains can cause the copper metal to warp and expand, while a light dust blast (which has little or no heavy impact on the surface) will scour without the tendency to warp.

Cast Iron

Cast iron provides a rather rough, porous substrate, as previously mentioned. The surface is porous to various gases, and unless these are removed, they can cause adhesion problems. Most cast iron surfaces are blasted at the foundry in order to remove the sand or core material from the surface, which adds to their surface roughness. However, cast iron is iron, and it has many of the same characteristics as steel when in a corrosive atmosphere.

Cast iron differs from steel in that it is more granular in nature and contains free graphite or carbon. Its surface is less homogeneous and dense than that of steel. Under soil or immersion conditions, the iron may be dissolved from around the carbon particles, causing a condition called *graphite corrosion*. Such a condition can exist where there are breaks or holidays in the topcoat. Where the cast iron surface is properly prepared, it provides a good substrate over which to apply protective coatings.

Lead

While it is not often that a coating is applied over a lead surface, they do tend to perform well since the lead provides an inert base for the coating. Lead is resistant to many acids and alkalies so that a variety of coatings can be used over it successfully. Lead provides a soft, smooth, and extremely dense surface and one which feels almost greasy. Because of this, some roughening of the surface is necessary. Good adhesion of organic coatings can be obtained over lead by lightly blasting the surface or lightly roughening the surface by power sanding or hand sanding.

Wood

Wood generally provides a smooth surface, depending on the way it is sawed and surfaced. It is not, however, a uniform surface. The smooth wood surface varies between the summer and winter wood, so that it can vary in density to a marked degree all across the surface (Figure 8.5).

The dark grain is winter wood and the light grain is summer wood. Each has a different density and physical structure. This is particularly true of plywood surfaces which have their own inherent characteristics caused by peeling the wood ply from the circumference of the log. This exaggerates the summer and winter grain.

Wood is moisture sensitive, and will swell during times of high humidity and shrink when conditions are dry. Also, it will tend to expand and contract at a different rate between the summer and the winter grain. This fact makes it difficult for many high-performance coatings to cover a wood surface, since they are not sufficiently flexible and extensible to withstand the continuing change in dimension. The change in both the wood and grain dimension can take place on a daily cycle, in addition to larger climatic changes which can take place over an extended period.

This difference in the winter and summer grains is particularly characteristic of woods such as Oregon pine, cedar, ash and many other open-grained woods. Maple, birch, and similar dense woods are considerably less of a

FIGURE 8.5 — Summer and winter grain on a wood surface.

The Substrate-Importance to Coating Life

problem from a grain standpoint. While all woods contain both types of surfaces, the hardwoods can tolerate a less flexible coating than that required for the softer woods. Interestingly enough, wood is quite resistant to many chemicals, particularly those which are acidic. Thus, coatings applied over wood exposed to an acidic atmosphere generally stand up well because of the resistance of the wood itself. Even some of the alkyd-type coatings will perform quite well in an acidic atmosphere.

So far, we have looked primarily at new, unused substrates over which to apply coatings. However, not only do different substrates create varying coating requirements, but contamination, corrosion, and corrosion residues multiply the surface conditions that can affect coating life. Under real-life conditions, most surfaces are either used or have been corroded or contaminated with some type of material, and this entirely changes the surface/coating relationship.

In industry, it can almost be stated as a rule-of-thumb that a surface over which a coating is to be applied has been contaminated in one way or another. As an example, any steel that has been stored near marine conditions or transported by barge or ship for even a short distance over ocean will have some chloride contamination. Pitting can occur and does occur prior to the use of such steel. Although the pits may be quite small, they are also focal points for coating breakdown, particularly if any chloride contamination remains in the pit itself.

Contamination, then, must be taken into consideration as a substrate condition whenever coatings are to be applied. The coating of contaminated or previously corroded surfaces can be considered a world-wide industrial and marine problem. Unless such a surface is given serious consideration, both as to proper surface preparation and proper coating selection, limited life may be expected. The various types of contaminants, because of their varied reaction with coatings, will be considered individually.

Types of Contamination

Rust

One of the most universally encountered contaminants is common iron oxide or rust. Rust is considered a contaminant of a steel surface and must be recognized as such, since areas of steel that previously have been rusted and then thoroughly cleaned by blasting will tend to rust again in the same areas where the rust formed the first time. As has been previously discussed, the corrosion of iron is electrochemical, and in order for iron to corrode, it first ionizes, after which it reacts with other ions in the area and with oxygen from the air, eventually changing to Fe_2O_3 or common red rust, as we know it. These general reactions are as follows.

$$\overset{Fe \ water}{\longrightarrow} Fe^{++} \tag{8.1}$$

$$Fe^{++} + 2OH^- \rightarrow Fe(OH)_2 \tag{8.2}$$

$$2Fe(OH)_2 + O_2 \rightarrow Fe_2O_3 + 2H_2O \tag{8.3}$$

Red Rust

The type of reaction involved in the formation of rust (*i.e.*, the formation of iron ions and their immediate reaction with any moisture to form ferrous hydroxide), is an intimate reaction on the steel surface. Anodic and cathodic areas are created. The grain structure of the steel is involved, as well as the grain boundaries. It is because of this complex microscopic reaction that it is difficult to clean the surface of the iron reaction products. The surface can be sandblasted to white metal, but if allowed to stand in a humid atmosphere, it is soon obvious that the previously corroded areas begin to corrode before other areas of the steel. Such a condition is shown in Figure 8.6*, where previously corroded steel was blasted and yet, within a matter of a few minutes, the pattern shown in the photograph had developed on the surface.

Chloride Contamination

Many different phases of chloride contamination have been previously discussed. Chloride contamination is most prevalent in marine environments where almost all surfaces are exposed. It is also found in the salt refining industry, chloralkali industry, coal mining, and any industry where soluble chlorides are used.

The problem with steel and chlorides is the formation of ferrous chloride wherever the two materials come in contact. For example, a chloride ion is absorbed into rust, and held on the steel surface. The iron ion, the chloride ion, and water form an iron chloride salt solution which is not only conductive and can aid in the electrochemical corrosion reaction, but is in itself a strong corrodant of the steel surface. The ferrous chloride oxidizes upon exposure to air to form ferric chloride ($FeCl_3 \cdot H_2O$). Ferric chloride is a hydroscopic salt which tends to draw moisture from humidity in the air, creating a ferric chloride solution on the steel surface. In this case, even though the steel is thoroughly sandblasted, even a few molecules of either ferric or ferrous chloride remaining on the surface or in the area of a pit tend to accumulate moisture from the air, causing a concentrated iron chloride solution.

This was once seen where a chloride-contaminated steel surface was sandblasted thoroughly. The areas of the pits were watched with a microscope and the water could be seen to condense from the air on the steel surface in the pit areas. Almost immediately the condensate turned to a light green, indicating a ferrous chloride solution. The absorption of the moisture continued and even in a relatively few minutes, the droplet of water that was forming was changed from a green to a light yellow, indicating a change in the ferric ion. The color change then continued with the solution becoming darker in color. Eventually, some of the iron in solution reacted with oxygen, forming a brown deposit. The last reaction involved the surface of the steel under and around the water condensation which become dark brown or black.

Even a surface which has been blasted two or three times can still show this type of a reaction in areas of previous corrosion. Organic coatings applied over such a

---

*See color insert.

surface would soon tend to fail by the iron salt drawing water through the coating by osmosis and creating an area of poor adhesion or blistering. This type of surface contamination has caused untold coating failures around the world and is one of the reasons why corroded areas in the marine industry are so different to recoat.

One solution to this problem is to pretreat the metal surface after the initial blast with a material that will form an insoluble iron compound, such as phosphoric acid. The iron phosphate is formed preferably to iron chloride so that when the surface is again blasted, if any iron contamination remains, it is in the form of an insoluble iron phosphate.

Chloride contamination of concrete can also be a problem, although of a much different nature than the chloride contamination of steel. Concrete itself is not chemically reactive with neutral salts such as sodium chloride. This does not mean, however, that there are not problems involved. Sodium chloride will penetrate into porous concrete, creating a hydroscopic condition. Coatings applied over such a chloride contaminated surface may well have poor adhesion to the concrete because of the hydroscopic nature of the sodium chloride.

One area where concrete is disintegrated or seriously affected by salt is in salt marshes where there is a constant wetting and drying of the surface. This action can build up a heavy crystalline deposit of sodium chloride on and in the concrete. These crystals form within the pores of the concrete, and as they grow, they swell and spall the concrete at and above the mudline. Concrete which has been exposed to chloride solutions provides a difficult surface over which to apply protective coatings.

Wood is affected by chlorides in a similar manner. The chlorides penetrate into the wood structure, but do little or no harm to the wood itself other than to provide a hydroscopic surface over which any coating would have poor adhesion. Also, as in the case of concrete, where wood is subject to a salt marsh condition or where there is constant wetting and drying of the surface, the salt will crystallize within the wood and tend to make it spall at and above the mudline.

Wood exposed to salt solutions presents a difficult surface over which to apply coatings. Lumber is often made from logs that have been floated in seawater. The moisture in the wood dries out slowly because of the hydroscopic salt, and when the wood is dry, salt crystals can still remain. Coatings applied over such a wood surface are often subject to blistering and poor adhesion.

Surfaces contaminated with chlorides present one of the most widespread coating problems found. All chloride salts are more or less soluble, many extremely so. Silver chloride is the only common exception. This almost universal solubility makes for conductive solutions which are seriously corrosive to many metals, as well as creating ideal conditions for osmosis, should a coating be applied over them.

Acid Contamination

Acid corrosion can be a catastrophic type of failure. It occurs wherever acids are manufactured or used, which is, for the most part, in the process industries. Acid salts should also be included in the acidic category since some of these materials are quite active and therefore extremely corrosive. All iron salts of the various strong acids are considered acid salts. Acid salts are the reaction product of a strong acid and a weak alkali. For instance, ferric chloride is the salt of a strong acid, hydrochloric acid, combined with a weak alkali such as iron hydroxide. These salts then hydrolyze to provide the acid, with the consequent corrosion of both steel and concrete surfaces.

We have discussed the chloride contamination of steel and noted that hydrochloric acid is one of the most troublesome of the acid contaminants. The problem is that once the hydrochloric acid is in contact with steel or iron, it leaves a chloride salt on the surface, either ferrous or ferric chloride. This has all of the characteristics of the chloride exposure previously described, in addition to being a relatively strong acid. Instances of such contamination illustrate the dramatic failure that can occur where coatings are applied.

In one particular case, a new hydrochloric acid plant was erected. However, because of the rush of construction, and in order to put the process into production, the plant was started prior to the initial coating of the structural steel. Even under the best operating conditions, some hydrochloric acid escapes from the process and the reaction of steel and hydrochloric acid is immediate, so that the steel surface was contaminated immediately after the plant was put into production. As might be expected, the hydrochloric acid reacted with the steel surface, forming ferric chloride.

Even though every effort was made to properly prepare the steel surface, it was still done during production and the coating applied during operations. Contamination was almost continuous. Within a matter of a few weeks, the coating was almost a complete failure. The chloride contamination on the surface was, of course, microscopic. However, because the deposit was on the surface, the coating tended to quickly blister in many areas. In the areas where the structural steel was most difficult to coat, holidays resulted in uncoated steel where the hydrochloric acid continued to react rapidly. In a matter of six months, the webs of the steel were actually penetrated so that major structural repairs were required.

Any plant that uses hydrochloric acids can have the same problem unless the steel is thoroughly and properly coated prior to any contamination of the structure. Hydrochloric acid pickling plants, titanium dioxide pigment plants, and some iron oxide pigment plants have contamination problems similar to the hydrochloric acid plant described. Every care must be taken in the coating of these plants to prevent the hydrochloric acid contamination of the steel. Once the contamination occurs, there is no easy method of removal.

Volatile Acids

Hydrochloric acid, in addition to being extremely reactive, is also volatile. Hydrogen chloride (HCl) is a gas, and the liquid hydrochloric acid is composed of water and the gas which combine to form a solution. Hydrochloric acid not only acts as a liquid acid, but also as a gaseous acid, and thus can penetrate any crack or crevice in the structure. The volatile HCl is also extremely hydroscopic so that, even in gaseous form, it rapidly picks up moisture from the air and can react on any metal surface where it makes contact. Surfaces, therefore, do not need to be close to splash or spray of the liquid acid to be quickly contaminated. Being a gas, it can be carried by air currents to any part of a plant and react with the steel in remote or inaccessible areas where surface preparation is difficult at best.

Other metals can also become contaminated. Zinc, copper, and aluminum all form soluble salts with hydrochloric acid. Once contaminated, such surfaces are as difficult to prepare and coat as is steel.

Nitric acid is another volatile acid with extremely strong penetrating powers. Both nitric and hydrochloric acids can cause contamination of the surface underneath certain types of coatings. Even some strongly acid-resistant coatings, when thinly applied, can be penetrated by these acids, so that in order to provide protection against them, the coating must not only have extremely strong adhesion and be acid inert, but it also must be applied in full thickness over the entire area. Thin spots, breaks, and holidays are all areas that are rapidly attacked and can become contaminated to the point where repair of the areas is very difficult. Iron salts of volatile acids can actually form beneath some coatings. In some cases, entire coatings have been penetrated, with the underlying substrate becoming contaminated to the point where complete removal of the coating was required prior to repair.

Nonvolatile Acids

Common examples of nonvolatile acids are sulfuric and phosphoric acid. Products such as ammonium sulfate, ammonium phosphate, and ammonium nitrate are found in general use throughout the fertilizer industry. These are all acid salts which hydrolyze to form the acid when under moist conditions. All of the iron salts which are formed from these acids are soluble, with the exception of iron phosphate, so that steel or metal surfaces are easily contaminated and easily corroded.

Concrete also can be severely contaminated with acid. Since concrete is a strongly alkaline material, the acids themselves are quickly neutralized by the cement, causing a disintegration of the concrete structure. Both hydrochloric and nitric acid form soluble cement reaction products which generally can be washed from the surface by thorough scrubbing with clean water. Unless the concrete itself has been badly disintegrated, coatings can be satisfactorily applied over such a surface once it is dry.

Sulfuric acid is quite a different problem since it does not form a soluble reaction product with cement, but instead forms a voluminous insoluble calcium sulfate. One of the areas most affected by sulfuric acid is that of concrete sewers and sewer structures. Here, the sulfuric acid is developed by bacteria from hydrogen sulfide. Droplets of condensate in the upper portion of sewers have been shown to have concentrations of sulfuric acid upwards of 10%. This provides a constant source of acid to react with the cement. Disintegration of unprotected sewer struc-

tures has been found to be as much as 2 inches per year. The formation of the voluminous calcium sulfate, which has a volume roughly 5 times that of the cement particle, causes physical disruption of the concrete, as well as the reaction of the sulfuric acid on the cement particles.

Sulfuric acid also tends to penetrate any imperfection in a coating. Since the concrete substrate is rapidly attacked, a pinhole in the coating can create a large tubercle due to the expansion of the calcium sulfate underneath the coating.

Figure 8.7 is an example of coatings applied to the interior of a concrete sewer chamber. The coating was penetrated through so many pinholes that it was almost completely lifted from the surface by the formation of calcium sulfate underneath. Such surfaces are difficult to repair as, in most cases, not only must the contaminated concrete be removed from the surface, but the concrete must be resurfaced prior to the application of any coating.

Phosphoric acid can aid in some areas of acid contamination on metal surfaces. As previously discussed, iron phosphate is insoluble so that it remains as a solid on the surface. Thus, it can more easily be removed from the surface by abrasive blasting than can the soluble salts. Minor residues of the iron phosphates, because of their insolubility, usually will not provide adhesion problems for a strongly adherent compatible primer.

FIGURE 8.7 — A coated concrete sewer structure seriously penetrated by sulfuric acid. Note the large sulfate tubercles formed underneath the coating due to the reactivity of the substrate.

Fatty Acids

Fatty acids are common throughout many industries, including the paper, food, and soap and detergent industries, as well as in the processing of grain. Surfaces contaminated with fatty acids are much more difficult to prepare and repair than are surfaces which are reacted with mineral acids. Many fatty acids react readily with metals to form insoluble soaps, with the fatty part of the molecule being the most difficult part of the contamination. These metal fatty acid salts are insoluble in water, difficult to remove from the grain boundaries of the steel surface, and, if they are overcoated, adhesion loss and blistering are quite common.

Organic acids and fatty acids range from formic acid, which is the lowest homolog of the series, to materials such as stearic acid, palmitic acid, and many similar long-carbon-chained acid products.

Fatty acids on a substrate, (e.g., steel, concrete, cast iron, copper, or others) react with the surface and create a metallic fatty acid salt such as iron stearate, copper stearate, calcium palmitate, etc. These metal fatty acid salts are also called soaps. Sodium stearate is the most common. The metal part of the metal fatty acid molecule is at one end of a long carbon chain. If this could be visualized, the metal would still be associated with the base metal. However, the long-chain fatty part of the molecule would be free, except at the metal attachment end. This makes for a chemically bonded molecule which creates a monomolecular hydrocarbon film on the surface.

A schematic drawing of such a situation is shown in Figure 8.8. Any coating which is not thoroughly compatible with such a metallo-organic film will have little or no adhesion. The calcium and aluminum atoms in cement act in the same way, tying the fatty molecule to the surface. Contaminated concrete can be an even worse substrate than steel since the fatty acid can penetrate into the porous concrete and react in-depth.

The food industry is one which has a major problem with fatty acid contamination. In this case, the preparation of contaminated surfaces is more difficult since abrasive blasting may not be possible or practical because of possible contamination of the foods involved. Fatty acid contamination has proven to be an extremely serious problem in the food industry. Only a thorough and meticulous cleaning of the surface can result in a satisfactory coating.

There is no practical means of removing fatty acid where it has penetrated concrete. Hot steam detergent washing removes some of the contamination from the surface, but does not remove that which has penetrated. Coatings applied over such a surface, even though the surface looks clean, invariably tend to fail from loss of adhesion. The only acceptable method of removing the contamination is abrasion of the concrete beyond the penetration of the fatty acids. In most cases, the concrete must then be resurfaced prior to being properly recoated. Some of the newer resinous concrete surfacing compounds do an excellent job in this area. They not only provide excellent adhesion to the abraded concrete surface, but also provide a smooth surface, and are resistant in themselves to fatty acid contamination.

Wood also presents a problem where fatty acid-type materials are found. The fatty acid does not necessarily react with the wood, but it does penetrate the wood and is extremely difficult to remove by any means. Thorough washing with alkaline detergents and hot water are a less than satisfactory practical answer. Following washing, drying, and resmoothing the wooden surface, where conditions permit, it is suggested that a primer with fatty acid compatibility be used (a medium oil alkyd).

Alkali Contamination

A common example of an alkali contaminant is caustic soda. The alkali grouping should also include alkali salts. By definition, an alkali salt is one which is a reaction product of a weak acid and a strong alkali. Sodium carbonate is an example. This is a reaction product of sodium hydroxide and carbonic acid. Alkali salts tend to be strongly alkaline and hydrolyze to a hydroxide when in contact with moisture. Many industries such as

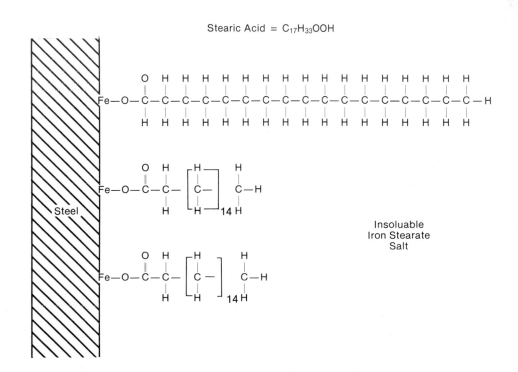

FIGURE 8.8 — Insoluble iron stearic acid salt (soap) as it might appear on a steel surface contaminated with a fatty acid.

The Substrate-Importance to Coating Life

the paper, rayon, and food industries, as well as refineries and many chemical companies use alkalies in their processing.

Surfaces which have been exposed to caustic or other alkalies are generally poor insofar as coating adhesion is concerned. Coatings which are based on an oil (*e.g.,* linseed oil, alkyds, or even some epoxy esters) may be expected to react with any caustic that might remain on the surface. Phenolic coatings would also tend to be adversely affected by any alkali that might be on the surface over which they were applied, resulting in lack of adhesion or possible disintegration.

Even vinyl resins, which are one of the most alkali-resistant of the synthetic resins, are definitely affected by alkali contamination of the surface over which they are applied. Where alkali dust or solution has been allowed to contaminate a steel surface, even though it is cleaned by normal procedures such as abrasive blasting, there is usually sufficient alkali remaining on the surface to neutralize the acidic groups that are attached to the vinyl resin molecule. These groups normally provide the adhesion for the coating. When these acidic groups are neutralized, the coating merely lies on the surface without the benefit of chemical or polar adhesion. The vinyl resins involved are the tripolymer vinyl chloride-vinyl acetate-maleic acid polymers.

The treatment of alkali-contaminated surfaces is fairly straightforward. The caustic contamination should be removed by washing the surface with water, followed by a dilute phosphoric acid wash prior to abrasive blasting. Phosphoric acid neutralizes the caustic that is on the surface, reacts with the steel to form an insoluble phosphate, and provides a slightly acidic rather than alkaline surface over which to apply the coating. This is a satisfactory base for almost any high-performance coating available today.

Concrete contaminated with alkali poses yet another problem. Many of the alkaline salts can react with the concrete, spalling the surface and creating a poor surface over which to apply any coating. The alkalies penetrate the concrete and tend to remain within the pore, even though the surface is thoroughly washed. Any severely contaminated concrete surface must be abraded down to solid concrete prior to the application of any coating. Alkali contamination that remains on the surface and does not necessarily react with the concrete can be removed by water washing, steam cleaning, and washing with dilute hydrochloric acid to neutralize the alkali. Once this has been done, the surface must be thoroughly washed and scrubbed free of the acid salt reaction product, and the surface thoroughly dried prior to the application of the coating.

Wood also presents a problem in times of alkali contamination. Wood is actually attacked by alkalies, breaking down the cellulose and providing a fibrous surface. In this case, all of the reacted and contaminated areas must be removed prior to the application of any coatings.

Sulfide Contamination

Sulfide represents a difficult type of steel contamination. Hydrogen sulfide is an insidious material, although sulfide contamination can come from other sources as well. It is a fairly widespread type of contamination, particularly in the marine industry and in the area of oil production and refining, as well as in chemical processing, such as the rayon industry. The marine industry can develop sulfide contamination through the breakdown of marine organisms on the surface of marine structures. When a marine organism, such as a barnacle, oyster, or other marine animal, dies, sulfides are formed immediately on, around, and underneath it. The sulfide then reacts with whatever iron or steel is available.

The production, transportation, storage, and processing of sour crude oil also create a sulfide problem. Sulfides can react directly with iron without the need for water; they can form a rather heavy iron sulfide scale which, when it has access to air, oxidizes quite rapidly to iron oxide. Iron sulfide is cathodic to steel, and therefore increases the corrosion problem where sulfide contamination exists. Iron sulfide oxidizes on exposure to air, but in so doing, also liberates sulfur or sulfide dioxide as reaction products.

In closed areas, this makes for a continuing contamination problem. Iron sulfide is sufficiently reactive with air so that it will ignite spontaneously if found in heavy concentration. A number of corrosion engineers have discovered this while taking samples of iron sulfide corrosion which they have wrapped in paper or cloth and stored in a pocket or briefcase. Sulfide is an extremely reactive material, and when ignited with a match, it burns with a blue sulfur flame.

When iron sulfide is present as contamination, it is also difficult to remove from the surface. Single sandblasting will not remove the hydrogen sulfide from the grain boundaries of the steel. The retention of sulfide is obvious in the oil industry where new plates have replaced old, corroded plates in a tank. When sandblasted, the new plates will remain bright and free of corrosion for a considerable period of time, while the old plates, blasted at the same time and to the same condition of surface preparation, will turn black in a short period. When such tanks have been coated, it has also been observed that, on the new plates, coatings will remain intact and unaffected by the high-sulfur crude and salt water conditions, while the coating on plates which have been previously exposed to sulfide will break down quite rapidly.

Sulfide contamination remains on steel, even though the steel is brought to a white metal by sandblasting. In order to remove the sulfide and to provide a proper surface for coating, the surface should first be blasted to white metal, then washed with water and allowed to rust, or washed with dilute phosphoric acid to form an iron phosphate salt on the surface. The surface then requires reblasting before a coating can be applied.

An indication of the surface contamination effect of hydrogen sulfide is shown in Figure 8.9*. Here, two areas of a clean steel panel were wetted with a dilute H_2S solution. The surface was allowed to dry and the plate then placed in the gelatine indicator solution previously de-

---

*See color insert.

scribed. Recall that the anode area turns blue and the cathode red. As can be seen, the H$_2$S-contaminated surface turned into a strong anode. When pure water was placed on the metal and placed in the same gelatine test, only a light anodic reaction took place.

Hydrogen sulfide is also developed by sulfate-reducing bacteria. These are found in many soils, particularly in boggy areas where there is plenty of decomposing organic matter. They are also found in some petroleum products and particularly in sewage. Here, they react not only on the organic matter in the sewer, but also on the sulfate that is in the water. This bacteria is found in a slime underneath the water level, and releases hydrogen sulfide into the water. The H$_2$S evaporates from the water into the air where it can react with steel directly, or, as we have previously seen, the aerobic bacteria on the surface of sewer structures above the water ingest the H$_2$S and form sulfuric acid.

This acid reacts rapidly with both concrete and steel. Any copper surface which is in such an atmosphere must be thoroughly protected, as it also reacts readily with both the sulfide and the sulfuric acid. Sulfide contamination is therefore much more widespread than most people realize. The sulfide-contaminated surface must be cleaned and completely freed of sulfide prior to the application of any coating if long-lasting protection is to be obtained.

Organic Contamination

Organic contamination is a prevalent problem throughout all industry, particularly where high-performance coatings are required. All industry uses oils, grease, and other similar lubricants. In addition, many industries use wax in some of their processes, and sugars as well as fats are used throughout the food industry. This type of contamination can be difficult since organic material (e.g., oils, greases, and waxes) tends to penetrate into the grain boundaries of metals; they wet the metal surface quite well and are therefore difficult to remove.

In oil production, for example, paraffin waxes, which are soluble in many of the oils, preferentially wet out a metal surface and accumulate in any area where the bare metal is exposed. Unlined steel tubing and pipe are often completely clogged with paraffin due to this preferential wetting and because of the decrease in solubility of the paraffin as the temperature of the crude oil is reduced.

Such surfaces, both on pipe and in tanks, are extremely difficult to prepare properly. The surfaces must be thoroughly steamed in order to melt the paraffin and allow it to be washed away from the surface. Even this does not completely eliminate the problem, because of the active wetting characteristics of the paraffin on the metal surface. Once the surface is thoroughly washed, it must be blasted and again washed with water in order to bring out any areas where paraffin may remain on the surface. These are readily recognized since they do not oxidize as rapidly as the surface, which is free of this type of contamination. Coatings that are applied over even a minute film of wax will not dry, but will remain tacky for several days. Even if they do eventually dry, they will have little

or no adhesion to the metal. Cast iron surfaces are even more difficult than steel because of the porosity of the cast iron, and the strong penetrating qualities of the wax.

In the early days of the wine industry, both steel and concrete tanks were coated with waxy compounds to prevent contact of the wine with either the metal or the concrete. As high-performance coatings became available, they were tested on both types of surfaces. They were singularly unsuccessful because of the wax contamination of the surface. Many of the early tank cars also were lined with wax, which was particularly unsatisfactory as a coating.

The expansion of the wine industry from local to worldwide distribution was a result of lining tanks with vinyls, epoxies, and similar high-performance coatings. However, it was only after it was discovered that it was the wax on the steel surfaces that was causing coating problems that they went to the trouble of properly preparing the surfaces by thoroughly removing the wax. This was only accomplished by the same procedure outlined for the petroleum tanks, i.e., a thorough steaming of the surface to melt and float the wax away. A thorough blasting of the surface, followed by an oxidation of the surface by a water wash in order to determine where any wax remained, and then a thorough reblast of the surface prior to the application of the coating. Unfortunately, there was not a less costly surface preparation method that would provide an uncontaminated surface over which to apply a coating.

Many other organic materials also present difficult coating problems. For instance, protein-contaminated metal or concrete surfaces are difficult to prepare for coating since organic coatings will not properly adhere where protein compounds remain on the substrate. Sugars are also difficult, particularly on concrete, since sugars penetrate and react with the concrete so that coatings applied over such surfaces do not have any adhesion.

Concrete surfaces contaminated with protein or sugar must be abraded to the point where the contaminants have ceased to penetrate. Such areas then require resurfacing. Organic resurfacing materials have proven quite satisfactory. Once applied over an uncontaminated surface, they provide good protection against further penetration into the concrete.

Where ordinary oils and greases are encountered, steel surfaces are reasonably easily prepared. This primarily involves steam cleaning the surface to remove the oil, followed by a thorough sandblast or acid pickle to prepare the surface for acceptance of the coating.

Concrete, on the other hand, poses a more difficult problem. Preparing and washing the surface with detergent and steam will not remove the oil and grease from the concrete to a sufficient depth to obtain satisfactory adhesion. Organic coatings normally contain solvents, and the solvents penetrate to the oil and grease which have accumulated within the concrete. They then dissolve the oil and grease and tend to bring it to the surface beneath the coating that has been applied. This not only makes for a porous coating where the oil and grease are not compatible, but it also produces poor adhesion of almost any applied coating. Again, if the contamination is serious, the concrete surface needs to be deeply abraded prior to the application of surfacers or coatings.

Much of the previously discussed contamination concerning steel also applies to cast iron. In the case of cast iron, however, because of its different crystalline structure and the fact that contamination of almost any type—chloride, acid, alkali, sulfide, or organic—is attracted to the surface, it is a much more difficult material to decontaminate than the denser steel surface.

Blasting of the cast iron is normal procedure, but this will not remove all of the contaminants. Heating the cast iron is a second procedure which often aids in the removal of contaminants from the surface. Heating must be carried to a relatively high temperature (*i.e.,* several hundred degrees) where such treatment is possible. The surface then requies a reblast prior to the application of coatings. It is also preferable to apply the coating to cast iron when the surface is still warm to the touch, which allows the coating to penetrate into the surface and to wet the cast iron surface to a better degree than if it were applied over a cold surface.

Aluminum, copper, brass, lead, and galvanized steel are all subject to the same organic contamination problems as is steel. Fortunately, they are all dense surfaces so that contamination does not tend to penetrate, unless the contaminant chemically reacts with the surface such as alkali with aluminum, sulfides with copper, or acidic contamination on zinc. Organic contamination of these surfaces, because of their density and smooth surface, is usually less severe than with steel and, particularly, cast iron. Nevertheless, similar surface preparation methods are required in order to eliminate all contamination prior to coating.

Summary

Contaminated substrates present a critical coating problem which must be recognized in order to obtain effective protection of those surfaces by coatings. Contaminated surfaces range from those that are merely rusty, to those exposed to chlorides, hydrogen sulfide, and organic contaminants. All must be considered important factors in the proper application of any coating.

Substrates have a major effect on coating performance and life. If the substrate prevents the coating from developing its full properties of adhesion and resistance, then the substrate cannot be satisfactorily protected. The substrate and the coating must be completely compatible before a coating can provide protection from any type of corrosive atmosphere.

A resistant substrate can improve coating life substantially as indicated by Larrabee,[1] Hudson,[2] and LaQue.[3] Reactive substrates also can improve coating life, *e.g.,* coatings applied over zinc in a marine atmosphere. Hard, strong substrates can provide a strong base and long coating life, while a soft substrate may mean a much shorter existence for the same coating because of reduced resistance to impact, abrasion, and physical abuse. Contaminated substrates even further complicate the problem by downgrading the properties of the original substrate which otherwise might have been a satisfactory base for the coating. A thorough understanding of the substrate and its condition must be developed before a proper coating can be selected and applied.

References

1. Copson, H. R. and Larrabee, C. P., Extra Durability of Paint on Low Alloy Steels, ASTM Bulletin, December, 1959.
2. Hudson, J. C., Protection of Steel by Painting Against Atmospheric Corrosion, Schweiger Archiv fur Augewnadte Wissenschaft und Tecknik, Vol. 2 (1958).
3. LaQue, F. L. and Boylan, J. A., Effect of Composition of Steel on the Performance of Organic Coatings in Atmospheric Exposure, Corrosion, Vol. 9 (1953).

9

Surface Preparation

Introduction

The objective of surface preparation is to create proper adhesion of a coating over the substrate. Adhesion is the key to coating effectiveness, and it determines whether the coating is merely a thin sheet of material lying on the substrate or whether it becomes an actual part of the substrate.

Adhesion becomes an even more critical condition for coatings applied in corrosive areas. Thus, proper surface preparation is vital to the long life and effectiveness of a coating applied in corrosive service.

One industry that has been especially concerned with proper adhesion is the automotive industry. Very seldom, if ever, does the original coating on the body of an automobile blister, peel, alligator, or otherwise delaminate from the surface. This represents a considerable achievement when one considers that there is probably no other piece of expensive equipment that is exposed to such varied conditions as the automobile. It is subject to rain, storms, snow, and dust, as well as industrial atmospheres and sodium chloride from deicing salts. In spite of this, there are very few failures of the exterior coating.

Meticulous preparation of the automotive body metal is mandatory to insure such comprehensive protection. The metal is mechanically moved through various baths, ranging from pretreatment washes to immersion in various metallic phosphate solutions, followed by additional washing and thorough drying to insure maximum surface cleanliness prior to coating application.

While many people have experienced catastrophic failures of automotive bodies due to corrosion, the greatest amount of this damage has occurred on the inner surface of the metal rather than the coated exterior surface. The excellence of automotive finishes and their resistance to severe weather and chemical atmospheres can be attributed to meticulous surface preparation and the outstanding surface adhesion thus provided.

It has been stated that paints fail in direct proportion to their lack of adhesion. Stating this in a positive way, paints (or anticorrosive coatings) are successful in direct proportion to their bond strength to the substrate. The purpose of surface preparation is to insure that the maximum bond strength will develop at the interface between the substrate and the coating. The weakest area across the coating should be within the adhesive or organic coating layer and not at the interface of the coating and the substrate. Failure within the coating is referred to as *cohesive failure;* failure at the interface between the substrate and the coating is referred to as *adhesion failure.*

If any failure is to be tolerated, the cohesive-type failure is preferable. A good demonstration of this is found in the vinyl resins. The vinyl chloride acetate resin is a strong polymer. When applied directly to a substrate, the polymer is sufficiently strong that the separation from the substrate is at the interface between the coating and the substrate. In other words, the vinyl coating made only with the vinyl chloride acetate resins fails at the adhesive bond between the two materials. On the other hand, the vinyl chloride acetate resin to which a small amount of maleic acid has been added and incorporated into the polymer, when applied to a substrate, has excellent adhesion. In this case, the vinyl coating itself will break before the bond between the coating and the substrate. This is the type of adhesion that is essential for coatings used in corrosive areas.

The goal of surface preparation should never be less than the level sufficient to insure cohesive failure of the coating. From a surface preparation standpoint, the ideal type of coating failure is 100% cohesive. Figure 9.1 indicates adhesive failure where the coating separates from the substrate cleanly and does not leave any coating attached to the substrate.

Figure 9.2 indicates cohesive failure in which the coating breaks within itself and leaves a continuous layer of coating on the substrate, even though the coating surface may be completely removed. Such cohesive failure is easy to understand with some of the soft coatings such as asphalt, coal tar, and similar materials where there is practically no cohesion within the coating. On the other hand, the majority of the coatings dealt with by corrosion engineers are hard, tough, and, in many cases, extremely strong; yet the adhesive bond must still be greater than the cohesive strength of the coating.

Another type of adhesion failure occurs where the substrate itself fails rather than the coating. Such failure is not uncommon on concrete. Epoxy coatings applied to concrete often will have a greater tensile strength and cohesion than the concrete. Figure 9.3 shows this type of substrate failure.

Types of Adhesion

The type of adhesive bond depends on both the substrate and the coating. The three types of adhesive bonds are: (1) chemical, (2) polar, and (3) mechanical. The chemical bond, which is created by a chemical reaction between the coating and the substrate, is undoubtedly the most effective bond. One example of this type of bond is hot dip galvanizing, where the steel and the zinc metal amalgamate or dissolve within themselves, as was indicated in Chapter 6 on inorganic coatings.

Inorganic zinc coatings are chemically bonded between the silicate molecule and the steel substrate. The vinyl wash primer reacts with the steel substrate to form an excellent adhesive bond. Such bonding is called *primary valence* bonding, where the chemical groups on the coating actually react across the interface with complimentary groups on the substrate, forming a chemical compound (Figure 9.4). An example of this is the oxygen bonding of the silicate matrix in an inorganic coating to the metal.

It is also possible that epoxy molecules are bonded to the metal surface by metal hydroxide groups through a condensation reaction (Figure 9.5). In both of these examples, a new chemical compound would have been formed, joining the coating and the substrate.

Polar adhesion is a more common type of adhesion than the previously described chemical adhesion, especially

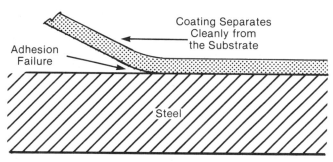

FIGURE 9.1 — Adhesive failure of a coating.

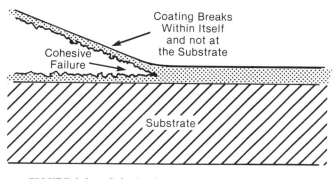

FIGURE 9.2 — Cohesive failure of a coating.

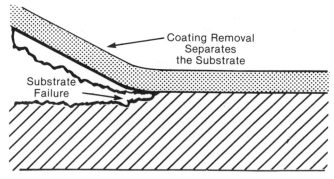

FIGURE 9.3 — Adhesion failure of the substrate.

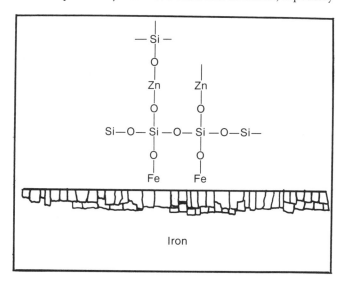

FIGURE 9.4 — Primary valence bonding.

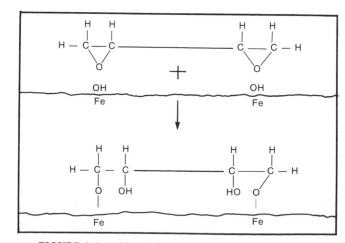

FIGURE 9.5 — Chemical adhesion of an epoxy coating to a metal substrate. (SOURCE: Corrosion and the Preparation of Metallic Surfaces for Painting, Unit 26, Federation Series on Paint Technology, Federation of Societies for Paint Technology, Philadelphia, PA, 1978.)

Corrosion Prevention by Protective Coatings

with organic coatings. *Polar adhesion* or *bonding* is the attraction of the resin molecule to the substrate. In this case, the resin acts like a weak magnet, with the north and south poles of the magnet attracting opposite groups on the substrate. This is the origin of the word "polar" to describe this type of adhesion. Actually, the polar groups are positively and negatively charged portions of the coating molecule that are attracted to oppositely charged areas on the metal or substrate.

Chemically, polar adhesion is considered secondary valence bonding where the adhesion occurs by way of physical physiochemical attractions between the resin molecules and the substrate surface molecular structure. In the case of vinyl chloride acetate materials, there is little or no polar bonding, and, as a result, the vinyl coating can be stripped from the substrate in sheets. However, when the maleic acid is incorporated as part of the molecule, this acidic group forms a polar molecule from the otherwise poorly polar molecule and allows it to be attracted to opposite poles that exist on the metal surface, with resultant adhesion. Adhesion is only possible when the two attracting polar groups from each surface approach each other, as described in *Corrosion and Preparation of Metallic Surfaces for Paint.*

> ...the strength of secondary valency attractions increases at a rate proportional to the sixth power of the intermolecular distance, but does not become effective until this distance is under 5 Å. As this distance is no more than three times the diameter of an oxygen atom, the importance of clean well-prepared surfaces in obtaining good adhesion can readily be appreciated. Grains of dirt, or even monomolecular films of oil, are considerably thicker than 5 Å, and will effectively nullify all adhesion.[1]

Figure 9.6 shows the secondary valence bonding of an hydroxylated coating, such as an epoxy, to the metal hydroxyl groups of a metallic surface by way of hydrogen bonds. It is not surprising, then, that the polar groups on the molecule significantly improve the adhesion of the coating.

The primary valence bonding, or the chemical bond (which is similar to the kind of bond that holds the polymer together), is significantly stronger than the secondary valence bonding. This is unfortunate, since most of the adhesion of organic coatings is of the polar or secondary valence type. Chemical bonding is the way in which conversion coatings work. (*Conversion coatings* are inorganic surface treatments that react by way of the primary valence bond producing an all-inorganic surface; *e.g.,* phosphate-treated steel.) These chemical surface treatments are more porous than the metal itself and provide considerably increased surface area for secondary or polar bonds. This is the type of surface treatment used on the body metal of automobiles and provides a combination of both chemical and polar bonding for the resulting excellent adhesion.

The wash primer is also a surface treatment, converting the metallic surface by way of both primary and secondary bonds to a surface that appears to be both inorganic and organic and making an easily paintable primary bonded organic coating on an inorganic substrate. (Coatings that have strong polar groups have excellent adhesion to the wash-primed surface.) The following is an explanation of this concept from *The Handbook of Surface Preparation.*

> The adhesive and cohesive forces are a combination of mechanical and chemical (or molecular). Mechanical forces result from interlocking of molecules or polymers with the asperities of the substrate surface, pigment or filler particles, and with the polymer chains themselves. The chemical or molecular forces include electrostatic forces, van der Waal's forces and ionic forces. These are related to the existing positive and negative sites that exist on all surfaces and in all molecules. Sufficient alignment of oppositely charged sites exists to result in strong cohesive and adhesive forces of attraction. Such forces operate in the film formation of surface coatings and affect the cohesion-adhesion balance.[2]

Mechanical adhesion is the type of adhesion that is associated with surface roughness or anchor pattern. *Anchor pattern* is the surface roughness formed by peaks and valleys on the substrate. These can vary over a relatively wide range of depth; however, of most importance to coatings is the number of hills and valleys which increase adhesion by the increase in surface area and by the actual roughness. Some coatings, because of the relatively poor adhesion within the coating itself or because of thickness, require good surface roughness and a deep anchor pattern in order to obtain adequate adhesion. Most high-performance coatings obtain adequate adhesion with an anchor pattern of from 1 to 2 mils in depth. Such a surface roughness substantially increases the surface area over which the coating has an opportunity to bond.

Concrete has a different type of mechanical adhesion. In this case, the surface is relatively porous, with many minute water and air pockets, surface checking, and other natural surface roughnesses which definitely aid in coating adhesion. Coatings for such types of surfaces preferably should be highly penetrating, so as to take advantage of the surface and subsurface roughness. Inorganic zinc coatings are relatively porous, allowing penetrating coatings to adhere by mechanical as well as by polar or chemical adhesion.

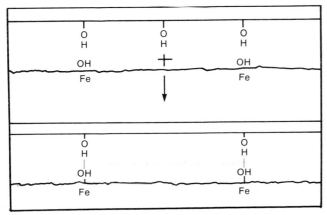

FIGURE 9.6 — Polar or secondary valence bonding of an hydroxylated coating to a metal substrate. (SOURCE: Corrosion and the Preparation of Metallic Surfaces for Painting, Unit 26, Federation Series on Paint Technology, Federation of Societies for Paint Technology, Philadelphia, PA, 1978.)

Surface Preparation Objectives

Strong adhesion is the key to coating performance and long life. If adhesion is weak, the coating will gradually fail by blistering, underfilm corrosion, or chipping and flaking. With strong adhesion, the coating can withstand most of the environmental resistance requirements that otherwise would affect its integrity. These include abrasion resistance, impact, flexure, moisture, moisture vapor transmission, humidity, salt conditions, resistance to corrosive chemicals, micro-organisms, and all of the many environmental conditions that affect the life of a coating. A coating with strong adhesion will have a longer life and greater resistance to corrosion than one which has marginal adhesion and is affected by all of the above atmospheric conditions.

The primary objective of surface preparation is to provide maximum coating adhesion. The actual mechanism of surface preparation has a twofold purpose. The first purpose is to remove any extraneous, loose materials from the surface of the substrate, as well as to eliminate chemically bonded scales, oxide films, and similar surface reaction products that cover active adhesion sites on the metal surface. The removal of such materials exposes the reactive sites so that the primers can have contact with them and develop the maximum adhesion possible.

The second purpose of surface preparation is to increase the surface area by increasing the roughness and anchor pattern of the surface. By this means, the actual exposed surface area per unit of actual area is greatly increased. By increasing this effective surface, many additional reactive sites on the metal surface are exposed, allowing for additional polar or chemical adhesion of the primer to the surface. This is extremely important, since increasing the opportunity for either primary or secondary valence bonding with the coating system is the key to the best possible adhesion of any coating.

In geometry, a straight line is the shortest distance between two points. The same principle is involved in increasing the surface area of metal. A sheet of cold-rolled steel, as an example, has a smooth surface and therefore a straight line surface. On the other hand, if a sheet of cold-rolled steel is sandblasted, the straight line changes to one that follows the peaks and valleys of the metal. The line therefore becomes much longer, thus the surface area has become much larger. The actual surface area, measured in square inches or square feet, is the same. However, the effective surface area can be doubled or tripled by roughening the surface (Figure 9.7).

Figure 9.8 shows the effect of even lightly contaminated and unprepared surfaces, and how the monomolecular film of coating adjacent to that surface has the polar sites on the molecule insulated from the reactive metal sites on the metal substrate. The metal reactive sites and the coating reactive sites are separated at some distance by the contamination, thus the full effect of either chemical or polar bonding is prevented.

Primers are applied directly to a surface. If this surface includes dirt, dust, scale, rust, oil, moisture, or other contamination, the adhesion of the protective coating to the surface can only be as good as the bond of the primer to the contamination and the bond of the contaminated

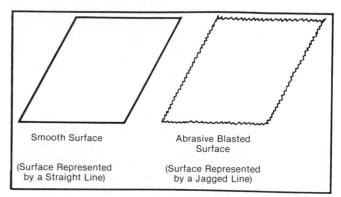

FIGURE 9.7 — An increase in surface area was achieved by increasing the roughness and anchor pattern of the surface.

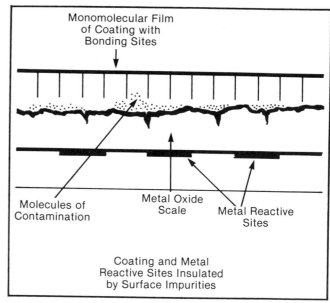

FIGURE 9.8 — An unprepared and lightly contaminated surface which prevents the full effect of either chemical or polar bonding. (SOURCE: Corrosion and the Preparation of Metallic Surfaces for Painting, Unit 26, Federation Series on Paint Technology, Federation of Societies for Paint Technology, Philadelphia, PA, 1978.)

material to the actual substrate. The contamination provides a secondary interface between the coating and the actual substrate. Primers, however, are formulated to adhere to metal, concrete, wood, and other structural surfaces rather than to the surface of contaminants.

Early painting systems were all brush applied, and the coating vehicle had a highly penetrating and strong wetting characteristic. The brushing action worked the primer into the contamination, and the highly penetrating and wetting characteristic of the vehicle allowed it to penetrate and actually contact the substrate below the contamination.

On the other hand, as the molecular weight of the coating resins increased and the application of the coatings changed from brushing to spraying, less of an opportunity existed for the large molecule to be worked into the contamination and down to the primary surface. Thus, scrupulous surface cleaning and preparation prior to priming became much more important factors in creating proper coating adhesion.

Figure 9.9 indicates how the polar areas on a coating molecule are attracted to and become attached to corresponding reactive sites on the clean metal surface. In this case, the surface is represented as a sandblasted or abrasive-blasted surface with relatively high peaks and valleys, which not only increase the surface area, but increase the number of bonding sites that become effective. This not only improves the mechanical adhesion of the coating, but improves the chemical or polar adhesion in a like amount.

It is important that the metal surface be entirely new in order for the maximum number of metal reactive sites to be available to the coating. It can be safely stated that the cleaner the metal surface, the better the bond or adhesion of the coating to that surface. Stated in a negative way, the adhesion of a coating is inversely proportional to the amount of contamination on the surface. The greater the contamination, the less adhesion the coating will have.

Painters of all types—industrial, marine, commercial, or even weekend painters—generally have one thing in common. That is, they tend to neglect the most important part of a coating job: surface preparation. The time and expense required to take this step is generally resented, creating a psychological barrier. Physical application of the coating seems to be a much more productive and therefore satisfying activity than sandblasting, chipping, scraping, or even thoroughly washing a surface. Unfortunately, if these activities are omitted (*i.e.,* proper surface preparation methods), the whole coating program is doomed to be a waste of both the time and expense involved.

Development of Surface Preparation Techniques

Proper surface preparation is historically a recent development. As was previously stated, surface sandblasting was not practiced until the late 1930s and, even then, was not of general interest until the Navy, during World War II, determined that they could leave their ships in service much longer if they sandblasted the surface and thus applied the coating over a thoroughly clean metal. Prior to this time, surface preparation of almost any surface consisted of chipping and scraping with hand tools. This was a costly and tedious process in which there was little interest.

This type of surface preparation was generally effective as long as oil-type coatings were in use. These had high wetting characteristics to penetrate the surface contamination. In addition, most coatings were brush applied, and the physical brushing action increased their penetration into the contamination. On new work, the mill scale was generally left on the surface so that under mild conditions and with thorough wetting and penetration of these surfaces by the oil coating, some long-term protection was obtained.

The advancement of surface preparation methods paralleled the development of high-performance protective coatings. First were the vinyls, which, in order to obtain optimal performance, needed a thoroughly cleaned surface. This was true for both steel and concrete. Inorganic zinc coatings added impetus to the proper surface preparation techniques, as well as the addition of epoxies, polyurethanes, and some of the heavy epoxy surfacers. It was also found, after surface preparation techniques were developed, that the old standby coatings, such as red lead and oil, asphalts, and coal tars also performed better and longer over a properly prepared surface.

The effect of surface preparation has been demonstrated in some dramatic ways. Figure 9.10 shows the same type of vinyl coating system applied over a sandblasted panel on the left, a rusty panel in the center, and a mill-scale panel on the right. These were exposed to a marine atmosphere for over nine years. In this case, the vinyl coating provided excellent protection for the sandblasted surface, while the rusty panel shows considerable corrosion activity, along with the mill scale covered panel which shows even more.

Figure 9.11 shows five panels: The first panel (left),

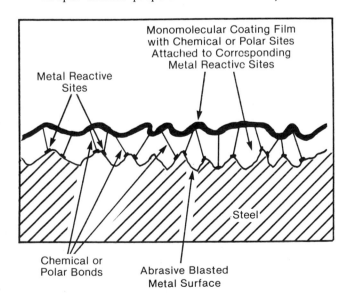

FIGURE 9.9 — An abrasive-blasted surface which shows how the polar areas on a coating molecule are attracted and become attached to corresponding reactive sites on the clean metal substrate. (SOURCE: Corrosion and the Preparation of Metallic Surfaces for Painting, Unit 26, Federation Series on Paint Technology, Federation of Societies for Paint Technology, Philadelphia, PA, 1978.)

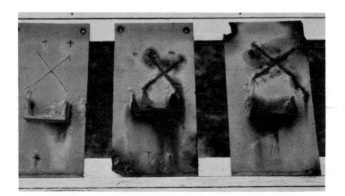

FIGURE 9.10 — Corrosion at scribes in the same coating (3-coat vinyl system) applied over sandblasted steel (left), rusted steel (middle), and intact mill scale (right), exposed for 9 years in a marine environment.

FIGURE 9.11 — Results of a test involving panels of, from left to right, galvanized steel, inorganic base coat, white sandblast, rust, and mill scale, exposed for 9 years in a marine environment with identical topcoats (3-coat vinyl system).

Surface Preparation	Durability, years	
	4-Coat Scheme (2 coats of red lead paint and 2 coats of red iron oxide paint)	2-Coat Scheme (2 coats of red iron oxide paint)
Intact Mill Scale	8.2[1]	3.0
Weathered and Wirebrushed	2.3	1.2
Pickled	9.5	4.6
Sandblasted	10.4	6.3[1]

[1]Indicates that at the most recent inspection the painting scheme had not failed on all the surfaces concerned.
[SOURCE: Hudson, J. C., Protection of Structural Steel Against Atmospheric Corrosion, Journal of the Iron and Steel Institute, Vol. 168, No. 6 (1951).]

TABLE 9.2 — Method of Surface Preparation for Painting

Painting Scheme (Two coats)	Life of Painting Scheme, years		
	Over Weathered Surface	Over Pickled Surface	Over Weathered and Heated Surface
Black Bituminous	2.6	13.1[1]	2.2
Lead Chromate	2.7	14.5[1]	7.5
Micaceous Iron Ore	0.6	15.0	0.6
Red Lead	1.1	13.1[1]	5.9
Red Oxide	1.8	8.1[1]	2.8
Red Oxide and Zinc Chromate	3.7	9.6[1]	5.3
White Lead	2.2	7.0	1.7
Average	2.1	11.5	3.7

[1]Back surface had not failed after fifteen years' exposure.
[SOURCE: Hudson, J. C., Subsidiary Paint Tests at Birmingham: Final Report, Journal of the Iron and Steel Institute, Oct. (1951).]

galvanized; the second panel, an inorganic base coat; the middle panel, a thorough white sandblast; the fourth panel, rusted; and the last panel (right), covered with mill scale. These panels were exposed in a marine atmosphere for over nine years. The topcoat on all panels was identical, consisting of a good quality vinyl coating applied in three coats. Results were similar to those shown on the panels in Figure 9.10 which were part of a dramatic test indicating the benefit of proper surface preparation.

Figures 9.10 and 9.11 give visual testimony to the impact of surface preparation. One of the original studies of the effect of surface preparation on coatings was done by J. C. Hudson of The British Iron and Steel Research Association. Many of his studies were conducted prior to the development of the highly corrosion-resistant coatings. The results of his tests, however, clearly indicated the benefits of a clean surface.

Table 9.1 gives the results of Hudson's extensive tests of varying types of steel surface preparation. According to these results, pickling or sandblasting dramatically lengthens the life of the coatings applied over rusted or weathered steel. They also show that a weathered and wire-brushed steel surface is an extremely poor surface over which to apply almost any coating. This includes those that have an affinity for and readily wet such a surface, such as the oil-base, red lead, and iron oxide coatings used in this particular instance.[3]

These tests also showed that the intact mill scale, when painted with these highly wetting coatings (particularly the red lead-base oil coating), lasted almost as long as the better prepared surface.[3] These results, however, would not apply to the usual scale available to industry. Partially rusted, loose, or broken mill scale would not perform in the same way.

In another article, Hudson reported on a series of tests that use a different series of coatings applied over weathered steel, pickled steel, and weathered and heated steel. These results are given in Table 9.2 and again indicate that the thoroughly cleaned steel, *i.e.*, a pickled surface, is many times superior to a weathered surface. The coatings involved, with the exception of the black bituminous coating, are all presumed to be oil-base coat-

ings pigmented with the materials shown in the table.[4]

The primary conclusion from Hudson's work, however, is that the properly prepared surface adds to the life of various types of materials. The weathered surface is presented as an extremely poor surface, with an average coating life of only 2.1 years, compared with an average life of 11.5 years over the pickled surface. Also, with the exception of the black bituminous coating, the micaceous iron ore-pigmented coating, and the white lead-pigmented coating, the life of the coating over the weathered and heated surface was much improved over the weathered-only surface.[4]

Experience also indicates that heating surfaces, particularly those that cannot be thoroughly prepared, improves the wetting of the surface by the coating and increases the life span of the coating itself. Coating a warm surface lowers the viscosity of the coating and increases its penetrating and wetting ability over either rusted or clean metal surfaces.

Types of Contamination

Mill Scale

A typical result of applying coatings over mill scale is shown in Figure 9.12, in which three different types of coatings were applied over hot-rolled steel containing mill scale. In this case, the mill scale was not entirely intact, although it was what could be considered good, clean mill scale without appreciable rusting. All three of the applied coatings had deteriorated badly after six years in the marine atmosphere.

Most of the structural steel that is used today for heavy construction, either in the marine industry or in chemical, petroleum, paper, and similar industries, is hot-rolled steel which contains varying amounts of mill scale on the surface. Some of the lighter shapes have a thin mill scale on the surface, while heavy plate and heavy shapes generally are covered with a heavy mill scale, amounting to 10 or more mils in thickness.

Mill scale forms on the hot-rolled steel because in the process of rolling, the steel is suitably heated below the melting point, yet at temperatures where the steel is plastic and can easily be shaped by the rolling mechanism. The steel is first heated in open furnaces where oxygen in the furnace atmosphere combines with the hot metal to form oxides on the iron surface. Much of the oxide formed in the furnace is broken off during the rolling process, and usually it is the oxide that is formed by the latent heat after the shape has been milled and is allowed to cool that remains on the surface. Since the rolls do not break it from the surface after the rolling is completed, it forms as a tight oxide on the surface, and the weight of the shape or the plate determines the thickness of the oxide coating.

Mill scale that is formed by this process is not uniform. The outer layer of the mill scale is primarily Fe_2O_3, or ferric oxide, and contains approximately 30% oxygen by weight. Beneath the outer layer is the layer that constitutes the majority of the scale. This has a general formula of Fe_3O_4, and contains approximately 28% oxygen. As the scale gets closer to the actual metal itself, the formula may approximate the formula of FeO, which contains some 22% oxygen. Underneath this is a layer of mixed oxygen and metal (scale binder) which has a still lower oxygen content. It is this scale binder that often causes sandblasters to conclude that the scale is embedded in the metal and cannot be removed. This, however, is not a proper evaluation and will not result in proper surface preparation.

The mill scale itself is quite brittle. It has a lower thermal expansion than steel and cracks on cooling. The cracks can extend through the scale and may be aggravated by any flexing of the steel during handling. The cracking of the scale allows access of oxygen and moisture through the scale down to the metal itself. Since the scale is strongly cathodic to the steel surface, rapid corrosion can take place even shortly after the plate or shape is rolled. This is particularly true if the metal is stored in the open or if it is transported by barge or rail through industrial or marine atmospheres.

An example of the pitting that can take place after a short period of time is shown in Figure 9.13, where the steel has been blasted free of mill scale. In this case, the steel was rather heavy plate and was allowed to stand for several weeks prior to its being fabricated. The pitting is the result of oxygen, water, and undoubtedly some salt penetrating the mill scale at the cracks and creating a corrosion cell at the bottom of the crack. Figure 9.14 is a schematic drawing of mill scale, indicating the various layers and their compositions.

When corrosion does occur in this manner, the corrosion products are of a considerably greater volume than the original metal. The increasing volume tends to create a wedge and pops the mill scale from the surface of the metal. This occurs rather easily when mill scale is weathered; however, it can also occur underneath a coating.

Remember that when a coating is applied over mill scale, the coating does not adhere to the steel itself. There-

FIGURE 9.12 — Results of a test involving three different types of coating applied over hot-rolled steel panels covered with mill scale. The panels were exposed in a marine environment for 6 years.

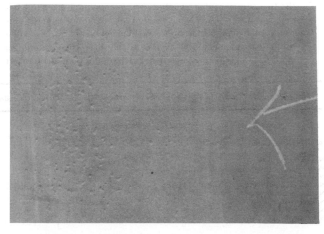

FIGURE 9.13 — Pitting of steel plate covered with mill scale after short atmospheric exposure (1 to 2 months). The plate was sandblasted to display pitting.

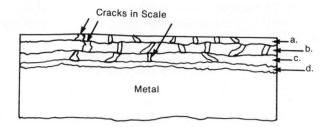

FIGURE 9.14 — Schematic drawing of mill scale, indicating the various layers and their compositions: (a) Fe_2O_3; (b) Fe_3O_4; (c) FeO; (d) FeO + Fe. (SOURCE: Good Painting Practice, Chapter 3, Vol. 1, Steel Structures Painting Manual, Steel Structures Painting Council, Pittsburg, PA, p. 50, 1955.)

fore, when corrosion takes place under the mill scale, the scale has no adhesion to the steel, allowing a loose opening between the coating and the steel surface. Moisture vapor then penetrates the coating and condenses in the area of loose adhesion. This creates blistering and eventual corrosion which loosens a greater amount of the mill scale as well as the coating. When a break occurs in the coating, rapid undercutting of the coating takes place by the corrosion working under the mill scale. Rust tubercules and heavy corrosion scale build up under the coating. This is a common occurrence in the marine industry and in other industries where chloride, sulfate, and similar ions are present.

Oil Contamination

Oil and grease should not be overlooked as serious factors in surface preparation simply because they are common materials used throughout most industries. Unit 26 of the *Federation Series on Paint Technology* comments on the coating problems created by oil and grease.

> Oil or grease on a metal surface can present a particularly difficult problem because most coating vehicles do not easily displace oil and thereby wet the surface. Generally, organic coatings have relatively low surface free energies (20-60 dynes per centimeter) and don't have much difficulty wetting metal and metal oxide surfaces with much higher energies (perhaps 400 dynes per centimeter).
>
> Oil contamination of a metal surface, however, can reduce the free energy of that metal surface to something like 20 dynes per centimeter, lower than the energy of most coatings. This would change the surface from one that can be wet by the coating into one that cannot. It is for this reason that, when designing a system for a poorly prepared metal surface, it is far better to select

a lower energy coating system having only moderate performance properties (*e.g.,* an oil paint), than to select a high performance coating of higher free energy (*e.g.,* a vinyl). A vinyl can give little protection if it cannot adhere to the substrate. Oil paint, being one of the lowest energy coatings systems, has a considerably better chance of adhering to the lower energy substrate. If, on the other hand, the oil can be removed and the surface properly prepared, then the vinyl system would undoubtedly be the better choice.[1]

Types of Surface Preparation

Various types of surface preparation mechanical equipment are used to clean the surface and thus provide proper coating adhesion. The Steel Structures Painting Council has given more attention to the various types of surface preparation than any other organization involved in coating work. NACE has also done extensive work through their technical committees in developing surface preparation standards; although their concentration has been primarily on surface cleaning through abrasive blasting.

Table 9.3 lists the surface preparation specifications outlined by both the Steel Structures Painting Council and NACE in a descending order of effectiveness. Each grade lower in the list allows a greater amount of contamination to be left on the surface prior to coating. This is extremely important since it is the degree of contamination that is the key to coating adhesion.

An NACE technical committee (T-6H-15 on Effects of Surface Preparation on Service Life of Protective Coatings) has run a series of tests covering 445 panels, in which 8 types of surface preparation were selected. Five types of protective coatings were tested over each method of surface preparation, and these were then exposed in seven different locations throughout the United States. Inspections were made on a yearly basis over a four-year period (Table 9.4).[5]

TABLE 9.3 — Surface Preparation in Descending Order of Effectiveness

1. White Sandblast — NACE #1, SSPC SP 5-63
2. Near-White Sandblast — NACE #2, SSPC SP 10-63
3. Commercial Blast — NACE #3, SSPC SP 6-63
4. Acid Pickling — SSPC SP 8-63
5. Brush Blast — NACE #4, SSPC SP 7-63
6. Flame Clean and Power Sanding — SSPC SP 4-63
7. Power Tool Cleaning — SSPC SP 3-63
8. Chip and Hand Wire Brush — SSPC SP 2-63
9. Solvent Wipe — SSPC SP 1-63

TABLE 9.4 — Surface Preparation Results Compiled from Tests Run by NACE Technical Committee T-6H-15

Surface Preparation Method	Rank[1]	Rust Rating Overall	Rank	Rust Rating[2] Local	Rank	Edge Perimeter Rating	Rank	Edge Penetration Maximum (Inches)	Rank	Edge Penetration Average (Inches)	Rank	Scribe Undercut Rating	Sum of Rank	Overall Rank	Overall Rank Adhesion Related
None Intact Mill Scale	3	93.8	5	83.3	8	38.1	8	1.87	8	1.26	8	21.9	40	7	8
Pickle Phosphate Treated	8	87	8	74.5	5	92.4	4	0.214	4	0.145	4	77.5	33	6	4
NACE 4 Brush Blast	7	90.8	6	82.9	7	81.0	7	0.539	7	0.325	7	64.9	41	8	7
NACE 3 Commercial Blast	5	93.3	2.5	84.2	3	94.4	3	0.207	3	0.137	3	79.8	19.5	2	3
NACE 2 Near-White Blast	6	91.0	7	82.0	2	98.1	2	0.131	2	0.090	2	83.2	21	3	2
NACE 1 White Blast	1	95.4	1	88.3	1	98.4	1	0.085	1	0.052	1	85.1	6	1	1
NACE 1 Wheelabrator-Grit	3	93.8	2.5	84.2	4	92.7	5	0.251	5	0.168	5	76.0	24.5	4	5
NACE 1 Wheelabrator-Shot	3	93.8	4	83.8	6	87.1	6	0.349	6	0.243	6	70.5	31	5	6

[1]Rank = Position within group of 8 surface preparation methods.
[2]Rating = Average of rating of all panels in specific surface preparation method: 100 = Perfect; 0 = 1% Failure.
[SOURCE: NACE, Effects of Surface Preparation on Service Life of Protective Coatings, Interim Statistical Report by NACE Technical Committee T-6H-15, Houston, TX, Dec. (1977).]

The ratings in Table 9.4 are a numerical average of all of the panels coated with the five different coating systems for each surface preparation method. The grading system used by the committee was 100 points for a perfect panel down to 0 points where failure had reached 1%. Therefore, in the rating shown, the highest figures are the best.[5]

This is not true, however, for the edge penetration ratings. These are shown in inches measured from the edge of the panel in to the area where the coating is intact. The overall rank of the surface preparation method that is shown in the last column of the table is almost classical in that the three grades of abrasive blasting which are most used throughout industry are rated 1, 2, and 3. Shot and grit blasting are shown as 4 and 5. Pickling is next in order, with brush blasting and intact mill scale with no surface preparation listed last.[5]

The results of these well-controlled tests were as expected on a more or less theoretical basis. The methods that allow the most contamination to remain on the surface show the poorest results. The results of the tests are actually closer to theoretical expectations if the first two ratings (*i.e.,* rust rating overall and rust rating local) are eliminated from the series. In many ways, these are ratings of overall appearance of the panel. On the other hand, if the edge perimeter rating, the edge penetration maximum, edge penetration average, and scribe undercut rating are used as the criteria for the effectiveness of the surface preparation (*i.e.,* the measure of the adhesion of the coating to the surface), then the ranking of the surface preparation methods is as follows:

1. NACE #1 white sand blast
2. NACE #2 near-white sand blast
3. NACE #3 commercial blast
4. Pickle, phosphate treated
5. NACE #1 grit
6. NACE #1 shot
7. NACE #4 brush blast
8. No surface preparation.

The amount of time, work, and effort required to achieve any particular degree of surface preparation depends to a great degree on the initial condition of the surface to be cleaned. It is necessary to take into consideration the amount of rust, old paint, contamination, and active corrosion or pitting on the surface to be protected. While there are many different initial conditions, they can be divided into approximately eight main classifications, as outlined by SSPC under Surface Preparation Specification.

New Construction

(A) Steel surface covered completely with adherent mill scale, with little, if any, rust.

(B) Steel surface which has begun to rust, and from which the mill scale has begun to flake.

(C) Steel surface from which most of the mill scale has rusted away or from which it can be scraped, but with little pitting visible to the naked eye.

(D) Steel surface where the mill scale has rusted away and where pitting is visible to the naked eye.

Maintenance

(E) Paint almost intact; some primer may show; rust covers less than one-tenth of one percent of the surface.

(F) Finish coat somewhat weathered; primer may show; slight staining or blistering; after stains are wiped off, less than one percent of area shows rust, blistering, loose mill scale, or loose paint film.

(G) Paint thoroughly weathered, blistered or stained; up to ten percent of surface is covered with rust, rust blisters, hard scale, or loose paint film; very little pitting visible to the naked eye.

(H) Large portion of surface is covered with rust, pits, rust nodules, and non-adherent paint. Pitting is visible to the naked eye.[6]

Each of these types of surfaces is more or less difficult to prepare; however, the methods of preparation outlined in Table 9.3 only consider the end result of the surface preparation method, irrespective of the difficulty with which that end result is achieved. This is because the end result of the surface preparation is the important factor since this dictates the degree of contamination that is left on the surface and therefore the ultimate degree of adhesion.

White Metal Blast

According to NACE and SSPC, a white metal blast is a surface with a gray-white (uniform metallic) color, slightly roughened to form a suitable anchor pattern for coating. This surface is free of all oil, grease, dirt, mill scale, rust, corrosion products, oxides, paint, and other foreign matter.[7,8]

White metal blasting is the highest degree of surface preparation recognized in industry for the protection and maintenance of large steel structures. This method may not be the ultimate choice from the standpoint of production-line surface preparation, *e.g.,* preparation of body steel by the automotive industry, but it is the best method from the standpoint of the corrosion engineer. A white metal blasted surface combines a clean, new metal surface with sufficient roughness to provide an enlarged surface area that allows the maximum mechanical and/or chemical and polar adhesion. This is well demonstrated in Table 9.4, where the NACE #1 white blast has the highest rating throughout the whole series of tests.

This does not mean that even with a well-blasted surface there cannot be contamination left on the surface. In fact, this is often the case where steel previously exposed to corrosive conditions is prepared for coating. In the marine industry, as discussed previously, minute quantities of chlorides can remain on the surface, particularly in rough and pitted areas, to the extent that within a short period of time after blasting in humid areas, the steel begins to rapidly change color because of corrosion from the retained chlorides or sulfates.

Figure 9.15 shows a white sandblasted surface that has been exposed to a few minutes of humid air. This steel previously had been in use under marine conditions and demonstrates the retention of minute amounts of chlorides, which are difficult to remove in a single cleaning of the surface, even down to a white sand blast condition. Any lesser degree of surface preparation would leave a considerable amount of additional contamination, with an even faster reaction to the marine atmosphere.

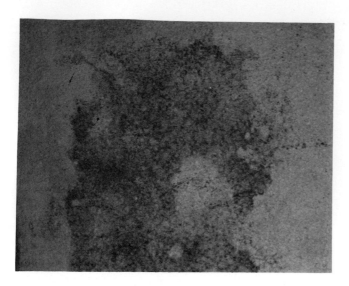

FIGURE 9.15 — Evidence of chloride contamination after steel that was previously corroded in a marine environment was white sandblasted and exposed to humid air for only a few minutes.

Near-White Sand Blast

The near-white sand blast (NACE #2, SSPC 10) is defined as a surface from which all oil, grease, dirt, mill scale, rust, corrosion products, oxides, paint, or other foreign matter has been removed, except for light shadows, streaks, or slight discolorations (or oxide bonded with metal). At least 95% of any given surface area has the appearance of white metal blast, and the remainder of that area is limited to slight discolorations.[7,8]

This is considered a practical degree of surface preparation because of the difficulty of removing the oxide binder (scale binder) from under mill scale. This is the material shown in Figure 9.14 as "d," and is indicated as containing a mixture of mostly metal with some iron oxide. It is darker than the light gray steel and may appear as dark streaks on the sandblasted surface. An area of steel that is rusted may also show a light streaking of the retained oxide. However, the amount remaining is impractical to remove without an excessive amount of effort. This is the reason for the near-white designation of a blast surface. While it is not perfect, it is substantially close. Nevertheless, as shown in Table 9.4, the near-white blasting is rated as a "2" from an adhesion standpoint, and the difference in the test ratings from NACE #1 to NACE #2 is significant.

Commercial Blast

Commercial blast (NACE #3, SSPC 6) is defined as a surface from which all oil, grease, dirt, rust scale, and foreign matter have been completely removed and all rust, mill scale, and old paint have been removed, except for slight shadows, streaks, or discolorations caused by rust stain or mill scale oxide binder. At least two thirds of the surface area should be free of all visible residues, and the remainder should be limited to the light discoloration, slight staining, or light residues mentioned earlier. If the surface is pitted, slight residues of rust or paint are found in the bottom of the pits.[7,8]

Commercial blast-cleaned surfaces are another practical approach to obtaining a degree of surface cleanliness which, under most conditions, can provide a satisfactory base for coating. Commercial blast is certainly a satisfactory degree of cleaning for many areas of relatively mild corrosion. Nevertheless, it is a degree lower in cleanliness than either NACE #1 or NACE #2, and this must be recognized by corrosion engineers who are applying coatings in relatively severe corrosion areas (*i.e.,* areas where there is high humidity, ions, and plenty of moisture condensation).

As can be seen in Table 9.4, the average rating for NACE #3 is considerably different from NACE #1. From a practical standpoint, the commercial blast should be the minimum considered for the application of high-performance coatings under even mild corrosion conditions.

Two other items that were tested and reported on in Table 9.4 are NACE #1 centrifugal blast grit and NACE #1 shot. While these are rated just below the commercial blast in the overall adhesion rating, they do not perform as well as the surfaces blasted by sand or mineral grit. The reason appears to be that, while a heavy profile can be obtained by either of these materials (grit or shot), the surface is not as clean as one which has been scoured by sand or mineral grit. Steel grit and shot may actually drive surface contamination into the surface because of the impact of the grit or shot particles. The grit particles do cut and make a high profile on the steel surface. Since steel grit is formed by crushing steel shot, there are still areas on the steel grit that are rounded and tend to peen the metal surface in a way similar to that of steel shot.

Both materials, when used as a surface preparation method, leave more contamination on the surface than sand or mineral grit. This does not mean that the method is not satisfactory for many purposes. It does mean, however, that where critical coating applications are to be made, this fact should be considered to prevent possible accelerated failure from occurring.

In Table 9.4, the actual distance of edge penetration on surfaces prepared by centrifugal blast shot compared to NACE #1 white sand blast is 4 to 5 times greater, indicating that the adhesion developed on such a surface, with the same type of coating used on both, is considerably less.

Acid Pickling

The definition given in the SSPC specifications for acid pickling (SSPC 8) is as follows: *pickling* is a method of preparing metal surfaces for painting by completely removing all mill scale, rust, and rust scale by chemical reaction, by electrolysis, or by both, and it is intended that the pickled surface shall be completely free of all scale, rust, and foreign matter. Furthermore, the surface shall be free of unreacted or harmful acid, alkali, or smut.[8]

Pickling is an in-plant operation and cannot be used on existing structures or on plants or equipment that has been erected. Pickling is restricted to steel objects that can be immersed in a pickle bath, so that the size of the bath determines the size of the object that can be cleaned in this way. In many steel mills, they have continuous steel baths in which steel sheet or coils are continuously passed through the pickle bath in order to remove all of the iron oxide scale from the surface. This is the principle of the pickle bath, *i.e.,* removing the oxide scale on the original hot-rolled metal in order to open up and free the steel surface of the oxide.

Pickling is the principal method by which objects to be galvanized are treated. In this case, steel plate, small tanks, or fabricated objects are pickled in the galvanizing plant and, immediately after washing, dipped into the molten zinc so as to obtain the maximum amalgamation of the zinc with the steel surface. If there is any contamination on the surface after it emerges from the pickle bath, the zinc will not adhere and the object will require retreatment. The same is true for plate that is to be used for coatings. Many steel fabricators also have pickling equipment into which they dip objects that are to be coated later with organic or inorganic coatings.

The general treatment of the steel during the pickling process is first to clean the steel and remove from the steel surface any materials that would prevent the pickling acid from contacting the surface and from penetrating and removing the scale. The biggest problem is oil or grease. These can be removed with solvents by any convenient means, such as rubbing with rags.

A thin film of oil is usually not a problem since the object goes from this stage into a cleaning bath. This can be an alkali bath which helps remove most of the contamination from the surface. Other harmful surface contamination is paint used for mill markings, wax pencil, crayon marks, and similar contamination. The hot alkali bath can remove most of these. The plate is then moved directly from the alkali bath into the hot pickling tank where it is left a sufficient amount of time to remove all of the scale and rust spots from the surface. Once this is accomplished, it is removed from this bath and moved into a water rinse (cold, warm, or hot). The water rinse removes the pickling acid and salts from the surface of the steel. However, at this point, the steel is quite reactive and must be prevented from immediately rusting. This is done through the use of a weak alkali solution of sodium carbonate or trisodium phosphate following the water rinse.

An alkaline surface does not rust rapidly; however, as previously noted, paint does not adhere well to an alkaline surface. For best results, the pH of the surface should be slightly on the acid side so that, in many cases, the pickled steel goes from the wash into a solution of phosphoric acid or phosphoric, chromic acid. As it comes from this bath, the steel may still be hot enough to immediately be given a prime coat, thus taking advantage of the dry surface and the fact that the primer penetrates and dries more rapidly over a warm surface.

In some cases, where there is still contamination showing on the surface, these areas have been given a quick brush blast in order to remove the contamination prior to applying the primer. This not only removes any residual contamination, but increases the anchor pattern as well.

Pickling is a good method of surface preparation for high-performance coatings since it assures that the surface over which the coating is applied is clean. However, it does have one drawback: The roughness of the steel surface coming from a pickling bath is considerably less than that of one which has been mechanically blasted. This means that the surface area is not as great so that there is little advantage derived from mechanical adhesion. However, coatings that do not require mechanical adhesion can be applied over a pickled surface with excellent results. As can be seen from the adhesion-related ratings in Table 9.4, pickling is ranked #4. Only the mineral grit surface preparation methods appear superior.

While the pickling method of surface preparation has proven itself, it is not entirely reliable since pickling solutions and rinsing baths can become contaminated. If improper care is taken, contaminants will collect on the surface of the liquids and be redeposited on the steel when it is withdrawn. This is not an isolated instance, since many pickling operators overlook the contamination, not realizing its effect on the coatings that may be applied. Therefore, if coating adhesion is critical and exposures are to be severe (constant immersion), acid pickling (unless the process is very well controlled) would not normally be the preferred means of metal preparation. In such cases, blasting would represent the better recommendation.

Brush Blasting

Brush blasting is defined by NACE as a surface from which oil, grease, dirt, loose rust scale, loose mill scale, and loose paint are removed, but tightly adhering mill scale, rust, paint, and coatings are permitted to remain if they have been exposed to the abrasive blast pattern, so that flecks of the underlying metal are uniformly distributed over the entire surface.[7,8]

Brush blasting is usually a field method of cleaning and is not generally used in fabricating shops or blasting plants. It is a method of cleaning steel, whether new or previously coated, that is fast and has a low cost. Other than the previous three techniques of blasting or pickling, brush blasting is undoubtedly the least costly and the best method of preparing field surfaces. This does not, however, mean that it should be used in highly corrosive areas or as a base for highly corrosion-resistant coatings.

Brush blasting is a good, low-cost procedure where tank exteriors or structural steel requires only the removal of surface contamination and loose paint or scale. It has the advantage of removing such materials rather readily, and the blasting operator can pick up areas that are badly corroded rather easily and remove the coating or heavy scale from those spots by concentrating the blast abrasive at that spot for a longer time than on the other part of the surface. Also, since abrasives are used, the surface is roughened to some extent, thus allowing mechanical adhesion both over previously applied coatings and over rusty areas. When recoating materials, such as polyurethanes and epoxies, the aged surface usually requires some roughening prior to the application of additional coats, and brush blasting is a good way to accomplish this.

The brush blasting procedure is one where the gun is held at a considerably greater distance from the surface than the other methods of blasting, and the surface is swept with the blast stream.

Where possible, brush blasting is preferred over other methods of hand cleaning (described later).

Flame Cleaning of New Steel

As defined in an SSPC specification (SSPC 4), flame cleaning of new steel is a method of preparing unpainted metal surfaces for painting by passing high-temperature, high-velocity oxyacetylene flames over the entire surface and then wire brushing to remove loosened scale and rust. It is intended that all unbonded scale, rust, and other detrimental foreign material be removed by this process, leaving a warm, dry surface to which the prime coat of paint is applied before the surface has cooled.[8]

Flame cleaning at one time had a number of advocates who felt that it was a good solution to metal cleaning problems in fabricating plants and was an economical and satisfactory surface preparation. Although actual practice has not necessarily verified these results, flame cleaning does have some advantages over other hand methods of surface preparation.

The principle of flame cleaning is to apply a very high temperature to the surface and pass it over the surface rapidly in order to expand the scale and to pop it off from the cooler metal surface. The thermal difference between the two materials, scale and steel, and the difference in rate of expansion, cause the scale to pop. Also, in case there is some water vapor within the scale or beneath it, the high temperature explodes the scale by the formation of steam and again pops it from the surface. However, tightly adhering mill scale cannot be removed by flame cleaning, making it necessary to follow the flame cleaning by mechanical wire brushing in order to remove the loose scale from the surface. Also, where the hot flame is applied over rusty surfaces, it dehydrates the rust, forming a dark powder.

The greatest advantage of flame cleaning is the application of the primer over the steel surface while that surface is thoroughly dry and warm. In this case, the viscosity of the primer is reduced and it is applied over a dry surface. The chemical activity of the primer is increased so drying occurs rapidly. Coatings which have been applied to steel that has been flame cleaned and then allowed to cool and absorb moisture before painting, generally perform about the same as those applied over plain wire brushed steel.

In an era of less sophistication in coating application and surface preparation, flame cleaning salesmen, in demonstrating their process, used to claim that the hot flame was driving the moisture out of the steel. This claim was based on the condensation of moisture ahead of the flame on the cool steel. However, this was actually the moisture from the flame reaction condensing on the steel ahead of it, rather than driving any moisture from the steel surface.

Power Tool Cleaning

The SSPC definition of power tool cleaning (SSPC 3) is a method of preparing metal surfaces for painting by removing loose mill scale, loose rust, and loose paint with power wire brushes, power impact tools, power grinders, power sanders, or by a combination of these methods. It is not intended that all mill scale, rust, and paint be removed by this process, but loose mill scale, rust, paint, and other detrimental foreign material shall be removed.[8]

More contamination is left on the surface by power tool cleaning than by any of the methods previously discussed. Nevertheless, there are areas where power tool cleaning is the only method possible. Presently, it is used primarily for the repair of damaged or undercut coatings where the damaged areas are not extremely large. While power wire brushes and power sanders have been used on large surfaces, brush blasting, where possible, is a much faster and better method of surface preparation.

There are a number of the newer impact tools that can effectively descale a surface and remove all paint and old rust. One of these is the pneumatic needle gun in which a group of needles impact the surface very sharply, breaking up any scale, rust, and paint that may be on the surface. There are also rotary impact tools which, in essence, flail the surface with a series of small, hardened wires or hammers. Here again, the coating and the rust or mill scale is removed by this means. Power sanding also tends to remove paint and rust as well as mill scale by grinding the surface free of these materials. In this case, the abrasive on the rotary sander cuts the metal and in this way increases the surface area so that some mechanical adhesion is derived as well. Some of the tools used in power tool cleaning are shown in Figure 9.16.

The poorest method of surface preparation is by power wire brush which, in many cases, tends to spread contamination on the surface and polish the surface of scale and rust rather than actually remove it. Oil or grease that is on the surface, and over which a rotary wire brush passes, is spread over the surface in a much wider area than originally existed. Prior to wire brushing or power tool cleaning, all such areas of oil and grease should be removed by solvent wiping.

Hand Tool Cleaning

According to SSPC specifications, hand tool cleaning (SSPC 2) is a method of preparing metal surfaces for painting by removing loose mill scale, loose rust, and loose paint by hand brushing, hand sanding, hand scraping, hand chipping, or other hand impact tools or by a combination of these methods. It is not intended that all mill scale, rust, and paint be removed by this process, but loose mill scale, loose rust, loose paint, and other detrimental foreign matter present should be removed.[8]

Hand cleaning is one of the oldest processes in use for preparing or cleaning surfaces prior to painting. Prior to the advent of high-performance coatings, it was a daily event for sailors on tankers and other marine equipment to be chipping and painting whenever the weather permitted. Actually, in many cases, a crew of several men was kept on board ship for this particular purpose. It was an ineffective way of preventing corrosion of the steel, primarily because of the amount of contamination that is left on the surface by this method of preparation.

Chipping hammers and other similar impact tools can crack heavy rust scale from the surface; however, much of what cannot be removed by this means remains

a ▲ ▲ b
c ▼ ▼ d

FIGURE 9.16 — Typical power equipment used in surface preparation: (a) a nonwoven abrasive cup wheel in use on a vertical power tool, (b) an electric tool which used a flap loading of heavy duty rotary peening to remove mill scale from carbon steel, (c) straight or in-line air tools, and (d) air-powered vertical or right-angle power tools. (Photos (a) and (b) courtesy of: 3M Company, St. Paul, MN, and (c) and (d) courtesy of: ARO Corp., Bryan, OH. Reprinted from Good Painting Practice, Chapter 2, Vol. 1, Steel Structures Painting Manual, Steel Structures Painting Council, Pittsburgh, PA, pp. 72, 73, 1982. Reprinted with permission from original sources and Steel Structures Painting Council.)

in the bottoms of pits. As a general rule, hand cleaning is used only when power-operated equipment or other surface preparation equipment is not available, where the job is inaccessible to power tools, or where the repair job on a coating is too small to warrant the use of power tools. The hand tools usually used are wire brushes, scrapers, chisels, knives, chipping hammers, and emery paper or sandpaper. Once the surface has been chipped free of heavy rust, scale, paint, or loose mill scale, going over the surface with a medium grade emery cloth aids materially in removing the contamination from the surface. It also

slightly cuts into the metal so that in some small proportion of the area, bare steel is exposed to the primer.

In applications where hand or power tools are used, a highly penetrating coating with a strong wetting action for steel and iron oxide should be used. Such materials are usually oil-based products, since these have the ability to penetrate and wet the metal surface to a greater degree than high-polymer high-performance coatings. Hand tool cleaning is actually a high-cost method of preparing a surface, since only a few square feet per hour can be satisfactorily cleaned by this method.

Solvent Cleaning

The SSPC definition for solvent cleaning (SSPC 1) is a procedure for removing detrimental foreign matter such as oil, grease, soil, drawing and cutting compounds, and other contaminants from steel surfaces by the use of solvents, emulsions, cleaning compounds, steam cleaning, or similar methods which involve solvent or cleaning action.[8]

Solvent cleaning is a form of surface preparation used specifically for the removal of oils and greases. It does nothing to remove rust, rust scale, mill scale, or old coating residues from the surface. If any of these are found on the surface and solvent cleaning is called for, the presence of the oil and grease may contaminate these other surfaces to the point where their actual receptivity for coating is worse than before the solvent cleaning.

Solvent cleaning can also spread contamination to an otherwise clean surface because of the oil or grease dissolved in the solvent. As soon as the solvent evaporates, the grease remains as a thin film over the entire wiped surface. As previously discussed, if a thin film of oil remains on the surface, it prevents the active groups both on the steel and in the coating from contacting and developing any proper adhesion. Oil and grease contamination is, however, extremely common and must therefore be removed before the application of any coating. Solvent degreasing is the only practical removal process and may be used in connection with any of the other methods of surface preparation as a precleaning procedure prior to such operations as sandblasting, acid pickling, etc.

The SSPC painting manual clearly outlines, as follows, the advantages and disadvantages of solvent cleaning in its chapter on surface preparation.

ADVANTAGES AND DISADVANTAGES OF SOLVENT CLEANING

A. Advantages. The following advantages may be listed for solvent cleaning (including vapor degreasing):

1. Solvents remove oils and greases readily.

2. They are easy to apply. Solvents remove oil and grease rapidly and easily. Cleaning equipment requires a minimum of floor space. In vapor degreasing, the work comes from the degreaser free of oil, warm, dry, and ready for any subsequent finishing operation.

B. Disadvantages. Unfortunately, there are some serious disadvantages inherent in solvent cleaning which impose limitations on its use in the cleaning of structural steel:

1. Both solvents and applicators are soon contaminated with oil and therefore instead of removing oil completely, only redistribute it.

2. Solvent cleaning is expensive if carried out properly. It involves considerable hand labor and is usually slow. In most solvent cleaning, except vapor degreasing, there is considerable loss of the solvent by evaporation, drag-out, and spillage.

3. In general (with the exception of vapor degreasing), solvent cleaning constitutes a fire hazard.

4. Only oils and greases are removed. Solvent cleaning is useless for the elimination of scale or rust. Rust stimulators, soaps, salts, and other water soluble materials remain on the surface, and should be removed or neutralized.

5. The fumes given off in solvent cleaning are, in many cases, toxic [and/or explosive].

6. Some chlorinated solvents are slightly decomposed by heat in contact with water and metal, forming hydrochloric acid which rapidly attacks the equipment and causes rusting of cleaned parts. This can be controlled by using solvents stabilized with volatile bases which tend to neutralize any acid which forms.[9]

One method of solvent cleaning (vapor degreasing) is effective where it is possible to place the item in a vapor degreasing unit. Vapor degreasing consists of removing oil, grease, wax, and other soluble materials by suspending the object in a vapor of trichlorethylene or other chlorinated solvent. The solvent vapor condenses on the surface of the object, and the contaminant and the solvent run off of the object and drip down into the reservoir. Since the solvent vapors are evaporated from the reservoir, they are completely free of contamination and the surface is therefore washed continually by a clean solvent. The clean solvent condenses on all surfaces so that every part of the object is subject to the cleaning action. The parts are not only cleaned by the solvent, but are also heated by the vapor temperature so that they quickly dry upon removal from the solvent degreaser.

A second method that has improved the solvent degreasing procedure is to use some commercial mixtures that contain a strong solvent, such as xylene or high flash naptha, with a strong emulsifying agent. This mixture is wiped or scrubbed over the surface to be cleaned, the grease or oil is picked up by the solvent, and once this has taken place, the solvent and the grease can be washed from the surface by water. A water jet is usually used. A thorough mixing of the solvent, the contaminant, the emulsifying agent, and the water takes place, with the contamination being washed away from the surface by the water.

This procedure is much superior for removing oil and grease from surfaces compared with the use of solvent alone. This can only be used where a water wash is possible. As has been previously stated, solvent cleaning is a specific method of surface preparation and is used primarily as a pretreatment for other surface preparation methods.

Dust Blasting

Dust blasting is defined as a cleaning of the surface through the use of very fine abrasive through a sand blast mechanism. Such an abrasive can be very fine siliceous or mineral abrasives, 80 to 100 mesh, or it can be fine, reused siliceous or mineral abrasives. The purpose of this method of surface preparation is to clean sensitive surfaces, such as aluminum, copper, lead, or galvanizing, without making a heavy etch and without stretching the metal surface and warping the metal, as would be the case with heavier abrasives. This procedure applies a fine etch to the metal surface, cleans the surface of any contamination or oxide, and provides a clean surface with some additional surface area over which coatings can be applied. It has proven effective for the surface preparation of aluminum, copper, and galvanized surfaces.

Metal Pretreatments

A number of metal pretreatments exist. While some are beneficial, others sport extravagant claims that are generally unsubstantiated by actual experience (particularly those advertised to eliminate the need for surface preparation). The Steel Structures Painting Council has several pretreatment specifications and methods worth considering.[10]

Wetting Oils

SSPC Pretreatment 1 (SSPC-PTI-64) is defined as wetting oils or penetrating oils that oxidize and air dry to solid, water insoluble protective films. Wetting oil treatment is a method of initially saturating the surface of rusty and scaled steel with wetting oil that is compatible with the priming paint, thus improving the adhesion and performance of the paint system to be applied.[10]

This pretreatment is not intended to take the place of a coat of paint or eliminate the necessity for proper surface preparation. Such wetting oil pretreatments generally are based on linseed oil, either raw linseed oil, raw linseed oil plus metallic dryers, or boiled linseed oil thinned to the point of being highly penetrating. Any surface preparation must be done before the application of the oil treatment. The wetting oil is applied freely to the surface by brush, spray, or other methods. The amount applied must be enough to saturate the rust or scale remaining on the surface. There should not be any excess of the oil treatment left on the surface. Any excess should be wiped from the surface, leaving only a very thin film that can quickly oxidize to a solid.

The wetting oil pretreatment can only be used where compatible topcoats will be applied over it. These are usually oil-base products, since most of the high-performance high molecular weight coatings will not adhere satisfactorily to the oil surface. Also, the high-performance coatings may contain solvents that can swell or otherwise damage the linseed oil pretreatment. This is a relatively specific pretreatment and is usually only effective where a coating is applied over a rusty surface. It can add to the life of the coating applied over such a surface.

Cold Phosphate

SSPC Pretreatment 2 (SSPC-PT2-64), best defined as cold phosphate surface treatment, is a method of converting the surface of steel to insoluble salts of phosphoric acid for the purpose of inhibiting corrosion and improving the adhesion and performance of paints. This type of pretreatment is often used as a light rust remover on production items. Otherwise, its use is specifically for the phosphate conversion of cleaned surfaces and for the prevention of immediate rusting of some cleaned steel surfaces, such as those that have been pickled. It is not intended to be used for removing rust or mill scale, nor is it satisfactory for use on rusty steel, since test results have shown that it actually degrades the performance of coatings over rusty surfaces.[10]

There are many proprietary-type materials of this nature available which make a variety of claims. Any claims of rust and scale removal should be carefully examined since, as stated earlier, rusty surfaces treated in this manner may have a degrading effect on the coating applied over the surface. The phosphate pretreatment has, however, proven effective for the neutralization of the highly alkaline weld flux that may remain on a welded surface. In this case, water washing after the treatment with the acid solution is recommended. Its effectiveness is primarily due to the neutralization of the alkaline flux, providing a slightly acidic surface over which the coating is applied.

The usual cold phosphate pretreatment is a combination of concentrated phosphoric acid in a water soluble solvent, such as butyl alcohol, ethyl alcohol, or other material. Treatment of clean steel surfaces with such phosphate solutions can inhibit the rusting of the surface for a considerable period of time (i.e., several days). On the other hand, many of the high-performance coatings, such as vinyls, have poorer adhesion to such surfaces than where the coating is applied directly over the clean steel surface.

An example of the use of phosphate treatments such as this occurred in a shipyard in Japan which developed a method for the treatment of entire block sections of tankers using a phosphoric acid wash after the block section had been completed. Prior to the steel being incorporated into the block section, it was thoroughly grit blasted to remove all the mill scale; however, during the forming of the block section, some surface rusting occurred (as might be expected). Considerable surface contamination also accumulated during the fabrication process.

The original theory was that by automatically washing the entire unit with a phosphoric acid solution, they could eliminate any of the minor rusting that had taken place, as well as wash away any contamination on the surface. A very large unit was constructed with many spray heads to cover the entire block section. Unfortunately, the complexity of the block sections did not allow complete contact of the phosphoric acid by the sprays in all areas, and other areas accumulated the acidic liquid, causing a buildup of large iron phosphate deposits in those areas which had to be recleaned from the surface.

It was concluded that a commercial sand blast of the block section would have been cheaper than the overall labor involved in removing the heavy phosphate deposits and cleaning the areas that did not come in contact with the acid. The project was thus abandoned.

There was considerable testing done, however, to determine the adhesion characteristics of various organic and inorganic coatings over the phosphate-treated steel. Most of the coatings applied over such a surface were not satisfactory after a test of several months in seawater where adhesion rapidly deteriorated. The best coating combination tested was a coating of water-base inorganic zinc followed by an epoxy primer and coal tar epoxy topcoats. Excellent adhesion over the entire duration of the test was obtained with this system. Figure 9.17 shows one of the block sections in the phosphate treating area.

Vinyl Butyral Wash Primer

SSPC Pre-treatment 3 (SSPC-PT3-64) consists of a basic zinc chromate vinyl butyral wash coat, sometimes referred to as wash primer. It is a pretreatment for metals which reacts with the metal and at the same time forms a

FIGURE 9.17 — Tanker block section in acid cleaning area showing spray nozzles.

protective vinyl film that contains inhibitive pigment to help prevent rusting.

This wash coat is supplied as two components which are mixed together just prior to use. The base contains an alcohol solution of polyvinyl butyral resin pigmented with basic zinc chromate. The diluent contains an alcohol solution of phosphoric acid which reacts with the vinyl resin, the pigment, and the steel. This pretreatment is intended to be used primarily on clean steel, *i.e.*, free of rust and scale, or on clean galvanized metal.

The vinyl butyral wash primer was developed by Union Carbide Corporation early in World War II in order to provide a proper primer for their vinyl resins. The wash primer was only part of the coating system; the entire system consisted of the primer, a second coat of vinyl alcohol polymer and red lead, and topcoats formed from vinyl chloride acetate copolymer resins. The system was a good one and is still in use today per Navy specifications.

As used by the Navy during the war, it was applied over a sandblasted steel surface and gave good corrosion protection to steel surfaces in a marine atmosphere. It may also be used over pickled steel and on galvanized surfaces.

On the other hand, the wash primer pretreatment should not be used over steel that has been pretreated in any other way. In addition to its use as a part of a vinyl coating system, it also has been found to be a good pretreatment for metal coated with alkyds, epoxies, and similar coatings. Where vinyls are used, the red lead intermediate coat is usually necessary in order to obtain proper adhesion of the topcoats over the wash coat primer. Therefore, it is not recommended to use the wash coat and then apply vinyl topcoats directly over the sur-

face. It may also be used over inorganic zinc base coats as a primer for other materials.

However, the material should not be considered a cure-all primer for all surfaces. There have been many instances of intercoat delamination, in addition to the development of osmotic blistering due to the basic zinc chromate used in the system. The government specifications for the wash primer are Bureau of Ships Formula 117 or MIL-C-15328B.

Hot Phosphate

SSPC Pre-treatment 4 (SSPC-PT4-64) is a hot phosphate surface treatment that converts the surface of steel to a heavy crystalline layer of insoluble salts of phosphoric acid. Its purpose is to inhibit corrosion and improve the adhesion and performance of the paint to be applied. These pretreatments have proven beneficial for steel or galvanized surfaces that are free of rust, scale, dirt, paint, or white rust preventatives.

This type of surface pretreatment is primarily used on production items, and there are a number of proprietary methods used in production plants for the treatment of parts and metal objects prior to the application of coatings. The pretreatment used on automotive body steel is this type of pretreatment which improves both the corrosion resistance of the steel and the adhesion of the topcoats applied. It is primarily useful where a very well-controlled metal treatment system can be used. Table 9.5 gives a summary of preparation techniques.

Water Blasting

Water blasting relies on the use of water at very high pressure, *i.e.,* 6000 to 10,000 psi or more, producing a velocity of 1300 feet per second. This kind of pressure rapidly removes most contaminants from a surface and is particularly effective in the removal of heavy mastic-type materials that have failed and under which corrosion exists. It is effective in removing accumulated salts, dirt, grease, and other similar contaminating materials from surfaces.

While water blasting will not produce an anchor pattern, it can remove the majority of heavy rust scale where tubercles have formed. It can be used to wash old coatings and remove contamination from tightly adhering coatings as well as inorganic base coats. The process of water blasting is considered several times faster than mechanical cleaning tools. It is also considered a better cleaning method than mechanical tools for deformed steel plate floors, expanded metal gratings, and similar areas.

The advantages of water blasting are use of relatively simple equipment, low use of noncritical material (water), and much less contamination of the area than where sandblasting is used. Water blasting is relatively easy to use due to the light hose and the fact that there is no dust, no spark, and that there is an immediate pressure drop should a hose break. One hazard involved, however, is that serious physical injuries can occur if a person is hit by the high-velocity water at short range.

Water blasting is a good surface preparation method for many areas where other systems are impossible, impractical, and slow. However, it does not produce a prac-

Name	SSPC Designation	NACE Designation	Description
Solvent Cleaning	SSPC-SP 1-63	None	Complete removal of oil, grease, wax, and other contaminants by cleaning with solvents, vapors, emulsions, alkalies, or steam. Interior environments of low humidity only.
Hand Tool Cleaning	SSPC-SP 2-63	None	Removal of loose rust, loose mill scale, and paint by manual labor with wire brushes, hand scrapers, sanders, etc.
Power Tool Cleaning	SSPC-SP 3-63	None	Much the same as above, but utilizing powered tools such as chippers, descalers, grinders, etc.
Flame Cleaning	SSPC-SP 4-63	None	High temperature flame dehydrates and removes rust, loose mill scale, and some tight mill scale. Usually followed by wire brushing or blasting.
White Metal Blasting	SSPC-SP 5-63	NACE #1	Complete removal of all visible rust, paint, mill scale, and foreign material by wheel or pressure blistering using (wet or dry) sand, grit, or shot. Suitable for all of the most severe environments, including immersion service
Commercial Blast	SSPC-SP 6-63	NACE #3	Sandblasted until at least two-thirds of each element of surface area is free of all visible residues.
Brush-Off Blast	SSPC-SP 7-63	NACE #4	Blast cleaning of all except tightly adhering residues of mill scale, rust, and old coatings, exposing numerous evenly distributed areas of underlying metal.
Pickling	SSPC-SP 8-63	None	Complete removal of rust and mill scale by acid pickling, duplex pickling, or electrolyte pickling.
Weathering, followed by Blast Cleaning	SSPC-SP 9-63	None	All or part of the mill scale is removed by allowing the steel to weather, followed by one of the blast cleaning standards.
Near-White Blast	SSPC-SP 10-63T	NACE #2	Blast clean until at least 95% of each element of surface is free of all visible residues.
Dust Blast	None	None	Complete removal of surface contamination from aluminum, copper, and zinc, without metal distortion or warpage. Abrasion of coating surfaces prior to recoating.
Wetting Oil Treatment	SSPC-PT 1-64	None	Saturation of rusty surface and scaled steel with a wetting oil that is compatible with primer.
Cold Phosphate Treatment	SSPC-PT 2-64	None	Conversion of clean surface of steel to insoluble phosphates to provide improved corrosion-resistant coating performance.
Wash Primer	SSPC-PT 3-64	None	A metal pretreatment which reacts with the metal and at the same time forms a vinyl film over which other primers may be applied.
Hot Phosphate Treatment	SSPC-PT 4-64	None	Application of a heavy crystalline layer of insoluble phosphate to the metal surface to inhibit corrosion and improve adhesion.

tical surface for most high-performance coatings in corrosive areas where tooth, cleanliness, and reactive areas on the steel are needed for proper adhesion.

Table 9.6 provides a list of coating types, together with the minimum surface preparation needed to provide good coating performance. Stress should be placed on the word "minimum" since any of these coatings will perform better over a better grade of surface preparation. This was shown rather conclusively by Hudson in Tables 9.1 and 2.

Abrasive Blasting

The preferred method for preparing steel for the ap-
plication of high-performance coatings is abrasive blast cleaning. It not only provides a clean surface and removes rust, scale, oil, paint, and similar contaminating materials, but it also roughens the surface and provides mechanical as well as chemical and polar adhesion for the coating. Abrasive blast cleaning consists essentially of impacting the surface with high-velocity abrasive particles to such an extent that the contamination on the surface is removed and a clean, active metal surface is obtained.

There are essentially three methods by which the abrasive particles can be accelerated to obtain sufficient impact to cut the metal surface. These are air blasting, water blasting (in which abrasive particles are included in the water stream), and mechanical rotary blasting (where the abrasive is discharged from a rapidly revolving paddle wheel, throwing the abrasive against the metal surface). The first two methods are hand methods of blasting, while the third is a mechanical method and is primarily an in-plant operation. The two hand methods may be used to prepare surfaces on location or wherever abrasive blasting is permitted.

Air Blasting

Air blasting has been the most common method of surface preparation since its inception in the 1930s. Many different abrasives may be used with this blasting procedure, and it may be used for blasting ships, industrial structural steel, concrete, and many other different surfaces. It is the most versatile type of surface preparation and undoubtedly involves the lowest cost of any surface preparation method besides the rotary blast method. It is also the most effective method of surface preparation, particularly for coatings that are to be used in highly corrosive areas.

Air pressure blasting may be used as an in-plant method of surface preparation for the cleaning and preparing of moveable tanks, structural steel, or small parts. It also may be used in large, stationary installations with workmen blasting in large, enclosed spaces where the air is rapidly changed and the abrasive cleaned and reused. Many shipyards now have fully enclosed and essentially air-conditioned blasting and coating facilities that will accommodate large block sections of ships. For new ship construction, this is a distinct advantage since it eliminates many weather problems which pose difficulties in various parts of the world.

Air pressure blasting may also be done wherever an air source is located or can be brought in. Portability, in fact, is probably its greatest advantage. Any place where compressors and sand blast pots can be moved, air blasting can be accomplished. Figure 9.18 is a typical view of an on-site blasting operation.

The field blasting installation shown in Figure 9.18 is a relatively small operation typical of those found at plant sites almost anywhere in the world. Figure 9.19 is a view of a large portable operation. It shows three sand hoppers, located on the deck of a large tanker, which feed many blast nozzles down in the ship's tanks. This figure shows only a small portion of the equipment used, which includes sand conveyors, hoppers, ventilation equipment, vacuum sand reclamation, dehumidifiers, etc., all of

TABLE 9.6 — Minimum[1] Surface Preparation Requirements for Steel with Commonly Used Types of Coating

Coating Type	Minimum Surface Preparation
Drying Oil	Hand or power tool cleaning (SSPC-SP2 or 3)
Alkyd	Commercial blast (NACE 3, SSPC-SP6)
Oleoresinous Phenolic	Commercial blast (NACE 3, SSPC-SP6)
Coal Tar (Emulsion or Cutback)	Commercial blast (NACE 3, SSPC-SP6)
Asphaltic (Emulsion or Cutback)	Near-white or commercial blast (NACE 2 or 3, SSPC-SP10 or 6)
Vinyl	Near-white or commercial blast (NACE 2 or 3, SSPC-SP10 or 6)
Chlorinated Rubber	Near-white or commercial blast (NACE 2 or 3, SSPC-SP10 or 6)
Epoxy	Near-white or commercial blast (NACE 2 or 3, SSPC-SP10 or 6)
Coal Tar Epoxy	Near-white or commercial blast (NACE 2 or 3, SSPC-SP10 or 6)
Urethane	Near-white or commercial blast (NACE 2 or 3, SSPC-SP10 or 6)
Organic Zinc	Near-white or commercial blast (NACE 2 or 3, SSPC-SP10 or 6)
Inorganic Zinc	White or near-white (NACE 1 or 2, SSPC-SP5 or 10)

[1]Listed coatings should not be used unless minimum surface preparation requirements can be met.

FIGURE 19.18 — Field sandblasting installation.

FIGURE 9.19 — Sand hopper installation on deck of a large super tanker which allows blasting operations in several tanks at the same time.

AIR PRESSURE SANDBLASTING

FIGURE 9.20 — The components of a good air pressure sandblasting system are: (1) the compressor giving an adequate and efficient supply of air; (2) an air hose, couplings, and valves of ample size; (3) a portable, high-production sandblast machine; (4) the correct size antistatic sandblast hose (with externally fitted quick couplings); (5) a high-production venturi nozzle; (6) a pneumatic remote control valve for safety and cost savings; (7) a moisture separator; (8) high air pressure at nozzle; (9) the correct type and size of abrasive; (10) an air-fed helmet and air purifier (in good working order); and (11) a well-trained operator. (SOURCE: Clemco-Clementina, Ltd., San Mateo, CA.)

which are placed on dock alongside the ship or on the deck. Many of the key components needed for proper air blasting of metal are graphically illustrated in Figure 9.20.

The principal advantage of an air-blast system is that the blasted surface is dry. While it can be blasted to a number of different stages, depending on cleanliness, the active sites on the metal are exposed to accept the coating polar groups for adhesion. Unless there is considerable humidity in the area, the surface will remain dry until the coating is applied, which is usually done on the same day.

Another advantage of air blasting is that the residue that may be left on the surface is simply dust and thus can be easily removed by blowing with clean compressed air, dusting with brushes, or by removal with a vacuum. When the dust contamination is removed, the surface is ready to coat. Air blasting also gives good production and a good profile; but above all, it can provide a good, clean,

Corrosion Prevention by Protective Coatings

dry surface over which a coating may be applied.

Figure 9.21 shows the actual blasting process as carried on in the interior of a large tanker tank. Note the white or near-white surface obtained. In this installation, all of the air in the tanks is dehumidified to prevent the clean surface from rusting. Without the dehumidification, the surface would begin to rust within minutes.

The disadvantage of dry sandblasting is primarily the dust which forms because of the breakup of the particles of sand or grit. The Environmental Protection Agency has taken a dim view of this, even though the contamination is primarily particulate and soon settles out of the air. The dust not only is objectionable to the agencies, but it is a contaminating influence to areas where sandblasting takes place. Motors, bearings, and other pieces of mechanical equipment are endangered because of the dust. The dust is also harmful to the individual who may breathe it. This is particularly true of silicate dust. While silica is most damaging because of silicosis, other dusts from nonsiliceous sources do not carry this threat. Nevertheless, no one should work in a dusty area of any kind without proper protection.

Because of the air pollution problem created by sandblasting, there recently has been more and more wet abrasive blasting. There are two methods of wet abrasive blasting, one uses a high-pressure water blast which includes a small amount of abrasive, while the other more common method simply adds water at the nozzle of a dry blasting operation. In both methods, the surface can be brought to a white blast.

Water Abrasive Blasting

The distinct advantage of water blasting (*i.e.,* water containing an abrasive) is the lack of dust during the operation. In place of dust, however, there is considerable buildup of water in the area, which also can be a hazard. The surface produced by a proper water blast unit using high-pressure water and a relatively small amount of abrasive is good and provides a good profile. The environmental agencies are much happier with a water blast setup than a dry blast. The high-pressure water blast cleans by the action of both the high-pressure water and the contained abrasive propelled by the high-velocity water. There is also a much greater water-to-abrasive ratio than in the wet blast procedure. Nevertheless, with the splash resulting during the blasting operation, some abrasive accumulates on the surface around the blast area, so that an after-blast washdown is necessary.

This can, however, be particularly troublesome in pocket areas where both the water and the abrasive accumulate. Unless thoroughly cleaned, these pockets form future focal points for corrosion and coating failure.

In this method, the blast surface is wet, which is also a distinct disadvantage. The wet surface can oxidize quickly, leaving a less-than-optimum surface for coating acceptance. To prevent oxidation, inhibitors, which in themselves can cause coating difficulties, must be used in the water. Some coatings have less than the best surface adhesion when used over an inhibited surface. However, most of the disadvantages of water-abrasive blasting are related to the water used to carry the abrasive. There are a few safety problems in the use of such high-pressure water. An individual hit at short range with high-pressure water containing abrasives would not only be injured by the high-pressure water, but the abrasives would be embedded in any body area in the path of the water.

Wet Blasting

The water blasting already described should not be confused with wet blasting, which is an incorporation of a small amount of water into a dry blast operation in order to reduce the dust and dirt. The wet blast system uses essentially the dry blast methods, equipment, and techniques, except that there is a modified nozzle where water is injected into the blast stream (Figure 9.22). There is a much smaller water-to-abrasive ratio in this wet sandblast method than in the water-blast system described earlier.

Because of the smaller water content in the blast

FIGURE 9.21 — Sandblasting on the interior of a tanker tank, showing a pitted steel surface after blasting.

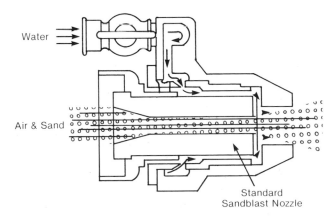

Water

Air & Sand

Standard
Sandblast Nozzle

FIGURE 9.22 — Cross section of a wet blast attachment which sucks water from any water supply for wet blast cleaning. (Drawing courtesy of: Sanstorm-Bowen Tools, Inc., Houston, TX. Reprinted from Good Painting Practice, Chapter 2, Vol. 1, Steel Structures Painting Manual, Steel Structures Painting Council, Pittsburgh, PA, pp. 19, 20, 28, 29, 1955. Reprinted with permission from original source and Steel Structures Painting Council.)

stream, considerable amounts of the abrasive can accumulate on the surface, making it necessary to wash the surface before the application of any coatings. Again, rusting takes place rapidly unless inhibitors are used both in the wash and in the blasting operation. Where inhibitors are involved and high-performance coatings are needed, every care must be taken to determine whether the coating will satisfactorily adhere to the inhibited surface. Many inhibitors interfere with maximum coating adhesion.

In most cases, it is best to allow the surface to dry as rapidly as possible and apply the coating directly over the uninhibited, dry surface. Under these circumstances, if at all possible, brushing of the prime coat is recommended in order to physically work it into the surface. Rolling the prime coat also aids in working the coating into the surface. Figure 9.23 shows the wet blasting of a ship hull.

FIGURE 9.23 — Wet blasting of a ship's hull.

With the present air pollution laws, wet blasting seems to be increasingly necessary, in spite of the adhesion problems it creates. However, use is still under study because of the unknown adhesion factors. NACE Technical Committee T-6G-26 on Performance of Organic Coatings Over Wet Sandblasted Steel addressed this, as follows, in a preliminary report:

> An obvious area of concern would be the effects of corrosion inhibitors upon the organic coatings which are used over them, with possible determination of the optimum inhibitor. Preliminary test data indicate insignificant differences in performance of organic coating systems over the inhibitors mentioned previously in this report, and only slight differences in prevention of flash rusting have been noticed. There also appears to be no clear trend in the performance of various types of organic coatings over the various inhibitors. Types of coatings on which test data are available include vinyls, epoxy coal tar, amine cured epoxy, and chlorinated rubber.
>
> The major area of concern, of course, is whether wet abrasive blasting involves some compromise in the performance of the organic coatings. Preliminary indications are that this is not the case, but it must be emphasized that this is based upon short term laboratory tests.[11]

In a series of tests outlined in a special API report on the use of inorganic coatings over a wet blasted surface, it concludes:

> ...if field tests confirm results obtained in the laboratory, we can meet government regulations and minimize pollution by using wet abrasive blast cleaning without chemically inhibiting the wet blasted surface and still achieve desired results—a long life coating system using inorganic zinc primer.[12]

While wet blasting is being recommended and used in order to overcome some of the environmental protection objections to dry blasting, it is not as good as the dry blasting procedure for organic coatings. Its use represents a definite compromise in order to satisfy air pollution regulations, since it results in a lightly contaminated surface compared with a thoroughly clean, dry surface. Unfortunately, until a better practical method of surface preparation is available, the compromise may be the best surface preparation that can be obtained under many circumstances.

Rotary or Wheel Blasting

Rotary blasting is entirely an in-plant operation where stationary units of rotors can be used to throw abrasives at various steel surfaces. The principal advantage of the rotary blast is that it is dry. The metal can actually be warm before going into the operation, and a warm, dry, clean surface is a definite advantage for the application of coatings.

The rotary blast system is also of relatively low cost. Since materials are handled mechanically, the production can be fast, using either metallic grit or shot. In many rotary blast setups, the prime coat is also automatically applied within a few feet of the blast unit so that the steel is coated almost immediately following the surface preparation, which is also an advantage. Because of the closed blasting chamber, any dust or air pollution from such units is automatically eliminated through suction blowers and precipitators so that no contaminants escape into the atmosphere.

The disadvantage of the rotary blasting process is that, for the most part, the steel must be blasted in its original state and not after fabrication. While some fabricated sections can be blasted in this way, any complicated unit requires hand blasting to properly clean all surfaces. Unfinished steel leaves welds and other areas unprepared. If a prime coat is not applied within a reasonable time after blasting, surface oxidation takes place and repreparation of the surface is necessary.

While a white rotary blast is considered to be entirely satisfactory from a coating application standpoint, the surface is not as satisfactory as one which is prepared by dry sandblasting. This is confirmed by the NACE T-6H-15 tests reported previously. Surface contaminants can be driven into the surface, which generally does not contain as many reactive sites for coating adhesion as would the sandblasted surface.

The mechanism, operation, and advantages of centrifugal blasting (or wheel blasting) equipment are well described by Arno J. Liebman, one of the best-known advocates of proper surface preparation.

Corrosion Prevention by Protective Coatings

....The many advantages of wheel blast cleaning over air or nozzle blast cleaning are sufficiently great to warrant the installation of extensive wheel blast cleaning machines where feasible. Because of the size and complexity of the installation, it is used only in shops, and particularly in shops where there is sufficient demand for blast cleaning to make the installation economical. The nature and shape of the work has a great effect upon the practicality of the operation. Highly irregular, large surfaces such as shop fabricated beams are difficult to clean in the usual set-up. These readily lend themselves to nozzle blast cleaning where sufficient space is available to install facilities. For production use, the wheel blast equipment is difficult to surpass in the low cleaning costs it achieves.

One of the big advantages of using wheel blast cleaning equipment is the elimination of air compressors and pipelines and of the attendant labor. Another advantage is in the compactness and self sufficiency of the unit. Many other advantages are associated with the use of this equipment, for example, the ease of starting, the simplicity of power supply, etc. The principal disadvantages of this type of equipment are the high initial cost, and the high maintenance cost and shut down time for repair and maintenance. Due to the nature of the operation, high wear is associated with the moving parts of the equipment; since the equipment is mechanically complex, it is more difficult to keep in operation than the conventional blast equipment. In spite of this, where the equipment can be used and adequate demand results in high percentage of time in service, the equipment will achieve low cleaning costs.

The principle of operation is illustrated in Figure [9.24]. Two general types of wheels are in service; the batter type and the slider type. In the batter type, abrasive is propelled by impact when it comes in contact with the edge of the vanes; this type of wheel is not in extensive use in this country. The commonly used type of wheel, which is known as the slider type, is shown in the illustration; in this type abrasive is charged through the hub of the wheel and slides on the vanes to the edge of the wheel where it is projected at high velocity towards the work. The blast pattern is not too good as the distribution of the abrasive across the pattern may vary greatly.

A typical wheel is approximately 2½ inches wide by perhaps 20 inches in diameter; it rotates at a speed of approximately 2000 r.p.m. One interesting attribute of the method of discharging the abrasive is the fact that all of the abrasive particles develop approximately the same velocity. This velocity will range from 5000 to 14,000 feet per minute. This is in comparison to nozzle blasting in which velocities of the abrasives vary with the particle size and the weight of the abrasive. Average velocities of abrasives in direct pressure blasting at customary high pressures are in the same region as for wheel blasting. A wheel pattern can be developed which is approximately 30 inches long and 6½ inches wide, but only an inner zone approximately 4 inches wide by 12 inches long attains a high intensity of blast cleaning. As a result, wheel type blast cleaning installations must be carefully designed, and in some cases a number of wheels must be installed in order to obtain sufficient coverage of the work.

The type of abrasive used with the wheel blast cleaning equipment is usually the metallic; that is, iron shot, grit, cut wire abrasives, malleable iron and steel abrasives. The synthetic nonmetallic abrasives are used occasionally on some work and sands are not used because of the excessive wear on the equipment and high maintenance costs that result. A typical wheel installation requires from 15 to 20 horsepower to drive the wheel. Such a wheel will project roughly 25,000 lbs. of metallic abrasive per hour. In order to throw this quantity of abrasive by direct pressure nozzle blasting, five 3/8 inch diameter nozzles requiring 200 horsepower would be necessary. About 1/10 the horsepower is necessary to operate wheel blast cleaning equipment as compared to direct pressure nozzle blast cleaning equipment. Cost figures, however, are not this attractive due to high depreciation and maintenance costs on the wheel equipment.[5]

Figure 9.25 shows a typical centrifugal blasting plant arrangement with several different wheels to apply abrasive streams at different angles. Such units are common for the automatic blasting of plate and shapes before fabrication.

Vacuum Blasting

Vacuum blasting, also called *closed recirculating blast*

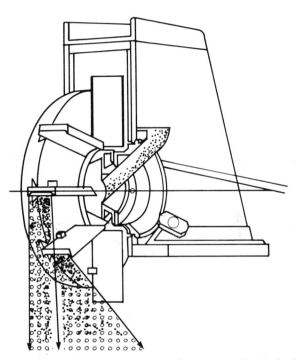

FIGURE 9.24 — Schematic drawing of the centrifugal wheel blast cleaning operation. (Drawing courtesy of: Pangborn Co., Hagerstown, MD. Reprinted from Good Painting Practice, Chapter 2, Vol. 1, Steel Structures Painting Manual, Steel Structures Painting Council, Pittsburgh, PA, pp. 19, 20, 28, 29, 1955. Reprinted with permission from original source and Steel Structures Painting Council.)

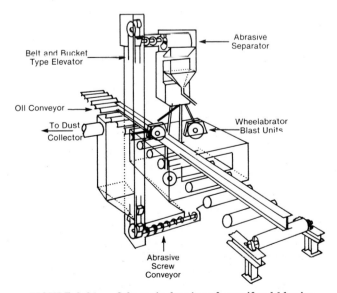

FIGURE 9.25 — Schematic drawing of centrifugal blasting operation for structural shapes. Cleaning area is enclosed to confine high-velocity abrasives so they can be recycled. (Drawing courtesy of: Wheelabrator-Frye, Inc., Mishawaka, IN.)

systems, is a method to do away with the dust and dirt of an open blast system. The results obtainable with a vacuum blast system can be equivalent to the best surface preparation by air or centrifugal blasting.

In vacuum blast systems, the blast nozzle is enclosed in a rubber cap or a brush which completely surrounds the nozzle and which prevents the abrasive from flying in all directions. This brush or cup that surrounds the nozzle is connected to a high-efficiency vacuum mechanism which pulls more air than is used by the blasting nozzle; therefore, as soon as the abrasive strikes the surface and bounces away, it is picked up by the vacuum system and carried back to the blast pot where it is cleaned before being automatically recirculated through the mechanism. This type of blasting can be used in areas where there is sensitive equipment, since no dust or abrasive is allowed to escape if properly operated. For the most part, the abrasives used in this type of equipment are metallic or aluminum oxide since they can be reused a number of times, while most sands are impractical due to their high breakdown rate.

In order for the vacuum blast method to properly operate, the brush or rubber cup around the blast nozzle must seal with the surface being blasted. While this is relatively easy on flat surfaces, when blasting structural shapes and configurations, specially shaped brush seals are necessary to prevent dust and abrasive from escaping outside the unit.

Unfortunately, this method of blasting is generally difficult, awkward, and slow for use with fabricated objects and can only be used effectively in special circumstances. In some cases, higher costs must be absorbed in order to eliminate the hazards of dust and abrasive in an area. Where the additional cost can be justified, this type of equipment can provide good surface preparation for all types of coatings.

This principle has been tried on the flat bottoms of ships using some rather large units. These units were mobile and self contained, and used a centrifugal wheel in place of air in order to blast the surface. The seal and the vacuum system were the same. The unit was self propelled, and where there was adequate area on the underside of the flat bottom, these units could move back and forth under the hull, providing an adequate surface preparation without dust or loss of abrasive.

In all abrasive blast operations, the actual blast cleaning operation depends on the surface to be cleaned, the type of structure, the equipment available, whether the work is being done in the shop or in the field, local environmental conditions, and many other factors. Only experience will serve as a guide to the best methods of carrying out the operations.

The efficiency of the blasting operation also depends on a number of conditions. It should be apparent that two of the items requiring careful attention are the distance of the nozzle or wheel from the work, and the angle of the abrasive stream to the work. Theoretically, maximum impact is obtained when the abrasive particles strike the surface perpendicularly. In practice, this results in inefficient operation because of rebounding abrasives slowing down the abrasive emerging from the nozzle. Thus, the best cleaning is obtained when the blast path is directed slightly away from the perpendicular angle. The exact angle at which the abrasive stream should be held will vary with the type of work. In some cases, *e.g.,* cleaning old paint, the blast path is directed at an angle of about 45 degrees from the surface to undercut the material to be removed.

The distance that the nozzle or wheel is held from the surface should be decided for each individual job. The closer that the abrasive driving force is to the surface, the greater the impact of the particle and the more concentrated the blast stream. As the source is moved away from the surface, the blast pattern widens and a greater area is covered. At the same time, the abrasive particles waste much of their energy in overcoming the resistance of the air and in expanding the blast stream. Each particular job will dictate the optimum distance the blast source should be held from the surface, the rate at which it must be traversed over the surface to obtain the degree of cleanliness desired, and the surface anchor pattern that is specified.

All hand blasting operations should be done in a consistent manner. While in many cases light and visibility are problems that must be overcome, the only practical method of blasting is to mark out a section of an area and blast it with an even motion of the blast nozzle until the area is completed to the desired amount of surface preparation. Following this, a second section can be marked off and the same procedure followed. In this way, the blaster can regulate his work, and the entire surface will be consistently and efficiently cleaned.

This procedure, however, is not always followed. Many blasters tend to wave the nozzle in several directions, sweeping over the surface any number of times and still not maintaining a consistent blast pattern. This often necessitates reblasting an area after inspection.

The consistent use of the block method and the careful blasting of each block prior to moving on to new surfaces provide the best and most economical procedure, as well as the most uniform surface.

Figure 9.26* illustrates the blasting of a ship's bottom. While a clean surface is desired, only a sweep blast is actually being obtained since the blaster is such a long distance from the surface. In this way, reaching even a commercial blast surface would require extensive blast time, and then may not even be as clean as desired. Such blasting procedures are common, although quite inefficient.

Abrasive Materials

There are a number of significant factors involved in the abrasive blasting of steel. Of particular importance from the standpoint of coating application and effectiveness, is the material or blasting media used to clean the surface. These can include any number of various grades of sand (*e.g.,* river or quartz sand), as well as synthetic grits or those made from refractory slags and steel grit or shot. Each of these materials will clean the surface in a different way and to a different degree, and will provide a different surface profile.

---

*See color insert.

Table 9.7 lists some of the commonly used abrasives. The hardest material in each group is listed at the top of the group; the others follow in order of decreasing hardness. While not all of the abrasives available for abrasive blasting are represented, these are sufficient to demonstrate a range of materials and abrasive characteristics.

Sands range from garnet, which is the hardest of the natural abrasives, to river sand, which may be a combination of a variety of materials. Garnet and quartz sand, being the hardest, provide the sharpest profile and have the heaviest cutting characteristics. The softer abrasives, often found in natural sands, range from those that clean well to those that contain sufficient dirt and clays to form a tremendous amount of dust. These are not only slow cutting, but may leave the surface heavily contaminated with dust and dirt. One of the difficulties with the natural abrasives is that they are usually siliceous and therefore considered hazardous. This, however, depends on the safety equipment and ventilation provided.

The synthetic grits or abrasives can range from the extremely hard silicon carbide, to steel slag or furnace slag that is crushed into the proper grading of abrasive. The Black copper slag abrasives are available from a number of sources in the United States, and have good cleaning characteristics as well as the ability to provide a good profile.

Figure 9.27 illustrates a good grade of silica sand. It is mostly crystalline silica of an even grade with little extraneous foreign material (*e.g.*, clay). Such a sand should be a good blast abrasive and one that would provide the best possible surface.

Figure 9.28 shows a typical black grit. This is made from boiler slag which is crushed and then graded to the proper size for a satisfactory blast medium. Note the sharp angular edges of this abrasive, as well as the lack of dust or dirt. This type of abrasive should provide the best possible surface.

Steel grit and shot are available in any number of different size ranges. The grit ranges in size from G10, which is the largest grit available, to G325, which is extremely fine. Shot also varies in similar ways; S1320 is a large size shot, while S70 is small. The usual sizes that are used in rotary blasting equipment are G16 to G50, and S330 and S230. These are medium-sized abrasives and are the type used for most coating work. Other sizes are used for different types of abrasive cleaning.

Figure 9.29 shows a typical steel grit. Grit is usually made by crushing metal shot to form the angular metal grit particles. The steel grit is angular, with a number of sharp corners for cutting, chipping, and cleaning. It is well graded and makes a good blasting medium.

Figure 9.30 shows typical steel shot of the size commonly used in rotary blasting equipment. Each of the particles is essentially a steel ball, which makes a good abrasive for cleaning mill scale from heavy steel plate or shapes.

Surface Profile

Profile is one of the measures of the use of abrasives and is determined by the type of abrasive, the hardness of the abrasive, the size of the abrasive, the velocity of impact, and the time that the surface is exposed to the

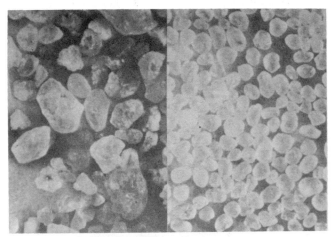

FIGURE 9.27 — Silica sand abrasive, magnification 8X. (SOURCE: Good Painting Practice, Chapter 2, Vol. 1, Steel Structures Painting Manual, Steel Structures Painting Council, Pittsburgh, PA, p. 48, 1982.)

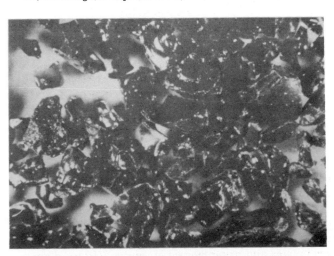

FIGURE 9.28 — Coal fired, boiler bottom ash slag, magnification 8X. (SOURCE: Good Painting Practice, Chapter 2, Vol. 1, Steel Structures Painting Manual, Steel Structures Painting Council, Pittsburgh, PA, p. 47, 1982.)

TABLE 9.7 — Commonly Used Abrasives

Metallic	Synthetic Nonmetallic Silica-Free	Siliceous
Chilled Cast Iron	Silicon Carbide	Garnet
Cast Steel	Aluminum Oxide	Quartz
Malleable Iron	Refractory Slag	Silica
Crushed Steel	Rock wool By-Products	Decomposed Rock
Cut Steel Wire		
Aluminum Shot		
Brass Shot		
Copper Shot		

[SOURCE: Steel Structures Painting Council, Steel Structures Painting Manual, Chapter 2, Good Painting Practice, Vol. 1, Pittsburgh, PA, pp. 12, 19, 20, 28, 29 (1955).]

FIGURE 9.29 — Steel grit G-50 for automatic blasting. (SOURCE: Ervin Industries, Inc., Ann Arbor, MI. Reprinted from Good Painting Practice, Chapter 2, Vol. 1, Steel Structures Painting Manual, Steel Structures Painting Council, Pittsburgh, PA, p. 40, 1982. Reprinted with permission from original source and Steel Structures Painting Council.)

FIGURE 9.30 — Shot S-230. (Source: Ervin Industries, Inc., Ann Arbor, MI. Reprinted from Good Painting Practice, Chapter 2, Vol. 1, Steel Structures Painting Manual, Steel Structures Painting Council, Pittsburgh, PA, p. 39, 1982. Reprinted with permission from original source and Steel Structures Painting Council.)

blasting process. With this number of variables, it can be easily seen that many different kinds of profiles can be obtained, even in the use of similar abrasives.

Table 9.8 gives a comparative maximum height of the profile obtained with a number of common abrasives. This is done using direct pressure blast cleaning of mild steel plates and using 80 psig air and a 5/16 in.-diameter nozzle. Different pressure and nozzle size could easily change these results. Nevertheless, this is an indication of the type of profile obtained from various materials.

The profile heights produced by various abrasives and their effect on the adhesion of coatings have been a subject for discussion dating back to the first use of blasting for surface preparation. It is undisputed that some coatings are seriously influenced in their adhesion and performance by the profile of the substrate. On the other hand, recent work on surface preparation by several groups indicated that profile may be secondary to other factors, such as type of abrasive used and degree of cleanliness.[13]

Table 9.9 shows the relationship of the various profile heights to the average bonding strength of the six coatings used in a study by Schwab and Drisko at the Naval Civil Engineering Laboratory on surface profile

TABLE 9.8 — Comparative Maximum Heights of Profile Obtained with Various Abrasives in Direct Pressure Blast Cleaning of Mild Steel Plates Using 80 psig Air and 5/16 in. Diameter Nozzle

Abrasive	Size[1]	Max. Height of Profile in mils
Large River Sand	Through U.S. 12, on U.S. 50	2.8
Medium Ottawa Silica Sand	Through U.S. 18, on U.S. 40	2.5
Fine Ottawa Silica Sand	Through U.S. 30, on U.S. 80	2.0
Very Fine Ottawa Silica Sand	Through U.S. 50, 80% through U.S. 100	1.5
Black Beauty (crushed slag)	Estimated at minus 80 mesh	1.3
Crushed Iron Grit	G-50	3.3
Crushed Iron Grit	G-40	3.6
Crushed Iron Grit	G-25	4.0
Crushed Iron Grit	G-16	8.0
Chilled Iron Shot	S-230	3.0
Chilled Iron Shot	S-330	3.3
Chilled Iron Shot	S-390	3.6

[1]Sizes listed are U.S. Seive Series Screen sizes or SAE grit or shot sizes.
[SOURCE: Steel Structures Painting Council, Steel Structures Painting Manual, Chapter 2, Good Painting Practice, Vol. 1, Pittsburg, PA, pp. 12, 19, 20, 28, 29 (1955).]

TABLE 9.9 — Ranking of Bonding Strengths Associated with Four Profile Heights on White Metal Surfaces

Rank	Profile Height	Abrasive	Average Bonding Strength (kg/sq cm)
1	High	Black Beauty 4016	108
2	Low	Flintshot	99
3	Medium	Steelgrit G40	99
4	Very High	Steelgrit G14	82

[SOURCE: Schwab, Lee K. and Drisko, Richard W., Relation of Steel Surface Profile and Cleanliness to Bonding of Coatings (Paper 116), CORROSION/80, NACE, Houston, TX, 1980.]

and coating performance. The black copper slag, which has a reasonably high profile, gave the best bonding strength. However, the steel grit G14, also with a high profile, had the poorest average bond strength. The low and the medium profile abrasives provided equal bond strength at a level between the two high profiles.[13] This indicates the variability of the surface provided by different abrasives and its effect on the bond strengths of coatings applied over it, rather than the effect of the surface profile.

This same study reveals some interesting information regarding the average bond strength of six different coatings to various types of abrasive blasted steel. As shown in Table 9.10, the abrasive itself can create a variable with respect to the bonding of coatings in addition to the profile and cleanliness of the surface.

Table 9.11 also indicates some differences in the level of adhesion between a white metal finish and a commercial finish. The average bond strength of the six coatings to a white metal finish was superior to that of the commer-

cial blast finish. The cleanliness variable is related to the initial surface, the type of abrasive, the speed of impact, and the time involved in cleaning the steel surface.

SSPC, in its work on surface profile for anticorrosion paints, summarizes the effects of blast cleaning conditions on profile as follows.

1. Profile height increases as the abrasive size increases.
2. Profile height also increases as the degree of cleaning is improved from commercial blast to white metal.
3. Profile height increases as the angle of abrasive impingement increases from oblique to perpendicular.
4. In general, profile obtained with metallic abrasives tends to be higher and less "disturbed" than that obtained with sand or other non-metallics.
5. Steel thickness has a relatively small effect on profile height.
6. The newly discovered phenomenon of hackles may, at least in some instances, overshadow the effects of profile height on paint performance.[14]

Actual graphs of profiles of steel surfaces obtained by the use of various abrasives are shown in Figures 9.31 and 9.32. The graph scale is 1/1000 in. vertically and 5/1000 in. horizontally, which provides a good indication of the actual surface left by the various abrasives used. The metal abrasives did not provide as deep or as sharp a profile as the ungraded silica sand and the black mineral slag.

The surface effects produced with various abrasives can range from deep cutting to gentle wiping or scouring of the surface. Selection of abrasives should not be done in a haphazard manner due to the number of variables possible. Some of the important factors that help to determine the abrasives to be used are:

1. Type of metal to be cleaned
2. Shape of the structure
3. Type of material to be removed
4. Coating surface finish desired
5. Profile of the steel to be coated and coating thickness
6. Amount of abrasives that will be lost during blasting
7. Breakdown rate of the abrasive
8. Reclamation of the abrasive
9. Hazards associated with the use of the abrasive
10. Area where the abrasive will be used and its danger to surrounding equipment.

The types of available abrasives vary from one part of the country or the world to another. The general categories are: shot, metal grit, or mineral abrasives. Each type of abrasive cleans in a different way and leaves a somewhat different surface from the other. The following section will discuss some of these surface differences and the cleaning characteristics obtained from different types of abrasive.

TABLE 9.10 — Ranking of Bonding Strength Associated with Different Abrasives on White Metal Surfaces

Rank	Abrasive	Average Bonding Strength (kg/sq cm)
1	Black Beauty 4016	108
2	Flintshot	99
3	Steelgrit G40	99
4	Steelshot S280	92
5	Black Beauty 400	91
6	Polygrit 80	87
7	Polygrit 40	86
8	Steelgrit G14	82

[SOURCE: Schwab, Lee K. and Drisko, Richard W., Relation of Steel Surface Profile and Cleanliness to Bonding of Coatings, CORROSION/80, Preprint No. 116, National Association of Corrosion Engineers, Houston, TX, 1980.]

TABLE 9.11 — Relating Level of Cleanliness to Bonding Strength

Rank	Level of Cleanliness	Average Bonding Strength (kg/sq cm)
1	White Metal Finish	97
2	Commercial Finish	90

[SOURCE: Schwab, Lee K. and Drisko, Richard W., Relation of Steel Surface Profile and Cleanliness to Bonding of Coatings, CORROSION/80, Preprint No. 116, National Association of Corrosion Engineers, Houston, TX, 1980.]

Types of Abrasives

Sand or Mineral Abrasive. One of the advantages of sand-type abrasives and the black mineral grits such as black copper slag are that they tend to scour in addition to cutting the surface. The scouring action is due to the fracturing of the nonmetallic particle as it hits the surface rather than having a direct impact and falling away, as

Metal Abrasive Profiles

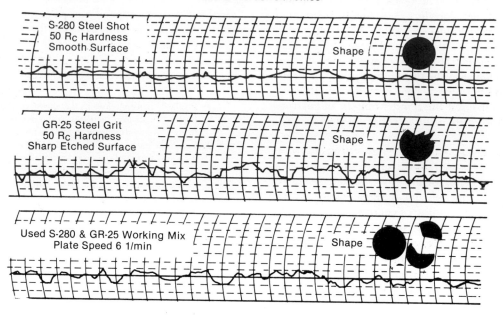

FIGURE 9.31 — Graph of the profiles of steel surfaces obtained by use of steel shot, steel grit, and working mix abrasives. (SOURCE: Surface Condition and Profile Produced by Various Abrasives Used in Abrasive Blast Cleaning, Preliminary Reports, NACE Technical Committee T-6G-25, National Association of Corrosion Engineers, Houston, TX, work currently in progress (1984).]

Surface Profile - Mild Steel - Silica

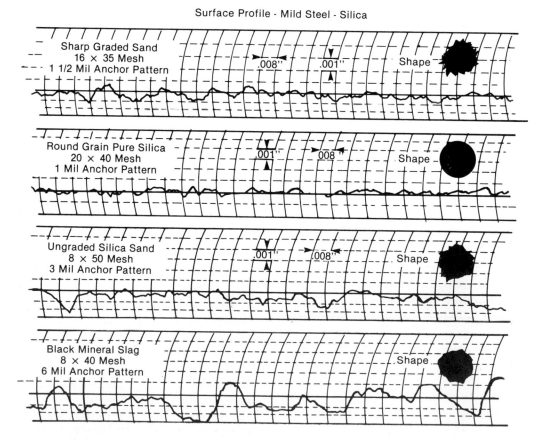

FIGURE 9.32 — Graph of the profiles of steel surfaces obtained by use of sand, silica, silica sand, and mineral slag abrasives. (SOURCE: Surface Condition and Profile Produced by Various Abrasives Used in Abrasive Blast Cleaning, Preliminary Reports, NACE Technical Committee T-6G-25, National Association of Corrosion Engineers, Houston, TX, work currently in progress (1984).]

with steel shot. This is one way in which a thoroughly clean, white, sandblasted surface is obtained. It is also effective in exposing the greatest number of reactive sites on the metal surface so that maximum adhesion of high-performance coatings can be obtained.

The exploding particle of sand or mineral grit has a different cleaning action from either steel shot or grit. It is hypothesized that the breaking of the sand particle into one or many pieces on impact tends to send the broken particles speeding away from the point of impact in a direction somewhat parallel to the metal surface.

This breaking of the particle and the abrupt change in direction of the pieces is the cause of the scouring action, which is in addition to the original impact (Figure 9.33). This action is essential in cleaning rusty areas and freeing pits of all contamination. The scouring action removes the fine rust and corrosion from flat areas, as well as from pits, much more effectively than shot or grit, which does not break and shatter on impact. The cleaning action of sand is less effective on heavy rust or mill scale than metal shot or grit because of its lesser impact energy. However, once the heavy scale has been popped off, the cleaning action of sand is superior.

The scouring action of sand is even more important in pits than on flat surfaces. As sand impacts the pit, the broken particles fly in all directions within the pit and finally exit at the opening (Figure 9.34). This scours the pit clean, even on the sidewalls. Such action is not possible with grit or shot since they do not break up, but merely rebound.

Steel Shot. By contrast, the action of steel shot is one of impact alone. Results are similar to striking the surface with a ball peen hammer. Shot will peen and hammer the surface (Figure 9.35), which is an advantage when heavy brittle deposits (*i.e.*, mill scale) must be removed from the surface. The energy of the heavy metal particle hitting the surface effectively cracks and pops the heavy brittle rust and mill scale from surfaces. It is not, however, as efficient in removing surface residues (*e.g.*, mill scale binder and surface rust) since these may be pounded into the surface by the peening action. The peening action on the metal both compresses the surface and stretches the metal, so that care must be taken in shot blasting thin metal sections or light steel plate. The stretching of the metal surface can cause excessive deformation and warpage.

Shot blasting is usually most effective on heavier steel plate and shapes which can absorb the impact of the shot and the surface compression without excessive warpage. The compression of the steel surface is also a factor in the adhesion of coatings over a shot-blasted surface. The compression increases the surface density and reduces the effective reactive surface sites that are necessary for coating adhesion.

Steel Grit. Steel grit is usually formed by crushing and cracking steel shot. Its blasting action, because of the sharp edges, is much more of a cutting action than either sand or steel shot. This is especially true when the grit is new. The sharp edges cut into the steel, forming sharp peaks and valleys. There is also some peening action by particles that hit with the blunt side. As the grit is used

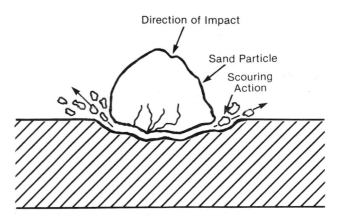

FIGURE 9.33 — Scouring action of sand or mineral abrasives.

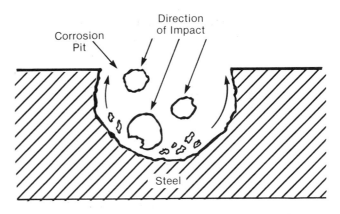

FIGURE 9.34 — Cleaning pits through use of sand-type abrasives.

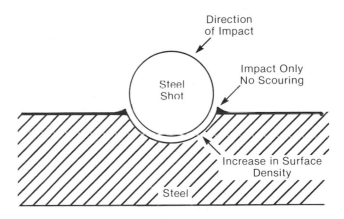

FIGURE 9.35 — Peening action of steel shot abrasive.

and reused, it becomes more rounded and blunt, which increases the peening action. The cutting action of grit opens new steel surfaces and somewhat increases the reactive sites for coating attachment. This increases the adhesion potential of the coatings applied over the grit-blasted surface. There is also less surface compaction than on surfaces blasted with shot alone, although the possibility of warpage should be considered in grit blasting thin steel cross sections.

Figure 9.36 illustrates the type of cutting that might be expected with steel grit. When it is considered that these sharp metal particles are spinning and twisting as they are thrown at the surface, it is easy to visualize sharp hackles or pinnacles of metal being created during the grit blasting process. An unpublished SSPC report, *Surface Profile for Anti-Corrosion Paints,* discusses this problem.[14] Excessive hackling was experienced on large diameter steel pipe blasted in a rotary blast setup. Since new grit was used, there were so many steel pinnacles or hackles on the surface that it was necessary to hand sand the surface with emery cloth before a coating could be applied successfully.

Types of Surfaces Produced by Abrasives

The three types of blasting abrasives present three rather distinct types of surfaces for use with coatings: (1) the compacted and peened surface from shot; (2) the sharp, angular cut surface from steel grit; and (3) the more finely cut, abraded, and scoured surface of sand or mineral grit. These three types of surfaces are shown in scanning electron microscope (SEM) photos developed by SSPC as part of their surface profile studies. Note the considerable differences in the appearance of the surfaces prepared by the three types of abrasive in Figure 9.37.

As might be expected, these three surfaces have different effects on the adhesion of the coatings applied over them. General observations of field results have been made. The behavior of high-performance coatings appears best over a sand or nonmetallic grit-blasted surface; the performance and adhesion is somewhat less over a metal grit-blasted surface; and the performance over a metal shot-blasted surface is somewhat less than either of the other two. This is also indicated in SSPC reports. One of their conclusions on the three types of surface preparation is as follows:

> Sand Versus Shot Versus Grit—Early outdoor exposures are verifying salt fog conclusions that small but measurable differences exist in the performance of typical coatings over surfaces cleaned with sand, shot, and grit. Use of sand and other non-metallic abrasives resulted in consistently good coatings performance equal to or better than with metallic shot or grit. No clear superiority was shown in comparing shot-blasted versus grit-blasted surfaces. However, with a number of paint/environment combinations, grit-blasted surfaces have resulted in better paint performance than shot, especially in the vicinity of damaged (scribed) areas. (It appears that these effects are related to the differences in surface textures achieved by these three differing cleaning mechanisms as described in the report section on scanning electron microscopy.)[14]

It must be noted here that all three types of surface are adequate and practical for high-performance coatings under most corrosive conditions. Nevertheless, corrosion engineers should understand that performance differences do exist and that where a critical exposure is involved, a sandblasted surface may provide a better substrate for a coating than grit- or shot-blasted surfaces.

The abrasive can also vary in terms of bond strength, profile height, and cleanliness. The type of abrasive used is also a major factor in the speed of cleaning the surface. Table 9.12 gives the comparative cleaning rate of a number of abrasives. The table shows that the Ottawa

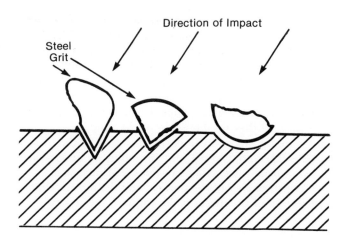

FIGURE 9.36 — Blasting action of steel grit abrasive.

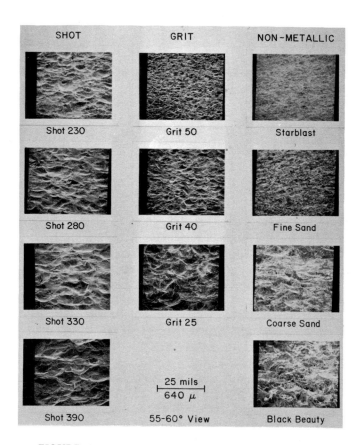

FIGURE 9.37 — SEM micrographs showing qualitative features of some blast-cleaned surfaces (near-white blast cleaned, SSPC-SP10). (SOURCE: Keane, J. D., Bruno, J. A., and Weaver, R. E. F., Steel Structures Painting Council report, unpublished.)

silica sand, which is a good sandblasting abrasive, has a cleaning rate of almost twice that of the usual combination of steel grits, and several times that of the various grades of steel shot. This is an important variable in terms of cost, but it is also important because of the additional cleanliness and better surface profile obtained by it use.

When the abrasive used is steel grit or steel shot, blasting is ordinarily confined to a closed blasting unit, where it is done either manually or by rotary wheels. For

TABLE 9.12 — Comparative Cleaning Rates of Various Sized Abrasives in Direct Pressure Blast Cleaning of Flat Plates Using 80 psig Air and 5/16 in. Diameter Nozzle

	Abrasive	Size[1]	Cleaning Rate Sq. ft. min.
1.	Ottawa Silica Sand	Through U.S. 30, on U.S. 80	2.14
2.	Crushed Iron Grit	Average U.S. 40 G50 (new)	1.56
3.	Crushed Iron Grit	10% G-50 and 90% misc. small grit	1.38
4.	Crushed Iron Grit	10% G-40 and 90% abrasive from 3	1.28
5.	Crushed Iron Grit	G-40 (new)	1.09
6.	Crushed iron Grit	10% G 25 90% abrasive from 4	1.05
7.	Crushed Iron Grit	10% G 16 and 90% abrasive from 6	0.90
8.	Chilled Iron Shot	S-330 (new)	0.72
9.	Chilled Iron Shot	S-390 (new)	0.67
10.	Crushed Iron Grit	G-25 (new)	0.66
11.	Crushed Iron Grit	G-16 (new)	0.48

[1]Sizes listed are U.S. Seive Series Screen sizes or SAE grit or shot sizes. [SOURCE: Steel Structures Painting Council, Steel Structures Painting Manual, Chapter 2, Good Painting Practice, Vol. 1, Pittsburg, PA, pp. 12, 19, 20, 28, 29 (1955).]

instance, metal abrasives were used outside in a relatively unrestricted area to blast the interior of the block sections of some large ships in a dry climate. In this case, the blasting was done manually and the grit was collected on asphalt pavement, later to be picked up by a vacuum cleaner.

There are several reasons for not using steel grit or shot for field blasting. The cost of the shot itself is quite high, and the shot must be reclaimed since it can be reused hundreds of times before being reduced to the point of being ineffective. The actual abrasive cost in an automatic centrifugal blast unit is low because of the number of times that the abrasive can be reused. On the other hand, in field or on-site blasting, sand or refractory abrasives are more generally used since they are not reusable due to the difficulty of reclamation and the breakup of the sand or nonmetallic grit.

Reclamation of the abrasive or its collection and disposal represents one of the most significant costs involved in blasting. In tanker holds, all of the sand must be brought to the surface by large vacuum units or brought to the deck by buckets. Either process is costly, and, for the most part, the abrasive is only used once. This is true of a great deal of interior tank work as well. Abrasive may or may not have to be collected when blasting is done in the outdoors, depending on the area where the work is done.

Concrete Surfaces

Another type of surface which requires effective preparation before the use of high-performance coatings is concrete. Concrete is commonly used throughout most industry, with the exception of marine and offshore structures.

Concrete is an extremely variable material. Not only are various mixtures of concrete used, sometimes on the same construction project, but there are also many different types of concrete surfaces. These range from floors that can be hard-troweled, wood-floated, or broomed, to walls that can be cast against plywood, steel, or an absorbent liner. These cast surfaces may also be plastered, sacked, or stoned; they may be covered with cement, stucco, or lime plaster; or walls may be made of concrete block of different sizes and shapes. Overhead concrete is usually cast against forms, producing a surface similar to concrete cast into walls. There are also various concrete pipe surfaces, ranging from mechanically made pipe to concrete pipes that are centrifugally cast with an extremely dense, smooth surface, usually covered with concrete laitance. Machine-made concrete pipe can be quite porous or it can be comparatively dense on the inner surface. Each of these surfaces generally receives differing types of surface preparation prior to the application of a coating.

Concrete also varies in the way in which it is cast or placed. The majority of concrete is cast, whether in floors or walls. One of the best placed concretes, and one of the most dense, is made by a process called *cast stone*. In this case, a carefully controlled concrete mix is carefully placed and compacted, using impact vibrators as well as vibrating tables. Very high-strength, durable concrete objects are made in this manner. At the other end of the scale, there are walls cast using very fluid concrete and poor compacting techniques which result in a wall with maximum porosity and rock pockets.

Concrete

Concrete is formed by a mixture of portland cement (which is finely ground powder), sand, aggregates of various sizes, possibly additives to aid in the placing of the concrete, and most of all, water. Water not only hydrates the cement powder, but it also provides the required fluidity for the concrete to take the required shape.

Water is one of the most important ingredients in concrete and can create or reduce the amount of surface preparation problems encountered. Concrete with too little water is difficult to place, even in a simple form, and rock pockets are easily formed unless the concrete is thoroughly vibrated and compacted. The addition of too much water in concrete makes it very fluid and creates a porous and relatively weak concrete structure. It also creates an excessive amount of water and air pockets in the concrete surface, all of which have a serious effect on the coating over such a surface.

Cement

Cement is a key ingredient in concrete. Portland cement is a finely pulverized powder which essentially consists of hydraulic calcium silicates. Chemically, portland cement is principally tricalcium silicate ($3CAO.SIO_2$) and beta dicalcium silicate ($B-2CAO.SIO_2$), together with lesser and variable quantities of tricalcium aluminate ($3CAO.AL_2O_3$). There also may be minor amounts of iron, magnesium, and possibly free lime. This combination is obtained by calcining limestone and clay or similar materials at temperatures of approximately 1500 C, and then grinding them to a fine powder. This fine powder is mixed with water and aggregates to form concrete.

When cement and water are mixed, a saturated solution of calcium hydroxide is rapidly formed. The hydrates of the various silicates form somewhat more slowly. The principal bonding agent is a colloidal gel of calcium silicate hydrates, which is the binding material for the sand and rock, and which, with the cement, form concrete.

There are several types of cement available. Portland cement, Type 1 (regular) is used for general concrete construction. Portland cement, Type 2 is used for general concrete construction where resistance to moderate sulfate action or where moderate hydration is required. Portland cement, Type 3 is a high early strength cement. This material is used where it is necessary for the concrete to harden and cure more rapidly than is possible with Type 1 or 2. Portland cement, Type 4 is for use where a low heat of hydration is required, such as in the heavy concrete masses found in dams or in atomic energy foundations. Portland cement, Type 5 is used where resistance to sulfates is required.

White portland cement is commonly found in areas where a white surface is required. This material is similar to portland cement, Type 1, however, it contains a low quantity of iron in order to maintain the white color. There are also air-entrainment cements which are portland cements with small quantities of air-entraining materials added, such as greases, tallows, or pozzolanic materials. These incorporate air into the cement while it is being formed into concrete and are thought to improve its durability in northern climates.

Most of these materials have similar properties where coatings are concerned. In this light, there is little difference in the surface formed by any of the portland cements. However, in some cases, there are additives used in the concrete (e.g., chlorides, sodium silicate, iron filings, and similar materials) which can and do affect the proper application of the coating over the surface. Whenever concrete is known to contain any of these materials, extra care should be required in both surface preparation and coating.

The aggregate, which is one of the largest ingredients in concrete, is also important to the surface obtained from the concrete. Any number of variations in sand and rock ratios can be found. Usually, where a large structure is involved and good compressive strengths are required, the rock aggregate is relatively large, ranging up to as much as 2 inches in diameter. The usual is 3/4 to 1 in. in diameter, combined with sand to form as dense a structure as possible in order to obtain maximum strength.

In other words, the area between the rocks is filled with sand and cement in order to obtain a dense structure. Where sand and rock combinations are not included in the right proportion, a porous, sandy structure with little strength can result, or a porous structure with many voids may result where insufficient sand is added to the mix. Neither of these situations is satisfactory for either strength or coating application.

Concrete Surfaces

Concrete surfaces are anything but uniform. Pours on the same day by the same crew using the same raw materials can vary greatly in physical and chemical characteristics, depending on the amount of compaction and the amount of puddling or vibration used in placing the concrete. Hot weather conditions make concrete set more rapidly, and, therefore, there is a greater possibility of rock pockets and voids. In cold weather, the concrete will not set as rapidly, creating different surface conditions. Under field conditions, concrete must be considered a nonhomogeneous material as far as surface preparation and coating application are concerned.

The most typical concrete surface of the greatest interest to corrosion engineers is hard-troweled concrete. This is the type of surface obtained on sidewalks and concrete floors. Normal troweling of concrete is obtained where the concrete is troweled smooth, but is not polished to the hard-troweled state. Wood-floated concrete is obtained where a wood float is used to smooth the concrete, leaving it slightly granular. Broomed and swept concrete is obtained when concrete is smoothed with a wood float and then swept with a broom. Maximum surface roughness would be of the type obtained on a concrete highway or driveway.

Concrete poured against wood forms creates a common type of surface, which is usually quite porous with many pinholes and air and water pockets. Concrete poured against steel forms creates a much smoother surface, but still contains many pinholes and water and air pockets. Concrete poured against an absorptive liner is an attempt to eliminate pinholes and water and air pockets. While this is successful to a considerable extent, the absorptive materials may leave some fibers in the concrete surface that could create coating difficulties. Poured concrete of most types tends to have offsets, form tie holes, and other types of surface disruptions which must be dealt with during surface preparation.

Gunited concrete is an extremely dense concrete. It usually has a rather uneven, granular surface which makes coating application difficult. However, when gunited surfaces are troweled smooth, they provide an excellent surface for coating application.

Regular concrete block has a hard, relatively granular surface with few openings caused by rocks or aggregates. Thus, it provides a relatively good surface over which to apply coatings.

Cinder block, however, is a quite different situation. This type of concrete block is extremely porous, with holes in the surface that may go completely through the block wall. This type of concrete block, then, requires considerable surface treatment.

Hard-Troweled Concrete

The hard-troweled surface provides a polished surface such as that found on hard-finished floors and sidewalks (Figure 9.38). It is usually formed by the application of a mortar that has a high cement content, combined with sand over the surface. This forms a smooth troweling mixture over the rougher poured surface. When this surfacing material is smooth and is set almost to the point of its final initial set, it is then troweled to a smooth surface that is quite hard and dense. Oftentimes, there is a thin layer of laitance brought to the surface during the troweling procedure. This type of hard-troweled surface provides a good surface over which to apply coatings since it is relatively nonporous, hard, and strong.

Normal-Troweled Concrete

Normal-troweled surfaces are those where the sand, cement and rock mixture is placed and then troweled smooth (Figure 9.39), without the use of the high-ratio cement mortar (where hard troweling is to be obtained). These surfaces are not usually troweled after the concrete has almost fully set, but are troweled while it is still relatively workable and smoothed to a uniform surface. This surface contains considerably more laitance than the hard-troweled concrete which must be removed before coatings can be applied. These surfaces are weaker than the hard-troweled surfaces and are less dense and somewhat more porous.

Both the hard-troweled and the normal-troweled surfaces can be found on cement plaster concrete walls. Such surfacing is usually done over a poured concrete surface in order to make it smooth, either as a tank lining or a smooth-surfaced wall or structure.

Wood-Floated Concrete

Wood-floated surfaces have been smoothed with a wood trowel after the concrete is poured (Figure 9.40). However, because of the relatively rough surface of the wood float, sand grains are brought to the surface, forming a granular surface. In this case, the cement surface also contains considerable laitance. The surface is relatively rough, more porous, and not quite as strong as those which are troweled with steel trowels. Because of its rough nature, a wood-floated surface is not suitable for the application of thin coatings.

Swept Concrete

Broomed or swept concrete surfaces are a step beyond the wood-float method in making the surface somewhat rougher and in bringing additional sand grains to the surface. In this case, actual grooves may be formed in the concrete, which increase the surface area (Figure 9.41). If such surfaces are to be coated, a heavy, thick coating must be used in order to overcome the surface irregularities. Nevertheless, a swept surface is dense with few pinholes. The surface is not as strong as a troweled surface and may contain more laitance. A penetrating primer is the most desirable coating for this type of surface.

Surfaces Poured Against Forms

As previously discussed, there are essentially three different types of surfaces which are poured against forms (Figure 9.42). The surfaces poured against plywood or steel forms are smooth and generally have the same surface characteristics. These surfaces generally are covered with water and air pockets and pinholes, all of which create coating problems.

FIGURE 9.38 — A hard-troweled concrete surface.

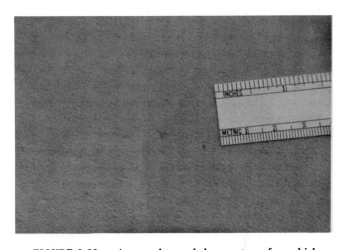

FIGURE 9.39 — A normal-troweled concrete surface which is slightly more granular and less dense than the hard-troweled surface.

FIGURE 9.40 — Wood-floated concrete which leaves an even, granular surface.

FIGURE 9.41 — Swept or broomed concrete which leaves an even-textured surface.

Many of the pinholes in such surfaces are merely small openings into a much larger opening under the surface of the concrete. Such a subsurface cavity is shown in Figure 9.43, which demonstrates the problem that exists with poured surfaces. Such cavities are a usual, rather than an isolated condition. It is oftentimes necessary to break these areas open in order to satisfactorily prepare the surface for coating.

Many of the hidden cavities have thin overhangs in the area of the pinhole which are frequently broken by a blast of air over the surface. They are not sufficiently thin, however, for a coating to break the surface and enter the cavity. A lightly sandblasted surface is shown in Figure 9.44. Note how many of the small openings have opened into larger cavities when the surface of the concrete was blasted. The large unopened cavities can cause considerable difficulty during coating application.

The pinholes, air pockets, and cavities are typical of a concrete poured against a form; whether the form is wood, steel, or an absorptive material makes little difference. The absorptive lining does reduce the number of cavities by allowing the air to escape into the absorptive form rather than remain on the surface, as it does with the other types of forms. The action of air in concrete when in contact with wood, steel, or lacquered forms is similar to the formation of bubbles on the surface of a glass containing soda water. The difference is primarily that the concrete is quite thixotropic and, when once set, does not allow the bubbles to move, but holds them in place on the surface. This results in the cavity forming just under the surface of concrete which is against the form. In order to properly coat this type of surface, considerable effort is necessary in order to fill all of the voids, cavities, and pinholes.

Gunite. As previously discussed, the gunite surface is extremely rough due to the guniting method of placing the concrete. This method forms an extremely dense surface without any pinholes, air pockets, or subsurface cavities. However, the surface is rough because of the sand aggregate used in the dry mixture as it is thrown against the surface by either compressed air or by mechanical means. The surface is not practical for coating unless that coating is extremely thick. Relatively thin sealers have been used on gunite surfaces which do prevent the penetration of some ions. However, a continuous coating is almost impossible to obtain on a gunited surface unless it is sufficiently thick so that the general roughness of the surface does not interfere with the continuity of the coating.

Troweled gunited surfaces are similar to other troweled surfaces and are therefore practical for the application of coatings.

Concrete Block. Concrete blocks are a standard building material and, more often than not, are coated for decorative purposes rather than as a protection against corrosion. They are, however, used quite generally in a number of industries, such as the nuclear power and chemical industries, where corrosion is a problem. The concrete block must be sufficiently coated in these areas so that there is no penetration of acids, chemicals, or

FIGURE 9.42 — A concrete surface poured against a form.

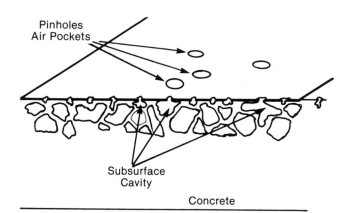

FIGURE 9.43 — A subsurface cavity in poured concrete.

FIGURE 9.44 — A lightly sandblasted, poured concrete surface showing opened cavities that were under the original surface.

decontamination materials. The surface of ordinary concrete block is quite dense, although it is not smooth. There are some depressions and cavities in the surface which require filling if an impervious coating is to be applied.

Cinder Block. Cinder block is a much more porous material (Figure 9.45). It may even have air passageways

Corrosion Prevention by Protective Coatings

through the entire block thickness. It is a coarse mixture and therefore quite porous with many surface cavities. These must be thoroughly filled prior to the application of any coating.

Concrete blocks of all types also have an additional surface problem, *i.e.,* the joints between the block. If the concrete surface is to be coated, the joints must be flush with the surface and the joint area must be treated in a similar fashion to the remainder of the block in order to fill any void. Joints that extend beyond the surface of the block must be removed. Many poured concrete surfaces have offsets at the junction between sections of plywood, steel forms, or joints between pours. These are areas that also must be leveled so that a smooth, continuous coating can be applied. Any concrete that extends beyond the general surface of the structure must be removed so that the entire surface is level and not subject to ridges or similar discontinuities. Figure 9.46 shows the offset caused by the junction of two plywood forms.

Most poured concrete also has form tie holes. These are areas where the forms are tied together in order to keep them from bulging due to the weight of the concrete. Form tie holes may be as much as 1 in. in diameter and 1 1/2 to 2 in. deep. These must be completely filled with whatever surface treatment is used in order to make the surface smooth and available for coating application. Figure 9.47 shows a typical form tie hole in a poured concrete wall.

In any description of a concrete surface, it is necessary to include the various methods of treating forms before the concrete is poured against them. There are a number of treatments; one of the common ones is the use of oil sprayed on the forms prior to placing the forms and before the concrete is poured. Oil is used in order to prevent the concrete from sticking to the form. However, the oil remains on the surface of the concrete and must be removed before any coating can be applied.

Waxes also are often compounded into form release agents. These are even worse than the oil since they are less affected by the concrete itself and remain on the surface. These also must be removed before coatings can be applied or even surface treatments applied prior to the application of coatings.

Lacquered forms are often used, as they possess an advantage from a coating standpoint. This is because the lacquer adheres well to the wood or steel surface and provides a break between the form and the concrete so that the form easily separates. In addition, lacquered forms can be used several times before requiring resurfacing. The lacquer does not leave any residue on the concrete surface so that coatings can be directly applied.

Preparing Concrete Surfaces

In the past, there were essentially four possibilities for making the surface of a poured concrete structure or a concrete block wall sufficiently impervious to accept coatings. These methods were those using cement or cement and sand to fill the surface. The methods consisted of sacking, stoning, floating, and plastering the surface.

FIGURE 9.45 — Typical cinder block surface.

FIGURE 9.46 — The offset in a concrete surface caused by the junction of two plywood forms.

FIGURE 9.47 — Typical form tie hole in a poured concrete wall.

Sacking

The sacking procedure essentially is the scrubbing of a buttery mix of cement mortar over the concrete using a cement sack, a gunny sack, or a sponge rubber float. In order to obtain proper surface preparation by sacking, the sacking procedure must be started as soon as possible after the concrete is poured and the forms removed. This is important since the mortar applied by sacking must be cured at the same time as the concrete wall if it is to thoroughly adhere to the surface of the wall and provide a solid base for the application of high-performance coatings.

If high-performance materials, such as the epoxies, are to be applied and are to be applied in any reasonable thickness for full protection of the surface, the sacked area or surface of the concrete must be as strong as the wall itself, or the epoxy coating will tend to pull the sacking away from the concrete surface. If the poured concrete is allowed to cure for too long before the forms have been removed, the sacking may disbond from the surface. Thus, the curing of the sacked surface is extremely important to coating effectiveness.

To start the sacking process, the cement wall should be thoroughly wetted with water prior to any of the filling. This prevents the concrete in the wall from sucking all of the water out of the concrete mortar used for sacking and thus making the surface too dry. The cement mortar that is used for sacking usually consists of a one-to-one volume of portland cement and fine sand or a one-to-two-part mix. Sufficient water is added to make the mortar a thick, creamy product. This is applied to the concrete in any reasonable way, or it may even be applied to the surface of the sack and then applied to the wall.

Once the mortar is on the wall, the cement sack or rubber float is then rubbed over the surface in a circular manner. This rubs the mortar into the concrete air and water pockets, thoroughly filling them with the buttery cement-sand mixture. The objective is to fill the pits and to leave as little of the sacking mix on the surface as possible.

Oftentimes, once the original mortar has been scrubbed into the surface with a sack, some dry cement is added to the surface and is again scrubbed with the sack in a circular motion in order to add the dry cement to the material in the water pockets and pinholes. This adds cement to the mortar in the water pockets and thus tends to dry the mortar, holding it in place and reducing shrinkage. Once the surface is nearly dry, it is gone over again with a dry sack, removing as much of the sacking material from the surface as possible. Since this is difficult to do, there is often a layer of mortar left on the surface. If it is not thoroughly cured with the body of the poured concrete, it can cause coating difficulties.

Figure 9.48 shows a typical sacked concrete surface. The concrete surface was poured on plywood forms, and some of the wood grain can be seen in the concrete. Also, the sacked mortar can be seen filling the roughness left by the grain of the wood. While all of the visible water pockets and pinholes appear filled, after the application of the first coat of coating or concrete primer, some pinholes will become obvious and will require filling.

Many millions of square feet of concrete surface have been treated by the sacking process, and it is still used as a surfacing method for many poured concrete surfaces. The original surfaces in nuclear power plants were sacked. Also, many of the concrete tanks used by the Navy for the storage of fuel oil and diesel oil have been treated by this method, as well as the concrete ships that were built during World War II.

All of these surfaces were overcoated with high-performance coatings in order to prevent any contamination of the underlying concrete surface or to prevent any penetration of the petroleum products into the concrete. In each of these cases, if any pinholes had remained, penetration would have taken place and leakage of petroleum products could have occurred or radioactivity could have formed in the pinholes, creating an untenable situation in the chemical areas of the nuclear processing unit.

Stoning

Stoning is another method by which concrete surfaces are prepared, and in may ways it is similar to the sacking method. Using the stoning procedure, the same type of cement mortar, a buttery mixture, is applied to concrete. The mortar is then ground into the surface with a Carborundum[1] brick using circular motions to move the mortar over the surface of the concrete. The brick grinds down any imperfections on the surface, opens up the pores of the surface, and thoroughly works the mortar into the concrete pores or cavities.

In many ways, this is better than the sacking process since it opens some of the pores and pinholes and removes any roughness or projections from the concrete surface. On the other hand, the stoning process does not provide as smooth a surface as the sacked surface unless, after the stoning process has been completed, the surface is rubbed with a sack in order to smooth it.

Figure 9.49 shows a stoned concrete surface that has not been rubbed by a sack. Note that Carborundum brick has ground the surface, and that the water pockets and pinholes are filled. The surface is generally smooth, although somewhat more granular than the sacked surface. The finer the Carborundum brick, the smoother the surface will be and the closer it comes to the smoothness of the sacked surface. Many architectural concrete surfaces are stoned with fine Carborundum stone in order to both smooth and fill the surface. Many of the surfaces of bridge

FIGURE 9.48 — Typical sacked concrete surface.

---

[1]A tradename of Carborundum Co., Niagara Falls, NY.

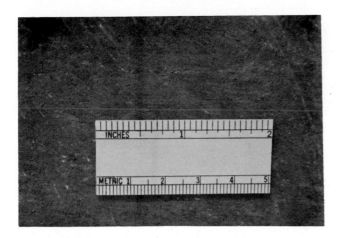

FIGURE 9.49 — A poured concrete surface stoned with a carburundum brick.

structures have been stoned or sacked. Also, many of the surfaces in nuclear energy plants have been stoned in order to assure a smooth, filled surface.

Wood Floating

A wood-floated surface is another method of preparing poured concrete surfaces. In this case, the same type of buttery cement mortar is used on the surface; however, in place of a stone or a sack, a wood float is used, again moving it over the surface in a circular motion and filling the pinholes and water pockets with mortar.

In many ways, this is not as effective as either sacking or stoning since the wood float stays on the surface of the concrete and leaves a considerable amount of mortar on the surface. The remaining mortar has to be thoroughly cured with the body of the concrete in order for this method to be effective. Also, the wood float leaves the surface with a sandy, granular finish (Figure 9.50). If an impervious-type coating is to be applied, this surface is less satisfactory than those that are smoother.

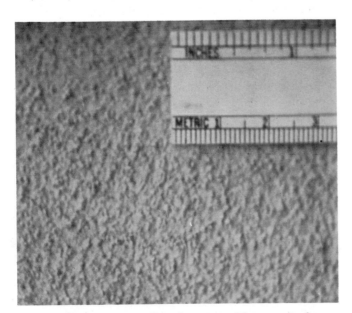

FIGURE 9.50 — Wood-floated concrete with an even surface that is too granular for anything but heavy mastic coatings.

Steel Trowel

The steel trowel method of preparing a poured concrete surface consists of applying a layer of cement plaster over the poured concrete surface by the use of a steel trowel (Figure 9.51). This is more or less a standard plastering technique and it allows from 1/8 to 1/4 in. of plaster to remain over the poured concrete surface. The steel trowel is moved over the surface, and as the trowel is held at a small angle, the cement mortar is forced into the pinholes and water pockets, filling and covering the imperfections in the poured surface. Where rock pockets are a problem, the troweling of mortar over the surface can fill the large as well as the small openings. In many ways, steel troweling is a preferred method of treating a poured concrete surface, since it completely covers the surface with sufficient mortar to provide a smooth, even surface over the entire area. This surface is satisfactory for many coatings and provides a reasonably pore-free surface over which to apply high-performance corrosion-resistant coatings.

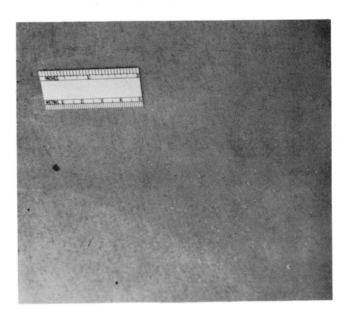

FIGURE 9.51 — Steel-troweled concrete which produced a smooth but not hard-troweled surface.

Hard Troweling

The plaster which is used in hard troweling is a cement-sand plaster mix. The cement plaster usually does not contain lime, as do many plasters which are used on the interior walls of homes and offices. Many walls in nuclear energy units and nuclear power plants use a hard-troweled surface over poured concrete. In this case, the mortar mix is considerably higher in cement and is generally applied in a thinner layer over the poured concrete. The mortar is applied as a relatively stiff mix and the surface is worked over with the trowel until the concrete surface almost becomes black. This forms an extremely dense surface and one which is hard and smooth.

Figure 9.52 shows a portion of a concrete wall which is troweled by this hard troweling method. It also shows the type of poured concrete over which it was used. Note the smooth, nongranular surface of the hard troweling.

There are some dangers in the hard troweling method in that, when the cement is worked in this manner, it also becomes brittle. Thus, unless it is thoroughly cured with the concrete wall itself, it can spall from the surface. This may not be discovered until after a coating has been applied, when the coating, upon shrinking and curing, tends to pull the hard-troweled concrete off the concrete wall surface. This has happened in many cases, primarily due to lack of adequate cure once the troweled coating was applied.

The smooth-troweled coatings, which are applied in somewhat greater thickness, also have this problem, although to a lesser degree. Nevertheless, in both cases, the cement plaster should be applied over the poured concrete surface as soon as possible after the forms have been removed. In fact, it is preferable to apply the coating the same day the form is removed. Following the application of the concrete plaster or hard-troweled cement, the walls must be maintained in a damp condition for several days prior to drying and coating application.

FIGURE 9.53 — A poured concrete wall that was lightly blasted to open the air pockets in the surface.

FIGURE 9.52 — Hard-troweled cement surfacer on a poured concrete wall which produced a smooth, dense surface.

Synthetic Concrete Surfacers

All of the methods discussed are reasonably costly. In fact, in some of the nuclear processing plants, the preparation of concrete by the troweling method has cost $5.00 a square foot or more for surface preparation alone. This expense has led to the use of synthetic concrete surfacers, which are a combination of the first coat of a coating and a filler to fill the pinholes and air pockets in the poured concrete surface. Often, during the application of this type of concrete surface preparation, the concrete is given a light blast in order to remove any concrete laitance on the surface, as well as to open all of the pinholes and water pockets that may be in the surface.

Figure 9.53 shows a sandblasted poured concrete surface prior to the application of the surfacer. The concrete is roughened and all of the water pockets and pinholes are opened so that they provide a broad open area for the penetration of the synthetic surfacer.

Sandblasting can also be used for any of the other

types of surface preparation (*e.g.*, sacking and stoning); however, it generally is an additional step that is not used except under special circumstances. This is not true with the synthetic surfacer, as sandblasting is often used to open the surface prior to the application of the material.

There are two general types of resinous surfacers. One is the thin type in which the surfacer is a thin thixotropic material that may be applied either by spray or, more usually, by squeegee. The material is sufficiently thixotropic to enter large air pockets, remain in place, and still provide a smooth surface without sagging out of the opening on a vertical wall. Some are sufficiently thixotropic so that they can even be applied into form tie holes without sagging. Most such materials are made from epoxy resins and are applied as thin as possible with the squeegee, trying only to fill the holes in the surface without leaving a heavy layer on the surface.

Such an application is shown in Figure 9.54 where the thin epoxy surfacer has been squeegeed over the surface, and has been removed as much as possible, except for the surface imperfections in the concrete. Figure 9.55 shows a cross section of a concrete surface after the application of the thixotropic synthetic surfacer. Note how the surfacer has penetrated into the water pockets to a depth of as much as 5 millimeters.

The second surfacer is a thick type in which the resinous surfacer is troweled over the poured concrete surface leaving 1/16 to 1/8 in. of synthetic resin mortar on the surface. While this is actually a surface preparation for the concrete, it also acts as the first application of a heavy coating over the concrete itself. These materials also are usually epoxy-based mortars. They may be applied directly by troweling or by the use of a gunite-like spray unit in which the material is sprayed over the surface and then troweled smooth. Figure 9.56 shows the smoothing of the surface by trowel after the heavy epoxy mortar has been applied directly to the poured concrete surface.

With either of these resinous-type surfacers, the sur-

FIGURE 9.54 — A thin epoxy surface squeegeed over a poured concrete wall, leaving a smooth finish and filling the air pockets and the wood pattern left by the forms on the concrete.

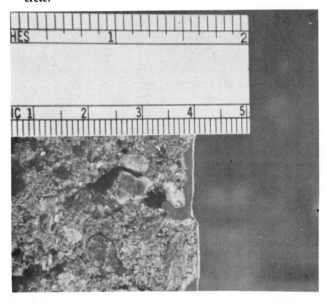

FIGURE 9.55 — A cross section of concrete illustrating the filling action of an epoxy surfacer, with only a small amount of surfacer remaining in the surface.

FIGURE 9.56 — Troweling of a heavy epoxy surfacer directly over a poured concrete wall, which may be followed by the application of any desired topcoats.

faced concrete is smooth and the high-performance coatings may be applied over the top of the surfacers. An excellent bond to the concrete surface is obtained as the epoxy resins tend to penetrate into the concrete itself, providing a tough mechanical bond to the concrete, which many times is stronger than the concrete itself. These concepts have been used in place of the sacking and stoning approach to surface preparation in critical areas where a completely corrosion-resistant or contamination-resistant coating is required.

Floor Troweling

In addition to the poured concrete surface, the next most common type is that of the troweled floor surface. This may be either the smooth steel-troweled surface or the sidewalk-type finish often used for floors. Both procedures are common for floors, and the surface preparation is generally similar.

The sidewalk finish provides a harder and somewhat denser surface over which to apply a coating. The smooth, troweled surface is a common floor surface; however, it is not as hard as the sidewalk finish. In both of these cases, there is some concrete laitance left on the surface, which must be removed prior to coating. Also, it is preferable to slightly open the troweled surface in order to allow the high-performance coatings to penetrate the concrete surface and obtain a physical bond. Many times, this is done by giving such surfaces a light sandblast. Where it can be done, this is the best and least costly method of preparing such surfaces.

Another surface where coatings are often applied is the interior of centrifugally cast concrete pipe. In this case, there is a heavy cement laitance on the surface and the only satisfactory method of removing it is by light sandblast. This breaks the surface of the concrete and allows coatings to have physical adhesion; however, it does not roughen the surface sufficiently to create a roughness problem where a smooth coating is desired.

Figure 9.57 shows a poured concrete slab which was smooth-troweled and lightly blasted to remove the surface laitance. Many of the pieces of aggregate are close to the surface and the blasting has removed the concrete laitance so that they show on the surface. Such a surface can be a good one over which to apply corrosion-resistant coatings. There is one imperfection in the surface that appeared during blasting: an area where several pieces of aggregate came together created a small pit that will need filling prior to coating.

Concrete cannot be heavily blasted. Where blasting a surface, the sand should preferably be very fine and the surface should only be etched without eating into the surface (as where heavy abrasive is used). Also, if the concrete is blasted heavily, it exposes the aggregate to too great a degree and forms a rough surface that is not practical for most impervious coatings. Blasting of the concrete surface is more of a sweep blast than a normal sandblast procedure for metal surfaces. Holding the blast gun too close to the surface can actually dig holes that may be much too deep to coat and thus must be patched prior to any coating application.

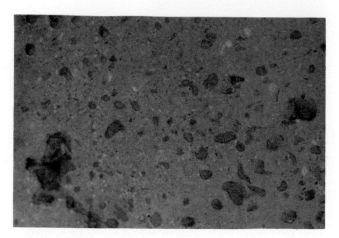

FIGURE 9.57 — Troweled concrete slab lightly sandblasted prior to coating.

Acid Etching

The other common treatment for troweled concrete surfaces such as floors is acid etching. While a sandblasted surface is preferable, in many areas sand is not a practical material because of the damage that might be caused to other equipment and other surfaces. Acid etching will remove the concrete laitance and will open the concrete surface so that coatings can obtain mechanical adhesion. The opening of the concrete surface is extremely important for most coatings, particularly those that have a high molecular weight and do not have high penetrating powers (*e.g.,* vinyl coatings). Unless the coating can penetrate the concrete, it does not obtain the adhesion necessary for a tight bond over the surface.

The procedure for acid etching is to use a relatively dilute acid, apply it to the concrete surface, allow it to react with the calcium compounds in the cement, and then wash the surface thoroughly to remove the acid salts. The most common acids used for this purpose are hydrochloric or muriatic acid. The concentrated acid, diluted, with three or four parts of water, is poured on the concrete, spread evenly, and allowed to react until all bubbling of the surface ceases. At this point, the acid has been neutralized by the calcium compounds in the cement surface. The reaction product is calcium chloride, which is soluble in water and can be easily washed away from the surface. It is generally preferable to scrub the surface with a broom after acid etching in order to be sure that all of the soluble products are removed.

Other acids may be used, but they are generally less effective than hydrochloric acid. Sulfuric acid will react with the calcium products and form calcium sulfate, which is an insoluble material and is much more difficult to remove from the surface. Phosphoric acid is another acid that can be used. It also reacts with the calcium products in the cement to form insoluble calcium phosphate materials. Nitric acid, because it represents a personnel hazard, is seldom used; on the other hand, the reaction products with the cement form soluble calcium nitrates which can be washed from the surface. All in all, the hydrochloric (muriatic) acids are the most practical and the most commonly used material.

There are some hazards involved in acid etching. The hydrochloric acid is a volatile material, and while it is the same acid that exists in the stomach, it can be toxic when the fumes are in significant concentration. Also, being volatile, it will attack metal surfaces in the area so that wherever sensitive metal objects are involved, the hydrochloric acid etching process should not be used. For instance, some expensive machine tools were installed on a concrete floor prior to coating. The concrete floor was acid etched with hydrochloric acid, and the following day, the machine tool surfaces were covered with rust. Needless to say, the damage was both extensive and expensive. In such areas, a nonvolatile acid, such as sulfuric or phosphoric, should be used and the floor should be thoroughly scrubbed after the action of the acid in order to remove any of the nonsoluble materials.

Any of these acids is extremely strong and should not be used in contact with skin or clothing. They will rapidly disintegrate clothing, and wherever acid may splash on cotton surfaces, a hole will soon form. Goggles, rubber gloves, and rubber boots should be used where acid etching is in progress.

Figure 9.58 shows a section of a concrete floor acid etched in three different degrees. In the lower right-hand corner, the unetched concrete is shown. In the upper right-hand corner is a light etch. The upper left-hand corner shows a somewhat stronger etch, and the lower left-hand corner is a strong etch. The preferable surface condition is shown in the upper right-hand corner, where the surface of the concrete is merely opened and does not expose an excessive amount of aggregate, as does the heavy etch in the lower left-hand corner.

Swept Concrete

Many concrete surfaces are swept after they have been smoothed by wood floating or similar means. This provides a rather granular surface only suitable for heavy coatings. Nevertheless, in order to prepare the surface for coating, it should also be lightly blasted or lightly acid etched. Figure 9.59 shows a lightly swept granular concrete surface. This surface has not been etched or blasted; however, the surface does not change appreciably other

FIGURE 9.58 — A section of concrete floor acid etched in three different degrees. The upper right-hand section demonstrates the preferable degree of etching.

FIGURE 9.59 — A lightly swept granular concrete surface.

than to remove the concrete laitance and to open the surface slightly for some physical adhesion. Heavily swept surfaces are not recommended for the application of coatings.

Concrete Block

Concrete block presents a surface similar in many ways to poured concrete. The surface of concrete block contains many cavities and deep holes, all of which must be filled prior to applying a corrosion-resistant coating. Concrete block walls can be plastered in order to provide an impervious surface; on the other hand, the two most effective ways for treating concrete block structures where they are exposed to corrosive problems is to use the resinous surfacers that have been described for poured concrete surfaces. The thin surfacer, when squeegeed over the surface of block, penetrates deeply into the depressions and locks into the block, forming a smooth, thin surface over the entire block area that can later be covered with corrosion-resistant coatings.

The thick type of resinous surfacer is also used to a considerable extent in nuclear energy areas. In this case, the block is sprayed with the heavy resinous mortar and then troweled smooth for later application of the corrosion-resistant topcoat. It has been found that the application of the surfacers to heavy concrete blocks that are filled with heavy concrete sometimes blister because the volatile materials in the surfacers expand within the pores of the concrete block. However, where the concrete blocks have not been filled with concrete, the volatile materials in the coating are able to pass through the block without developing any back pressure. The application of a coating to unfilled block is thus more satisfactory than the application of a coating after the block has been filled with concrete.

The joints between concrete blocks present another area where thick resinous surfacers greatly aid in obtaining a smooth surface. Areas of the joint that are sharp and uneven are also easily filled with resinous surfacers. Concrete block joints can also be filled with latex fillers. These

are quite satisfactory for architectural purposes, but are not satisfactory for areas where the block will be subject to chemical corrosion or immersion. The cement latex surfacers are easy to apply and do a good job of sealing the block against moisture penetration due to driving rains and weather. They can be applied to surface by brush, trowel, or squeegee; however, before application of the surfacing material, the surface should be cleaned and the holes and voids opened by blasting, power wire brushing, or power standing.

Concrete Curing Compounds

The concrete curing compounds are, for the most part, waxy-type materials applied to prevent the moisture in the concrete from evaporating too rapidly. In this way, the moisture is held in the concrete for reaction during the concrete curing process. Where concrete curing compounds have been used, it is always hazardous to apply coatings. Such surfaces should be blasted free of the curing compound and then treated, as previously described, by sacking, the use of resinous surfacers, or similar means. Blasting is by far the best procedure to use in eliminating the concrete curing compounds, since it breaks the surface of the concrete and blows the curing compound away with a small amount of the concrete surface. There is no other completely satisfactory means for eliminating such materials.

In some cases, the concrete curing compounds will weather away. This is particularly true where they are used on concrete highways; however, the coatings applied under such conditions are not critical. Complete removal of concrete curing agents is essential, primarily wherever high-performance coatings are to be applied.

Another way in which concrete can be cured is through the application of a prime coat over the uncured concrete prior to its curing. This has been done quite satisfactorily in a number of cases. For instance, when concrete ships were built during World War II, the placement of the concrete was extremely critical. Thus, there were few imperfections (*e.g.*, rock or air pockets) in the concrete surface. As soon as the forms were removed, a vinyl coating, thinned approximately 50% with a moisture compatible ketone solvent, was applied to the surface. As soon as this had dried, a full coat of the vinyl coating was applied, which then acted as the concrete curing coating. This procedure was used for a number of concrete ships, none of which had any topcoat adhesion problems due to the application of the vinyl coating. In more recent years, some of the epoxy polyamide coatings have to be used in a similar way to eliminate concrete curing compounds in critical coating areas.

Concrete hardeners are used to increase the surface hardness of concrete floors. These materials are many times incorporated into the concrete mix. For the most part, however, they are materials such as sodium silicate, magnesium fluorosilicate, zinc fluorosilicate, or similar materials applied in dilute solution over the new concrete surface. These materials react with the cement chemicals and combine to form the hard, impervious surface. Where coatings are to be applied, however (and this is true for almost any coating that may be desired on a concrete floor), concrete hardeners cannot be applied. Most of the high-performance coatings for concrete will not adhere satisfactorily to the dense silicate surface created by such hardeners.

Where such hardeners have been applied to the concrete, sandblasting is the only satisfactory treatment for the surface. The surface must be broken up and the silicate treatment eliminated from the surface. Acid etching is not entirely satisfactory for such surfaces since the silicate reaction product is reasonably acid resistant. Concrete scarifiers may be used to break the concrete surface and to remove a thin layer of the concrete surface containing the silicate hardener. These mechanical surface preparation devices, however, usually remove too much of the concrete and make it too rough for a thin coating to be applied satisfactorily. Only heavy epoxy floor toppings can be satisfactorily used over sandblasted or scarified surfaces.

Contaminated Concrete

Contaminated concrete is always a problem, both to prepare and to coat satisfactorily. There are many types of contaminants; oils and greases are among the most common. Food products and waxes are some of the most difficult. There are also many types of chemicals that can contaminate concrete and thus require removal during the surface preparation period. Acids are also considered difficult concrete contaminants because of their rapid reaction with the chemicals in cement. Since the cement chemicals are highly alkaline, the acids are actually neutralized by the concrete itself. However, voluminous end products of the acid-concrete reaction do pose difficulties and therefore must be removed prior to coating application.

One of the most difficult types of food contamination is protein materials found in meat and meat products, many vegetables, eggs, and similar foods. The protein materials penetrate the concrete and react in some measure with the concrete chemicals. The resulting surface, which is covered with dried protein materials, is insoluble in most coating solvents and allows little, if any, adhesion. Because the protein-type material also penetrates the surface, it is necessary to remove the surface of the concrete either by mechanical means, *e.g.*, a concrete scarifier, rotary impact chipper, or by sandblasting. Sandblasting is undoubtedly the preferable way to remove such materials from a concrete surface. Alkaline washes, acid etching, or other such procedures are not satisfactory.

Oils and greases must always be thoroughly removed from the concrete prior to coating. If the concrete has been deeply penetrated, sandblasting or removal of the surface concrete is the only satisfactory way to eliminate the penetrated oil and grease. Steam cleaning, solvent wash, or use of trisodium phosphate are often recommended in this case. However, concrete badly contaminated with oil usually will not be sufficiently cleaned by such methods to accept high-performance coatings properly.

Lubricating oils and greases are less difficult to remove than fats and oils used in food processing. Most of the fatty materials are slightly acidic; many may become rancid or acidic due to exposure to air and, in so doing, become reactive with the calcium chemicals in the cement. They therefore become a permanent fixture on the surface, so that the actual surface must be removed in order to apply a satisfactory coating. Acid etching, alkaline cleaners, or similar procedures usually will not effectively prepare the concrete where such contamination is present.

While wax is another problem contaminent, it is somewhat preferable to the fatty oils and greases in food products. Wax is a relatively high molecular weight material that will, for the most part, remain on the surface, unless the concrete and wax have been heated to the softening point of the wax. Cold wax does not penetrate the concrete easily so that under these conditions the surface will not need to be deeply removed. Sandblasting or scarifying are the more practical methods of preparing such surfaces.

Chemicals, such as alkalies, may also contaminate concrete surfaces. Many of these are in a dry salt form and thus do not actually penetrate the concrete. A thorough washing of the surface with clean water, or steam cleaning the surface will remove the majority of such contamination. The concrete can then be acid etched, which not only neutralizes any alkaline material that may be on the surface, but also tends to etch the surface of the concrete and thus prepare it for coating.

Acid contamination of concrete is more difficult, usually because the acid has eaten away the cement area of the concrete, leaving the aggregate exposed. Also, in the case of sulfuric acid, the reaction products of the acid and the concrete are sufficiently voluminous so that the concrete may be soft for some depth. In this case, the entire reaction product must be removed, including the aggregate, in order to arrive at a solid concrete surface. A sulfuric acid reaction with concrete is found particularly in sewer and sewer structures. In many cases, the reaction has penetrated to a depth of one or more inches, all of which must be removed before the concrete can be replaced or prior to the application of a coating. In many cases, the concrete surface must be replaced, followed by the procedures previously described for either poured or troweled concrete.

Concrete surfaces that have been exposed to various salts, particularly those that are nonreactive with concrete, merely require washing from the surface by water blasting, steam cleaning, or similar methods. Where some reaction has taken place, as in the case with acid salts, the surface must be removed down to solid concrete prior to the application of the coating. This can be done by sandblasting, the use of scarifiers, or even acid etching, if the salts have not penetrated to a major degree.

Other Influences on Surface Preparation Selection

Environment

Environment has a significant bearing on surface preparation. Where coatings are to be immersed, or where they are to be used in areas where they are continually wet with water, moisture, or chemical solutions, there should be no compromise on surface preparation. These areas are difficult to maintain, and if the coating fails, the cost of repair can be several times that of the original coating cost.

Tank linings and immersion coatings should never be applied over a surface preparation that is less than NACE 2 or SSPC 10. Tank linings, in addition to preventing corrosion to the tank itself, are often used to protect the contents of the tank from iron contamination. Thus, coating failure may not only endanger the tank surface, but it may also contaminate a sizable quantity of valuable product. A wine tank coating, for example, must maintain a perfect barrier, since any iron ions that may come in contact with the wine will create a metallic taste and a blue-black precipitate in the wine.

Atmospheric Conditions

There are three general atmospheric conditions where coatings are used: (1) marine, (2) industrial, and (3) rural. Marine conditions dictate that anything less than the best surface preparation is usually poor economy. Even with a white or near-white sand blast, marine surfaces that are contaminated with chloride and sulfate ions can create immediate surface reaction problems, as previously noted. The difficulty of repair also must be taken into consideration. Repair on offshore structures is extremely costly, and coating failure can seriously compromise the safety of such structures. Economic costs alone would dictate that the additional cost of the best surface preparation is a small insurance premium to pay to protect against coating failure.

Industrial atmospheres include a wide range of conditions. Plants which are nonchemically oriented and in a relatively rural atmosphere would not require the degree of surface preparation of those industries which are involved with acids, alkalies, and salts. Where these strong contaminants and corrosive ions are found, the best surface preparation is, again, not too good and is still considered a low premium to pay for insurance against coating failure.

Underground pipelines are another area where corrosive conditions can be considered serious. Here again, the best surface preparation should be used since maximum adhesion is necessary to resist moisture, salts, earth movement, and undercutting of the coating where damage may occur during laying, from rock point pressure, and under similar conditions.

Rural atmospheres present less of a surface preparation problem because of the fewer corrosive conditions that exist. Serious corrosion of steel in this type of atmosphere may take a long period of time. Thus, coatings such as oil paint, alkyds, or similar products may adequately protect the surface over a practical period of time, even though the original surface was rusty. Many of the coatings applied in rural areas are for decorative purposes rather than for resistance to severe corrosion. Nevertheless, a coating showing spot rusting is still not a satisfactory one, even from an appearance standpoint.

All surface preparation procedures are aimed at providing a surface over which a coating will have strong adhesion. Surface preparation, by providing a surface over which a coating can strongly adhere, is the key to long and effective coating life. Without a proper surface, high-performance coatings cannot provide the corrosion resistance for which they were intended.

References

1. Corrosion and the Preparation of Metallic Surfaces for Painting, Unit 26, Federation Series on Paint Technology, Federation of Societies for Coatings Technology, Philadelphia, PA, 1978.
2. Snogren, Richard C., Handbook of Surface Preparation, Palmerton Publishing Co., p. 37, 1974.
3. Hudson, J. C., Protection of Structural Steel Against Atmospheric Corrosion, Journal of the Iron and Steel Institute, Vol. 168, June (1951).
4. Hudson, J. C., Subsidiary Paint Tests at Birmingham: Final Report, Journal of the Iron and Steel Institute, Oct. (1951).
5. Effects of Surface Preparation on Service Life of Protective Coatings, Interim Statistical Report by NACE Tech. Committee T-6H-15, National Association of Corrosion Engineers, Houston, TX, Dec. (1977).
6. Surface Preparation Guide, Steel Structures Painting Manual, Vol. 2, 2nd Ed., Steel Structures Painting Council, Pittsburg, PA, p. 36, 1955.
7. NACE Coatings and Linings Handbook, Part 2, Atmospheric Coatings, Sec. 4, Surface Preparation, National Association of Corrosion Engineers, Houston, TX.
8. Good Painting Practice, Chapter 2, Steel Structures Painting Manual, Vol. 2, Steel Structures Painting Council, Pittsburgh, PA, pp. 33-45, 1982.
9. Good Painting Practice, Chapter 3, Steel Structures Painting Manual, Vol. 1, Steel Structures Painting Council, Pittsburg, PA, p. 39, 1966.
10. Good Painting Practice, Chapter 2, Steel Structures Painting Manual, Vol. 2, Steel Structures Painting Council, Pittsburg, PA, pp. 89, 91, 95, 99, 1973.
11. Performance of Organic Coatings Over Wet Sandblasted Steel, NACE Proposed Tech. Committee Report, T-6G-26, National Association of Corrosion Engineers, Houston, TX, p. 3, Aug. (1980), unpublished.
12. Garroutte, J. D. and Hutchinson, M., How to Use Inorganic Zinc Primer Over Wet Abrasive Cleaned Steel, Special API Report, Hydrocarbon Processing, p. 95, May (1975).
13. Schwab, Lee K. and Drisko, Richard W., Relation of Steel Surface Profile and Cleanliness to Bonding of Coatings, CORROSION/80, Preprint No. 116, National Association of Corrosion Engineers, Houston, TX, 1980.
14. Keane, J. D., Bruno, J. A., and Weaver, R. E. F., Surface Profile for Anti-Corrosion Paints, Steel Structures Painting Council Report, Pittsburgh, PA.

10

Application of Coatings

A coating is a unique product. It is manufactured, placed in a container, and sold as a packaged unit, which is what the consumer buys. Yet in this state, as merely a liquid in a can, the coating is valueless. It is only after application that the coating becomes valuable and useful. The manufacturer can be as scientific and as careful as possible in producing the liquid coating, yet the final variable in the product's usefulness lies in the hands of a third party, *i.e.,* the applicator. That is why proper and careful application is often stressed as the key to the success of any coating. A less effective yet well-applied coating can provide better and longer lasting protection than the best coating material poorly applied.

The application of a coating is actually only one of the keys to proper coating protection. Coating protection is like a tent supported by three poles: (1) the material, (2) the surface preparation, and (3) the application. If any one of the poles is weak and breaks, the whole tent collapses and the protection it provided is gone.

This concept is illustrated in Figure 10.1. If the material is ineffective, the coating production is nil. If the surface preparation is improper, the protection provided by the coating is short-lived. Finally, if the application is poor or careless, coating protection will not be achieved, regardless of the strength of the material and surface preparation. Application, then, must be considered one of the important factors in arriving at effective coating protection.

The purpose of coating application is first to develop a protective layer of material over the substrate. Secondly, this layer must provide a continuous film over the surface. Thirdly, the continuous film must be of a relatively constant and even thickness. Fourthly, the film must adhere tightly to the substrate. The overall purpose, then, is to develop a continuous, highly adherent film of an even thickness over the substrate. The achievement of this purpose must take into consideration many additional factors which will be discussed individually,

The Type of Coating

The application characteristics of the various coating types is one of these factors. Each coating type has its own application characteristics; in fact, many of the individual coatings have unique characteristics. However, because of the large number of coatings available for corrosion-resistant applications, it is only possible to describe the unique characteristics of the coating types. While there may be individual differences within the coating type, such as the differences in two coatings supplied by two different manufacturers from the same base material, they nevertheless would have some general properties which

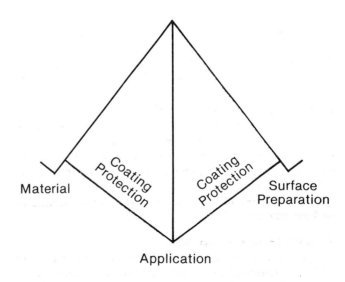

FIGURE 10.1 — The three keys to coating protection.

TABLE 10.1 — Application Characteristics of Various Coating Types

Application Characteristics	Oil-Base	Lacquers	Coreactive Coatings	Water-Base (Emulsion)	High Solids	Inorganic
Application by:						
Brush	Excellent	Poor	Fair	Good	Poor	Poor
Roller	Excellent	Poor	Good	Excellent	Fair-Poor	Poor
Spray	Good	Good	Good	Excellent	Good	Good
Film Thickness	Low	Low	Med.-High	Low	High	Medium
Flow	Excellent	Fair	Good	Poor	Good	Fair-Good
Drying Time	Slow	Fast	Med.-Slow	Fast	Fast-Med.	Fast
Wetting of Surface	Excellent	Poor	Fair-Good	Poor	Poor-Fair	Fair-Good
Film Build/Coat	Low	Medium	High	Low	High	Medium
Consistency	Oily		Sticky	Thixotropic	High Viscosity	Thixotropic
Pigment/Vehicle Ratio	Medium	Low-High	Med.-High	Low	Low-Med.	High
Curing Method	Air	Solvent Evaporation	Internal	Evaporation	Internal or Melt	Air
General Application Characteristics	Excellent	Good	Good	Easy	Fair	Good

are common to both. These are the properties that will be described in the following sections (Table 10.1).

Oil-Base Coatings

Oil-base coatings include materials that are based entirely on oil, such as linseed oil-base products, alkyds, alkyd enamels, oil-base varnishes, and similar materials. One of the principal characteristics of this type of coating is that it is generally applied in thin films. If a thick, overall coating is desired, the oil-base material must be applied in several coats. In fact, these materials produce the best results when they are applied as thin films in several coats. This is because they react with oxygen from the air, which is external to the coating. The oxygen must penetrate through the coating and react throughout the coating in order to provide a strong, resistant film. Thickly applied coats of these materials tend to react on the surface, which may cause a number of coating problems, *e.g.*, wrinkling, checking, or cracking.

A second important characteristic is that these materials generally have good workability. Since they are based on oil, they have good lubricating characteristics, which makes application by brush much easier than most other protective coatings. Another outstanding characteristic is their good wettability of most surfaces. Due to the oil content, they wet both metal and wood surfaces easily and effectively. This is also true of their application as repair materials; they easily solvate previous coatings of the same type. Because of their good wetting characteristics, oil-base coatings also flow well. In many cases, it is necessary to design some thixotropic properties into the coating to prevent the rapid formation of runs and sags on a vertical surface.

Good workability is the overriding characteristic of oil-base coatings.

Lacquer-Base Coatings

Lacquer-base coatings are based on synthetic resins which are dissolved in solvents and do not change in properties upon application. One of their best application characteristics is that they generally dry rapidly. This is dependent on the solvent structure of the coating. However, even if the solvents are relatively slow, the drying characteristics are reasonably fast, since it only requires the evaporation of the solvents for the film to be formed.

Materials included in the lacquer category are nitrocellulose, vinyl, and chlorinated rubber. These materials usually have a relatively low solids content compared with coreactive materials, and therefore result in thin films once the coating is applied and has dried. Lacquer-base materials are principally composed of relatively large molecular weight resins. Thus, the coatings have relatively poor wetting characteristics, since the large molecules tend to remain on the surface and therefore do not wet it to the same degree as the oil-base products. The large molecular weight resins also make for difficult workability.

Thus, lacquer-base coatings are usually applied by spray. Even in this case, however, the characteristic of difficult workability can be a disadvantage, particularly when combined with relatively fast solvents. Since these coatings easily dry before they reach the surface, overspray and similar application-related difficulties sometimes result.

The most important characteristic of lacquer-base coatings is the ability to form a fast-drying coating over the substrate.

Co-Reactive Coatings

Co-reactive coatings include epoxies and urethanes. One of their principal application characteristics is that since they can leave a relatively high solids content in the film, they can be applied at a greater thickness per coat than is possible with the much higher molecular weight lacquer-type products.

Co-reactive coatings begin at a relatively low molecular weight, and upon application react with a catalyst or other reactive resin which increases the molecular weight and forms the film. The film formation is generally an internal reaction of the coating; therefore, coating thickness is less critical than for coatings which are primarily oxygen reactive. These coatings have good build characteristics, and thick films can be applied which still retain the resistant characteristics of the basic materials.

One of the disadvantages of these coatings is that they are relatively sticky. This means that even though they are of an intermittent molecular weight, they are not easy to brush or sometimes to apply by roller because of the sticky characteristic of the resins.

One factor that must be taken into consideration during the application of co-reactive coatings in their curing. Because of their internal cure and because chemical reactivity is related to temperature, many of these materials must be applied and cured at temperatures above 60 F. Some of the urethane materials may be reacted at lower temperatures. Nevertheless, the curing and surface temperature is something that must be taken into consideration as a possible application problem.

Water-Base or Emulsion-Type Coatings

Water-base or emulsion-type coatings are based on vinyls, acrylics, epoxies, or other materials dispersed in water in the form of an emulsion. Undoubtedly, the primary application characteristic of these materials is that of easy workability. Because of the dispersion of the resin as small individual particles in water, these materials are more thixotropic than viscous. The resin particles are therefore relatively easy to move, making for easy brushing and spraying. Because of their thixotropic nature, these coatings are buttery and have a relatively limited flow. Also, due to the nature of the emulsion, even though the water easily wets the surface, there is no particular penetration of the resin into the underlying surface.

One important factor in the application of this type of coating is its water content. In order for these materials to be effective, the water must evaporate out of the emulsion in a relatively even, regulated fashion. The best film formation occurs at moderate humidities and temperatures. If the humidity is high and the temperature is relatively low, evaporation of the water is extremely slow, which can cause the formation of poor films and poor adhesion. If the temperature is too high and the humidity low, the water evaporates from the coating too rapidly often causing the formation of mud cracks or checks. Temperature and humidity are therefore extremely important in the application of these materials.

Easy workability is undoubtedly the outstanding characteristic of the water-base materials.

Very High or 100% Solids Coatings

Very high or 100% solids materials vary from hot-melt coatings such as asphalts and coal tars, to epoxies and polyurethanes. One of the principal application characteristics of this type of coating is poor wetting. This is primarily due to their high solids content. While some of them may be liquid (*e.g.*, liquid epoxy), they must be catalyzed in order to form a solid coating. The reaction is often rather rapid, requiring mixing of the catalyst and the resin at the time of application and not before. Many application problems arise when materials are catalyzed

at the spray gun and then applied to a surface. Because of the rapid reaction, wetting of the surface is extremely poor and results in a complete lack of adhesion in many cases.

Hot melts, on the other hand, such as the asphalts and coal tars, also have wetting difficulties, since they often are applied as a hot melt to relatively cold surfaces, so that the phase of the coating changes from a liquid to a solid in a very short period of time. Much lower solid and molecular weight primers are usually required along with these materials in order to obtain effective adhesion.

While the hot-melt coatings may be applied by flowing on the surface, or by daubing in the case of hot asphalt or coal tar, most of the high solids coatings are applied by spray. This is the most effective way of distributing the high solids coatings over the surface in a relatively thin, yet continuous film. Some coatings also require in-line heaters to reduce the viscosity sufficiently for easy application.

High solids materials generally have more critical application characteristics than many of the other coating types.

Inorganic Coatings

Inorganic coatings primarily refer to the zinc coatings of either a water or solvent base. The principal application characteristic that must be considered in the use of these materials is their high pigment:vehicle ratio. This characteristic makes it somewhat more difficult for them to flow through hoses and gun orifices, and in certain instances permits the high zinc loading to build up and create uneven flow conditions. Due to the type of vehicle and the high pigment content, these materials do not have the lubricating characteristics of organic coatings, and are therefore much more difficult to apply by brush or roller.

The principal method of application for inorganic zinc coatings is by spray. The inorganic zinc coatings have a base of either water or solvent, which affects application characteristics. The cure is one of the key elements. Water-base materials are not satisfactory under cold, highly humid conditions. Solvent-base materials are difficult to apply under hot, dry conditions because of the moisture cure requirement for the ethyl silicate-base inorganic products. If the humidity is too low, these materials will not cure. Hot, dry conditions also tend to increase the incidence of mud cracking and overspray.

Preparation for Coating Application

Coatings, as they are packaged at the manufacturing plant, are thoroughly dispersed, with the pigments fully suspended and of a uniform consistency in terms of both texture and color. Unfortunately, very few coatings are applied within a short time after manufacture. They may be placed in inventory at the manufacturing plant or sent to a distribution point where they will be held for a period of time. Also, the coating may be purchased several months before its actual use and again stored under any number of different conditions. Thus, coatings generally must be remixed and redispersed prior to actual applications.

A pigment, which is usually heavier than the vehicle,

tends to settle and may even cake in the bottom of the container. Coatings vary to a wide degree in this particular charcteristic. Some may stay suspended for many years; others settle out hard in the bottom of the container. While this makes for difficult redispersion on the job, it does not necessarily mean that the coating is poorly manufactured or that it will be ineffective. On the other hand, coatings which have gelled in the container or in which the pigment *livered* (*i.e.,* became thick and rubbery) are not satisfactory for use and cannot be practically redispersed.

The purpose of remixing and redispersion is to make the coating completely homogeneous, so that upon application the pigment and vehicle can produce the film that was intended by the manufacturer. In certain cases, particularly in oil-type vehicles, there may be skins on the surface of the liquid. These should be removed before redispersion, since they will not redissolve into the vehicle.

Mixing

The mixing process is never particularly easy, even if the coating has not settled hard. It is a process that is often neglected by applicators, particularly in coatings which have settled rather solidly. There have been examples of coatings that were applied with at least half of the pigment remaining in the bottom of the container undispersed and later thrown away with the container. This procedure does not allow for the maximum performance of coating properties and normally leads to rapid coating failure.

Mechanical Mixing

It is always best to use a mechanical mixer of some type, since mechanical mixing always produces a more uniform coating and does so much more rapidly than manual mixing. Manual mixing should only be done under unavoidable circumstances and only in containers with the maximum of a 5-gallon capacity.

There are many types of mechanical mixers available. These range from propeller attachments for hand drills and mechanical vibrating shakers, to large portable mixing units which can be used to mix as much as 55 gallons of coating at one time. Except for the vibrating shakers, most of these attachments are the propeller type that can be driven either by electric motor or by air-powered motors. Even where the coating has settled rather hard, the propeller-type agitater can break it up and redisperse it to a point which is closely equivalent to its original manufacture. Nevertheless, care should be taken in the mixing operation, particularly to ensure that the material on the bottom and lower sides of the container has been well separated from the container and redispersed. Some materials form a soft sediment which clings to both the sides of the container and the bottom, making it necessary to scrape these off before they can be properly dispersed. This is usually a manual operation.

The mixing should be done in such a manner that splashing is avoided. Splashing not only makes for a messy mixing area, but is also a fire hazard to both the person mixing the coating and the surrounding area. The speed of a mechanical mixer should be as low as possible

in order to obtain the redispersion of the pigment in the vehicle. The coating should have a slight vortex at the surface. A large vortex tends to mix air into the coating, which can cause pinholes and air bubbles during application. This is particularly true of latex-type coatings. Air in a thixotropic material such as a latex is extremely difficult to eliminate.

The mixed coating should appear uniform throughout the container, and should not show any separation pigment or color streaks on the surface.

Manual Mixing

If manual mixing is necessary, the liquid portion of the coating should be separated into a clean container. The lower, thicker part of the coating can then be more readily mixed into a heavy paste, including the material which is clinging to the sides of the container. Once the heavier material is mixed into a smooth paste, the remainder of the liquid from the second container can be remixed into the original container with the heavy material, making sure that the two are thoroughly mixed into a uniform coating. One way to do this is to pour the material back and forth between the two containers. This is called *boxing*. The materials should be poured back and forth several times to assure complete uniformity (Figure 10.2).

Two-Component Coatings

In the case of two-component coatings, there are two materials that must be checked to determine whether or not they are properly dispersed prior to being mixed together. Two-component coatings are extremely common at the present time. They include numerous kinds of epoxy coatings, coal tar epoxy coatings, polyurethane coatings, and inorganic zinc coatings. With two-component coatings, it is essential that the two components be separately and thoroughly mixed. Two-component materials are designed to react chemically, so that if they are not thoroughly mixed, the chemical reaction may not take place properly. Mechanical blending of the two components is recommended to obtain a thoroughly mixed product. The two component materials often are different colors so that a satisfactory mix can be readily determined. The fully mixed coating should have a uniform color and consistency.

Mixing Dry Powder and Liquid

The primary purpose in mixing dry powder and liquid components is the use of inorganic zinc coatings. Inorganic zinc coatings are made from a liquid component and dry powdered zinc. The first step is to determine whether or not the liquid component is thoroughly mixed and dispersed to a completely homogeneous liquid. This usually is not difficult since most liquid components are lightly pigmented.

Second, stir the total contents of the powder slowly into the total content of the liquid until it is a well-dispersed, semithixotropic material. In the case of the inorganic zinc coatings, the manufacturers supply the liquid and the powder in two different containers in the exact amount that should be mixed. It is essential that the total

Pour off thin portion into a clean container.

Stir the settled paste, breaking up the lumps, if any. For gallon cans, use a paddle about 1½ inches wide.

Mix thoroughly, using a figure 8 motion. Then follow with a lifting and heating motion.

Continue stirring while gradually returning the poured-off portion to the original container.

Box paint by pouring several times back and forth from one container to the other until uniform.

FIGURE 10.2 — The manual mixing procedure. (SOURCE: Department of Defense, Paints and Protective Coatings, Chap. 4, Army TM5-618.)

powder and total liquid be used in order to obtain the desired final coating. Mixing small portions of zinc and liquid is not recommended, since correct proportions are seldom measured under field conditions.

Materials other than inorganic zinc often are mixed in the reverse order; *i.e.,* a small amount of liquid is added to the powder until a paste is formed, thoroughly dispersing the dry powder with the liquid. The remainder of the liquid is then added to the paste. Some mastics, caulking materials, and floor compounds are mixed in this manner.

Thinning

Most manufacturing procedures today produce coatings with the correct proportion of liquid and pigment so that no thinning is necessary. Therefore, do not thin unless it is recommended by the manufacturer or where local conditions (*e.g.,* cold or hot temperatures) require thinning for proper spray. In this case, the amount of thinning should be only that recommended by the manu-

facturer and using only the manufacturer's thinner for that particular product. Applicators too often think that any thinner will work with any coating. This is not the case, and a number of severe failures resulting from improper thinning have proven it.

The thinner should be of the same composition as the solvents in the coating, which is normally only known by the manufacturer. In one instance, kerosene was added to a vinyl coating. While some kerosene can be added to a vinyl coating without precipitating the resins, once the coating was applied and because the kerosene was the slowest solvent in the system and incompatible with the resins, it was squeezed out of the coating onto the surface being coated. The vinyl coating came off in sheets almost as large as the structure being coated. Other solvent combinations in other coatings may precipitate the resins or may coagulate the pigment so that an uneven coating is applied to the surface.

Do not use just any thinner with any coating, and do not overthin. Thinning reduces the total solids content of the film, and therefore reduces the thickness of the film applied to the surface. If the coating is thinned 50%, then only 50% of the pigment and vehicle will be applied to the surface, and it will be half as thick as the coating should be. Thinning should be discouraged, except where recommended by the manufacturer's representative or in accordance with the manufacturer's instructions.

Straining

Most coatings are thoroughly strained prior to being placed in their container. When the container is opened, if the contents have not settled to a hard deposit in the bottom, straining in the field may not be necessary. On the other hand, if the pigment has settled hard, if used buckets are used in the field for boxing or mixing, if the coating has a skin on the surface, or if the product is a material such as inorganic zinc, straining is recommended. Straining prior to spraying often eliminates considerable downtime due to gun clogging by small particles that restrict the orifice in the gun.

Straining can be done with a fine fly screen or through a nylon stocking. A nylon stocking does not contain any lint and is a very fine mesh that most coating materials can readily pass through. However, the nylon can stretch or become bonded and burst. Mosquito netting or similar materials also are used, although they often contain some lint which can cause problems.

Keeping the material clean in the field is essential, not only to prevent gun stoppages, but also to prevent imperfections in the coating due to particles of dirt that were allowed into the coating material.

Temperature

The temperature of the coating material is a factor that must be considered during any application. Coating materials often are stored in a cold warehouse. As the temperature drops, the coating tends to become thicker, more viscous, and even, in some cases, turn to a gel. If the coating has been stored for a long period of time in cold temperatures, the coating should be brought back to normal temperatures [*i.e.,* somewhere between 65 and 85 F

(18 and 30 C)] at least 24 hours before use. If this is not done and the temperature is not stabilized, then the cold liquid will be sufficiently thick to make it difficult to apply, and it may require excessive thinning in order to be applicable at all. As previously stated, any thinning reduces the solids content of the film, thus reducing its thickness.

High temperatures also can cause difficulties. In some cases, such as with epoxies, it may cause the epoxy materials to react too rapidly and even gel once they are mixed. With some amine-type epoxies, if the liquid is at elevated temperatures and the catalyst is added, they may react violently and boil up out of the container. Other coatings may be reduced in viscosity to the point where they are so liquid that a thin film is applied to the surface or excessive running and curtaining takes place. Thus, the temperature of the coating material should be taken into consideration at the beginning of any application.

Application Methods

There are a number of application methods by which coatings can be applied. The three principal methods are by brush, roller, or spray. Other methods consist of paint pad applicators, paint mitts, electrostatic spray, electrocoating, dipping, and fluidized bed coating. The latter methods, from electrostatic spray on, are primarily for in-plant application and generally are used for product finishes rather than for the field application of coatings.

Of the three principal methods, brushing is the slowest and spraying is by far the fastest. The relative coverage per day is indicated in Table 10.2.

The actual choice of application methods depends on a number of factors. The first is the type of coating. Most oil-base coatings can be easily applied by brush; however, for large, flat surfaces, roller or spray application is much faster. For small and relatively intricate areas, brushing probably would be the recommended method. The repair of small areas or touch-up coating often is done by brush because the areas are small and brushing allows better control of the application. Fast-drying coatings, *e.g.,* lacquers or vinyl lacquers, are difficult to uniformly apply by brush or roller. Thus, spraying is the preferred method for these materials. Slower drying materials usually can be applied by any of the three methods.

The type of surface is also a factor to be taken into consideration. As stated previously, brushing is probably the best method for small areas and odd shapes. Also, if the surface is used and pitted, brushing of the first coat is recommended. Rolling is practical for large, flat areas, *e.g.,* tanks, side walls, or flat tank roofs. However, spraying would be an even faster method in such cases. Spraying usually is preferred on large areas and is not limited to flat surfaces. It can be used on curved surfaces, such as pipe, or irregular surfaces, such as structural steel shapes.

Environmental problems also should be considered. Brushing and rolling can be done in almost any area, since the liquid coating is transferred from the brush or roller directly to the surface. Spraying, however, represents problems with toxic solvents, as well as a possible fire hazard due to fume buildup. Spraying in small, enclosed areas usually is not suggested if either brushing or

TABLE 10.2 — Application Methods

| Method | Relative Coverage Per Day | |
	Sq. Ft.	Sq. Meter
Brush	650	93
Roller	1200-2600	186-372
Air Spray	2000-6000	372-744
Airless Spray	3000-8000	744-1115

(SOURCE: Levinson, Sidney B. and Spindel, Saul, Paint Application, Chapter 5, Steel Structures Painting Manual, Vol. 1, Steel Structures Painting Council, p. 160, 1982.)

rolling can be done. Also, areas close to parked cars or areas where welding is being done should be avoided when spraying. When spraying in the open, overspray may be a problem since slow-drying liquid coatings can be carried for a long distance under windy conditions. Spraying where there are parked cars or where coating contamination may occur on adjacent structures should be limited to fast-drying coatings (*e.g.,* vinyl lacquers), since any overspray would dry before hitting the other surfaces.

Cleanup is also a factor to consider. Cleaning a brush is the least difficult procedure, cleaning a roller is second, and cleanup of spray equipment is the most time-consuming and the most complicated procedure. Where a small area is involved, brushing or rolling would be preferable because of easier cleaning procedures.

Brush Application

Brush application is not an outmoded method of applying coatings. This may seem to be the case because many of the high-performance corrosion-resistant type of coatings are preferably applied by spray. Nevertheless, there are areas and conditions where brushing is a valuable method of application for even these types of materials. In many instances, it is recommended to brush difficult areas, *e.g.,* edges, around rivets, in corners, around boltheads, along welds, and similar areas, prior to the application of a general spray coat. Also, brushing is useful in obtaining improved wetting of primers, particularly on surfaces which are difficult to coat, such as those just mentioned. Brushing is worthwhile for blending in edges and corners prior to spraying the general area. Brush application definitely should not be overlooked in the application of high-performance coatings.

There are a number of types of brushes. The most common type, which is used for structural steel and similar areas, is the conventional wall-type brush, which may vary in width from 3 to 6 inches and with varying lengths of bristles up to 7 inches. Oval brushes often are used for structural steel and marine applications. These can be used for the application of coating around irregular surfaces, such as rivets, boltheads, and similar areas, as well as piping, railings, etc. A list of the various types of brushes and their general uses is given in Table 10.3.

The most commonly used brush is the conventional

TABLE 10.3 — Paint Brushes

Width, in.	Name	Apply	On
6-8	Calcimine, flatting	Water-thinned paint	—
3 1/2-5 (4 in. usual)	Wall	Flat, semigloss, primers, industrial paint	Walls, siding, structural steel
2-3	Enamel, varnish	Enamel, varnish, metallic paint	Woodwork, small areas, trim rolled areas
1-1 1/2	Sash and trim	Enamel, varnish, metallic paint	Narrow woodwork, radiators
1-2	Angular sash	Enamel, varnish, metallic paint	Window sash
No. 2 to No. 12	Oval	Enamel, varnish, metallic paint	Irregular surfaces, piping, railings

(SOURCE: Levinson, Sidney B., Facilities and Plant Engineering Handbook - Painting, McGraw-Hill Book Co., New York, NY.)

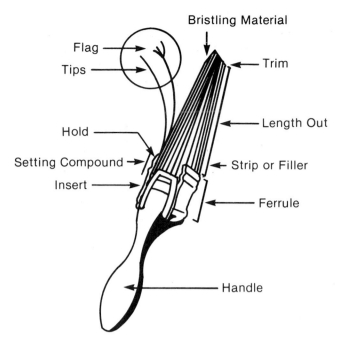

FIGURE 10.3 — Typical standard paint brush. (SOURCE: Department of Defense, Paints and Protective Coatings, Chap. 4, Army TM5-618.)

type with a 4-in. width. Larger sizes can be used, e.g., up to 6 inches. However, these usually are only satisfactory for the most easily worked coatings, such as latex-type wall and masonry coatings. Other materials have sufficient pull so that the 4-in. brush is the optimum for easy handling. Figure 10.3 is a cutaway view of a typical standard type of brush. All good brushes would have this or a similar type of construction.

The flat brush is a new type of brush that generally is made up of short nylon bristles of 1 to 1½ inches long, attached to a flat base of about 4 × 7 inches. This provides much wider coverage on the surface, and this type of brush can hold a considerably larger quantity of coating than the conventional brush. It also generally applies the coating to flat surfaces more rapidly; however, it cannot be used effectively on complex surfaces. It is effective in forcing coating into cracks in wood, into pitted surfaces, and into similar uneven areas on flat surfaces. This brush requires the use of a roller tray in order to fill the brush, since its size does not allow it to be practically placed in a container.

There are generally two types of bristles used in paint brushes. Synthetic fibers are satisfactory for most brushes. However, where they are used in lacquers which contain powerful solvents such as ketones, the synthetic fibers may be affected by the solvents. Nylon is an excellent synthetic fiber. It is not affected by most solvents, although strong lacquers may affect the nylon bristle. Hog bristle used to be the primary material used for paint brushes. It is still the best quality bristle and gives better leveling for most coatings. Hog bristle brushes are expensive, since most of the hog bristle comes from China. They are water sensitive and should not be used for water-thinned coatings, but can be satisfactorily used for lacquers. Nylon can be used for almost all coatings, with the limited exception of very strong solvents. It is less expensive than the hog bristle and is good with water-thinned coatings. Nylon also has excellent abrasion resistance where coatings are applied to rough, uneven steel, concrete, and masonry surfaces.

The proper application of a coating by brush depends on the proper handling of the brush. The brush should be held with the fingers, much like holding a pen-

cil. It should not be held in the fist as with a club. Proper handling techniques, as shown in Figure 10.4, give a lighter brush touch, provide better control, leave fewer brush marks, and cause less fatigue than other brush handling methods.

Brushes should not be dipped deep into the paint. Dipping into the paint approximately 1 inch is adequate for most applications. This keeps the liquid out of the heel of the brush and prevents dripping and running of the coating onto the handle. The proper technique of loading a brush is shown in Figure 10.5.

The coating should be spread over the surface holding the brush at an angle of approximately 45 degrees to the work. The coating should be spread evenly and quickly by a number of light strokes, using the wrist and arm to spread the brush-loaded coating on the surface. Once the coating is evenly spread, the coating should be smoothed by light parallel strokes of the brush over the surface to eliminate any irregularities in the coating. Do not push down on the brush during application; this only makes for hard work and does not improve the brushing procedure. Pushing the brush too hard, especially on the finish coats, can create brush marks which can be focal points for eventual coating failure.

When the next brush load of coating is applied, the final smoothing strokes should be from the latest applied coating into the previous brush load in order to spread the overlap between the two areas of coating and to provide a smooth, uniform coating over the whole surface. Always brush the final strokes from the last application into the previous one, sweeping the coating into the wet edge of the previous application to prevent lap marks. It is often best to use cross strokes for both spreading and smoothing the coating; however, the final strokes should all be parallel and in one direction.

Application of Coatings

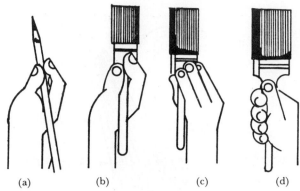

(a)   (b)   (c)   (d)

FIGURE 10.4 — Brush handling techniques: (a) and (b) grasping brush with pencil grip; (c) grip used for painting walls and floors; and (d) simple grip with all fingers around brush handle, suitable for use when painting ceilings. (SOURCE: Department of Defense, Paints and Protective Coatings, Chap. 4, Army TM5-618.)

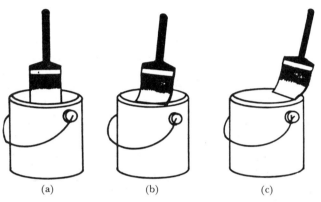

(a)   (b)   (c)

FIGURE 10.5 — Loading a paint brush: (a) The brush should not be dipped into the paint more than half the length of the bristles and (b) excess paint should be removed from brush by gently tapping against the side of the can as shown at left, and (c) not by wiping brush across top of can.

In applying fast-drying coatings (*e.g.,* vinyl lacquers) by brush, the brush should be dipped and filled. The coating should be applied to the surface as rapidly as possible and then left undisturbed. Going back over such coated surfaces only results in piling up the coating and roughening the surface. Use a few rapid strokes to apply the coating to a small area and then leave it alone. Apply the next brush load adjacent to it in the same manner, always brushing the lapped area from the new into the wet edge of the previous brush load. Where there are edges to be coated, brush out over the edge, pulling the coating out over the edge. This helps build the coating on this critical area (Figure 10.6).

Do not brush against an edge, as this pulls the coating away, causing a thin area on the edge. This is particularly true when brushing chlorinated rubber, vinyl, and some lacquer-type coatings, since they are permanently soluble in their own solvents. The rubbing of the brush on the sharp corner will dissolve any previous coat which might be there.

It must be stressed again not to overload the brush. A brush with a heel full of coating, *i.e.,* coating up to the handle, will cause dripping, runs, and continual application problems. A brush which is loaded to the heel is much

FIGURE 10.6 — Brushing out over an edge rather than against it, helps build the coating in the critical area.

more difficult to clean than a brush that is kept properly loaded. When the coating dries in the heel area, the brush loses its flexibility, the bristles tend to stand out, and the brush usually is destroyed.

Roller Application

Rollers are excellent for large, flat areas, *e.g.,* tank tops, decks, ship hulls, side walls of tanks, walls, ceilings, etc., and wherever application does not require the skill needed for brush or spray application. Roller application is more rapid than brush, but not quite as rapid as spray. The roller generally holds considerably more coating than a brush, so that a much larger area can be covered with one roller load of coating.

Rollers are much more satisfactory on smooth surfaces as compared with rougher, difficult-to-wet surfaces such as hand-cleaned, rusted, or pitted steel, or even blasted steel that is badly pitted. Brushing or a good, wet spray coat would be preferred on such surfaces.

The paint roller consists of a simple roller with a cover which slides over the roller mechanism and to which a handle is attached (Figure 10.7). The roller covers are 1½ to 2 inches in diameter, and may vary from a narrow roller to one as much as 18 inches long. The average length is approximately 9 inches, and the cover consists of lamb's wool, mohair, or a synthetic fabric attached to an impregnated paper, plastic, or wire mesh holder. The length of the fiber on the cover or the pile ranges from ¼ to 1¼ inches, depending on the use of the roller. For most smooth surfaces, the shorter fiber is preferred. The very long fiber is only used for very rough surfaces or for coating wire fencing. A very long fiber will hold a great deal of coating. However, for smooth surfaces, it will not apply as smooth a coat as the shorter fiber.

The roller must be used with a tray which holds the coating. The tray has a sloping ramp on which the roller is cleared of excess coating before applying it to the surface. There is also a roller grid, which is used in 5-gallon cans of coating and attaches to the side of the can (Figure 10.8). The roller is worked over this surface in order to remove the excess coating.

Corrosion Prevention by Protective Coatings

FIGURE 10.7 — Industrial paint roller and tray. (Photo courtesy of: Arsco Paint Rollers, Inc.; reprinted from Good Painting Practice, Steel Structures Painting Manual, Vol. 1, Steel Structures Painting Council, Pittsburgh, PA with permission.)

FIGURE 10.8 — Roller and grid screen in a 5-gallon bucket.

Rollers also come with extension handles up to 16 feet in length. Extension handles and poles enable the relatively easy coating of high rank surfaces, walls, and ceilings without the use of ladders. The extension handle with the roller is much easier than using a ladder or scaffold to reach high smooth areas. It is a decided advantage on high broad, flat surfaces such as a ship hull. The types of rollers available are shown in Table 10.4.

Some special roller equipment also is available. For example, a pipe roller is made up of narrow roller segments on a flexible spindle which can be made to conform to curved surfaces such as piping (Figure 10.9). The size of the pipe determines the number of segments. However, only ¼ or ⅓ of the round surface should be coated with a

TABLE 10.4 — Paint Rollers

	Type of Surface		
	Smooth	Medium	Rough, wire fence
Roller Cover:			
Dynel (D)	1/4-3/8 in.	3/8-3/4 in.	1-1 1/4 in.
Lamb's Wool (L)	1/4-3/8 in.	1/2-3/4 in.	1-1 1/4 in.
Mohair (M)	3/16-1/4 in.		
Paint:			
Solvent-Thinned:			
Enamel	D or M	D or L	D or L
Flat	D	D or L	D or L
Oil Paint	D	D or L	D or L
Floor Paint	D or M	D	
Varnish	D or M	D	
Water-Thinned	D	D	D
Lacquer	M	L	L
Two-Package	M	L	L

(SOURCE: Levinson, Sidney B., Facilities and Plant Engineering Handbook—Painting, McGraw-Hill Book Co., New York, NY.)

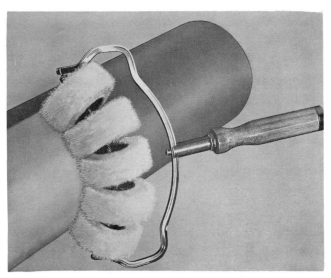

FIGURE 10.9 — Pipe paint roller. (Photo courtesy of: EZ Painter Corp., reprinted from Good Painting Practice, Steel Structures Painting Manual, Vol. 1, Steel Structures Painting Council, Pittsburgh, PA with permission.)

roller at any one time. This is a good method for applying coatings to round or curved surfaces.

There are also rollers for use on fencing. In this case, the roller has a very long nap which surrounds the wire in such a manner that it can coat both sides at the same time. While a wire fence can be brushed, the only practical way of applying the coating to an open mesh fence is by such a fence roller (Figure 10.10).

Pressure rollers also are available. The coating is pressurized in a normal coating container through a hose to the handle of the roller. The amount of coating is controlled by a fingertip valve on the handle, allowing a continuous rolling procedure over very large surfaces.

Use of the roller requires less skill than either brushing or spraying. Nevertheless, rolling should not be considered a poor application method. It can be effectively used with many high-performance coatings, particularly those which have some leeway in drying time so that the roller can always be worked into a wet edge.

FIGURE 10.10 — Showing one of the many ways rollers can be used. In coating chain link fence, material is applied by one roller, in which a second roller is used on the other side to distribute material evenly. (Photo courtesy of: Arsco Paint Rollers, Inc., Hialeah, FL.)

The proper procedure for using the roller is to immerse the roller in the coating tray or bucket. The roller should then be rolled back and forth on the tray ramp or on the grid on the interior of the bucket to remove the excess of coating from the roller so that excessive drip and spatter will not result. Apply the roller to the surface to be coated, and spread the coating liquid in the roller over the surface by forming a "W" or an "M" in the area that the amount of coating in the roller will cover. The amount of area to be covered depends on the coating used. However, generally a 9 × 2 1/2-in. roller with a 3/8-in. nap will cover about 10 sq. ft. After the initial spread of the coating, fill in the area by rolling the roller back and forth over the entire area to be covered. The finished rolling of the coating should all be in one direction so that the appearance will be uniform.

When one block of coating is completed, start another block adjacent to that one and repeat the procedure. Roll the new area into the previous one so that there is a continuous film over the surface. The roller should be rolled evenly, smoothly, and relatively slowly, so that the nap on the roller will not spatter the coating. Spatter is particularly a problem on ceilings and on vertical structures. The finish of the rolled coating will not be as smooth as by brush or by spray because the nap on the roller will make a slightly stippled effect on the surface. The thicker the nap on the roller, the greater the stippling effect on the surface.

A coating with good flow characteristics will flow smoothly. Faster drying coatings will show a greater amount of stipple than thixotropic coatings. It has been stated that fast-drying coatings such as lacquers are difficult to roll. While this usually is true, there has been an instance where the entire hull of a ship was rolled with a vinyl coating, applying three coats to all damaged areas and an overall coat over the entire surface. In this case, the temperature of the steel was somewhat below freezing, and the outdoor temperature was close to freezing. The

surface was perfectly dry and with the cold temperatures, the vinyl coating rolled on the surface very easily, spread evenly, did not become sticky, and produced excellent results for several years following the application. While this is an exceptional case, it does demonstrate the ability of a roller to be used with almost any coating under certain circumstances.

When rolling materials like chlorinated rubber or vinyl, which is permanently soluble in its own solvents, the roller performs much like a brush. The application of the second coat may very well pick up the first coat. In other words, if a white vinyl is to be applied over a red primer, the finish may very well be a pink or a rose-colored topcoat rather than white. Either brushing or rolling tends to do this, which is another reason why spray application of lacquer-type materials is preferable.

Spray Application

The largest volume of all maintenance and high-performance coatings is applied by spray. This method of application is not only faster than any of the alternative methods, but the resulting coating films are more uniform in thickness, more evenly applied, and contain fewer imperfections than with other methods.

The spray method of coating utilizes a stream of highly atomized coating particles which are directed to the surface in a uniform pattern. The particles then flow together to form a continuous and even film. Obtaining the finely atomized spray is the key to spray application, along with directing the spray toward the target so that a continuous film is formed. This is done by essentially two types of equipment: (1) the conventional air spray gun, and (2) the airless spray equipment which uses hydraulic pressure to form the fine mist.

Air Spray

Air spray is a process where compressed air and the coating liquid are brought together in a way that forms a fine spray. The spray gun is the key element in the spray system and is the mechanism which brings the air and the liquid coating together. This is accomplished in such a way that the liquid is broken up into a fine spray, which is also directed toward the surface to be coated by the air stream.

There are two types of spray guns. The first and most common is the external mix gun. In this case, the liquid and the air first contact each other outside of the spray gun. The jets of air impinge on the stream of liquid coating coming out of the gun, dispersing it and forming the fine spray. The air is directed toward the liquid stream through a specially designed air cap, which may have from only two to as many as eight or ten air openings to help break up the liquid stream.

There are two adjustments on most spray guns. One regulates the amount of fluid which passes through the gun when the trigger is operated, and the second controls the amount of air passing through the gun, also actuated by the trigger mechanism. In most production-type spray guns, the trigger operates both the liquid and the air; however, there are some spray guns where the air flows in a continuous stream and only the liquid is controlled. The design of the air cap, with the number of holes involved, determines the fan or the stream of dispersed coating particles which is directed toward the surface.

In conventional air-spray equipment systems compressed air is used to perform two functions. The first, as previously discussed, is to atomize the coating at the tip of the gun. The second is to apply pressure to the liquid coating and force it through the liquid orifice in the gun. The pressure of the air going through the gun, as well as the pressure on the liquid, have a great influence on the type of air spray and the type of fan developed. If there is too little air pressure at the gun, the liquid stream will not be broken up satisfactorily and the spray will merely be spattered when it reaches the surface. This does not provide a satisfactory coating film. If the air pressure on the liquid is too high in comparison with the air pressure at the gun, the same thing can occur.

If the air pressure at the gun is too high in comparison with the liquid pressure, the coating will be over-dispersed, with a great deal of turbulence that will not apply the liquid coating to the surface, but will disperse it into the surrounding atmosphere. Also, if the liquid pressure is too low in comparison with the air pressure, turbulence will also occur, forming an unsatisfactory film on the surface. The latter type of gun adjustment often creates what is called *overspray,* and, with fast-drying coatings, will form a rough, uneven film on the surface. The principle of external air spray is shown in Figure 10.11.

The external type of spray system is one of the simplest and most versatile of all of the spraying units. It consists essentially of an air source applying air at a constant pressure of 100 pounds in a volume of approximately 25 cu. ft./min. The compressed air in the pressure tank for the liquid coating forces the liquid from the tank to the gun. It also supplies the air hose carrying air to the gun, which brings the air and the liquid coating together.

The system has been widely used over the past 50 years. During this time, the gun, nozzles, regulating devices, etc., have been developed and refined to the point of making them practical for almost every conceivable type of coating material and use. The conventional air spray provides a wide variety of spray patterns, degree of atomization, and wetness of the coat applied to the surface. It is therefore the most practical solution to problems where the above factors are required on the job. This type of spray equipment, using the proper liquid orifice and tip, can spray everything from heavy mastics to water-thin coatings by adjusting the variables of air and liquid pressure. Because of the many variables, follow the coating manufacturer's recommendations as to the proper spray gun, air cap, fluid tip, and air and liquid pressure for the application of any specific material. Using the manufacturer's recommendations, the adjustment of the gun for both air and liquid pressure then allows the specific coating to be applied under optimum conditions.

Figure 10.12 shows the application of a heavy epoxy-type mastic coating to the leg of an offshore platform. Note the even, smooth finish obtained even with the heavy coating.

The second method of applying coatings by air spray

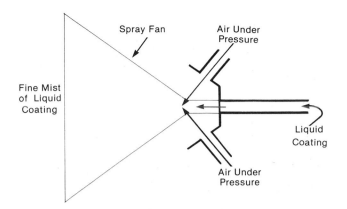

FIGURE 10.11 — The external air spray principle.

FIGURE 10.12 — Spraying a heavy epoxy mastic with a typical air spray pole gun with external atomization.

is by use of the internal air spray gun. In this case, rather than the air impinging on the stream of liquid coating outside the spray gun proper, the air within the spray gun surrounds the liquid coating nozzle and the two are mixed right at the tip of the liquid tube (Figure 10.13). The actual fan, shape, and size is determined by the shape of the air cap. It is claimed by advocates of this air spray principle that a much finer dispersion of the liquid coating can be obtained by this method. While this is true for many coatings, one of the disadvantages of this type of air spray is that, with relatively fast-drying coatings, the coating may build up in the head itself and cause gun stoppages.

Hot Spray

The hot spray procedure is primarily used with conventional spray, although it also can be used with airless spray. With hot spray, the same type of application equipment is used, including the cap and tip of the gun. However, the difference lies in the heating unit, usually located in the feed line between the pressure pot and gun. This can offer several advantages with certain types of coatings and under certain operational conditions.

Most organic coatings will become more fluid at higher temperatures. In other words, as the temperature goes up, the viscosity of the solution goes down, making a viscous coating into one which will spray and be dispersed by the air-spray mechanism. However, it is also possible, using the hot-spray system, to apply higher solids content

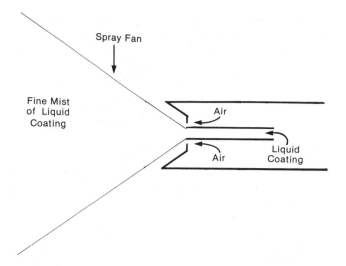

FIGURE 10.13 — The internal air spray principle.

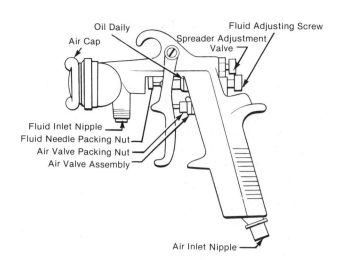

FIGURE 10.14 — A typical air spray gun. (SOURCE: Corr. Control Principles and Methods, Sect. 7, Ameron Inc., Monterey Park, CA.)

organic coatings. This, of course, produces a heavier coating and reduces the amount of solvent that is required to evaporate from the coating. With the heater in the line, the warm liquid coating has better atomization at lower air pressure, and under many circumstances can reduce the amount of overspray produced. Better flow-out of the coating may be obtained, and, under some conditions, pinholing of some coatings is effectively reduced. The temperature of the liquid coating is normally raised to between 130 and 160 F (55 and 70 C), although it may be lower or higher depending on the liquid coating being sprayed.

In general, the advantages of hot spray are as follows:

1. Faster application due to lower viscosity.
2. Lower pressure required.
3. A satisfactory spray at below ambient temperatures.
4. Less overspray.
5. Less solvent fumes, and no added solvent to reduce viscosity.
6. Increased thickness per coat.
7. Somewhat faster drying overall.
8. A possible smoother finish because of faster flow-out of the warm coating.
9. Less atomization air pressure required.

Conventional Spray Equipment

The application of coatings is similar to many other operations, *e.g.,* sandblasting. Many people use the wrong spray equipment and then cannot understand the reason for their poor results. Good spray equipment is like a good tool of any kind; it is essential for the proper application of coatings.

Spray equipment consists primarily of an air compressor, pressure spray pot and fluid hose, spray gun, ample air supply, respirator, air mask, and air transformer. A diagram of a typical conventional air spray gun is given in Figure 10.14.

There are two methods for bringing fluid to the gun: suction- or pressure-feed.

Suction-Feed Gun. For suction-feed, the gun is usually fitted with a 1-quart cup that holds the fluid to be sprayed. When the trigger is pulled, suction is developed at the tip of the gun, drawing fluid out of the cup and up to the nozzle where it is sprayed. This type of equipment has severe limitations; *e.g.,* the gun works best when only pointing horizontally, it will not spray viscous material, spraying is relatively slow, and the cup must be refilled frequently.

Pressure-Feed Gun. With the pressure-feed, the fluid to be sprayed is forced to the gun under pressure. The pressure type has several advantages; *e.g.,* material of higher viscosity can be sprayed, a heavier coat can be applied, and spraying is much faster. Although a 1-quart pressure cup is sometimes used for small applications, the majority of pressure-fed spray operations are conducted with a separate pressure pot. This system insures better pressure control and permits faster spraying.

Pressure Pot. The pressure pot is a closed chamber of a usually 2 to 10-gallon capacity which contains the fluid to be sprayed. A large air hose, preferably 3/4-in. ID or better, connects the pressure pot to the air source. A fluid hose and an air hose connect the pot to the spray gun. Air is directed into the pot through a regulator which maintains the proper pressure upon the fluid. The fluid is forced under pressure out through the fluid hose to the gun. The adjustment of the pressure regulator on the pot determines the amount of fluid available to the spray gun. Air is also bypassed from the main source into the air hose leading to the gun. Figure 10.15 shows the main components of a paint spray pot.

Air and Fluid Hose. It is extremely important that the correct size fluid and air hoses are used. For best results, the gun should be equipped with a 1/2-in. ID fluid hose and a 5/16-in. air hose. Smaller air hoses should be avoided since they cause excessive pressure losses. For example, if a DeVilbiss MBC gun with 765 air cap is connected by 50 feet of 1/4-in. ID hose to an air supply at 90 lbs. pressure, when the trigger is opened, the actual pressure at the gun will drop to 50 lbs. This is not enough

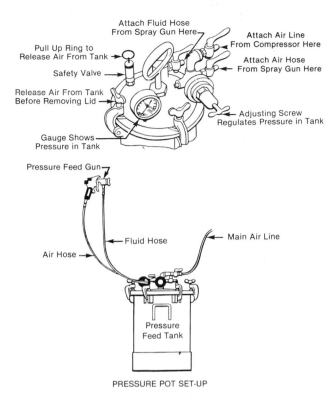

TYPICAL PRESSURE POT FITTINGS

Attach Fluid Hose From Spray Gun Here

Pull Up Ring to Release Air From Tank

Safety Valve

Release Air From Tank Before Removing Lid

Gauge Shows Pressure in Tank

Attach Air Line From Compressor Here

Attach Air Hose From Spray Gun Here

Adjusting Screw Regulates Pressure in Tank

Pressure Feed Gun

Fluid Hose

Air Hose

Main Air Line

Pressure Feed Tank

PRESSURE POT SET-UP

FIGURE 10.15 — The main components of a paint spray pot, (a) pressure pot fittings and (b) pressure pot setup. (SOURCE: Corr. Control Principles and Methods, Sect. 7, Ameron Inc., Monterey Park, CA.)

pressure to spray most coatings. However, when 50 feet of the recommended 5/16-in. air hose is used, under the same conditions, the pressure will only drop to 75 lbs., which is satisfactory in most cases.

It is extremely important from an efficiency standpoint that threads on all hoses, spray guns, regulators, and pressure pots be standardized so that any gun or hose will fit any other gun, hose, or pot on the job. This used to be a greater problem than it is at the present time, although threads of different dimensions may still be found on equipment produced by various manufacturers. This point cannot be overemphasized, since a great deal of time can be lost obtaining and installing proper adapters. All fluid hose should be of the special paint-grade type made for materials containing strong solvents. The powerful solvents used in vinyls will dissolve the lining of some types of hose. This not only leads to coating contamination, but the lining in the hose swells and prevents the proper material flow through the hose. The hose should be made with a working pressure of 150 psi.

It should be emphasized that all material hoses, pots, and guns should be perfectly clean. If one has been used for oil-base coatings or alkyds, it should not be used for lacquers. Lacquer solvent is sufficiently strong that it can dissolve any trapped or remaining oil-base coating. Most epoxy solvents are also strong and may do the same thing. Internally reacted, coal tar epoxy, epoxy, and urethane coatings also can leave residues in the hoses if they are

not properly and promptly cleaned after use. Subsequent use may cause these to separate from the hose and clog the gun or, worse, contaminate the coating. Clean material hoses are essential. In fact, it is recommended that on any large, new application, new hoses be used. This can save enough downtime to more than pay for the new hose.

Water and Oil Separator. A water and oil separator is an essential piece of equipment and may be part of the air transformer. This is installed in the main air line prior to the pressure pot so as to remove all oil, grease, water, and similar contamination that may be carried through the air line from the compressor. There are many coating problems that can result from such contaminating materials getting into the liquid coating; *e.g.,* loss of adhesion, blistering, delamination, cratering, bug eyes, etc.

Air Supply. There should be a compressor capable of supplying a minimum of 25 cu. ft./min. of free air at 100 psi to each spray gun. Most industrial spray guns work well with this quantity of air. If an air mask is used for spraying on the interior of tanks and similar vessels, an additional 5 to 10 cu. ft./min. should be added for each operator.

Spray applications which take place in enclosed space or which involve tank linings may require additional equipment. In confined areas, the worker should wear an air mask such as Devilbiss P-MPH 527 and MPH 529. Forced air, supplied to the mask through a ¼-in. hose and discharged in front of the eyes, provides clear visibility and freedom from fumes.

Miscellaneous Equipment. In confined areas it is essential that fans or blowers be provided to remove solvent fumes. If fumes are allowed to accumulate, they cause acute discomfort to the workers and create a serious fire hazard. For safe application, the concentration of fumes must be kept below 0.05% by volume of air.

To best accomplish this, the blower should suck air from inside and exhaust it outside. Blowing into the tank normally results in stirring up rather than removing the vapors. Since the solvents used in most maintenance coatings are heavier than air, it is necessary to attach a conduit or tube to the blower to draw the vapors up from the bottom of the tank.

When used for continuous immersion, many tank lining materials should be force-dried. Small coated units should be dried in an oven with circulating air at 140 to 200 F (60 to 93 C).

For large tanks, a heater is required to circulate warm air through the tank. One of the most practical systems uses a gas- or butane-fired heater with a blower to force air through the heater and into the tank. Explosion-proof and spark-proof equipment should always be provided for coating operations in enclosed areas. This includes motors, junction boxes, lights, and any other equipment which could generate a spark or flame.

Setting Up Spray Equipment

Cleaning. Spray equipment first must be cleaned with a suitable solvent. If the equipment is not thoroughly cleaned, old, dried paint from the interior of equipment and hoses may be removed by the new solutions, causing gun stoppage and unsatisfactory results. If possible, new fluid hoses should be used.

Adjusting Fluid Pressure. Most manufacturers' application instructions give approximate pressures for spraying a particular coating. However, the exact pressure depends on temperature, size of hose, length of hose, etc. The equipment should be adjusted as follows:

1. Pour the mixed coating into the pressure pot and clamp on the lid.

2. Turn on the air supply to the pot.

3. Turn on the fluid hose to the gun.

4. Turn off the air hose to the gun.

5. Remove the air cap from the spray gun.

6. Open the fluid-adjusting screw on the spray gun several turns and pull the trigger all the way back.

7. Holding the gun waist high, pointing out in a horizontal position, screw the pressure regulator on the pot until the fluid is forced out in a stream that travels 4 to 6 feet before hitting the ground. This is the proper fluid pressure.

8. Replace air cap.

Adjusting Air Pressure. The air pressure at the gun is the pressure of the air source. It is not controlled by the fluid pressure regulator on the pot. The pressure at the source, *i.e.*, line pressure, usually should be 90 to 100 psi. (Using 50 feet of 5/16-in. air hose to the gun, this will result in about 75 to 80 psi at the gun.) After fluid pressure is adjusted as described above, test for proper air pressure as follows:

1. Turn on air to the gun (fluid also turned on).

2. Pull trigger back.

3. Adjust fan to about 6 to 10 inches by turning fan adjustment screw on gun.

4. Spray a test pattern, holding gun about 10 inches from surface, and make a rapid pass to produce a thin coat. If droplets of liquid hitting the surface appear to be less than 1/16 in. across, air pressure is adequate. If coating is spattered with drops appearing 1/8 in. or larger, coating is not being atomized and air pressure must be increased.

Adjustment at Gun. Adjust the fan to produce a spray pattern 6 to 10 inches wide. Adjust the fluid control at the gun so that a smooth, wet film is produced on the surface when the gun is moved at the desired speed. Figure 10.16 shows some typical fan adjustments possible with standard air spray guns.

Spraying Procedure

The most important point to keep in mind in spraying high-performance coatings of almost any type, particularly the relatively fast-drying synthetic resin-based materials, is that they must be applied in an even, heavy, wet film. The film should not be so heavy that it will run on a vertical surface. However, if the coating is allowed to dry before it hits the surface or thin spots occur in the film, the protective value of the coating is greatly reduced. The proper spraying technique is essential to the success of the job, and the following points should be observed when using air-spray equipment.

1. Hold the gun not more than 10 inches from surface.

2. Use a normal fan, as shown in Figure 10.17. It is

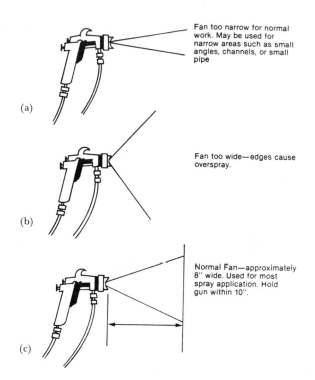

FIGURE 10.16 — Typical fan adjustments for air spray guns: (a) fan too narrow for normal work, may be used for narrow areas such as small angles, channels, or small pipe; (b) fan too wide as the edges cause overspray; and (c) normal fan is approximately 8 in. wide with the gun held within 10 in. of surface and is used for most spray applications. (**SOURCE:** Corr. Control Principles and Methods, Sect. 7, Ameron Inc., Monterey Park, CA.)

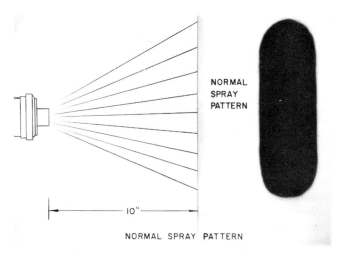

FIGURE 10.17 — A normal fan and spray pattern.

possible to check this by holding the gun the proper distance from the surface and pulling the trigger quickly. This will leave a spot on the vertical surface which indicates the type of fan you will obtain with the gun setting. Figure 10.17 shows the normal fan. With a normal spray pattern, the spray area will be an even, smooth oblong with no heavy center or dry spray at the ends.

3. Figure 10.18 shows the type of fan you would obtain by the above procedure if the pressure on the liquid is too high and the air pressure too low. This makes a heavy

liquid cream in the center of the fan with too little liquid at the ends. Air may be blown into this heavy liquid area, creating bubbles which become pinholes upon drying. This type of gun adjustment generally will cause pinholes in the wide area of the fan.

Figure 10.19 shows the type of fan which is obtained when the gun adjustment is such that the liquid pressure is too low and the air pressure is too high. The high air pressure tends to split the fan, causing a thin center and dry spray at the ends. The liquid-starved spray will cause overspray with fast-drying coatings and a spatter coat with relatively slow-drying materials.

5. Figure 10.20 demonstrates the type of fan which may be obtained if the gun has a problem. This might be caused by a nick in the needle seat, partially clogged tip, or the tip of the needle could be bent slightly. In this case, the fan is teardrop-shaped, with too little liquid at the top which may cause overspray, and too much liquid at the bottom of the fan which may cause running or curtaining. If such a fan is obtained, the gun should be thoroughly inspected to determine the difficulty.

6. The gun should be pointed perpendicular to the surface. The tip of the gun should not be pointed either up or down from the perpendicular angle, as the spray will again be uneven (Figure 10.21). In that case, the shape of the fan would be heavy at one end and very light at the other end, resulting in either spatter or overspray at one end and running at the other.

7. Keep the gun parallel to the surface at all times. *Do not swing the gun in an arc or shoot it off at an angle at the end of a pass.* The gun should be kept an even distance from the surface, so that the fan is maintained at an optimum width. Work should be done with straight, uniform strokes, moving backward and forward across the surface so that the spray pattern will overlap the previous pass by 50%. Overlapping each pass is essential in preventing holidays and missed areas.

The preferred method of spraying is to cover an area approximately 3 to 4 feet wide with one pass, triggering the gun at the start and the end of each pass. Do not hold the trigger open at the end of the pass and then make a return pass. This builds up heavy coating at the end of the pass, which causes uneven coating thickness and runs and other coating problems in a heavy area. Let up on the trigger of the gun at the end of the pass; however, continue to go past that point with your stroke. In returning,

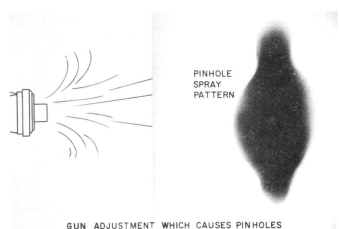

GUN ADJUSTMENT WHICH CAUSES PINHOLES

FIGURE 10.18 — A gun adjustment where the liquid pressure is too high and the air pressure is too low causes pinholes.

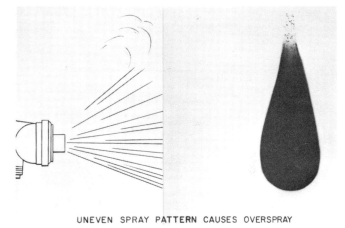

UNEVEN SPRAY PATTERN CAUSES OVERSPRAY

FIGURE 10.20 — An uneven spray pattern caused by gun difficulties results in overspray.

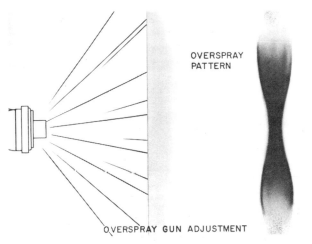

OVERSPRAY GUN ADJUSTMENT

FIGURE 10.19 — An overspray gun adjustment where the liquid pressure is too low and the air pressure is too high.

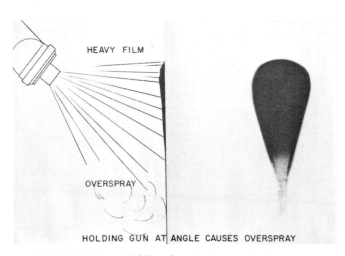

HOLDING GUN AT ANGLE CAUSES OVERSPRAY

FIGURE 10.21 — Holding the gun at an angle causes overspray.

Application of Coatings

261

trigger the gun at the same point that the previous spray pattern was stopped. Continue this type of operation until a practical area has been blocked out.

Correct and incorrect movement of the gun is shown in Figure 10.22. Figure 10.23 shows the proper procedure to use in the starting and stopping of the stroke and in overlapping each pass. Figure 10.24 shows the procedure to use in blocking out sections and in joining sections where a large areas is being sprayed.

Banding is a method often used by some spray operators to block out the sections on which they are working. Some believe that it reduces overspray at the end of each stroke. This method is demonstrated in Figure 10.25. The single vertical stroke at the end of each panel assures complete coverage and eliminates the waste of coating which results from trying to spray right up to a vertical edge. During the usual horizontal stroke at the top and bottom of the panel, the gun is aimed at the edge and is automatically a banding stroke. This procedure is more often used in spraying smaller surfaces than it is where there are a number of sections to coat on a very large surface.

8. Apply an extra coat to all corners, welds, rivets, bolts, or other sharp edges. This is similar to banding in that the edges are given a pass prior to the body of the surface. This is important in order to assure full coverage of the above areas. This is demonstrated on an edge in

Figure 10.26. As shown in the "wrong" procedure in the figure, the spray pattern is tangent to the edge of the corner and there is no direct spray impingement on the corner using such a procedure. The proper procedure is to point the spray gun straight at the corner and make a stroke covering not only the corner, but a small area on each side of the corner. When the body of the surface is then sprayed, each pass is carried onto the previous corner area.

An inside corner is sprayed in a similar way, spraying directly into the corner and moving the spray gun linearly with the corner. While this produces a somewhat heavier coat on the sides of the corner, for most work it is a satisfactory method of covering a corner area (Figure 10.27).

A second method is to apply a band of coating linearly with the corner, not spraying directly into the corner. In this case, a band is sprayed on each side of the corner,

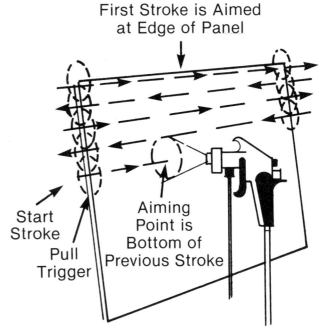

FIGURE 10.23 — When spraying a panel use alternate right and left strokes, triggering the gun at the beginning and end of each stroke. The spray pattern should overlap one-half the previous stroke for smooth coverage without strokes. (SOURCE: Industrial Maintenance Painting, National Association of Corrosion Engineers, Houston, TX, p. 88, 1973.)

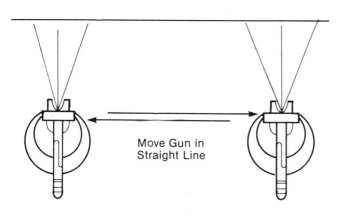

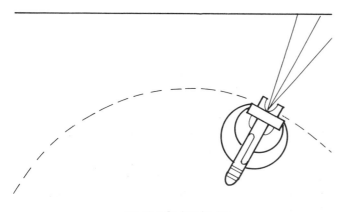

Do Not Swing in Arc

FIGURE 10.22 — Proper and improper gun movement. (SOURCE: Corr. Control Principles and Methods, Sect. 7, Ameron Inc., Monterey Park, CA.)

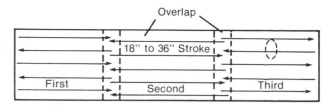

FIGURE 10.24 — Long work is sprayed in sections of convenient length, each section overlapping the previous section by 4 inches. (SOURCE: Industrial Maintenance Painting, National Association of Corrosion Engineers, Houston, TX, p. 88, 1973.)

Corrosion Prevention by Protective Coatings

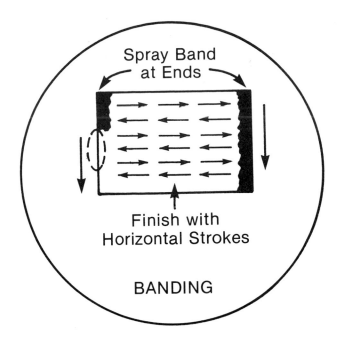

FIGURE 10.25 — Vertical bands sprayed at the ends of a panel prevent over-spray from horizontal strokes. (SOURCE: Industrial Maintenance Painting, National Association of Corrosion Engineers, Houston, TX, p. 88, 1973.)

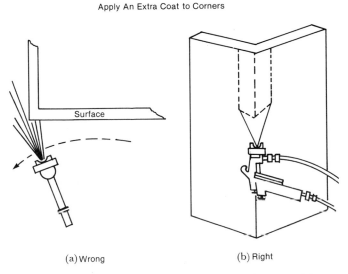

FIGURE 10.26 — Applying an extra coat to corners: (a) wrong procedure and (b) right procedure. (SOURCE: Corr. Control Principles and Methods, Sect. 7, Ameron Inc., Monterey Park, CA.)

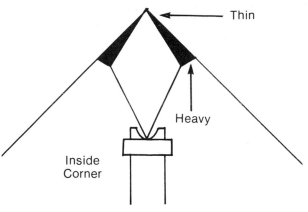

FIGURE 10.27 — Spraying directly into the corner gives an uneven coating but is satisfactory for most work. (SOURCE: Industrial Maintenance Painting, National Association of Corrosion Engineers, Houston, TX, p. 88, 1973.)

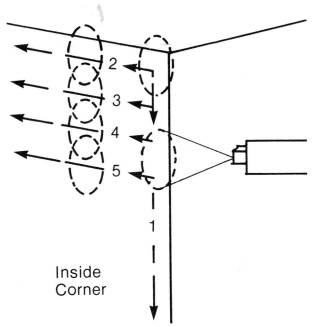

FIGURE 10.28 — Spraying each side of the corner separately gives an even coating. Use a vertical spray pattern. (SOURCE: Industrial Maintenance Painting, National Association of Corrosion Engineers, Houston, TX, p. 88, 1973.)

overlapping into the corner in order to fully cover it (Figure 10.28).

In either case, the spray passes which are used for the adjacent area to the corner should be triggered at the corner, carrying the coating into the previously coated area of the corner.

Spraying of a corner, followed by spraying of the flat area adjacent to the corner, is shown in Figure 10.29. In this photograph, an H-beam is being coated. Note the first pass of coating in the corner, with the second on the flat area overlapping the original corner spray. Such a procedure insures full coverage of the area.

Cross Spraying. *Cross spraying* is a technique which is used by many applicators to obtain a coating with an even thickness and no holidays. The procedure involves applying a relatively thin spray pass over a blocked-out area, as described earlier, and then applying a second pass over the same area at right angles to the first pass. Each pass should be properly lapped (50% overlap) and should fully cover the blocked-out area. It can easily be seen that when the second pass is applied at right angles, a full, even coverage of the area results (Figure 10.30). Cross spraying is a good technique to use with high-performance coatings, since continuous holiday-free films are necessary for full corrosion protection. A properly applied cross-spray coat will result in an even film which is not necessarily overly thick.

Round Work. It is generally assumed that cylindrical objects, *e.g.,* pipe or tanks, are the easiest type of surface to coat because they are planar surfaces with no sharp edges or corners. While this is true for large tanks, it is not necessarily true for smaller surfaces. For instance, small-diameter pipe usually must be sprayed in a linear fashion, overlapping each pass at least 50%. This is necessary because a considerable amount of the spray is tangent to the round surface, and therefore either passes out into the air or does not impact the surface directly (Figure 10.31). It is these areas which cause holidays and thin spots in the coating, and are the reason why care must be taken in overlapping each spray pass.

Lack of proper overlapping of each spray pass is the cause of many coating failures on pipe. The holidays are usually linear with the pipe, indicating that the coating had not been overlapped sufficiently during application. Such a holiday is shown in Figure 10.32. The area of the holiday did not receive any direct spray, only coating applied at a very long angle or tangent to the pipe surface.

The diameter of the pipe will determine the number of passes necessary to fully cover it. It must be kept in mind, however, that even though the surface may appear covered, tangent spray is not a continuous film and corrosion will soon begin in these areas. Figure 10.33 shows the proper procedure for coating cylindrical objects.

The coating of handrails, uprights on handrail grills, and similar slender-type work is difficult at best. The rule in this case is to make sure the spray pattern fits the job, by the adjustment of the fan on the gun. The gun fan may be adjusted from a very slender, almost cylindrical pattern to a very wide one. A wide fan should not be used. The objective is to cover the narrow work without excessive overspray and loss of material. Such work can be

FIGURE 10.29 — Coating an H-beam using air spray.

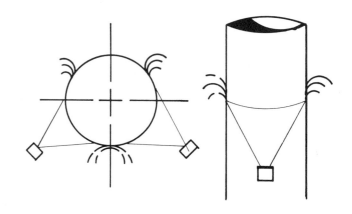

FIGURE 10.31 — Tangent spray can cause holidays in pipe coatings.

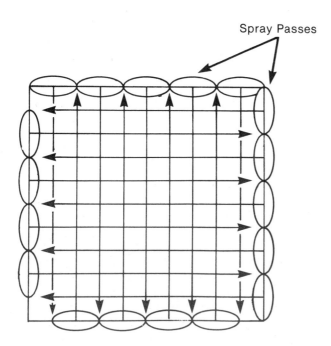

FIGURE 10.30 — Cross spraying a blocked-out area with spray passes at right angles.

FIGURE 10.32 — A holiday caused by improper overlapping of the spray pattern on a cylindrical surface.

Corrosion Prevention by Protective Coatings

done by the use of a brush; however, in most circumstances it is preferable to apply coatings to such work by spray because there is a better chance of fully covering the edges, corners, and similar areas than there is by other means. Figure 10.34 provides an example of the way such small work should be coated.

Open Grill Work. It is often necessary to apply coatings to open grill work (*e.g.,* steps, platforms, and walkways), due to the failure of galvanizing in an industrial or marine atmosphere. This is a difficult job and must be done carefully if the coating is to be successful. Figure 10.35 illustrates an open-work step which is typical of this type of application. Not only is the application of the coating difficult, but the surface preparation is difficult as well. Such a grill must be coated from at least four different angles, *i.e.,* two from the top and two from the bottom, in order to fully cover the surface. Figure 10.36 shows an open-work bridge deck which was coated with

one coat of inorganic zinc and showed no corrosion after 10 years of service.

Expanded metal is used extensively in industry and is also difficult to coat due to the many metal directions and sharp edges. In this case, coating applications must be made from several different directions. Figure 10.37 shows expanded metal partially coated with a high-build vinyl. This example shows good coverage of the sharp edges and corners by the heavy-bodied material.

Figure 10.38 shows the spraying of two types of open work. Note the angles of the spray used to cover as much area as possible. The same surface must be coated from the opposite angle, as well as from two angles on the opposite side of the work. Note also the shield used to produce spray bounce-back and reduce overspray.

Each application is unique, and the spray applicator must adjust his gun for each application in a way which will cover the work in the most efficient fashion and in a

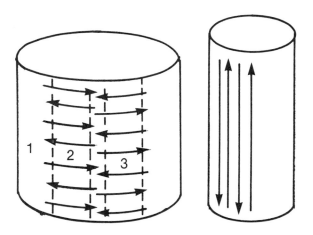

FIGURE 10.33 — Round work should be treated to curve the strokes to conform with the surface. (SOURCE: Industrial Maintenance Painting, National Association of Corrosion Engineers, Houston, TX, p. 92, 1973.)

FIGURE 10.35 — Typical open work grating step on an industrial structure. Note the depth of the grating and the difficulty of obtaining full coating coverage.

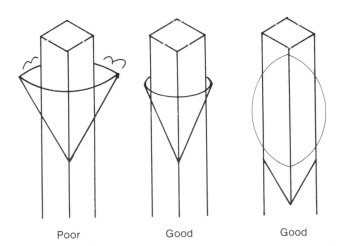

Poor   Good   Good

FIGURE 10.34 — Procedure for coating slender surfaces. Avoid excessive overspray. Adjust the spray pattern to fit the job. The center illustration shows the best method for most work. (SOURCE: Industrial Maintenance Painting, National Association of Corrosion Engineers, Houston, TX, p. 91, 1973.)

FIGURE 10.36 — Open work grating used as bridge deck.

Application of Coatings

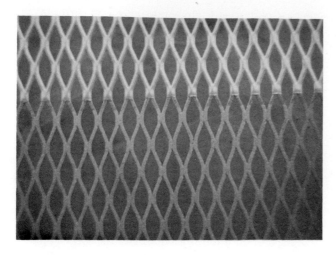

FIGURE 10.37 — Expanded metal partially coated with a high build vinyl.

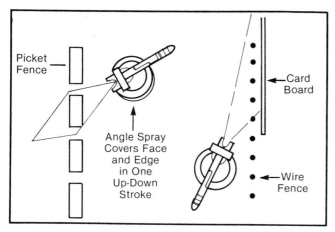

FIGURE 10.38 — The spraying of two types of open work from four different angles (two on each side) to cover as much area as possible. (SOURCE: Corrosion Prevention by Coatings, Course 5, National Association of Corrosion Engineers, Houston, TX.)

way that will best eliminate pinholes, holidays, and coating loss due to overspray.

Spray Difficulties

There are a number of difficulties involved with an air-spray setup. The primary cause of most difficulties is dirty equipment and poor equipment maintenance. The Steel Structures Painting Council provides a list of many of the difficulties and their remedies, which is given in Table 10.5.

Airless (Hydraulic) Spray

The word *airless* is used in connection with this type of spray equipment since there is no air used in the atomization of the liquid coating. The liquid is forced through a very fine orifice by hydraulic pressure and forms a spray pattern as it expands into the atmosphere. The combination of the pressure and the small orifice breaks the liquid coating up into a very fine mist which is forced onto the surface to be coated. In this way, its operation is similar to that of a garden sprayer. In a garden sprayer, the spray is

of very low viscosity and therefore does not require much pressure. Pressure is built up within the container by a simple hand pump. The fan of the atomized liquid may be anything from a cone to a flat fan, depending on the orifice. The process is extremely simple and quite effective.

The gun for airless spraying is a simple device, consisting of not much more than a handle used to direct the spray in the proper direction. As previously mentioned, the fan is developed by the shape of the orifice through which the high-pressure liquid is directed. This is the most critical part of the airless spray gun, and the spray tips must be carefully made and machined to obtain the proper expansion of the liquid and the formation of the fan.

There are essentially no adjustments to be made on an airless spray gun. In order to change the fan size or shape, it is necessary to change the orifice tip. It is also necessary to change the orifice. depending on the type of coating that is being sprayed. Very thin liquid will require a very small orifice, while a heavy, viscous liquid will require a larger opening.

Because of the adjustments needed due to the use of different tips, a number of tips are essential. These vary from an approximately 0.010-in. opening, up to one which may be over 0.040 inches. The larger the orifice, the greater the flow of material and the greater the speed and coverage that can be obtained. On the other hand, if a relatively large opening is used with a low-viscosity material, the rate of discharge may be such that it is impossible to keep up with the speed of the flow. As a general rule, the best orifice is the smallest orifice that will allow the coating material to pass through without plugging.

As with the garden sprayer, a number of different spray patterns also can be obtained. These vary from almost cone-shaped patterns to fans with various widths. These can vary from approximately a 25-degree angle up to a 65-degree angle. The finer angles would be used for narrower work, such as on relatively small dimensional structural steel, while the wide patterns would be used for large, broad, flat surfaces. Figure 10.39 provides a schematic drawing of the airless spray gun tip and its operation.

Pressure Pumps

In order to supply the pressure on the liquid, hydraulic-type pressure pumps are used and may be operated by either air or an electric motor. A small gasoline-driven compressor can be used for portable operations. Most of the hydraulic pumps are actuated by an air cylinder. These have an air cylinder to hydraulic cylinder ratio of from 8:1 up to about 45:1. The hydraulic pressure can be regulated by the pressure on the air cylinder. Generally, it is better to have a pump with a higher ratio than required, rather than one with a ratio which is too low. This seems rather obvious, since too low a pressure would not provide the liquid dispersion required for proper application. On the other hand, the lowest air pressure into the pump which will produce the pressure required to form a uniform pattern is all that is required. Higher

Corrosion Prevention by Protective Coatings

TABLE 10.5 — Air Spray Painting Faults and How to Remedy Them

Trouble	Possible Causes	Suggested Remedies
Sags	1. Dirty air cap and fluid tip (distorted spray pattern).	1. Remove air cap and clean tip and air cap carefully.
	2. Gun stroked too close to the surface.	2. Stroke the gun 6 to 10 in. from surface.
	3. Trigger not released at end of stroke (when stroke does not go beyond object).	3. Operator should release the trigger after every stroke.
	4. Gun stroked at wrong angle to surface.	4. Gun should be stroked at right angles to surface.
	5. Paint too cold.	5. Heat paint in an approved paint heater.
	6. Paint piled on too heavy.	6. Learn to calculate depth of wet film of paint.
	7. Paint thinned out too much.	7. Add the correct amount of solvent by measure.
Streaks	1. Dirty air cap and fluid tip (distorted spray pattern).	1. Remove air cap and clean tip and air cap carefully.
	2. Insufficient or incorrect overlapping of strokes.	2. Follow the previous stroke accurately. Deposit a wet coat.
	3. Gun stroked too rapidly ("dusting" of the paint).	3. Avoid "whipping." Take deliberate slow stroke.
	4. Gun stroked at wrong angle to surface.	4. Gun should be stroked at right angles to surface.
	5. Stroking too far from surface.	5. Stroke 6 to 10 in. from surface.
	6. Too much air pressure.	6. Use least air pressure necessary.
	7. Split spray.	7. Clean the fluid tip and air cap.
	8. Paint too cold.	8. Heat paint to get good flow out.
"Orange-Peel"	1. Paint not thinned out sufficiently.	1. Add the correct amount of solvent by measure.
	2. Paint too cold.	2. Heat paint to get good flow out.
	3. Not depositing a wet coat.	3. Check solvent. Use correct speed and overlap of stroke.
	4. Gun stroked too rapidly ("dusting" the paint).	4. Avoid "whipping." Take deliberate slow strokes.
	5. Insufficient air pressure.	5. Increase air pressure or reduce fluid pressure.
	6. Using wrong air cap or fluid nozzle.	6. Select correct air cap and nozzle for the material and feed.
	7. Gun stroked too far from surface.	7. Stroke the gun 6 to 10 in. from surface.
	8. Overspray striking a previously sprayed surface.	8. Spray detail parts first. End with a wet coat.
Excessive Paint Loss	1. Not "triggering" the gun at each stroke.	1. It should be a habit to release trigger after every stroke.
	2. Stroking at wrong angle to surface.	2. Gun should be stroked at right angles to surface.
	3. Stroking gun too far from the surface.	3. Stroke the gun 6 to 10 in. from the surface.
	4. Wrong air cap or fluid tip.	4. Ascertain and use correct setup.
	5. Depositing a paint film of irregular thickness.	5. Learn to calculate the depth of wet film of finish.
	6. Air pressure too high.	6. Use the least amount of air necessary.
	7. Fluid pressure too high.	7. Reduce pressure. If pressure keeps climbing, clean regulator on pressure tank.
	8. Paint too cold.	8. Heat paint to reduce air pressure.
Excessive Spray Fog	1. Too high air pressure.	1. Use least amount of compressed air necessary.
	2. Spraying past surface of the product.	2. Release trigger when gun passes target.
	3. Wrong air cap or fluid tip.	3. Ascertain and use correct setup.
	4. Gun stroked too far from the surface.	4. Stroke the gun 6 to 10 in. from surface.
	5. Material thinned out too much.	5. Add the correct amount of solvent by measure.
	6. Too little material at gun.	6. Adjust pot pressure or gun or both.
Paint Won't Come From Spray Gun	1. Out of paint (gun begins to sputter).	1. Add paint, correctly thinned out and strained.
	2. Settled, caked, pigment blocking gun tip.	2. Remove obstruction; stir paint thoroughly.
	3. Grit, dirt, paint skins, etc., blocking gun tip, fluid valve, or strainer.	3. Clean your spray gun thoroughly, and strain the paint. Always strain paint before using it.
Paint Won't Come From Pressure Tank	1. Lack of proper air pressure in the pressure tank.	1. Check for leaks or lack of air entry.
	2. Air intake opening, inside of pressure tank lid, clogged by dried-up paint.	2. This is a common trouble. Clean the opening periodically.
	3. Leaking gaskets on tank cover.	3. Replace with a new gasket.
Paint Won't Come From Suction Cup	1. Dirty fluid tip and air cap.	1. Remove air cap and clean tip and air cap carefully.
	2. Clogged air vent on cup cover.	2. Remove the obstruction.
	3. You may be using the wrong air cap.	3. Ascertain and use correct setup.
	4. Leaky connections on fluid tube or nozzle.	4. Check for leaks under water, and repair.
Gun Sputters Constantly	1. Fluid nozzle not tightened to spray gun.	1. Tighten securely, using a good gasket.
	2. Leaky connection on fluid tube or needle packing (suction gun).	2. Tighten connections; lubricate packing.
	3. Fluid pipe not tightened to the pressure tank lid.	3. Tighten. Check for defective threads.
Paint Leaks From Spray Gun	1. Fluid needle packing nut too tight.	1. Loosen nut, lubricate packing.
	2. Packing for fluid needle dry.	2. Lubricate this part daily.
	3. Foreign particle blocks fluid up.	3. Remove tip and clean.
	4. Damaged fluid tip or needle.	4. Replace both tip and needle.

(SOURCE: Steel Structures Painting Council, Good Painting Practice, Steel Structures Painting Manual, Vol. 1, p. 106, 1966.)

pressure is no advantage. Insufficient pressure causes streaks or fingers in the fan pattern. Adjustment of the air pressure usually eliminates these streaks or fingers. When the correct operating pressure for the particular coating liquid is obtained, the spray pattern evens out the smooth fan.

A pump with a ratio of 20:1, 28:1, or higher is commonly used to apply the pressure required. With air pressure of 100 psi, a 28:1 pump would provide a pressure on the liquid coating material of 2800 psi. If the ratio was 20:1, an air pressure of approximately 50 psi would provide 1000 psi pressure on the coating liquid. The pump pressure must be regulated to meet the application characteristics of the various materials, even as the fluid tip is selected for the material being applied.

Figure 10.40 demonstrates a properly adjusted airless spray gun. In this operation, a 100% solids epoxy coating was being sprayed. Note the even spray pattern with no fingers or streaks in the fan. Note also the distance of the spray gun from the surface (12 to 14 in-

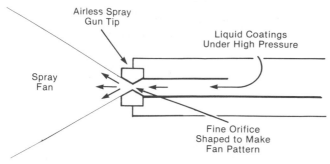

FIGURE 10.39 — Operation of an airless spray gun tip.

FIGURE 10.41 — A heavy-duty airless pump on wheels.

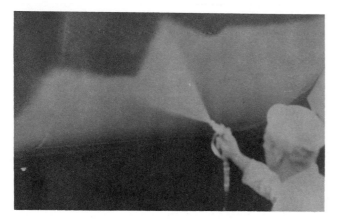

FIGURE 10.40 — Airless spray application with proper spray width and gun distance from the surface.

ches) and the width of the fan. Both are optimum for the coating being sprayed.

The hydraulic pump used to pressurize the liquid is a mechanism with two cylinders. The air actuates the large cylinder and applies pressure to the smaller cylinder which pumps the liquid coating under the higher pressure. The ratio of the air pressure to the fluid pressure is approximately the inverse ratio of the square of the respective cylinder diameters.

Hydraulic pumps vary in size from ones which can be used with a gallon of liquid, to those which can be used in very large, centralized coating areas, with the pumps operating out of 50-gallon drums or even larger containers. Some pumps can fit directly on the top of a 5-gallon can; others come equipped with wheels, so that a container of liquid is placed on the wheeled vehicle, with the hydraulic pump dropping down into the liquid. This makes for a very portable unit (Figure 10.41).

The spray gun and pump are the two most important items in airless spray equipment. The third is the high-pressure fluid line. The hose is usually fluorcarbon or nylon lined, which will withstand a working pressure of 2500 psi or more. Fluid hose is usually a 1/4 or 3/8-in. ID hose and, if lined with a fluorcarbon, has a very low coefficient of friction. Generally, any air source may be used for the air-actuated hydraulic pumps. A small volume of air is required to operate the pump, and a 3/8 or 1/2-in. hose is all that is required from the air compressor to the liquid pump. Table 10.6 provides a list of operating

equipment, sizes, and conditions for two types of organic coatings.

An essential element in airless spraying is to have not only scrupulously clean equipment, but the coating liquid completely free of any small particles which might clog the tiny spray gun orifice. The clogging of spray gun openings can cause a great deal of lost time and annoyance. Thus, any material used should be thoroughly screened or filtered to eliminate this kind of problem. It occurs often enough to make having several extra nozzles or tips on hand a good practice, so that they can be changed quickly when they become clogged. There are reversible tips which can be turned around and cleaned by blowing the fluid through the tip in a reverse manner, forcing out the clogging material.

Cleaning the gun also can be hazardous due to the very high pressure of the liquid. If the spray gun is close to a person's body and is accidently triggered, liquid coating can actually be forced into the flesh of the arm or hand. This has happened on some occasions. The safety precautions necessary for the operation of the gun may easily be forgotten while attempting to clean out small particles of contamination.

While the actual operation of an airless spray gun is similar to that of a conventional gun, there are some important differences. The principal one is that there is no air sprayed with the coating in airless spray equipment. The high pressure atomizes the coating, and there is no compressed air involved to cause air rebound from the surface to which the coating is applied. This aids materially in coating the interior of corners, corrosion pits, angles, and many other complex areas found where structural steel is fabricated. Since there is no compressed air to cause a cushion in these areas, the coating tends not to rebound, but stays on the surface.

This is a decided advantage, since a properly operated airless spray gun can apply the coating without bubbles or pinholes and yet obtain a smooth, wet film. In order to do this, however, the spray gun must be held some distance from the surface, *i.e.*, several inches farther from the surface than conventional spray guns, so that the coating is applied to the surface with maximum dispersion. If the airless spray gun is held too close to the surface, too much liquid coating is applied, causing runs and, in many cases, bubbling and pinholing. The dis-

Corrosion Prevention by Protective Coatings

TABLE 10.6 — Typical Airless Spray Equipment and Operating Conditions

	Fluid Pump	Gun Orifice Size	Foot Valve Filter	Fluid Pressure psi	Fluid Hose I.D.	Air Hose I.D.	Spray Pattern Width (inches)	Gun Distance from Surface (rushes)
High-Viscosity Coatings	30-1	0.019 to 0.026	None to Course Mesh	2400 to 3200	1/4 in.	1/2 in.	10-16	14-16
Low-Medium High-Viscosity Coatings	30-1	0.015 to 0.026	#50 Mesh	2400 to 3200	1/4 in.	1/2 in.	10-16	12-14

tance may vary with the coating being applied; however, it generally should be approximately 14 to 16 inches (Figure 10.40).

Overspray is also reduced with airless spray, since there is no air intermixed with the coating as it is being applied. The volume of coating applied to the surface also can be somewhat greater with airless spray equipment. The spray is designed for the production application of coatings to large areas, so that a higher liquid flow is an advantage. For large surfaces, airless spray is generally faster, cleaner, and more economical than conventional air spray.

Figures 10.42 and 10.43 show the advantages of airless spray, e.g., a high volume of materials going on a very large surface. These figures also demonstrate poor coating technique. The gun is too far from the surface, with most of the coating never reaching the surface and falling onto the dock. (Note the red coloration of the dock adjacent to the spray operation.) Even though the application was improper, this project could not have been done at all without airless spray equipment.

While most coating materials can be applied by either air or airless spray, some coating materials are specially formulated for airless application.

In airless spray, more coating generally is applied to the surface and less lost into the surrounding air. Solvent consumption is also generally reduced because higher viscosity coating materials can be applied by airless spray. These usually are high solids content materials with a reduced amount of solvent as compared with conventional air-spray materials. In many cases, thicker film builds can be obtained by airless application. This depends on the material being applied. However, if the coating does not run readily, heavy single-pass applications can be made by airless spray. Table 10.7 provides a comparison of the two application methods, indicating the positive characteristics of each.

Heated Spray

Heating the coating liquid prior to application has been used with both air and airless spray applications. This is accomplished through the use of an in-line heater which heats the liquid prior to the time that it passes

FIGURE 10.42 — Airless spraying of an epoxy hull coating at an excessive distance from the surface, using poor application techniques.

FIGURE 10.43 — Coating by airless spray using poor application procedures.

through the spray gun. The increased temperature of the liquid generally reduces viscosity, allowing the coatings to be applied more readily and at a somewhat faster speed. It also has the reverse advantage. As the coating is propelled into the air, it is cooled rapidly and increases in viscosity shortly after it hits the surface. This helps the coating remain in place and reduces the chance of sags and curtains, generally making a smoother, more uniform film.

The low-boiling solvents in a coating liquid tend to evaporate readily between the gun and the surface. This leaves a higher quantity of the higher boiling solvents in the material to aid in leveling the coating and providing a smooth, even film. Heating a coating formulated for hot

TABLE 10.7 — Airless Versus Conventional Spray

	Conventional	Airless
Overspray	10-30%	Below 10%
Pinholing	Possible	Unusual
Material Loss on Application	Considerable	Up to 35% less
Penetration of Corners and Voids	Fair	Good
Paint Clogging Problems	Slight	Considerable
Safety During Cleaning	Excellent	Fair
Coating Contamination from Air Source	Possible	None
Wind Loss	Considerable	Negligible
Gun Distance from Surface	6-10 in.	14-16 in.
Film Build per Coat	Lower	Higher
Versatility	More	Less
Thinning Before Spray	Usual	Sometimes
Hoses to Spray Gun	2	1
Compressors	Large	Small
Portability	Fair	Excellent
Masking	Considerable	Moderate
Coverage 1000/sq. ft./day	4 to 8	8 to 12

application also allows higher solids and higher viscosity materials to be sprayed, which permits more material to be applied per coat without the danger of runs or sags. Heating of the coating liquid not only lowers the viscosity, but also tends to provide a more uniform viscosity material for application.

There are a number of suppliers of coating heaters, which come in various sizes, shapes, and forms. They generally preheat the coating from 120 F (50 C) to possibly 180 to 200 F (77 to 93 C). Many of the heaters are portable and attach to the applicator's belt. They are heated with electricity and therefore require an electric cord to the heater. This, along with the weight of the heater itself, makes the equipment somewhat more cumbersome for the applicator. The coating also may pass through a stationary heater between the pump and the gun. There can be considerable heat loss using this method, depending on the length of the hose.

As mentioned previously, heated coatings can be applied by either airless or air spray, and the benefits are approximately the same for both. Many product finish application units use airless spray and in-line heaters as a standard procedure. This equipment is less commonly used on large structural steel units, offshore rigs, large storage tanks, etc.

Some of the advantages of heated spray are as follows:

1. The coating tends to have a more uniform viscosity.

2. Even coverage reduces runs and sags.

3. Drying time is often reduced.

4. Faster applications are possible due to the lower viscosity.

5. Coatings can be sprayed readily at lower ambient temperatures.

6. There is less overspray and bounce-back from corners or enclosed areas.

7. There are less solvent fumes because of no added solvents.

8. There may be an increased thickness of coating applied per coat.

9. There is finer atomization with lower application pressures.

10. With reduced viscosity, there is the possiblity of improved coating adhesion.

Electrostatic Spray

Electrostatic spray may be used with either air or airless spray equipment. Airless spray equipment undoubtedly has some considerable advantages where electrostatic procedures are used. The objective of electrostatic spraying is to apply a rather high electrostatic charge to the coating particles as they pass away from the spray gun. An 80,000-volt static charge may be applied to the coating particles, with the object to be coated having an opposite charge. Generally, the particles are negatively charged and the object to be coated is positively charged or postively grounded. When the coating passes through the spray gun and into the air, the opposite charges attract each other, and the finely divided coating liquid is attracted to all sides, edges, corners, and some recesses of the object to be coated. The opposite side of the object also may be coated, provided the distance to be traveled is reasonable.

The principle of electrostatic spraying is shown in Figure 10.44. In this case, the coating particles pick up the negative charge as they evolve from the gun and carry the charge to the positively grounded surface. Almost all of the coating to be applied arrives on the surface to be coated with very little loss.

There is portable electrostatic spray equipment available, although for the most part this method of coating is

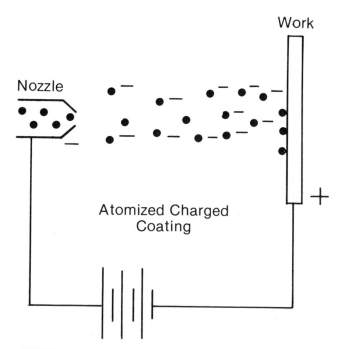

FIGURE 10.44 — A high-voltage DC power supply. (SOURCE: Corrosion Control Principles and Methods, Section 7, Ameron, Inc., Monterey Park, CA.)

Corrosion Prevention by Protective Coatings

used for product finish operations. One of the most common uses for portable electrostatic spray is on fencing (Figure 10.45).

The advantages of electrostatic airless spray are: (1) complete coverage of odd shapes, (2) complete use of the liquid coating, (3) almost no coating loss, (4) reduction in overspray to almost zero, and (5) very uniform film build.

There are also some substantial disadvantages in the use of electrostatic spray.

1. Not all coatings can be applied by electrostatic means.

2. The formulation of each coating is critical; *i.e.*, the coating must be designed for electrostatic spray.

3. The electrostatic procedure cannot be used on a nonconductive surface; it can only be used on bare metal or conductive coatings.

4. Only one coat of the material can be applied, because of the insulating characteristics of the original material.

5. Because of the very high voltage used, there is some hazard of electric shock during application.

6. The equipment is considerably more expensive than either air or airless spray setups.

The electrostatic method also can be used for dry powder coatings. In this case, the particles of the powder are negatively charged and the object to which they are to be applied is positively charged. The same electrostatic action draws the powder particles to the article to be coated, and a uniform layer of the powder is applied to the metal surface. The powder can then be fused in place, in a normal powder coating procedure.

Powder Coating

The powder coating procedure is generally an in-plant production-line procedure where a finely divided coating material is applied to a metal substrate in a fluidized bed (or by electrostatic spray, as discussed previously). The coating particles are dry and are completely formulated, including pigment. In a fluidized bed, they are floated in an air stream so that they are in con-

FIGURE 10.45 — Electrostatic spray gun is being used to coat industrial fence. (Photo courtesy of: Ransburg Electro-Coating Corp., Indianapolis, Indiana.)

stant movement. The object to be coated is usually heated to the fusion temperature of the coating and dipped into the fluidized bed, so that the coating particles fuse on the surface. The thickness of the coating can be controlled by both the heat and the time of exposure in the fluidized bed. After it is removed from the fluidized bed, the coated object then goes into a heating oven where the coating is further fused to the object until it flows into a continuous film.

Due to environmental restrictions, mainly on solvents, more coating applications are being made in this manner wherever such a procedure is practical. Many different types of coatings can be applied, including the thermoplastic materials (*e.g.,* vinyls and acrylics) and epoxies. All powder coatings must be specially formulated for this procedure. Up to the present time, high-performance coatings have not been developed for practical application to existing or new large structural steel objects.

Dip Coating

Dip coating is another method of application which, in many ways, is similar to powder coating. However, in the dipping procedure, the coating is in a liquid form and the part to be coated is dipped into the fluid bath. Again, the coating material must be properly formulated to provide an even-thickness film over all parts of the object to be coated. In dip coating, one of the principal requirements is to remove the part from the bath at exactly the same speed as the coating flows from the surface. In this way, an even coating is formed without drips, tears, or heavy coating accumulation on the lower part of the work.

Dip coating and powder coating procedures usually are used where only one coat is applied. However, there is nothing to prevent additional dip coats from being applied to the object using the same procedure. Dip coating is an in-plant procedure and is usually part of a production line for the application of product finishes to fairly complicated parts or ones which will receive only one coat. The process is not applicable to high-performance, multiple-coat coating systems.

Electrocoating

Electrocoating is a modified dipping procedure where a low-voltage current is passed from a negative electrode to the part to be coated, which is the positive electrode. Very little active current is used in electrocoating, making the process similar to electroplating in character (Figure 10.46). The coating must be carefully and specially formulated for use in this process since the liquid coating must be conductive. The parts to be coated are usually on a conveyor line passing through the electrocoating bath. The coating is partially precipitated on the part in a uniform coat. Automotive frames and similar parts have been primed in this manner. This is a specialized process and can only be used on parts which are coated in-plant.

Application of Multiple-Component Systems

With the increased emphasis on 100% solids coatings and high-solids coatings, the use of fast-acting catalyst coating systems has been considered, and some success-

Electrocoating

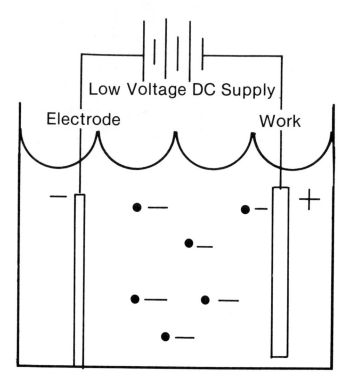

FIGURE 10.46 — A schematic illustration of the electro-coating process.

ful ones have been developed. In multiple-component systems, the procedure is to mix the reactive components of the coating within minutes, or even a few seconds, of their application. To do this, special equipment is required so that the exact amount of each coating component can be mixed and then quickly applied to the surface.

One of the first of the multiple-component systems was the catalyzed polyester. These coatings were glass reinforced and applied to the interior of many tank bottoms. In this case, the catalyst and the polyester were mixed at the tip of the gun, applied to the surface, and within a matter of seconds were reacting in place. From this relatively crude beginning, more systems are coming into existence, along with the proportioning equipment and mixing equipment to go along with them. The general procedure is described by Donald J. Wisdorf in *Modern Paint and Coatings,* as follows:

> Proportioning Base and catalyst are proportioned by air operated piston pump, a gear pump or a diaphragm pump. Piston and diaphragm pumps tend to be more widely used, partially because they are less complicated, have proved to be effective and are well known. The positive displacement principle tends to insure greater accuracy. Gear pumps tend to be sensitive to changes in viscosity so that speed adjustment is necessary. Slippage or leakage is common because of the metal to metal interface.
>
> The proportioning pump meters the base and catalyst in the prescribed ratio (one to one, ten to one, or other) within whatever accuracy is judged appropriate for the application.
>
> Pot life or reaction time of the material will significantly in-

fluence the design of the system. Pot lives of one to five minutes and longer will require one type of equipment. A five second pot life will require a different, somewhat more elaborate design. For example, five seconds will require a manual or automatic flush for the unit to stay in operation. Short pot life coatings can be accommodated in existing equipment...100% solid plural component coatings have been mixed and sprayed successfully for more than ten years in shipyards, aircraft manufacturing and in the petrochemical industry.[1]

Once the proportioning of the ingredients has been accomplished, the actual application of the coating can be done by conventional, airless, or electrostatic spray. The key is to mix the catalysts and the components thoroughly, just prior to their being applied. More and more of this type of equipment is being developed as more and more 100% solids and very high-solids coatings are being used. Multiple-component polyurethane coatings are being used on aircraft; epoxy materials are being used on the interior of tankers; and polyester materials are being applied to the interior of tanks and similar equipment.

The advantages of this multiple-component procedure are obvious. They involve fast-curing coatings, with high solids and therefore a low solvent base with little air contamination. Also, heavy coating thicknesses can be applied in one coat. The multiple-component system generally is not used for maintenance coatings in plants where corrosion engineers are most involved. On the other hand, sophisticated computerized equipment is being developed which will allow broader use of this type of system.

Drying or Curing

The drying or curing of a coating after application may be as important as the application itself, depending on the use of the coating.

Exterior Applications

The curing of coatings on the exterior of structural steel, tanks, and similar surfaces is considerably less critical than where the coating is applied in areas of splash or spray of chemicals or where the coating is used as a lining for the interior of a tank. Nevertheless, curing is important. Oil-base coatings and medium-oil or long-oil alkyds require a longer period of drying time than do some of the other high-performance coatings. Because of this, weather conditions can create problems, particularly where such coatings would be applied to tank tops and similar areas where water can accumulate from either rain or heavy dew. If the coating is exposed to water prior to its taking at least its initial set, which may be 3 or 4 hours, coating problems can occur.

Lacquer-type coatings are undoubtedly the easiest materials to handle on exterior surfaces. This is because they dry rapidly, *i.e.*, within a matter of minutes, and exposure to rain at that point is not detrimental. The same is true for chemical fumes or chemical splash and spray. Lacquer-type coatings are considerably more resistant once the solvents have been removed from the film. Nevertheless, the coating is substantially resistant to the environment after the initial evaporation of the solvents.

Even though cold temperatures reduce the evaporation rate of the solvents from lacquer-type coatings, the initial drying of the coating usually takes place within a half-hour. Epoxies and polyurethanes, since they are internally cured (particularly those used for corrosion resistance), are very sensitive to temperature. This must be taken into consideration during application. Polyurethanes are much less sensitive to temperature than the epoxies; nevertheless, the chemical reactivity is temperature sensitive. Epoxies, including coal tar epoxies, can cure rather rapidly at temperatures above 70 F (21 C). However, if the temperature goes below 70 F (21 C), the curing time increases substantially, even to the point that the coating may remain tacky for from 12 to 24 hours.

Curing is difficult under these conditions, and both atmospheric conditions and chemical fumes, splash, or spray can damage the coating prior to proper cure. Polyurethane coatings, particularly those which react with moisture, act rather well even though the temperature is cool, as long as the humidity is sufficiently high to supply moisture for curing the coating. On the other hand, hot, dry conditions can retard the cure of moisture-cure polyurethanes, making them susceptible to solvent and chemical action.

Water-base coatings cure by the evaporation of water, and subsequently by the coalescence of the coating particles which were dispersed in the water. Proper coalescence of the water-dispersed resin particles is essential for the proper life of water-base materials. If the atmosphere is hot and dry, the coating may dry sufficiently rapidly so that the particles never are able to properly coalesce. If atmospheric conditions are cold and damp, the water may evaporate sufficiently slowly so that the resin particles agglomerate without proper coalescence. Again, the coating is improperly cured and will not be effective.

Temperature and weather conditions for the proper cure of water-base materials are somewhere between 50 and 80 F (10 and 47 C), with a relative humidity in the median range (i.e., 30 to 60%). Under these conditions, the water evaporates sufficiently slowly so that the resin particles are moved in place in the film in such a way that the particles coalesce and form a solid film. Exterior exposure may not be as critical from a total resistance standpoint for exterior coatings as for immersion coatings. On the other hand, curing is important, and unless it takes place within a reasonable time, coating problems can be expected.

Curing of all of the above coatings is much easier under controlled drying conditions, such as those that exist in product finishing operations or where the coating is done under controlled interior conditions. In product finishing operations, there is usually a forced drying or curing part of the total coating application process. If this is the case, proper curing conditions are established, and the maximum resistance of the coatings can be quickly obtained.

Interior Coating Applications

Coatings applied to the interior of tanks and similar areas where the coating will be subject to immersion, water, or chemicals, are in a much more critical situation from a curing standpoint. In this case, the optimum curing conditions must exist for the coating to be fully resistant.

Vinyl coatings and some other lacquer-type materials are known to retain solvents. This is caused by a mutual solubility of the solvents and resins. While the resins are soluble in the solvents, the solvents are soluble in the resins, which is where solvent retention comes in. With the solvents dissolved in the resins, it is much more difficult for them to evaporate. Many are held for long periods of time and can almost be considered plasticizers. Unless it is force-dried, 2 to 4% of the solvent can be retained in the coating film almost permanently. Force-drying increases the vapor pressure of the solvent to the point where it escapes from the resin which is retaining it.

There have been instances with vinyl coatings where the solvent retention in the coating lasted for many months or a year or more. Inclusion of solvents in the coating, even though it is hard and abrasion resistant, tends to reduce the resistance of the coating for immersion conditions. In order to remove the remaining solvent from within the vinyl coating, it is necessary to increase the temperature and provide good air movement to carry the solvent away from the surface. Without the air movement, it is possible for the solvent in the air and in the coating to come to equilibrium when in a closed space. This would result in solvent remaining in the coating.

Force-drying does not mean that the coatings would be baked, but that the temperature would be raised for a period of time from 130 to 160 F (55 to 70 C) in order to drive the residual solvents out of the coating. This may take a number of hours, during which the hot air is circulated in order to remove the solvents from the area. Such treatment can make a measurable difference in the resistance of vinyl coatings used for immersion purposes.

Chlorinated rubber coatings, even though they are lacquer-type coatings, do not have strong solvent retention. The chlorinated rubber resin does not have the same mutual solubility characteristics as the vinyls, and these coatings are known to "through dry" rather quickly.

Epoxy coatings also react well to force-drying conditions. Although the epoxy coatings are internally cured, the chemical reaction of the cure proceeds at a much more rapid pace at elevated temperatures. In the same temperature range as used for force-drying vinyls, curing takes place more rapidly and more completely, making the epoxy coatings that much more solvent and chemical resistant.

Epoxy phenolic coatings often require some additional heat to reach their maximum cure and resistance. Generally, they are slower to react than the unmodified epoxy coatings, and some measure of force-curing improves their resistance. This is also true of phenolic coatings, which require a high bake in order to develop their resistance to water and chemicals. In this case, the heat-curing is part of the coating process and is necessary to develop the properties of the coating. It is believed that force-curing tends to improve almost all coatings, by increasing their adhesion and chemical and solvent resistance, and by reducing the chance of chemical contamination of the film prior to its reaching a full cure.

All coatings can be overcured to the point where their properties are diminished. In many cases, it is better to have a somewhat overcured condition than to have an undercured condition; *i.e.,* overcuring to a point just before the coating begins to break down. The coating will then have its maximum properties in terms of adhesion and resistance. In the case of the thermosetting coatings, however, undercuring leads to poor adhesion, poor water resistance, and poor resistance to solvents and chemicals Therefore, a somewhat overcured coating is preferable.

This is particularly true where coatings are to be used for immersion, for areas of heavy condensation, or where they are subject to constant chemical contact. Even some of the water-base organic coatings are improved by heat treatment. The original inorganic zinc coatings based on a high alkali silicate were cured by heat at around 350 to 400 F (180 to 205 C). The coatings reached their maximum resistance within approximately ½ hour and maintained these properties over a period of many years. Even now, while the heating process generally is not used for such coatings, if a rapid cure is required, heating of the water-base inorganic coating can substantially increase the speed of cure for quick handling and exposure.

Weather Conditions

Weather conditions can have a substantial effect on coating application. Weather conditions are less critical where coatings are applied to interior surfaces, although even in this case problems can occur. The ideal conditions for applying almost all coatings include a temperature of 70 F (21 C), little or no wind, and a humidity within 50 to 60%. All coatings respond well to such conditions. Moving away from the ideal, problems can occur which have a decided effect on the resistance of the coating.

Humidity

Humidity is one of the weather conditions which is an almost universal problem. This is because humidity may change rapidly, *e.g.,* from relatively low to very high during a single working day. It is the consensus of most coating authorities that coatings should be applied at temperatures no lower than 5 F (3 C) above the dewpoint. This may be difficult for the actual applicator to determine, since condensation is difficult to see. Wherever high humidities are encountered, particularly in coastal and marine areas, coating application must be carefully controlled as the temperature of the substrate approaches the dewpoint. Coatings applied over a surface covered with even a light film of condensation are almost always doomed to failure.

Table 10.8 shows the percent relative humidity above which moisture will condense on metal surfaces that are not insulated. This chart provides good guidelines as to when coating applications should be restricted. For example, if the metal temperature is 60 F (15 C) and the surrounding air temperature is 70 F (21 C) (which are reasonably normal conditions), there is a good chance of surface condensation if the humidity is 56% or above.

Table 10.9 shows how to determine the dewpoint if the air temperature and relative humidity are known. Both of these charts are helpful in assuring proper application conditions.

Hot, Dry Weather

Hot and dry weather conditions may or may not be good coating conditions. Under hot, dry conditions, overspray is much more of a problem, and coatings tend to dry much faster, often resulting in dry spray. For example, unless lacquer-type coatings are properly formulated for hot, dry conditions, excessive overspray can result when applied under such conditions. This is even more likely where there is even a mild wind condition. Dry spray also may result during the application of ethyl silicate-based inorganic coatings. Also, because these coatings require moisture to cure, they may not cure satisfactorily under such conditions. The solvents in epoxies may even dry so rapidly that the coating develops an uneven, orange peel-type surface. Moisture-cured polyurethanes also may not cure properly under such condi-

TABLE 10.8 — Percent Relative Humidity Above Which Moisture Will Condense on Metal Surfaces Not Insulated

Metal Surface Temp.	4 C 40 F	7 C 45 F	10 C 50 F	13 C 55 F	15 C 60 F	18 C 65 F	21 C 70 F	24 C 75 F	27 C 80 F	30 C 85 F	32 C 90 F	35 C 95 F	38 C 100 F	40 C 105 F	43 C 110 F	46 C 115 F	49 C 120 F
1 C(35 F)	60	33	11														
4 C(40 F)		69	39	20	8												
7 C(45 F)			69	45	27	14											
10 C(50 F)				71	49	32	20	11									
13 C(55 F)					73	53	38	26									
15 C(60 F)						75	56	41	17	9							
18 C(65 F)							78	59	30	21	14	9					
21 C(70 F)								79	45	34	25	18	13				
24 C(75 F)									61	48	37	29	22	16	13		
27 C(80 F)									80	64	50	40	32	25	20	15	
30 C(85 F)										81	66	53	43	35	29	22	16
32 C(90 F)											81	68	55	46	37	30	25
35 C(95 F)												82	69	58	49	40	32
38 C(100 F)							% OF RELATIVE HUMIDITY						83	70	58	50	40
40 C(105 F)														84	70	61	50
43 C(110 F)															85	71	61
46 C(115 F)																85	72
49 C(120 F)																	86

(SOURCE: Ameron Co., Corrosion Control Principles and Methods, Section 7.)

Corrosion Prevention by Protective Coatings

TABLE 10.9 — Dewpoint Determination

Air Temp °C	Dewpoints (°C) at Various Relative Humidities							
	30%	40%	50%	60%	70%	80%	90%	100%
−1	—	—	—	—	−6.5	−4	−2	−1
4	—	−6.5	−4	−2	0.5	1.5	3.5	4.5
10	−6.5	−3.5	0.5	2	3.5	5.5	8.5	10
15.5	0	2	4	8	10	11.5	14	15.5
21	3	6.5	10	13	15	18	19.5	21
26.5	7	12	15.5	19	21	23.5	25	26.5
32	13	16.5	20.5	24	25.5	28.5	30.5	32
38	18	22	25.5	29	31	33.5	36	38

NOTE: It is essential to ensure that no condensation occurs on blasted steel or between coats during painting.

Air at a given temperature can only contain a certain (maximum) amount of water vapor. This proportion is lower at lower tempertures.

The dewpoint is the temperature of a given air-water vapor mixture at which condensation starts, since at that temperature its maximum water content (saturation) is reached.

In practice a safety margin must be kept, whereby the substrate temperature is at least 3 °C above dewpoint.

tions. In addition, water-base water emulsion coatings may dry so rapidly under these conditions that the coating does not properly coalesce, but may mud crack or craze, thereby resulting in an unsatisfactory coating.

Hot, Humid Weather

Hot, humid weather comprises another set of conditions to be considered. High humidity can create condensation problems, particularly as the late afternoon or evening approaches. This means that a constant check of the surface is necessary to make sure that coatings are not being applied over such condensation.

Warm, moist conditions are good for moisture-cured coatings such as ethyl silicate inorganics and polyurethanes. However, they are not as satisfactory for vinyls, since many solvents evaporate rapidly due to the temperature and, because of the rapid evaporation, the temperature of the coating may be reduced below the dewpoint so that progressive blushing can easily occur. This is also a problem with most lacquer-type coatings under humid conditions. *Blushing* is the condensation of moisture on the surface of the semiliquid coating, causing it to whiten. This blushing is caused by the precipitation of the coating resin on the surface due to the water, and it remains even after the coating is dry. A blushed coating is not satisfactory from either an appearance or resistance standpoint.

Cool, Dry Weather

Cold or cool, dry conditions [*i.e.,* above 40 F (5 C)] are usually good conditions for the application of lacquer-type or water-base coatings such as water-base inorganics, acrylics, or similar products. Under these conditions, the water dries from the coating relatively slowly, allowing the coating to make a good strong film. Silicate dispersion and lithium silicate-type inorganic zinc coatings are particularly good under such conditions. On the other hand, moisture cure inorganics may not cure for a long period of time under conditions which are cold and dry. This also may be true of moisture-cured urethanes.

Cold, Humid Weather

Cold, humid conditions are more satisfactory for moisture-cured coatings such as the ethyl silicate-base inorganics and moisture-cured polyurethanes. They are not as satisfactory for water-base coatings such as alkali-base inorganics or the water-base organics. It can be so cold and humid that the water will not evaporate from the coating over a period of many hours. Under such conditions, neither the water-base inorganics nor the organic water-base products will make a proper film on the surface. A poor coating with little resistance to corrosive conditions results. Epoxy coatings are particularly subject to poor curing during cold conditions. Increased humidity may be reactive with amine-type curing agents, causing further coating resistance problems.

Weather conditions are critical to the application of coatings. In fact, it often is necessary to select the type of coating that may be applied under the particular weather conditions which exist at the time.

Coating Coverage

Coating coverage can be defined in a number of ways. It can be the area which a gallon of coating will cover, the thickness of the film covering the surface, or the amount of coating required for a given area at a given thickness. Coverage, as it normally is described, is considered in an NACE Technical Committee Report entitled *Determination of Theoretical Coverage Rates of Inorganic and Organic Protective Coatings.*

The "coverage rate" of a coating is defined as the area of a surface that can be covered at a uniform specified dry film thickness by a specified unit volume of a coating. In metric units, the coverage rate is expressed in sq. meters/liter at a dry film thickness of 100 microns (0.1 mm). In U.S. units, the coverage rate is expressed in sq. ft/U.S. gallon at a dry film thickness of 1 mil. In British units, the coverage rate is expressed in sq. ft/Imperial gallon.

The following table [Table 10.10] is presented as an example of the theoretical yield of a 100% solid (no volatiles) coating spread at a specified unit film thickness assuming 100% utilization. In this example, the area covered by one liter of 100% solid coating at a thickness of 100 microns is 10 sq. meters.[2]

TABLE 10.10 — Theoretical Yield for a Unit Volume of 100% Solid Coating at Specified Unit of Film Thickness

Measuring System	Unit Volume	Unit Dry Film Thickness	Area
Metric	1 Liter	100 microns	10 sq. meters
U.S.	1 U.S. Gallon	0.001 inch	1604 sq. ft.
British	1 Imperial Gallon	0.001 inch	1925 sq. ft.

Since coverage is an important factor in any application, the above description indicates only theoretical coverage. A more detailed discussion of coverage is given in Koppers, *Protective Coatings,* Technical Data Sheet.

A gallon measure contains 231 cubic inches which when spread out covers an area of 231 square inches one inch thick (231 sq. in. equals 1.604 sq.ft.). Since one mil is 1/1000 of an inch one gallon of solid material, not containing any volatiles will cover a thousand times 1,604 or 1604 square feet one mil thick. For simplicity we drop 4 square feet and state one gallon of solid material of any kind will produce a one mil thickness on 1600 square feet. This is strictly a question of volume and has nothing to do with the weight of the gallon or the kind of material contained in the gallon. If more than one mil is needed the coverage will of course be proportionally less and it is easy to see that one gallon will produce 2 mils on 800 square feet, (½ of 1600 square feet), 3 mils on 533 square feet (⅓ of 1600 square feet), 4 miles on 400 square feet (¼ of 1600 square feet) and so on. On these facts all the following computations are based.

If all coatings would contain only solid material, all coating regardless of type would give the same mil thickness per square foot. Most paints, however, contain solvents to make application possible. The amount of solvent used is governed by the method of application and the inherent viscosity of the solid content of the paint and it may vary greatly. These solvents go out of the paint film after application and do not contribute to the final dry mil thickness of the coating. They do contribute however to the wet mil thickness.

For example: one gallon of paint containing 50% solid material by volume, (also called nonvolatile by volume), and 50% solvent by volume will cover 1600 square feet 1 mil wet or 1600 square feet ½ mil dry. Dry thickness is less because one half of the volume is made up by solvent which will leave the film and dissipate. Expressed differently one gallon of paint having 50% nonvolatile volume will cover 800 square feet one mil dry. All mil thicknesses mentioned will be dry mil thicknesses if not otherwise stated. Mathematically if you multiply 1600 by the nonvolatile volume percentage of the paint you get the coverage at one mil dry thickness.

$$1600 \times 50/100 = 800 \text{ square feet 1 mil thick.}[3]$$

While theoretical coverage rates are never attainable in actual practice, since they do not take into account additional solvent evaporation, application losses, wastage, surface complexities, etc., they can be used in an impor-

tant way. Applicators can make appropriate deductions from the theoretical values to compute estimated practical coverage rates. The factors that must be taken into consideration include nonuniform dry film thickness; application wastage (*e.g.,* overspray); material retained in containers, spray pots, and lines; surface profile developed by surface preparation procedures; and the effect of pitted or irregular surfaces.

These elements of material loss and increased consumption are beyond the control of the coating manufacturer. Consequently, manufacturers cannot be held accountable for the practical coverage rates obtained at the job site. However, they can be held accountable for total volume solids and theoretical coverage rates of products as furnished.

When all factors are taken into consideration, the practical coverage amounts to considerably less than the theoretical coverage. The average reduction in coverage is from 10 to 30%. The practical coverage may become even lower on heavily pitted steel, rough concrete, or while spraying in the wind.

It is not possible for even the best applicator to hit the mil thickness right on the nose; thus, there should be a reasonable leeway. The unavoidable error, however, lies on the high side of the required thickness, since most specifications call for a minimum mil thickness. Wet thickness gages are commercially available and easy to use.

Mild atmospheric conditions call for a thickness of at least 3.5 mils. In industrial environments, where the conditions are more severe, 5 mils are recommended. Submerged surfaces need a thickness of from 5 to 15 mils, depending on the corrosiveness of the liquid and the coating system used. Heavy bituminous coatings are best applied in 18 to 20-mil thicknesses and in some instances as high as 1/32 in.

As stated previously, coating coverage and thickness are closely related and are both important in obtaining proper coating performance. The importance of film thickness is discussed by John D. Keane in a paper entitled *Minimum Paint Film Thickness for Economical Protection of Hot-Rolled Steel Against Corrosion.*

A ten-year paint exposure study was conducted to determine whether or not there is a minimum or optimum paint film thickness—such as 5 mils—for the economical protection of hot-rolled steel. In all work to date, each additional mil of paint thickness has been accompanied by an increase of about 20 months in paint life.

It was shown that there was indeed an economic initial paint film thickness above which long-term protection was afforded throughout the entire study period. This minimum effectual thickness, however, was not a constant, being thinner for the more durable and less permeable paints. In addition, this thickness tended to be lower in milder environments, and to an unexpectedly limited extent, lower on well-prepared surfaces. It was found far more economical to apply sufficiently heavy films at the beginning rather than to undertake more frequent subsequent maintenance painting.[4]

Coverage also implies coating continuity. In any application of any high-performance coating to prevent corrosion, coating continuity is extremely important. Without continuity, the coverage (*i.e.,* coating thickness) means very little. Continuity, on the other hand, should mean that the coating is applied evenly over the entire surface with no coating defects. This concept is important when using any anticorrosive coating, but it is even more important where coatings are applied to the interior of containers to protect against corrosion or to protect the contents against contamination.

In order to assure a proper coating job, it is essential that quality controls be enforced during and following the application. Each coat of a coating application should be carefully inspected before the succeeding coating is applied. Overspray or dust should be removed by light sanding and dusting; loose brush hairs in the prime coating should be removed; and any holidays should be reprimed.

Thickness and continuity are two of the most important items which should be considered in application quality control checks. One procedure that is recommended by most manufacturers, is to assure proper coverage by alternating the color of each succeeding coat. If the prime coat is red, the second coat may be black or gray, the third coat an alternate of the black or gray, and the final coat can be whatever is selected as being in harmony with the rest of the structure. Alternating colors between coats not only helps in the inspection of the coating during the application, but also aids in obtaining the proper thickness and continuity of the coating. It also aids the applicator in being able to visually determine when an area has been thoroughly covered. In any tank lining job, visibility is less than ideal and alternating the color of each coat helps assure good judgments.

Inspection

Specifications for the finished coating system usually call for a minimum film thickness. A variety of instruments are available for such measurements. The simplest and least accurate of these is the wet film gage. By sinking this instrument into the wet coating film, a reading can be obtained. Naturally, coating thickness will diminish as the solvent in the coating evaporates. Therefore, the wet film gage will nearly always indicate a greater film thickness than that which actually exists. This is why it is important to take a reading immediately after coating application. A second drawback in the use of this instrument is the inability to measure total film thickness in a multiple coat application, since only wet coat readings are obtainable. A third problem is when lacquers are measured. Since the previous coat is soluble in the following one, the wet film gage will sink into the previous coating, giving a greater-than-actual reading for the second coat.

At best, the wet film gage can give the inspector a rough idea of the thickness of each coat as it is being applied. Its areas of greatest use are in spot checking the coating film as it is being sprayed, which allows the applicator to immediately correct any application defects. It is also possible with this instrument to measure coating films which are being applied over smooth concrete and wooden surfaces.

Dry film thickness readings over steel are usually taken with nondestructive gages which function by permanent magnet-calibrated spring mechanisms to give readings on a graduated scale. The most popular of these gages are the Mikrotest,[1] the Inspector,[2] and the Elcometer.[3] These instruments are small enough to be carried in a pocket, and should be readily available to every inspector.

A hand microscope device, the Tooke Gauge,[4] requires cutting the film and can be used for dry film thickness measurements over nonferrous substrates. It is also useful over steel, particularly in measuring the thickness of each coat in a system.

It is extremely important that the dried coating film contain no bare spots or holidays. Such areas often are too small to be noticed with the naked eye. Therefore, to insure 100% coverage, *holiday detectors* are provided. These electrically operated instruments are capable of detecting any pinholes or bare areas in the coating and may be used to inspect any nonconductive coatings applied over a conductive surface. A typical example of such a device would consist of a rod, which is electrically connected to a power supply, holding a wet sponge. This power supply also is connected to the metallic surface over which the coating is applied. The coating film acts as an insulating barrier, thereby preventing completion of the electrical circuit. If, however, there is a holiday in the coating, the circuit will be completed, current will flow, and some type of signaling device such as a bell or buzzer will alert the inspector.

This method of inspection is extremely important where the coating is to be continuously immersed in a highly corrosive solution or used to prevent liquid contamination. Holidays, if not patched, will become focal points for corrosion, thereby resulting in deep pitting of the underlying surfaces.

The completed coating should also be inspected for runs, overspray, and roughness. Any areas which show these or other signs of improper application should be repaired or recoated.

An important part of any application job is the cleaning of the equipment upon completion of a day's work. Dirty or improperly cleaned equipment can cause more downtime and coating difficulties than almost any other part of the application procedure. The cleaning of the coating tools should be done immediately after use, before the coating can become hard, because once it reaches this state, many of the high-performance coatings are insoluble and extremely difficult to remove from either interior or exterior surfaces.

The first step in the cleaning procedure is to remove as much of the liquid coating as possible from brushes, rollers, paint pots, pumps, air lines, and guns, and clean these tools at least two and perhaps three times in the correct solvent until no coating is noticeable in the last cleaning. If possible, the solvent should be somewhat stronger than the solvent within the coating itself, in order to insure that the coating is dissolved from the interior of brushes, hoses, spray guns, etc. The brushes, pots, and coating containers should be washed and then wiped thoroughly dry.

After solvent cleaning, brushes and rollers should be flushed with a good detergent and warm water. Household dishwashing or clothes washing detergent is good for this purpose. It is surprising, even after using the solvent, how much additional material can be removed through the use of a detergent and warm water. The brushes and rollers should then be rinsed in clear water, the excess water removed, and in the case of brushes, the bristles should be brushed straight with a short-fiber wire brush. This not only helps to straighten the bristles out, but also tends to remove any of the coating which may have dried on the outside bristles adjacent to the handle of the brush. Any coating in this area should be removed to prevent the brush from becoming stiff and unusable elsewhere in a short period of time. Rollers should be stood on end and allowed to dry. Brushes should be laid flat so the bristles will not warp or become fuzzy during the drying period.

Also, use extreme caution when cleaning airless spray guns. The high pressures involved can be very hazardous, particularly when the spray head is removed. The airless spray must be pointed away from any person or part of the body. The high pressure can drive the solvent or coating into the flesh, making a wound difficult to heal. With good care, all coating tools and equipment will last much longer and will always be in proper condition for additional use. Clean equipment is also insurance against contamination of the applied coating.

Application Problem Areas

Chapter 7 on structural design indicated some of the coating problems that exist with much of the structural design that is common today. Since most of these are application problem areas, they will be discussed in greater length in the following sections.

Edges and Corners

The coating of sharp edges and exterior corners has been discussed previously in this chapter. However, due to the importance of these areas, this application should be stressed. Figure 10.47 provides a good example of sharp edges and corners as they often exist. The photograph is a view of the interior of a block section showing the bottom reinforcing members in a large tanker. While this is a complex area, all of the edges and corners shown must be thoroughly coated in order to provide the corrosion protection required. Such areas should be double coated, with the first coating applied directly at the corner edge and the second application overlapping the corner or edge in the flat area. Where possible, it is best to stripe the corners and edges by brush, flowing the coating out over the sharp edges in order to build up maximum coating.

Figure 10.47 also shows the number of inside corners which are common in such structures. In this case, the first coating should be directed into the corner, followed by additional passes on the flat area adjacent to the corner, and carrying the spray pattern into the corner each time. In coating inside corners, there is the problem of applying too much coating under certain conditions. Since coatings tend to shrink, if too much coating is applied into a corner (particularly where drying of the coating is relatively slow), the coating may shrink away from the inside corner. This causes a void underneath the coating, as shown in Figure 10.48.

Thus, sufficient coating must be applied to thoroughly cover the inside corner with coating thickness.

FIGURE 10.47 — A block section of the bottom of a tanker tank showing the many sharp edges which require special coating attention.

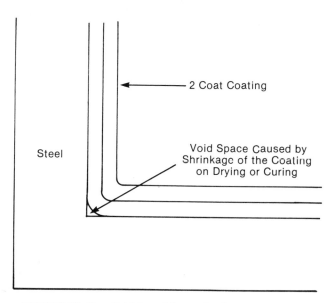

FIGURE 10.48 — Bridging of the coating in a square corner.

However, the coating should not be applied so thickly that a void occurs in the corner upon shrinking. This can only be accomplished through extreme care on the part of the coating applicator, since only he can control the coating thickness. Coating of the inside corner also may be complicated by welds in that area. In this case, additional care must be taken to assure application of the proper coating thickness over the weld and corner area.

Welds

Welds nearly always represent an application problem, since they usually are rougher than the adjacent steel surfaces (Figure 10.49). The coating must be applied over such a weld so that all of the rough areas, pits, under-

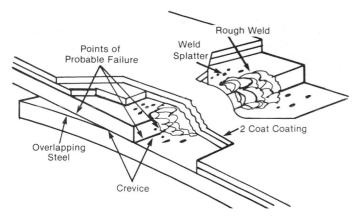

FIGURE 10.49 — Coating difficulties on welded plates.

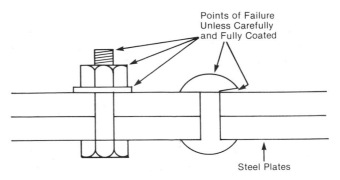

FIGURE 10.50 — Coating difficulties on bolts, nuts, threads, and rivets.

FIGURE 10.51 — Corrosion resulting from spraying rivets from only one direction.

cutting, and sharp areas are thoroughly coated. In this case, it is always good practice to apply at least the first coat over the weld by brush as opposed to spray. Using a brush, the coating can be worked into the rough areas to make sure that they are properly filled and without holidays.

If spraying of such areas is required, it is recommended that welds be striped with a coat before coating any adjacent areas. When the adjacent areas are coated, the coating should be carried out over the weld area, providing a double coat. Where this is properly done, the weld areas normally will outlast the flat areas in a corrosive atmosphere. This is due to the additional coating thickness, along with the extra care taken during the application.

Bolts, Nuts, and Rivets

While bolts, nuts, and rivets provide the most common way of fastening structural steel besides welding, they pose obvious coating difficulties. In this case, it also is good coating practice to brush-coat these areas prior to coating the flat area. With the brush, the coating can be worked into all crevice and corner areas, building the coating up and making sure that no holidays remain (Figure 10.50).

Figure 10.51 shows the effect of spraying a riveted area. While the entire area may have seemed to be covered during the spray operation, it can be seen that after some exposure, the left-hand area of the rivets is rusty, while the remainder of the rivet is well protected. This is caused by the applicator spraying from one direction only, with the surface away from the spray being shaded from the coating (Figure 10.52).

Such application deficiencies are common on steel structures. Rivets or bolts should be sprayed from at least four different angles to prevent holidays from occurring adjacent to the rivets or bolts. Rivet and bolt areas should be sprayed first and thoroughly covered, and then recoated during the coating application to the adjacent flat areas.

There are many coating difficulties that are commonly encountered during application. While most of these are under the control of the applicator, anyone supervising an application job should be on the look-out for them.

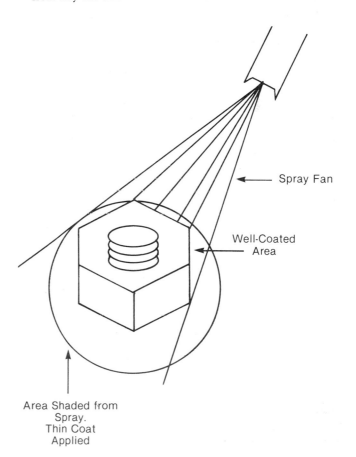

FIGURE 10.52 — Incorrect spraying procedure from only one direction leaves a shaded area of thin coating.

Brush Marks

Brush marks pose a common problem in brushing operations, particularly when using many of the more thixotropic formulations which are available in the heavier bodied coatings. A coating that flows very well tends not to have brush marks, since the marks will flow out to a smooth film. With the heavier coatings, however, the coating should be thoroughly brushed to cover the area, making sure that full coverage is obtained. Following this, very light strokes of the brush should be applied all in one direction to minimize any heavy brush marks that might be left in the original pass (Figure 10.53). The light stroking of the brush as the area is completed vastly improves the problem of brush marks with heavier materials.

FIGURE 10.53 — Typical brush marks in a heavy-bodied coating.

Stippling or Orange Peel

A stippling or orange peel type of surface is common when using thixotropic coatings with a roller, particularly a roller with a heavy nap. In many ways, the orange peel is similar to the brush marks in a coating in that there are heavy and light areas, even though the film may be continuous. They are more common where the roller is pressed heavily against the surface and rolled rapidly. While they cannot be completely overcome in thixotropic coatings, they can be improved considerably by a light, slow rolling of the surface once the entire area has been covered. This light rolling should be in all one direction in order to provide a uniform texture. Coatings that have considerable flow-out are much less of a problem in this way, and usually a thinner film is applied to the surface.

Orange peel also can result from the spraying process. Again, it is more prevalent with a thixotropic-type coating and usually results when the spray gun is not adjusted properly for maximum dispersion in the spray. Where this is a problem, the spray should be adjusted so that it is as fine as possible without causing overspray.

Overspray

Overspray usually is caused by gun adjustment, although there are some coatings which have a greater tendency to overspray than others. Chlorinated rubbers and vinyls are examples of these. Some ethyl silicate in-

organic zinc coatings also tend to overspray more than others. The problem is caused by the fast-drying coating losing sufficient solvent from the time that it leaves the gun until it hits the surface, so that a dry particle of coating sticks to the surface (Figure 10.54). This often is the case where one area of coating overlaps another or at the end of the spray pass.

The gun adjustment, which causes most overspray, is where there is too little material passing through the gun or where there is an excessive amount of air being used to disperse the coating. Reducing the amount of air and increasing the pressure on the liquid, so that both are optimum, will reduce overspray to a minimum. Holding the spray gun too far away from the surface is also another cause of overspray. Tilting the gun when spraying either high or low can cause excessive overspray. For instance, during the spraying of some tanks, the applicator tried to spray above and below a scaffold platform, with excessive overspray in both areas. These areas failed quickly and were extremely costly to repair. Thus, care should be taken to assure that the coating is applied to the surface as a good, wet film without excessive dry spray at the edges of the fan.

Overspray can cause coating failure by creating areas of pinholes surrounding the dry spray that sticks to the surface. If overspray occurs on the surface, it should be removed by a light sanding of the surface prior to the application of additional coats. Figure 10.55 shows the reason why corrosion may begin where a coating is applied over overspray. The overspray particles actually create a shadow for additional spray over the surface, creating a small bare spot in and around the shaded side of the overspray particle. This is a common occurrence and should not be overlooked during an application. Once these bare areas around the overspray particles exist, it is extremely difficult to obtain a continuous coating over the area, and pinpoint corrosion begins very quickly (Figure 10.56).

Overspray is more commonly associated with the standard air spray equipment than it is with airless spray. Since no air is incorporated into the coating during the airless spraying process, this problem is largely eliminated.

FIGURE 10.54 — A typical overspray condition on a coating.

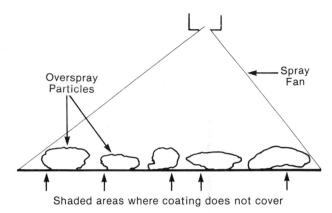

FIGURE 10.55 — Pinhole corrosion due to nonuniform coverage of overspray.

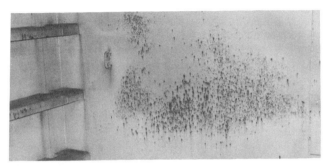

FIGURE 10.56 — Typical example of overspray and the resulting pinpoint corrosion.

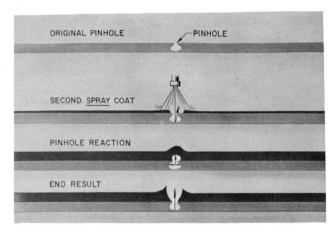

FIGURE 10.57 — The mechanism of pinhole formation.

Pinholing

Pinholing is another problem that can be caused by improper gun adjustment. This usually occurs where there is too much pressure on the coating liquid and too little air pressure to disperse the coating. The pinholes usually occur in the center of the fan where the coating is heavy and where air has been blown into the coating. These small bubbles break and cause the pinhole in the coating. Coatings with excellent flow-out often flow sufficiently well so the pinholes are removed. On the other hand, most of the high-performance coatings are sufficiently heavy-bodied so that the pinhole, once formed, tends to remain.

Once formed in a coating, pinholes are extremely difficult to remove since a second coat usually will not cover the pinhole, as indicated in Figure 10.57. In this case, the pinhole is formed in the original coating, and when the second spray coat is applied, it covers the pinhole, leaving a small bubble underneath the coating. The solvent vapor pressure in the bubble tends to cause it to expand and break, again leaving a hole clear through the entire coating.

Pinholing can be a serious problem wherever corrosion-resistant coatings occur. Gun adjustment must be watched very carefully to see that this is not the cause of the trouble. Pinholing can also occur because the gun is held too close to the surface, not allowing the coating to be sufficiently broken up before it hits the surface.

Pinholing may not be as prevalent with airless spray as it is with air spray, although it can occur with airless

spray as well. In this case, it usually occurs where the gun is held too close to the surface, creating bubbles in the coating. The airless spray should be held a sufficient distance from the surface so that the spray is entirely broken up and as fine as possible as it hits the surface. A wet, even film is the answer to the problem.

Figure 10.58 shows a typical surface with fine pinholes throughout. Figure 10.59 illustrates a pinholed coating magnified to show the actual openings in the film.

Spatter Coat

Spatter coat is a condition that can occur at the end of a spray stroke or where each pass of the spray is not sufficiently overlapped. Spatter coat is where the individual particles of coating from the spray hit the surface, but the coating is not continuous. There is a small area between the individual coating droplets which is bare, even though during the application the applicator may think the surface is covered. Spatter coat is a major cause of holidays in a coating. The area may look as though it is fully covered because of an even coloration; however, insufficient coating has been applied to create a continuous coating of even thickness (Figure 10.60).

Cratering

Cratering is also called *fish-eyeing,* or *bug-eyeing.* There are a number of causes for this kind of coating difficulty. It is usually associated with phenolic or epoxy coatings, and, in some cases, can be traced back to improper coating formulation. The solvent balance within the coating can create this condition. A typical example is shown in Figure 10.61.

Cratering also can be caused by very fine oil droplets in the air used to spray the coating. In this case, the coating does not properly wet the surface and tends to pull away from the fine droplet of oil. It also can be caused by oil in the sandblast air, which is left in a small spot on the surface. These are extremely fine droplets and ordinarily cannot be seen on the sandblasted surface.

Another cause of this problem is particles from the air which have the same charge as the coating. The static condition of the particle causes the coating to be repelled from that area, causing a small crater (Figure 10.62). Since this problem usually is caused by contamination,

FIGURE 10.58 — A pinholed coating.

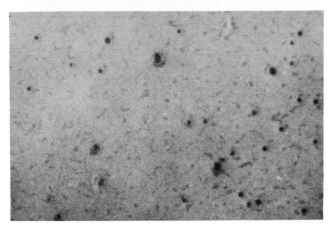

FIGURE 10.59 — A pinholed coating magnified to demonstrate actual pinhole openings. (SOURCE: Steel Structures Painting Manual, Vol. 1, Good Painting Practice, Steel Structures Painting Council, Pittsburgh, PA, p. 106, 1966.)

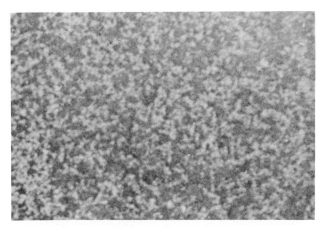

FIGURE 10.60 — Spattercoating.

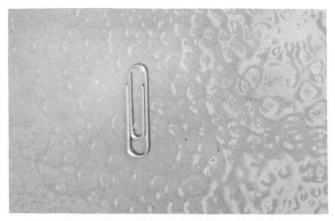

FIGURE 10.61 — Typical cratering or "bug-eyeing" of a coating.

Spot of Contamination

FIGURE 10.62 — Cratering due to the static condition of the particle of contamination which causes the coating to be repelled from that area.

the solution to the problem is to eliminate the contamination. This should be done as soon as there is evidence that the coating is tending to crawl or crater. If the coating is applied and cured in this condition, it is usually very difficult to apply a second coat over such a surface. Contamination still exists and the cratering or bug-eyeing often recurs at the same spot. If the coating has dried in this condition, it is necessary to hand or power sand the surface thoroughly prior to applying the second coat.

If the cratering condition is caused by improper solvent balance, there is little the applicator can do to aid the situation. In this case, the coating supplier should be called in to overcome the problem.

Runs or Sags

Runs or sags are also called *curtains,* and generally are caused by the heavy application of a coating in a certain area, so that, on a vertical surface, the weight of the coating causes it to run down the surface and create runs or curtains (Figures 10.63 and 10.64). This often leaves a thin area above the run or sag. The area above the sag also may contain pinholes. The sag or run itself also may contain bubbles, which can break, causing voids in the coating.

Runs and sags usually are in the direct control of the applicator. If the gun accidentally goes too close to the surface and a run or sag is formed, it should be removed from the area before the coating is dry and the coating reapplied at that point. Where they are left on the surface and the coating dries, these areas should be thoroughly sanded to level out the coating and remove voids or pinholes. Coating should then be reapplied to the area.

Solvent Blistering

Solvent blistering is a condition that can occur in almost any coating which contains a solvent. It may occur more readily with lacquer-type coatings, which tend to dry on the surface rapidly and trap solvent underneath. This does not necessarily mean that it cannot occur with epoxies and other similar coatings which contain solvent and which also may dry on the surface faster than the solvent evaporates from the body of the coating. Solvent blistering usually occurs in areas where the sun hits the

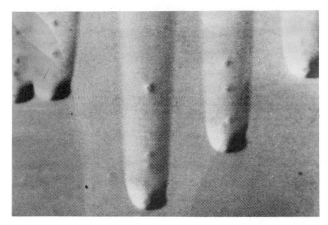

FIGURE 10.63 — Runs in a coating. (SOURCE: Steel Structures Painting Manual, Vol. 1, Good Painting Practice, Steel Structures Painting Council, Pittsburgh, PA, p. 106, 1966.)

FIGURE 10.64 — A coating sag. (SOURCE: Steel Structures Painting Manual, Vol. 1, Good Painting Practice, Steel Structures Painting Council, Pittsburgh, PA, p. 106, 1966.)

coating during the application, causing the surface to become quite warm and evaporating the solvent rapidly. It often occurs with heavy-bodied coatings because of the greater coating thickness.

Should solvent blistering become a problem, one answer to the situation is to apply the coating in the afternoon, after the sun has ceased to create such hot, drying conditions. If this is impossible, the coating should be applied in several thinner coats rather than in one thick coat, since the solvent can evaporate more readily out of the thin coating. This requires additional passes over the surface in order to obtain the proper coating thickness. However, it can eliminate the problem of solvent blistering and the creation of a weak spot in the coating at that point. An example of solvent blistering is shown in Figure 10.65.

Blistering of this type can also occur over inorganic coatings which contain some porosity. In this case, the solvent expands within the porous inorganic coating, causing sufficient pressure to cause blistering of the topcoat. There are many formulations of topcoats available which do not tend to blister over inorganics. These materials usually contain higher boiling solvents which lessen the problem, unless the coating is applied on a very hot surface. Again, if this is a problem, coating in the afternoon can often overcome it. Thin coats and more coats can also help in this situation.

The same condition can also exist over concrete.

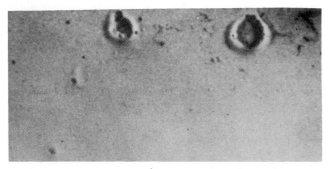

FIGURE 10.65 — Solvent blistering. (SOURCE: Steel Structures Painting Manual, Vol. 1, Good Painting Practice, Steel Structures Painting Council, Pittsburgh, PA, p. 106, 1966.)

Even when hard-troweled, concrete is sufficiently porous that coatings applied over it in the sun may tend to solvent blister. The reason for the difficulty is solvent expansion in the porosity of the concrete, creating a pressure that is sufficient to raise a blister. Coating late in the day, or in several thin coats, can also remedy this condition.

Pigment Separation

Pigment separation is not a common problem. However, one of the areas where it has been noted is where water-base inorganic zinc coatings are applied under cold conditions, i.e., sufficiently cold so that the water does not evaporate from the coating rapidly. If the coating is applied on a vertical surface, the weight of the zinc in the coating tends to separate from the liquid in the coating, causing small curtains where the pigment is concentrated. This leaves areas without pigment, which can then corrode.

Lead-containing coatings applied heavily under slow-drying conditions can also have pigment separation. This situation also can be caused by improper solvent balance. It is usually caused by thinning the coating with a solvent which is marginal, thus causing the pigment to agglomerate. When this occurs, the coating has an uneven appearance, with streaks of the pigment occurring as the coating dries.

In the case of the coating applied where the temperature is too cold, this can be overcome by warming the surface or by waiting for changing weather conditions. In the case of the improper solvent balance, a change in the thinning procedure should be started as soon as it is noticed.

Bleeding

Bleeding is a condition which usually exists in lacquer-type coatings where the first coat is one color and the second coat a different color. If the primer is red and the second coat a lacquer gray, and if the coating is brushed or rolled, the red pigment in the first coat may bleed into the second. This is usually less of a problem where the coatings are sprayed.

Bleeding can also occur where a dye has been used as a pigment in one coat, followed by a second coat which has a solvent that will attack the dye and dissolve it. In this case, the dye then bleeds into the second coat, causing discoloration. In the latter case, there have been all types of remedies suggested for a bleeding surface prior to the application of another coat. These include a coat of aluminum-pigmented coating, a coat of a material dissolved in alcohol (*e.g.,* shellac), or a coat of a thermoset-type coating.

These remedies are usually satisfactory for decorative-type paints. In working with high-performance and corrosion-resistant coatings, such procedures are not recommended since they change the nature of the coating system and may cause failure after a period of time. Such pigment types usually are not used in corrosion-resistant coatings, and therefore there is a limited chance of it occurring. On the other hand, vinyls and chlorinated rubbers are standard for corrosion-resistant coatings and the first situation mentioned can easily occur.

In order to overcome the problem, spraying of the second coat is recommended. A light film should be applied so that the solvent in the second coat does not have time to strike into the first coat and allow the pigment to bleed through to the surface. Once the light second coat is applied, and is dry, a full coat can then be applied over the surface, usually without additional difficulty.

Dirt in the Film

The problem of dirt in the film is not at all uncommon. Applicators have actually been known to coat over a pile of sand on a deck. This is strictly the result of a lack of attention, care, and good application procedures. Dirt and contamination is a problem in many corners and inside angles, and extra care needs to be taken to remove the dirt and contamination before the coating is applied (Figure 10.66). If the coating is applied over such contamination, rapid corrosion may occur since the coating will not be continuous. Where such contamination is overcoated and the coating dried, the area should be scraped and sanded smooth prior to applying additional coats.

Lifting of Previous Coatings

The lifting of previous coatings usually occurs during coating repair and where one coating is applied over a previously dried coating. The previous coating may be

FIGURE 10.66 — Airborne contaminants in a coating. (SOURCE: Steel Structures Painting Manual, Vol. 1, Good Painting Practice, Steel Structures Painting Council, Pittsburgh, PA, p. 106, 1966.)

corroded or damaged, and if a coating of a different type is applied (particularly one which has a rather strong solvent base), it may swell the previous coating around the break and cause it to lift and curl. The application of a coating with a strong solvent over an oil-base material is often the cause of this difficulty. In this case, the second coat should be thoroughly compatible with the first so that such swelling does not occur. A good example is the application of a vinyl over an old alkyd. While this can be done, where the alkyd is thoroughly dried and if there is a break in the original coating, the coating with the strong solvent usually causes the undercoating to swell and curl. This also can occur where the first coating is the same as the repair coat. In this case, solvents in the repair coat strike into the first coat, causing it to swell and lift from the surface.

The answer to these situations is to thoroughly sand the coating smooth around the damaged area, feathering the broken edge sufficiently far from the damaged area so that the coating adhesion is sound. Apply a thin coating so that it can dry quickly. This can be followed by several other thin coats, each well dried before another is applied, to build up the required thickness. Thick coats can cause the undercoat to wrinkle and swell and generally lose adhesion to the surface.

There also is a possible related reaction. When one coat is applied over another after a considerable period of time (months or years), there may be a tendency for the first coat to lose adhesion, although not to the extent of swelling, wrinkling, or peeling from the surface. Nevertheless, the adhesion is weakened, and, if used for immersion or in a critical corrosion service, blistering or other failure can occur. This often happens in repair areas and where an overall refresher coat is applied. Under immersion conditions, the area surrounding the repair often fails due to the loss of adhesion of the initial coating. This has occurred when vinyl and other water immersion coatings have been repaired. It is related to the effect of the solvent in the latest coat on the original material. Some swelling and softening will undoubtedly occur, causing a physical strain on the original bond. This causes some degree of adhesion release of the original coating, which later results in coating failure.

Not all coatings are affected in this manner. Oil-base coatings, alkyds, etc., are less affected since they are strongly cross-linked and their solvents are very mild. High-performance coatings, however, particularly when used for immersion or when constantly wet, can react in this manner. When high-performance coatings are to be repaired and have been used for immersion, it is best to test the adhesion of the original coating before and after it is overcoated. A small patch is usually enough to determine any adhesion change. If there is some change for the worse, it may be more cost effective to remove all of the old coating and start anew, than to repair and risk an additional failure.

Mud Cracking

Mud cracking is a problem that is more common with inorganic zinc coatings than with organic coatings. On the other hand, it does occur with water-base coatings that have a high pigment loading. Mud cracking is usually the result of the coating being applied too heavily so that, as the coating dries, it shrinks sufficiently to cause the coating to crack (Figure 10.67). This generally occurs where there is a high pigment loading in the coating or where there is a rapid drying of the coating in addition to its being applied in a thick film. The usual answer to the problem is to apply the coating under less rapid drying conditions and in thinner coats, allowing the thin coats to dry before applying an additional coat.

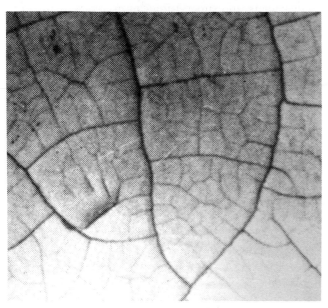

FIGURE 10.67 — Mudcracking of a coating. (SOURCE: Steel Structures Painting Manual, Vol. 1, Good Painting Practice, Steel Structures Painting Council, Pittsburgh, PA, p. 106, 1966.)

Cost of Application

Since the beginning of their widespread use in the 1930s, there has been continuing discussion on the cost of the application of high-performance coatings. All kinds of analyses have been made to justify the cost of high-performance coatings, and, under almost all conditions, the analyses have indicated that high-performance coatings will do the job, last longer, and be more economical than conventional coatings under corrosive conditions. Nevertheless, in the context of this book, it is difficult to establish any specific costs for application, since every application has its unique problems and requirements. No two applications of protective coatings are the same, unless they are being applied to a piece of equipment on a production-line basis. In general, anticorrosive coatings or maintenance coatings are specific for the particular job on which they are being applied.

The exterior of a tank, for example, is a large, broad, generally flat area and therefore the application costs involved are probably at a minimum. On the other hand, the cost of applying a coating to a structural steel installation would be quite different, due to the many edges, corners, bolts, and other complex areas that require coating. Therefore, there is no rule-of-thumb as to the application cost on any particular job.

Also, when considering inflation, there is no way of calculating exact costs which would be applicable today and for a few months in the future. Percentages, however, can provide some continuing basis for application costs. Sol T. Sprayer, in an article entitled *Smart Money Painting*, gives some interesting information from an industry standpoint.

> The chemical industry alone is said to spend more than $600 million a year to fight plant and equipment deterioration resulting from corrosion. On the average, annual painting costs for a chemical plant are said to approximate 0.5 per cent of the capital investment.
>
> Thus, the annual cost of painting a $10 million plant is calculated at around $50,000 per year.
>
> It is interesting to note that the relationship of cost of maintenance painting to capital investment varies with the industry involved. In the heavy chemicals industry, as already noted, the average cost of painting is 0.5 per cent of the capital investment. In the plastics industry it is somewhat lower, 0.42 per cent. The pharmaceutical industry spends the same amount as the plastics industry, whereas the metallurgical industry spends the very low figure of 0.21 per cent.
>
> On the other hand, plants manufacturing organic intermediates spend 0.77 per cent per year of their capital investment for painting, whereas the corresponding figure for the petrochemical industry is 0.62 per cent.[5]

To be somewhat more specific, Table 10.11 provides some percentage costs of the three major components of a coating application. These figures are derived from three different types of coatings, ranging from a three-coat alkyd to an inorganic zinc coat and two coats of high-performance topcoat. The average of the costs for the three different types of coatings was 44% for surface preparation, 36% for application labor, and 20% for materials. While the figures for each of these three components vary for each coating, as is shown in the table, the average figures are indicative of the range of costs for each one of the items.

Probably the most important point in the average is the relatively low cost for the material in an overall application. It only averages 20% for three different coatings. The cost of material, therefore, is small compared with the other costs involved in an application. The importance of this is that even the best coating, or coating system, is still only a fraction of the cost of the surface preparation and application labor. Thus, it should be emphasized that the cost of material is secondary to the other costs involved and that the best possible material is low-cost insurance for the long-time corrosion protection of a valuable structure.

TABLE 10.11 — Cost of Three Major Components of a Coating Application

	Inorganic Zinc and 2 Coats	3-Coat Alkyd	3-Coat Organic High Performance	Average
Surface Preparation	42.6%	40%	49%	44%
Labor	32.0%	46%	33%	36%
Material	25.4%	14%	19%	20%

The material cost is even lower when the total cost of the application is considered. These three major components do not include the other miscellaneous items (*e.g.,* scaffolding, clean-up materials, general overhead, and profit) which would be included in an overall contract. When considering the material in relation to the entire cost, it is an even less significant factor than when comparing these three major components of an application.

There are a number of hidden costs in maintenance coatings which generally do not occur at the time of the application, but are obscured by one, two, three, or more layers of coating and which require time to develop to a point of failure. These are such items as contamination left during surface preparation, exterior contamination of the surface prior to the application of coatings, the character of the underlying surface itself, and imperfections in the application of the coating (*e.g.,* holidays, pinholes, brush marks, or overspray).

These often are not recognized or even cause concern until the point of failure is reached. At this time, the solution or elimination of the problem is often much more costly than the original application. For example, if the original application cost is $1.00/sq. ft., the repair of a failed coating due to some of the hidden deficiencies could cost $2.00, $3.00, $5.00, or even $10.00/sq. ft., depending on the complexity of the structure and its location. The economics of maintenance coatings and their application are, therefore, extremely complex.

While the variables involved are innumerable, they are, however, primarily people-oriented. If an improper coating is selected, if the character of the surface is disregarded, if the design of the structure for corrosion resistance is improper, if the specifications have been written in a vague manner, if the applicator applies the coating to minimum specification, and if the inspection on the overall application job is cursory, it is generally too late for the management of a company to do anything but live with less than the job they paid for. On the other hand, since these variables are all people-oriented, proper management and supervision can overcome many of the costly application problems. This is particularly true when management is aware of the problems and has the help of corrosion engineers who can intelligently recognize and appraise the various factors involved in the application of a coating.

References

1. Wisdorf, Donald J., New Generation of Plural-Component Systems, Modern Paint and Coatings, December (1979).
2. NACE, Determination of Theoretical Coverage Rates of Inorganic and Organic Protective Coatings, NACE Publication 6A184, Houston, TX, 1981.
3. NACE, Koppers Co. Treatise on Protective Coats, Corrosion Prevention by Coats, Course 5, pp. 2-61.
4. Keane, John D., Wettack, William, and Bosch, Wouter, Minimum Paint Film Thickness for Economical Protection of Hot Rolled Steel, Journal of Paint Technology, Vol. 41, No. 533, June (1969).
5. Sprayer, Sol T., Smart Money Paint.

11

Coatings for Concrete

Introduction

Concrete may provide the largest surface area of all construction materials. While major emphasis has been placed on the use of steel as a surface for corrosion-resistant coatings, it is believed that the total area of concrete surfaces may be even greater, although there is a substantial portion of concrete surfaces that do not require coatings. Concrete in itself is considered a corrosion-resistant substance and is often applied over steel to prevent its corrosion. In fact, where concrete is well-placed and dense, it is one of the best corrosion-resistant coatings available for steel. It provides a thick, dense, water-resistant barrier, as well as creating an inhibitive atmosphere that prevents steel from corroding.

Most of the major water lines around the world are either reinforced concrete, cast iron, or steel-lined with a cement mortar coating. This concrete coating has maintained its properties and prevented steel from corroding under water pipe operating conditions for up to 100 years. There are few other coatings that can make an equivalent claim. Concrete provides the best protection when exposed to fresh water and to the atmosphere. Unfortunately, because of its reactive characteristics, concrete requires protection from many common conditions found in industrial and marine areas throughout the world. Thus, coatings are needed to provide this protection.

The amount of concrete requiring coatings is quite extensive because of its use as a primary construction material. For example, it was estimated that approximately 80 million tons of cement were produced in the United States which was capable of producing 250 million tons of concrete and some 30 billion cement blocks.[1] Worldwide, many times this amount is produced. In many countries, concrete is the preferred building material, since steel is generally costly and in short supply.

Concrete has been in use in one form or another since the time of the Romans, some 2000 years ago. Some of the aquaducts of this ancient period are still in use. Many concrete structures that are hundreds of years old include dams, canals, buildings, wharfs, and bridges. This type of durability is one reason for its wide useage. The following are some of the properties that lend concrete its durability.

Properties of Concrete

Concrete is inorganic and is therefore essentially a rock. Very few organisms, such as fungi or bacteria, attack it as they do organic materials. It does not "rot" in the common sense of the term. It is generally unaffected by sunlight, weather, moisture, dryness, or other similar conditions.

Concrete is also hard and does not wear away easily. Its abrasion resistance is determined by the aggregate used. The use of hard, durable granitic aggregate makes it very abrasion-resistant, even though hydrated cement alone is not a highly abrasion-resistant material.

Concrete has good compressive strength, which is its most outstanding property. Few normally occurring conditions, except earthquakes, can cause it to fail in compression.

Concrete also improves with age. Crystallization takes place under water over a long period, thus increasing its hardness and compressive strength. In many cases, such crystallization will actually heal minor cracks in a concrete structure. Since concrete originally contains considerable lime, it reacts with carbon dioxide from the air to form calcium carbonate or limestone. This also increases its hardness and compressive strength.

Concrete has proven a durable building material under most common environmental conditions. However, it also is a very reactive material to many chemical exposures, primarily acidic ones. Even pure flowing water will dissolve concrete under certain conditions. Thus, many concrete deionized water tanks have been lined to prevent water contamination by calcium and metal ions.

Composition of Concrete

In order to better understand the requirements for a coating used for concrete, the basic chemical composition of cement should be discussed. Portland cement consists principally of tricalcium silicate ($3CaO \cdot SiO_2$) and beta dicalcium silicate (B-$2CaO \cdot SiO_2$), together with lesser and variable quantities of tricalcium aluminate ($3CaO \cdot Al_2O_2$), tetracalcium alumino ferrite ($4CaO \cdot Al_2O_3 \cdot Fe_2O_3$), or some solid solution of the iron phase, periclase (MgO), free lime (CaO), and trace amounts of many other compounds.[2]

To produce concrete from portland cement, hydration is required. The chemical reactions which result in the formation of the hard, durable concrete surfaces are shown in Table 11.1. The structural strength of concrete depends on the conversion of the anhydrous calcium silicates to hydrated calcium silicates, as shown in the table. It will be noted in the hydration reaction of dicalcium and tricalcium silicates that a sizable amount of lime is formed, up to as much as 20% of the hydrated product. It is largely this lime content which gives rise to the high alkalinity in concrete, which can be in the neighborhood of a pH of 13 when concrete is saturated with water.

This high alkalinity provides the corrosion resistance to steel that is coated with concrete. Steel is passivated at an alkalinity of approximately 11.5. Saturated concrete, then, provides a good media for the encapsulation of steel and for maintaining the steel in a corrosion-resistant state. However, this strong alkaline condition can also cause problems with the coatings applied over a concrete surface.

The hydration of the concrete or cement particles or the curing of concrete can require some period of time. As a matter of fact, cement continues to cure or continues the hydration process for many years. However, it does develop its primary physical properties within 28 days, if the concrete is maintained with a proper moisture content. Twenty-eight days is the time usually used by engineers to indicate that the concrete has cured to its proper physical state.

This process may be speeded up to a considerable extent by steam-curing concrete. In this process, the concrete is heated to 140 to 160 F in a steam atmosphere, which increases the chemical reactions, and in a few hours the cement will be fully hydrated and the concrete will reach its full strength. Many manufactured products, such as cement-asbestos pipe or concrete pipe, are treated in this manner to speed the curing process. Such treatment is difficult in the field and therefore seldom used.

The hydration process is one which definitely bears on the ability of concrete to be coated properly. In addition to the air and water pockets that have been discussed in Chapter 9, the cement particles and their curing reactions also create porosity. This is explained well in an article entitled *Coatings for Concrete and Other Masonry Substrates* by Joseph E. Ilaria. His explanation is as follows:

The amount of mixing water added is usually at least two

TABLE 11.1 — Hydration Reaction of Portland Cement

$$2(3CaO - SiO_2) + 6\,H_2O \rightarrow 3\,CaO \cdot 2SiO_2 \cdot 3\,H_2O + 3\,Ca(OH)_2$$
(Tricalcium Silicate) (Tobermorite Gel) (Lime)

$$2(2CaO \cdot SiO_2) + 4\,H_2O \rightarrow 3\,CaO \cdot 2SiO_2 \cdot 3\,H_2O + Ca(OH)_2$$
(Di Calcium Silicate) (Tobermorite Gel) (Lime)

$$4\,CaO \cdot AL_2O_3 \cdot Fe_2O_3 + 10\,H_2O + Ca(OH)_2 \rightarrow 6\,CaO \cdot Al_2O_3 \cdot Fe_2O_3 \cdot 12\,H_2O$$
(Tetra Calcium Alumino Ferrite) (Calcium Alumino Ferrite Hydrate)

$$3\,CaO \cdot Al_2O_3 + 12\,H_2O + Ca(OH)_2 \rightarrow 3\,CaO \cdot Al_2O_3 \cdot CA(OH)_2 \cdot 12\,H_2O$$
(Tricalcium Aluminate) (Tetracalcium Aluminate Hydrate)

$$3CaO \cdot Al_2O_3 + 10\,H_2O + CaSO_4 \cdot 2H_2O \rightarrow 3\,CaO \cdot AL_2O_3 \cdot CaSO_4 \cdot 12\,H_2O$$
(Tricalcium Aluminate) (Calcium Mono Sulfo Aluminate)

[SOURCE: Ilaria, Jos. E., Coatings for Concrete and Other Masonry Substrates, 12th Annual Liberty Bell Corrosion Course, Philadelphia, PA (1974).]

times that needed to satisfy the anhydrous cement. However, this excess water is needed especially in poured inplace concrete to make the mix workable and allow it to take the shape of its mold. The excess water evaporates off, in time, giving rise to voids or air spaces within the cement gel.

Figure [11.1] illustrates the physical aspects of cement hydration. The dark center of the grain is surrounded by cement gel in which the hydrates precipitate. When the excess water evaporates off, there is within the gel, pores known as gel pores. These pores may be of any shape but assuming them to be rectangular slits, their size would be in the 2 × 4 micron range. The excess water between grains, due to its relatively low specific gravity compared to the cement compound (1.0 vs 2.5), will also have a tendency to migrate giving rise to channels within the mass. These channels are called capillary pores. Keeping in mind the relative sizes in the cement hardening process (if we magnified one hydrated particle to one inch, a piece of sand of the same scale would have a 10 foot diameter). You can probably visualize the amount of free void space in a hydrated dry cement product. Absorption measurements on high strength concrete (5-600 lbs. cement/cubic yard) which is a measure of the % weight of water a predried (212°F.) concrete specimen will hold can vary between four and six % weight which converts to 10-15% by volume.

The total porosity of cementiferous materials depends on the efficiency of the mix (the sand, stone, cement and water), the type product, the method of curing and the use of certain chemical admixtures.[1]

Much of the detail of cement porosity, pinholes, and water and air pockets, as well as the general character of concrete is covered in Chapter 9 under *Surface Preparation of Concrete*. Proper surface preparation is vital to the effectiveness of a coating, and concrete preparation is much more critical than that of steel. While that information will not be reiterated here, reference should be made to Chapter 9 whenever concrete is to be coated in order to develop the best method of preparing the concrete surface. In this chapter's discussions, all concrete surfaces will be assumed to have been properly prepared in accordance with Chapter 9.

There are numerous types of concrete that may be used with coatings. Some of these are cement-asbestos products such as transite board, cement-asbestos shingles, corrugated cement asbestos roofing or siding, in addition

Corrosion Prevention by Protective Coatings

to the more conventional nonreinforced concrete, reinforced concrete, prestressed concrete, gunited concrete, and centrifugally cast concrete. Concretes may be made with all types of aggregate, ranging from river sand to granitic aggregates, limestone aggregates, and various fibrous types of aggregates such as glass and asbestos. Each of the types of concrete and the different aggregate mixes may present a different surface over which a coating is to be applied, which affects the selection of a coating.

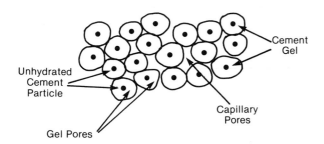

FIGURE 11.1 — Physical aspects of cement hydration. (SOURCE: Ilaria, Jos. E., Coatings for Concrete and Other Masonry Substrates, 12th Annual Liberty Bell Corrosion Course, Philadelphia, PA, 1974.)

Problems in Coating Concrete

There are some physical and chemical properties of concrete that have a direct bearing on coatings. Concrete performs best when under compression, and it has a relatively high compressive strength, *i.e.,* 2000 to 6000 psi. In this state, it resists cracking and checking, and the surface does not appreciably change, even with time. This makes a good, constant surface over which to apply coatings. Unfortunately, the opposite property—tensile strength—is low, so that concrete in tension can crack, making the surface difficult to protect. Cracks can occur due to foundation settling, earth movement (as in the joints of concrete pipe), the vibration of machinery, and many other reasons. Whenever concrete changes from a state of compression to one of tension, cracks appear. Concrete is also brittle, which adds to the cracking problem and often causes spalling.

One of the ways in which cracking is reduced is to prestress concrete or place it in permanent compression. Many concrete tanks, both above and below ground, are of a prestress design which means that the reinforcing is in sufficient tension to maintain the concrete under compression. Cracking, however, is not completely eliminated, as some prestressed cylindrical tanks which are wound with high tensile wire may develop a few cracks parallel to the reinforcing wire. Much concrete pressure pipe is also made using a prestressed design. Each of these physical conditions can create coating problems where concrete is used under corrosive conditions.

Moisture Vapor Transmission

As discussed previously, concrete is porous. It also contains a permanent amount of free moisture that can evaporate or be absorbed, depending on the humidity to which the concrete is exposed. Concrete is saturated when immersed in water, while the moisture content goes down when exposed to low humidity and increases rapidly when the humidity rises above 80%. Concrete can be formed in such a way that it will not transmit liquid water; on the other hand, concrete is not impervious to moisture vapor or moisture vapor transfer, as indicated by its change in moisture content due to humidity.

The fact that water, in the form of vapor, can readily pass through the concrete structure is extremely important to any coating that is applied. For example, in a coated concrete vault which is kept at a constant humidity of 50%, if there is no waterproofing on the exterior, and the earth around the vault is damp, there is a constant positive vapor pressure on the concrete side of the coating. Unless the coating has excellent adhesion, blistering and coating failure will occur. This is not an isolated example. Subterranian structures, basements, subways, and electrical vaults all have problems of this sort. The cause is the vapor pressure passing through the concrete and constantly pushing on the back of the coating. This may be the most important problem encountered in obtaining a proper coating over a concrete surface.

It is difficult to apply a coating on the interior of a concrete surface when the exterior is subject to moisture conditions, and to maintain the interior of the structure free of moisture. During World War II, there were literally hundreds of large underground tanks built to contain petroleum products. Most were of a prestressed design to reduce cracking; however, most did not have any exterior water barrier. They were coated on the interior with a petroleum-resistant coating and many coating blistering problems as well as actual leaks developed. The coating was applied to a generally dry surface, while well points kept the ground water down. Once the well points were removed, it was obvious that the moisture was against the back of the coating. Wherever coatings are to be applied to underground concrete structures, there should be an exterior waterproofing membrane applied.

The alkalinity of concrete also comes into play under such conditions where the moisture is being transferred from the exterior of a structure to the interior. The alkaline salts within the cement can be brought to the concrete surface where the calcium oxide then reacts with carbon dioxide in the air to form rather voluminous white crystalline deposits. In many cases, where somewhat porous coatings are applied, these are actually pushed off of the surface by the formation of the calcium salts which are brought to the cement surface from the interior of the wall. Any time that moisture conditions such as these exist, it is very difficult, without the very best penetration and adhesion, to obtain coatings that will satisfactorily adhere over long periods of time.

In summary, the properties of concrete structures that are made with portland cement are generally as follows:

1. The compressive strength is relatively high, from 2000 to 6000 psi.

2. The product has a relatively low tensile strength,

in the neighborhood of 500 psi.

3. The concrete is alkaline, developing a pH of from 12 to 13.

4. Concrete is porous; when dry, its void volume can vary from 10 to 25%.

5. Concrete contains free moisture and holds it within the structure even under severe drying conditions. The amount of free moisture in a cured concrete structure depends on the relative humidity in which the structure exists.

6. A concrete structure can prevent liquid water from transferring through the concrete. On the other hand, concrete is porous to moisture vapor and will transfer moisture vapor through a concrete structure from a side of high humidity to one of lower humidity.

There are a number of other difficulties or problems which are associated with concrete structures, several of which have a direct bearing on the application of coatings to concrete surfaces.

Cracking

One characteristic of concrete is that it may crack in various ways. Cracks are the result of the low tensile strength of concrete, and may be due to the shrinkage of concrete during the curing process.

Shrinkage Cracks

In general, there are two types of cracking. The first is the cracking caused by shrinkage; such cracks are normally found during the initial curing period of the concrete and many of them are only surface cracks. These cracks are not working cracks and are only subject to stress as the concrete is subject to temperature expansion and contraction. Shrinkage cracks, for the most part, can be mended during the surface preparation procedure. Large shrinkage cracks can be filled; small surface shrinkage cracks can be opened sufficiently to allow the coating to penetrate and seal the crack.

Figure 11.2 shows shrinkage cracks in hand-troweled, sidewalk-finished concrete. They are only surface cracks and thus do not penetrate deeply. Nevertheless, they do represent weaknesses in the surface, which can allow water to be absorbed more readily than into a solid surface, and which must be filled during surface preparation or by the coating.

Figure 11.3 is another type of shrinkage crack. Such cracks often form on walls and similar areas that are plastered with cement plaster. Many times such areas require a high-performance coating, and cracks such as this can cause difficulties such as blistering, lack of coating penetration, etc. If the coating penetrates readily and fills the crack, the area is usually stronger than without the coating. These usually are not working cracks, and proper filling during surface preparation or coating makes a satisfactory surface.

Structural or Working Cracks

Structural or working cracks are those which form due to movement in the structure, ground movement, earthquakes, and similar causes. They can appear at almost any time during the life of the structure. Such

FIGURE 11.2 — Typical shrinkage cracks in troweled concrete.

FIGURE 11.3 — Shrinkage crack in cement-plastered wall.

cracks often take place in areas where the structure changes direction, such as on inside corners. They may also be diagonal across a structure wall due to a slight shift in the foundation. Such cracks are working cracks, which means that as the structure moves, the cracks expand and contract.

Figure 11.4 shows a structural crack in a concrete foundation due to ground movement. It is a working crack and thus moves periodically. Structural (working) cracks usually are much wider and longer than shrinkage cracks, and spalled concrete may exist along the crack proper, as is the case in Figure 11.4.

If a coating has been applied over a structure where a working crack develops, this will crack the coating and will cause corrosion or other difficulties at that spot. Once these cracks have developed, it is difficult to seal them with coating since they periodically move. When this takes place, the coating will tend to crack along with the structure.

Engineers often talk about coatings that will bridge cracks, and many kinds of highly flexible and extensible materials have been tried. Generally, however, there are few materials that adhere tightly enough to concrete to bridge a working crack. The reason for this is that when the coating is applied over the surface of the structure and cracking occurs, there is infinite expansion at the crack almost irrespective of the size of the crack. The crack starts from a point of zero and then expands to 1/32, 1/16, 1/8 in. or more. Even in a 1/32-in. crack, the stress applied on the coating at the point of the crack is much more than the tensile strength and extensibility of most materials (Figure 11.5).

One procedure used to repair cracks is to vee them out and fill the ''V'' with additional mortar or similar materials. If this is a working crack, sooner or later the area will crack again, and if coated over will crack the coating as well.

There are some coatings that have a better chance than others of bridging very small cracks. These are coatings with appreciable thickness, usually 20 to 40 mils or more, as well as some extensibility. In this case, the stress is spread over the total thickness of the coating, which allows the coating closest to the surface to be subject to considerably less stress. Heavy rubber linings, 1/8 in. or so in thickness, which are tightly adherent to the surface, can accomodate some cracking. On the other hand, because of the stress put into the coating across the crack, it is a weak area in the coating film, and while it may not crack immediately, it can do so on aging.

Another procedure is the use of a fairly heavy extensible primer, followed by a strong, tough coating. When cracking, the major stress is accepted by the rubbery primer and is not transmitted to the stronger topcoat (Figure 11.6). It must be stressed, however, that such coatings can only bridge small, narrow cracks or checks and not cracks of any major proportion.

From the previous examples it becomes obvious that thin coatings are not satisfactory to bridge cracks. This is only possible through use of thick coatings and then only for small, so-called ''hair'' cracks. Coatings to bridge cracks are largely a myth where even mild corrosive environments exist.

Cracking of concrete is a serious problem, and one which can cause severe corrosion to either the concrete or the reinforcing within the concrete. A tightly adherent coating does not provide complete protection. Only a material that will spread the strain over a considerable area is practical.

One type of lining material has been designed to

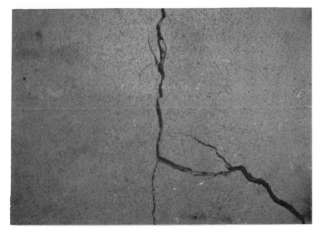

FIGURE 11.4 — Typical structural crack in a concrete foundation resulting from earth movement.

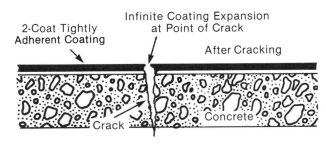

FIGURE 11.5 — Cracking of coated concrete.

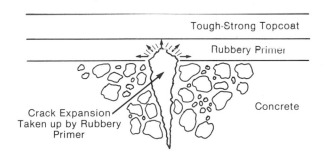

FIGURE 11.6 — Cracking of concrete coated with a heavy extensible primer, followed by a strong topcoat.

resist strong corrosive solutions and to overcome possible cracking problems in a concrete structure. This is a plasticized P.V.C. sheet which has ''T'' extensions on the back. The sheet is placed on the concrete form with the T extensions facing into the concrete. The concrete is then cast around the T's as the plastic sheet is held against the concrete form. In this case, the T's are spaced at 50-mm

(2-in.) to 63-mm (2 1/2-in.) intervals and there is no adhesive holding the sheet tightly against the concrete surface. The principle of this lining material is shown in Figure 11.7.

Should a crack occur, the sheet lining is only tied down at intervals and is free from the surface of the concrete in between. This allows the extension of the concrete at the crack to be taken up by the 63 mm (2 1/2 in.) of sheet between the tees. This places little strain on the lining and it remains intact over the cracked area. A concrete pipe lined in this manner was once dropped from a truck. A 1/2-in. crack occurred under the lining. The pipe was dammed up and a strong sulfuric acid solution kept in contact with the lining over the crack for several years. There was no evidence of acid penetration in spite of the excessive crack in the concrete. This type of lining has been in service for 30 to 35 years in many highly corrosive areas such as sewers, copper leaching tanks, and ferric chloride tanks without any break due to cracks in the concrete structure.

An engineering design made to overcome thermal expansion and contraction and shrinkage cracking is the use of expansion joints. Expansion joints have proven to be satisfactory under many conditions. However, where there is concrete movement, cracking between expansion joints can also take place. From a coating standpoint, there are also many problems in making expansion joints completely corrosion-free, so that each installation requires an individual solution.

Alkalinity

The alkalinity of concrete is an inherent problem, particularly where coatings are concerned. Any coating that is not thoroughly alkali-resistant will eventually break down and be destroyed by the alkali content in the concrete. There is a substantial amount of alkali in concrete which is particularly available whenever moisture is present. When a coating is applied over the surface of concrete, it seals the surface to the internal moisture, and a pH of 12 to 13 is available at the moist concrete/coating interface.

One action of the alkalinity in concrete has already been mentioned; i.e., the formation of efflorescence or a white crystalline deposit on the concrete. This is caused by the moisture passing through the concrete and carrying the soluble concrete salts with it to the surface. The moisture evaporates, leaving the alkaline salts (primarily calcium hydroxide) on the surface. This reacts with carbon dioxide from the air, creating a fluffy, white crystalline deposit on the concrete surface.

When coatings are somewhat porous to moisture or porous to carbon dioxide, this reaction can take place underneath the coating, causing it to blister, with the white, crystalline deposit pushing the coating off the surface. At the same time, some of the concrete surface may also be softened and destroyed by the crystal action. This is not an isolated problem, but takes place wherever there is a source of moisture in the concrete, such as ground water, or moisture coming from cracks, waterproofing, flashing, and similar structural areas.

Figure 11.8 demonstrates the formation of the white

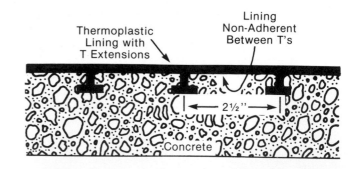

FIGURE 11.7 — A cast-in-place lining for concrete structures.

FIGURE 11.8 — Moisture-developed efflorescence, causing damage to both the concrete and the coating.

salts underneath a coating applied over a concrete block building. In this case, the moisture was coming from a crack in the roofing of the building, getting into the concrete block and coming to the surface in the area shown.

The alkalinity in concrete is also a primary cause for the chemical activity of concrete. Being highly alkaline, it is subject to attack by any material that is even slightly acidic. In the case of a material like hydrochloric acid, the calcium salts and hydration products that make up the concrete are attacked and become soluble. These can easily be washed away, leaving the aggregate without any binder.

In the case of sulfuric acid attack, such as would be the case in a chemical plant, a concrete sewer outfall, or a sewage treatment plant, the acid again attacks the calcium products in the concrete, creating calcium sulfate. The calcium sulfate is not soluble like the calcium chloride; however, its volume is several times that of the original hydrated cement. As the reaction takes place, the calcium sulfate increases in volume. It is a soft product, but because of the expansion in volume, it tends to spall and break the concrete. It pushes off the coatings, and in sewers the expansion has been strong enough to push cast-in-place, tile lining off the surface, and crack the tile lining. In this case, it is a combination of both a chemical and a strong physical attack. Even very slightly acidic products, such as pure water, will attack the calcium compounds in concrete and dissolve them. Concrete flumes bringing pure snow water from the mountains have been found corroded to a depth of an inch or more, purely by the solvent action of the snow water.

Concrete, in spite of its durability to ordinary atmospheric conditions, is a reactive material to many chemical products. This must be taken into consideration when coatings are selected for concrete surfaces.

Freezing and Thawing

The moisture content within concrete can be a problem. As discussed previously, moisture in concrete can cause efflorescence. Water is also responsible for a much more catastrophic concrete failure where concrete is subject to freezing and thawing. During the freezing cycle, the ice that is formed from the moisture in the concrete expands approximately 10%, causing a tremendous physical pressure within the concrete. Because of the relatively low tensile strength of the concrete, it cracks and spalls. This is a major problem with all types of concrete structures that are located within areas where freezing weather can occur. The structures subject to such damage are bridges, buildings, foundations, concrete pipelines, highways, sidewalks, and similar areas. One of the answers to this problem is a proper coating, *i.e.*, one that will seal the concrete against absorption of moisture from the surface and keep the humidity within the concrete as low as possible.

Porosity and Soluble Salts

Also as previously discussed, concrete is porous to moisture. It is also porous to various gases and chemical products. Some of these are the various soluble chloride salts. These occur in seawater, in deicing salts, and wherever the concrete might be subject to sodium chloride. Sodium chloride and calcium chloride, which are the two common soluble chloride salts, are very soluble in water and can be carried into the concrete by the water on the surface and the moisture within the concrete itself.

These salts do not necessarily leave, but are held within the concrete. When the concrete dries, the salts form crystals and spalling of the concrete can occur. When the concrete is re-exposed to moisture, the salts redissolve and then recrystallize on drying, thus starting the disintegration all over again. This cycle of dissolving and recrystalling can be quite damaging to concrete and is a common phenomenon on concrete highways and bridges where deicing salts are used. The problem is also common in chemical plants and is quite prevalent in areas where concrete piers are used to cross salt marshes. The concrete spalls readily and rapidly at the marsh/concrete interface.

The answer to the problem is a coating that will not allow penetration of the sodium, calcium, or chloride ions. Figure 11.9* shows the bow section of a concrete ship after 25 years of exposure to seawater. This area was originally coated with three to four coats of a vinyl coating which was not repaired during the entire time. There is no spalling of the concrete at or directly above the water line due to the vinyl coating sealing the surface against ionic penetration, in spite of the long exposure time.

The sealing of concrete surfaces against salt penetration is not as easy as it appears. It requires a careful and uniform application of a coating, usually of some thickness, in order to prevent pinholes or imperfections from allowing the salt to penetrate and crystallize. Many different kinds of coatings have been used to prevent such ionic penetration, from the vinyls (Figure 11.9) to coal tar epoxy and coal tar cutbacks.

Corrosion of Steel in Concrete

More of a problem than crystallizing and spalling is the problem of chloride and sulfate ions and oxygen reaching the reinforcing within reinforced concrete structures and setting up a corrosion cell. When this occurs, the steel corrodes, forming a voluminous rust scale which expands several times over the original steel volume. With the relatively low tensile strength of the concrete, this expansion creates internal pressure, rupturing the concrete at the point of the salt penetration.

Some such rupturing can be seen in Figure 11.9 on the upper area of the bow where the original wooden bumper was located (part of the bumper still hangs on the surface). The concrete in these concrete ships was extremely dense and very heavily reinforced. The concrete was a minimum depth over the reinforcing, so that any salt penetration would have caused extensive damage.

A similar problem is shown in Figure 11.10* where a reinforced concrete pipe was located in a difficult soil condition. After 15 years, there was sufficient ion and oxygen penetration to set up a corrosion cell, as shown. The concrete over the reinforcing broke away because of the internal pressure of the iron oxide formed due to the corrosion cell.

The more porous the concrete, the more readily this reaction takes place. This is particularly damaging to concrete used for pipe, piling, roadways, bridges, and similar structures. The action of chloride salts combined with freeze-thaw conditions create a serious, continuing problem with any concrete structure.

As indicated in the section above, coatings on concrete subject to salt crystallization, can help eliminate the corrosion problems of concrete-coated steel or reinforced concrete. Coatings can prevent chloride and other ions from penetrating into concrete of high porosity. Coatings, however, cannot prevent chlorides from penetrating into exposed, cracked areas of the concrete. Cracks make easy access both to water and to the chlorides, creating substantial corrosion problems.

Reactive Aggregates

In certain areas, there are concrete aggregates which react with the moisture and alkali in the concrete to cause disruption of the concrete. Normally, the total aggregate is not reactive, since aggregates are usually closely examined for incorporation into a concrete structure. On the other hand, pieces of the reactive aggregate can get in-

---

*See color insert.

*See color insert.

to the structure, and areas have been seen where the aggregate has expanded, spalling the concrete, and rising above the surface between 1/4 and 1/2 in. The expanded aggregate actually extended beyond the concrete surface by that amount.

The possibility of reactive aggregate should be kept in mind wherever there is a critical concrete structure to be coated. The aggregate should be thoroughly examined prior to use in the structure. There is little that can be done if reactive aggregate is present, except to physically remove the aggregate and to repair the concrete in the area. Such aggregate will push coatings off of the concrete wherever the reaction occurs.

Properties Required for Coatings Used on Concrete

There are any number of properties needed by a coating to be effective on a concrete surface. The following are some of the most pertinent.

Coating Penetration of Concrete

The surface preparation of concrete surfaces has been well covered in Chapter 9, and the general application of coatings has been covered in Chapter 10. While these subjects need not be reiterated in this chapter, there is one point worth covering. An area where concrete is quite different from steel is in the penetration of the concrete surface by the coating. When applying a coating to concrete, penetration of the surface is extremely important for the best adhesion and for the best resistance and life of the coating.

Penetration is to concrete what surface profile is to steel. In both cases, maximum adhesion is the goal. Coatings which penetrate concrete have much improved adhesion and increased long-time performance as compared with those that remain on the surface and rely on adhesion alone. One good example of this is the penetration of some liquid epoxy coatings into a well-prepared troweled concrete surface. There are instances where the resin has penetrated as much as 1/4 in. below the concrete surface. This not only makes for maximum adhesion, but increases the density of the concrete surface as well. Not all epoxy coatings will do this. Those based on higher molecular weight solid resins will only penetrate as far as the solvent will carry the resin into the surface. While this is very limited penetration, the more dilute the coating, the greater the penetration.

Penetration is limited by the molecular size of the resin. Some liquid epoxies have a sufficiently low molecular weight and an excellent concrete wetting characteristic so that in-depth penetration is achieved. This penetration is also aided by highly wetting solvents. In this case, the epoxy actually impregnates the surface and then reacts to a solid, making the penetrated area much stronger than the remainder of the concrete. If the concrete is shattered, the penetrated area will usually remain with the coating.

Figure 11.11 is a photo of a cross section of concrete with a liquid epoxy primer (unpigmented) applied to the surface. Note in this case that the primer has primarily penetrated into the concrete, with little of the primer re-maining on the surface. The concrete was sawed to provide the cross section. The sawn surface was etched with acid to bring out the depth of penetration of the primer. The penetration in this case was approximately 3 mm (1/8 in.). Both halves of the concrete are shown in the figure; the upper half has not been acid etched, the lower half has been etched. The penetration can be seen in both sections; however, it is more obvious in the lower section.

It is this penetration principle which has been used for many of the epoxy coatings that have been developed especially for concrete. Many of both the spray-applied coatings and the trowel-applied, heavy-bodied coatings for nuclear power plant reactor areas have been designed to penetrate, and have thus reduced to a minimum blistering due to deionized water immersion and D.B.A. (design basic accident) testing. The D.B.A. test is a combination of steam, high-temperature, pressure, and vacuum and is a very difficult test of coating adhesion.[3] Coatings with a strong penetration quality perform best under such test conditions and on continuous immersion.

Vinyl coatings were the first of the high-performance coatings for concrete, and penetration of the vinyl polymer was the initial method of obtaining adequate adhesion. Vinyl coatings are good concrete coatings because of their chemical properties, and a small penetration can be obtained by applying the vinyl primer as a very dilute solution. On a well-etched concrete surface, the dilute primer will penetrate to some degree (not nearly to the extent of a liquid epoxy) and will improve the vinyl coating adhesion to a marked degree. One of the standard procedures in the early nuclear energy programs where vinyl coatings were used was to apply a thin, clear vinyl resin solution to the surface as a preliminary step before applying the heavier coating. Good adhesion resulted even under immersion conditions.

FIGURE 11.11 — A concrete surface penetrated by an epoxy primer.

Corrosion Prevention by Protective Coatings

The surface of concrete that is best for penetration is a surface which has been either blasted or acid etched to a fine sandpaper finish. In this case, all of the concrete laitance is removed from the surface, thus allowing for maximum wetting and maximum penetration of the coating into the concrete. All of the small hair cracks, checks, and pores are opened by this surface preparation, and there is usually easy penetration of the coating into these areas. Following the penetrating coat (primer), the surface should be fully sealed by the application of additional coats.

The recommended application procedure for almost any coating on concrete is to dilute the first coat and apply a very wet film. The coating should not run or curtain, but should be as heavy as possible without running, to allow the liquid to penetrate the surface. The application can be by brush, roller, or spray, as long as a sufficient amount of the liquid coating is applied and allowed to penetrate.

Penetration of the concrete surface provides physical adhesion. On steel, the reactive sites on the metal provide chemical or polar adhesion. In both cases, substantial improvement in adhesion results.

Low Molecular Weight

Low molecular weight is an extremely important property of any coating that must be highly penetrating. The molecular size of the resin molecules should be sufficiently low so that there is easy penetration between sand grains and cement pores, as well as into the various minor cracks, checks, and openings in a concrete surface. This allows the coating to penetrate as deeply as possible beneath the concrete surface.

There are many low molecular weight materials. Many of the oils have a relatively low molecular weight and are very penetrating. On the other hand, the oils are soon turned to soap because of the high alkalinity of the concrete, and they are therefore not satisfactory for this purpose. Many low molecular weight materials also remain liquid and therefore cannot develop into a strong binder. Some of the vinyl and chlorinated rubber coatings contain relatively low molecular weight additive resins. These lower molecular weight materials aid in the penetration of concrete surfaces and help to bind the higher molecular weight vinyl and chlorinated molecules into the surface. Polyester resins can begin as low molecular weight materials and solidify on curing. On the other hand, these materials are subject to attack by alkali, and, while more resistant than the drying oils, would still be somewhat questionable where strong corrosion-resistant surfaces are required.

Of all the materials that fit into this category, the catalyzed liquid epoxies are undoubtedly the most effective. Even with the epoxies, there are materials with lower molecular weight and greater penetrating power than others. The liquid epoxy resins combined with liquid curing agents and strongly penetrating solvents provide the best answer.

Wetting the Surface

In many ways, the property of wetting goes along with the low molecular weight. Generally, most low molecular weight materials will wet and penetrate the concrete surface well. To properly wet a surface, the coating should not bead or stand up on the surface, but should readily flow out over the surface, spreading itself. If a drop is applied to the concrete surface, the drop should spread out over the surface without any additional physical effort required. Many solvents aid in this. It is preferable to use somewhat slower evaporating solvents rather than fast drying materials to develop the maximum wetting of the concrete surface. The liquid epoxy coatings, because of their excellent compatibility with cement and their resistance to alkali, wet concrete surfaces very well.

Alkali Resistance

Any coating that is to be applied over a concrete surface must have good chemical resistance, particularly on the alkaline side. With a possible pH of 12 to 13, any tendency to react with alkali will create a deficiency in the adhesion of the coating. In many cases, lack of alkali resistance can go beyond adhesion to the point of actually saponifying the coating and causing it to disintegrate.

The two best known alkali-resistant coatings are those based on vinyl and epoxy resins, and both make good coatings for concrete. The vinyl coatings have broader weather and general chemical resistance than the epoxies, while the epoxies have the properties of penetration and wetting which gives them maximum adhesion.

Increased Density of the Surface

Increased density is a result of the penetration and wetting of the coating. There is only one coating type that really improves the density of a concrete surface, and that is the catalyzed liquid epoxy coating. Even without overcoating, the penetration of the liquid resin can flow into checks, cracks, and between sand grains and increase the density to a point where there is little water absorption or chemical activity. Figure 11.11 demonstrates this property, showing the depth of the penetration into the concrete porosity on the surface. This not only improves the adhesion of whatever coatings are applied over the top, but it increases the resistance of the surface to both chemical and physical attack as well.

Appreciable Thickness

Coatings applied to concrete surfaces should have appreciable thickness. It is more important to have a thick coating over concrete than it is over a smooth steel surface. Not only should the coating penetrate the concrete, adding to its thickness, but it should have a substantial thickness over the surface to cover any imperfections and irregularities of the concrete. Appreciable thickness also helps in the minor shrinkage cracks which take place on the concrete surface. Without some thickness, there is no resistance to the stresses involved. As previously discussed, coatings highly adherent to concrete will not resist working cracks or even large shrinkage cracks. On the other hand, coatings in the 10 to 20 mil category should be capable of bridging hair cracks and checks, particularly if the cracks are in the concrete surface before the coating is applied.

The connotation of bridging cracks is somewhat different than that of the application of coating over an existing crack. When the term *bridging* is used by engineers in a specification, it means that if the concrete cracks after the coating has been applied, the coating will stretch over the crack and act as a solid bridge, preventing liquid penetration or corrosion from occurring. As mentioned, this is impossible with thin coatings, and very difficult, if not impossible, with coatings of some thickness. It is only the specially designed coatings and linings described previously that will perform satisfactorily under most cracking situations.

Resilience - Flexibility

Coatings for concrete need to have some resilience, flexibility, and extensibility. Coatings for steel do not require these properties to the same extent as those for concrete since steel is a solid base. If the coatings have good adhesion, they need only slight resilience to perform well. Concrete, on the other hand, is a much softer and brittle base, so that coatings for concrete need to be somewhat resilient, extensible, and flexible in order to follow any shrinkage or expansion and resist the impact to which concrete is frequently subjected.

This is particularly true on floors subject to traffic. The coating should not be rubbery, but should be tough and resilient. When this property is combined with penetration into the surface so that strong adhesion results, the coating will have good wear resistance.

Adhesion

Adhesion is undoubtedly the primary property required for concrete coatings, and many of the previously mentioned properties aid in obtaining the necessary adhesion. It is one thing for the coating to have adhesion and to retain the adhesion when the concrete is dry; it is another problem when it is subject to moisture vapor pressure or even to liquid water within the concrete and directly underneath the coating. There are many cases where this condition prevails, particularly on subsurface structures such as basements, underground tanks, and similar areas. The coating should have the penetrating power described, and a thorough wetting characteristic for the concrete surface; it should retain the penetration and the adhesion, even though there is water within the concrete adjacent to the coating.

In addition to adhesion, the coating needs to have appreciable thickness and be physically strong in order to resist the back pressure of the water. If it did not have this property, it would not be satisfactory for any area where water pressure can develop within the concrete and underneath the coating. This is a common problem, and blistering of coatings on concrete where such back pressure exists or where the coating is immersed is also common. A coating can maintain its integrity only with a strong bond to the surface. The concrete also needs to be strong (in tension), since most coatings have a higher tensile strength than concrete and with back pressure can actually rupture the concrete surface in the blistering proc-

ess. A weak concrete surface, or one containing concrete laitance, will easily cause a coating to blister.

Abrasion Resistance

All of the properties discussed in the previous sections have a bearing on abrasion resistance. Abrasion resistance is important with most concrete coatings since most concrete surfaces are subject to hard usage, impact, and wear. This is especially true of floors where walking, sliding of heavy objects, and wheeled traffic are normal conditions. Any coating subject to these conditions should wear out rather than chip and peel. The coating systems that have the greatest abrasion resistance are liquid epoxy-type floor toppings which contain hard abrasion-resistant aggregates, or a polyurethane coating system using a liquid epoxy primer for penetration and adhesion, followed by abrasion-resistant urethane topcoats.

Reasons for Coating Concrete

If concrete is such a durable material itself, then why should it be coated? Its various properties, which have been previously described, provide the reasons.

Decoration

Concrete is a rather drab, dull, gray material without much overall attraction other than its utilitarian aspect. Decoration, therefore, is the reason for coating the majority of concrete surfaces. Almost any alkali-resistant, weather durable, color coating can satisfy this requirement. The coatings used range from stucco color coats to many different vinyl and acrylic water- and solvent-base materials. These materials are usually not the high-performance coatings with which we are most familiar; however, this does not mean that the high-performance materials should not contribute to the decoration of the concrete structure in addition to their corrosion-resistant purposes.

Waterproofing

The prevention of water penetration into concrete structures, either above or below ground, is also a major reason for coating concrete. Without coating or waterproofing, interior surfaces can be damp and efflorescent. With coatings that do not provide waterproofing, the surface can be peeled, blistered, or thoroughly disrupted by the moisture. Waterproof coatings can overcome these problems by preventing the water or moisture from passing through the concrete, and are thus essential for most subterranean structures.

Chemical Resistance

Chemical resistance is another significant reason for coating concrete. Since concrete is such a reactive material, it is essential to keep it separated from other reactive materials, either to prevent corrosion to the concrete or to prevent contamination of the chemical solution. Most food processing plants coat their concrete surfaces, not only to prevent food acids from attacking the concrete, but to maintain sanitary conditions in the plant and to prevent food contamination.

Corrosion Resistance

Corrosion resistance for concrete is another common reason that coatings are needed. Chemicals, acids, food acids, carbonic acid solutions, pure water—all can take their toll on uncoated concrete.

In order for a coating to provide full corrosion protection, it must form and maintain a continuous film over the surface. This is often difficult because of the variable surface presented by the concrete. To provide a continuous film, the coating must have sufficient thickness and film strength to thoroughly cover the entire surface and all its imperfections. Any lesser coating could only be used for decorative purposes.

Freeze-Thaw Cycles

Concrete, because of its moisture content, is susceptible to damage by continuing freeze-thaw conditions. The physical forces of ice are greater than the concrete's strength, thus causing the concrete to spall and shatter. The only prevention method is to maintain the concrete with as little contained free water as possible, and coatings seem to be the only practical way to do this.

Reinforcing Steel

Reinforcing steel in all kinds of concrete structures can be seriously corroded if the concrete is porous to chloride, sulfate, other less common ions, or oxygen. Most of these materials, if left unchecked, will cause corrosion cells to form on the reinforcing steel, thus breaking and spalling the concrete in the corrosion cell area. High-performance coatings, being resistant to the passage of ions and oxygen, provide a practical answer for the protection of concrete and the reinforcing steel imbedded in it.

Types of Coatings for Concrete

As previously outlined, the reasons for coating concrete are quite numerous. The concrete installations normally coated include floors, buried pipe, exteriors of structures and tanks, tank interiors, nuclear power contaminent vessels, interiors of sewer pipe, water reclamation equipment, etc. These structures need protection against a variety of conditions such as food products, food chemicals, fats, oils, sewage, corrosive soils, chemical plant waste and fumes, radioactive chemicals, brines, and even fungus and bacteria growth. The following are some of the more common coatings used for these purposes.

Bituminous Cutbacks

Bituminous cutback coatings are solvent solutions of coal tar or asphalt. The word bituminous includes both, and both have been extensively used on concrete. The coal tar cutbacks have better chemical resistance and better water impermeability than the asphalts. On the other hand, the asphalt cutbacks have better weather resistance and sunlight resistance than the coal tars. The coal tars tend to alligator and check when applied where they are unprotected from the weather. Asphalt cutbacks are used to line concrete water pipe and water storage tanks where no odor or taste can be tolerated. A thin coat is generally applied as a primer, followed by heavier coats over the surface to resist water penetration.

Heavy and rather thick-bodied coal tar cutbacks do an excellent job in protecting concrete from chemical attack and water absorption and in aiding in the flow rate of concrete pipe. These materials generally wet concrete well, making for good adhesion without extensive surface preparation. They are not attacked by ground bacteria. They have been used on concrete sewer pipe, catch basins, filter tanks, and underground storage basins, and they make excellent waterproofing for the exterior of foundations and structures. As a waterproofing for concrete structures, they may be applied as a single coat, although more often they are installed as a built-up membrane of several coats, including fiberglass mat for reinforcing. Both coal tar and asphalt may be used in this way.

Some bituminous cutbacks are used as concrete penetrants. In this case, they are applied as a thin material and allowed to soak into the concrete, increasing the surface density and helping to minimize water penetration and concrete spalling due to freezing. Bituminous coatings are also available as water emulsions. These materials have many of the properties described above. However, they do not have the penetration characteristics of the coal tar cutback, nor are they as effective for continuous immersion or wet conditions. Their properties are better suited for spillage or spray, or for relatively dry areas where a black, reasonably chemical-resistant coating is satisfactory.

Prior to applying emulsions to concrete, the surface should be dampened, but not running wet or puddled. This effectively thins the material at the material/concrete interface, allowing deeper penetration and greater bond or adhesion. This also precludes the water in the emulsion from being "sucked" suddenly into a dry concrete "sponge." If this later occurs, there is a risk of "sucking" resin into the concrete, leaving a powdery, chalky layer of pigment on the surface. Later, subsequent coat adhesion problems could develop.

Emulsions cure by water evaporation and resin coalescence, leaving a breathing-type film. This film will allow moisture vapor to pass through without lifting the film, but will not permit water to pass because of surface tension. This characteristic is only important if the total coating system is emulsion based.

Chlorinated Rubber Coatings

Chlorinated rubber coatings have the water and chemical resistance, as well as the adhesion required for concrete coatings, including the all-important alkali resistance. They are lacquer-type and therefore dry relatively rapidly to form a good resistant film on the surface. They have been used extensively for the coating of concrete water tanks and swimming pools, and, where properly and carefully applied, have done an excellent job. They are specifically formulated for tough, abrasion-resistant concrete floor enamels. They can be supplied in many colors and offer excellent adhesion. They perform well under conditions of high humidity and they can be recoated with relative ease once the previous surface has been thoroughly cleaned. Good adhesion results from re-coating. Since these are solvent-sensitive coatings, the new coating tends to bite in and thus satisfactorily adhere to the previous coating.

Chlorinated rubber coatings are not satisfactory for sewer coatings or sewer linings, as they are not resistant to animal and vegetable oils and greases, and are rather quickly softened when these materials deposit on the surface. The chlorinated rubber coatings for use on concrete should not be modified with drying oils or alkyds, but with low molecular weight, chemical-resistant, permanently plasticizing-type resins. This is important, since the chemical resistance of chlorinated rubber is reduced by the amount of any oil-type modifiers.

Vinyl Coatings

Vinyl coatings have been used for many years as a coating for concrete of all types, from tank linings to nuclear energy chemical installations and concrete floors in food plants.

Figure 11.12 shows large keyed concrete sections used as radiation shields at a nuclear energy chemical extraction plant. The heavy vinyl coating was subject to heavy radiation densities, radioactive materials, and decontamination chemicals. Any penetration of chemicals into the concrete would have been both dangerous and damaging.

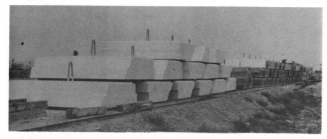

FIGURE 11.12 — Keyed concrete radiation shields coated with a vinyl coating.

Vinyl coatings have excellent chemical resistance to both acids and alkalies, as well as being resilient and flexible. Vinyl coatings, because of their high molecular weight, should be preceded by a thin primer for maximum penetration into concrete surfaces. Heavy-bodied, thick, viscous vinyl coatings do not provide adequate penetration. However, when combined with adequately thinned vinyl primers, they have proven to have excellent resistance and adhesion over a period of many years.

Vinyl coatings are lacquer-type materials which dry rapidly. The primer should be applied as a heavy, wet coat in order to obtain maximum penetration into the concrete. Since they dry rapidly, additional coats often can be applied within minutes, which is considered an advantage. However, there are also disadvantages in this fast drying process. One is the possibility of overspray. The second is surface drying of the film while solvents remain within the pores of the concrete surface. These solvents are reasonably volatile, and if the concrete surface is warm, as from exposure to the sun, solvent blistering of the vinyl coating may occur. This can be avoided by application during a cooler time of day.

Vinyl coatings generally have relatively low solid content, and this may require additional coats in order to build the desired thickness. Again, thickness is important, particularly in areas of abrasion and where the vinyl coatings are used as tank linings. For exterior purposes, vinyl acrylic topcoatings provide outstanding weather resistance.

Epoxy Coatings

Epoxy coatings have been well proven in concrete service. While all epoxy coatings adhere well to concrete surfaces, the ones that are outstanding are those formulated with liquid epoxy resins, liquid curing agents, and highly penetrating solvents. This combination allows the epoxy resins to penetrate deeply into the concrete surface, react, and increase the density as well as the strength of the surface of the concrete. There are many of these epoxy coatings available, and, where properly formulated, they will provide a coating with excellent chemical resistance and maximum adhesion to the concrete surface.

Epoxy coatings have been used extensively throughout all industry as coatings for concrete areas such as floors, tank linings, pump bases, etc., as well as in nuclear power plants. Here, they have been used as linings for the interior of chemical process areas, nuclear fuel storage areas, waste fuel storage areas, and many areas where coatings are not only subjected to high radiation chemicals, but to continuous immersion in ionized water. One such application is shown in Figure 11.13 where an epoxy coating was applied to the interior of a small concrete nuclear test reactor.

FIGURE 11.13 — A concrete nuclear test reactor coated with a multiple-coat epoxy coating system.

Another known application of an epoxy coating was on concrete surfaces in a chemical reaction area of a nuclear extraction plant. The coating was dry, but subject to high radiation. This concrete reaction chamber was broken into after 17 years of service, and the epoxy coating was completely intact, adherent, glossy, and, except for a slight yellowing, appeared to be entirely unchanged.

There are several types of epoxy formulations, each

of which has its place in coating concrete:

1. Epoxy coatings, relatively thin, applied over concrete surfaces which are prepared as outlined in Chapter 9. These are usually solvent-based epoxies using relatively high molecular weight resins and are quite similar, if not identical, to many epoxies used for steel.

2. Thin thixotropic liquid epoxy-based primers which can be applied to the original concrete surface and provide both surface preparation and a base for other epoxy topcoats.

3. A thick epoxy surfacer applied by trowel, or a combination of spray and trowel. This is applied directly to a clean, but otherwise unprepared concrete surface. It also fills the concrete surface imperfections and can be used alone or with additional coats of conventional epoxy topcoats.

The conventional epoxy coatings are applicable to and have been used extensively on concrete surfaces. They have the necessary alkali resistance, are fully compatible, and have good adhesion to concrete. The adhesion is benefited by some thinning of the primer for as much penetration as possible.

These coatings may be either the amine-cured or polyamide-type epoxy and should be applied over smooth, filled concrete which has been etched or blasted to a fine sandpaper finish. These are normally relatively thin coatings, 8 to 15 mils in thickness, and a smooth concrete surface is necessary to obtain a continous film. They are used in chemical plants, food plants, paper plants, and nuclear plants as a general purpose, chemical-resistant coating. They are generally used in areas of fumes, chemical dusts, splash, or areas requiring possible decontamination, but not for immersion in difficult chemicals. Figure 11.14 is an example of a large concrete surface in a nuclear chemical processing plant which was being prepared for application of this type of epoxy coating.

FIGURE 11.14 — The interior of a nuclear chemical processing area prior to application of a conventional epoxy-type coating.

The second type of epoxy coating is one which is a combination of a thin thixotropic base coat (primer) with the conventional epoxy topcoats described above. The base coat is a combination of liquid epoxy resins and highly penetrating solvents which penetrate and have excellent adhesion to concrete. This may be applied by spray over a clean, but otherwise unprepared concrete surface, and then squeegeed over the surface to fill all of the imperfections common to poured concrete or concrete block. As little as possible of the base material is left on the surface.

Figure 11.15 shows a concrete surface with the base material applied and with all of the normal concrete imperfections filled, including the form tie hole in the center. Notice also the dark color of the concrete, indicating the thorough wetting of the surface by the epoxy base coat.

Figure 11.16 is a closer view of a concrete surface, showing both the base coat and the conventional epoxy topcoat. These coatings are not exceptionally thick, but do provide a dense, impervious surface over even exceptionally rough concrete. Excellent adhesion to concrete is one of this coating system's primary properties.

FIGURE 11.15 — Poured concrete surface coated with a thixotropic base coat which provides excellent filling and wetting of the surface, including the form tie hole in the center.

FIGURE 11.16 — Poured concrete surface coated with the thixotropic epoxy base coat (top), followed by the conventional epoxy topcoat (bottom).

Such concrete coating systems have been used to a considerable extent by the nuclear power industry in critical areas which are exposed to radiation and which may require decontamination. Even concrete water storage areas and retention basins for used nuclear fuel have been coated. Such coatings may, however, be used on any concrete surface subject to water immersion, immersion in mild chemicals, or in areas of dusts, splash, or spray.

The third system is a combination of a relatively heavy concrete surfacing material, which may or may not be followed by conventional epoxy topcoats. The concrete surfacer is a mixture of liquid epoxy resins and catalysts with finely graded aggregate to form a material that may be applied by spray or trowel. There is sufficient liquid epoxy present so that it can effectively wet and penetrate the concrete surface to obtain maximum adhesion. The surfacer is applied to the surface in a sufficiently thick layer so that it not only fills the imperfection in the concrete surface, but provides a surface layer with a thickness of 1.6 mm (1/16 in.) to 3.2 mm (1/8 in.).

Figure 11.17 shows a cross section of concrete with an epoxy surfacer, followed by a conventional epoxy topcoat. The surface imperfections in the concrete are well filled and the surface is smooth and dense.

Figure 11.18 shows the spray application of a concrete surfacer to a drainage trench in a plating plant. Such a trench is exposed to plating chemicals, including strong acids such as chromic and sulfuric. Drainage areas such as this are extremely difficult to protect from corrosion. However, when carefully and properly applied, the epoxy surfacers do an excellent job.

Figure 11.19 shows an epoxy surfacer being applied by trowel to a concrete wall. Trowel application is usually, but not always preceded by spray application of the heavy surfacer. The troweling then brings the surfacer to a smooth, even, dense surface. This procedure, when followed by one or more coats of epoxy topcoats, makes a smooth, dense, chemical-resistant surface for any concrete structure subject to corrosion and abrasion.

Concrete surfacers such as this are used extensively as a coating for concrete floors that are subject to abrasion and corrosive chemicals. The finely graded aggregate in the surfacer is tightly held in the epoxy matrix and makes a highly abrasion-resistant surface. The combination of strong adhesion to concrete, abrasion resistance, chemical resistance, and density makes it well suited for difficult floor conditions. These same properties make it applicable for immersion in many chemical solutions (except strong acids).

The three types of epoxy coatings provide a versatile answer to the coating of concrete under many different exposure conditions. These coatings are manufactured by several high-performance coating producers who specialize in coatings for severe exposure conditions. While each manufacturer has his own proprietary system, they generally follow the description of the three epoxy coating types.

Coal Tar Epoxies

Coal tar epoxies might be classed as a fourth type of epoxy coating that is uniquely applicable to concrete.

FIGURE 11.17 — Cross section of concrete treated with an epoxy surfacer and coated with a conventional epoxy topcoat.

FIGURE 11.18 — The spray application of a concrete surfacer to a chemical drainage trench in a plating plant.

These materials combine the useful properties of both the coal tar and the epoxy into a particularly effective coating for concrete surfaces. They require application over a properly prepared concrete surface, as they can be classed as a comparatively thin coating: two coats at 15 to 20 mils thickness. Thicker applications can be made under certain circumstances. The coal tar epoxy coatings have good adhesion to concrete, are chemical-resistant, abrasion-resistant, and relatively easy to apply by roller or conventional or airless spray. They can be used wherever a black coating is acceptable.

An amine-cured coal tar epoxy coating is one of the few coatings that will effectively withstand the action of sewer corrosion, bacteria, H₂S, and acid condensate. Therefore, they have been used extensively as coatings for concrete structures in this field.

FIGURE 11.19 — The trowel application of an epoxy surfacer to a concrete wall.

Special Materials

There are some concrete structures that are subject to sufficiently strong corrosion conditions that require specially designed lining materials. One of these areas is that of a septic sewer where corrosive sewer gases develop. Several conditions combine to create the corrosion problem in sewers. These are a relatively warm atmospheric condition, high-sulfate water, and a relatively long retention time or life of the effluent. These conditions are most common in the more southern areas of the country, the west coast, the middle south, and the southern east coast. The more northern area also may have problems, depending on the sewage effluent; however, since the average temperature of the sewage is lower, there is less overall bacterial action, creating less H₂S and less sulfuric acid.

While the subject of sewer corrosion has been discussed previously, it is worth reiterating due to the increasing need for efficient waste water systems and treatment plants throughout the United States and the world. These plants and the concrete pipe and structures that go along with them, represent large areas of concrete. Wherever there is a serious septic sewage problem, corrosion of concrete surfaces occurs. The problem is the chemical reactivity of the concrete, which creates a need for a strong, impervious coating or lining. There have been literally hundreds of coating materials tested under

sewage conditions, and very few, if any, have been effective where sewer corrosion has been severe. There is no known sprayed-on coating that has been effective on concrete under severe weather conditions for any economical length of time.

The cause of the reactions, as discussed in previous chapters, is a biological one. The bacteria and the liquid sewage react with the sulfate and sulfide products that are in the sewage, creating hydrogen sulfide. The hydrogen sulfide is consumed by the bacteria that are in the crown of the sewer or in the area of the treatment plant exposed to air. These bacteria form a slime on the concrete surface and absorb the hydrogen sulfide, oxidizing it to sulfuric acid. The sulfuric acid content on the exposed areas has been measured to as high as 10%.

The corrosion to concrete is thus quite severe, with a constant source of such an acid on the surface. Not only is the acid constantly being renewed, but it is freshly formed on the surface, making it extremely penetrating. It will penetrate even the most minute imperfection in a coating. The newly formed acid penetrates and attacks the calcium products in the concrete, creating calcium sulfate. Calcium sulfate occupies several times the volume of the original calcium compounds. This creates a physical problem in addition to the chemical one. Figure 11.20 shows the interior of a flowing sewer that is unprotected. Note the exposed aggregate above the water line in the concrete sewer.

In many areas where there are such serious corrosion problems, concrete will disintegrate at rates as high as an inch or two a year, rapidly exposing the reinforcing steel to sulfuric acid attack. Figure 11.21 shows the interior of a sewer structure. In this case, the structure was previously coated. However, note the complete coating failure, with the heavy attack on the concrete surface. The pile of calcium sulfate debris on the shelf above the pipe shows the

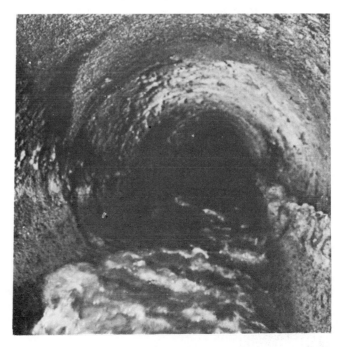

FIGURE 11.20 — The interior of an unlined concrete sewer showing exposed aggregate above the water line.

extent of the reaction.

Many coatings have been applied to such concrete sewer structures and piping. Almost all have resulted in a complete failure similar to that shown in Figure 11.21. The serious nature of this type of corrosion has not been understood, and it was generally believed that any coating that would withstand sulfuric acid would solve the problem. Unfortunately, because of the nature of the concrete, such coatings are subject to imperfections, and even a minute imperfection will cause the concrete to disintegrate under the coating, pushing it from that particular area.

Most of these structures are also buried and are thus subject to earth movement. This is particularly true in piping. Where movement occurs, some cracking, particularly at the pipe joints, may also occur, which opens the concrete to immediate attack. Many early concrete sewers were tile lined. However, under such conditions, they were completely spalled from the concrete surface within a few years due to the penetration of the acid into the joints between the tile.

FIGURE 11.21 — The interior of a previously coated concrete sewer structure showing complete coating failure and the heavy accumulation of calcium sulfate that formed and then broke away from the surface.

Because of the harsh corrosive conditions in sewers, a different approach was developed for the protection of formed concrete. A cast-in-place molded or extruded plastic lining which could be heat welded to itself was formed with "T" extensions on the back of the sheet to hold the sheet in physical contact with the concrete. The principle of this lining has been described and shown previously in Figure 11.7.

The sheet lining is formed into sheets of the size required by heat welding the sheets together. The sheets are then placed on the form and the concrete poured in a normal fashion. Many millions of square feet of surface have been coated this way in sewer applications. However, the same principle has also been extended to chemical areas such as the sulfuric acid leaching of copper ores. Joints between the plastic sheets can be heat welded by hand fusing the two plastic surfaces together, forming a complete and impervious plastic lining for the entire surface.

Coatings such as this for concrete structures have been in use since the late 1940s and, under sewage conditions, have been in continuous use since that time. The key to the success of such a coating has been: (1) its ability to bridge structural cracks which may occur; and (2) its appreciable thickness that can withstand the tensile loads that may be found should some damage occur. Because the lining is either molded or extruded, it can be made completely impervious. While there is the possibility of human error in the welding, this can be overcome with proper training. A continuously plastic-lined tank or structure can be obtained in this manner.

The waterproofing of large structures, where waterproofing is a very critical item, also has followed the same plastic lining procedure as just described. The extensions on the back of the plastic sheet are small knobs, but the principle of their use is the same. Cracks can be bridged without damage to the sheet. In this case, the plastic lining is either cast on the exterior of the structure or is cast as a lining between two sections of concrete to prevent water or moisture vapor from reaching the interior of the structure.

Figure 11.22 shows a large plastic waterproofing membrane or liner being held in place prior to casting a heavy structure against it. In this particular case, it was a large, heavy concrete mass in a nuclear power plant. By heat welding the joints, complete waterproofing of the structure is possible.

The coating and lining of concrete is a more complicated and difficult procedure than providing the same type of protection to steel. Steel is a smooth, impervious, uniform, dense surface, while concrete is soft, brittle, subject to cracking, filled with air and water pockets, porous, water permeable, and chemically active. It is not surprising then that special surface preparation, product design, and application methods are required to coat and protect concrete effectively from the corrosive conditions in which it is used.

FIGURE 11.22 — Plastic waterproofing sheet held in place prior to casting concrete against it. Knob-like extensions are visible on the concrete side of the sheet.

References

1. Ilaria, Jos. E., Coatings for Concrete and Other Masonry Sub-Strates, 12th Annual Liberty Bell Corrosion Course, Philadelphia, PA, 1974.

2. Clark and Hawley, The Encyclopedia of Chemistry, Reinhold Publishing, New York, NY, 1957.

12

Coating Selection

Introduction

The selection of a proper coating for use on a structure, marine installation, or plant equipment is a difficult decision. In fact, it may well be the most important decision involved in the whole coating process. Not only must the coating resist the demands of the corrosive atmosphere, but the coating determines the type and quality of surface preparation, as well as the method of application. As Sidney Levinson states in *Selecting Paints:*

> To the architect, the cost of painting various parts of a structure he designs amounts to a very small part of the total cost. Yet this thin film of paint, rarely more than 5 or 6 thousandths of an inch thick and often less, is the surface most visible to the eye. Furthermore, it may be the shield which must protect the substrate underneath from injury or even destruction by man or by the elements; by moisture, sunlight, rapid temperature changes, dirt, chemical fumes or abrasion. And it must retain its appearance for many years even when exposed to these sources of destruction.[1]

Robert Baldwin, in *Selecting Coatings for Plant Painting Projects,* also states:

> Picking the right coating for plant painting projects might better be described as a process of elimination, rather than selection.... The plant engineering approach begins by eliminating coatings that require more surface preparation than is feasible. For example, a coating that must be applied over a surface sandblasted to white metal should not be considered unless such surface preparation can be provided. Likewise, slow drying or slow curing coatings cannot be considered when the project must be completed over the weekend. The coating would be subjected to severe service conditions before it was cured, inviting premature failure.
>
> Regardless of the process used, selecting the right coating is very difficult when the full range of proprietary products—dozens from each of hundreds of manufacturers—must be considered.[2]

Coating selection is recognized, then, as a difficult yet important process. It is important in that the coating must satisfy the specific conditions under which it will be used.

It is difficult because of the many variables involved, ranging from the substrate to surface preparation, from rural atmospheres to a complex industrial condition, from immersion to air exposure, or from wall paint to an inorganic coating. Also, there are literally thousands of different coatings manufactured and sold, from which one must be selected to overcome the specific conditions where it is to be used. This adds up to hundreds of variables, all of which affect the life or effectiveness of the coating selected. Actually, the number of variables and conditions that affect coating selection is the reason that so many coatings exist.

Since there are many variable coating properties, as well as many individual coatings from which to select one, some method of reducing the number of selections is needed. One procedure is to list the generic coatings and their advantages and limitations (Table 12.1). While a coating cannot be selected simply by looking at such a chart, it does provide a guide for the selection of one, two, or three generic coatings which *may* satisfy the coating requirements. With this information, further, more specific data can be obtained from literature, laboratory, or field tests which provide a basis for a more definitive selection.

There are two key points to remember in the selection of a coating:

1. There is no such thing as a universal coating that can be used for all conditions. If there were, the corrosion engineer's life would be much easier. A few decades ago, linseed oil paints came as close as any coating to being the universal answer to all coating problems. They could be applied to wooden surfaces, as well as to metal surfaces, and provided reasonable protection to each. As a matter of fact, prior to the 1930s, most of the industrial, railroad, and so-called corrosion-resistant coatings were made from a linseed oil base.

2. All coating selections represent compromises

TABLE 12.1 — Evaluation of Coating Systems

High Chemical Resistance (Immersion or Splash & Spillage)	Outstanding Advantages	Chief Limitations	Typical Applications
Converted Epoxy	1. Alkali Resistance 2. Abrasion Resistance 3. Recoatability 4. Surface Moisture Tolerance 5. Adhesion 6. Solvent Resistance 7. Water Resistance	1. Two Package 2. Pot Life 3. Yellowing 4. Exterior Chalking 5. Exterior Fading 6. Apply over 50 F	1. Chemical Plants 2. Oil Refineries 3. Tank Lining 4. Floor Finish 5. Marine Applications 6. Machinery 7. Paper Mills 8. Water & Sewage Plants
Epoxy Polyester	1. Abrasion Resistance 2. Stain Resistance 3. Acid Resistance 4. Ext. Gloss Retention 5. Water Resistance 6. High Film Build 7. Ext. Color Retention	1. Two Package 2. Pot Life 3. Alkali Resistance 4. Apply over 50 F 5. Limited Flexibility	1. Wall Finish 2. Machinery Finish 3. Kitchens & Bakeries
Moisture-Cured Polyurethane	1. Acid Resistance 2. Abrasion Resistance 3. Single-Package Converted Coating 4. Water Resistance	1. Package Stability after Opening 2. Yellowing 3. Exterior Chalking 4. Exterior Fading 5. Recoatability 6. Adhesion to Metal	1. Floor Finish 2. Tank Lining 3. Machinery 4. Textile Mills
Vinyl	1. Acid Resistance 2. Recoatability 3. Ext. Color Retention 4. Ext. Gloss Retention 5. Water Resistance	1. Spray Application Only 2. Low Solids 3. Low Flash Point 4. Careful Surface Preparation 5. Heat Resistance	1. Tank Linings 2. Chemical Plants 3. Oil Refineries 4. Marine Applications 5. Machinery 6. Potable Water Tanks
Intermediate Chemical Resistance (Chiefly Splash & Spillage)			
Chemical Resistance Latex	1. Chemical Resistance 2. Ext. Color Retention 3. Ext. Gloss Retention 4. Ease of Application 5. Water Clean-up 6. Nonflammable 7. Surface Preparation	1. Abrasion Resistance 2. Application Temperature 3. Heat Resistance	1. Storage Tanks 2. Chemical Plant Equipment 3. Oil Refineries 4. Marine Superstructure
Chlorinated Rubber	1. Chemical Resistance 2. Water Resistance (immersion) 3. Abrasion Resistance	1. Solvent Resistance 2. Heat Resistance	1. Chemical Plants 2. Water Immersion 3. Machinery 4. Floor Finish
Epoxy Ester	1. Chemical Resistance 2. Abrasion Resistance 3. Surface Preparation 4. Ease of Application 5. Water Resistance	1. Ext. Gloss Retention 2. Ext. Color Retention 3. Film Build	1. General Interior Plant Use 2. Floor Finish 3. Primers
Phenolics (Oil Modified)	1. Water Resistance 2. Acid Resistance 3. Abrasion Resistance 4. Hardness 5. Ease of Application	1. Yellowing 2. Recoatability (Intercoat Adhesion) 3. Exterior Fading 4. Exterior Chalking	1. Marine Applications 2. Floor Finishes 3. Water & Sewage Plants 4. Machinery

TABLE 12.1 — Evaluation of Coating Systems

Intermediate Chemical Resistance (Chiefly Splash & Spillage)	Outstanding Advantages	Chief Limitations	Typical Applications
Tar Epoxy	1. Water Resistance 2. Chemical Resistance 3. High Film Build	1. Bleeding 2. Ext. Color Retention 3. Ext. Gloss Retention 4. Two Package 5. Pot Life 6. Dark Colors	1. Crude Oil Tank Lining 2. Sewage Disposal Plants 3. Pipe Lining or Exterior Coating
Low Chemical Resistance (General Purpose Products)			
Oil, Linseed	1. Ease of Application 2. Minimum Surface Preparation 3. Flexibility 4. Excellent Adhesion	1. Slow Dry 2. Soft Film 3. Low Chemical Resistance 4. Poor Solvent Resistance 5. Water Resistance 6. Abrasion Resistance	1. Wooden Buildings (Exterior) 2. Metal Surfaces (Exterior)
Alkyd	1. Ease of Application 2. Surface Preparation 3. Low Cost 4. Good One Coat Hiding 5. Durability 6. Gloss Retention	1. Chemical Resistance 2. Water Resistance	1. Tank Exteriors 2. Structural Metal 3. Machinery 4. Plant Equipment 5. Interior or Exterior, Wood or Metal
Exterior Latex Products	1. Water Reducible 2. Nonflammable 3. Ease of Application 4. Blister Resistance	1. Freeze-Thaw Limitations 2. Heat Resistance	1. Exterior Wood 2. Exterior Concrete, Stucco & Masonry 3. Exterior Metal
Silicone (Masonry Water Repellent)	1. Colorless 2. Invisible 3. Effective for 10 Years 4. Prevents Staining	1. Not for Use on Limestone 2. Use Only on New Masonry	1. New Brisk, Mortar, Sandstone, & Poured Concrete
High-Temperature Coatings			
Silicones	1. Air-Dry Tack-Free (Some Formulations) 2. Service Temperature up to 1000 F 3. Interior or Exterior Service	1. High Cost 2. Low Film Build 3. Limited Solvent Resistance 4. Limited Chemical Resistance	1. High-Temperature Stacks 2. Boilers & Boiler Breeches 3. Exhaust Lines, Manifolds & Mufflers
Special Products			
Zinc-Rich Coatings (Organic)	1. Tolerant of Surface Preparation 2. High-Temp. Tolerance 3. High Film Build 4. Abrasion Resistance 5. Primer and Finish Coat	1. Solvent Resistance 2. High Cost 3. Poor Acid or Alkali Resistance	1. Rust Inhibitive Primer 2. Exterior Equipment in Petroleum and Chemical Plants 3. High-Temp. Stacks
Zinc-Rich Coatings (Inorganic)	1. Solvent Resistance 2. High-Temp. Tolerance 3. High Film Build 4. Abrasion Resistance 5. Primer and Finish Coat	1. Critical of Surface Preparation 2. High Coat 3. Poor Acid and Alkali Resistance	1. Solvent Tank Lining 2. Exterior Equipment in Petroleum and Chemical Plants 3. High-Temp. Stacks

(SOURCE: NACE, Course #5, Corrosion Prevention by Coatings, Houston, TX.)

made to satisfy all of the variables involved. The properties of coating ingredients are balanced to produce the best results for specific requirements at a reasonable cost and for an estimated service life. Again, using linseed oil-base paints as an example, most of the industrial and corrosion-resistant paints were based on red lead pigmentation for the prime coat. Some coatings contained as many as 26 pounds of red lead to the gallon. However, because of the heavy pigmentation and the fact that red lead offered no advantage to coatings for wood, the red lead was replaced by zinc oxides, iron oxide, or other pigments to satisfy the requirements of wood applications.

Because of the much more complex corrosion problems present today, many more compromises must be made in order to satisfy specific conditions. The same coating formulation used for marine immersion ordinarily cannot be used for a marine topside application. Compromises or changes to meet the specific conditions are both necessary and advisable in order for the coating to meet the specific conditions for which it is being applied.

Just as there is no universal coating that fits all conditions, there also is no specific set of conditions for which a particular coating can be entirely satisfactory. Each coating job is unique, even if it is for an identical process in a different plant. This is because environmental conditions can make a substantial difference. For example, a coating for refinery tanks in Galveston, Texas would require an inorganic zinc base coat and weather-resistant topcoats to assure protection against the Galveston marine environment. On the other hand, refinery tanks in Illinois or Michigan could be adequately protected by a top grade alkyd primer and topcoats. Therefore, each coating job must be individually considered, and specific conditions surrounding the job must be thoroughly evaluated in order to assure proper protection by the selected coating.

Considerations in Coating Selection

There are so many factors to be considered when selecting a coating that all of them cannot be listed in relation to the specific nature of coating use. Table 12.2, however, provides a general list of considerations for coating selection, including most of the essential information needed to determine the proper coating material. Some of these points are discussed in more detail in the following sections.

One of the most important questions, and the first that must be asked, concerns the purpose or the reason for coating application. Why is the coating going to be used? Is the coating for corrosion resistance? Is it for resistance to atmospheric conditions? Are the atmospheric conditions rural, marine, or otherwise? Is the coating's purpose to be decorative? Is it to be applied on new construction or is it a maintenance coating? Once the reason for coating has been determined, the specific requirements for the coating can be more accurately outlined.

One of the first points to be considered is whether the coating will be used on new surfaces or whether the coating will be applied over previously coated surfaces. If the surfaces are new, such as new plant facilities which have not been previously exposed to serious corrosion condi-

tions, an unequalled opportunity for coating selection is present. Every effort should be made to take advantage of this prime opportunity in starting with new, clean, uncontaminated surfaces that can be properly prepared. Essentially, there is only one time in the life of a plant or structure that a coating job can be done both properly and economically. This is during plant construction. If a poor material selection is made at this point, because of cost cutting or ignorance of operating conditions, the cost of in-plant maintenance can be both high and continuous during the life of the plant.

Another often-made mistake is to begin startup operations before the entire coating system has been applied. This could be particularly serious where corrosive conditions exist. There are several things which affect early coating failure when such a procedure is used:

1. The primer may not be sufficiently resistant to existing conditions to prevent corrosion from starting prior to the time the topcoats are applied.

2. Once the decision is made to apply the topcoats, even though the primer may be in good condition, there still may be enough surface contamination to require an excessive amount of work for its proper removal. In many

---

TABLE 12.2 — Considerations in Coating Selection

A. Coating Properties

   1. Abrasion resistance
   2. Flexibility
   3. Color and gloss retention
   4. Temperature range
   5. Drying time
   6. Mildew resistance
   7. Appearance
   8. Water or fuel resistance
   9. Wettability

B. Nature of Substrate (wood, masonry, steel, old coating, etc.)

C. Basic Function of Coating on Substrate

   1. Deterioration control (corrosion, weathering, fire, etc.)
   2. Waterproofing
   3. Temperature control
   4. Marking
   5. Appearance (color, texture, gloss, etc.)

D. Accessibility (time or space) and Equipment Available for Satisfactory Surface Preparation and Application

E. Environmental Factors

   1. Temperature (extremes and variations)
   2. Humidity (dry, damp, immersed, marine, etc.)
   3. Chemical contact (fumes, acids, alkalies, solvents, etc.)
   4. Solar radiation
   5. Biological problems (fouling, mildew, etc.)

F. Life Cycle Costs

(SOURCE: Drisko, Richard W., Chemistry and Technology of Coatings, Civil Engineering Laboratory, Port Hueneme, CA.)

Corrosion Prevention by Protective Coatings

cases, this means reblasting the steel and starting over.

3. With some contamination present, particularly in the difficult-to-coat areas of a steel structure, blistering of topcoats is highly likely, ultimately resulting in coating breakdown and severe corrosion.

The best opportunity to coat a structure properly where corrosive conditions will be involved, is when the structure is new, uncontaminated, and the coating work can be scheduled under proper conditions so that maximum benefit from the coating can be obtained.

When the structure is already in place and maintenance becomes the main concern, coating selection becomes somewhat more difficult. This is primarily because the substrate in this instance is not the base material (*e.g.*, concrete or steel); rather, a previous coating now forms the base. Many times the kind of coating previously applied is unknown, which makes the selection of the new maintenance product even more difficult. Any maintenance procedure must take into consideration the type of coating that presently exists on the surface. If this is not done, the new coating may be completely incompatible with the old, resulting in an unsatisfactory coating system that leads to early failure. Without some record of the original coating application, it is often extremely difficult to determine the type of the previous coating.

Table 12.3 indicates the common coating types and their areas of use. However, this can only serve as a guide, since there are a number of different coatings that may be used under the same conditions.

Evaluating Operating Conditions

Probably the most important step in coating selection is to evaluate the conditions under which the coating must operate. This cannot be a superficial evaluation, but must take into consideration all of the conditions that may exist. Even small and seemingly irrelevent factors should be taken into consideration.

For example, a tank was to be lined for the purpose of holding slightly acidic waste water. All other conditions were reasonably innocuous, including ambient temperature. The overall water solid content was low, and a vinyl coating was selected for the tank lining. However, one factor had been overlooked during the evaluation. A small amount of high-boiling aromatic organic liquid contaminated the water. This contaminant was not soluble in the water, but floated on the surface. Although it was insignificant in volume, it accumulated on the coating surface as the liquid level changed, causing the coating to swell, lose adhesion, and blister. This action allowed the acidic water to attack the steel substrate, causing severe corrosion. Since the small amount of contamination was overlooked during the selection of the coating, a costly repair job was required.

Compatibility

If the existing coating is unknown, a practical determination can be made by cleaning a small area (usually several square feet) of the surface to be coated. Prepare the surface as dictated by the coating and the type and degree of failure. Then apply the selected coating to the test surface. When this is done, it is possible to determine:

TABLE 12.3 — Common Coating Types and Areas of Use

Area of Use	Coating Type	Environment
Exterior Iron and Steel	Oil Alkyd Urethane	Rural-Mild Industrial
	Oil-Based Phenolic	High Humidity-Mild Industrial
	Inorganic Zinc Vinyl Epoxy	Corrosive Industrial - Marine- Severe Weather
Interior Iron and Steel-Metal	Alkyd Urethane Epoxy Ester	Mild Industrial
	Vinyl Epoxy Chlorinated Rubber	Acid or Other Corrosive Fumes or Dust Fallout
Interior Storage Tanks, Chemical Solutions	Vinyl Epoxy Phenolic Epoxy	Salt Solutions Dilute Acid Alkalies
Interior Water Tanks	Vinyl Epoxy Chlorinated Rubber Coal Tar Epoxy Coal Tar Asphalt	Water Immersion, High Humidity
Interior Concrete Tanks or Reservoirs	Vinyl Epoxy Chlorinated Rubber	Water Immersion, High Humidity
Interior Solvent Storage Tanks	Inorganic Zinc Epoxy Phenolic Epoxy	Solvent Immersion Fumes - Condensation
Exterior Concrete, Plaster - Masonry	Cement Acrylic Vinyl Acetate	Mild - Industrial
	Vinyl Epoxy Chlorinated Rubber	Corrosive Humidity
Exterior Wood	Oil Alkyd	Rural - Mild Industrial
Interior Plaster or Wall Board	Acrylic Alkyd Chlorinated Rubber	Mild Industrial
Wooden Floors	Alkyd Polyurethane	Mild Industrial
Concrete Floors	Epoxy Ester Polyurethane	Mild Industrial
	Chlorinated Rubber Vinyl Epoxy	Spillage Water - Dilute Chemicals

[SOURCE: Munger, C. G., Repairing Protective Coatings—Effect of Coating Type, Plant Engineering, Feb. 17 (1977).]

(1) if the new coating is going to lift the previous one; *i.e.*, swelling the previous coating and causing it to lose adhesion; (2) whether or not the new coating strikes into the old one; *i.e.*, the immediate absorption of the new coating

by the old coating, which often causes a loss of adhesion that may only become apparent at some later date; and (3) peeling, in which the new coating completely loses adhesion to the old coating. Generally, this can be determined within a reasonable period of time after all coats of the new coating have been applied; *e.g.,* after an overnight dry. This simple application test will not necessarily indicate the exact type of the existing coating. However, it will indicate the general compatibility of the new coat with the old one.

A procedure for a more definitive identification of previous coatings has also been developed. It consists of a series of tests designed by the Naval Civil Engineering Laboratory at Port Hueneme, California and is published in one of their technical data sheets.[3] A field test kit was developed for the sole purpose of determining the type of coating over which a new maintenance coating was to be applied. Such a laboratory kit is available commercially.[4]

The Substrate

Where a new coating is to be applied over an old one, the substrate must also be taken into consideration. It is relatively unimportant if the previous coating is continuous and unbroken, retains strong adhesion, or requires coating only to add thickness for additional life. These conditions rarely exist so that for the most part, the substrate is exposed and corroded. In this case, the nature of the substrate is extremely important to the repair of the existing coating. The new coat must not only be compatible with the old coating, but it must be compatible with the substrate and substrate corrosion as well. The principal substrates encountered are steel, cast iron, galvanized metal, aluminum, and concrete. Each of these corrodes differently and each must be treated in a different manner in order to obtain the proper repair coating.

The substrate is equally important if the structure is new. However, the problem of selecting a coating for new surfaces is less complicated than for the old ones. A thorough discussion of the substrate and the problems involved is provided in a previous chapter (Chapter 8). Nevertheless, the substrate is one of the conditions that must be given serious consideration during the selection of any coating for a corrosive environment. Table 12.4 lists some common substrates, together with the types of coatings that may be applied.

The Environment

The environment is an extremely important part of

TABLE 12.4 — Types of Coatings Commonly Used on Different Substrates

Substrate	Paint	Comments
Interior Wood	Oil	Generally slow drying and relatively soft.
	Alkyd	May be hard or soft.
	Latex (Vinyl or Acrylic)	Can be applied over oil, alkyd, or latex primer.
Exterior Wood	Oil	Good wetting of weathered wood and paint chalk; slow drying; soft.
	Alkyd	Good wetting; variations give variety of properties.
	Silicone Alkyd	Good wetting; good gloss.
	Latex (Vinyl or Acrylic)	Poor wetting or weathered wood and paint chalk; easily applied.
Interior Masonry, Plaster, and Wall Board	Acrylic Latex	Easily applied; brushing is good on coarse surfaces; must remove all loose chalk.
	Vinyl Latex	Same as Acrylic Latex.
	Chlorinated Rubber	Good for waterproofing.
Exterior Concrete and Masonry	Acrylic Latex	Fill coats of these materials will reduce water penetration.
	Vinyl Latex	Same as Acrylic Latex.
	Chlorinated Rubber	Good for waterproofing.
	Vinyl	For concrete in very corrosive environments.
Interior Iron and Steel	Alkyd	Never on continuously damp or immersed environments.
	Vinyl	Good resistance to water; poor resistance to strong solvents.
	Epoxy	Good durability and chemical resistance.
	Urethane	Good durability and chemical resistance.
Exterior Iron and Steel	Oil	For mild environments only.
	Alkyd	For mild environments only.
	Silicone Alkyd	For mild environments only; good gloss.
	Inorganic Zinc	Very abrasion resistant; limited life in seawater without topcoat.
	Vinyl	Good durability; easily touched up.
	Epoxy	Good durability and chemical resistance, but chalks in sunlight.
	Urethane	Aliphatic type has good weathering over epoxy primer.

(SOURCE: Drisko, Richard W., Chemistry and Technology of Coatings, Civil Engineering Laboratory, Port Hueneme, CA.)

Corrosion Prevention by Protective Coatings

any coating selection. It makes a great deal of difference as to whether the environment is dry or moist; whether it is rural, industrial, heavy industrial, or marine; and whether the coating is to be used on the interior or exterior of a structure. The environment for a coating to be used in a rural area is much less severe than one which will be used in a marine area. On the other hand, there are difficulties in a rural atmosphere as well. The atmosphere of a plant in the southern part of the United States is generally much more corrosive than that in the northern area of the continent. This is because the temperature and solar radiation conditions differ significantly between the two locations. Coatings that will readily and quickly chalk in southern areas may last for a substantial period in northern areas.

Industrial atmospheres may have any number of different chemical components in addition to the normal atmospheric conditions in the same area. Marine corrosive conditions are generally the same the world over, with the exception of areas that are in the southern portion of the Northern Hemisphere, which are generally more corrosive because of the higher temperature and higher reactivity of the corrosion process due to temperature.

Soil Problems

There are many protective coatings that are used to coat pipe in underground and underwater conditions. The type of soil is extremely important in any underground installation. Soil in contact with pipe coating can severely damage the coating over a period of time because of earth movement caused by the swelling and shrinking of the soil, due to periodic wetting and drying. The damage to the coating can also include penetration by rock, improper backfill, and damage during handling and laying. Any coating selected for underground conditions must have properties that will withstand the physical problems involved in pipe installation as well as the corrosion problems encountered in soil conditions.

Internal Surfaces

There are a number of different factors that influence the selection of a coating for the interior of a plant, vessel, or tank. On the interior of a plant, the coating need not be developed or designed for exterior atmospheric exposure, since solar radiation has little effect on the interior. On the other hand, fumes, condensation, dampness, and spillage all create conditions that affect the coating to be selected.

For example, an epoxy ester might make an excellent coating for equipment inside a plant that is subject to mild chemical conditions. However, the same coating used outside of the plant and exposed to essentially the same chemical conditions, would not be satisfactory due to its rapid degradation by solar radiation. In the early days of the epoxy esters, because of their relatively easy application and good appearance, many were applied to exterior areas subject to mild chemical conditions. Such coatings were not only a disappointment, but a failure because of their rapid chalking and reaction to solar radiation conditions.

Coatings for plant interiors may also be formulated with different kinds of pigments from those used on the exterior. Interior pigments need not be particularly chalk resistant since they are not subject to weathering conditions, but should be pigmented for water, fume, and chemical resistance. Clear coatings that can be used for interior work would be completely unacceptable for exterior exposure.

Tank linings require a different kind of selection process. Generally, the coating must be selected for its specific properties with relation to specific chemicals or solutions where it must either protect the tank surface, or it must protect the contained material from the tank itself. Since tanks are generally single-purpose units, one of the prime considerations for the coatings is the contained liquid. In a closed tank, if the coating will satisfactorily resist the liquid and protect the tank surface, then other properties, such as resistance to solar radiation or weather resistance, are of little importance.

Corrosive Conditions

Since most of the coatings with which corrosion engineers are directly involved are high-performance coatings, one of the primary considerations for coating selection must be the principal corrosive condition that exists or against which the coating must protect. There are essentially three corrosive conditions that affect the coating: (1) fumes, (2) splash and spillage, and (3) total immersion in the corrosive material. The selection of a coating may be quite different for any one of these three conditions.

Fumes and Splash or Spillage

For fumes, such as those in a galvanizing plant or a sulfuric acid leaching facility undercover, the coating would need to have strong resistance to acidic conditions, condensation, or precipitation of the acidic fumes on the coating surface. High humidities would be a considerable influence on the general corrosive condition, and one might expect high-humidity conditions under the examples given.

Nevertheless, the conditions in the fume area might be considerably less of a problem than those in the area where there was splash, spillage, or leakage from piping and pumps. In the case of fumes, the coating required would not need to have any abnormal thickness, but would need to form a continuous film over the surface. Where there was spillage, dripping, and similar liquid conditions, the coating would need to have greater physical properties and general thickness than for fumes alone.

Immersion

A coating for immersion service is unique in that, generally, it must have outstanding adhesion, as well as an excellent specific resistance to the product in which it is immersed. In this case, the concentration of the product in solution is essential, as is the temperature of the liquid. A 10% caustic solution in some ways can be more difficult than a 50% caustic solution at 150 F. In this example, the 10% solution could allow some moisture vapor to pass into the coating and cause blistering, while the 50% solution is sufficiently hydroscopic so that it would tend to remove any moisture that might be in the coating and thus reduce any blistering tendency. This has been proven under actual service conditions.

The reverse situation can also be true. In the case of a volatile acid, such as hydrochloric acid, the higher the concentration, the greater the tendency to penetrate the coating.

In tank coatings, some minor side effects also must be considered. For example, a ship's tank was coated with an epoxy coating, and plans were made for it to transport molasses from Hawaii to Los Angeles. The tank was a dedicated tank, so that it returned from Los Angeles to Hawaii essentially empty. The coating effectively withstood the immersion in molasses, both in laboratory tests and on the first trip from Hawaii to Los Angeles.

On the second trip, however, it was noticed that the coating was blistered, and on the third trip, it was blistered to an even greater extent. What was not taken into consideration in the initial coating selection, was that once the molasses was removed from the tank, there was sufficient water and condensation in the tank to cause the remaining molasses on the surface to ferment. While the coating was thoroughly resistant to molasses, it was not resistant to acetic acid, alcohol, and the other fermentation products which developed on the return trip. Thus, it is often a minor element in an evaluation or coating selection that creates a problem, and not the principal conditions under which the coating must stand up.

Product Contamination

In tankage, contamination of the contained product must also be one of the considerations in coating selection. In many cases, the prime reason for the coating is to prevent contamination of the contained liquid. The coating therefore should be evaluated from this standpoint, as well as its resistance to the liquid itself.

A coating for a liquid sugar tank is an example. If the liquid sugar is crude sugar and will be further refined, then the lining in the tank merely requires resistance to the sugar solution in order to protect the tank. On the other hand, if it is a refined sugar solution or glucose, which then may be used in sensitive food products, any taste transfer or other contamination to the sugar solution will cause serious problems in the end product. If the lining in the tank contributed an improper odor or taste and was not thoroughly cured or thoroughly free of solvent, serious sugar contamination would occur.

Coating Curing

The curing of a coating used as a tank lining is also an important consideration in the selection of lining materials. If high-temperature cured linings are to be used, the tank must be designed so that the proper temperature can be obtained over the entire coated area. There have been many failures experienced where tanks were designed in such a way that there were cold sections in the tank. These are areas where the tank could not be properly insulated or where supports or reinforced areas acted as a heat sink and could never be brought up to the proper curing temperature. The poorly cured lining in such spots soon failed in service. Where such conditions exist, and there are many of them, a coating with a longer curing time at lower temperature should be selected in order to develop equal resistance over the whole tank area.

Surface Preparation

Another factor in the selection of a coating is the degree of possible surface preparation. In many instances, particularly on new construction, the best surface preparation possible is available for the coating. On the other hand, on the interior of a plant that is operating and yet where a coating is required, sandblasting may be a total impossibility. Many times blasting can be done during a shutdown period; however, with sensitive equipment, products, or reagents involved, blasting may still be impossible. This means that a coating which will tolerate less surface preparation must be applied. Under these conditions, a coating that will provide reasonable resistance as well as adhesion over a hand-cleaned surface would have to be selected. It must be realized, however, that in this case the coating would have a shorter life and would require continued maintenance in order to provide the necessary corrosion protection.

Table 12.5 shows the minimum surface preparation recommended for a number of different generic types of coatings. In the example used above, the chlorinated rubber or epoxy ester varnish might have been the coating selected because of its toleration of less than the best surface preparation.

Timing

One other factor involved in the selection of a coating is the length of time available for coating application. This is not so much of a problem on new work, since it is usually scheduled along with the construction job. However, the application of a coating in an existing plant can be a problem, where application must be scheduled during a shutdown period, sandwiched in-between plant operations, or completed over a weekend.

If it is an exterior application, the weather during the application is also an extremely important factor. If it has to be done during a hot, dry time of the year, one type of coating might be selected, while if it is done in the cold, damp time of the year, another would be preferable. Usually, any selected coating would have to dry and cure rapidly. A vinyl or a chlorinated rubber might satisfy this condition. Some coating application properties, including approximate drying times of various generic type coatings, are shown in Table 12.6.

There are also other application problems that are dictated by time. These would be such things as wind and dust, high temperatures, fallout from adjacent plants, moisture condensation, and similar difficulties. For example, a solvent-base inorganic zinc coating was applied to the wind girder of a large oil storage tank. The application took place in the fall of the year and in the relatively late afternoon after the blasting of the wind ring had been completed. The tank was located in a marine area. During the time that the coating was being applied, the

TABLE 12.5 — Coating Comparison Guide—
Surface Preparation

Coatings Formulated as Primers for Steel	Minimum Surface Preparation
Acrylic Lacquer	Normally used as topcoat.
Alkyd Varnish	Tolerant of hand and power tool cleaning. (Better coatings normally used with blast cleaning.)
Bituminous Cutback	Commercial blast.
Bituminous Emulsion	Commercial blast.
Chlorinated Rubber	Commercial blast. (Tolerant of hand and power tool cleaning.) White metal blast for immersion service.
Epoxy Ester Varnish	Commercial blast. (Tolerant of hand and power tool cleaning.)
Epoxy-Amine (2-part)	Commercial blast. Near-white blast for immersion service.
Epoxy-Polyamide (2-part)	Commercial blast. Near-white blast for immersion service.
Epoxy-Coal Tar (2-part)	Commercial blast. Near-white blast for immersion service.
Latex (water-base)	Commercial blast (acrylic latex). Normally used as topcoat.
Oil-Base	Hand or power tool cleaning. (Better coatings normally used with blast cleaning.)
Phenolic Varnish	Commercial blast. (Tolerant of power tool cleaning.)
Polyester-Epoxy	Normally used as topcoat.
Polyurethane (oil-modified)	Normally used as topcoat.
Urethane (moisture-cured)	Not normally used on steel.
Urethane (2-part)	Normally used as topcoat.
Vinyl Lacquer	White metal blast. Normally used as topcoat.
Zinc-Inorganic	Commercial blast. White metal blast for immersion service.
Zinc-Organic	Commercial blast. White metal blast for immersion service.

[SOURCE: Baldwin, Robert, Selecting Coatings for Plant Painting Projects, Plant Engineering, Feb. 6 (1975).]

humidity changed sufficiently so that the wind girder condensed moisture shortly after the coating had been applied. This resulted in rapid curing of the coating because of the moisture present on the surface. The coating checked and cracked over nearly the entire surface. This problem would not have occurred had the coating been applied during the middle of the day or during a somewhat warmer and less humid time of the year.

Safety

Another consideration in coating selection is the safety problems that may be encountered. These vary from coating to coating and from one type of application to another. For instance, solvents in a coating create possible hazardous conditions, such as: (1) explosion, (2) fire, and (3) personal contact with the solvent or a fume.

While safety during coating application will be discussed in more detail in a later chapter, there is one rule-of-thumb concerning use of solvents that is worth remembering: Keep the volume of solvent vapor in air below a concentration of 1%. Actually, it should be kept far below 1% to maintain the best safety conditions. This can be accomplished by proper ventilation which eliminates the hazard of an explosion, since the explosive limit of most solvents is well above 1%. Personal contact with solvents can be avoided through use of proper clothing and air masks. Catastrophic problems such as fire or explosion can be greatly reduced by the rule-of-thumb mentioned above.

Previous Experience

Previous coating experience is an excellent criterion for coating selection. If the experience has been good and the conditions are reasonably similar, then this experience lends confidence in making the coating selection. If the coating has not performed satisfactorily, then it is certainly a reason for eliminating the coating from consideration.

In order for previous experience with a coating to be valid for coating selection, the experience should be personal and not based on hearsay from someone else in the field or from coating salespeople. Personal observation of the application of a coating, followed by evaluation of the performance of the coating under actual operating conditions, makes for first-hand knowledge that cannot be equaled by any second- or third-hand comments. Outside sources cannot evaluate the coating or coating potential with the same objectivity as someone who has made first-hand observations. However, this is not always possible, and outside reports and third-party evaluations may be the best information available. Nevertheless, wherever possible, personal inspections of existing coating installations are suggested whenever a critical coating selection is being made.

As previously stated in this chapter, no two applications are identical, and, therefore, a coating application in one plant can be considerably different from the coating in another. Nevertheless, the experience of vendors in the application and use of their coatings supplies important information and should also be taken into consideration during coating selection. Well-documented information on the life of a coating under similar conditions should be given serious consideration in the selection of any coating. Robert Baldwin, in his article, *Selecting Coatings for Plant Painting Projects,* suggests:

TABLE 12.6 — Coating Comparison

Coating	Film Forming Solids, % by volume	Recommended Dry Film Thickness, mils/coat	Number of Topcoats Over Primer (if req.)	Minimum Application Temperature, °F	Flash Point, °F (open cup)	Drying Time To Touch, hours	To Recoat, hours	Final Cure, days
Acrylic Lacquer	25-37	1.5-2.5	1-2	40	53-85	1/4-3/4	12	1-7
Alkyd Varnish	43-56	1-2	2	40	50-100	2-12	3-48	4 days - several weeks
Bituminous Cutback	32-74	12-15	1-3	40	80	4-12	24-72	3-14
Bituminous Emulsion	10	12-15	1-3	40	—	1-2	12-24	10-15
Chlorinated Rubber	29-46	1.2-4	1-3	40	50-100	1/4-4	1/4-24	7
Epoxy Ester Varnish	34-44	1.5-2.5	2	50	85-100	2	6-72	4-14
Epoxy Amine (2-part)	37-60	2-8	1-3	50	60-90	2	12-18	3-10
Epoxy-Polyamide (2-part)	45-60	2-5	1-3	50	60-90	1-3	3-40	5-7
Epoxy-Coal Tar (2-part)	60-86	8-10	1-2	60	60-96	1/2-2	4-96	1-7
Latex (water-base)	35-50	1.5-4.5	1-2	40	—	1/4-2	1-12	5-30
Oil-Base	70-90	3	1-2	40	60-105	8	8-48	several weeks
Phenolic Varnish	32-43	1.5-4	1-3	50	60-105	1/2	8-48	7-12
Polyester-Epoxy	51-61	2.5-3.5	1-2	60	90-100	2-6	8-48	10-14
Polyurethane (oil-modified)	30-50	1-3	1-2	40	60-100	1/2-6	4-16	2-14
Urethane (moisture-cured)	30-50	1-4	1-3	40	60-90	1/2-12	3-48	2-7
Urethane (2-part)	30-100	2-40	1-3	40	60-90	1/2-12	3-48	2-14
Vinyl Lacquer	4-28	1-5	1-3	35	28-95	1/4-1	1/2-8	3-7
Zinc-Inorganic	4-71	1.5-5	1-2	40	50-70	1/4-1/2	1/4-1/2	1-7
Zinc-Organic	4-40	2-3	1-2	35	28-100	1/4-1/2	2-24	7

[SOURCE: Baldwin, Robert, Selecting Coatings for Plant Painting Projects, Plant Engineering, Feb. 6 (1975).]

Once the requirements and properties of various generic coatings have been compared to circumstances existing in the plant, the choices should be sufficiently narrowed so that specific formulations can be investigated. If it has not already been done, now is the time to involve the technical representatives of paint companies manufacturing all the generic types under consideration. They are intimately familiar with application and performance characteristics of their lines.

The following information should be obtained for each proprietary product:
* Resistance to corrosive chemicals
* Estimated life or recoating cycles
* Solids content by volume
* Mil feet per gallon
* Film thickness per coat
* Theoretical coverage per gallon
* Total applied cost of coating system
* Cost per square foot per year of service
* Names and addresses of users with exposure conditions similar to proposed plant project
* Ease of repairing damaged areas
* Recommended surface preparation
* Dry time and curing time
* Ease of application.

Each of the coatings under consideration should be compared item by item under these categories.[2]

Coating Cost

The cost of a coating is always given serious consideration in coating selection. There is no question that cost is a serious factor in coating selection. In fact, under some circumstances, it may be the only factor. Coating cost, however, in most circumstances should be less important than the coating properties that provide the basis for long-time effective coating protection.

Cost should really only come into consideration once the coating or coatings have been selected that will satisfactorily overcome the corrosion problem for which they will be used. During the selection process, perhaps two or three coatings may all have the necessary qualifications. Cost, then, can enter the picture to determine which of the coatings that entirely satisfy the conditions involved will be the most economical choice.

It has been stated that the proper corrosion control system is the lowest cost system that will retain the qualities of the structure needed to fulfill the objectives of esthetics, safety, and performance throughout the service life of that structure. Retaining the structure qualities needed to fulfill the objectives is the key to this statement, and if a coating does not do this, for any reason whatsoever, cost becomes secondary.

Frequently coatings are selected by persons who are not entirely knowledgeable in the field and lack basic facts on which to make a sound decision. Statements like, "We can't afford to sandblast or to use exotic protective coatings like vinyl," are often made. However, a second-rate coating system used over less than the best surface preparation causes the company considerable losses because continual, costly maintenance is required to keep the structure from corroding to the point of becoming a safety hazard.

There have been numerous articles written on cost and how it should be related. Cost per square foot per year is a common way of comparing coatings and, generally, a good one. Cost studies can go from this point to discounted cashflow and the time value of money, all of which can become fairly complicated, even though they may present an accurate cost comparison.

It is not intended to include the details of this kind of cost comparison in this chapter. The NACE standard, *Direct Calculation of Economic Appraisals of Corrosion Control Measures,* is an excellent reference if a detailed cost study is to be made.[5] Another excellent cost comparison reference is a paper entitled, *The Simplified Cost Calculations and Comparisons of Paint and Protection Coating Systems, Expected Life and Economic Justification.*[6] The information given in these two references should provide the details needed to begin an effective cost study.

Coating Properties

The properties of the coating itself are some of the most important considerations in coating selection. Many of the properties of various high-performance coatings are summarized in Table 12.7. This table provides a comparison of the generic types of coatings and their properties. However, this information alone should not be used to make the final decision as to what coating to select for any specific job. Once it is determined which generic type of coating needs to be used, then there is a direct comparison of the various coatings within that generic type. Without this further comparison, it is difficult to make a proper coating selection.

Generic Comparisons

It may be somewhat difficult to understand why it is necessary to test individual coatings in a generic type, even though they supposedly have been manufactured for similar exposures. Actually, there are hundreds of different formulations for any specific generic type of coating. Some manufacturers will take a standard formulation published either in a governmental specification or by a supplier, and manufacture the coating according to that standard. Other manufacturers will use those standards as a starting point, and then spend considerable research time determining the best combination of ingredients within that generic type of material that will do the best job.

For instance, they make a formulation that is easier to apply or that has less of a tendency toward problems like pinholing, they may add reinforcing pigments to the coating to make it tougher, they may add plate-type pigments such as mica to reduce the moisture vapor transformation rate, and they may add specific pigments to improve both the workability and the resistance of the coating. Since there can be such substantial differences between generic types of coatings, comparative testing is the only way to select the best material.

Comparative Tests

The best way to select a coating for a specific pur-

pose, after determining the characteristics necessary, is to make comparative tests, both in the laboratory and under actual working conditions. While such a procedure is somewhat costly and requires the time of some technical personnel, where a large application job is involved, it can easily save many times the original coating cost before the job is completed or during the life of the coating.

The material cost of a coating is the smallest part of the cost of the total applied material. Thus, there is very little applied cost difference between a cheaper, less effective coating and the best coating that can be obtained. The difference, which amounts to a small percentage of the total job, is merely good insurance of adequate protection of the structure being coated.

Laboratory coating tests are, of course, the quickest and easiest to run. They can, however, be less than definitive. The testing of coatings under actual operating conditions is the most effective way of comparing the value of one coating over another. Figure 12.1* is an example of this. While the five inorganic systems tested were supposedly comparable, a field test in a marine atmosphere demonstrated the benefit of using this type of test for coating selection.

Figure 12.2 is an example of the actual field testing of various types of coating systems under actual operating conditions. In this case, a petroleum production tank in an area of high hydrogen sulfide fumes required a better-than-average coating system to provide reasonable protection to the metal surface. The coatings involved in this test were inorganic zinc (top), inorganic zinc and polyamide epoxy topcoats (center), inorganic zinc and an inorganic white topcoat (bottom), and an epoxy polyamide primer and epoxy polyamide topcoats (left side of tank).

FIGURE 12.2 — Field testing of various coatings exposed to actual operating conditions, which serves as an excellent method of coating comparison when time permits.

---

*See color insert.

TABLE 12.7 — Comparison of Resistant Coating Uses

Uses of Resistant Coatings	LACQUER COATINGS				COREACTING COATINGS		
	Vinyl Chloride-Acetate Copolymer	Vinyl-Acrylate Copolymer	Chlorinated Rubber	Coal Tar Pitch (Hot Melt)	Epoxy-Amine	Epoxy Polyamide	Coal Tar Epoxy
Abrasion Resistance	Good	Good	Good	Fair	Good	Good	Good
Bacteria & Fungus Resistance	Excellent	Good	Good	Good	Good	NR	Good
Chemical Resistance	Broad Spectrum Resistance	Good	Broad Spectrum Resistance	Good	Good	Good	Good
Acid-Oxidizing	Spray or Fumes	Fumes	Spray or Fumes	NR	NR	NR	NR
Nonoxidizing	Spray or Fumes	Fumes	Spray or Fumes	Fumes	Spray or Fumes	Fumes	Fumes
Organic	Immersion Fatty Acids	Fumes	Dissolves in Fatty Acids	NR	Spillage-Fumes	Fumes	NR
Alkali	Good	Dusts	Good	Good	Good	Good	Good
Salts: Oxidizing	Splash-Spray	Fumes or Dusts	Fumes or Dusts	NR	Fumes	Fumes	Fumes
Nonoxidizing	Immersion OK	Fumes or Dusts	Immersion OK	Seawater OK	Immersion OK	Immersion OK	Immersion OK
Solvent: Aliphatic	Excellent	Fair	Good	NR	Excellent	Good	Good
Aromatic	Swells	Dissolves	Dissolves	NR	Good	NR	NR
Oxygenated	Dissolves	Dissolves	Dissolves	NR	NR	NR	NR
Water:	Very Good Immersion	Good	Very Good Immersion	Excellent	Very Good Immersion	Very Good Immersion	Very Good Immersion
Moisture Permeability	Low	—	Low	Low	Low	Low	Low
Contamination of Contacting Materials							
Food	OK-Odorless-Tasteless-Nontoxic	—	Water OK	Water OK	Good	Good	NR
Chemical	Very Good	—	Good	—	Good	Good	Fair
Decontamination	Very Good	—	—	—	Very Good	Very Good	NR
Friction Resistance (Faying Surfaces)	—	—	—	—	—	—	—
Heat Resistance							
Wet	48 C (120 F)	38 C (100 F)	38 C (100 F)	48 C (120 F)	48 C (120 F)	48 C (120 F)	48 C (120 F)
Dry	65 C (150 F)	65 C (150 F)	60 C (140 F)	65 C (150 F)	95 C (200 F)	95 C (200 F)	95 C (200 F)
Radiation Resistance	10^8R	—	1×10^8R	—	1×10^9R	1×10^{10}F	5×10^8R
Soil Resistance	—	—	—	Excellent	—	—	Good
Weather & Light Resistant	Good-Properly Pigmented	Excellent	Good-Properly Pigmented	NR	Heavy Chalking	Fair-Good	Chalking
Principal Hazard-Application	Solvent Fumes	Solvent Fumes	Solvent Fumes	Coal Tar Fumes[1]	Dermatitis Solvent Fumes	Solvent Fumes	Dermatitis Solvent Fumes

Uses of Resistant Coatings	COREACTING COATINGS			CONDENSING COATINGS			INORGANIC COATING		
	Urethane 2-Package	Urethane Moisture Cure	Urethane-Aliphatic Isocyanate Cure	Baked Phenolic	Epoxy Phenolic	Epoxy Powder	Water-Base	Solvent-Base	Fused Ceramic
Abrasion Resistance	Outstanding	Outstanding	Outstanding	Good-Brittle	Good	Good	Very Good, Metallic	Very Good, Metallic	Very Good, Brittle
Bacteria & Fungus Resistance	—	—	—	Excellent	Excellent	Good	—	—	Excellent
Chemical Resistance	Good	Good	Good	Very Good	Very Good	Good	NR	NR	Very Good
Acid-Oxidizing	Spray or Fumes	Fumes	Spray or Fumes	NR	Splash-Fumes	NR	NR	NR	Immersion OK
Nonoxidizing	Spray or Fumes	Fumes	Spray or Fumes	OK, Ambient Temp.	Splash-Fumes	Splash Fumes	NR	NR	Immersion OK
Organic	Spray or Fumes	Fumes	Spray or Fumes	OK, Ambient Temp.	Splash-Fumes	Splash Fumes	NR	NR	Immersion OK
Alkali	Spray or Dust	Dust	Spray or Dust	NR	Excellent	Good	NR	NR	NR
Salts: Oxidizing	NR	NR	NR	NR	Fumes	Fumes	NR	NR	Immersion OK
Nonoxidizing	Immersion OK Good-Primer Required	Spray	Spray	Immersion OK	Immersion OK	Splash-Fumes	OK Marine Splash & Spray	OK Marine Splash & Spray	Immersion OK
Solvent: Aliphatic	Excellent	Very Good	Excellent	Excellent	Excellent	Excellent	Excellent	Excellent	Excellent
Aromatic	Excellent	Very Good	Excellent	Excellent	Excellent	Excellent	Excellent	Excellent	Excellent
Oxygenated	NR	NR	NR	Excellent	Alcohols-OK, NR-Ketones	NR	Excellent	Excellent	Excellent
Water:	Good-Primer Required	—	Good-Primer Required	Excellent	Excellent	OK Immersion	Good	Good	Excellent
Moisture Permeability	Medium-Low	Medium-Low	Medium-Low	Low	Low	Low	—	—	Low
Contamination of Contacting Materials									
Food	—	—	—	OK-Odorless, Tasteless, Nontoxic	OK-Nontoxic	—	—	—	OK, Odorless, Tasteless, Nontoxic Good
Chemical	—	—	—		Good	—	—	—	
Decontamination	Very Good	Good	Very Good	Very Good	—	—	—	—	
Friction Resistance (Faying Surfaces)	—	—	—	—	—	—	Excellent Coef. of Friction 0.47	Excellent Coef. of Friction 0.52	Good
Heat Resistance									
Wet	38 C (100 F)	—	38 C (100 F)	82 C (180 F)	82 C (180 F)	48 C (120 F)	—	—	95 C (200 F)
Dry	120 C (250 F)	120 C (250 F)	120 C (250 F)	120 C (250 F)	120 C (250 F)	95 C (200 F)	370 C (700 F)	315 C (600 F)	260 C (500 F)
Radiation Resistance	5×10^8R	5×10^8R	5×10^8R	—	1×10^{10}R	—	1×10^{10}R	1×10^{10}R	1×10^{10}R
Soil Resistance	—	—	—	—	—	Thin Film-Care Required for Backfill	NR	NR	NR
Weather & Light Resistant	Good, Yellows	Good	Excellent, Color & Gloss Retention	NR	—	Good-Chalks	Excellent 20 + Years	Excellent 20 + Years	Excellent 20 + Years
Principal Hazard-Application	Solvent Fumes	Solvent Fumes	Solvent Fumes	Phenol & Solvent Fumes	Solvent Fumes	None	None-Water Base	Solvent Fumes	None

[1]Critical for immersion.
NR = Not Recommended.
[SOURCE: Munger, C. G., Coatings-Resistant, Kirk-Othmer: Encyclopedia of Chemical Technology, 3rd Ed., Vol. 6, John Wiley & Sons, New York, NY (1979).]

The testing of coatings under marine conditions is a common practice. Panels may be exposed to a marine atmosphere, such as is illustrated in Figure 12.3*, marine immersion conditions where panels are placed in tidal zones, or actual full immersion conditions. Such a marine test installation is shown in Figure 12.4*.

A comparison of a series of antifouling coatings is shown in Figure 12.5*. There are a number of commercial testing laboratories which have marine testing facilities. One of the best known is the LaQue Center for Corrosion Technology, Inc. (LCCT, Inc.) at Wrightsville Beach, North Carolina. Many thousands of tests have been conducted at this station, comparing materials, different coatings, different coating systems, and for many different companies throughout the United States. These locations have become recognized standard marine test areas.

Many individual companies also maintain private panel testing stations in their plants at particularly corrosive areas. Such testing areas make for an easy comparison of coatings at relatively low cost. Once the panel preparation, which represents the greatest cost, has been completed, actual testing is merely a matter of time. Selection of a coating on the basis of such tests can be quite definitive with reasonably predictable results for large-scale applications.

Figure 12.6 shows the comparative testing of two zinc-rich coatings under plant operating conditions. Both were scribed down the center of the panel in equivalent manners. This test was of relatively short duration (a few weeks), and yet there was a decided difference between the protection provided by the two coatings.

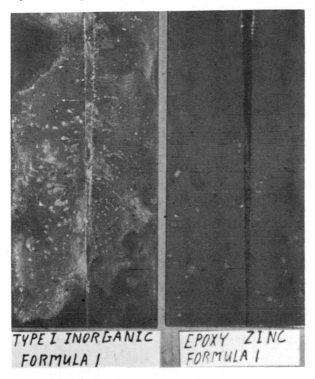

FIGURE 12.6 — Comparison of two zinc-rich formulations after exposure to plant environmental conditions.

*See color insert.

Laboratory Tests. Laboratory coating tests can also provide some interesting comparative results. There are a number of laboratory tests that can be used essentially for screening purposes to indicate which coating can be most effective for a particular purpose. Laboratory tests can be quite valuable in determining a proper coating for an area where there is a single factor of corrosion or contamination involved. For example, a laboratory test can be definitive when comparing various coatings for exposure to 50% sodium hydroxide at 180 F. This type of exposure can be easily duplicated in the laboratory. There are many similar types of exposures where laboratory tests would be equally effective.

There are, however, a great many exposures that include a number of variables, *e.g.,* atmospheric corrosion in a plant. In cases such as these, laboratory tests can, at best, only be comparative between the materials or coatings being tested, and are only indicative of the resistance of the various materials to the actual plant exposure. Where there are many variables involved, it is preferable to test coatings under the actual exposure conditions. Small plant test racks holding only a few panels can be set up quite inexpensively.

Several organizations have established standards and methods for the laboratory testing of coatings. NACE Standard TM-01-74, *Laboratory Methods for the Evaluation of Protective Coatings Used as Lining Materials in Immersion Service*[7] outlines several comparative test methods for differentiating between coating materials. ASTM also has a number of standard tests to compare coating materials for various purposes. Table 12.8 lists some of the testing methods and standards established for coating comparison.

Salt Spray Tests. Laboratory comparison tests are used throughout industry to evaluate various coating materials. The salt spray test is one of the common laboratory tests that has been used by both government and industry for many years to compare the resistance of coatings. It consists of a fine mist of salt solution which precipitates on the coating and keeps it continually wet with the dilute salt solution. The spray chamber is also highly aerated so that adequate oxygen is present for active corrosion.

It can be used as a test to compare coating adhesion, and as a test to compare resistance to the transfer of ions (*e.g.,* sodium and chloride) through the coating. When a panel is scribed, it can be used as a test to determine the corrosion resistance of the coating and its resistance to undercutting. ASTM, in their description of the test, states:

> The test has been used to a considerable extent for the purpose of comparing different materials or finishes. It should be noted that there is seldom a direct relation between salt spray (fog) resistance and resistance to corrosion in other media because the chemistry of the reactions, including the formation of films and their protective value, frequency varies greatly with the precise conditions encountered....Method B 117 is considered to be most useful in estimating the relative behavior of closely related materials in marine atmospheres, since it simulates the basic conditions with some acceleration due to either wetness or temperature or both.[8]

Figure 12.7 shows a typical salt spray cabinet, to-

TABLE 12.8 — Standard Methods of Testing Coatings and Linings

NACE Standard TM-01-74, Laboratory Methods for the Evaluation of Protective Coatings Used as Lining Materials in Immersion Service

ASTM Standard B-117-73, Salt Spray (Fog) Testing

ASTM Standard B-287-74, Acetic Acid-Salt Spray (Fog) Testing

ASTM Standard B-368-68 (1973), Copper Accelerated Acetic Acid-Salt Spray (Fog) Testing (CASS Test)

ASTM Standard D-610-68 (1974), Preparation of Steel Panel for Testing Paint, Varnish, Lacquer, and Related Products

ASTM Standard D-611-64 (1973), Evaluating Degree of Rusting on Painted Steel Surfaces

ASTM Standard D-659-74, Evaluating Degree of Chalking of Exterior Paints

ASTM Standard D-660-44 (1970), Evaluating Degree of Checking of Exterior Paints

ASTM Standard D-661-44 (1970), Evaluating Degree of Cracking of Exterior Paints

ASTM Standard D-662-44 (1970), Evaluating Degree of Erosion of Exterior Paints

ASTM Standard 714-56 (1974), Evaluating Degree of Blistering of Exterior Paints

ASTM Standard 772-47 (1970), Evaluating Degree of Flaking (Scaling) of Exterior Paints

ASTM Standard D-822-60 (1973), Recommended Practice for Operating A Light-Water-Exposure Apparatus (Carbon Arc Type) for Testing Paint, Varnish, Lacquer, and Related Products

ASTM Standard D-823-53 (1970), Producing Films of Uniform Thickness of Paint, Varnish, Lacquer, and Related Products on Test Panels

ASTM Standard D-870-54 (1973), Water Immersion Test of Organic Coatings on Steel

ASTM Standard D-968-51 (1972), Test for Abrasion Resistance of Coatings of Paint, Varnish, Lacquer, and Related Products by the Falling Sand Method

ASTM Standard D-1005-51 (1972), Measuring the Dry Film Thickness of Organic Coatings

ASTM Standard D-1014-66 (1973), Conducting Exterior Exposure Tests of Paints on Steel

ASTM Standard D-1186-53 (1973), Measuring of Dry Film Thickness of Nonmagnetic Organic Coatings on a Magnetic Base

ASTM Standard D-1212-70, Measurement of Wet Film Thickness of Organic Coatings

ASTM Standard D-1400-67, Measurement of Dry Film Thickness of Nonmetallic Coatings of Paint, Varnish, Lacquer, and Related Products Applied on a Nonmagnetic Metal Base

ASTM Standard D-1653-72, Test for Moisture Vapor Permeability of Organic Coating Films

ASTM Standard D-1654-61 (1974), Evaluation of Painted or Coated Specimens Subject to Corrosive Environments

ASTM Standard D-1735-62 (1973), Water Fog Testing of Organic Coatings

ASTM Standard D-2197-68 (1973), Test for Adhesion of Organic Coatings

ASTM Standard D-2247-68 (1973), Testing Coated Metal Specimens at 100% Relative Humidity

ASTM Standard D-3258-73, Test for Porosity of Paint Films

ASTM Standard D-3276-73, Recommended Guide for Paint Inspection (this is an outline for inspection and testing procedures)

ASTM Standard D-3359-74, Measuring Adhesion by Tape Test

ASTM Standard D-3361-74, Recommended Practice for Operating Light and Water Exposure Apparatus (unfiltered carbon arc type) for Testing Paint, Varnish, Lacquer, and Related Products Using the Dew Cycle

ASTM Standard D-3363-74, Test for Film Hardness by the Pencil Test

ASTM Standard E-376-69, Recommended Practice for Measuring Coating Thickness by Magnetic Field or Eddy Current (electromagnetic test methods)

ASTM Standard G-8-72, Test for Cathodic Disbonding of Pipe Line Coatings

ASTM Standard G-12-72, Nondestructive Measurement of Film Thickness of Pipe Line Coatings for Steel

[SOURCE: Munger, C. G., Coatings-Resistant, Kirk-Othmer: Encyclopedia of Chemical Technology, 3rd Ed., Vol. 6, John Wiley & Sons, New York, NY (1979).]

gether with the pertinent working parts. This is taken from ASTM Standard B 117-73 which fully describes the conditions, requirements, and workings of salt spray (fog) testing equipment.[8] If a salt spray test is considered for evaluating coatings, it is recommended that the ASTM Standard be followed.

Again, it should be reiterated that the salt spray test, as well as the other tests that will be described later, should be used only for comparison between coatings and should not be used as a tool to establish the life of a coating under actual operating conditions. Many years ago Frank LaQue said, "The salt spray test only deter-

mines the life of a coating in the salt spray cabinet." The salt spray test, in particular, has been used to try to make such life comparisons, with people claiming that each hundred hours in a salt spray is equal to a year of life under actual conditions. Such life forecasts have never been accurately correlated between the laboratory test and actual life under service conditions. Any correlation which has been attempted has worked out poorly.

The most that can be said is that a coating which withstands salt spray for 1000 to 2000 hours without corrosion to the panel or damage to the coating itself, *should* have good resistance to moist salt air conditions in the

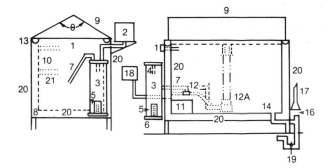

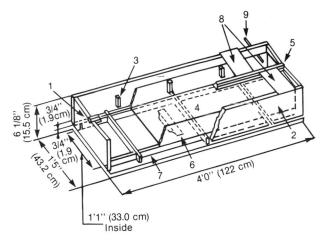

θ—Angle of lid, 90 to 125 deg
1—Thermometer and thermostat for controlling heater (Item 8) in base
2—Automatic water levelling device
3—Humidifying tower
4—Automatic temperature regulator for controlling heater (Item 5)
5—Immersion heater, non-rusting
6—Air inlet, multiple openings
7—Air tube to spray nozzle
8—Strip heater in base
9—Hinged top, hydraulically operated or counterbalanced
10—Brackets for rods supporting specimens or test table
11—Internal reservoir
12—Spray nozzle above reservoir, suitably designed, located, and baffled
12A—Spray nozzle housed in dispersion tower located preferably in center of cabinet
13—Water seal
14—Combination drain and exhaust. Exhaust at opposite side of test space from spray nozzle (Item 12), but preferably in combination with drain, waste trap, and forced draft waste pipe (Items 16, 17, and 19).
16—Complete separation between forced draft waste pipe (Item 17) and combination drain and exhaust (Items 14 and 19) to avoid undesirable suction or back pressure.
17—Forced draft waste pipe
18—Automatic levelling device for reservoir
19—Waste trap
20—Air space or water jacket
21—Test table or rack, well below roof area

FIGURE 12.7 — Typical salt spray (fog) cabinet. [SOURCE: ASTM, Salt Spray (Fog) Testing, Standard B-117-73, Philadelphia, PA, 1973.]

1—End supports, 3/4 by 6 1/8 in. by 1 ft 3 1/2 in. (1.9 by 15.5 by 39.4 cm) (slanted 1/4 in./8 in. (0.64 cm/20.3 cm) toward center).
2—Raised side supports, 3/4 by 4 7/8 in. by 4.0 ft. (1.9 by 12.4 by 122 cm).
3—Tank spacers, 1/2 by 3/8 by 2 1/2 in. (1.3 by 0.95 by 6.4 cm).
4—Water tank, 14 1/2 by 46 1/2 by 3 in. (36.8 by 118 by 7.6 cm) (distilled water and stainless steel preferred).
5—Depressed central panel support (rigid insulating type, for example, methylmethacrylate resin).
6—Six strip heaters (attached to bottom of tank) spaced equally for uniform heat, rheostat-controlled (115-V, 150-W steel sheath strip heaters and 7 1/2-A capacity rheostat).
7—Tank support and heat barrier, 3/4 by 1 3/8 in. (1.9 by 3.5-cm) stock.
8—Typical 8-in. (20.3-cm) long test panels or blanks (all positions on cabinet must be occupied) (blanks may be glass).
9—Thermometer (bulb 1 in. (2.5 cm) below test surface).

NOTE 1—Arrangement of 2 and 3 permits access of fresh air.
NOTE 2—A satisfactory construction material is 3/4-in. (1.9-cm) exterior grade plywood coated with a polyamide crosslinked epoxy.

FIGURE 12.8 — Cleveland condensing humidity cabinet. (SOURCE: ASTM, Coated Metal Specimens at 100% Relative Humidity, Standard D-2247-68, reapproved 1973, Philadelphia, PA, 1968.)

field. Because of the many variables in field conditions, no measure of life can be accurately inferred from such a test.

Humidity Test. There are several types of humidity tests. The ASTM standard for testing *Coated Metal Specimens at 100% Relative Humidity* is #D 2247-68 (reapproved 1973). There are essentially two methods shown. The first is for a humidity cabinet similar to a salt spray cabinet in which panels are exposed to 100% humidity. The entire panel is exposed to condensing moisture.[9]

The second is an apparatus where only one side of the panel is subject to humidity and condensation.[9] This is the preferable procedure since it creates a temperature differential across the panel which makes a more severe test. This unit was developed by the Cleveland Society of the Federation of Societies of Paint Technology and is referred to as *The Cleveland Test.* The apparatus used is the Cleveland Condensing Type Humidity Cabinet. Details of the cabinet are shown in Figure 12.8.

The Cleveland Test is one where coatings are compared for resistance to moisture condensation on the surface. The test mechanism has a storage area for water, which is kept warm and at a constant temperature by electric heaters. The panels are suspended on an angle above the water in such a manner that the moisture will constantly condense on the underside of the panel. The

panels are held at a slight angle so that the condensed water runs off and is replaced by new condensation on a continual basis. If a coating has questionable adhesion, it will show up quickly in this test by the formation of blistering. Any coating that tends to release its adhesion under moist conditions will do so in this test.

As under most condensation conditions, the interior of the coating, or the coating that is exposed to the interior of this unit, is at a higher temperature than the surface of the panel on the exterior. Thus, there is a temperature gradient across the coating from the warm, moist interior to the cooler interface between the panel and the coating. Such conditions are much more severe than where a coating is merely immersed in a liquid with the temperature remaining the same over the entire panel.

This testing device can also be set up with an alternate wetting and drying condition, as well as with ultraviolet radiation. It is believed, however, that the primary value of this test is its ability to compare the water resistance and adhesion of various coatings.

The Weatherometer Test. The weatherometer is a mechanical piece of test equipment in which there is an attempt to simulate weather conditions on an accelerated basis. The coating in most weatherometers is subject to cycles of warm water spray for a period of time. This is followed by a drying period, during which time high-intensity ultraviolet radiation is directed against the coating. These cycles of heating, cooling, wetting, drying, and radiation are accurately controlled. The panels are rotated around the light source to obtain even exposure, and the resistance of the coating is measured in weatherometer hours. While this is a good comparative test, it should not be used as a definitive measure of the coating life under actual operating conditions.

ASTM has several standards related to the weatherometer:

1. Recommended practice G-23 for operating light- and water-exposure apparatus (carbon arc type) for exposure of nonmetallic materials.

2. Recommended practice D-822-60 (reapproved 1973) for operating light- and water-exposure apparatus for testing paint, varnish, lacquer, and related products.

3. Recommended practice D-3361-64 for operating light- and water-exposure apparatus (unfiltered carbon arc type) for testing paint, varnish, lacquer, and related products using the sea cycle.[10]

The weatherometer test compares the coatings for resistance to ultraviolet and simulated solar radiation, and can measure the comparative resistance of the coating to chalking, checking, cracking, rusting, blistering, and similar breakdown. It also can be used to measure gloss retention and color change. The wetting and drying cycle also has some merit in testing the adhesion of a coating, since such treatment causes accelerated stress within the coating similar to what it might receive in ordinary weather conditions. This type of testing can cause rapid failure in the coating film, and good control panels are required to obtain a proper comparison.

The above three tests, the salt spray, humidity tester, and weatherometer, make a good series of screening tests for coatings to withstand atmospheric exposure; either rural, mild industrial, or marine. A coating which stands up well under all three should also stand up well under actual exposure. However, no interpolation of actual life should be made from these tests.

J. A. Burgbacher, in his paper *Laboratory Evaluation of Protective Coatings for Marine Environments,* outlines a series of tests which were developed to determine the marine resistance of coatings for use on offshore platforms. He used a series of three tests: continuous immersion, salt spray, and intermittent immersion in seawater. His summary of the findings of these tests on comparing a large number of different types of coatings is interesting and might apply as well to the salt spray, humidity, and weatherometer tests.

> In summary, a series of tests has been devised to screen protective coatings for marine applications. These tests are so severe, poorer formulations fail in two weeks. Variations in performance of coating systems of any given generic type suggest that a testing program can be extremely profitable.
>
> Coatings that perform poorly in the test will, in all probability, give substandard performance offshore. Conversely, outstanding performance should be expected from the system that showed least deterioration on tests. Somewhere between these extremes the corrosion engineer will have to establish the cutoff as to which coatings merit consideration and this decision will have to be based largely on practical experience.[11]

Control Panels. Control panels, or specimens, are extremely important in all laboratory testing. It is only by controls that the tests can be compared with a standard. For example, if a series of coatings is tested in a humidity test and there is no control, the only knowledge gained is the resistance of the coatings in comparison with each other. If a control is included, *e.g.,* a coating which is known to have good humidity resistance, the information obtained is not only comparative within the coatings tested, but also provides a comparison with a proven coating. This provides a standard of comparison on which to make the coating selection.

Exposure test racks, both for weathering and for marine exposures, are a valuable way of comparing coatings. However, because the coatings are usually exposed to an extremely severe weather or marine atmosphere, actual applications may not follow the same pattern as indicated in the test.

Actual operating condition tests are also valuable. This is the exposure of test panels directly on a structure where the coating is to be used. Where the structure is large and the atmosphere is relatively constant around the structure, this is a good comparison procedure and should determine which coating would be most satisfactory in that exposure. However, there are some problems involved as well. If panels are applied directly to the structure, problems may result from that direct contact. It is always recommended that panels be insulated from both the structure and each other so that there is no outside influence on any particular test.

An example of this was the installation of a series of inorganic zinc panels on the interior of a tanker. The test was to determine whether the inorganic zincs would stand such an exposure. The panels were welded directly to the steel in the tanker tank, and since the inorganic zinc is anodic to steel, the panels attempted to protect the entire tank and thus failed rather rapidly. On the other hand, when this was discovered and the panels were insulated from the tank, a good comparison of the various inorganic zinc coatings resulted.

Test Panels. There have been many different types of panels recommended for testing in various corrosive atmospheres. One of these is the Ken Tator Panel or KTA panel.[4] This is specially constructed in an attempt to illustrate all of the variables found in a steel structure, including angles, corners, crevices, welds, etc. This has been a popular panel, particularly for testing directly on structures in a plant atmosphere. However, because of the panel complexity, variation in application may cause a variation in the results obtained. Whenever such panels are used, the application variable should be taken into consideration.

Figure 12.9 shows two coated KTA panels which had been exposed in an acidic atmosphere on a rack attached to the plant structure. The comparison of the two coating systems involved is very good.

Flat, rectangular panels are generally preferred for the comparative testing of coatings. These can vary in size, depending on the desires of the person testing, and can range from 2 × 6-in. panels to 12 × 12 in. or larger.

A preferred size is 6 × 8 in. or 6 × 12 in. Such a panel provides sufficient area to reduce the edge effects (corrosion and undercutting on the panel edges), is small enough so that a number of panels can be placed on one rack, and is large enough that a uniform coating can be applied. The coatings also can be scribed to test the corrosion resistance of the coating.

The flat panel has the advantage that most of the application variable is eliminated when the samples are carefully prepared, making for a more consistent comparison of the coatings being tested. Figure 12.10 shows three different size exposure panels: 4 × 12 in., 6 × 12 in., and 12 × 12 in. This is a typical exposure rack for atmospheric testing.

Immersion Tests. The selection of coatings or linings for immersion service is often somewhat easier than one for general atmospheric service. This is because most immersion service is rather specific and therefore a laboratory test can be set up which can reasonably duplicate the conditions under which the coating must operate. For example, in testing for marine immersion, coatings can be immersed in 3% salt solution which provides many of the conditions that exist in marine immersion. The one thing that it does not include is the high oxygen content of the water, particularly just below the splash zone. This can be accomplished by adding air to the static immersion test. In this case, the air is merely bubbled through the salt solution, saturating it with oxygen. This makes a simple and yet effective immersion test for many coatings. Again, while the test itself is not entirely definitive, the comparison of coatings in this test does indicate those which will perform better in actual service.

An even better method is to use actual seawater in this immersion test, also bubbling air through the liquid. However, this is not necessarily an accelerated test. Nevertheless, it is one that can be done in the laboratory without the necessity of traveling back and forth to a marine test station.

There are many other simple immersion tests used to determine the resistance of a coating. Some specific ones could be sugar syrup at 82 C (180 F), 50% caustic soda at 65 C (150 F), or crude petroleum at 70 C (158 F). Each of the above could be the condition on the interior of a storage tank, so that duplication of the conditions in a laboratory is relatively easy, and the results of the test can be quite definitive.

A recommended series of screening tests is shown in Table 12.9. There are few coatings that will withstand all of these reagents. On the other hand, when determining the general resistance of a series of coatings, it does indicate the resistance area of the various coatings being tested. All of these are simple immersion tests and require only a closed container of sufficient size that a sample can be placed within the liquid. Usually, it is good to have the sample half in and half out of the liquid, since many times the adhesion of the coating will change at the water line. When the coating is tested for adhesion across that area, the line can easily be detected. This is an important part of the test, since some of the reagents may very well cause the coating to lose adhesion without any other obvious reaction.

FIGURE 12.9 — Typical coated KTA panels exposed under plant conditions.

FIGURE 12.10 — Typical atmospheric exposure rack showing three different sized panels, most of which are scribed.

TABLE 12.9 — Screening Test Reagents for Coatings and Linings

---

Simple Immersion

Water	
Neutral Salt	Sodium Chloride
Acid Salt	Ferric Chloride
Alkaline Salt	Sodium Carbonate
Oxidizing Salt	Sodium Hypochlorite
Acid	10% Sulfuric Acid
Volatile Acid	10% Hydrochloric Acid
Oxidizing Acid	10% Nitric Acid
Organic Acid	Acetic Acid
Alkali	10% Sodium Hydroxide
Aliphatic Hydrocarbon	Gasoline - VM&P Naphtha
Aromatic Hydrocarbon	Toluol
Vegetable Oil	Corn Oil
Alcohol	Ethyl Alcohol
Chlorinated Hydrocarbon	Trichlorethylene
Ester	Butyl Acetate
Ketone	Methyl Isobutyl Ketones

Other simple tests that may be made on these immersion samples would be hardness (before and after immersion), swelling and softness of the coating as compared with its original state, and possibly an increase or decrease in weight of the sample. There are also many other obvious results, such as blistering, discoloration, and actual dissolution, all of which can provide some valuable information regarding the resistance of the coating. Even where some specific requirements are necessary for a coating, this series of tests can give a good indication of a coating's overall resistance, even though the coating composition may be unknown prior to the test.

There is another, more specific yet valuable test for the evaluation of coatings and linings. This is the so-called *one-side test,* or tests using Atlas cells. This test consists of applying a coating to one side of a plate and fastening it into the test apparatus with the coating on the inside and the bare steel on the outside. This not only tests the coating for specific resistance to the liquid contained in the test container, but where there is a temperature differential from the test liquid to the outside ambient air, it also tests the ability of the coating to withstand the temperature differential between the internal and the external surfaces of the coating. This is the normal condition of a coating on the interior of an uninsulated tank. This is a practical test and is an excellent way to determine the adhesion of a lining to a container. Most internal coatings and linings are subject to this differential temperature.

In this apparatus, there are generally two plates, one for each side of the container. This is illustrated in Figure 12.11, which shows the various parts of the test apparatus. Figure 12.12 indicates the apparatus in use, with all of the electrical connections, condensers, monitors, and other equipment that can be used, particularly when the cells are at elevated temperatures.

These last two series of tests, the simple immersion test and the one-side test, are well described in a NACE Standard entitled, *Laboratory Methods for the Evaluation of Protective Coatings Used as Lining Materials in Immersion Service.*[7] Selection of a lining is greatly simplified if the lining

FIGURE 12.12 — Illustration of a test cell for one-side testing designed to approximate the volume-to-surface ratio found in a standard 8000-gallon tank car. (SOURCE: Laboratory Methods for the Evaluation of Protective Coatings Used as Lining Materials in Immersion Service, NACE Standard TM-01-74, NACE, Houston, TX, 1974.)

material performs satisfactorily in these one-side and simple immersion tests.

Summary

As discussed, there are many methods of selecting

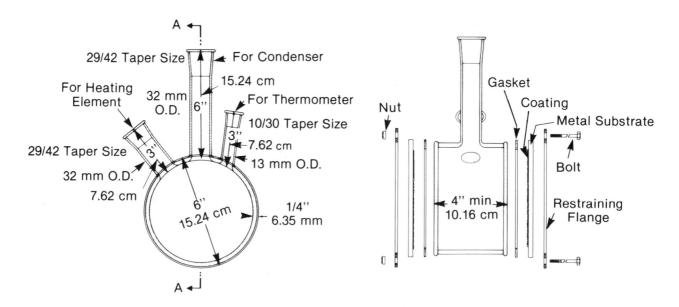

FIGURE 12.11 — Front and side views of a test cell for the exposure of one side of two separate panels which are mounted on the cell. This glass cell has openings for a thermometer, a condenser, and a heating element. (SOURCE: Laboratory Methods for the Evaluation of Protective Coatings Used as Lining Materials in Immersion Service, NACE, Standard TM-01-74, NACE, Houston, TX, 1974.)

Corrosion Prevention by Protective Coatings

coatings for a particular use. Some of the most common ones are:

1. Select a coating based on cost. This is not recommended.

2. Select coatings from general charts of generic coatings. This provides information on the general properties of coatings.

3. Compare literature on coatings from various manufacturers. This gives information on what coatings are available.

4. Request coatings information from various vendors. This provides information on the coatings that are available.

5. Make laboratory comparison tests of coatings from several manufacturers. This provides specific information on the resistance of various coatings.

6. Obtain a recommendation for a coating from the supplier with the best reputation in the field. This provides an outside source of information based on experience.

7. Make a selection of a coating from field tests. This indicates a coating that is the best one tested.

8. Select a coating based on personal experience. This may be the best criterion if the experience is similar to the proposed coating exposure.

The above methods of coating selection are generally in order of increasing value. Actually, the best procedure is to use all of the methods, combining the information developed. The question of coating material cost should only be considered after all other selection methods have been explored. If the above procedure is followed, a good coating selection can be made for almost any exposure.

References

1. Levinson, Sidney B., Selecting Paints, Architectural & Engineering News, March (1970).
2. Baldwin, Robert, Selecting Coatings for Plant Painting Projects, Plant Engineering, Feb. 6 (1975).
3. Vind, H. P. and Drisko, R. W., Field Identification of Weathered Paints, Naval Civil Engineering Laboratory Technical Report R-766, April, 1972.
4. Tinker and Rasor, San Gabriel, CA.
5. NACE, Direct Calculation of Economic Appraisals of Corrosion Control Measures, RP-02-72, Houston, TX, 1972.
6. Brevoort, Gordon H. and Roebuck, A. H., The Simplified Cost Calculations and Comparisons of Paint and Protective Coating Systems, Expanded Life and Economic Justification, CORROSION/79, Preprint No. 37, National Association of Corrosion Engineers, Houston, TX, 1979.
7. NACE, Laboratory Methods for the Evaluation of Protective Coatings Used as Lining Materials in Immersion Service, Standard TM-01-74, Houston, TX, 1974.
8. ASTM, Salt Spray (Fog) Testing, Standard B-117-73, Philadelphia, PA, 1973.
9. ASTM, Coated Metal Specimens at 100% Relative Humidity, Standard D-2247-68, Reapproved 1973, Philadelphia, PA, 1968.
10. ASTM, Paint—Tests for Formulated Products and Applied Coatings, Part 27, Book of Standards, Weatherometer Recommended Practices, Philadelphia, PA.
11. Burgbacher, J. A., Laboratory Evaluation of Protective Coatings for Marine Environments.

13

Coatings and Cathodic Protection

Introduction

Coatings and cathodic protection have often stood on opposite sides of the fence as exclusive and opposing approaches to corrosion protection. Proponents of coatings are often on one side discounting the advantages of cathodic protection and claiming that a good, well-applied coating is the only necessary protection for steel. On the other side are proponents of cathodic protection who often claim that any immersed or buried metal structure can best be protected by a well-engineered cathodic protection installation. Under many conditions, both sides may be correct in their assertions. However, under many more commonly occurring conditions, the ideal corrosion protection is actually a combination of both protection concepts.

Cathodic protection and coatings are both engineering disciplines with the primary purpose of mitigating and preventing corrosion.[1] Each process is different: cathodic protection prevents corrosion by introducing electrical currents from external sources to counteract the normal electrochemical corrosion reactions; coatings form a barrier to prevent the flow of corrosion current between the naturally occurring anodes and cathodes or within galvanic couples. Each of these processes has been successful in its own right. Coatings by far represent the most widespread method of general corrosion prevention. Cathodic protection, however, has protected hundreds of thousands of miles of pipe and acres of steel surface subject to buried or immersion conditions.

As corrosion protection has become more critical, and all types of metal structures more valuable, a marriage of the two corrosion prevention systems has naturally occurred. Experience has shown that damage to organically coated structures is almost unavoidable during construction and service. Breaks or holidays in coatings expose metal surfaces to corrosion, particularly in underground or immersion service. Attempts to eliminate all coating holidays drastically increase costs and are usually unwarranted. Thus, under many conditions, the combination of both systems actually provides better, more reliable, and, in many cases, less costly corrosion protection.

The concept of corrosion protection by coatings has been well described in earlier chapters. In the case of coatings for use with cathodic protection, it is necessary to use those coatings that are based on the impervious coating concept. This is because wherever coatings are used in connection with cathodic protection, they generally must be either immersed or in moist underground conditions. Without immersion or moisture, the cathodic protection mechanism will not properly operate.

As discussed in a previous chapter, the impervious coating concept requires the coating to be relatively impervious to the transfer of moisture, oxygen, air, and the various ions that may be in contact with the coating. Also, it must be resistant to the passage of electrons or electric current so that it forms an impervious film over the surface to be protected. As previously noted, the coating is not fully impervious to moisture vapor. Each coating material has a moisture vapor transfer rate characteristic of that particular material. Nevertheless, the coatings used in connection with cathodic protection are much more impervious than those types of coatings used in atmospheric exposures. Figure 13.1 is an example of a coating that should be satisfactory with cathodic protection.

Cathodic Protection

Cathodic protection is a technique used to reduce the corrosion of a metal surface by providing it with enough cathodic current to make its anodic dissolution rate become negligible. The cathodic protection concept operates by extinguishing the potential difference between the local anodic and cathodic surfaces through the application of sufficient current to polarize the cathodes to the potential

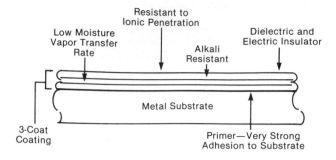

FIGURE 13.1 — An example of a coating that is compatible with cathodic protection.

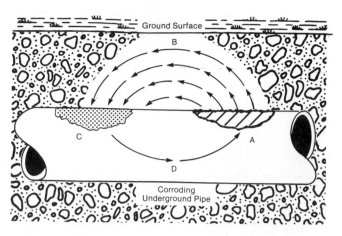

FIGURE 13.2 — Direct current flow on a typical corroding structure: (A) anodic area, where current leaves the steel to enter the surrounding earth—steel is being corroded here; (B) current flow through earth *from* anodic area *to* cathodic area; (C) cathodic area, no corrosion at pipe surface; and (D) current flows through the pipe steel from cathodic area back to anodic area to complete the circuit. (**SOURCE: Corrosion Basics - An Introduction, Chapter 5, Cathodic Protection, NACE, Houston, TX, p. 179, 1984.**)

of the anodes. In other words, the effect of applying cathodic currents is to reduce the area that continues to act as an anode, rather than reduce the rate of corrosion of such remaining anodes. Complete protection is achieved when all of the anodes have been extinguished.

From an electrochemical standpoint, this indicates that sufficient electrons have been supplied to the metal to be protected, so that any tendency for the metal to ionize or go into solution has been neutralized. Frank LaQue summarized this when he said:

> The most probable mechanism of cathodic protection of steel in seawater is a sufficient number of electrons from a preferred external source to accomodate a cathodic reaction, such as oxygen reduction or hydrogen evolution, over the whole surface of the metal being protected. Without cathodic protection, the electrons reacting with oxygen at the cathodic surfaces must be supplied by corrosion at the anodic areas. As additional electrons are supplied from an external source, the oxygen reduction reaction is accomodated by these additional electrons and fewer are required from the original anodes, some of which are converted to cathodes. The current reaching the cathodic surfaces from the remaining anodes decreases as the external current increases so that the total cathodic current density does not change substantially until all of the anodes are extinguished and the current density increases on the whole of the metal surfaces.[2]

An understanding of cathodic protection can be complicated by the existence of two different concepts concerning the direction of current flow in a corrosion cell. The *engineering approach,* which is used by most cathodic protection engineers, shows the current flow from the area of corroding metal through the electrolyte to the cathode (Figure 13.2). As indicated, the circuit is completed by the current returning from the cathode to the anode through the metal.

The *scientific concept* of the corrosion cell is that the iron changes from metallic to ionic iron at the area of corrosion, while simultaneously releasing electrons. These electrons pass through the metal to the cathode area where the cathodic reactions, such as oxygen reduction and formation of hydrogen gas, occur. The circuit is completed through the electrolyte by the transfer of positive charges, derived from the positive ions coming from the anode (*i.e.,* corroding area), in the general direction of the cathode. It is this neutralization of the electrons that is the controlling factor in the corrosion that exists at the anode.

The reason for the existence of two different concepts is that theory on the direction of electric current flow was

presented before the electron theory had been fully developed. The electron flow concept (Figure 13.3) is easier to understand than the electric current flow concept, although the latter is the one most commonly used in electrical engineering.

The purpose of cathodic protection is to reverse the current flow or electron flow so that an excess of electrons is present on the metal to be protected. With an excess of electrons, corrosion is reduced to nil. In other words, when there is an excess of electrons on a steel surface, any tendency for the iron to go into solution as a positive ion is reduced to zero. As long as there are negative electrons present in excess, all of the positive iron ions are held in the metal surface (Figure 13.4).

Coatings

The current or electron flow is the key to any corrosion that takes place at the anode. The total amount of corrosion that will occur on the anode is directly proportional to the total value of current flow in the circuit, while the penetration rate is a function of the current density per unit area and is directly related to the anode:cathode area ratios.[3] This area relationship is extremely important since, as previously discussed, if there is, for instance, a

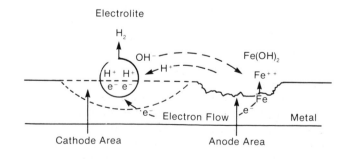

FIGURE 13.3 — Electron flow concept of a corrosion cell.

Corrosion Prevention by Protective Coatings

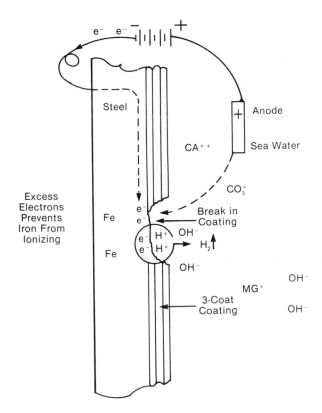

FIGURE 13.4 — Cathodic protection reverses the current flow so that an excess of electrons is present on the metal to be protected. [**SOURCE:** Munger, C. G. and Robinson, R. C., Coatings and Cathodic Protection, Materials Performance, July (1981).]

large anode with an area of 100 coupled to a small cathode with an area of 1, the corrosion on the anode will be only slight. On the other hand, if the area ratios are reversed so that the cathode area is 100 and the anode area is 1, the corrosion at the anode will be severe. This is why coatings become important in reducing the size of any cathodic area to a minimum and reducing the current flow to a small quantity. This is indicated in Table 13.1, which shows the approximate current requirements for cathodic protection of steel in various environments.

As can be seen from Table 13.1, the cathodic protection required varies over a wide range, depending on the area involved. The table also shows that a poorly coated area of steel in water requires substantially less cathodic protection than bare steel under the same conditions. A well-coated area of steel under the same conditions requires less than 1/1000 of the current required for even a poorly coated area of steel. Reducing the amount of area where cathodic protection is required, and providing a dam or impervious coating over the surface clearly reduces the cathodic protection requirements of the substrate while still providing full protection.

The coating mechanism of providing a dam which prevents any current flow to or from the surface of the metal is quite complex. This is due to the many variables that exist in a coating, many of which can be directly affected by reactions at the anode or cathode or by the mechanism of cathodic protection.[1]

TABLE 13.1 — Approximate Current Requirements for Cathodic Protection of Steel

Environment	mA/sq. ft.	mA/sq. meter
Sulfuric Acid Pickle (Hot)	35000	380.000
Seawater: Cook Inlet	35-40	380-430
North Sea	8-15	90-160
Persian Gulf	7-10	80-110
U.S., West Coast	7-8	80-90
Gulf of Mexico	5-6	50-60
Indonesia	5-6	50-60
Soil	1-3	10-30
Poorly Coated Steel in Soil or Water	0.1	1.0
Well-Coated Steel in Soil or Water	0.003	0.03
Very Well-Coated Steel in Soil or Water	0.0003 or less	0.003 or less

(SOURCE: Treseder, R. S., Ed., NACE Corrosion Engineers Reference Book, National Association of Corrosion Engineers, Houston, TX, p. 61, 1980.)

Coating Characteristics

Where coatings are used with cathodic protection, they must have several essential physical and chemical properties in order to be effective. Most of these are illustrated in Figure 13.1

Dielectric Strength

A coating used with cathodic protection must have good dielectric strength. Its dielectric characteristics must be sufficient so that both cathodic protection potentials and current flows would not affect its ability to act as a corrosion prevention membrane. When a cathodic protection potential is applied to a steel surface, the coating over that surface must have sufficient dielectric strength to resist the potential and to prevent any current flow through the coating. Coatings with a low dielectric strength, or those that will allow some current flow, often allow the buildup of cathodic deposits on the surface or under the coating, causing coating breakdown. This is not an uncommon occurrence where coatings contain metallic pigments.

Adhesion

Any coating, in order to be effective under immersion conditions, and particularly when combined with cathodic protection, must have strong adhesion to the surface over which it is applied. Marginal basic adhesion, or marginal adhesion caused by the application of the coating over less than properly prepared surfaces, will cause early coating breakdown when subject to cathodic protection potentials. If there is marginal adhesion and if there are breaks in the coating, which cathodic protection is designed to protect, delamination of the coating from the surface, blistering, or both, may occur.

Electroendosmosis

Adhesion is also important where electroendosmosis may be encountered. *Electroendosmosis* is defined as the forcing of water through a semipermeable membrane by an electrical potential in the direction of the pole that has the same charge as the coating membrane. Most coatings are **negatively** charged, and under cathodic protection the **cathode** has an excess of negative electrons, making it **negatively** charged. Thus, under cathodic protection, **coatings** with a high moisture vapor transfer rate and/or **questionable** adhesion would be more subject to damage **and** blistering by cathodic potentials.

This has often occurred under actual cathodic protection exposures where extensive blistering of pipeline coatings has occurred. The blisters were all filled with liquid, indicating the electroendosmotic reaction. Similar conditions have occurred in water tanks where adhesion was poor prior to the application of the cathodic current. While there is no corrosion to the steel underneath the coating, there is usually a substantial increase in current demand.

Alkali Resistance

It is essential that the coating be chemical-resistant, particularly toward alkalies. One of the key chemical reactions that takes place at the cathode is the development of a strong concentration of hydroxyl ions. This is due to the removal of the hydrogen ions in the area by the excess of electrons that are present on the cathode. This is illustrated in the chemical diagrams for the formation of molecular hydrogen.

$$2H_2O \rightarrow 2H^+ \quad + 2OH^-$$
$$\text{Water} \quad \text{Hydrogen} \quad \text{Hydroxyl}$$
$$\text{Ion} \quad \text{Ion} \quad (13.1)$$
$$\text{Water Ionization}$$

$$2H^+ \quad + 2e^- \quad \rightarrow H_2$$
$$\text{Hydrogen} \quad \text{Electrons} \quad \text{Molecular}$$
$$\text{Hydrogen} \quad (13.2)$$
$$\text{Cathode Reaction}$$

As shown, each molecule of hydrogen that is formed on the cathode leaves two free hydroxyl ions in the area which create the strong alkaline condition. If the coating is not sufficiently alkali-resistant, it can be completely disintegrated by this strong alkaline reaction. Even the current flow between local anode and cathode areas on a piece of steel can develop sufficient alkali to disintegrate some coatings in the cathode area. Coatings that contain aluminum pigment and are subject to cathodic protection often fail because of the reaction of the alkali with the metallic aluminum.

Ionic Resistance

The coating also must be resistant to the passage of ions. While this may be considered a part of chemical resistance, there are many variations in the susceptibility of coatings to ionic transfer based on coating composition. When combined with cathodic protection, it can be readily understood that a coating which had some tendency to transmit either positive or negative ions might soon be forced from the surface by the action of the cathodic current. Its electrical resistance would be affected, as well as its resistance to the passage of moisture. Coatings have, in fact, been forced from the surface by the formation of calcareous salts underneath the coating due to cathodic protection currents.

Electron Resistance

Resistance to the passage of electrons is equally important for coatings that are used with cathodic protection. Apparently, some coatings readily transmit electrons, since large buildups of calcareous deposits have been seen on the surface of coatings exposed to cathodic currents. Inorganic coatings that are not dielectric can allow calcareous deposits to form. In fact, this reaction has occurred with some epoxy coatings.

Coating Thickness

Coating thickness is an extremely important characteristic, particularly where coatings are to be combined with cathodic protection. Several critical properties are directly affected by thickness. Electrical resistance is improved as coating thicknesses increase. The moisture vapor transfer rate is directly proportional to coating thickness. Ionic transfer is reduced as the coating becomes thicker, as is electron movement.

Therefore, to be practical, when used with cathodic protection, the coating should have an *optimum* thickness for the coating type involved. Coating thickness may have a negative effect on adhesion under certain circumstances. For example, unreinforced epoxy coatings applied at too great a thickness may continue to cure, shrink, crack, and, in many cases, have been known to actually pull themselves off from the steel substrate. Therefore, the thickness of the coating should be as great as is practical, based on its composition.

Moisture Vapor Transfer Rate

The coating, to be effective under immersion conditions while also subject to electric potentials, should have a low moisture vapor transfer rate, as well as low moisture absorption. Moisture absorption is the molecular moisture that is absorbed into and held within the molecular structure of the coating. This property is important to coating effectiveness only where the moisture absorption might be sufficiently high to lower the dielectric characteristics of the coating and therefore offer less resistance to the passage of the current.

Moisture vapor transfer, on the other hand, is important, not only to a coating used for immersion service, but even more so to one exposed to an exterior source of current, such as cathodic protection potentials. Coatings with a high moisture vapor transfer rate are much more susceptible to electroendosmosis than are tightly adherent coatings with a low moisture vapor transfer rate. It is a general rule-of-thumb, all other properties being equal, that the lower the moisture vapor transfer rate, the more effective the coating.

Inorganic Zinc Coatings

The coating properties described above are generally for organic coatings, and the characteristics described are particularly important for organic coatings. The inorganic zinc coatings, which are also important for corrosion prevention, do not necessarily have these same characteristics. On the other hand, inorganic zinc coatings, including galvanizing, can be used in connection with cathodic protection, and excellent results can be obtained.

As previously discussed in Chapter 6 on inorganic zinc coatings, these materials are designed to provide cathodic protection and to produce a potential that is close to that of freely corroding zinc. In the case of galvanized steel, the potential would be the same as that of zinc. These coatings are also conductive in order to provide cathodic protection so that any impressed cathodic current can freely pass through the coating with little or no effect on it.

This has been proven in a number of areas, particularly in tidal zones where inorganic zinc coatings alone have been applied to piling or offshore structures down to below the mean low tide level. The remainder of the area was protected by cathodic protection so that the tidal area of the inorganic zinc coating received cathodic protection during the time when the surface was covered with water. In these instances, the cathodic protection applied to the structure merely augmented the protection offered by the inorganic zinc without harm to the inorganic coating. As a matter of fact, the electrons impressed on the surface by the cathodic protection current prevented either the underlying steel or the zinc from going into solution during the time that the area was covered with water, while the inorganic zinc coating protected the area during exposure to the atmosphere.

The inorganic zinc coatings, because of their excellent adhesion, combining both mechanical and chemical adhesion to the steel, have made an excellent base over which to apply organic coatings and to reduce both the undercutting and blistering that might occur where the organic coating was applied to the steel surface alone. This same characteristic of providing an excellent base coat for organic coatings also applies to many organic coatings subject to cathodic protection. An inorganic zinc base coat provides the strong adhesion needed by the organic materials to resist the cathodic protection current. Inorganic zinc coatings can, then, be used in connection with organic coatings to improve coating effectiveness when combined with cathodic protection. This has been proven through actual service, as well as by extensive laboratory tests.

Chemical Reactions

When considering the use of cathodic protection and coatings, the chemical reactions that take place at the anode and cathode areas are very important. Some of these reactions were briefly considered earlier in the chapter. However, due to their significance, a full description of the processes in the order they occur is warranted.

Anodic Reaction

The anodic reaction should be considered first. While this reaction is extremely important in the corrosion process, it is much less important where the cathodic protection process is involved. When metallic iron is placed in an electrolyte, an almost immediate solution of the iron takes place at specific areas on the surface. These areas are usually corners, edges, sharp protrusions, or any area where the metal has been damaged. The iron goes into solution as the ferrous ion, and in so doing releases electrons into the metal substrate. Iron ions going into solution quickly react with the hydroxyl ions in the area, forming ferrous hydroxide. This can then continue to react with oxygen, forming the black oxide of iron, and continuing even further to the red oxide of iron. The insolubilization of the iron ion as soon as it goes into solution creates a continuing demand for the iron ion, so that as long as there is an outlet for the electrons that are released, the iron will continue to go into solution (Figure 13.5).

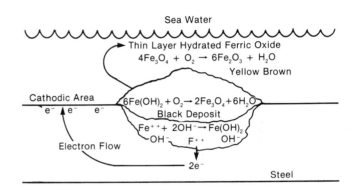

FIGURE 13.5 — Anodic reactions on iron.

Cathodic Reactions

The most important reactions shown in Figure 13.5 are those of the iron going into solution as ions and the release of the electrons. The action of cathodic protection on this freely corroding iron anode is to reverse the process by creating an excess of the electrons on the iron surface. The excess of electrons, from whatever outside source, reduces the tendency or ability of the iron to form ions, so that no iron goes into the solution. It can be said that the solution potential of the iron has been completely neutralized by the excess electrons. When this occurs, the anodic area, or the pit, becomes cathodic and thus prevents the iron from going into solution.

The chemical reactions at the cathode are the most important since as they take over under cathodic protection, the entire surface becomes cathodic. Thus, any coating applied over such a surface is subject only to the cathodic chemical reactions. As the metal surface is saturated with electrons from whatever exterior source and the water adjacent to the surface is ionized, the hydrogen ions from the water react with the electrons on the surface, forming atomic hydrogen which can then react with the oxygen that is dissolved in the water to form additional water.

The second reaction that occurs is for water and oxygen to absorb electrons, thus forming hydroxyl ions. These two reaction absorb electrons from the surface which need to be continually replaced in order for cathodic protection to be effective. They are continually replaced from an outside source of electrons, either an impressed current or current from a galvanic anode. With a coating over the surface, these two reactions occur only at a break in the coating, as was shown in Figure 13.4. The remainder of the surface is insulated from the liquid, and if the coating is impervious, as it should be, no reaction occurs underneath or on the coating.

There are two other reactions at the cathode, involving calcium and magnesium ions. Seawater, or even fresh water, it must be remembered, contains calcium and magnesium ions in solution. As the hydrogen ions are removed from the metal surface, hydroxyl ions are left in an excess close to the metal surface. Hydroxyl ions are also generated by the reaction of the electrons with water and oxygen. This creates a heavy concentration of the negative hydroxyl ions, which is easily demonstrated with zinc and an iron panel connected by an exterior wire in a phenolthalein indicator solution. When there are hydroxyl ions present, the phenolthalein becomes bright red.

Figure 13.6* shows the iron-zinc couple. Note the bright red coloration of the iron panel which is being cathodically protected by the zinc panel. This concentration of hydroxyl ions is sufficient to disintegrate many alkali reactive coatings that might be used under cathodic protection conditions.

This excess of hydroxyl ions, which creates an unstable situation in the water layer adjacent to the steel, is demonstrated in Figure 13.7. This causes the calcium and magnesium ions to react with the hydroxyl ions and the carbon dioxide from the air, forming magnesium hydroxide and calcium carbonate. These materials precipitate on the steel surface and can build up a substantial calcium deposit. This reaction is verified by F. W. Fink in *The Corrosion of Metals in a Marine Environment* in which he says:

> The application of cathodic current promotes the formation of hydroxyl ions on the cathodic areas. Also, the concentration of calcium and magnesium ions tends to increase in the film of sea water over the cathodes. As a result of these changes, the solubility of calcium carbonate and magnesium hydroxide is exceeded and the calcareous coating is deposited.[4]

All of these reactions are shown in Figure 13.7. That these calcium deposits are built up is indicated in Figure 13.8*, which shows a center pole of a coated domestic water tank with linear holidays in the coating. Cathodic protection was also used on the tank, and the calcareous deposits build up wherever a coating holiday occurred.

Consequences of Poor Coating Selection

The cathodic reactions that occur when cathodic protection is applied to a steel surface have a definite bearing on the effectiveness of the coating applied over the steel surface. The coatings which become a part of a coating and cathodic protection system must be designed to overcome any of these chemical reactions. The following are some examples of the effects of cathodic protection on coatings when the coatings are not properly selected or formulated.

*See color insert.

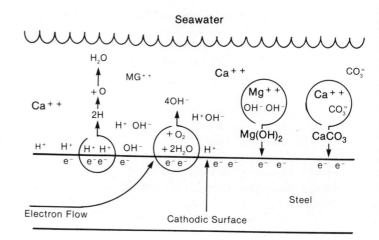

FIGURE 13.7 — Chemical reactions at the cathode under cathodic protection. [SOURCE: Munger, C. G., and Robinson, R. C., Coatings and Cathodic Protection, Materials Performance, July (1981).]

Hydrogen Blistering

A coating which has high water absorption or a high moisture vapor transfer rate will have moisture available at the steel coating interface and, as such, the electrons that are present there can react with the ionized hydrogen ions, forming hydrogen gas. If the adhesion of the coating is questionable, hydrogen blisters can form due to the hydrogen produced from the moisture. This is not an uncommon occurrence, even though the hydrogen has to develop at a sufficient rate to build up pressure underneath the coating.

In some cases, coatings are sufficiently porous to hydrogen so that any of the gas produced would pass through the coating rather than produce blisters. Of course, when this occurs there are excess hydroxyl ions that develop and create a strongly alkaline condition underneath the coating, which can also act to release the coating adhesion and to react with any acidic pigments or other acidic materials that may exist within the coating itself.

Reactive Pigments

Oftentimes metallic pigments, such as aluminum, are incorporated into a coating to help reduce the moisture vapor transfer and to make the coating more impervious to water. Aluminum pigments are very reactive with alkalies. Particularly under an excess of cathodic protection, coatings containing aluminum pigments may disintegrate because of the reaction of the aluminum and hydroxyl ions.

Delamination

Coatings that allow some ionic penetration or are sufficiently affected by the hydroxyl ion buildup at the coating/steel interface to allow ionic penetration, may actually be pushed off from the surface by the precipitation of the calcareous deposit underneath the coating. In this case, the coating is pushed off from the surface and usually breaks and spalls away, leaving a large area to be cathodically protected.

Corrosion Prevention by Protective Coatings

This has been shown to take place in tests of asphaltic-base coatings which were softened by the reaction of the hydroxyl ions and were then pushed off from the surface. Figure 13.9 shows an offshore platform that was taken out of service. The coal tar wrap coating was damaged, and cathodic protection deposits had built up under the coating at the damaged area, pushing it off from the surface. In this instance, the white deposits are the calcium deposits from the cathodic reaction.

FIGURE 13.9 — A coal tar wrap on an offshore platform where cathodic protection built calcareous deposits under the coating and stripped it from the surface.

Calcareous Deposits

Some coatings seem to be permeable to electrons. In this case, the electrons pass through the organic coating and are neutralized on the surface of the coating. When this occurs, the calcareous deposit can build up on the exterior of the coating. In fact, with only a small amount of excess cathodic current, quite a heavy layer can build up. While this has been demonstrated in tests, it is not common in actual practice.

Special Problems

Inorganic Coatings. Inorganic coatings are, of course, porous to electrons. As a matter of fact, the inorganic zinc coating concept is based on the solution of the zinc in the coating and the development of electrons to provide cathodic protection. When the cathodic protection current is at or no more than the potential developed by the iron-zinc couple, little or no reaction occurs within the coating. In this case, the zinc in the coating does not go into solution, and the coating acts as an effective film.

On the other hand, if an excess of electrons is forced onto the surface by an impressed current system with a higher potential than the zinc-iron potential, excess alkali can build up and possible breakdown of the inorganic zinc coating can occur. Hydrogen can also be produced, and while the inorganic zinc coating is porous, if sufficient hydrogen is produced within the coating, blistering can occur. It is also possible for the calcareous deposits to form on the outside of the zinc coating so that all of the cathodic reactions would have some effect, providing the cathodic current is applied in excess of that needed for corrosion protection.

Vinyl Coatings. Vinyl coatings generally have excellent alkali resistance and are resistant to the transfer of ions, as well as having a relatively low moisture vapor transfer rate. They generally perform well under cathodic protection, as long as the cathodic potential is not increased above the 1-volt limit. Most vinyl coatings rely on acid groups built into the molecule for adhesion to metal surfaces. Many of the primers for vinyl resins are formed from a tripolymer which contains maleic acid. While this has a bond to the steel, it may still react with excess hydroxyl ions that may be formed by absorbed moisture in the film. This reaction with the acid groups on the vinyl resin molecule tends to release the coating adhesion and create blistering.

All of these are examples of negative reactions of cathodic protection with the use of coatings, and are primarily demonstrations of what can happen if the wrong coating is selected or the cathodic protection system is improperly designed for the coating that has been selected. This is to emphasize the importance of considering the cathodic protection and the coating as parts of an overall protective system which must be designed *together* in order to obtain proper effectiveness. There are actually millions of square feet of surface that are protected by both coatings and cathodic protection, and they are generally performing better than either one of the protective systems alone.

Testing

With the number of coating systems available today, there have been several tests established to determine the comparative resistance of various coatings combined with cathodic protection. The object of these tests is to determine which coating systems perform best under cathodic protection conditions.

Early Tests

There were some early tests run at the Harbor Island Test Station at Wrightsville Beach, North Carolina, as well as some tests established at the North Florida Test Station of the Battelle Memorial Institute at Daytona Beach, Florida. These tests were conducted in the early 1950s, and the principal coatings which appeared satisfactory for marine immersion at that time were the vinyl, epoxy, asphalt, and early inorganic zinc coatings. Each of these coating types was tested, using a range of cathodic protection potentials varying from zero for the control and beginning with −0.8 volts and increasing the potential in increments to −1.4 volts to determine the effect of such potentials on the various coatings. Duplicate panels of each coating were used, one scribed and one unscribed. The tests were conducted in natural seawater, with the tide changes creating a flow of seawater across the panels at all times.

This series of tests was one of the first, if not the first, well-controlled tests to determine the effect of cathodic protection on coatings. In these tests, all of the coatings performed reasonably satisfactorily below a 1-volt negative potential, and each of the coatings failed rather characteristically at negative potentials above 1 volt. Tests were conducted over a 1-year period, keeping the potentials as constant as possible.

Where the potential was above 1 volt, each of the coatings reacted differently. For example, the vinyl coating tended to blister along the scribed area, with calcareous deposits forming in the scribe. The asphalt coating became generally soft and mushy, with the greatest effect occurring in the area of the scribe. This was attributed to the high concentration of alkaline hydroxyl ions reacting with the ingredients in the asphalt coating.

The epoxy coating that was used, an amine-cured epoxy, reacted quite differently. In this case, there was a fine, general blistering of the coating overall, indicating the possible formation of hydrogen on the steel underneath the coating. In addition to this, under the higher potentials, calcareous deposits formed on the surface of the coating in a heavy, thick layer. This indicated that there must have been transfer of electrons through the coating, causing the precipitation of the calcareous salts over the entire surface. In this case, the precipitation of calcium deposits was not at the scribe alone, but extended out over the entire panel. Thickness appeared to be a controlling factor, as well as the general coating resistance.

The vinyl coatings that were applied in thicknesses of only 2 mils showed some effect on the coating at all potentials, while the vinyl coatings that were applied at 8 mils showed little effect at potentials over 1 volt in the 1-year test period. The vinyl coatings were also overcoated with an antifouling paint and were unaffected even at the direct iron-magnesium potential. The antifouling coating continued to be effective, even at these potentials.

The early inorganic zinc coatings were also included in this test. Surprisingly, at voltages above 1 volt, some blisters formed in the coating, indicating considerable hydrogen formation and the formation of calcareous deposits over the surface as well.

The conclusions drawn from these early tests were:

1. Coatings, to be effective, must be dielectric so as to prevent the passage of cathodic current through the coating.

2. The coating must be resistant to moisture vapor transfer in order to prevent hydrogen blistering.

3. Coatings must be highly resistant to alkalies so that they are not disintegrated, or the adhesive bond affected by the high hydroxyl concentration.

4. Coatings must be of sufficient thickness to prevent the passage of current and to maintain a dielectric condition.

5. None of the coatings were appreciably affected at the negative potential of -0.9 volts.

These early tests were among the first controlled tests of coatings and cathodic protection and they helped establish the -850 to -800 mV range of potentials for the effective use of coatings with cathodic protection.

Later Tests

There were later tests established to determine the effect of cathodic current on coatings. One test series was set up to determine the compatibility of impressed current cathodic protection in connection with inorganic zinc coatings, both alone and overcoated.[5] The samples tested were bare steel; zinc anode material; steel with zinc primer only; steel with a complete paint system, including inorganic zinc primer; 8 mils of epoxy coating plus 3 mils of an antifouling; and the above epoxy system without the zinc primer. The tests were designed to answer five basic questions:

1. At what polarization levels and current density will hydrogen gas begin to evolve from bare steel?

2. At what polarization level and current density will hydrogen gas begin to evolve from bare zinc anode material?

3. Will hydrogen gas evolution begin at the same level when inorganic zinc primer is cathodically protected?

4. How will topcoats affect the activity of the zinc primer?

5. How does a system containing zinc primer compare with the same system without zinc primer?

The answers to these questions were:

1. A polarization level at which hydrogen gas evolution begins on bare steel is -996 millivolts and a current density of 35 milliamps per square foot.

2. A polarization at which hydrogen gas begins to evolve from the surface of the zinc anode material is -1.33 volts at a current density of 67 milliamps per square foot.

3. Polarization levels at which hydrogen evolution began on the inorganic zinc primer were about -1.31 volts, and a current density of 64 milliamps per square foot.

4. A polarization level of the zinc primer plus the topcoat was similar to the zinc primer only.

5. Without the zinc primer, the epoxy coating polarized and began hydrogen evolution at a slightly higher potential than that of bare steel.

The conclusions from this series of tests were:

1. The current density and the polarization levels at which hydrogen gas begins to evolve from the surface of the inorganic zinc primer are higher than that for similar coating systems without the zinc primer.

2. Any potential and current density which is safe with the epoxy system without the zinc primer, is also safe for the systems including the zinc primer.

3. In order to avoid damage to either coating system, the polarization level of 1 volt referenced to a silver chloride half-cell must not be exceeded.

Cathodic Disbonding Tests

There are numerous underground coatings that are often combined with cathodic protection. There have also been any number of tests to determine the most effective combination. One of these is the cathodic disbonding test where coated pipe samples are subject to cathodic current to determine their adhesion retention and resistance to cathodic potentials and current flow. One such test was carried out with four different pipe coatings. This was conducted on 4-inch pipe, 24 inches long, in accordance with ASTM-72 procedures (Figure 13.10).

The results of these cathodic disbonding tests are shown in Table 13.2, with the pipe coatings tested listed in the order of increasing disbonding of the coating at the holiday. The two polyethylene-coated pipes, extruded and double wrapped, were unaffected during the tests. This is due to the high dielectric strength, excellent adhesion, and coating thickness (40 mils). The coal tar epoxy showed only minor disbonding after approximately one year's exposure, while the powdered epoxy coating showed a disbonded area of more than ten times that of the coal tar epoxy.

Three of the pipe sections are shown in Figure 13.11, with the polyethylene tape showing no disbonding, the coal tar epoxy showing a minor undercutting at the 1/4-in. drilled hole, while the powdered epoxy shows a relatively large area (the area marked out in black) of coating disruption. Basic adhesion, water resistance, and coating thickness can account for the different performances of the coal tar epoxy and the powdered epoxy. All coatings were subject to a direct couple to a magnesium anode with approximately 1.5 volts potential.

General Testing

In 1975 Ameron, Inc. began an extensive series of tests to evaluate a number of coating combinations under cathodic protection, representing one of the most carefully controlled and extensive series known.[6] There were 23 coating systems put into test and subject to potentials of 0.0, −850, 1100, and 1200 millivolts, compared to a copper-copper sulfate reference cell. The coating systems are listed in Table 13.3. Actually, three different test series were initiated during the 5-year period, since many of the coating systems failed quickly and were replaced by other systems. The testing is still in progress and is scheduled to continue with newer coatings as they become available.

The test apparatus consists of four 8-in. × 7-in. × 9-ft long wooden cells filled with synthetic seawater and equipped with water circulating and aeration devices, makeup and renewal systems, necessary electrodes, potentiostats, wiring, shunts, and monitoring instruments. To provide impressed-current cathodic protection, a continuous platinized-titanium electrode was installed along the longitudinal center axis of a cell and connected to a potentiostat. To control and monitor potential levels, two continuous silver/silver chloride electrodes were installed symmetrically, one on each side of the center electrode. Figure 13.12 shows some of the test cells during the testing program.

FIGURE 13.10 — Cathodic disbonding test for pipe coatings with the four magnesium anodes at the corners of the salt tank.

TABLE 13.2 — Cathodic Disbonding Determinations for Coated Steel Pipe Specimens[1]

Specimen Coating Description	Exposure Time, Days	Determinations[2] Unsealed[3] Area, Sq Cm	Potential[4] Volts to CSE	Cathodic Current mA
Extruded Polyethylene	0	0.3	−1.18	1.00
Butyl Rubber Primer	365	0.3	−1.19	0.68
Avg 45 Mils DFT	699	0.3	−1.17	0.30
Polyethylene Tape	0	0.3	−1.18	1.65
Double Wrapped	365	0.3	−1.18	0.85
Avg 40 Mils DFT	699	0.3	−1.15	0.40
Coal Tar Epoxy	0	0.3	−1.16	1.35
No Primer	84	0.3	−1.18	0.72
Avg 24 Mils DFT	334	5.4	−1.18	0.47
Powdered Epoxy	0	0.3	−1.19	1.53
Avg 19 Mils DFT	299	25.8	−1.20	1.78
	365	67.9	−1.20	2.65

[1]Specimens were 4-inch diameter by 24-inch long pipe with one end plugged.
[2]Tests were conducted using ASTM G8-72 procedures.
[3]One 1/4-inch diameter holiday was drilled through coating prior to test.
[4]Pipe potential to copper sulfate electrode (CSE) determined after magnesium anode disconnected.

FIGURE 13.11 — Cathodic action at the small holiday in each pipe: Pipe 1—polyethylene tape, double wrapped (40 mils); Pipe 2—coal tar epoxy (24 mils); and Pipe 3—powdered epoxy (19 mils). (SOURCE: Ameron Corp., Monterey Park, CA.)

The performance of the 23 coating systems is shown in Table 13.4. Blistering was the most common type of failure because of the adhesion and undercutting related failures. The only coating systems with a significantly different failure were the aluminum-pigmented coating systems. Solution of the aluminum by the strong alkali build-up caused this failure.

First coating blisters developed at the most vulnerable locations, either adjacent to damaged coating at the

TABLE 13.3 — Coating Systems Tested

Number	Description	Total Average Dry Film Thickness (mils)
1	Inhibitive Epoxy Primer (2 mils) Epoxy Polyamide Topcoat (6 mils)	8.6
2	Inhibitive Epoxy Primer (2 mils) Epoxy Polyamide Topcoat (6 mils) (Panels Unscribed)	8.8
3	Epoxy Polyamide Topcoat (4 mils) Epoxy Polyamide Topcoat (4 mils)	7.6
4	Inhibitive Epoxy Primer (2 mils) Epoxy Polyamide Topcoat (6 mils) Epoxy Polyamide Topcoat (6 mils)	13.9
5	Salt-Contaminated Surface Inhibitive Epoxy Primer (2 mils) Epoxy Polyamide Topcoat (6 mils) Copper/Rosin Antifoulant (4 mils)	11.3
6	Inorganic Zinc Primer (0.75 mil) Inhibitive Epoxy Primer (2 mils) Epoxy Polyamide Topcoat (6 mils)	9.4
7	Inorganic Zinc Primer (0.75 mil) Inhibitive Epoxy Primer (2 mils) Epoxy Polyamide Topcoat (6 mils) Copper/Rosin Antifoulant (5 mils)	13.8
8	Salt-Contaminated Surface Inorganic Zinc Primer (0.75 mil) Inhibitive Epoxy Primer (2 mils) Epoxy Polyamide Topcoat (6 mils) Copper/Rosin Antifoulant (5 mils)	14.7
9	Epoxy Amine Primer (5 mils) Epoxy Amine Topcoat (6 mils)	11.4
10	High Solids Epoxy Amine	10.3
11	100% Solids Epoxy Cladding	188.0
12	One Coat Coal Tar Epoxy Polyamide	18.2
13	Inhibitive Epoxy Primer (2 mils) Coal Tar Epoxy Polyamide (10 mils)	13.3
14	Inhibitive Epoxy Primer (2 mils) Coal Tar Epoxy Polyamide (18 mils)	21.6
15	Inorganic Zinc Primer (3 mils) Inhibitive Epoxy Primer (2 mils) Coal Tar Epoxy Polyamide (16 mils)	22.7
16	Coal Tar Epoxy Amine (9 mils) Coal Tar Epoxy Amine (9 mils)	17.9
17	Inhibitive Epoxy Primer (2 mils) Coal Tar Epoxy Amine (9 mils)	11.6
18	Inhibitive Epoxy Primer (2 mils) Coal Tar Epoxy Amine (9 mils) Coal Tar Epoxy Amine (9 mils)	19.3
19	Aluminum Chlorinated Rubber Primer (9 mils) Chlorinated Rubber Topcoat (6 mils)	9.8
20	Aluminum Chlorinated Rubber Primer (3 mils) Aluminum Chlorinated Rubber Primer (3 mils)	6.5
21	Aluminum Chlorinated Rubber Primer (3 mils) Chlorinated Rubber Topcoat (6 mils)	9.0
22	Chlorinated Rubber Primer (3 mils) Chlorinated Rubber Topcoat (3 mils) Chlorinated Rubber Topcoat (3 mils)	9.1
23	Aluminum Pigmented Hydrocarbon Coating (2.5 mils) Aluminum Pigmented Hydrocarbon Coating (2.5 mils) Aluminum Pigmented Hydrocarbon Coating (2.5 mils)	7.0

FIGURE 13.12 — Apparatus for testing samples of four specimens of each coating system without and with cathodic protection at −850, −1100, and −1200 mV protection levels. (SOURCE: Ameron Corp., Monterey Park, CA.)

scribes or at inadequately coated edges. Blistering at these locations is not truly representative of deterioration under normal service conditions and leads to increasing protective current requirements of the specimens and accelerated coating deterioration. For these reasons, blisters closer than about 1/4 in. from the scribes were neglected, and occurrence of first blisters on the plate surface was used to assess coating performance.

Although there were some minor differences in blistering performance of a coating system without cathodic protection and with −850 mV protection, it became significant only with further increase in the protection level. In all cases, worst blistering occurred at the highest protection level, and the scribed sides of the specimens deteriorated most.

With −1100 and −1200 mV protection, many of the coating systems had medium to dense blistering and few systems had none. Testing of chlorinated rubber systems numbers 21 and 22 was terminated after 5 months exposure due to excessive blistering and deterioration. Aluminum-based systems numbers 20 and 23 started to dissolve around the scribe in about 45 days. Photographs of specimens from these systems are shown in Figures 13.13 and 13.14.

Figure 13.15 shows system number 10, a high-solids epoxy amine coating. The blistering shown is typical of many of the amine epoxy formulations and is similar to the results obtained with the epoxy amine coatings in the early cathodic protection tests.

The best performing system in terms of lack of blistering and lowest current demand is system number 15. This is the polyamide-cured coal tar epoxy used in conjunction with an inorganic zinc-rich primer and epoxy inhibitive primer.

Figure 13.16 shows the excellent performance of this system after 4 years of continuous exposure. After a period of approximately 20 days, the current demand dropped to 10 microamperes per square foot. This is an extremely small value, considering that the system specimens had 1% bare steel area at the scribes. A completely bare steel specimen under similar circumstances would be expected to require about 5 milliamperes per square foot

Corrosion Prevention by Protective Coatings

TABLE 13.4 — Test Performance of Coating Systems

Coating Number	Thickness/Mils	0.0 mV Days to First Blisters		−850 mV Days to First Blisters		−1100 mV Days to First Blisters		−1200 mV Days to First Blisters	
		At Scribe	On Plate	At Scribe	On Plate	At Scribe	On Plate	At Scribe	On Plate
Epoxy Series									
1	8.6	20	20	20	20	20	27	20	20
2	8.8	—	>532	—	>532	—	532	—	>532
3	7.6	8	20	8	27	13	20	8	13
4	13.9	20	48	20	41	20	34	13	34
5	11.3	3	3	3	3	3	3	3	3
6	9.4	13	365	309	365	69	111	13	55
7	13.8	13	64	13	167	55	62	34	69
8	14.7	13	41	20	55	13	34	13	34
9	11.4	69	199	44	58	44	58	22	37
10	10.3	22	22	22	22	22	22	13	13
11	188.0	>1484	>1484	>1484	>1484	>1484	>1484	>1484	>1484
Coal Tar Epoxy Series									
12	18.2	60	60	60	152	60	60	41	60
13	13.3	174	667	199	667	120	260	293	293
14	21.6	667	>878	590	>878	174	321	351	351
15	22.7	>1484	>1484	>1484	>1484	>1484	>1484	>1184	>1184
16	17.9	321	667	260	667	293	293	120	260
17	11.6	260	260	146	146	120	120	91	91
18	19.3	293	>878	779	>878	351	392	321	321
Chlorinated Rubber Series									
19	9.8	524	>532	13	532	13	41	10	245
20	6.5	27	245	200	365	10	152	10	20
21	9.0	>156	>156	>156	>156	69	69	2	13
22	9.1	>156	>156	>156	>156	91	91	7	7
Hydrocarbon Coating									
23	7.0	27	152	20	110	20	41	20	27

after the initial 20 days polarization. Thus, the current needed to protect system number 15 is about 1/500th of the current needed to protect bare steel.

This same coating system has shown similar good results in other seawater tests and has been used extensively on large mobile offshore drilling equipment. Figure 13.17 shows the performance of the above epoxy system (inhibitive polyamide primer and coal tar epoxy polyamide topcoats) without the inorganic zinc base coat. This system is also showing excellent resistance after 3 years exposure at the 1200 mV potential. Excellent adhesion and appreciable thickness, combined with the basic excellent water resistance of the coal tar epoxy are major contributors to its success.

To determine the effect of primers on the performance of a coating system, primed and unprimed specimens with the same topcoats were tested. In all cases, systems with primers had superior performance. For coating systems with zinc primers which inhibit corrosion at damaged areas, the service life is substantially increased over systems without primers. The differences in performance were greater with polyamide coal tar epoxy topcoats, for which the primers increased the time to first blisters two to three times.

Thickness is of real importance where coatings are to be used with cathodic protection, as has been stressed before. This series of tests also demonstrates this in Table 13.4, where the thicker coatings generally have shown the best resistance. Other factors also enter into these results; nevertheless, thickness remains an important factor. It is no coincidence that the two most effective coatings in the test were also the thickest (coatings numbers 11 and 15). This information is confirmed by Alexander in his paper, *Variables which Influence Cathodic Disbonding Test Results.* He states:

> Coating thickness was studied over a range of 20 to 70 mils. The effect of coating thickness is quite pronounced. Disbonding increased almost tenfold as the thickness decreased from the 65 mil range to the 20 mil range.[7]

Further data is given in Table 13.5.

Coating number 11 is a thick, tough, resistant epoxy coating designed for the splash zone and tidal area of offshore platforms. It is the one coating type tested which has been designed for this service where it is exposed to the most corrosive conditions as well as heavy abrasion and impact. As such, it should be combined with cathodic protection for the best overall corrosion protection.

Its thickness (188 mils) is a key factor in its effectiveness, combined with good adhesion and overall electrical and chemical resistance. This coating type is recognized as an effective answer to the tidal area problems on offshore structures. Its performance in this test

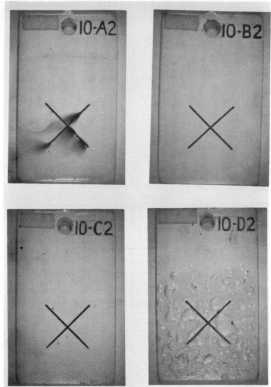

FIGURE 13.13 — Coating system number 21 of a chlorinated rubber primer and topcoat (9.0 mils) after 5 months exposure testing: (a) no cathodic protection; (b) – 850 mV; (c) – 1100 mV; and (d) – 1200 mV. (SOURCE: Ameron Corp., Monterey Park, CA.)

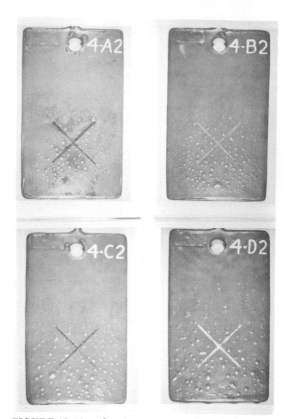

FIGURE 13.15 — Coating system number 10 of a high-solids epoxy amine (10.3 mils) after 5 months exposure testing: (a) no cathodic protection; (b) – 850 mV; (c) – 1100 mV; and (d) – 1200 mV. (SOURCE: Ameron Corp., Monterey Park, CA.)

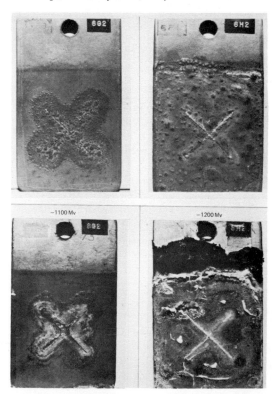

FIGURE 13.14 — Coating system number 20 [(a) and (b)] of aluminum-pigmented chlorinated rubber (6.5 mils): (a) – 1100 mV; and (b) – 1200 mV. Coating system number 23 [(c) and (d)] of aluminum-pigmented hydrocarbon (7.0 mils): (c) – 1100 mV; and (d) – 1200 mV. Both systems are shown after 10 months exposure testing. (SOURCE: Ameron Corp., Monterey Park, CA.)

FIGURE 13.16 — Coating system 15 of inorganic zinc primer/epoxy primer and coal tar epoxy polyamide topcoat (22.7 mils) after 4 years exposure testing: (a) no cathodic protection; (b) – 850 mV; (c) – 1100 mV; and (d) – 1200 mV. (SOURCE: Ameron Corp., Monterey Park, CA.)

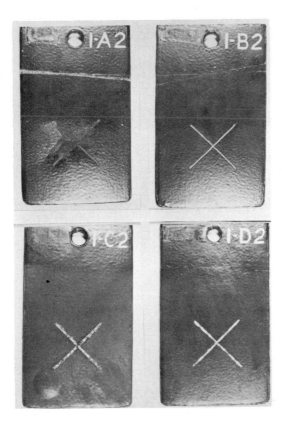

FIGURE 13.17 — Coating system number 14 of an epoxy primer and coal tar epoxy polyamide topcoat (21.6 mils) after 3 years exposure: (a) no cathodic protection; (b) – 850 mV; (c) – 1100 mV; and (d) – 1200 mV. (SOURCE: Ameron Corp., Monterey Park, CA.)

TABLE 13.5 — Effect of Coating Thickness on Cathodic Disbonding

Coating Thickness, Mils	Disbonded Area, In.^{2}	
	Each Panel	Average
20	0.42	
20	0.58	0.500
26	0.12[1]	
46	0.07	
37	0.10	0.090
40	0.10	
71	0.04	
65	0.07	0.057
58	0.06	

[1]This value not used in calculating the average. It is less than one-half the value of the average of the other two values.

[SOURCE: Alexander, Stephen H., "Variables Which Influence Cathodic Disbonding Test Results," National Association of Corrosion Engineers 25th Annual Conference, Preprint No. 13, 1969, National Association of Corrosion Engineers, Houston, TX.]

confirms this. Figure 13.18 shows its excellent resistance to cathodic protection after 4 years in tests.

The effect of the cathodic protection potential is shown in Table 13.4. Almost all of the coatings tested (compared to the controls) showed little effect at the preferred protection potential of – 850 mV. Some even showed better performance (longer time to the first blister) at the – 850 mV protection level than the control panels of the same coating without cathodic protection. At the – 1100 mV level, almost all of the thinner and more questionable coatings blistered in shorter intervals, while the time to first blister at the – 1200 mV level for most coatings reduced substantially over the controls at the – 850 mV potential.

Only the thick, tidal-area epoxy and the coal tar epoxy polyamide coating over an inorganic base coat showed no blistering over the duration of the test. The latter coating was the best performing coating in terms of lack of blistering and lowest current demand; in fact, both of the latter systems (systems numbers 11 and 15) had the lowest current demand throughout the test period, and the current demand after polarization remained practically constant.

Each of the coating systems used in this series of tests performed in a different way—most failing by blistering in a reasonably short period of time. The coating type which performed best overall, with the exception of the heavy coating for tidal exposure, was the coal tar epoxy.

FIGURE 13.18 — Coating system number 11 of 100% solids epoxy cladding (188 mils) after 4 years exposure testing with no undercutting at the scribes: (a) no cathodic protection; (b) – 850 mV; (c) – 100 mV; and (d) – 1200 mV. (SOURCE: Ameron Corp., Monterey Park, CA.)

When combined with an inorganic base coat, it was the outstanding coating system.

The coal tar epoxy also performed well in the salt box cathodic disbonding test and has proven superior in many actual cathodic protection installations on offshore struc-

FIGURE 13.19 — A coal tar epoxy coating used for an anode shield on a large tanker with no failure in evidence.

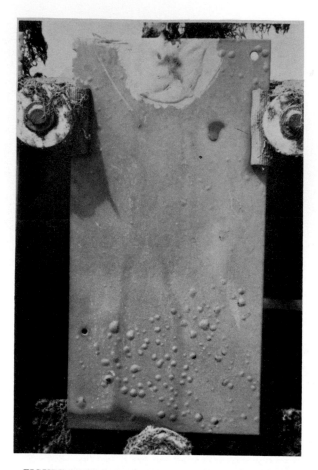

FIGURE 13.20 — An organic zinc-rich coating with typical blistering failure after exposure for 114 days in seawater and subject to an average current of 0.120 mA.

tures, ship hulls, water tanks, and similar exposures. In fact, it is commonly used as a shield around impressed current cathodic protection anodes on ship hulls and other structures because of its resistance to cathodic disbonding. In this case, it is subject both to heavy current flow and maximum potential. Figure 13.19 shows a coal tar epoxy coating used for an anode shield on a large tanker. No blistering or other failure is in evidence, even after sweep blasting.

Coating Failure

Blistering is one of the most common forms of failure of coatings under cathodic protection. Blistering can be caused by a lack of one of the essential coating properties outlined earlier in this chapter. Any coating lacking resistance to alkali, electroendosmosis, or which has some basic adhesion problems, will eventually fail by blistering under cathodic protection. Figure 13.20 shows an organic, zinc-rich coating with typical blistering failure after exposure for 114 days in seawater and subject to an average current potential of -1034 mV (SCE) and an average current of 0.120 mA..

An even more dramatic example is the blistering on the bottom of a large tanker where the impressed current protection system went out of control and applied excessive potential to the hull. The coating was an epoxy system (inorganic zinc preconstruction primer, epoxy primer, epoxy topcoat, and antifouling) applied over a well-sandblasted surface. The blistering shown in Figure 13.21 was present after several months service, and cathodic deposits can be seen where the blisters were broken.

In this case, the blisters were filled with a yellow-brown liquid having a pH of 11.3 and a chloride content of 12.4 grams/liter. The blistering was between the inorganic preconstruction primer and the epoxy. There was no corrosion under any of the unbroken blisters. This indicates that there was a basic adhesion deficiency between the inorganic zinc and the epoxy aggravated by the cathodic current, with the water passing through the epoxy into the area of poor adhesion by electroendosmosis. The

FIGURE 13.21 — Blistering on the bottom of a large tanker due to overprotection by cathodic protection.

Corrosion Prevention by Protective Coatings

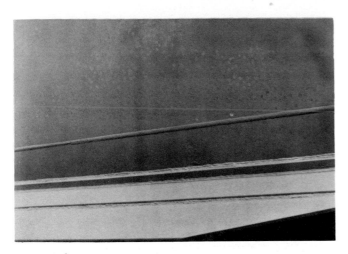

FIGURE 13.22 — A blistered hull after sweep blasting showing tight adherent coating between blisters.

high pH of the liquid occurred because of the normal cathodic reaction on the surface of the zinc primer

$$4e + 2H_2O + O_2 \rightarrow 4OH^- \qquad (13.3)$$

Liquid under blisters of this type is not uncommon, and generally the pH of the liquid is above that of the surrounding liquid. Once the hull coating was sweep blasted and the blisters removed, it appeared as shown in Figure 13.22. The inorganic base coat appeared unaffected. Normally, there is little corrosion under blisters of this type, even without the inorganic base coat. Where the coating is over a sandblasted surface, the original surface is usually bright and the blasting profile readily visible. Where no inorganic zinc is present, corrosion begins as soon as the blister is broken.

Most Common Type of Coating

The coatings with by far the greatest use in connection with cathodic protection are the hot-applied coal tar pitch or coal tar enamel coatings. These can be applied as the hot melted pitch alone or as a hot melt combined with fiberglass and other reinforcing material to form a built-up coal tar coating system. These coatings are usually of substantial thickness, *i.e.,* 125 mils or greater. Under most conditions the steel is blasted, a coat of coal tar primer applied, followed by the application of the coal tar hot melt, either reinforced or unreinforced. Under most conditions at the present time, the coal tar coatings for pipe are reinforced with glass fiber matte or cloth. While there have been difficulties with the coal tar hot melt systems, by far the great majority of the pipe coated in this manner has performed well with no great blistering, even under higher than normal potentials.

Where these coatings have failed in connection with cathodic protection, it has been primarily a problem of adhesion to the steel surface, and with some questionable adhesion, electroendosmotic blistering occurred, pushing the coating from the steel surface. All in all, many hundreds of millions of square feet have been coated in this manner and protected with cathodic protection as well.

The properties that make this coating type successful are its generally good adhesion to steel surfaces, its resistance to water and moisture penetration, and its appreciable thickness, which is much greater than usual for spray-applied coatings such as epoxies or coal tar epoxies.

Most Common Areas of Use

There are numerous areas where coatings and cathodic protection are used; however, there are a few where the combination of the two systems makes a much more effective corrosion-resistant system than can be obtained by either one alone. These areas take the best properties of both systems and combine them to provide the best type of corrosion protection for such areas. Some of these are as follows.

Tidal Areas

Under tidal conditions, immersion, moisture, and even drying conditions exist. Tidal changes occur cyclically on approximately a 12-hour basis. There are the effects of temperature and humidity at the lower tide level and the effects of immersion and an excess of oxygen in the water at high tide level and in the splash zone. All of these create serious corrosion conditions. When the tide is high, the area is fully immersed, and under these conditions can be protected by cathodic current in a manner similar to protection of a structure that is constantly immersed. At low tide, the area is completely exposed to the atmosphere and usually remains somewhat damp with full oxygen exposure from the air. Under these conditions, severe corrosion takes place in the tidal area where cathodic protection is only partially effective.

On the other hand, this tidal area can be covered with a protective coating and the area protected from the atmosphere and from the corrosion that normally occurs. Such corrosion is shown in Figure 13.23, which is the area

FIGURE 13.23 — Corrosion of piling on an offshore structure in a tidal zone.

of an offshore structure shown at low tide. In this case, the original coating was not adequate, and after a short period of time, corrosion occurred.

The tidal area is one of the most corrosive conditions found in a marine environment. This type of corrosion is shown in Figure 13.24, which is the corrosion profile of an offshore structure after 8 years of service in the Gulf area. In this case, the steel piling was nearly perforated in the tidal area.

The benefit of the combined system is shown in the following two figures. Figure 13.25 shows the corrosion profile of piling under seawater conditions when protected by either zinc or aluminum anodes. As can be seen, the tidal area still is an active area of corrosion, even though the corrosion rate for the fully immersed metal is quite low. Figure 13.26 shows the profile of a piling that was coated with coal tar epoxy and also protected with zinc anodes. Here, the atmospheric and tidal areas are fully protected, as well as the area that is fully immersed. The piling from which this data was derived was part of a test conducted by the U.S. Army Construction Engineering Research Laboratory. One of their conclusions from this 5-year test was:

> Coal tar epoxy coating with zinc znodes for cathodic protection provides protection to the atmospheric zone and the immersed zone and has an added capability of protecting the steel in the immersed zone should damage to the coating occur.[8]

Water Storage Tanks

An area that produces many of the same conditions is the interior of water storage tanks. In this case, the underside of the roof of the tank along with the area a short distance down the shell is constantly exposed to a highly humid atmosphere. The shell of the tank, for most of the distance from the roof to the bottom, is periodically exposed to immersion and then to the highly humid atmosphere on the interior of the tank.

An added factor in the case of water storage tanks is the temperature that is created by the sun on the exterior of the tank. When the water level in the tank is low, the coating which was previously subject to immersion is heated quite rapidly from the exterior, creating a vapor pressure. If the coating has poor adhesion characteristics, the pressure between the coating and the steel will cause blistering. Unfortunately, many such tanks have been coated with coal tar coatings. Since the coal tar is thermoplastic, even to the point of flowing when heated, warm

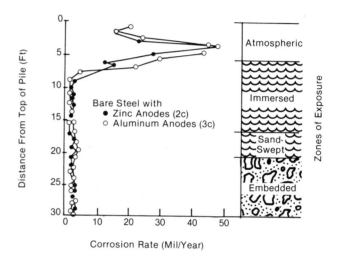

FIGURE 13.25 — Corrosion profile of cathodically protected bare carbon steel. [SOURCE: Kusnar, A. and Wittmer, D., Coatings and Cathodic Protection of Pilings in Sea Water: Results of 5 Years Exposure, Materials Performance, December (1979).]

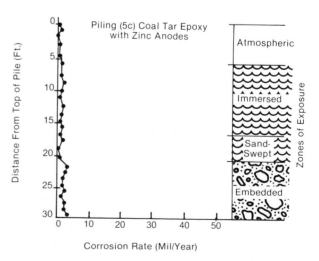

FIGURE 13.26 — Corrosion profile of coal tar epoxy-coated, cathodically protected piling. Steel conversion factors: 1 foot = 0.3048 m; 1 mil = 0.0254 mm. [SOURCE: Kusnar, A. and Wittmer, D., Coatings and Cathodic Protection of Pilings in Sea Water: Results of 5 Years Exposure, Materials Performance, December (1979).]

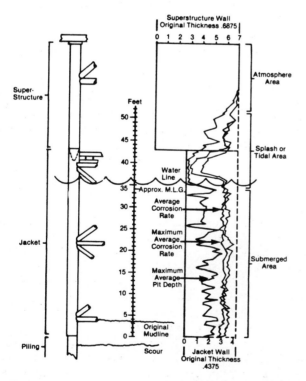

FIGURE 13.24 — Corrosion on an offshore structure after 8 years service in the Gulf of Mexico. (SOURCE: Grosz, O. L., Important Methods of Corrosion Control in Offshore Operations, Chevron USA, New Orleans, LA. Reprinted from Offshore, June, July, 1958.)

Corrosion Prevention by Protective Coatings

areas of the tank pose serious problems.

Thermoset coatings, such as a coal tar epoxy, have less of a problem in this way. However, they, too, are subject to the vapor pressure caused by the warm exterior of steel and the moisture that may be absorbed into the coating. While the coating protects this area and the above-water area in the tank, cathodic protection provides coverage for any damaged areas on the bottom of the tank or on the side walls where damage is most likely. On the side walls, should the coating blister and the blisters break, the part-time cathodic protection will materially reduce the corrosion caused by the break in the coating.

Table 13.6 shows a comparison of the coating performance of three steel water tanks. These 750,000-gallon tanks contain Colorado River water which has a high conductivity and is therefore more corrosive than most. Each of the three tanks was coated with a 50-mil hot-applied coal tar enamel. In this case, the enamel was applied by the daubing process, which is the reason the coating appears so uneven. Also, this type of application, where brush marks are a factor, may show linear steaks of corrosion because of thinning in the brush-marked areas. Two tanks were not protected by cathodic protection for ten years, while the third tank had cathodic protection installed after approximately seven years.

Figures 13.27 and 13.28 show the interior of one of the unprotected tanks and some of the corrosion that had started over the 10-year period. The coating definitely had started to deteriorate over this period, with blistering occurring and rusting showing wherever there was a pinhole in the coating.

The third tank, which had a combination of the same coating and cathodic protection, showed no steel corrosion within the same period of time. The coating in this tank appeared sound. While there were some white calcareous deposits present on the inside surface, indicating some pinholes or imperfections in the original coating, there was no coating adhesion failure at these points, even though the system was operating at a high current level for most coatings. The good adhesion, water resistance, and thickness of the coating enabled it to satisfactorily withstand these conditions.

Often, where tanks of this type are overprotected,

FIGURE 13.27 — Corrosion on the interior of a coal tar enamel-coated water tank with no cathodic protection after 10 years service.

FIGURE 13.28 — Corrosion in pinholes on a coal tar-enameled tank with no cathodic protection after 10 years service.

electroendosmosis can occur with blistering of the coating. In this case, the coating withstood the excess potential. However, if there had been an area of questionable adhesion, the water would have been driven through the coating to the metal/coating interface with resultant blistering.

Boottopping

The boottopping area on a ship has a similar combination of wetting and drying conditions, with the coating being fully immersed, possibly weeks at a time, followed by an equal period where the boottopping may be fully exposed to the atmosphere. This is particularly true on the large supertankers which often have a boottopping as wide as 30 to 50 feet.

An example of such a tanker is shown in Figure 13.29*. While this ship is not completely empty, an extensive area of boottopping can be seen. Many of the large crude carriers and ships of this magnitude have installed impressed current cathodic protection systems. Where

---

TABLE 13.6 — Comparison of Coated Water Tank Performance With and Without Cathodic Protection

Description	Duration of Cathodic Protection	Tank Potential, Volts to CSE[3]	Interior Surface Observations
North Tank[1]	None for 10 Years	−0.67 to −0.71	Coating Deteriorated. Inside Rusted and Pitted. Tank Repairs Needed.
Middle Tank[1]	None for 10 Years	−0.65 to −0.71	Coating Partially Deteriorated. Some Rusting and Pitting. Tank Repairs Needed.
South Tank[1,2]	Continuous for 3 Years	−1.10 to −1.18	Coating Sound. No Steel Corrosion Noted. Protection System Adjusted.

[1]Steel tank 3/4 MG capacity with 50 mil enamel coating.
[2]Impressed current cathodic protection system rated 12 volt, 5 amp d-c.
[3]Copper sulfate electrode (CSE) placed against immersed tank surface.

*See color insert.

Coatings and Cathodic Protection

such systems have been properly maintained and controlled, the combination of coatings and cathodic protection have effectively prevented corrosion to this area of the ship, as well as to the bottom.

On the other hand, as previously seen, there have been a number of instances where excessive impressed current has caused extensive damage to the coating of the bottom and the lower boottopping area. Coating in the vicinity of the impressed current anodes is often exposed to excessive electroendosmotic blistering, as well as the breakdown of the coating. Calcareous deposits form underneath and around any coating break. This has been overcome by a combination of heavy polyester glass flake coatings surrounding the electrode or coal tar epoxy coatings applied in the same area. The combination of firmly adherent, corrosion-resistant coating and cathodic protection has provided excellent corrosion resistance to many of these large ships.

Undersea Storage Tanks

With the advent of offshore petroleum production throughout the world, there have been many undersea storage tanks installed for the crude petroleum produced by the offshore wells. These are often massive structures that use the flotation principle; *i.e.,* the crude is floated on the surface of the seawater in the tanks, seawater is pumped out, the oil enters, and the seawater is pumped in to empty the tank of the oil. In almost all of these cases, a combination of galvanic anodes and coal tar epoxy coatings has been used for the protection of these massive structures, both on the interior and exterior.

Figure 13.30 shows such a large undersea storage tank being towed into position. Note the galvanic anodes attached at intervals to the exterior of the structure.

Offshore Platforms

Much was said previously regarding the corrosion of offshore platforms. Except for some of the early installations, most have used a combination of cathodic protection and coatings, although not for the complete structure. Most of the structures are coated down to a level slightly below the mean low tide line, while until now the remainder of the underwater structure has been protected either by galvanic anodes or impressed current.

The tidal area has been previously discussed as the most corrosive area on an offshore platform. In addition to being subject to considerable physical damage, it is an area where maximum corrosion protection is needed. Thus, the area is best protected when the coating is augmented with cathodic protection.

Figure 13.31 illustrates an offshore structure that is protected in the tidal area by both a coating and cathodic protection. In this case, the coating was inorganic zinc. In the center of the photograph, a vertical wire is shown which is the wire lead to a zinc anode providing galvanic cathodic protection to this area. Even though exposed for several years, this area of the platform shows little or no corrosion.

As the offshore platforms have become more sophisticated and costly, they have been used in greater and greater depths of water. More of the new platforms are being entirely coated to augment the cathodic protection in the immersed area. Not only are these areas relatively inaccessible for the life of the platform, the weight and quantity of the anodes required for galvanic protection are significant. The massive uncoated steel area demands a high current density in order to be polarized to an effective protective potential. In addition, the protection range of the anodes is limited, oftentimes, to a line of sight, and

FIGURE 13.30 — Large undersea storage tank with galvanic anodes attached at intervals to the exterior of the structure. (SOURCE: Chicago Bridge & Iron, Oakbrook, IL.)

FIGURE 13.31 — Tidal area of an offshore drilling platform coated with inorganic zinc to below the low tide area and cathodically protected with zinc anodes after 2 years service.

many anodes must be attached to provide complete protection to all crevices, corners, and shielded areas.

The galvanic anodes have a life which is usually substantially less than the design life of the structure, and the complete replacement of the spent or damaged anodes is extremely costly. If the cathodic protection system fails for any particular reason, there is no backup system to protect the steel from attack by the seawater environment. The combination of coatings and cathodic protection reduces the number of anodes, increases anode life, and provides a dual system to protect the structure in case there is damage to either the cathodic protection system or the coating.

High Water Velocity

The combination of the two protective systems can be highly beneficial where relatively high seawater velocities are encountered. As shown in Figure 13.32, comparing the applied current density in milliamps per square decimeter with water velocity in feet per second, the cathodic current required at 10 ft/sec. is 4 to 5 times as great as that at 1 ft/sec. This is illustrated by an example cited in Frank LaQue's book, *Marine Corrosion.*

> The current required for protection of uncoated submerged surfaces of offshore structures such as oil drilling rigs is influenced by temperature, dissolved oxygen content and especially the flow velocity in the sea water. The high velocity of ocean currents and the high oxygen content of water off the Alaskan coast have required much higher current density as in other locations such as the Gulf of Mexico.[2]

He also states:

> Practical experience with cathodic protection of ships has indicated that a ship at rest with a new intact paint system can be protected by a current density of about 1/10th milliamp per square foot. Underway, the current increases from one to three

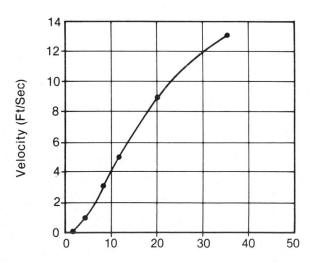

FIGURE 13.32 — Effect of seawater velocity on current density requirements for protection of bare steel. (SOURCE: LaQue, Frank L., Marine Corrosion, A Wiley-Interscience Publication, John Wiley and Sons, New York, NY, p. 205, 1975.)

milliamps per square foot depending on the condition in the bottom paint.[2]

Where such conditions exist, the combination of both systems not only provides for a better current distribution, but also lowers the amount of cathodic current required over that for an uncoated structure.

Improved Current Distribution

Distribution of cathodic protection currents over large structures is important, and is made possible by the insulating characteristics or dielectric properties of protective coatings. In the case of a bare structure, because of the conductivity of the liquid or soil, considerable driving force is required to distribute the current evenly over the surface. On the other hand, with protective coatings on the surface, the current is distributed only to bare areas in the coating. It makes a much more even distribution, reduces the amount of current required, and provides full protection for any area which is uncoated, or where the coating is broken.

Ships such as fishing vessels and even larger ships use this principle to protect the underwater hull from corrosion. Figure 13.33 shows the location of zinc anodes on the bottom of a large tuna clipper. In this case, the keel is protected as well as the hull of the boat up to the heavy load line. Zinc anodes operate only if a break in the coating occurs, and the flow of electrons from the zinc anodes is distributed through the hull to the break.

FIGURE 13.33 — Zinc anodes on the bottom of a fishing vessel to provide hull protection.

Galvanic Corrosion

The barrier principle of the coating combined with cathodic protection team up to prevent galvanic corrosion. The best example of this is the use of large bronze propellers on ships, which create a serious galvanic couple between the steel hull and the bronze propeller. In this case, with the propeller being strongly cathodic to the hull, if there is any break in the coating (and there is considerable abrasion on the leading edge of the rudder), deep pitting can quickly occur.

On the other hand, with the combination of cathodic protection and a strong coating, such corrosion can be prevented. It is standard practice aboard ship to apply the best coatings possible in the best manner possible in the

FIGURE 13.34 — Zinc anodes installed on the stern area and rudder adjacent to the bronze propeller on a large ship.

FIGURE 13.35 — Portion of a pontoon of a large semisubmersible drilling platform protected by a coal tar epoxy coating and zinc anodes.

stern area of the ship, on the stern frame, and on the rudder. These areas are subject to excessive turbulence, as well as the galvanic cell established between the propeller and the steel hull. In order to provide both cathodic protection and the coating, zinc anodes are generally applied to the rudder and the stern area of the ship in order to offset the action of the galvanic cell. Installation of zinc anodes is shown in Figure 13.34. Zinc anodes can be seen attached at intervals to the stern area of the ship and the rudder surrounding the propeller.

Semisubmersible Structures

Much has been said of offshore drilling platforms and the problems which exist there. The same problems exist, only in some cases to a greater extent, on the large semisubmersible drilling rigs that are presently being used. Most of these are extremely large, costly, and complicated structures. To assure full protection of the entire critical structure, coatings as well as cathodic protection are the most standard procedure.

For example, one of the largest semisubmersible drilling platforms in the world was coated with the following coating system. An inorganic zinc preconstruction primer was applied to the entire steel area prior to erection of the structure. When the structure was complete, a polyamide epoxy primer was applied over the preconstruction primer. Following the epoxy primer, a coat of heavy-bodied polyamide coal tar epoxy was applied, using a total dry film thickness of 20 to 25 mils.

The system was then augmented with zinc anodes attached at regular intervals over the complex structure. In this case, the structure measured approximately 100 yards square. The underside of the drilling platform was over 150 feet above the top of the immersible pontoon area. In addition, the platform was able to be self-propelled for placement and for holding in place during the drilling operation. The structure, when complete, was so large that there were no dry docks in the world capable of repairing the structure. Thus, it was assumed that the coating and the cathodic protection system would have to maintain the structure corrosion-free for the life of the structure.

Figure 13.35 shows a section of the pontoon area following the application of the final coat of the coal tar epoxy as well as the zinc anodes. The zinc anodes can be seen at intervals on the structure. The structures, because of their complexity and sophistication, require the best corrosion protection available, and it is for this reason that the combination of both coatings and cathodic protection was selected.

Cargo Ballast Tanks

The protection of the interior of cargo ballast tanks aboard tankers has been a subject for discussion for a considerable period of time. There has been some success obtained by the use of anodes in dedicated seawater ballast tanks. The Zinc Institute published an article, *Installation of Zinc Anodes in Cargo Ballast Tanks,*[9] on two ultra-large cruise carriers. The zinc anodes in this particular case were used as supplementary protection to selectively applied epoxy-type coatings in these tanks. In the case of these two vessels, almost 1,000,000 pounds of zinc anodes were required for the combined protection.

Ballast tanks have also been well protected by the use of zinc anodes alone. On the other hand, the combination of the coating and the zinc anodes reduces the number of anodes required and provides backup and much longer term protection for these tanks, which seriously corrode without the best protection possible.

Many companies have attempted to protect the cargo ballast tanks with galvanic anodes as well. In many cases, this has resulted in less than the best protection to the tanks, particularly where the ships are crude carriers. In the cargo ballast tanks, progressive pitting has resulted, with numerous cases of such pitting occurring within an inch of the anode position. Ships carrying sulfur crude with considerable hydrogen sulfide dissolved in the crude have been particularly troublesome where cathodic protection alone has been attempted. Figure 13.36 shows the pitting which resulted in a sulfur crude ballast tank. The photograph was taken on one of the horizontal stiffeners on the interior of the tank and the zinc anodes can be seen on the horizontal surface.

Figure 13.37 is a close-up of a zinc anode located on

FIGURE 13.36 — Zinc anodes on a horizontal stiffener in a cargo ballast tank. There is pitting adjacent to the anodes, and the light surface coloration is a white sulfur precipitate and not a calcareous cathodic protection deposit.

FIGURE 13.37 — Zinc anodes on the edges of steel plate in an uncoated cargo ballast tank, showing corrosion under and close to the zinc anodes.

such a steel stiffener. Corrosion of the steel can be seen closely adjacent to the zinc anode. In this case, hydrogen sulfide not only reacted with the zinc, but it reacted with the iron as well. It can do this regardless of any moisture being present. Therefore, the sulfide reaction can occur during the time that the crude petroleum covers both the iron and the zinc surface. Sulfide is strongly cathodic to steel and the zinc anode was deactivated by the zinc sulfide formed on its surface. Deep pitting of the steel surface was the result.

Generally, where chemical corrosion such as this is a problem, cathodic protection alone is not a complete answer. In several of the large crude carriers, the interior of the tanks were thoroughly cleaned by blasting, and a retrofit application of an epoxy primer and coal tar epoxy coating was made in addition to the zinc anodes in order to overcome the deep pitting problem.

Underground Structures

Underground structures are not unlike structures that are fully immersed. As a matter of fact, many underground structures are beneath the underground water level and, in essence, are under immersion conditions. In any case, the soil is usually damp at the level of the structures so that reasonable conductivity exists at most times. In underground pipelines running cross-country, there are many different areas of soil conductivity, and anodic and cathodic areas develop between these areas of differing activity. Most underground pipelines are therefore protected by both cathodic protection and coatings, with the combination providing good corrosion protection to the pipe.

There are several problems that necessitate the use of the combined corrosion protection systems, *e.g.,* holidays in the coating. Damage to the coating on the exterior of the pipe due to backfill operations is another example. The weight of the pipeline over a long period also can cause penetration of the coating by rock or similar hard materials in the backfill. Such coating breaks and discontinuities are almost impossible to detect once the pipe is in the ditch and the backfilling started.

Most pipelines are under pressure as well. Pitting of the pipe can cause stress problems from either water or high-pressure gas. The pipeline is generally inaccessible once it is in place, so that a minimum of maintenance and repair is desirable without extreme cost. Thus, the combination of both coatings and cathodic protection has provided the best answer to pipeline corrosion protection.

Also, many underground tanks, particularly in gas stations, were given a light coat of an asphalt coating and were placed in the ground without any other protection. Large quantities of gasoline and oil have been lost because of such poor protection, and at the present time, many municipalities have prohibited metal tanks from being used. Fiberglass-reinforced plastic tanks have replaced the metal tanks in many areas, although metal tanks are still used in some places. In these cases, coatings are applied and are usually combined with cathodic protection to eliminate the loss of fuel, as well as to prevent the contamination of ground water.

Summary

To obtain maximum corrosion resistance from the combination of coatings and cathodic protection, a number of factors, which have a basic influence on their combined effectiveness, must be taken into consideration.

1. Application of coatings needs to be done in the best possible manner, with excellent surface preparation. Poor application and poor surface preparation can only lead to coating failure under the influence of cathodic current by either hydrogen or electroendosmotic blistering.

2. For the coatings to be effective, they must be highly dielectric and must maintain such dielectric properties over their entire life.

3. Coatings must be highly chemical-resistant, particularly toward alkalies, because of the strong buildup of hydroxyl ions by the cathodic current. Any coatings which have even a small tendency to saponify, or which contain pigments that are alkali reactive, should not be used.

4. Coatings should have a low moisture absorption and a low moisture transfer rate, in addition to being dielectric, in order to prevent hydrogen blistering under the cathodic current.

5. Optimum coating thickness is essential in order to act as a long-time barrier and insulator for the cathodic current. Each coating has its own optimum thickness. However, in general, it can be stated that the thicker the coating (based on its own inherent properties), the better the results when used with cathodic protection.

6. It is indicated that antifouling coatings using alkali-resistant resins and pigments, applied over a proper dielectric coating, are not affected by the application of cathodic protection potentials.

7. From all evidence, indicated from actual tests as well as field installation, the cathodic potential, where used in connection with coatings, should be below − 1 volt, and the optimum potential is − 0.085 volt.

If these factors are taken into consideration, the results of both laboratory testing and actual use of the combined system indicate that maximum corrosion protection can be obtained through the use of coatings *and* cathodic protection.

References

1. Munger, C. G. and Robinson, R. C., Coatings and Cathodic Protection, Materials Performance, July (1981).
2. LaQue, Frank L., Marine Corrosion, A Wiley-Interscience Publication, John Wiley and Sons, New York, NY, p. 205, 1975.
3. Dean, Roy O., Coatings for Underground Environment and Cathodic Protection, Presented at short course 9/12-15/66, Surface Coating for Metals in Aggressive Environments, Univ. of California.
4. Fink, F. W., The Corrosion of Metals in Marine Environments, National Technical Information Service #AD-712 585-S, May (1970).
5. Dietle, B. and Gleason, J. D., Compatibility of Impressed Cathodic Protection with Paint Systems, CORROSION/71, National Association of Corrosion Engineers, Preprint No. 66, Houston, TX (1971).
6. Simpson, D. V. P. and Robinson, R. C., Experimental Studies Relate Effect of Cathodic Protection with Certain Generic Coating Systems, Presented at the 12th Annual Offshore Technology Conference, May, 1980.
7. Alexander, Stephen H., Variables Which Influence Cathodic Disbonding Test Results, National Association of Corrosion Engineers 25th Annual Conference, Preprint No. 13, 1969, National Association of Corrosion Engineers, Houston, TX.
8. Kusnar, A. and Wittmer, D., Coatings and Cathodic Protection of Pilings in Sea Water: Results of 5 Years Exposure, Materials Performance, December (1979).
9. Zinc Institute, Installation of Zinc Anodes in Cargo Ballast Tanks, New York, NY, 1979.

14

Coating Failures

Introduction

Coatings are much like human, animal, or plant life. Sooner or later failure occurs and "life" ceases. According to this concept, all coatings are doomed to failure. However, the coating failures described in this chapter are those that occur quickly in the life cycle and thus are not anticipated. As G. W. Seagren says,

> Paint is an expendable material which is used to protect or decorate another material more costly to replace; and the best paint for any given application is the one which will give optimum service during the longest period of service. Eventual failure of any paint is inevitable; but rapid deterioration of a coating is undesirable and can be prevented by careful observance of sound painting practice.
>
> Paint failure may be defined as deterioration of the paint system, or corrosion of the coated structure, more rapidly than would normally be expected, under the service conditions.[1]

Coating failure is also affected by the attitudes of those involved in coating application. During new construction, the application of coatings is usually the last job to be completed. By this time, the plant or equipment is ready to operate, at least for test runs; workers are anxious to finish the job; management is pushing to get into production; and the painters are pushed by all concerned (the owner, the general contractor, and their supervisors) to complete the work as rapidly as possible. The same is true for coating repair, only perhaps more so since repair time is always limited. The resulting attitude is thus "get the job done fast." This operating philosophy can only result in lack of proper care during application and thus eventual coating failures.

It has been estimated from past experience that 70% of all coating failures have resulted from poor or inadequate surface preparation.[2] This percentage only increases when a lack of care in application of the coating is evidenced. Thus, care in surface preparation, application, and inspection is the best insurance against coating failure. While such close attention to the physical application of the coating may initially seem costly, it is much less costly than continuing repair over the life of the structure.

A coating is a complex material, made up of a whole series of interacting ingredients, *e.g.*, resins, plasticizers, pigments, extenders, catalysts, fungicides, solvents, etc. They are applied as a very thin film of a few microns or thousandths of an inch thick. Solvents must evaporate. The nonvolatile portion must deposit a continuous film over the surface, react with and adhere to the surface, react with internal curing agents or with oxygen from the air to become insoluble, and provide a good-looking finish as well. This thin film must then withstand wind, rain, sun, humidity, cold, heat, oxygen, physical abuse, chemicals, biodegradation, and other forces.[3] It is understandable, then, that, as resistant as the modern protective coatings are to corrosive environments, they are still subject to many destructive variables and mechanisms. Obviously, with as complex and sophisticated materials as protective coatings, there can be many causes for coating failures, including the following:

1. A number of types of failures can occur due to the basic formulation of a coating. Such failures can be chalking, checking, cracking, discoloration, and similar phenomena.

2. Many coating failures are due to improper coating selection. A coating that was designed for steel surfaces is often extended out over a concrete or a wood surface, rapidly failing over these other surfaces. A coating that may be excellent for the exterior of a ship can be a complete failure on a chemical reaction vessel. Thus, careful selection according to the particular exposure is extremely important.

3. Many coating failures also result from the nature

of the substrate; *i.e.,* a coating can be incompatible with the surface over which it is applied. In this case, a chemical reaction may occur between the surface and the coating. In addition, the density or smoothness of the surface may cause the coating to have poor adhesion. Thus, the substrate directly affects the performance of a coating.

4. As previously discussed, a large majority of coating failures can be traced back to improper or poor surface preparation. In this case, lack of or poor surface profile, surface contamination, condensation, mill scale, or many other surface-oriented difficulties can influence the performance of a coating.

5. Application and surface preparation together form a substantial part of the difficulties which create coating failure. Inadequate thickness, pinholes, overspray, improper drying, improper curing—all are causes for rapid coating failure.

6. Adhesion is related to a number of the other causes for coating failure. In fact, adhesion-related failures are numerous and generally catastrophic. Such failures include blistering, flaking, peeling, and intercoat contamination.

7. As previously noted, the design of a structure itself is often the cause for severe failure. Sharp edges, crevices, skip welds, back-to-back angles—all are focal points for failure.

8. Exterior forces are forces such as chemical exposure, abrasion, reverse impact, severe weathering—all of which can cause rapid coating failure.

The above are all relatively common, general causes for coating failure, each of which will be individually discussed.

Formulation-Related Failures

There are some coating failures over which the corrosion engineer has little or no control. Those which have to do with the formulation of the coating itself are one of these general types. If the coating is formulated poorly and the corrosion engineer selects that coating, the coating will fail in spite of any effort that is made to insure its proper application.

Failures do occur as a result of the basic ingredients used and their combination in the coating, the resins used, the pigment used, or even improper solvent formulation. The formulation-related types of coating failure are as follows.

Chalking

Chalking was probably one of the most difficult types of coating failure that the automotive industry has had to contend with over the years. The early natural cellulose lacquer coatings often failed rather rapidly by chalking, and it was not until the advent of some of the alkyd-base materials that pastel colors, which are so popular today, became possible. Even then, over a period of time, the color on a car would change from a light blue to a light gray. When this occurred, it was necessary to abrade the surface, using a body cleaning compound, to remove the chalk and return the automobile to its original color. This same type of action takes place on all kinds of industrial

structures; however, some coating materials are superior to others in this regard.

The mechanism of *chalking* is essentially one where the coating binder tends to gradually disintegrate, leaving the surface covered with the pigments that have been held on the surface by the binder. This process continues until the surface coating is worn through, at which time the primer is visible or corrosion begins to occur on the substrate. Action such as this is shown in Figure 14.1, which shows the deck of a ship where an epoxy coating chalked so rapidly that the entire topcoat eroded down to the primer.

Chalking is strictly a surface phenomenon. While in many cases it takes a period of months or years to chalk down to the state shown in Figure 14.1, it can also be a rather rapid process, with an entire topcoating several mils thick eroding within a year. At the time that the automotive coatings failed by chalking, it was common practice to polish the coating using a very fine abrasive, followed by a coat of wax to protect the binder from weather exposure. This, to a substantial degree, aided in reducing the chalking. On the other hand, it was a temporary solution, making it necessary to wax a car every few months.

FIGURE 14.1 — Rapid chalking of the coating on a ship's deck.

Powdering or chalking of a coating is due to the exposure of a coating to the actinic rays of the sun and the action of the radiation on the organic binder. Shaded areas seldom tend to chalk. All of the airborne reactants, *e.g.,* humidity, oxygen, and air pollution, play a part in the chalking action. These react with the resins in the binder, causing it to disintegrate, thus leaving the pigments free on the surface.

One example is amine epoxy coatings or epoxy ester coatings which chalk very rapidly. Epoxy ester coatings are generally used only for interior work because of their rapid chalking tendency. Other resins, such as the acrylic or acrylic-modified resins modified with other resins (such as vinyls, epoxies, and alkyds), now have excellent weathering and chalk-resistant properties so that they are only slightly affected by the sun's radiation. Several years exposure is usually required before an appreciable surface reaction occurs. The same is true for some of the new aliphatic isocyanate-catalized polyurethane resins. These materials not only have a high gloss, but retain the gloss without chalking over a period of several years.

Pigments

The pigmentation of coatings is extremely important in the chalking reaction. The original white pigments, including the anatase titanium dioxide, tended to chalk readily and rapidly in almost any coating vehicle. On the other hand, when the rutile titanium dioxide pigment became available, it had a major impact on the chalking properties of coatings. These materials made possible the use of pastel colors and the mixing of colored pigments with white in order to obtain pastel shades. The combination of the chalk-resistant resins and pigments has made possible the type of automotive finishes that are available today, which last for several years and still maintain their appearance without appreciable attention.

Many coating pigments will tend to catalyze the chalking reaction. There are some black pigments which will remain unaffected in a coating for long periods of time, while others will tend to chalk in a short period, with the black surface becoming dull and easily rubbed away. Blue pigments also tend to bronze and then continue on to the chalking stage. Many other pigments tend to change color on the surface. Figure 14.2* shows the effect of sunlight on a red pigmented coating, turning it white and chalking it over a relatively short period of time.

Pigments also have a positive influence on chalking, as they tend to reduce the chalking by shielding the resin from the sun's rays. This is the action of some of the black pigments and was one of the reasons that the Ford automobiles, when black was the only color available, maintained their appearance over a relatively long period of time. The shading effect is also demonstrated to an even greater degree when coatings are pigmented with aluminum. This is an excellent example of shielding the sun's rays from the binder. The shingle effect of the aluminum plate pigment prevents any of the sun's rays from contacting the vehicle binder. This is the reason that the weather resistance of almost any coating vehicle is improved when the coating is formulated with aluminum pigments.

Figure 14.3 schematically shows the type of orientation developed by a leafing aluminum pigment in a coating film. The overlapping metal flakes provide an opaque shield for the binder resin. Other opaque pigments also shield the resin from the sun, but not as effectively as the orientation of the metal flake.

The chalking reaction has also been used in a positive manner by a number of coating manufacturers. These manufacturers have formulated coatings with a controlled chalking rate, so that the surface erodes slowly, but still at a sufficient rate for the surface to remain white and clean. This reaction is regulated through the proper formulation of the coating binder and the pigments, and it reduces the effect of grime and dirt which otherwise might accumulate and adhere tightly to the coated surfaces. In this case, the white remains much more attractive over a longer period of time. In all of these cases, the reduction in the thickness of the coating is controlled so that the thickness reduces slowly, thus allowing the coating to provide the protection needed and still maintain its appearance over a number of years.

Chalking is also considered one of the least objectionable types of coating failure. It is preferred because a clean chalked surface that is free of corrosion provides a good base for additional topcoats which can bring the coating back to its original thickness and effectiveness.

In order to evaluate the extent of chalking, ASTM Standard D 659-80 provides a method for determining the degree of chalking for exterior coatings. Figure 14.4 shows the various degrees of chalking failure outlined in this standard. The procedure used here is a simple one—merely wrap a piece of black felt or velvet on a thumb or finger, rotating the finger through a 180-degree angle when pressed against the coating. The cloth can then be removed and the degree of chalking checked against the standard shown. A dark black felt is used for white or pastel coatings, while a white cloth is used for colored coatings.[4] It is a simple yet very effective test for comparing the chalking characteristics of various coatings.

In order to reduce chalking to a minimum, formulations for topcoats based on the acrylics, the acrylic-modified vinyls, the acrylic-modified epoxies or alkyds, and the aliphatic polyurethanes combined with chalk-resistant pigments and such coating additives as ultraviolet absorbers, will provide resistance to chalking or reduction in gloss for a period of several years.

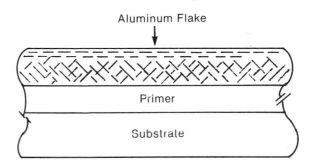

FIGURE 14.3 — Orientation developed by a leafing aluminum pigment in a coating film.

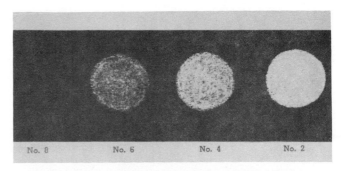

FIGURE 14.4 — Standards showing the degrees of chalking failure. (SOURCE: Federation of Societies for Coating Technology, Pictorial Standards of Coating Defects, Chalking Resistance, Philadelphia, PA.)

*See color insert.

Erosion

Erosion is also a surface reaction usually associated with chalking. It is often seen in brushed coatings where the high ridges of the brush marks show greater erosion than the valley areas. It can also be caused or aggravated by the surface being exposed to heavy rainfall, hail, high winds, or a combination of high winds and rain. *Erosion* is essentially a wearing away of the coating surface in a gradual manner similar to the chalking mechanism.

There are various types of erosion mechanisms, *e.g.,* sand erosion, which is very common in the western deserts. Automobiles caught in sand storms can have the entire coating on one side of the car eroded by wind-blown sand within a short period of time. This is because the high winds and sand create a condition much like sandblasting. Sand erosion is also caused by winds on beaches, where the constant movement of the sand against the coating on piling or similar structures will wear it away in the area of impact. Sand erosion by wave action can also take a toll on coated piling or bulkheads.

Generally, however, erosion, as it is considered in the coating industry, is a reaction similar to chalking, *i.e.,* the gradual wearing away of the coating due to weathering action. The resistance to this type of coating failure is also similar to that of chalking, and the use of the same type of resistant resins and pigments substantially reduces the erosion mechanism. Figure 14.5 shows an eroded coating. In this case, the coating has been brushed, and erosion has taken place on the high spots in the coating.

Checking

Checking of a coating can be described as small breaks in a coating surface that are formed as the coating ages and becomes harder and more brittle. Checking is a surface phenomenon and does not penetrate the full depth of the coating. This is important, since in many cases a coating may check on the surface soon after it is applied, yet remain in a stable condition, corrosion-free, for a long period thereafter. Some checking of a coating is almost invisible to the naked eye and can only be readily seen

under a low power magnifying glass or microscope. On the other hand, other coatings may check sufficiently to be readily visible. Figure 14.6 shows macrochecking on a highway sign. In this case, the topcoat has checked, but the failure does not penetrate through to the substrate.

Checking, for the most part, is a formulation-related reaction. It is, again, a combination of the resins and pigments which are not properly combined, so that, as the coating dries or continues to react, the surface becomes hard and brittle and surface stresses develop in the coating. Certain pigments will catalyze or cause checking to occur. This has been proven by using the same vehicle yet formulating with different pigments and having various checking patterns develop from the different pigments. This can range all the way from no checking at all, to severe checking due to the pigmentation.

It was originally stated that checks in the coating do not penetrate to the substrate. However, if the coating is not maintained, it will eventually break down to a point where it cracks, exposing the underlying surface. The surface stress in the coating is aggravated by wetting and drying, heating, cooling, and solar radiation—all contributing to the reactions on the coating and creating a more brittle and less expansive surface.

Since checking is primarily a formulation problem, its prevention is a matter of proper coating selection. The coating should be formulated with very weather-resistant resins, with nonreactive pigments that do not contribute to checking, with some permanent plasticizer, and with reinforcing pigments that reduce the stress in the surface of the coating. Figure 14.7 shows an example of fine yet severe checking of a coating film. ASTM Standard D 660-44 (reapproved in 1981) provides a method for the evaluation of the degree of checking in exterior coatings.[5] Such a method can be quite valuable in comparing coatings prior to selection for a given project.

Alligatoring

Alligatoring can also be considered a checking reaction in which the surface of a coating hardens and shrinks at a much faster rate than the body of the coating itself. Actually, it can be considered a macrochecking type of failure caused by stresses set up on the surface of the coating. As in the case of checking, alligatoring usually does

FIGURE 14.5 — Erosion by weathering of a brushed coating. The tops of the ridges of the brush marks are worn away first. (SOURCE: Federation of Societies for Coating Technology, Pictorial Standards of Coating Defects, Erosion Resistance, Philadelphia, PA.)

FIGURE 14.6 — Macrochecking of the topcoat on an interstate sign.

Corrosion Prevention by Protective Coatings

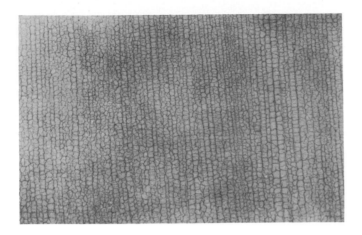

FIGURE 14.7 — Severe, fine checking of a coating film.

FIGURE 14.8 — Alligatoring of a coating.

not penetrate through the coating, but is a surface reaction. The reaction occurs where a hard, tough coating may be applied over a softer, extensible coating. The harder material tends to shrink and float on the surface of the underlying material, with the surface cracking in large segments (Figure 14.8).

Coal tar coatings, in particular, tend to alligator when exposed to the weather and sunlight. In this case, the coal tar coating hardens on the surface, while the underlying area remains soft. It is a typical coal tar reaction, particularly where the coating is applied rather thickly. It can also occur where harder coatings, particularly oxidizing types, are applied over an asphalt surface. Here again, the harder material, such as an alkyd, may tend to shrink sufficiently to pull it into an alligator pattern over the softer asphalt material.

Alligatoring has also happened where some air drying or chemically cured coatings are applied over a cold surface and then heated from the coated surface side, with a rapid temperature rise curing the surface much more rapidly than the underlying material. *Alligatoring,* then, is the shrinkage of the surface while at the same time the body of the coating does not change in the same way. Figure 14.9 is a schematic drawing of an alligatored surface, also showing a cross section of the coating. As can be seen, the surface of the coating ruptures due to shrinkage, but the underlying softer material merely pulls while not penetrating to the substrate.

The thickness of the coating has much to do with alligatoring, and, while the thickness is reduced in the checks of the alligator pattern, sufficient thickness usually remains to protect the substrate from corrosion for a considerable period of time. Eventually, failure will occur in these areas due to the thinner coating.

Alligatoring is a formulation-related failure so that prevention is also a matter of coating selection. A coating should be selected which does not have a soft primer under a harder topcoat. The coating, in general, should be reinforced with reinforcing pigments or fibers, and thin coats should be applied and cured before the application of the second coat. It should be a rule-of-thumb never to apply a hard coating, which oxidizes or requires polymerization, over a permanently softer or more rubbery undercoat.

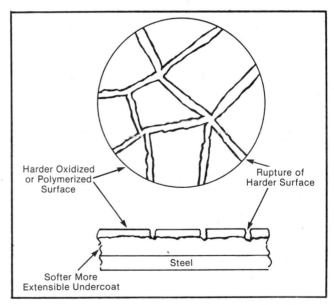

FIGURE 14.9 — A schematic of an alligatored surface, showing a cross section of the coating.

Cracking

Under most conditions, *cracking* is a formulation-related failure and one due to aging and weathering. It is different from checking and alligatoring in that it is not a surface problem, but one where the breaks in the coating extend from the surface through to the substrate. Thus, it is a much more serious type of failure than checking. Both checking and cracking are the result of stresses in the coating film which exceed the strength of the coating. Checking results from stresses on the surface, while cracking is caused by stresses throughout the film and between the film and the substrate.

Cracking can take several different forms. Figure 14.10 shows the cracking of a coating where the coating was applied heavily on the interior of a steel channel. In this case, it is nonlinear, taking a rather circular formation. Figure 14.11 shows a completely different type of cracking where, again, a coating is applied rather heavily on the interior of a steel channel. In this case, the cracking is in relatively short cracks in a heterogeneous pattern over the surface. The difference in configuration is undoubtedly due to the different resin and pigment formulations of the two coatings.

Coatings on wood, etc., usually are much more susceptible to cracking than are coatings on metals. This is due to the wood grain. When wood expands and contracts, it does so much more across than along the grain. Therefore, the stress in the coating is greatest across the grain, with the cracking being linear with the grain.

Severe cracking of a coating on wood is shown in Figure 14.12. In this case, there was severe shrinkage of the coating over the wood surface, creating wide cracks

between the sections of coating. A more normal cracking over a wood surface is shown in Figure 14.13. Here, the cracking generally follows the wood grain and is caused by the differential expansion and contraction of the hard and soft grains.

Cracking occurs as the coating ages, as it is subject to expansion and contraction, wetting and drying, absorption and desorption of moisture, etc. Epoxy coatings (amine-cured) have been known to crack on steel, which is subject to expansion and contraction due to temperature changes. This is especially true where the coating may be applied in a relatively thick film. The thicker the film, the greater the stress within the film as curing continues.

For instance, an epoxy coating was applied on the exterior of a cylindrical tank which was alternately heated and cooled. The epoxy, as it cured, developed sufficient internal stresses to crack and pull itself away from the surface in large pieces. Lack of adhesion undoubtedly contributed to the failure, even though the coating was in tension due to its outside position on the cylinder.

Cracking can be relieved by the use of proper resins, plasticizers, and pigments. Reinforcing pigments which are fibrous or acicular aid materially in reinforcing the coating against cracking action. The addition of plasticizing resins or plasticizers also aids in softening the coating, allowing it to be more extensible and elastic, and thus tends to reduce the stresses caused by temperature

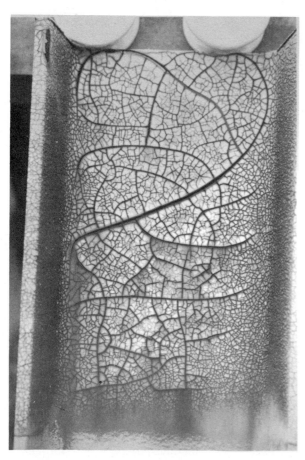

FIGURE 14.10 — General cracking of a coating, showing larger cracks in heavily coated areas.

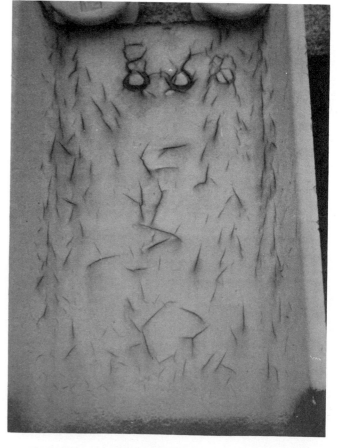

FIGURE 14.11 — Cracking of a coating, showing short, non-continuous cracks.

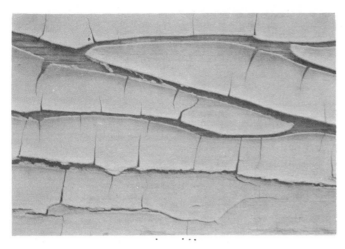

FIGURE 14.12 — Severe cracking of a heavy coating film over a wooden surface.

FIGURE 14.13 — Cracking of a coating film over a wooden surface, with cracks generally following the wood grain.

changes, weathering, and continued polymerization. Coatings which continue to polymerize, or which oxidize over a period of time (*e.g.,* oils and alkyds), are more susceptible to cracking failure than are fully polymerized vinyls and chlorinated rubbers.

In order to evaluate the extent of cracking in a coating, reference to ASTM Standard D 661-44 (reapproved 1981) is suggested.[6] This provides a standard for the evaluation of the degree of cracking so that coatings can be directly compared where cracking failure is a problem.

Mud Cracking

Mud cracking and alligatoring often seem similar. However, they are caused by quite different sets of circumstances. While alligatoring penetrates only to the more flexible underlying coating, mud cracking goes directly through to the substrate and therefore is a source of immediate corrosion, along with possible chipping and flaking of the coating from the surface. *Mud cracking* occurs when highly filled coatings are applied rather heavily, which may be the case in runs and sags in the coating.

This type of failure is usually found in overly thick areas of coatings. Water-base coatings that are highly filled often fail by mud cracking.

Mud cracking is also a rather immediate reaction, occurring as soon as the solvent or water begins to dry out of the coating. The more rapidly the solvent or water evaporates, the greater the chance that mud cracking will occur. Water-base coatings are usually emulsions or dispersions, and, in this case, the resin is not a continuous phase, but is broken up into small particles or droplets. When these materials dry rapidly, the resin particles do not properly coalesce and the pigment acts like mud in a dried out pond.

Zinc-rich coatings often tend to mud crack, even though manufacturers have spent much time in formulating to overcome the problem. The reason is the high pigment:vehicle ratio and the nature of the zinc particles. Being spherical, they have poor reinforcing characteristics, and, when this is combined with a high pigment loading and a low tensile binder, an increased rate of shrinkage occurs upon rapid drying, leaving a mud-cracked surface.

The prevention of mud cracking is a combination of coating selection and proper application. Avoid highly filled water-base coatings if fast drying conditions exist. Coatings should have a proper pigment:vehicle ratio for the type and volume of pigment being used. If possible, reinforcing pigments should be included in the coating. Coatings should be applied during drying conditions that are moderate, with the elimination of sags or puddles which can contribute to mud cracking. Apply coats as thinly as possible, using a second pass to develop the required thickness.

Wrinkling

Wrinkling is a peculiar phenomenon and is usually associated with coatings that are applied at too great a thickness, many times occurring in sags and puddles. In place of shrinkage, as is the case with checking and cracking, *wrinkling* is the result of the swelling of a coating where the surface of the coating expands more rapidly during the drying period than the body of the coating. Wrinkling occurs most often with oil-base coatings, alkyds, and similar materials. It is often seen where a spill may have occurred with linseed oil and the coating is allowed to dry somewhat before cleanup. The surface of the linseed oil expands as it absorbs oxygen from the air, and this causes it to wrinkle over the unoxidized body of the coating.

Many oil-base materials also contain driers which serve to increase the drying speed. Some driers are used to speed the surface cure and others are used as through-drying materials. If a coating contains an excess of surface drier, wrinkling may occur wherever the coating is slightly thicker than normal. Cobalt materials or cobalt driers are a common type of surface drier. Zinc and lead compounds are classified as the through-drying driers. All of these reactions depend on oxidation, air temperature, surface temperature, and coating thickness.

Temperature has a considerable influence on the problem of wrinkling. Coatings that cure without wrinkling at ordinary application temperatures may wrinkle badly if the curing is accelerated by baking. An increase in temperature tends to cure the surface much more rapidly than the body of the coating. This is particularly true if the coating is formulated with a balance of surface and body driers for curing at a lower temperature. A substantial increase in surface temperature would cause the surface to cure more rapidly and therefore have a greater tendency to wrinkle. Wrinkling can also occur in cold weather when the thickened coating is applied so that a heavy film develops, or in hot weather in the sun when the topcoat dries quickly, but the coating underneath remains wet (Figure 14.14).

Wrinkling is unsightly; however, it may not severely reduce the corrosion resistance of a coating, since the surface swells rather than shrinks. If the coating can eventually through-dry and retain its adhesion, the wrinkling of the surface should not appreciably reduce its effectiveness. As a matter of fact, wrinkle finishes are designed and used as decorative finishes for many small pieces of office equipment or similar metal objects where the need for corrosion resistance is not great and yet a reasonably long-lasting coating is required.

In coatings which have a tendency to wrinkle, such as oil-base materials and alkyd coatings, care should be taken during the application to make sure that several even, thin coatings are applied and that there are no areas of excessive thickness, such as sags or puddles. The elimination of all heavy areas in a coating application materially reduces the possibility of wrinkling.

Biological Failure

Bacteria and fungi are the primary microorganisms that can act on coatings. There are two types of action. One is the activity of a microorganism due to dirt and contamination on the coating. In this case, the bacteria or fungi merely live on the surface of the coating and do not necessarily affect its resistance. The second type is where the microorganisms actually use the coating for food and derive their energy from it. Under certain conditions, coatings can be rapidly disintegrated by this type of action. These latter coatings are all organic and usually of the oil type, *e.g.*, alkyds, polyamide epoxies, and coatings which use biodegradable plasticizers. In these cases, the portion of the coating which is used for food is the cause of the difficulty, even though the resin portion of the coating might be inert under other circumstances.

One example of this is polyamide coatings, which are used under sewage conditions or under other conditions where moisture, bacteria, and fungi are active. Coal tar epoxy coatings are particularly effective under sewage conditions. However, when polyamide curing agents are used, they tend to fail quite rapidly. If amine-type curing agents are used, the coal tar epoxies have excellent resistance to sewage conditions.

The area where fungus attack is most obvious is on coatings that are exposed to a highly humid atmosphere, particularly in the northern hemisphere on the north sides of tank surfaces and buildings. These areas remain damp for much longer periods of time than the sunny side, and, if the coating can be used for nutrients, ugly, dark, blotchy areas can form on the coating which make it very unsightly. Sooner or later, the coating's corrosion resistance is affected because of the breakdown of the film by the fungus attack (Figure 14.15).

When observed under a microscope, the black discoloration on the coating can be seen as black fungus colonies, and, unlike surface dirt, they are difficult to wash from the surface. Figure 14.16 is a photomicrograph of the fungus colonies on the surface of a coating. Note how the organism grows from a central spot out over the surface.

Such a failure does not occur when coatings based on very inert resins (*e.g.*, vinyls, chlorinated rubbers, and acrylics) are used. Even with these materials, if they are plasticized, the plasticizers should be inert, nonbiodegradable plasticizers as well. Where other types of coatings are used, the fungus growth can be largely or completely eliminated by the use of zinc oxide as part of the pigmentation, together with the addition of fungicides, bactericides, or a combination of all three in the basic formulation.

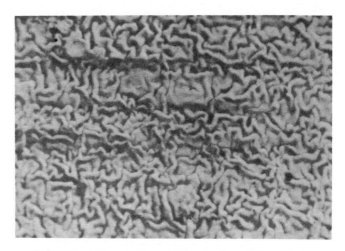

FIGURE 14.14 — Wrinkling of a coating. (SOURCE: Department of Defense, Paint and Protective Coatings, Army TM5-618, Figure 10.5, p. 5-26).

FIGURE 14.15 — Fungus growth on a coating film, with fungus colonies covering most of the surface. (SOURCE: Federation of Societies for Coating Technology, Pictorial Standards of Coating Defects, Mildew Resistance, Philadelphia, PA.)

Corrosion Prevention by Protective Coatings

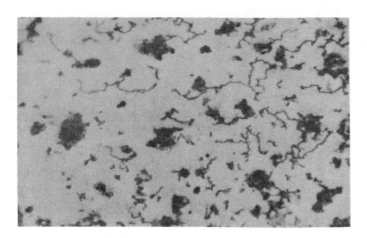

FIGURE 14.16 — Photomicrograph of fungus growth on a coating. (SOURCE: Federation of Societies for Coating Technology, Pictorial Standards of Coating Defects, Mildew Resistance, Philadelphia, PA.)

Where such a condition exists, the fungus or mildew should be removed from the surface of the coating by abrading the surface, by scraping, by power sanding, or by a light blast. In order to prevent the fungus from continuing its growth, the surface should then be washed with a solution composed of trisodium phosphate, a detergent, and sodium hypochlorite. The hypochlorite will both kill and bleach the fungus growth. Following such treatment, the area should be fully rinsed with clear water and thoroughly dried before recoating.

Discoloration

The discoloration of a coating may not seem like a major failure mechanism; however, appearance is a part of the coating's purpose. If its appearance is poor, the coating is generally considered unsatisfactory. Thus, coatings that fade, discolor, or become unsightly a short time after application, can be considered to have failed. This is primarily a formulation-type failure. On the other hand, under certain conditions, particularly in recoating a surface, the previous material may bleed through the new topcoat, also creating a poor appearance. Materials such as wood treated with creosote preservatives, copper naphthanate, or previous coal tar or asphalt coatings, can easily bleed into the new topcoats. If such materials exist, care should be taken to apply a coating that will not dissolve these materials and allow them to bleed through the new material.

Some dyes and pigments will also cause bleeding. More often, however, the resins that are used to formulate the coatings are the cause of discoloration. Some epoxy resins tend to yellow in the dark, as will linseed oil coatings. Polyurethane resins tend to yellow in the light. Vinyl chloride acetate coatings, for example, if improperly pigmented, may turn yellow or brown. Many clear coatings tend to discolor rather rapidly. This is due to the resins containing photosensitive groups on the molecules which, when exposed to strong sunlight radiation, become dark, yellow, or otherwise discolor.

Pigments also cause color changes. Many colors, especially the brighter ones, fade and turn dull with time.

Tinted coatings may chalk and become light colored. Orange pigments may be color-reactive and darken to a dull brown. Other yellow pigments may tend to gray or whiten. The lead pigments are susceptible to darkening in sulfide atmospheres and may even turn very dark or black. Some black and blue pigments tend to bronze and turn brown. It is obvious that pigments such as these should not be used in formulations for high-performance coatings, where they are to be exposed to severe weather conditions. Proper inert pigments of many colors and shades are actually beneficial and will help shade the resin used for the coating and prevent discoloration. For the most part, the best preventive solution to discoloration is experience based on actual operating conditions.

Inorganic Coatings

Up to this point, the discussion of formulation-related failures has been primarily oriented toward organic coatings, which is where most such failures occur. On the other hand, inorganic coatings are also susceptible to some formulation-related failures.

One of the principal causes of formulation problems with inorganic zinc coatings is the fact that they are made with silicate vehicles, either alkali silicate or organic silicate such as ethyl silicate. Neither of these materials is a good film former in the sense of film formation by organic materials. When unpigmented, these materials usually form a brash, clear deposit on the surface which tends to check and crack and to have little coating strength. It is only after the reaction with the zinc pigmentation that these materials form a film which has both adhesion and strength. Under these conditions, it is obvious that formulation is critical, and small variations in additives or pigmentation can cause the coating to be soft, poorly adherent, or to have generally poor film characteristics.

Checking

Because of the character of inorganic zinc coatings, they are particularly subject to checking if the formulation is not proper. In addition to the fact that the silicates in themselves are poor film formers, inorganic zinc coatings are highly filled with powdered metallic zinc and other pigments. The ratio of pigments to binder is high, and under these conditions, particularly where drying is rapid, fine surface checking can occur.

Formulations often include fibrous pigments to help prevent the checking, as well as the incorporation of slower solvents and mixtures of silicates to aid in controlling the drying rate and the checking. As with organic coatings, the checking is usually a surface reaction, and as such, with otherwise properly formulated inorganic zinc coatings, the small checking imperfection can heal due to the continuing curing reactions that occur within the zinc silicate film. Checking can also be prevented by proper application techniques. Thin, multiple-pass coats are often helpful. Application of the coating under less severe drying conditions, or under cover in order to keep the surface cool, improves the checking problem.

Cracking

Cracking or mud cracking also occurs with inorganic

zinc coatings. It often occurs rapidly and, again, is related to the rapid drying of the coating as well as the coating thickness. On the other hand, the high pigment:vehicle ratio is the basic cause. Where the coating is applied heavily and allowed to sag or puddle, mud cracking easily occurs, particularly where drying conditions are rapid. Its appearance is similar to that of the mud in a puddle of water which begins drying out. As the water evaporates, the finely divided soil particles contract and form mud cracks, which is where the name of this reaction originated.

Mud cracking usually occurs prior to the time that any chemical action of the silicate and the zinc occurs, so that when the coating is heavily applied, it is much like a heavy slurry. As it dries, it shrinks and the film cracks. Many times, the cracked area is curved up from the surface in chips or flakes. On the other hand, inorganic zinc coatings have been applied as much as 1/8 in. thick and still maintained excellent film characteristics and adhesion without any evidence of checking or mud cracking. It must be stressed, however, that in this case the curing conditions of the coating were ideal, with just the right humidity, temperature, and evaporation rate for the water or solvent.

Primarily, the elimination or prevention of this problem is in the application of the coating i.e., applying a relatively thin film, with additional passes if necessary, in order to obtain the thickness recommended by the manufacturer.

Pinpoint Rusting

Pinpoint rusting usually occurs where a coating has been in service for a long time and is nearing the end of its useful life. However, if a coating is improperly formulated, pinpoint rusting can begin even a few days after application. This has occurred in a number of instances with some of the zinc coatings with a low zinc loading. (These are the lower priced coatings, intended to be very competitive.) It has also occurred in some of the coatings where pigments have been added to provide a pastel color to the inorganic zinc coating.

In one instance, the zinc was removed and colored pigment was added in sufficient volume to mask the zinc so that the zinc became ineffective as an anticorrosive pigment. Figure 14.17 shows the condition of such a zinc coating after only a few weeks in service. In such cases, the only remedy is proper coating selection. Personal knowledge of the coating or testing prior to use is the best procedure.

A summary of formulation-related failures giving the failure appearance, general cause of the failure and the usual remedy for the failure appears in Table 14.1.

Failures Due to Coating Selection

Since Chapter 12 deals exclusively with coating selection, most of the information that might be covered here is covered in that chapter. However, since failures due to improper coating selection can be catastrophic, some discussion of failure due to improper coating selection seems worthwhile.

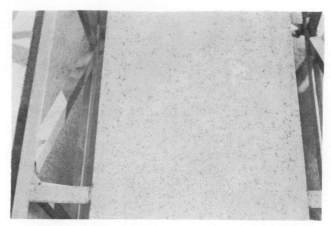

FIGURE 14.17 — Pinpoint rusting of an inorganic zinc coating shortly after application.

Mixed Surfaces

One area where coating selection can be difficult is where a coating extends from one type of surface to another, e.g., structural steel in a concrete foundation. The coating selected for the steel surface in the particular atmosphere might be entirely adequate. However, because of the entirely different surface presented by the concrete, the coating could fail quite rapidly over the concrete in a severe corrosive condition. The corrosive liquid or atmosphere could then attack the steel at the concrete/steel interface. In this case, the selection of a coating for one surface would not be adequate for the second surface. Where situations like this occur, it is best to consider both surfaces, and to select either a coating which is adequate over both surfaces or a coating for each surface, extending one or the other out over the coating on the other surface in order to thoroughly seal the joint between the two surfaces.

Chemical Exposure

Selection of a coating that is inadequate for the subsequent chemical exposure of the film is a common occurrence. Many times, this is due to minor chemical contaminants in the atmosphere or solution which were not taken into consideration during the coating selection process. One example occurred in an alcoholic beverage rectifying plant, which formulates alcoholic beverages such as sloe gin. The coating that was applied to one of the large mixing vessels was tested and found to be thoroughly resistant to alcohol, which is the primary ingredient. Shortly after the mixing tank was placed in service, it was noticed that on one side of the container the coating had disappeared.

Upon investigation, it was found that the area where the coating had disappeared was the area where the workers added the special additives and flavorings to the alcoholic solution. In this case, they were adding a rather high boiling aldehyde which was very penetrating and which actually dissolved the coating in the area where it ran down the tank. Obviously, all of the conditions of the atmosphere in which the coating was to operate had not been taken into consideration during the coating selection.

TABLE 14.1 — Formulation-Related Failures

Organic Coating Failure	Failure Appearance	Cause of Failure	Remedy
1. Chalking	Surface soft and powdery. Easily removed by wiping surface.	Surface disintegration by actinic rays of sun on the organic resin binder; improper pigmentation.	Select coatings formulated with radiation-resistant resins (acrylics) and noncatalytic, nonchalking pigments.
2. Erosion	Similar to chalking. Surface removed on high spots and brush marks to base coating or primer.	Chalking mechanism with coating surface removed by weathering.	Select chalk-resistant coating with good flow out to a smooth film.
3. Checking	Surface phenomenon—uneven, small, non-continuous fissures in coating which do not penetrate to the substrate.	Surface stresses caused by shrinkage due to weathering and continued surface polymerization and oxidation.	Select coating formulated with weather-resistant resins and inert reinforcing pigments in addition to noncatalytic colored pigments.
4. Alligatoring	Very large macrochecking, generally cross-hatched pattern.	Internal stresses where surface shrinks more rapidly than body of coating. Hard topcoat applied over soft undercoat.	Apply thin coats and thoroughly dry before adding additional coats. Never apply hard topcoats (epoxy) over soft undercoats (asphalt).
5. Cracking	Small breaks in coating to substrate. May be linear, cross-hatched, or curved. Cracks may or may not be continuous.	Stress set up in coating due to continued polymerization and oxidation; improper pigmentation.	Select coating formulated from nonreactive weather-resistant resins, reinforcing pigments, and nonreactive colored pigments.
6. Mud Cracking	Large macrocracking. Coating may curl at cracks and lose adhesion.	Rapid drying of highly filled coatings, especially water-based materials (water emulsion paints).	Use coatings with strong adhesion. Apply coatings under proper drying conditions and prevent sags, puddles, or areas of excess thickness.
7. Wrinkling	Furrows and ridges in coating surface. May be linear or random pattern. Wrinkle may be fine or quite large.	Surface reaction where surface of coating expands more rapidly during drying than does the body of the film.	Choose coatings with even, thorough drying characteristics. Apply evenly; avoid excessive thickness.
8. Biological Failure	Softening or slime reaction of coating. Blotchy brown or black spots on coating surface causing poor, dirty appearance.	The biodegradation of the coating by bacteria or fungii. The coating is used as a source of nourishment.	Select oil-base coating which contains permanent fungicides or bacteriacides. Nonoil coatings should use nonbiodegradable modifiers.
9. Discoloration	Yellowing, greying, or darkening of coating.	Resin or pigment color change due to weather or chemical action.	Select coating formulated with both color stable resins and pigments.

Inorganic Coating Failure	Failure Appearance	Cause of Failure	Remedy
1. Checking	Usually fine visible or microscopic checks. Do not penetrate to the substrate.	The zinc pigment to binder ratio is high; rapid drying conditions cause surface checking.	Formulation should include reinforcing pigments. Apply coating as thin as recommended. Second coat, if necessary. Apply under favorable drying conditions.
2. Mud Cracking	Fine to fairly large segments (1/4 in.) flaking from surface.	Application of coating too heavy. Rapid drying conditions.	Apply coating at no more than recommended thickness. Apply under favorable drying conditions.
4. Pinpoint Rusting	Pinpoint spots of corrosion progressing from a few per square feet to almost continuous. Early failure can be catastrophic.	Zinc pigment mask by other pigmentation or improper zinc/binder ratio. Uneven coating thickness; thin coated areas show first failure.	Usual remedy: remove coating and reapply more satisfactory zinc coating. Apply maintenance coat at first sign of pinpoint failure.

(SOURCE: Steel Structures Painting Council, Causes and Prevention of Paint Failure, Chapter 23, Good Painting Practice, Vol. 1, Steel Structures Painting Manual, 1982.)

Tank Linings

Another area where there have been failures due to coating selection has been in tank linings, tank car linings, and tanker ship linings, where combined service causes difficulties. Tank cars and tanker ships are frequently subject to alternate cargoes of quite different materials. While a coating may be entirely satisfactory for one of the cargoes, when a second loading follows the first, the coating may fail. This has actually happened many times under alternate loading conditions. In one instance a vinyl coating was used as a lining for a refined oil tank in a tanker. The coating stood up well to gasolines, diesel fuels, lube oils, and similar products. However, when toluol was included as a lading, the entire tank lining blistered. In fact, the blistering was sufficient for the lining to be considered a total failure. In this particular case, however, the ship went back into gasoline and diesel service. After 10 years of service, the lining had still prevented any corrosion in the tank, in spite of the blisters which still remained.

Tank cars and tanker ships are also often steamed or washed with hot water after unloading one material and prior to loading another. While the coating may stand up well under the original material, if there is some absorption of the material by the coating, the hot water and steaming can often cause blistering as well. Alternate ladings are a difficult service, because of the varying kinds of chemicals to which the coating may be subjected from one trip to the next, and coating selection is critical in making sure that any coating used will not only be satisfactory for the various ladings, but will withstand many of the tank washing procedures as well.

Repairing and Recoating

Coating selection is even more of a problem when recoating is being considered. In this case, not only is it necessary to determine the properties of the coating from the standpoint of the exterior exposure, but also the coating must be satisfactory over the existing coating on the structure. Many failures have occurred due to a lack of attention to the properties of the undercoat and adhesion characteristics of the new repair coating. Table 14.2 shows some of the compatibility and adhesion characteristics of coatings when one material is applied over another. This table, developed by NACE Unit Committee T-6H as part of a proposed state-of-the-art report which has not yet been published, is a good general reference for the compatibility of some generic materials.[7] If there is ever any question, a test area of the new material should be applied on the old in order to make sure of the adhesion and compatibility of the new material, as well as the retention of adhesion of the old material after recoating.

TABLE 14.2 — Compatibility and Adhesion Properties of Coatings and Paints[1]

Existing Painted Surface (Aged 12 mos. minimum)	Long Oil	Med. Oil Alkyd	Short Oil Alkyd	Silicone Alkyd	Oil Modified	Phenolic Modified (Epoxy Ester)	Amine	Polyamide (Catalyzed Epoxy)	Coal Tar Epoxy	Chlorinated Rubber	Solution Vinyl	Moisture Cured (Urethane)	Catalyst Cured (Urethane)	Coal Tar (Urethane)	Asphalt	Latex
Long Oil	G	G	F	G	G	G	X	X	X	X	X	X	X	P	G	G
Med. Oil/Alkyd	G	G	F	G	G	G	X	X	X	X	X	G	F	F	G	G
Short Oil/Alkyd	G	G	G	G	G	G	X	X	X	X	X	G	F	F	G	G
Silicone Alkyd	G	G	G	G	G	G	X	X	X	X	X	F	F	F	G	G
Epoxy Ester Oil Modified	G	G	G	G	G	G	X	X	X	X	X	G	F	F	G	G
Phenolic Modified	G	G	F	G	G	G	X	X	X	X	X	G	G	F	F	G
Catalyzed Epoxy Amine	X	X	X	X	X	X	X	X	X	X	X	X	G	X	X	X
Polyamide	G	G	G	G	G	G	G	G	G	G	G	X	G	G	G	G
Coal Tar Epoxy	X	X	X	X	X	X	G(2)	G(2)	G(2)	F(2)	X	X	F	G	G	G
Chlorinated Rubber	G	G	G	G	G	G	X	G	X	G	G	F(2)	F	G	G	G
Solution Vinyl	G	G	G	G	G	P	X	G	X	G	G	P	P	P	G	G
Zinc Rich Inorganic	X	X	X	X	X	X	G	G	F	G	G	F(2)	G(2)	G	G	G
Organic	X	X	X	X	X	X	G	G	(3)	G	G	G	G	G	G	G
Coal Tar	X	X	X	X	X	X	X	X	X	X	X	X	X	G	P	G
Asphalt	G	G	G	G	G	G	X	X	X	X	X	X	X	X	G	G
Latex	G	G	G	G	G	G	G	G	G	G	G	X	X	G	G	G
Urethane Moisture Cured - Type II	P	P	P	P	P	P	X	X	X	X	X	P	G	X	X	X
Catalyst Cured - Type IV	G	G	G	G	G	G	G	G	G	G	F(2)	G	G	G	G	G

G = Good-Always work. F = Fair-Will work-Some exceptions. P = Poor-Will not work-Some exceptions. X = Not recommended.

(1)This table should be used only as a guide, and as stated in the text, a test patch must be used to confirm satisfactory performance of the intended coating or coating system.
(2)Application critical.
(3)Properties will be same as for binder shown for existing paint.
(SOURCE: Proposed NACE Technical Committee Report, "Combating Adhesion Problems When Applying New Onto Existing Finish Coats of Paint," T-6H-27 1982 Draft, subject to change, NACE, Houston, TX.

Corrosion Prevention by Protective Coatings

Substrate-Related Failures

Chapter 8 stressed the importance of the substrate in its relation to coatings. However, some additional points should be emphasized from the standpoint of failures due to substrate reaction.

Steel

Steel is probably the most important substrate because of the volume of steel surface which requires coating. While steel may be the easiest surface to coat because of its dense, impervious nature, steel is also reactive to many different environmental conditions, and can thus become a significant factor in coating failures.

There are also many different types of steel surfaces over which coatings may be applied. Hot-rolled steel with its mill scale surface is certainly one of these, and when overcoated, mill scale can be a major cause of coating failure due to its delamination from the steel surface and the ability of corrosion to undercut a mill-scaled surface. Mill scale is also cathodic to bare steel, and as such aggravates the coating breakdown through the action of a massive cathode and a small anode. Previously corroded steel also presents a very different surface from new, clean steel. When rust has occurred and coatings are subsequently applied, even though the surface is blasted clean, the coatings will fail more rapidly in these areas than over the virgin metal due to contamination.

Cold-Rolled Steel

Cold-rolled steel is vitally important because of the many millions of square feet of surface that are coated. This includes most appliances, automobiles, and many other similar pieces of equipment. Cold-rolled steel is particularly difficult to coat with an organic material because of its smooth character. There are many coatings that will adhere to it; however, the same coatings will adhere and retain their adhesion much better if the surface is lighty blasted to increase the tooth.

One way in which cold-rolled steel is improved is to treat the surface with a zinc phosphate. Most automotive and appliance surfaces are treated with zinc phosphate to reduce underfilm corrosion of the applied coatings. Robert Iezzi and Henry Leidheiser, in a paper on the treatment of cold-rolled steel, indicate that carbon contamination of the steel surface is a major factor in the adhesion and effectiveness of the zinc phosphate treatment. They state:

> Samples with low surface contamination have a high electrochemical reactivity and a high surface area available for zinc phosphate nucleation. This combination results in nearly complete zinc phosphate coverage with low porosity and good adhesion to the steel surface.
>
> Samples with high surface contamination have a less electrochemically active steel surface and a reduced area for zinc phosphate nucleation. Nucleation of zinc phosphate crystals grow laterally over contamination, but coverage is incomplete and the bond is poor. Samples with high surface contamination fail much more rapidly than those with low surface contamination because of the lower bonding area of the zinc phosphate layer to the steel

surface and the consequent higher oxygen and moisture availability to the steel surface due to the incomplete coverage of the zinc phosphate layer, *i.e.*, porosity. This greater oxygen and moisture access to the steel surface promotes the cathodic delamination reactions and accelerates paint failure.[8]

This is a considerably different surface condition than exists for most structural steel. However, it demonstrates the different types of surfaces that steel can present as a substrate for coatings.

We've previously seen how the same coating can act very differently on one surface as compared with another. This is an important finding, since there is a tendency, when a coating works satisfactorily in an atmosphere and over a specific surface, to then apply the same coating over other surfaces that may be in a similar atmosphere.

A good example of this is the use of coal tar epoxy coatings subject to a sewer atmosphere. Coal tar epoxy applied over steel provides a good corrosion-resistant material, and will fully protect the steel surface if it is properly applied. When the same coating is applied over concrete, because of the inherent porosity of the concrete itself and the difficulty of obtaining a continuous film, the coal tar epoxy coating can fail quite readily.

Going a step further, if the coal tar epoxy is applied over wood, early failure may be expected because of the continual contraction and expansion of the wood due to changes in atmospheric moisture. In this case, since the coal tar epoxy is somewhat brittle, it would soon crack over the wood surface, allowing penetration of whatever materials are in the atmosphere.

Aluminum

Aluminum is also becoming a much more common metal surface over which to apply coatings. In many ways it is a good coating surface, particularly where the metal surface is broken up by a light abrasive. Vinyl coatings that have been applied to aluminum surfaces have remained on the surface for more than 35 years, maintaining satisfactory protection and excellent adhesion.

Aluminum is often attached to steel in order to provide a lightweight superstructure. While such a combination of metals may be good from an engineering standpoint, it may create difficulties from a coating standpoint. Unless the junction of the steel and aluminum is insulated so that there is no actual bimetallic contact, a galvanic corrosion cell is easily established. This is particularly true if the coating has a tendency to crack at the junction of the two materials.

Lightweight deck houses have been seen on steel ships where the coating on the aluminum was undercut for several inches away from the junction of the two metals due to the penetration of seawater at the junction and the rapid corrosion of the aluminum in that area. Also, the adhesion to the aluminum was not as effective as it would have been had the aluminum itself been blasted prior to the application of the coating. While the steel substrate adjacent to the aluminum was well protected, the combination of seawater, aluminum salts, and alkali buildup at the cathode caused the coating to peel away from the junction of the two metals.

Zinc

Galvanized steel attached to ordinary steel can also cause similar problems under many corrosive conditions. Zinc, as previously seen, is an excellent surface over which to apply many coatings. On the other hand, other coatings tend to fail fairly rapidly over a zinc surface because of their reactivity with that surface. Oil coatings, alkyds, and other coating materials that contain drying oils, often react poorly over a galvanized surface or over an inorganic zinc surface due to the reactivity of these materials with the zinc salts that form between the coating and the substrate.

Figure 14.18 provides an example of the failure of a coating over a galvanized surface. In this case, an alkyd coating is cracking and spalling away from the zinc surface in large pieces. This indicates the bond between the zinc and the coating is deteriorating due to chemical reactions between the zinc and the coating. No corrosion exists because of this galvanized steel surface.

Zinc surfaces oxidize and react with the atmosphere quite readily, and many times the zinc surface is allowed to react with the atmosphere before the coating is applied. Any coating that goes over a zinc oxide or zinc carbonate surface must be thoroughly compatible with these zinc salts; otherwise, early adhesion problems develop.

FIGURE 14.18 — Coating failure over galvanized steel.

Used Surfaces

Used surfaces are much more difficult to coat than new ones, and the failure of a coating over a used surface is much more probable than one applied over a new original surface. Surfaces that have been subject to sulfides are a particular problem, and coatings have failed extensively over such surfaces. In this case, the metal surface, even though it apparently is blasted free of corrosion products, contains enough iron sulfide and iron oxide to create a substrate which is less than satisfactory for the application of coatings.

Figure 14.19 is a schematic drawing of a corroded area on a piece of steel where minute microscopic contamination remains even after blasting. That these areas exist after blasting has been demonstrated several times with both chlorides and sulfides being identified. After the

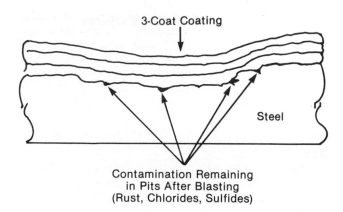

FIGURE 14.19 — Contamination remaining in a corroded area of steel after blasting.

surface has been blasted, and with high humidity, these minute areas which are invisible to the naked eye, soon become visible by absorbing moisture from the air. When this happens, these tiny areas of contamination turn dark and become pinpoint rusted areas. If these small areas are coated over, they become focal points for coating failure, and thus are one of the primary causes for continuing failure in the same area where coatings had failed previously and where corrosion existed.

Such a substrate is difficult for most organic coatings to cover and remain intact for long periods of time. (Chapter 8 on surface preparation covered this problem in more detail.) One of the options that has proven satisfactory in many areas is the application of inorganic zinc coatings. In this case, the iron ions which are present in the tiny spots of contamination are reacted into the coating, helping to nullify the contamination. The problems involved with used steel surfaces should not be underestimated in any way. Untold failures of coatings have occurred because the problem of prior use of a substrate has not been recognized.

Wood and Concrete

Similar problems are not necessarily found with wood or concrete surfaces, although they are definitely affected by previous use. The problem with these surfaces is that they absorb and retain soluble or reactive materials which are difficult to remove without removing part of the actual surface of the concrete or wood. The contamination which has been absorbed will cause blistering, lack of adhesion, and coating failure. Previously used surfaces of all types should be given extra care during their preparation in order to make a satisfactory substrate for the coating.

Cleanliness

Cleanliness of the surface is part of the substrate problem and is vitally important to the life of the coating. It can be stated as a rule-of-thumb that any coating applied over a perfectly clean surface will last longer and be more effective than the same coating applied over a less than clean surface. This not only applies to concrete and wood, but to all types of metal as well. The best coating available will fail rapidly if it is applied over a surface that

Corrosion Prevention by Protective Coatings

has not been properly prepared to receive the coating. This is true even when it can be proven that the coating material itself can resist the environment and protect the metal. The probability of coating failure over a surface is directly proportional to the amount of contaminant left on the surface.

The substrate is important, since otherwise there would be considerably less emphasis on surface preparation throughout the coating field. The problem of substrate preparation is recognized by several corrosion-related organizations which have established standards for the preparation of surfaces and the cleanliness of the substrate over which a coating is to be applied. Table 14.3 provides a summary of substrate-related failures.

Surface Preparation-Related Failures

It is little wonder that improper surface preparation is the cause of a sizable number of coating failures. There appears to be a psychological block with painters of all types when it comes to proper surface preparation. However, there is no substitute for proper surface preparation,

as previously discussed in Chapter 8, if long life is expected from a coating.

When applying a repair coat over a previous material, even greater care is necessary to make sure the surface is prepared properly to accept the repair coating. Compatibility of the repair material, as well as proper cleaning of the surface, is the key to a satisfactory repair job. Intercoat contamination as well as outside contamination of a properly prepared surface before coating application is often a cause for coating failure. Such contamination can come from many sources, e.g., from the chemical plant next door to the air used to sandblast the surface. Many air compressors which have had poor maintenance may pass small quantities of oil into the air, which can often contaminate an otherwise clean, blasted surface, causing the applied coating to blister wherever a minute oil droplet remains.

There are a number of methods of surface preparation available. The Steel Structures Painting Council has prepared a list of surface preparation specifications covering most of the common methods of preparing steel.[9] This

---

TABLE 14.3 — Substrate-Related Failures

Coating Failure	Failure Appearance	Cause of Failure	Remedy
1. Previously Used Steel	Blistering, rust, tubercles, loss of adhesion in areas where steel was previously exposed to corrosive conditions.	Retention of minute amounts of corrosion product or contaminant along grain boundaries of the steel surface, even though blasted to white metal.	Wash-blasted surface with water or dilute phosphoric acid solution and reblast. Apply an anticorrosive primer with strong adhesion. Where applicable, an inorganic zinc primer may provide a good base coat by reacting with the minute surface corrosion after the first blasting.
2. Galvanized or Metallic Zinc Surface	White zinc corrosion product forming under the coating or actually breaking through the coating.	Formation of zinc salts (oxide, sulfide, oxychloride, zinc soap) underneath coating.	Brush blast zinc surface or treat with commercial zinc treatment. Apply a nonoil base, inert, strongly adherent primer.
3. Aluminum	White corrosion product causing pinpoint failure in coating; loss of adhesion because of very smooth surface. Possible blistering.	The very smooth aluminum oxide surface. No physical adhesion.	Very lightly dust blast the aluminum surface, or where applicable, treat with commercial aluminum treatment. Apply a primer with known compatability and strong adhesion to aluminum surface.
4. Copper	Grey-green corrosion product; loss of adhesion.	Very smooth copper oxide surface. No physical adhesion.	Brush blast copper surface or etch with commercial copper treatment. Apply a primer with known high adhesion to copper.
5. Wood	Checking and cracking of coating. Flaking from hard winter grain. Dense coatings with a low MVT rate may blister due to absorption and evaporation of moisture from the wood.	Expansion and contraction of wood due to varying temperature and humidity. Differential expansion summer and winter grain. Very hard dense winter grain combined with soft porous summer grain, causing a variation in coating adhesion.	Start with a clean, newly sanded wood surface. Apply low molecular weight highly penetrating paint, preferably oil base, with sufficient elasticity to expand and contract with the wood surface. Paint should have relatively high moisture porosity to allow wood to breathe.
6. Concrete	Blistering of coating. Formation of calcium salts under coating, forcing coating from the surface. Loss of adhesion and peeling.	The chemical reactivity and moisture content of concrete. Its nonhomogeneous very porous structure. Pinholes, water, and air pockets in poured concrete surfaces.	The concrete should be clean and the surface dry. It may be acid etched or lightly blasted to obtain proper surface condition. Use a low molecular weight highly penetrating primer with strong alkali resistance (liquid epoxy). Primer should be heavy bodied and thixotropic to fill imperfections in concrete surface.

(SOURCE: Federation of Societies for Coating Technology, Pictorial Standards of Coating Defects, Chalking Resistance, Erosion Resistance, Mildew Resistance, Philadelphia, PA.)

list is given, in a descending order of effectiveness, in Table 14.4.

As one goes down the list in Table 14.4, the method of surface preparation leaves increasing amounts of contamination on the surface and therefore creates an increasing chance of coating failure, with the method at the bottom of the list (solvent cleaning) being almost a sure failure with any of the high-performance coatings. This method is effective only if used in connection with some of the other cleaning methods listed. A proper coating must be selected to go with each of these methods in order to obtain even reasonably good life from the coating. The high-performance coatings, which are those with which most corrosion engineers are familiar, should not be used with a primary surface preparation less effective than number 4 on this list.[9]

There are coatings which are specially formulated for application to rusty hand-cleaned surfaces. Some of these do a good job, but only where used in a light to very moderately corrosive atmosphere. Where a severe corrosive atmosphere exists or where the coating must withstand immersion, only the best surface preparation is satisfactory. Any techniques following acid pickling (Table 14.4) leave too much contamination on the surface to prevent coating failure. Even within the group of the first four, failures can easily occur because of the use of poor abrasives during blasting or sloppy pickling techniques.

TABLE 14.4 — SSPC Surface Preparation Specifications

Method of Surface Preparation	Specification Number
1. White Metal Blast Cleaning	SSPC SP 5
2. Near-White Blast Cleaning	SSPC SP 10
3. Commercial Blast Cleaning	SSPC SP 6
4. Acid Pickling	SSPC SP 8
5. Brush-Off Blast Cleaning	SSPC SP 7
6. Flame Cleaning and Power Sanding	SSPC SP 4
7. Power Sanding	SSPC SP 3
8. Power Wire Brushing	SSPC SP 3
9. Chipping andd Hand Wire Brushing	SSPC SP 1
10. Solvent Cleaning	SSPC SP 1

Abrasives

There are literally hundreds of abrasives used to prepare metal surfaces. These range from the very good silica materials, such as garnet, flint, silica sand, etc., to river sand. The unwashed river sands usually contain considerable amounts of clay. When this is impacted on the surface, it not only creates a large amount of dust, which makes for poor visibility, but when it is driven on the surface, the impact of the clay makes it tightly adhere to the surface. In fact, it is almost impossible to remove by blowing or brushing. Thus, application of a coating over such a surface is almost a sure failure. There is little that can be done with it other than to wash the surface with clean water and scrub it to remove the clay from the sur-

face profile. Even this is a less than satisfactory procedure.

The abrasives also have a broad range of particle size. The optimum abrasive size is believed to range from 16 to 30 mesh. This will produce the most uniform surface and surface profile, and is effective for the greatest number of coatings. While there are heavy-bodied coatings which require heavy surface profiles, they represent the exception more than the rule.

All of the different abrasives provide a different anchor pattern and profile on the surface. As mentioned above, it is believed that the silica sand using a 16 to 30 mesh sand produces a universally satisfactory profile. Steel shot and steel grit provide a completely different profile from that of the sand, and coatings which may be entirely satisfactory over the sandblasted surface may not perform as well and may fail faster over a steel shot- or grit-blasted surface. On the other hand, there are other coatings which perform equally well, if not better, over the cut profile of steel grit. There are dangers here, however, from too great a roughness of the surface caused by too great a size in either the sand or the steel grit.

For instance, the exterior of a steel pipe had such a heavy profile that it was necessary to go over the surface with an emery cloth prior to coating application. The surface in this case was essentially the same as a heavy rasp or file. There have also been ship bottoms blasted with silica sand with a grain size up to 1/8 to 3/16 in. In this case, the profile was extreme and was not satisfactory for most ship bottom coatings. There are other cases where there is insufficient surface profile or pattern on the surface where the blasting has occurred too rapidly so that there are many areas of plain surface in between the impact areas of the steel or shot. This has been seen particularly in automatic blasting units where the steel has passed under the unit at too rapid a pace. Blasting such as this can leave sizable particles of mill scale, as well as rust and contamination, all of which make for early coating failure.

Inhibitors

At the present time, with the increase in wet sandblasting, there are a number of different inhibitors used, along with a number of different methods used for treating steel surfaces to prevent them from changing color. Citric acid and some of the citrates have been used for this purpose. A number of tests have been made using such materials. In many cases, the adhesion of coatings over such treated surfaces was considerably less than a plain white metal blast when tested in seawater for several months. This was true for both organic and inorganic coatings.

There are also many phosphates used, with the phosphate being applied during water blasting or during and after blast wash. Again, it is suggested that these be used with caution, since many coatings do not adhere as well to such a treated surface as they do over a standard sandblast. Coating failures are more probable, therefore, when inhibitors are used than when the same coating is applied over a dry sandblasted surface. If a coating is to be immersed, or where it is used in an area of high humidity or condensation, metal surface inhibitors should be used with caution, and the coating thoroughly tested over such a surface prior to making a full-scale application.

Application-Related Failures

Chapter 10 was dedicated entirely to the importance of and techniques involved in coating application. Nevertheless, there are a number of failures that are a direct result of improper application, and thus should be addressed in any discussion of coating failures. Almost all application-related failures are due to carelessness and poor workmanship. For instance, the surface may not be sufficiently clean, with dirt, dust, and many more obvious types of surface contamination (*e.g.*, sand or steel grit left from blasting operations, gum wrappers, cigarette butts, tobacco juice, and even, on occasion, urine). All of these things have been known to be coated over during coating application, resulting in failure once the coating was put into service. Most of the above contaminants are not so much the result of poor workmanship as of plain carelessness, with the applicator and inspector overlooking items that will obviously cause prompt coating failure.

Poor workmanship refers to improper coating application by being too close to or too far away from the surface, or by spraying the surface at angles which cause coating imperfections. All application-related failures are due to lack of care, lack of an understanding of coating fundamentals, or poor workmanship, which may be the result of poor training in the application of high-performance coatings or little pride in proper workmanship. Figure 14.20 shows a general coating failure as a result of poor application procedures. Areas in the same tank where application was consistent showed no failure whatsoever.

Brush Marks

The failure which occurs when pronounced brush marks are left in a surface is due to the hills and valleys left in the coating by the brushing technique. There are many heavy bodied coatings that are difficult to apply and which leave pronounced brush marks. However, these can be overcome through the use of proper application techniques. The failure starts as pinpoint rusting in the

FIGURE 14.20 — General failure of an epoxy coating in a ship's tank due to poor and inconsistent coating application.

low, thin areas of the coating, which soon undercut, and, because the low areas of the coating are rather close together, rapid total failure results.

Runs and Sags

Failures which are due to runs and sags are usually a result of too great of a coating thickness. Many of the internally reacted or catalyzed coatings continue to cure over a considerable period of time, and, in areas where the coating is heavy (*e.g.*, runs and sags), they often check and crack, leaving the substrate open to rapid coating failure. Runs and sags are also a result of careless application. In brushing, it means that the material was not brushed out sufficiently to eliminate overly thick areas in the coating and that the curing was sufficiently slow that these areas sagged. With spraying, improper gun adjustment can cause sags or runs, or merely the overly heavy application of the coating at any one spot. This can be caused by the gun being held too close to the surface and not moved rapidly enough over the surface in order to obtain a uniform film.

Inorganic coatings often crack and scale in runs and sags. While corrosion may not begin immediately, it will proceed more rapidly in these areas than in the smooth, evenly coated areas. Also, if such areas are overcoated with organic coatings, these rough areas are focal points for the breakdown of the organic coating. Runs and sags should be eliminated soon after the application of the coating, and before the coating has dried sufficiently so that the sags cannot be removed.

Improper Coating Thickness

There are two types of failures which occur from improper coating thickness. One, as with runs and sags, is a result of the coating being too thick, and the other is a result of the coating being too thin. Where a coating is applied too thickly, particularly with a lacquer-type coating, the coating may not dry properly and may have an excessive retention of solvent in the coating. Retained solvents may cause blistering and poor adhesion to the substrate because the underlying coating is softer and because of the retained solvents close to the substrate. Where the coating is catalyzed or internally cured, the coating may have internal shrinkage because of the thickness, which causes checks, cracks, and even scaling. Where this occurs, rapid undercutting of the coating may result from the poor adhesion of the overly thick film.

In the case of thin coatings, failure is primarily due to rapid application which does not leave enough material on the surface. The type of failure which occurs in these areas is due to pinpoint rusting, with the pinpoints gradually becoming larger until the entire coated area is undercut in the thinnest spots. This is a common type of coating failure and can easily be overcome by careful workmanship on the part of the applicator, and by careful checking on the part of the inspector to insure that the proper thickness of the coating is applied.

Figure 14.21 is an example of improper coating thickness on the thin side. The area above the weld is corrosion-free, while the area of the pile below the weld was undoubtedly done at a different time by a different work-

FIGURE 14.21 — A thin coating area showing both pinpoint and general corrosion after a period of exposure.

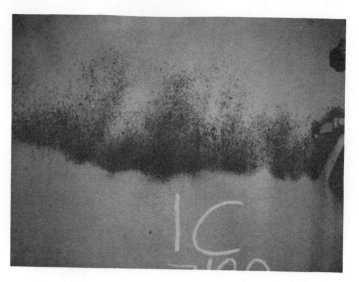

FIGURE 14.22 — A typical holiday which is readily visible only after the coating has been placed in service. Both pinpoint and general corrosion are evident.

er who applied an inconsistent and thin coat, resulting in both pinpoint and heavy rusting.

Where coatings are applied over wood, excessive coating thickness can result in cracking and scaling as the alkyd or oil coating continues to oxidize in the air. Again, this is a common type of failure and the remedy is simply the application of thin coats in order to obtain a thorough cure before application of additional coats.

Holidays

Holidays are also related to coating thickness. *Holidays* are areas where the applicator has missed coating the surface, or where an extremely thin spot remains after the coating is applied. These can be on plane surfaces where the coating was not overlapped sufficiently. They are more often found in difficult-to-coat areas, such as interior corners, along welds, around bolts and rivets, or wherever there are rapid changes in the direction of the surface. The type of failure resulting in these areas is an early pinpoint rusting of the area, followed by the formation of rust scale and the undercutting of the coating adjacent to the holiday. Again, this is a common type of failure and is due to lack of care during coating application. A 50% overlap of each pass of the coating is recommended in order to insure against holiday-type failure.

Figure 14.22 shows a typical holiday after the coating has been in servcie. The coating in this area was not overlapped sufficiently to make a continuous film. The failure ranges from pinpoint rusting to continuous rust in the thinnest area of the holiday.

Overspray

A failure because of overspray is due to the pinpoint rusting of the film in and around the overspray particles. When overspray is allowed to remain on the coating, and additional coats are applied, the coating generally will not penetrate around the overspray particles, but will leave imperfections in the coating at this point. Pinpoint rusting then occurs, with the rusting growing to where the coating is undercut, resulting in a complete failure of the area.

Overspray is a common type of coating failure, particularly where fast-drying coatings are used. However, even heavy-bodied epoxies and similar materials will cause overspray as a result of improper application. *Overspray,* in general, is a result of poor application techniques, a result of either poor gun adjustment or application of the coating at too great a distance from the surface and not applying an even, wet film to the surface. Overspray can be prevented by proper gun adjustment and careful application.

Figure 14.23 shows a typical overspray failure, *i.e.,* pinholing around or adjacent to the rough overspray particles. In this case, it can be seen that the lower part of the surface had been given a cursory wipe before the second coat was applied. There is no corrosion where even this minimum cleaning was done. On the other hand, above the wiped area, heavy pinhole rusting has occurred where the second coat was applied over the rough overspray of the first coat.

Pinholes

Pinholes in a coating can result from overspray. For the most part, however, they result from heavy application of a coating, with air being blown into the coating surface during the application. As with overspray, pinholing can be caused by improper gun adjustment, where the gun does not atomize properly and air is entrapped in the coating. It can also be caused by the application of the

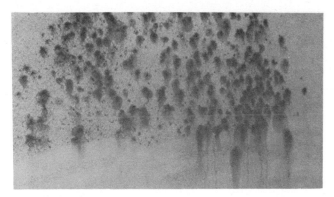

FIGURE 14.23 — Typical overspray failure, showing pinholing adjacent to the rough overspray particles.

coating too close to the surface, creating an area in the center of the fan where too much material is applied and where air bubbles are entrapped in the wet coating. The air bubbles remain, unless the coating is a very slow drying one, creating a pit which may go to the substrate or entirely through any one coat of the coating. In any case, the coating is thinner at these spots than in the body of the coating, and these are focal points for rapid coating failure. Pinpoint rusting occurs first, followed by undercutting of the coating around the pinhole. This is also a common type of coating failure caused by improper application and can be prevented by careful workmanship.

Figure 14.24 shows a typical pinholed coating after being in service. The one streak of pinholes is undoubtedly due to a pass of the gun which was too close to the surface, with air being blown into the wet coating. In this case, the coating was a heavy-bodied epoxy and was sufficiently thixotropic so that once the air bubbles were blown into the coating, they remained, creating the pinholes as shown.

Spatter Coat

Spatter coat is much like a holiday or other workmanship defect. It is an area where the coating is not applied as a continuous film, but is applied sufficiently fast and in a small enough volume so that the coating reaches the surface not as a wet film, but as a series of discrete droplets with bare areas in between. This is a common application problem, and is the basis of most "holidays." The coated area may look like it is covered, but on close inspection, the droplets are visible, as are the bare substrate areas in between. The coating is thin, which is easily detected by a thickness gage. If such areas occur during an application, they are easily repaired by the application of an additional coat. If the coating is placed in service without repair, spatter coat areas are the first to fail.

Cratering

Cratering (or bug-eyeing or crawling) may be either an application problem, a material problem, or both. Some materials which have a high surface tension have a greater tendency to behave in this manner than those with a lower tension. Much cratering is due to contamination and dirt that may be on the surface or that falls into the wet coating during application. When a vehicle has some

cratering tendency, minor contamination can seriously aggravate the problem. Oil in sandblast or atomization air is one source of such contamination and can be removed by proper oil and water filters on the application air line.

Cratering occurs before the coating dries, and as such can be eliminated during application. Often, the wet coating must be removed and reapplied. If the coating cures with the cratering in place, the coating surface, particularly in the cratered area, must be roughened sufficiently to break the surface before another coat is applied. The repair coat should be brushed to work the new coat into the previous one. The coatings most susceptible to this problem are those which thermoset, *e.g.*, epoxies, phenolics, etc.

The areas of a coating where a crater forms are thin and are therefore focal points for coating failure. Without repair during application, pinhole corrosion can occur. Figure 14.25 shows the cratering or crawling of an epoxy coating and the uneven appearance which results.

Application failures are due to human and not mechanical or material failures. Application failures are often blamed on the material, poor equipment, lack of

FIGURE 14.24 — Pinholing failure of a coating, with the pinholes being somewhat isolated in comparison with those found in overspray failure.

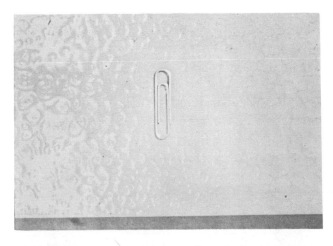

FIGURE 14.25 — Cratering or crawling of a thermosetting coating.

time, and many other similar excuses. Application failures, generally, are due to lack of care and poor application techniques on the part of the applicator and not to any other cause. Application-related failures are summarized in Table 14.5.

Design-Related Failures

Design and its effect on corrosion has been covered in several previous chapters. However, it is an important subject in terms of coating failures because it is difficult to change an existing design, and design does contribute to the corrosion resistance (or lack of resistance) of a structure.

Edges

The failure of coatings on edges is important since this is the area where corrosion usually first begins. The edge provides a sharp break in the coating and is also subject to more damage than flat areas. Edges invariably seem to be a problem, unless special care is taken during the application of the coating to these areas.

Figure 14.26 shows the failure of a coating on the edges of structural steel which obviously had not had any particular attention during application. Edges also represent critical areas of the structure, and continued failure can result in cracking from a decrease in steel thickness. Areas such as those shown in Figure 14.26 are quite typical, particularly in the marine area where structural steel shapes are welded and riveted together to form the overall structure. Note that almost all edges in the figure show some edge corrosion. In this particular case, the coating was only two years old.

TABLE 14.5 — Application-Related Failures

Coating Failure	Failure Appearance	Cause of Failure	Remedy
1. Poor Workmanship	Dirt, dust, grime on surface and in coatings. Holidays, overspray, pinholes, runs, sags.	Lack of care in application.	Better training for workmen. Instill sense of pride in work.
2. Runs, Sags, Curtains	Heavy areas in coating which flow down vertical surface in streaks or curtains.	Lack of care in application.	Remove runs and sags with a brush prior to initial set of coating. Smooth area with light spray coat.
3. Brush Marks	Linear hills & valleys in coating. Considerable difference in coating thickness from hills to valleys. Rusting starting in valleys.	Poor workmanship. Very heavy bodied (thixotropic) coating.	Train workmen to brush smoothly. Brush coating out well, finishing by light brushing in one direction.
4. Improper Coating Thickness	Areas of pinpoint corrosion between areas of solid coating. Where coating is over thick. Possible checking and cracking.	Thin areas, spatter coating, holidays. Runs, puddles, excessive number of spray passes in areas where coating is difficult.	Careful application. Even spray passes with each pass overlapped 50%. Use cross spray technique.
5. Overspray	Very rough coating surface. May appear like sand in coating. Some dry coating, like dust, on surface. Pinpoint corrosion throughout rough areas.	Improper spraying technique. Uneven spray passes with gun too far from surface.	Apply coating with care and with even wet spray passes overlapped 50%. If overspray occurs, remove before overcoating.
6. Pinholes	Small, visible holes in coating (1/32 in.). Holes generally appear in concentrations with a random distribution. Pinpoint corrosion in pinholes.	Improper spray technique. Spray gun too close to surface with air bubbles being forced into coating. Spray pot pressure too high with atomizing air pressure too low. Pinholes may exist in the substrate (concrete).	Apply coating with care with spray gun at the optimum distance from surface. Make sure gun is properly adjusted. If pinholes already exist, apply coating by brush, working it into surface.
7. Holidays	General corrosion in bare or thin areas of surface which were uncoated by the painter. Most often in difficult areas to coat.	Poor, inconsistent application. Lack of care.	Apply coating in careful, consistent manner, making certain that no areas remain uncoated. Overlap each pass 50%.
8. Spatter Coat	Pinpoint rusting in area of thin coating, usually at end of spray pass or around a complex section of structure. Small spots of coating which are noncontinuous over substrate. In poor light, may seem continuous.	Discrete coating droplets which are not continuous over surface. Inconsistent spray passes not overlapped 50%. Spray gun flipped at end of spray pass.	Apply coating with care. Use even, wet spray with each pass overlapped 50%. Use cross spray technique.
9. Cratering	Pinpoint rust forming in thin areas of bug eyes, fish eyes, or craters randomly dispersed over coated area. May be more prevalent in thicker sections.	Improper solvent mixture, oil in atomizing air, surface contamination, particulate fallout during application, high surface tension.	Once cratering occurs, sand or roughen crater area. Apply second coat by brush, working coating into cratered area.

(SOURCE: Steel Structures Painting Council, Causes and Prevention of Paint Failure, Chapter 23, Good Painting Practice, Vol. 1, Steel Structures Painting Manual, 1982.)

FIGURE 14.26 — Edge corrosion in a marine atmosphere after approximately 2 years of exposure, with nearly all edges showing corrosion and undercutting of the coating.

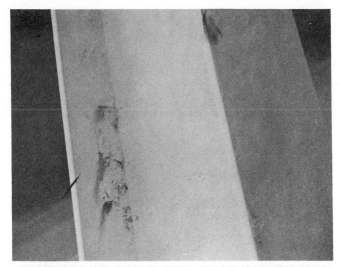

FIGURE 14.28 — Corrosion in the interior corner of an H-beam due to contamination which was allowed to remain on the surface and therefore caused a noncontinuous coating.

Figure 14.27* shows the typical edge effect on a panel exposed to a marine atmosphere. While the plane surface is unaffected and the coating is fully protecting the surface, the edges are seriously corroded, with the coating gradually being undercut. Coating failures on corners and edges can be expected, unless additional care is taken on these areas during application.

Corners

Exterior corners can be considered the same as edges, and the same corrosion occurs on exterior corners as would be expected on edges. This is shown on the corners of the box beams in Figure 14.26. Interior corners, however, are quite different. These are generally two types of failure that can be expected on interior corners. The first is the shrinkage of the coating away from the interior corner, causing a blistered area or an area where the coating bridges the corner. In this case, corrosion starts under the lifted area of the coating in the corner and spreads from that point out. The other common type of coating failure in a corner is due to dirt and trash accumulation in the corner which is not removed during the coating process. In this case, the coating is discontinuous over the contamination, and corrosion begins very quickly.

Figure 14.28 shows a column in which contamination was coated over and where early corrosion started after only a few months of exposure. In the case of this type of interior corner failure, the only remedy is to remove the contaminated coating and recoat. The best way to eliminate the failure is to improve the workmanship of the applicator.

Welds

The failure of coatings on welds is very common. There are several causes, the primary cause being that the weld is a rough and discontinuous surface area on a plane surface. On the other hand, many welds are tested for leaks, after which a soap solution is applied over the weld.

In many cases, the coating is applied over such a surface, with early failure of the coating resulting over the soaped surface. Many welds are cleaned less than perfectly, and weld slag and similar contamination can cause early coating failure. Figure 14.29 shows a typical weld failure on a plane surface. In this case, the weld was not given the proper attention and the coating failed over the rougher weld area in a short period of time.

Figure 14.30 shows another weld failure. Again, it is a change in direction of the surface, with the weld being considerably rougher than the plane surface adjacent to it. Extra attention was not given to such a weld, as recommended in previous chapters, and therefore the weld is the first area to fail.

Welds are almost always the first area in a structure to show failure. Main construction welds in tankers have been known to fail after one year, with the coating completely corroded across the weld area, even to the point of heavy scale developing on the weld. The cause was purely lack of attention to the weld area during the coating process. As stated in previous chapters, welds should be given proper attention and an extra pass for each coat applied.

On the other hand, weld areas in ships can turn out to be the best areas of the entire tank, primarily because of extra attention given the weld during coating application. In some cases, the primer is brushed on the weld and an extra pass applied to the weld during each coat on the flat surface.

Nuts, Bolts, and Rivets

Corrosion on nuts, bolts, rivets, and similar construction fasteners is almost inevitable, unless extra care is taken during the application of the coating. This is due to the many edges and corners that are present in a bolted structure. Figure 14.31* shows a critical bolted joint, with heavy corrosion occurring on the bolts and joint after only a few months of service. This is typical of many bolted

*See color insert.

*See color insert.

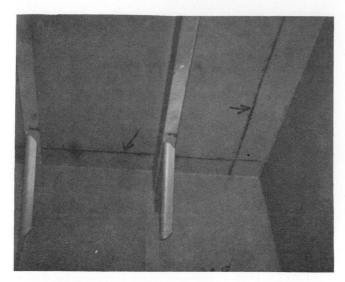

FIGURE 14.29 — Typical coating failure over a weld on a plane surface.

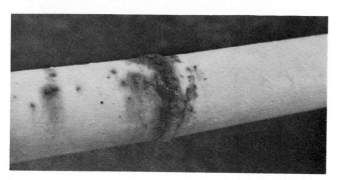

FIGURE 14.30 — Coating failure of the weld area on a pipe.

areas and only leads to the conclusion that extra care is necessary wherever such a bolted joint is used.

Boltheads also can be coated effectively, as shown in Figure 14.32.* Even in this case, damage occurred to some of the boltheads and edges, causing early corrorsion. Extra care in application is a requirement wherever bolted or riveted structures are used. Inorganic zinc coatings as a base coat for such areas have vastly improved the corrosion resistance of such joints.

Overlapping Joints or Plates

There are always critical corrosion areas whenever bolted or riveted structures are used. Failure of the coating in these areas is, in many cases, rapid and is due to the crevice between the two plates and the rapid change in direction of the coating in the joint. The coating in the joint often cracks, with moisture, salts, and similar contamination penetrating the overlap and causing rapid corrosion.

An excellent remedy is the use of inorganic zinc coatings on the faying surfaces (*i.e.,* the surfaces being joined) prior to bolting or riveting. In this way, the coating is carried underneath the plate and is protected for long periods

---

*See color insert.

of time. Figure 14.33 shows an overlapped joint and the corrosion which can occur when proper care is not taken in the application of coatings over such areas.

Construction Aids

Construction aids, hold downs, and similar areas are areas where corrosion is usually initiated and coating breakdown quite rapid. The cause is that these aids to construction are temporary and are usually welded rapidly, without particular care during the welding process. The welds are also usually quite rough. The metal pieces which are welded to the structure are rough cut and may be skip welded or only welded on one side. This allows easy access to moisture and salts with resulting corrosion. These temporary aids are too often left on the surface as well.

Figure 14.34 shows a construction aid which was left

FIGURE 14.33 — Typical overlapping joint, with corrosion beginning in the joint and with undercutting of the coating on each side of the overlap.

FIGURE 14.34 — Coating failure on a typical bracket or scaffolding support which served as a construction aid.

Corrosion Prevention by Protective Coatings

on a main piling of an offshore structure and the type of failure which can occur, even though the coating was applied over the construction aid during the original application. Unfortunately, application to these temporary aids is also usually less than the best. Corrosion starts at the weld and at the cut edges of the angle.

The remedy for such problems is to remove the construction aid once it has served its purpose, properly smoothing the cut edges, and, where the aid is removed, preparing the surface properly and applying the proper number of coatings in the repair area. While corrosion in such areas may seem trivial, once the coating has broken down and corrosion begins, pitting can occur, and if such aids are in a highly stressed area, failure of the structure can occur because of cracking beginning at the pit.

Pipe Structures

The failure of coatings on pipe structures is usually due to two principal causes. One is the welding of the pipe itself, with the failure of the coating occurring on the weld. The second is due to holidays left in the coating where the coating has not been properly overlapped during the coating process. In this case, the failure areas are usually longitudinal with the pipe and take the form of areas where there is first pinpoint rusting, followed by expansion of the corrosion over the whole thinner area, out to the point where the coating is applied at full thickness. These represent a typical type of failure on a pipe and are easily seen and recognized on pipe structures. The remedy is to overlap each pass of the coating as it is applied to the pipe by at least 50%, in order to eliminate such longitudinal holidays on welds. Add additional coats to the weld in order to overcome the uneven surface.

Adhesion-Related Failures

As reiterated throughout the discussions on coatings, adhesion is one of the key factors in the effectiveness of a coating. Conversely, lack of adhesion can be the cause of a number of major coating failures. Without adhesion, a coating is merely a film on the surface, subject to all types of disruption and damage caused by exterior forces. The coating must adhere to the substrate if the substrate is going to be protected from corrosion. Any factor which prevents or reduces coating adhesion will subsequently create coating failures of one type or another.

Blistering

One of the most common types of failure related to adhesion is that of blistering (Figure 14.35). *Blisters* can be large or small, although generally they are round, hemispherical projections of the coating from the surface and are either dry or liquid filled.

The size of the blister usually depends on the degree of adhesion of the coating to the surface and the internal pressure of either gas or liquid within the blister, stretching the coating to the point where the internal pressure is balanced against the degree of the coating adhesion. The usual cause for blistering is the penetration of moisture through the coating into areas of poor adhesion. The moisture vapor may condense and form a liquid blister, or the vapor pressure of the moisture at the interface be-

tween the coating and the surface may be sufficient to lift the coating away from the substrate. There are generally two types of blisters: (1) those that are formed from the substrate, with the coating separating from the substrate, and (2) those that are formed between coats and where the topcoat separates from an undercoat, forming circular blisters.

Figure 14.36 is an illustration of a topcoat blistering from the undercoat. In this case, the blisters were removed, showing the round areas where the blisters formed. There is little adhesion of one coat to the other, as indicated by the scratch between blisters. In this case, the topcoat chips and flakes away from the undercoat with no apparent adhesion between blisters. This type of failure indicates a complete lack of compatability between coats. To repair such a condition, the entire topcoat must be removed and recoated with a coating that is more compatible with the undercoat.

Both of the above types of blisters, whether they are from the substrate or between coats, may be either dry or filled with liquid. Blistering can be caused by a number of different conditions.

FIGURE 14.35 — A typical blistered surface, showing mostly hemispherical projections without breaks in the coating.

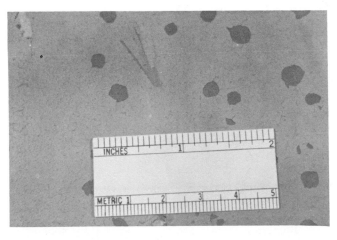

FIGURE 14.36 — Broken blisters in a topcoat that is incompatible with the base coat.

Soluble Pigments in the Primer

Soluble pigments in the primer have been a cause for many blistering failures. In this case, the soluble pigments absorb moisture vapor as it passes into the coating, creating a rather concentrated solution. At that point, the phenomenon of osmosis occurs, pulling water through the coating into the areas where the pigments have been dissolved. This is the reason why impervious types of coatings (previously described) are recommended for highly humid or immersion service. This same phenomenon can occur if there are soluble salts contaminating the substrate or contaminating the surface between coats of the coating. *Osmosis* is the transfer of moisture through the coating in the direction of the most concentrated solution. When there are soluble materials on or within the coating, blistering is almost inevitable. Blisters, in this case, would be liquid filled from the osmotic action.

Blistering can also be caused by contamination of the surface by materials (*e.g.,* oils, waxes, dust, etc.) that will not allow proper adhesion of the coating. Again, the moisture vapor transferring through the coating is the cause for the blistering. The moisture vapor tends to concentrate in these areas of low adhesion, with the vapor pressure causing the blister to form. In this case, the blisters are so-called "dry" blisters.

Shop Primers

Blistering can also be caused by the selection of an improper primer for the surface. This can often be the case when a shop primer is used on surfaces which are later overcoated with a high-performance coating. The primer does not have the necessary adhesion or physical properties to provide adequate adhesion for the high-performance coatings. Also, under some circumstances, the solvents in the high-performance materials will cause this shop primer to disbond from the surface.

Shop primers are always suspect where a high-performance material is to be applied over them. Unless the shop primer is designed as an undercoat for epoxies, vinyls, and similar coatings, it should be removed before the high-performance material is applied. Where primer failure is involved, the blistering is usually from the surface of the substrate.

Incompatability

Incompatible coatings often have poor adhesion which causes blistering between coats. Figure 14.37 shows an incompatible topcoat over an inorganic base coat. In this case, the topcoat had very poor adhesion, as indicated both by the blistering on the surface and the areas where the blisters have been broken, and by the area where the coating flaked away from the surface. The blistering occurred primarily in areas where a water layer was involved. Higher areas in the tank, exposed to oil only, were not affected in such a dramatic way, even though basic adhesion was poor.

Retained Solvents

Blistering failure also can be caused by poor or inadequate solvent release by the coating. Solvents retained by the coating act like plasticizers, making the coating

FIGURE 14.37 — Blistering in a water area of a petroleum tank, indicating the basic incompatibility of the topcoat with the inorganic zinc base coat.

softer and more flexible, and, depending on the water sensitivity of the solvent, can increase the water absorption and moisture vapor transmission of the coating. If the basic adhesion of the coating is at all marginal, the decrease in water resistance can cause adhesion release and blistering failure.

Where there is solvent retention and the coated surface changes in temperature, the solvent itself may create a sufficient vapor pressure to cause blisters to form in the coating. The most serious of these types of blisters are those caused by the moisture vapor passing into the coating and combining with the solvent, in some cases dissolving into the solvent, and causing a combined vapor pressure with resulting blistering of the coating. Solvent odor is usually connected with retained solvents, and under many conditions, solvents are detectable even after months and years of service. The blisters caused by solvent evaporation alone usually occur shortly after the coating has been applied, and, for the most part, can be relatively easily repaired.

Blisters of the solvent type may also occur where a topcoat is applied over a porous substrate or a porous undercoat. Such blistering often occurs where an organic coating is applied over an inorganic base coat and the temperature of the surface is quite warm. Blisters of these types rise rather quickly, often breaking and leaving a bare area where the blisters formed. This type of blistering can be prevented by a very penetrating prime coat flowing into the more porous substrate or base coat. The primer is usually applied in a relatively thin but wet film, so that is dries quickly yet has time to penetrate the surface. Changing the time of application to one when the surface is relatively cool may overcome this type of blistering as well.

Cathodic Protection

Blistering is often caused where cathodic protection

and coatings are used to supplement each other. This type of blistering is most often caused by hydrogen gas being formed on the metal substrate underneath the coating in sufficient volume so that the hydrogen vapor pressure pushes the coating off of the surface. Figure 14.38 shows blistering of this type, where excess cathodic protection caused blistering in the neighborhood of a scribe or damaged area of the coating. Cathodic disbonding is not unusual where coatings have less than the best adhesion.

Previously corroded surfaces can also cause blistering where salts or oxides remain in the surface pores of the substrate, creating focal points for rusting and blistering. Blistering can also be caused by a corrosive gas (*e.g.*, hydrogen sulfide, hydrogen chloride, or carbon dioxide) passing through the coating or being absorbed by it and pushing the coating from the surface. Such strongly active penetrating materials can react with the surface, causing a gas pressure with resultant blistering. Coatings used in fermentation tanks often blister because of the passage of carbon dioxide through the coating into an area where the coating adhesion is less than adequate.

Any blistering of a coating is serious. If it is localized from surface contamination, the coating can be removed and repaired. If it is general, the entire coating should be removed from the surface, the surface reprepared in order to obtain adequate adhesion, and if the coating selected is satisfactory for the service, it should be reapplied.

Peeling

Peeling is definitely an adhesion-related failure. It includes peeling from the surface or peeling between coats. It is generally caused where the tensile strength of the coating is higher than the adhesive strength or bond strength. This may be caused by poor surface preparation which reduces the bond of the coating to the surface. It can also be caused by contamination between coats, where the topcoat then peels from the surface; or it may occur when the thickness of the coating is too great, causing a film that is stronger than its adhesive characteristics.

Figure 14.39 shows the peeling of a topcoat from an undercoat, which is due either to surface contamination or to lack of compatibility of the topcoat with the undercoat. Figure 14.40 demonstrates an area of poor adhesion, where the coating originally blistered and the blisters were broken, after which the coating was easily peeled from the nonblistered surface. In this case, the coating was nonadherent to or incompatible with the substrate.

Flaking

Flaking is also an adhesion-related failure, and is much like peeling, with the exception that the flaked coating is usually hard and brittle and tends to pull itself away from the substrate. Once the coating cracks, the edges may tend to curl away from the surface, creating the flak-

FIGURE 14.39 — Peeling of a coating because of surface contamination or the incompatibility of the topcoat with the undercoat.

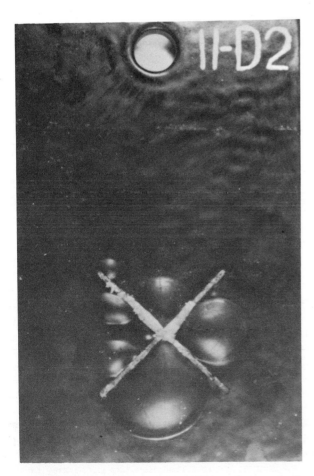

FIGURE 14.38 — Blistering of a coating subject to cathodic protection.

FIGURE 14.40 — Peeling due to poor adhesion.

ing tendency. Figure 14.41 shows a relatively heavy, hard coating flaking away from a metal substrate. Coatings applied over wood have a strong tendency to flake because of the working of the wood surface. Oftentimes, coatings are applied quite heavily over a wood surface, and, as it oxidizes and cures, the film becomes quite hard and brittle and tends to shrink. Figure 14.42 shows the coating as it is flaking away from a wood surface.

FIGURE 14.41 — Flaking of a heavy organic coating from a metal surface because of poor adhesion.

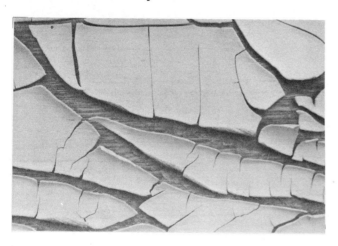

FIGURE 14.42 — Flaking of an oxidized organic coating from a wood surface due to the coating's obvious lack of penetration of the wood surface.

Intercoat Delamination

Delamination is the loss of adhesion between coats and is a common type of coating failure. The failure occurs most often where repair or maintenance coatings are being applied over existing coatings that have been in service for some period of time. Under these conditions, the original coating usually is chalky, has dirt embedded in the surface, or has oxidized or cured on the surface to the point of complete insolubility and impenetrability. Whatever the cause, the repair coating has difficulty wetting and establishing bond areas over the previous surface.

There are many causes for intercoat delamination. Incompatibility of two coating materials is one. It is always preferable to put the same kind of a topcoat or repair coat over a previous coat since there is less chance of incompatibility with similar materials. This is an area where mixing coating systems or mixing suppliers can cause some serious problems. Unless the two coatings are of the same generic type, or unless tests have been made to determine that the coatings are compatible, the mixing of coating systems is not a recommended procedure.

Contamination

The contamination of a surface is always a possible cause of delamination. Coatings which have been in service for some time and are being repaired accumulate dust, dirt, grease, chemicals, and similar contamination which must be thoroughly removed prior to the application of an additional coat. This can often occur quite quickly, even during the original application of a coating system. In areas where there is considerable atmospheric contamination from chemical fallout, dust, or similar materials, sufficient contamination can fall on the original coat to the point where a second coat may not adhere to the first coating. Although this is a difficult situation at best, the application of fast-drying, resistant coatings such as vinyls or chlorinated rubbers can help eliminate the rapid contamination between coats. In this case, the second coat should be applied as quickly as possible after the first coat, even within minutes.

Chalking

A chalky surface is a difficult one over which to apply any coating. Some are much more susceptible to delamination from chalking than others. On a chalky surface, it is recommended that it be thoroughly scrubbed prior to the application of any additional coats. Even with the substrate scrubbed to remove the chalking as much as possible, a coating that will wet the chalked surface is preferable, since it will penetrate any chalk and adhere to the original coating. Some materials are much more satisfactory for this type of exposure than others. Emulsion-type coatings are particularly poor for application over a chalked surface, and quick peeling or delamination can occur.

Overcured Surfaces

Overcured surfaces occur with many coatings, particularly those that react with air or are internally cured. Oil-base materials, such as linseed oil, alkyds, epoxy esters, and some similar materials, oxidize sufficiently so that over a period of time, if an additional coat is required, intercoat delamination may be a problem. It is always recommended that the surface of highly oxidized materials be broken or abraded by sanding or lightly blasting to allow mechanical adhesion, as well as some access of the new coat to the lesser oxidized body of the coating.

The same applies to epoxy coatings. However, in this case, they are internally cured, and many amine-type epoxies cure sufficiently hard so that the application of a second coat to the first coat that has cured may be difficult, making delamination possible. Coal tar epoxy coatings are even more difficult than epoxy coatings. The

Corrosion Prevention by Protective Coatings

combination of the two materials, coal tar and epoxy, apparently creates a coating that will cure rapidly on the surface, sufficiently so that even a few hours curing will cause delamination between coats.

As discussed previously, when coal tar epoxy coatings are exposed to sunlight, the sunlight will catalyze the curing of the surface to the point where additional coats will not adhere properly. Both amine- and polyamide-cured coal tar epoxies react in this way. This is not an uncommon problem, and has occurred in some extensive applications of coal tar epoxy. In fact, it has occurred on large applications where the coating was exposed to both sunlight and condensation. Massive delamination occurred even before the unit could be placed in service.

Another cause of delamination with coal tar epoxies is the exposure of the surface to moisture, such as condensation or dew, rain on the surface, and, of course, dust or dirt. In the case of Figure 14.43, the coal tar epoxy suffered massive failure, with extensive delamination covering many thousands of square feet, even though the coating was plant-applied and the second coat applied within 48 hours. The problem, therefore, can be a major one, and one which must be taken into consideration during the application of coal tar epoxy coatings. The one answer to this problem is the use of some of the newer high solids coal tar epoxy coatings that can be applied at a dry film thickness of 20 to 25 mils per coat. A single coat over a primer is often adequate and eliminates the delamination problem.

Polyurethane coatings are also highly polymerized, with a high-gloss surface. The combination often causes problems when applying additional coats. In every case where epoxy, coal tar epoxy, or polyurethane coatings are to be given more than one coat, the second coat should be applied prior to the time that the first coat has cured to insolubility on the surface. This may be within a few hours, under many circumstances. If this is not possible, or the original coating has aged, breaking the surface by a light brush blast, hand sanding, or light belt sanding is recommended to eliminate intercoat delamination.

The application of too great a thickness of a coating can also lead to delamination. It can cause delamination not only down to the substrate, but also between coats. The shrinkage stresses in a heavy coating can increase until they exceed the force of the adhesion between coats, and the heavy coat will actually pull itself apart and away from the surface. This is something that often occurs with heavy-bodied epoxy coatings where excessive thicknesses are applied. This is particularly true where there is a substantial temperature variation of the coated surface.

Chemical attack can also cause delamination. Volatile acids, such as acetic, hydrochloric, or nitric, often are sufficiently penetrating to cause delamination between coats. They also will often cause delamination from the substrate for the same reason. Where coatings are subject to strong solvents, the solvent may also separate the various coats, even though the coats appear tightly bonded to each other. If there is a distinct interface, solvents will often penetrate sufficiently so that delamination occurs.

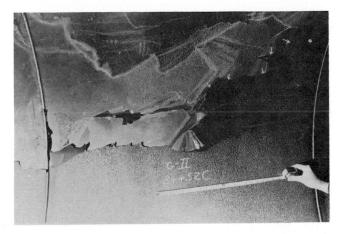

FIGURE 14.43 — Delamination of a coal tar epoxy coating on the interior of a large pipe.

As discussed, the oxidized or internally polymerized coatings are much more difficult from a delamination standpoint than materials that are permanently soluble in their own solvents such as lacquer-type coatings, vinyls, chlorinated rubbers, etc. Materials that are permanently soluble in their own solvents have less of a problem, since the solvents strike into the original coating and allow some penetration and mixing of the coating at the interface. This is one of the advantages of the vinyl and chlorinated rubber-type high-performance materials. Where delamination is a possibility as a result of exposure, aging, or contamination, vinyl and/or chlorinated rubber coatings should be considered.

Correcting delamination failures depends on a number of things. If the delamination is spotty and is limited to certain localized areas, repair of these areas is often possible. Where the delamination is extensive, it usually is preferable to remove the entire coating down to the substrate, starting out with a clean surface and reapplying the coating, while making certain that the factors causing the initial delamination have been eliminated.

Undercutting

Undercutting of a coating is the action of rust under the coating, usually forming around a small break in the coating. This precondition is not necessary, however, if the coating has been applied over a previously corroded surface or one that may be contaminated with chlorides. In the latter case, depending on the coating (particularly those that cure by oxidation), corrosion can occur underneath the coating, undercutting the coating prior to any actual break.

It is often difficult to see minor abrasions that may be sufficient to cause undercutting in a coating. Figure 14.44* is an example of the type of undercutting that occurs because of minor abrasions on the coated surface. Figure 14.45 is a close-up of a more massive coating failure, showing the growth of the corrosion under the coating. Each of the round, blister-type spots is a growth of iron

---

*See color insert.

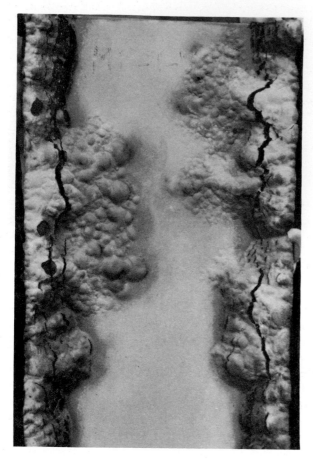

FIGURE 14.45 — Undercutting from the edges toward the center of a panel exposed to a marine atmosphere.

oxide underneath the coating. In both cases, corrosion has penetrated the coating and is building up under the coating at the interface between the coating and the metal. Corrosion occurs quite rapidly under such conditions, inasmuch as the corrosion product will retain moisture and/or salts, creating an ideal corrosion cell.

A coating that has been applied over mill scale is susceptible to undercutting, since once a break in the coating occurs, corrosion can begin between the mill scale surface and the steel, and corrosion can spread rapidly between the two surfaces. Moisture and oxygen both penetrate the coating and react at the scale/metal interface and at the coating interface, causing a loss of adhesion, with progressive corrosion beneath the mill scale and the coating.

Undercutting is also common where a coating is applied over a smooth surface, since the adhesion between the coating and the surface is marginal because of this condition. Here again, the corrosion, once started, acts as a wedge, pushing the coating away from the surface and building up corrosion product between the surface and the underside of the coating. Figure 14.46* is a good example of the undercutting of a coating from a break in the coating. Note that the panel on the right shows an extremely heavy corrosion product at the break and that the coating is actually being turned up at the sides of the

---
*See color insert.

break as the corrosion product works underneath.

Undercutting is a serious problem, and very deep pitting can result from the corrosion which builds up underneath a coating. Undercutting is definitely adhesion related. The greater the adhesion, the less tendency for the coating to undercut. This is one of the reasons why the use of an inorganic zinc base coat under organic coatings has been so successful in eliminating undercutting. In any area where there are general serious corrosive conditions, such as in marine areas, industrial plants, refineries, etc., an inorganic base coat should be seriously considered for any coating project so as to eliminate the possibility of undercutting and serious corrosion.

Failures Due to Exterior Forces

In many ways, it can be said that all types of failures are due to exterior forces, since it is the environment in which a coating operates that is the primary cause of coating failure. If the environment was always benign, there would be no coating failure, even though there are coating imperfections. Nevertheless, there are certain types of exterior forces that cause coating failure which are over and above the ordinary atmospheric types of exposure to which the coating is expected to be resistant.

Chemical exposure is one of the principal types of exterior forces that can cause coating failure. Not only is the chemical industry one of the largest industries in the world, but it is also of a broad scope, covering many different types of chemical exposure. Chemical attack on coatings can be by simple solution, where a solvent for the coating material dissolves the coating and removes it from the surface. Even though the solvent may not remove the coating, but only tend to swell it, in many cases it opens the coating to attack from the atmosphere by making it more porous.

Volatile acids are particularly damaging to coated surfaces since they are highly penetrating. In fact, acetic acid acts as both a solvent and an acid, so that easier and greater penetration of many coating materials is possible. Hydrochloric and nitric acids are very penetrating, and under many circumstances will pass through a coating as a vapor, attacking the substrate without appreciably damaging the coating itself. Once the penetration has occurred, the corrosion takes place underneath the coating, forcing it off from the surface and increasing the corrosion possibilities.

The caustic chlorine industry is a difficult one in terms of coatings. Not only is there strong caustic available, but chlorides are also present, and chlorine gas is a very penetrating oxidizing material as well. The combination makes this industry a difficult one for coatings, although there are a number which effectively resist the corrosion.

Chemical transport is another industry which is difficult from a coating standpoint. Not only are there many different chemicals involved, but there may be abrasion involved as well, so that breaks in the coating increase the possibility of coating failure in this service.

Reverse Impact

Reverse impact is an area which can be particularly

damaging to coatings (Figure 14.47). As can be seen in the figure, the coating is heavily star cracked because of the impact on the exterior of the coated surface. Reverse impact is a difficult problem for many coatings, and in some cases it is used as a test for coating adhesion. Actually, because the coating is heavily stressed, it is not only a test of adhesion, but a test of extensibility as well. Reverse impact is much more difficult for a coating to resist than is direct impact on the surface. There are coatings that are designed to have extremely good adhesion as well as extensibility and which will resist the reverse impact shown in Figure 14.47.

FIGURE 14.47 — Star cracks due to reverse impact.

Abrasion

Abrasion is another common cause of coating failure. Floors and similar surfaces are one of the major areas where abrasion is a problem. Wheel traffic, foot traffic, sliding of heavy objects—all cause abrasion on floors and severely damage the coating. In order to resist such forces, the coating must have the ability to resist abrasion, but, at the same time, have sufficiently strong adhesion to the substrate so that, as the coating is worn away, it does not flake or chip, but wears down to an even, feathered edge. Good adhesion is essential for an abrasion-resistant coating.

Ships and other marine structures are subject to considerable abrasion. Figure 14.48* shows the hull of a ship

*See color insert.

where the anchor chain has scraped over the surface. This type of abrasion is called *riding the anchor,* and coatings must be particularly abrasion-resistant and have good adhesion to reduce to a minimum the corrosion caused by such abrasion. Such abrasion actually cuts into the steel itself. However, with a coating with high adhesion and abrasion resistance, the damage is reduced to a minimum, and repair can be made with a minimum of effort during dry docking. Inorganic zinc base coats have been a major factor in holding this type of coating abrasion to a minimum as a result of their strong adhesion, malleability, and corrosion resistance.

Figure 14.49* shows the rudder of a ship and the abrasion which occurs directly behind the propeller. Notice, in this case, the zinc anodes that are used to help reduce the corrosion caused by the abrasion of the coating in this area. In this case also, a very abrasion-resistant and highly adherent corrosion-resistant coating is required in order to minimize the corrosion caused by the abrasion.

Faying Surfaces or Joint Movement

Failures can occur where joints in a steel structure are fastened by riveting or by the use of high-strength bolts. Such joints are common throughout industries of all types and are used on many steel structures, such as bridges. Faying surfaces are designed to resist movement. However, at this point there is a change in direction of the coating where cracking will often occur. When this happens, coating failure occurs at the junction of the two metal surfaces, resulting from excess moisture and atmospheric contamination. Most coatings are unsuitable for faying surfaces because of the fact that the coefficient of friction is not sufficient for the bonding of the joint. Fortunately, inorganic zinc coatings are satisfactory for use on these surfaces and are an answer in protecting the overlapping surfaces from corrosion.

Summary

Coating failures, then, may be due to many different factors, ranging from the characteristics of the coating itself to its formulation and the materials from which it is made. Failure can occur from the underlying surface when improperly prepared, by the application of the coating over the surface, by incomplete coverage of the surface, and because of the basic design of the structure itself.

There is no set manner in which to deal with coating failures. Each problem is a unique one and must be analyzed as such in order to overcome that particular type of failure. Once the failure has occurred, repair is usually possible. On the other hand, unless the repair is done very carefully, continuing corrosion can occur in the repaired area. Massive failures can only be repaired by removing the coating, repreparing the surface, and reapplying a coating that will be more resistant to the particular source of corrosion. Any generalization on the prevention of coating failure can be made only by proper analysis of the exposure and the structure, and specifying a material that will completely resist the exposure. No compromise should be made, either in the coating or with its applica-

*See color insert.

tion, since repair of failures are much more costly than proper application of the original material.

Once a material has been selected, a specification should be written, covering not only the material, but its application as well. The specification should be detailed, making certain that all possible causes of coating failure are taken into consideration and eliminated. Following the specification, a thorough inspection of the coating should take place during its installation, as well as upon its completion. There is no substitute for proper material selection, complete specifications for both the material and its installation, and a strong follow-up inspection to ensure that it has been properly applied. These three points can do more to prevent coating failure than any others known.

References

1. Seagren, G. W., Causes and Prevention of Paint Failure, Steel Structures Painting Manual, Ch. 18, Vol. 1, Good Painting Practice, Steel Structures Painting Council, Pittsburgh, PA, 1966.
2. NACE, Causes and Prevention of Paint Failure, Publication 6D170, Houston, TX.
3. Munger, C. G., Understanding Protective Coating Failures, Part 1, Plant Engineering, April 15 (1976).
4. ASTM, Standard Method of Evaluating Degree of Chalking of Exterior Paints, D-659-80, ASTM, Philadelphia, PA.
5. ASTM, Standard Method of Evaluating Degree of Checking of Exterior Paints, D-659-80, ASTM, Philadelphia, PA, 1982.
6. ASTM, Standard Method of Evaluating Degree of Cracking in Exterior Paints, D-661-44 (Reapproved 1981), ASTM, Philadelphia, PA, 1982.
7. NACE, Combating Adhesion Problems When Applying New Onto Existing Finish Coats of Paint, Proposed Technical Committee Report, 1982 Draft (subject to change), Houston, TX.
8. Iezzi, Robert A. and Leidheiser, Jr., H., Surface Characteristics of Cold Rolled Steel as They Affect Paint Performance, Corrosion, Vol. 37, No. 1 (1981).
9. SSPC, Causes and Prevention of Paint Failure, Steel Structures Painting Manual, Ch. 23, Vol. 1, Good Painting Practice, Pittsburgh, PA, 1982.

15

Coating Repair and Maintenance

Introduction

The maintenance and repair of coatings is one of the corrosion engineer's primary responsibilities. This is particularly true when dealing with critical structures where corrosion is a constant threat, *e.g.,* boats, ships, offshore structures, refineries, chemical plants, paper plants, and sewage treatment plants. Linings for the interior of tanks, chemical reaction vessels, leaching tanks, transportation equipment, and even food processing and storage vessels are also included, since maintenance is an important procedure in the protection of both the structure and the contained liquid. Since coatings represent the first line of defense in the protection of these structures, they must be kept in the best possible condition for the longest period of time to guarantee proper protection over their intended life span. Therefore, both coatings and linings need to be maintained and repaired on a continuing basis to assure their maximum usefulness.

A subtle difference exists between the terms *maintenance* and *repair*. *Maintenance* means preservation, conservation, safe-guarding, upkeep, or to hold or keep in a particular state. *Repair* means fix, recondition, renew, rebuild, patch, or restore to a sound state. As defined, maintenance is a much broader term referring to the protective management of the surface. This does not simply mean the periodic touch-up of a few damaged spots, but rather the management of the coating process from the beginning of the original application to its present as well as future state. It includes the inspection, record keeping, and observation of the coating and its condition. It also requires material studies and the establishment of limits of failure, beyond which repair and recoating are automatic procedures.

Repair, on the other hand, refers to the actual act of reconditioning, rebuilding, and restoring a coating to its proper condition and protection standard. Thus, proper repair is a part of good maintenance and coating manage-

ment. One such general maintenance procedure is described in a Department of Defense manual on paints and protective coatings.

Programmed painting is a systematic process for establishing when painting is required, what painting should be done, by whom, with what materials, at what time, and in what manner. Paint systems deteriorate and will lose their protective ability unless the film is intact. The principal objective of painting is to prevent deterioration of the substrate at a minimum cost per square foot per year. One procedure frequently used for providing protection has been to completely repaint after the original coating has failed. This failure results in an unsightly surface, expensive preparation before repainting, and possible deterioration of structural members. Another procedure is to completely repaint by applying two and even three coats at arbitrary intervals. This may be too late in cases where deterioration has already taken place, but completely unnecessary in others. Extensive surface preparation will be required in the first case, and film thickness will eventually become excessive in the latter case, leading to early failure by cracking and peeling. The most practical method of protection, therefore, is a continuous program of inspection, and painting as necessary. By means of programming and careful record keeping, a history of past performance is accumulated which aids materially in selecting the best paint systems and painting procedures.

Applied paint systems do not deteriorate uniformly. Even when they are applied by skilled painters, some pinholes, holidays and breaks at sharp edges or seams are often present. Left untouched, corrosion and deterioration will start at these points, eventually undermining the coating and then spreading to adjacent areas. Furthermore, as corrosion increases, it does so at an accelerated rate until large areas of the surface are left unprotected. Programmed painting enforces inspection and work scheduling to provide for relatively easy spot-painting of these minor breaks in the film long before any serious harm is done. Spot-painting describes the painting of only the small or localized areas in which the coating has begun to deteriorate. Not only does spot-painting save costly surface preparation and repainting of large areas, but the life expectancy of the paint system and structure can be extended considerably. Furthermore, when repainting is desired to achieve adequate film thickness or for uniform appearance, it can be accomplished economically with the minimum number of coats, since the surface will be in sound condition. An added advantage derived from preventive maintenance is the detection of faulty structural conditions or problems caused by leakage or moisture before they become serious due to oversight.[1]

The repair and proper maintenance of coatings is equally as important as application of the original coating. In fact, the repair of coatings is actually a more difficult procedure than coating a new structure or one that has been completely sandblasted. One of the principal reasons is that coating repair carries its own psychological barrier. People just generally don't like to do it. It might be different if the process could be done quickly and with no preparation. Unfortunately, any repair job done in this manner would not last long. The hang-up seems to be the surface preparation that is needed to do the job right. For example, how often have you put off painting the kitchen because you knew the ceiling would have to be washed first? The same problem exists on industrial jobs where a coating has failed on tank roofs or corroded structural steel. However, if these areas are not properly prepared, the repair effort will be wasted and have to be repeated after a short period of time. For example, ships have been repaired by giving them what is called a quick "shave and haircut," only to have the coating wash away from the surface when the ship was refloated in drydock.

The timing of any coating repair is critical. Recoating before it is required is uneconomical and eventually results in a heavy, thick film buildup which leads to deterioration of the coating system by delamination, cracking, chipping, etc. For instance, a ship was coated after every round trip of approximately two months in order to maintain the best possible appearance of the ship for public relations purposes. However, when the time came to remove the coatings and begin from scratch, it was found that various areas of the coating had delaminated, both between coats and from the surface, and in many areas the thickness of the coating was from 1/4 to 3/8 in. thick. On the other hand, recoating scheduled at overly long intervals results in costly surface preparation and may be responsible for damage to the structure which also may require extensive repairs.

Results from an NACE report on Industrial Maintenance Painting Programs shows that the frequency of inspection, and therefore of repainting in severe environments, was six months to a year in about half the plants reporting; about one-third of the plants with moderate or normal corrosion rates reported inspections at six months to a year intervals without necessarily requiring any repair. The tendency for longer intervals between inspections increased as the atmospheric corrosion rate diminished.[2]

Spot painting should be done before any appreciable amount of rust appears. Where inorganic zinc primers are used, spot painting should be done only when the second coat (primer, where used) is exposed and before the zinc primer is exposed. If repainting is done before rust appears, surface preparation costs can be avoided. If more than 10% of the surface needs spot painting, it is more economical to paint the entire surface. While more material will be used, less labor will be required. Labor for spot painting with a brush costs approximately six times as much as the material used.[3]

When is it time to recoat? The coating of equipment or structures that are already in poor condition should not necessarily receive top maintenance priority. Since such a coating is already disintegrated, the structure is corroded and the overall coating not salvageable. A few months more or less at this point would probably make little difference to the basic structure. On the other hand, a tank or structure that is weathered to the point where the primer may be showing through in a few spots, but where no active corrosion has occurred can yet be salvaged. Here, with proper cleaning and additional topcoats, a continued trouble-free life of many years can be expected. Once this surface has been brought up to an "as good as new" condition, the repair of the badly disintegrated coating can begin.

Primary Repair Considerations

Once the decision is made to repair a coating, there are a number of items that must be considered in making decisions as to the type of repair, the amount of the repair, the coating to be used, and the number of coats. Several of the most important considerations are:

1. Type of failure
2. Extent of failure
3. Adhesion of existing coating
4. Type of coating
5. Type of substrate
6. Area of coating use

Type of Failure

The preceding chapter on coating failures discussed the major types of failures. Nevertheless, since coating failure is the primary reason for maintenance or repair, a brief summary of failure types in this context follows.

Undercutting

Corrosion undercutting and buildup of corrosion beneath the coating is one of the most common causes of coating failure and deterioration. The cause of undercutting is inherently an adhesion problem stemming from the original coating or its application. This type of failure is most often found on structural steel, bridges, open steelwork, construction in the chemical industries, offshore structures, and in the marine industry in general. Actually, it is common to most industries to one extent or another.

Blistering

Blistering is certainly one of the more common types of coating failure that should be considered in relation to coating repair and maintenance. It occurs in immersion service, in areas of splash and spray, and in high humidity. Therefore, it is a broad range type of failure and may be found in almost any industry. However, it may be more common in the marine industry due to a greater amount of immersed surface and surface exposed to splash or spray and high humidity. Blistering is usually caused by poor adhesion, a lack of proper surface preparation, cathodic disbonding, and various types of surface contamination.

Pinpoint Rusting

Overall, pinpoint rusting is probably the most common type of coating failure. Where repair of a coating is contemplated, pinpoint rusting is the first sign of coating failure. It is usually found first where the coating is thin and not fully protecting the surface. It occurs in both organic and inorganic coatings, and is the first sign of thin areas or holidays that may have occurred during application. Pinpoint rusting is a type of failure found throughout most industries in which high-performance coatings are used. It is the aging type of failure for zinc coatings such as galvanizing, inorganic zinc, and even organic zinc-rich coatings. It is the type of failure that is often overlooked for a considerable period of time, to the point where it becomes sufficiently widespread to require major repair operations. However, pinpoint rusting is probably the easiest type of failure to repair, if observed in time and followed by prompt repair procedures.

Delamination

Coating delamination is not quite as common a problem as some of the previously described failures. Nevertheless, it can prove serious from a repair standpoint. Delamination is essentially a lack of adhesion between coats. While a small area may blister or otherwise be pushed from the surface, if that small spot is recoated, additional delamination usually occurs around the repair due to the original lack of adhesion between coats. This type of problem is usually found where thermosetting-type coatings are encountered. Even oil paints can be considered in this category since they are essentially thermoset once they are strongly oxidized. The amine- or polyamide-cured coal tar epoxy coating is one of the most common of this type of coating. These materials, as well as other epoxies and urethanes, cause delamination problems where overcured between coats by heat, sunlight, or water exposure. In this latter instance, the surface becomes dense and insoluble, making it difficult for additional coats of the same material to adhere. This problem can occur in almost all industries.

Chalking

Chalking is an extremely common type of failure for all except a few coatings. It is a weather-oriented surface difficulty; therefore, it is normally found on the exterior surfaces of buildings and structures and is common to all industries.

Checking

Checking is similar to chalking. It is a weather-oriented problem, and occurs on the surface only, rather than penetrating through the coating.

Extent of Failure

The extent of coating failure is one of the most im-portant items to determine when a coating is to be repaired. The extent of the failure determines to a considerable degree the method as well as the extent of the repair. In order to know the extent of the repair, it is necessary to rate the condition of the coated surface so as to know when and how to recoat. Weaver, in his book *Industrial Maintenance Painting,* describes four states of coating failure, as follows.

1. Thin Finish Coat. Some primer showing, rust negligible.
2. Thin Finish Coat. Considerable primer showing, 10% of surface containing rust, loose scale, and loose paint film.
3. Finish coat is thoroughly weathered and is blistered badly. Approximately 30% of the surface contains rust with pitting and hard scale.
4. Badly pitting surface with rust nodules.[3]

This series of coating ratings, however, is not adequate to describe properly the extent of failure on the surface. If 10% of the surface is showing rust, or other failure, repair is undoubtedly necessary. In most cases, the method would be to completely remove all of the coating, unless the corrosion was sufficiently localized to allow easy repair of the corroded areas while the remainder of the surface remained completely corrosion-free.

ASTM has attempted, by a series of standards for types of coating failure, to evaluate the extent of the failure numerically, in a way that can be reasonably duplicated by more than one person. The ASTM standards that are most useful in determining the extent of coating failure are as follows.

1. ASTM Standard D610-68 (1974), *Degree of Rusting on Painted Steel Surfaces, Evaluating.*
2. ASTM Standard D659-74, *Chalking of Exterior Paint, Evaluating Degree of.*
3. ASTM Standard D660-44 (1970), *Checking of Exterior Paint, Evaluating Degree of.*
4. ASTM Standard D661-44 (1970), *Cracking of Exterior Paint, Evaluating Degree of.*
5. ASTM Standard D662-44 (1970), *Erosion of Exterior Paint, Evaluating Degree of.*
6. ASTM Standard D714-56 (1974), *Blistering of Paint, Evaluating Degree of.*
7. ASTM Standard D772-47 (1970), *Flaking (Scaling) of Exterior Paint, Evaluating Degree of.*[4]

Each of these standards for evaluating coating failure includes a photographic standard which serves as a useful tool in rating the extent of failure. The ASTM Standards generally use a rating of from 10 to 0, with 10 being perfect and 0 being completely corroded, blistered, etc. This is considered a good method since the scale of 10 to 0 is discrete enough to accommodate accurate reporting of the condition. In some cases, where a more general scale is needed, it can be modified to proceed from 10 to 8, 6, 4, 2, and then 0, which allows more leeway between the numerical values.

Figure 15.1 shows the scale and gives a description of the rust grades outlined in ASTM Specification D610-68.[4] This ASTM Standard is designed to rate the condition of the coating as a percentage of the surface rusted. However, it can also be used for blistering, scaling, and similar types of failure.

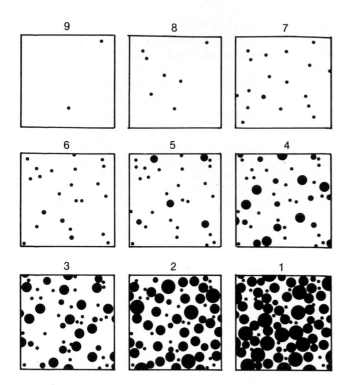

FIGURE 15.1 — Rating of painted steel surfaces as a function of area percent rusted. (SOURCE: ASTM-D610/ SSPC-Vis 2, ASTM, Philadelphia, PA.)

TABLE 15.1 — Scale and Description of Rust Grades

ASTM Rust Grades	ASTM Description	Alternative Description
10	No rusting or less than 0.01% of surface rusted	No corrosion
9	Minute rusting, less than 0.03% of surface rusted	1 to 2 pinpoints of rust per square foot
8	Few isolated rust spots, less than 0.1% of surface rusted	3 to 6 pinpoints of rust per square foot
7	Less than 0.3% of surface rusted	±20 pinpoints of rust per square foot
6	Extensive rust spots, but less than 1% of surface rusted	±30 pinpoints to 1/32 in. rust spots per square foot
5	Rusting to the extent of 3% of surface rusted	±30 pinpoint to 1/16 in. rust spots per square foot
4	Rusting to the extent of 10% of surface rusted	±50 1/32 to 1/8 in. rust spots per square foot
3	Approximately 1/6 of the surface rusted	Approximately 1/4 of surface rusted with 1/32 to 1/8 in. rust spots
2	Approximately 1/3 of surface rusted	Approximately 1/3 of surface rusted with 1/32 to 1/4 in. rust spots
1	Approximately 1/2 of surface rusted	Approximately 1/2 of surface rusted with 1/32 to 1/2 in. rust spots
0	Approximately 100% of surface rusted	Approximately 100% of surface rusted

(SOURCE: ASTM Description from ASTM Standard D-610-68, Evaluating Degree of Rusting on Painted Surfaces, Philadelphia, PA, 1982.)

While the percentage method is a good one, it can be simplified using the same approach as shown in Figure 15.1. If, for instance, in place of the percentage rating, the rating of the number of corrosion spots or blisters per square foot was used, results would be easier to visualize in a report. If each one of the squares in Figure 15.1 was considered to be a square foot, then the following rating would result. A rating of 9 would equal 1 to 2 pinpoints of rust per square foot. A rating of 8 would equal from 3 to 6 pinpoints of rust per square foot. A rating of 6 would equal plus or minus 30 pinpoints to 1/32 in. rust spots per square foot. A rating of 4 would equal plus or minus 50 pinpoints to 1/4-in. rust spots per square foot. This would then scale up rapidly to the point of complete failure at 0.

This alternate rating method is suggested because experience in reporting failures via written reports has led to the conclusion that it is easier to visualize 2 pinpoints of rust per square foot than it is 0.003% of surface rusted. For instance, in the case of a rating of 8, it is easier to visualize 3 to 6 rust spots per square foot as compared with a rating of less than one tenth of one percent of the surface rusted.

Table 15.1 shows the ASTM rust grades and their descriptions, combined with the alternative descriptions suggested above. These alternative descriptions can be related to Figure 15.1 for a visual evaluation of the rating. ASTM also has photographic standards corresponding to the various ratings, which can be obtained from them[4] or from the Federation of Societies on Coating Technology.[5]

The ASTM standard for evaluation of rust on painted steel surfaces provides the basis for the rating of the other types of coating failure that have been discussed. The evaluation for the degree of blistering, however, is somewhat different. The size of the blisters is rated on a 10-to-0 basis, with 10 being no blistering, 8 being pinpoint blisters, 6 being blisters up to 1/16 in. in diameter, 4 being blisters up to 1/8 in. in diameter, and 2 being blisters 3/8 in. in diameter or larger. In addition to the size of the blisters, the quantity of blisters is also rated. In this case, the rating is none for no blistering, few for one or more blisters per unit area, medium for several blisters per unit area, medium-dense for an area with many blisters but with some flat space between the blisters, and dense for continuous blisters over the surface.[4] Figure 15.2 shows blisters with a rating of number 2 dense. Other ASTM ratings for degree of coating failure are as follows.

1. ASTM-D659-44 (reapproved 1965), *Evaluating the*

FIGURE 15.2 — Complete blistering of coating (ASTM Rating 2, Dense). (SOURCE: Federation of Societies for Coating Technology, Pictorial Standards of Coatings Defects, Blistering, Philadelphia, PA.)

Degree of Resistance to Chalking of Exterior Panels. Chalking is the powdering of the surface of a coating, and again is evaluated as 10 where there is no chalking and 0 where the chalking is extremely heavy. The method of rating is to take a dark wool cloth, press it against the surface, and twist it under your thumb for 180 to 360 degrees. With a rating of 10, there is no discoloration on the dark cloth; 8 leaves a slight discoloration, while 2 results in a completely opaque white surface.

2. ASTM-D660-44 (reapproved 1970), *Evaluating the Degree of Resistance to Checking Exterior Paints.* In this case, checking means a break in the surface that does not penetrate to the substrate. This means checking is a surface condition. The rating system is from 10 to 0, with 10 being no checking, 9 being minor checking, 8 being a few checks, 6 being moderate checking, 4 being nearly continuous checking, down to 2 which is a completely checked surface.

3. ASTM-D661-44 (reapproved 1970), *Evaluating the Degree of Resistance to Cracking of Exterior Paints.* The differentiation between cracking and checking is that cracking is completely through the coating film down to the substrate. The appearance may be quite similar, but, cracking is a much more severe type of disintegration, with corrosion often occurring in the cracks. In other words, cracking is a step beyond checking from the standpoint of severe coating failure. The rating system is essentially the same as for the previous ASTM Standard on checking.

4. ASTM-D772-47 (reapproved 1970), *Evaluating the Degree of Flaking (Scaling) of Exterior Paint.* Flaking may be considered a continuation of the previous cracking failure. Flaking can be considered a cracked surface where the small areas between the cracks have lost adhesion and are curling or popping from the surface. The rating system is similar, with 10 meaning no scaling, 8 meaning 1 to a few small flakes per unit area, 6 meaning moderate flaking, 4 meaning 20 to 25% of the surface flaking, and 2 meaning 40 to 50% of the surface flaking.[4]

Table 15.2 summarizes the ratings for evaluating the above failures. In connection with the evaluation of coatings, it must be kept in mind that there are photographic

TABLE 15.2 — Standard Coating Ratings

Standard	Scale	Description
ASTM D714-56, Evaluating the Degree of Blistering Paints	Size of Blister	
	10	No blister
	8	Pinpoint
	6	Pinpoint to 1/16 in.
	4	1/8 in.
	2	3/8 in. or larger
	Frequency of Blisters	
	10	None
	8	Few
	6	Medium
	4	Medium-Dense
	2	Dense
ASTM D659-44, Evaluating Degree of Resistance to Chalking of Exterior Paints (wool cloth pressed on surface and turned 180 degrees)	10	No chalk or discolor on cloth
	8	Slight discoloration
	6	Light discoloration
	2	Completely opaque chalk
ASTM D660, Evaluating Degree of Resistance to Checking (checking is a break in the surface not penetrating to the substrate)	10	No checking
	9	Very minor checking
	8	Few checks
	6	Moderate
	4	Almost continuously checked
	2	Completely checked
ASTM D661, Evaluating the Degree of Resistance to Cracking of Exterior Paints (cracking extends through coating to substrate)	10	No cracking
	9	Very minor cracking
	8	Few cracks
	6	Moderate
	4	Almost continuously cracked
	2	Completely cracked
ASTM D772-47, Evaluating the Degree of Flaking (scaling) of Exterior Paint	10	No flaking
	8	Few flakes
	6	Moderate flaking
	4	20 to 25% of surface flaked
	2	40 to 50% of surface flaked

[SOURCE: Munger, C. G., Repairing Protective Coatings: Effect of Coating Types, Plant Engineering, Feb. 17 (1977).]

standards for the various degrees of failure, which can aid immeasurably in evaluating the type and amount of coating failure.

As previously stated, it is important to know the degree of coating failure in order to determine the type of repair that is needed. If the amount of failure is relatively minor, with a rating of 8, the surface can usually be repaired without completely removing the entire coating from the surface. If the rating is much less than 8, it is usually more economical to remove the entire coating and begin from scratch. While this may seem somewhat drastic, preparation of individual spots of failure is often much more difficult and time consuming than removal of the entire coating.

Adhesion

It is extremely important to know the adhesion of the original coating wherever repairs are to be made. The adhesion of the original material can make the difference between success and failure in the application of a repair coating over that surface. For example, in the case of the undercutting of a coating by corrosion products, if the undercutting is localized to a break in the coating or to an edge, and the adhesion beyond the undercutting is good and continuous, then the undercut areas can be satisfactorily prepared, the area patched with the original number of coats, and an overall coat provided over the entire surface.

On the other hand, if the adhesion is only marginal beyond the area of undercutting, it is quite probable that the addition of an overall topcoat over that surface may further decrease the adhesion and create areas for blistering or for further undercutting. Also, in the case of blistering, if the coating on the plane area beyond the blister has poor adhesion, then the entire coating must be removed. If the adhesion between the blisters is strong, the blistered areas often can be repaired with an overall coating over the remainder of the surface. Without the strong adhesion between the blisters, however, this would not be practical.

In the case of cracking, the coating may have strong adhesion except at the crack. If the coating is generally cracked, and corrosion has started on the substrate within the crack, then the only recommended procedure is to remove the entire coating from the surface. If the adhesion is good, and there is little or no corrosion starting, repair by additional coats may be practical. Since chalking and checking are surface coating phenomena, adhesion generally is not affected by either one of these types of coating failure. Thus, these types of coating failure can usually be remedied by the proper application of a topcoat over the existing material. With chipping or scaling, however, adhesion to the surface is usually questionable, so that unless the failure is restricted to local areas, it is generally preferable to begin repair from the substrate out by removing the existing coating.

Type of Coating

Any repair program must take into consideration the type of coating that already exists on the surface. Otherwise, the entire repair job may result in early failure. For instance, a topcoat that is incompatible with the existing coating can create adhesion problems, blistering, delamination, and other similar problems, any one of which can cause the repair material to fail within a short period of time.

It is often difficult to determine the type of the original coating. This is particularly true where no record has been kept of the original material and its application. Table 15.3 shows many of the common coating types and areas where they are commonly used. However, even

TABLE 15.3 — Common Coating Types and Areas of Use

Area of Use	Coating Type	Environment
Exterior Iron and Steel	Oil Alkyd Urethane	Rural-Mild Industrial
	Oil-Based Phenolic	High Humidity-Mild Industrial
	Inorganic Zinc Vinyl Epoxy	Corrosive Industrial - Marine - Severe Weather
Interior Iron and Steel-Metal	Alkyd Urethane Epoxy Ester	Mild - Industrial
	Vinyl Epoxy Chlorinated Rubber	Acid or Other Corrosive Fumes or Dust Fallout
Interior Storage Tanks Chemical Solutions	Vinyl Epoxy Phenolic Epoxy	Salt Solutions Dilute Acid Alkalies
Interior Water Tanks	Vinyl Epoxy Chlorinated Rubber Coal Tar Epoxy Coal Tar Asphalt	Water Immersion High Humidity
Interior Concrete Tanks or Reservoirs	Vinyl Epoxy Chlorinated Rubber	Water Immersion High Humidity
Interior Solvent Storage Tanks	Inorganic Zinc Epoxy Phenolic Epoxy	Solvent Immersion Fumes - Condensation
Exterior Concrete Plaster - Masonry	Cement Acrylic Vinyl Acetate	Mild - Industrial
	Vinyl Epoxy Chlorinatd Rubber	Corrosive Humidity
Exterior Wood	Oil Alkyd	Rural - Mild Industrial
Interior Plaster or or Wall Board	Acrylic Alkyd Chlorinated Rubber	Mild - Industrial
Wooden Floors	Alkyd Polyurethane	Mild - Industrial
Concrete Floors	Epoxy Ester Polyurethane	Mild - Industrial
	Chlorinated Rubber Vinyl Epoxy	Spillage Water - Dilute Chemicals

[SOURCE: Munger, C. G., Repairing Protective Coatings: Effect of Coating Types, Plant Engineering, Feb. 17 (1977).]

Corrosion Prevention by Protective Coatings

with this type of information, it is difficult to determine the character of the previous coating. The table does, however, help eliminate many of the possibilities that need to be explored.

For example, on the interior of storage tanks for some types of chemical solutions, the principal material on the surface may be either a vinyl, an epoxy, or an epoxy phenolic. With this information, and by knowing the particular characteristics of these three coatings, some deduction can be made as to the actual type of coating on the surface. A practical determination of compatibility can then be made by cleaning several square feet of the coated surface, treating the surface with the proper surface preparation, and then applying the repair coating material over that surface. In doing this, it will be possible to determine: (1) if the coating is going to damage the previous one (*i.e.,* cause it to swell, dissolve, or lose adhesion); and (2) whether or not the new topcoat will have adequate adhesion to the previous coating. If it does not, the new coating can generally be peeled from the older surface within a reasonable period of time after the new coating has been applied (*e.g.,* after an overnight dry). Even if the new topcoat cannot be physically peeled from the surface, if it shatters, pops, or scales from the surface when cut with a knife, then it is obvious that the new material is incompatible with the original coating.

Although the new topcoat may appear to adhere satisfactorily to the undercoat, it often tends to decrease the adhesion of the original coat to a substantial degree. Again, if a knife is used to cut the surface and the entire coating tends to pop, shatter, or scale from the surface, the basic adhesion of the previous coat has been damaged and future failure can be anticipated.

While such tests do not indicate the exact type of the existing coating, they do indicate the compatibility of the new topcoat over the previous coating. If the test shows none of the possible difficulties, there is a good chance that the two coatings are compatible and that the new topcoat will make a satisfactory repair over the previous coating.

Another procedure has been developed, however, which provides a much more definite identification of the previous coatings than the simple test indicated earlier. It consists of a series of tests developed at the Naval Civil Engineering Facility at Point Hueneme, California which provides a chemical identification of the unknown coating on the surface. When this procedure is used, a reasonably accurate identification can be made, and selection of the coating can proceed from that point.[6]

Table 15.4 lists a number of different common types of protective coatings and the areas where they are generally used. These may be combined into various general coating categories that usually can be treated in a similar manner when repair is necessary. While the various classes of coatings listed in Table 15.4 have all been described in previous chapters, they nevertheless should be reviewed in terms of their repair characteristics.

Solvent-Drying Lacquers

Solvent-drying lacquers are made from resins that are fully polymerized prior to being formed into a coat-

TABLE 15.4 — Common Coating Types—Maintenance Procedure

Class of Coating	Type of Vehicle	Maintenance
Air-Reactive Coating	Oil, Varnish, Enamel, Alkyds, Oil-Modified Phenolic, Epoxy Ester	(1) Remove all oil or grease. (2) Wash/scrub with clean water. (3) Power sand or light brush blast. (4) Brush first coat on repair areas. (NOTE) If general surface has heavy or medium chalk, sanding may be eliminated.
Internally Reactive Coatings (Catalyzed)	Epoxy Amine, Epoxy Polyamide, Epoxy Phenolic, Coal Tar Epoxy, Urethanes, Polyesters	(1) Remove all oil or grease. (2) Wash/scrub with clean water. (3) Power sand or light brush blast. (4) Brush first coat on repair areas. (NOTE) Surface of coating must be abraded for proper topcoat adhesion.
Moisture-Reactive Coating	Moisture Cure Urethane	(1) Same as Internally Reactive Coatings.
Solvent Drying Lacquers	Vinyl, Vinyl Acrylic, Coal Tar, Asphalt, Nitrocellulose	(1) Remove all oil and grease (2) Wash/scrub with clean water. (3) Reapply solvent dry coatings.
Water-Based Coatings	Vinyl Acetate P.V.A., Vinyl Acrylic, Acrylic, Coal Tar Emulsion, Asphalt Emulsion	(1) Remove all dirt, dust, and grease (2) Wash/scrub with clean water. (3) Reapply water-base coating.
Inorganic Zinc Coatings	Water-Based Silicate, Solvent-Based Silicate	(1) Remove all oil and grease. (2) Wash/scrub with clean water. (3) Light power sand or light sweep blast. (4) Reapply coating. (NOTE) Light abrasion of surface is necessary to open up new zinc. (NOTE) Inorganics will not adhere to any organic/oil-grease coating.

[SOURCE: Munger, C. G., Repairing Protective Coatings: Effect of Coating Types, Plant Engineering, Feb. 17 (1977).]

ing. There is no change in the resin after the coating is formed, and they remain permanently soluble in their original solvent. The drying process is merely one of the solvents evaporating from the coating, leaving the fully polymerized resins and pigment on the surface. These coatings are thermoplastic coatings, indicating that they are softened with heat and usually provide a tough, semi-rubbery, partially extensible film. Hard, internally polymerized coatings are therefore generally not recommended for repair of these materials.

The coatings that come under the classification of solvent-drying lacquers are generally single-package coatings, since they require no catalyst. They include vinyl coatings, chlorinated rubbers, cellulose lacquers, phenoxy coatings, vinyl acrylics, as well as coal tar and asphalt cutbacks. As discussed, all of these materials are soluble in their own solvents, even after several years of exterior exposure. Repair coatings of similar types, *e.g.*, a vinyl over a vinyl, will soften the previous undercoat and will allow some softening of the old material at the interface between the new and the old coating. Thus, the new material generally has good adhesion to the old undercoat. This is an important characteristic of the thermoplastic-type of coating since it eliminates many of the difficulties caused by the interface between the new and the old coatings. Intercoat adhesion is generally good because of the permanent solubility of these materials.

The general repair procedure suggested for the lacquer-type coatings is as follows.

1. Remove all oil and grease or similar contaminant with solvent.

2. Wash the surface of the coating thoroughly with water and detergents, scrubbing with brushes or sponges to remove the surface contamination physically.

3. Wash the surface with clean water.

4. Allow the surface to dry thoroughly after which a similar type of lacquer coating should be applied over the clean surface.

The repair coating thus applied should have all of the properties of the original coating, and the renewal will substantially increase the life span of the coating.

Thermoset, Internally Reactive Coatings

In the classification of thermoset, internally reactive coatings, all coatings revert to insolubility as a result of internal condensation or polymerization shortly after application. They are, for the most part, two-part coatings which have to be mixed just prior to application. They change from a liquid to a solid due to the internal polymerization reactions. Once formed, these materials are insoluble in most common solvents, and therefore the solvents in a repair coat would not necessarily strike into the previous coatings and allow adhesion in a way similar to lacquer-type coatings. Many coatings in this classification are of a high solids type. Nevertheless, there is usually a resin-rich layer on the surface which makes the penetration of the topcoat much more difficult. This surface layer is not necessarily any more highly polymerized than the remainder of the coating, but, because of the thin resin-rich surface, it makes penetration difficult. Thus, the topcoat must be completely compatible with this surface.

The coatings which are usually included in this classification are amine epoxies, polyamide epoxies, phenolic epoxies, coal tar epoxies, bitumen epoxies, two-component polyurethane coatings, and some polyester coatings. These coatings are mostly thermosetting, and provide a hard, somewhat brittle film over the surface. Polymerization may continue for some time with this type of coating, making it more resistant after it has been exposed for a long period of time.

The recoating of these internally reacted coatings usually involves the following procedure.

1. Remove all oil and grease and similar contaminants from the surface.

2. Wash the surface, using a sponge or brush, with clean water to remove the soluble contaminants from the surface, as well as any chalk or similar surface material.

3. Allow the surface to dry, and abrade it lightly by sand sweeping or by hand or power sanding. This step is necessary for this coating classification to break up the surface of the coating and allow some mechanical adhesion in addition to the normal adhesion of the compatible topcoat.

4. If possible, it is preferable to brush the first coat of the repair material over the surface since this aids in wetting the surface with the new coating.

Using this procedure, the topcoats of these coatings generally have good compatibility with the old coating being repaired, and a continuing life span of the total coating can be expected following repair.

Air-Reactive Coatings

Air-reactive coatings comprise a comparatively large group of coating materials due to the many variations available. Most of these materials are based on drying oils, or are resins combined with drying oils that have similar air-curing properties. Air-reactive coatings have the characteristic of hardening and polymerizing through reaction with oxygen from the air. This reaction first occurs on the surface, and then, as the oxygen penetrates the surfaces, passes on through the depth of the coating. Due to the surface reaction, the surface is often more highly polymerized and more highly insoluble than the underlying coating, which is less reacted. This surface layer is relatively insoluble, particularly in the mild solvents generally associated with these coatings, and therefore the repair coating must not only be compatible with, but it must also thoroughly wet the previous material in order to obtain satisfactory adhesion. It cannot rely on solvent softening of the previous coating for its adhesion. Coatings in this classification are enamels, linseed oil paint, medium- and long-oil alkyds (as well as alkyds that are modified with chlorinated rubber), silicones, acrylics, phenolics, and epoxies.

The repair recommendations for air-reactive coatings are as follows.

1. Thoroughly remove any oil, grease, or other similar contamination from the surface.

2. Wash the surface with clean water, while scrubbing with a brush or sponge. This not only removes soluble contamination from the surface, but also helps remove any chalk or dirt that has accumulated on the surface over

a period of years.

3. Lightly power- or hand-sand the surface of the coating in order to break up the surface and allow penetration of the new coat. The surface may also be lightly sweep blasted with fine sand to gently abrade the surface.

4. Apply the compatible topcoat material to the lightly abraded surface, preferably by brush.

These coating materials generally have good wetting characteristics. Nevertheless, the physical working of the coating into the old surface aids in that wetting property. Because these materials react with oxygen from the air, some time is required for them to cure sufficiently for additional coats to be applied. The proper drying of each coat after it is applied is important in the resistance of these materials.

Moisture-Reactive Coatings

The only common type of coating in the classification of moisture-reactive coatings is the moisture-cured polyurethanes. The main characteristic of this coating type is that it has a tightly crosslinked and polymerized surface. This surface is usually a glossy one, with the gloss being retained over long periods of time, indicating a resin-rich surface. This surface is very insoluble so that it is necessary to break up the surface and allow mechanical, as well as chemical adhesion. The general repair procedure with the urethane-type coatings is similar to that outlined for the thermoset, internally reactive coatings.

Water-Base Organic Coatings

Water-base organic coatings are usually emulsions of different resins. Some of them are water dispersions or emulsions of polyvinyl acetate, polyvinyl acrylic copolymers, acrylic resins, coal tars, or asphalts. They also include water-based thermosetting-type coatings, such as epoxies and coal tar epoxies. The thermoplastic-type materials have characteristics somewhat similar to the lacquer-base materials. The epoxy emulsion coating, and even some water-base alkyds, have some of the characteristics of the thermoset-type coatings. Although these characteristics are quite different, all of these materials must coalesce out of the water dispersion and form a film.

Water-base coatings are not as dense, nor do they have the same type of film as that formed from a solvent-based coating. They are usually rather heavily pigmented and have a rather flat low-gloss surface which aids in coating repair. Some of these materials do tend to chalk, and, because of their water base, tend not to wet the chalky surface as well as some other coating classifications. However, these coatings can usually be repaired and recoated with little difficulty. The repair procedure, as follows, is reasonably simple.

1. Remove oil and grease from the surface.

2. Scrub the surface with clean water to remove all of the soluble dirt and chalk.

3. Allow the surface to dry and then reapply the water-base coating.

Inorganic Zinc Coatings

Inorganic zinc materials are of a completely different class than the organic coatings. They may be water-based (composed of soluble sodium, potassium, or lithium silicates), or they may be the solvent type, which are solvent solutions of organic silicates such as ethyl silicate. When these coatings are cured to insolubility, there is little difference between the two types of films, except for the amount of zinc and the various fillers that may be added to the coating.

Either a water-based or a solvent-based inorganic zinc can be applied over either inorganic zinc-type coating. When applied without topcoats, these materials usually remain on the surface several years before requiring repair. The surface at that time is a hard, rocklike zinc surface that is no longer soluble in either water or the solvents of the solvent-base inorganic. As mentioned, these coatings continue to react over long periods of time with oxygen and carbon dioxide from the air. The carbon dioxide and oxygen reaction products gradually penetrate throughout the coating to form both zinc oxide and zinc carbonate on the surface and within the coating. The resulting product is hard, with the zinc particles encapsulated in a matrix of zinc silicate combined with zinc oxide and zinc carbonate.

It usually is not satisfactory to apply an inorganic zinc coating over an old inorganic zinc film without lightly abrading the surface to break up the zinc carbonate-zinc oxide film on the surface, and to expose the zinc in the old film so as to allow it to combine with the silicates in the new vehicle. In this way, the old or original zinc coating combines with the new repair coating, creating a chemical bond between the two. If the surface is properly abraded, good adhesion between the new and the old coatings usually results.

The recommended procedure for recoating the inorganic zinc coatings is as follows.

1. Remove oil and grease from the surface.

2. Wash the surface free of any contamination with clear water, using brushes or other means to release the contamination on the surface. (Any organic type of contamination will prevent the proper adhesion of the repair coating.)

3. Once the surface is free of contamination, it should be lightly abraded, preferably by sweep blasting the surface with a fine abrasive. The surface may also be abraded by hand or power sanding. Since inorganic zinc coatings are generally 3 to 4 mils thick, even after several years in the atmosphere, the abrasion should be light, yet sufficient to remove the carbonate film from the surface while not removing the zinc coating itself.

4. Once the surface of the coating is abraded, the new repair coat of inorganic zinc may be applied.

In the discussions of the various coating types, it has been assumed that the original coating is a continuous film with no general corrosion breaking through the surface. If there is corrosion on the surface, the repair procedure is much more complicated.

Type of Substrate

As previously discussed, the substrate is an important part of any coating system. This continues to be true where coatings need to be repaired, since the substrate may be exposed at any break that occurs in the coating. The substrate is relatively unimportant if the coating on the surface has good adhesion, is intact and unbroken, and if the repair involved is only an addition of another coat to improve appearance and add additional life to the coating. Unfortunately, this is seldom the case in any corrosive atmosphere. There usually are breaks in the coating where not only the coating substrate interface is exposed, but also where there is active corrosion and perhaps undercutting of the coating. In this case, the repair coating must not only be compatible with the previous coating, it must be compatible with the substrate and must thoroughly wet, penetrate, and adhere to any area of the substrate that is pitted or corroded.

The principal substrates encountered are steel, cast iron, galvanized steel, aluminum, wood, and concrete. Each of these corrodes or disintegrates in a different manner, and each presents a different surface over which the repair coating must be applied.

Steel

Steel is the most commonly coated substrate. Coatings are the primary method of protecting steel, so that much coating maintenance and repair involves steel surfaces. Where a steel surface has been coated, and where the coating has been in service for a considerable period of time, there is usually some degree of corrosion and coating failure taking place. Rust tubercles form wherever there are breaks in the coating, even at minor breaks such as pinholes. The rust spot may be small, even pinpoint in size, or it may cover an extensive area. Pitting corrosion may be undercutting the coating, with corrosion building underneath the coating. In each of these cases, the basic steel substrate will be pitted.

Where repair is necessary, the corrosion products at the breaks in the coating must be completely removed. The pit must be cleaned free of corrosion so that the repair coating can properly adhere to the steel surface. However, whether large or small, the coating surrounds each one of these areas. Thus, in order to make the repair, the corrosion product and the coating must be removed from the surface back to the point where the coating is solid, intact, and tightly adhering to the surface.

The removal of the coating around the corroded area is extremely important, since the adhesion of many coatings is released in the areas of corrosion, and unless the loose, poorly adhering coating is removed, failure will continue at the same spot. It is also essential that the coating be tapered from the corrosion area out to the solid coating, creating a feather edge at the junction between the steel and the old coating. This provides a transition area where the repair coating extends from the corroded steel surface out over the existing coating. This is not only necessary for proper repair, but also gives the repair job a much better appearance, since there are no jagged, rough edges to give the appearance of a patch job.

The repair of coated steel surfaces is a difficult and time-consuming task. Unfortunately, many plant managers, maintenance personnel, marine operators, and ap-

plicators too often take coating repair for granted. If there is coating damage (*e.g.,* blisters, rust, undercutting), the immediate reaction should involve chipping and scaling, or brush blasting to remove some of the corrosion products and failed coating, and to apply additional coats of the same material over these areas. In situations where there may be some areas of pinpoint rusting or undercutting while the remainder of the coating is in good condition, the repair is often a stopgap measure to provide some immediate improvement in appearance, while the primary problem may be aggravated by damage to the original coating from the repair procedures used. There are many difficulties and hazards in repairing coatings over steel surfaces which should be understood before any repair procedure is followed.

Many repair difficulties stem from the methods of repair used. Most of these methods go back to lessons learned about the application of coatings over original surfaces where the surface is cleaned to white or near-white metal and then the coating applied. Sandblasting or grit blasting is the preferred method of obtaining such a surface, so, when it comes to repair, the first thought is to sandblast the damaged areas of the coating and then reapply the original coating over the clean surface. Unfortunately, the preferred procedure for new work can prove damaging where a coating is to be repaired. First, the steel structure, tank, ship, or bridge now exists as a completed unit, so that blasting must be done by hand, using sand or synthetic grit, and usually from some sort of a scaffold. Steel grit or shot are impractical because of the loss of the materials. Spot hand-blasting is therefore included in many specifications where a coating is to be repaired.

There are, however, repair problems associated with spot blasting. First, the blasting procedure uses impact as the method of cleaning the surface and removing the coating. The impact of the sand grain, and the scouring of the sand grain as it disintegrates on impact, provides the cleaning process. Impact of the sand grain is one of the inherent problems of coating repair. The procedure often outlined in the specification is to spot blast the rusty area, feather the edge of the coating surrounding the area, and sweep blast the rest of the coating to remove dirt, chalk, and other surface contaminants, in addition to lightly abrading the surface of the coating. This light abrasion is necessary for alkyds, epoxies, and polyurethanes where the surface requires some tooth or mechanical advantage when the repair coating is applied. The blasting procedure to remove the area of rust is good. However, to be effective, all corrosion products must be removed (white metal) since the corroded area is more sensitive to continued corrosion than the original metal underneath the remaining solid coating.

The repair problems primarily occur at the area of the interface between the blasted areas and the old coating where the edge is supposed to be feathered and the coating beyond that slightly roughened (Figure 15.3). In these areas, the impact of the sandgrains (particularly where highly cross-linked and brittle coatings are involved) can cause fractures of the coating on impact and have a tendency to loosen the adhesion of the coating in the area

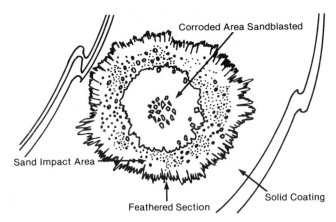

FIGURE 15.3 — Interface between blasted area of corrosion and solid coating, with feathered area between. (SOURCE: Munger, C. G., Practical Aspects of Coating Repair, Materials Performance, Vol. 19, No. 2, p. 46 (1980).]

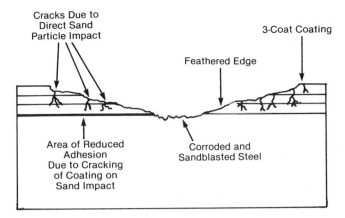

FIGURE 15.4 — Cross section of coating showing a corroded area that has been sandblasted. Note the tapered, feathered edge and the star cracks caused by heavy sand grain impacts. [SOURCE: Munger, C. G., Practical Aspects of Coating Repair, Materials Performance, Vol. 19, No. 2, p. 46 (1980).]

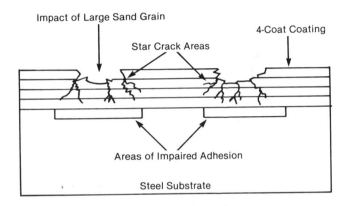

FIGURE 15.5 — Star cracks caused by the impact of heavy sand particle. [SOURCE: Munger, C. G., Practical Aspects of Coating Repair, Materials Performance, Vol. 19, No. 2, p. 46 (1980).]

of impact (Figure 15.4).

This poor adhesion is often evident when the repair coating is applied. The edges of the coating swell and curl, creating an area of questionable effectiveness, as well as one of poor appearance. The feathered edge is the most subject to this action, since it is the part of the coating most subject to sand action other than the corroded area itself. There also may be star cracks in the coating area just beyond the feathered edge from impact of the sand grain. This makes a porous spot in the coating and an area of questionable adhesion (Figure 15.5). These spots can then become focal points for the failure, as evidenced by the fact that many such repaired sections soon fail again; beginning, not where the previous rust occurred, but in a circle surrounding the area where the old coating was feathered (Figure 15.6).*

Much of the failure of organic coatings is due to the impact of the blasting media impairing the adhesion of the original coating. This is combined with the difficulty of the repair coating to sufficiently penetrate the original film at the star cracks, and in the feathered area to gain new ahesion to the substrate. The feathered area is also a transition zone between the sandblasted area and the solid coating beyond the feathered edge. As the coating dries and cures, stress may further reduce the adhesion in that particular transition zone, which often curls the edge of the old coating.

Considering these difficulties, the following repair procedure is recommended.

1. Use very fine abrasive, such as − 60-mesh sand, so as to eliminate the impact of large particles on the coating.

2. Blast the corroded area clean, preferably to white metal, by normal blasting procedures. Following this, however, back off the blast nozzle to 3 or 4 feet from the surface and blast the feathered edge from this distance at an oblique angle to the surface. This further reduces the impact of the sand grains on the coating and tends to wear the coating away in the feathered edge area.

_____
*See color insert.

3. To clean and slightly roughen the solid coating surrounding the repair, continue to use the very fine sand, sweeping the existing coating both at a distance from the surface and at an oblique angle to prevent direct impact of large sand grains. The *incorrect* procedure is shown in Figure 15.7. With the sand blast nozzle held close to the surface, any heavy sand grain can star crack the coating, causing a spot of incipient failure. The heavy impact is satisfactory for removing corrosion deposits. However, for feathering edges and slightly roughening existing coated surfaces, the blast nozzle should be removed to a distance of 3 to 4 feet from the surface, and the blasting done at an angle oblique to the surface, as shown in Figure 15.8.

4. Apply a very liquid, highly penetrating primer to the surface to penetrate any possible edge areas of poor adhesion in the feathered area, and any possible star cracks in the body of the coating. Such a primer should be, if possible, a low viscosity and low molecular weight material that is thoroughly compatible with the existing coating and steel surface, and which has high wetting characteristics for each.

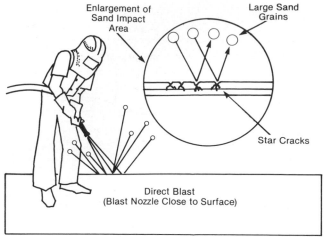

Incorrect Method of Sweep Blasting for Coating Repair

FIGURE 15.7 — Incorrect blasting procedure.

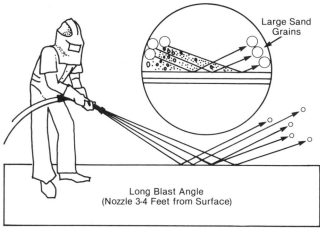

Correct Method of Sweep Blasting for Coating Repair

FIGURE 15.8 — Correct blasting procedure.

5. Mechanical application of the primer (preferably by brush) aids in the wetting and penetration of the primer into the porosity of the feathered section of the coating.

Depending on the severity of the coating breakdown, other methods of surface preparation can be used. If the coating has only started to show minor failure in isolated spots, blasting might cause more damage to the solid coating by sand ricochet and accidental impact than the benefit gained. Power sanders, while not leaving the metal surface as clean as sandblasting, can be used to effectively clean and feather isolated corrosion spots (*i.e.*, pinpoint rust spots or small coating breaks) without damage to the large surrounding areas of solid coating. A rotary power sander does an excellent job of feathering the edge of the coating around a break. Impact tools, such as needle guns, are effective for small complex corroded areas (*e.g.*, rivets, boltheads, welds, etc.), but are not as effective in feathering the coating edge as a rotary power sander. The combination of both tools may be more effective than either one used alone. Should the mechanical method of cleaning be used, it is essential that a highly

penetrating primer be brushed into the repair area for the most effective results.

One of the most common repair problems is that of cleaning and recoating old steel structures, tanks, ships, or offshore drilling or production platforms, where the coating has broken down on the edges of structural steel shapes or over welds, or where rust scale is forming under the plane areas of the coatings. This is a typical undercutting type of failure, where the corrosion begins at an edge or a break and gradually works its way under the coating, forming a hard black rust scale that can be as much as 6 millimeters (1/4 in.) or more in thickness.

Figure 15.9 shows typical edge undercutting of a coating in a marine atmosphere. This type of corrosion can be quite severe, completely eating away the steel under the coating. Figure 15.10* shows an edge where the steel has been completely corroded away, leaving only the coating intact.

Rust scale is primarily suboxides of FeO-OH, with some red Fe_2O_3 on the surface and in the fractures. The extent of the undercutting of an organic coating depends on the strength of the adhesion of the coating. The stronger the adhesion, the less undercutting and the easier the repair. Where the adhesion is marginal, the surface preparation necessary to clean the corroded area free of rust may damage the apparently good coating to the point where no repair other than complete renewal of the coating is practical. Where the coating adhesion is good, the repair procedures should be as described earlier, followed by mechanically working a highly penetrating primer into the cleaned and feathered area, building the topcoats to the required thickness, and, if the repair areas are extensive, applying a topcoat overall.

Heating the Repair Area. One procedure that should be mentioned during a discussion of repairing coated steel surfaces is the use of heat to warm the surface, to remove any moisture that may be on the surface, and to eliminate any volatile corrosive materials that might be present. In many industrial plants, particularly where chemicals are involved and the coating repairs are in areas where a thorough removal of the corrosion on the steel by sandblasting is impossible, the use of heat to warm the surface of the metal, after the best possible surface preparation, can be beneficial in increasing the coating resistance and adhesion.

Once the surface is prepared, the steel area can be heated with an acetylene torch or other heating device to the point where it is hot to the touch, but not to the point where the coating that surrounds the area is blistered. Once the surface has reached this temperature, the primer should be applied by brush, applying a full wet coat to the surface so that it will remain wet sufficiently long to penetrate into any of the irregularities of the surface. Heating of the steel in this manner reduces the viscosity of the coating and allows it to penetrate to a much greater degree than would otherwise be possible.

However, when the surface is hot, some care must be taken not to swell or blister the coating that is feathered

---

*See color insert.

Corrosion Prevention by Protective Coatings

FIGURE 15.9 — Typical edge corrosion and undercutting of a coating in a marine atmosphere.

around the area of corrosion. The warm surface may allow the topcoating to penetrate faster than would ordinarily be the case. Adhesion between the two coating surfaces is usually increased by the heating process, since the interface is thoroughly free of any dampness or moisture. On any critical area of corrosion where such a heating procedure is possible, the procedure is recommended since it does increase the adhesion and coating resistance over the corroded steel surface.

Galvanized Steel

Galvanized surfaces that have been coated, while usually not severely corroded, often require repair because the coating has started to chip and scale from the galvanized surface. The primary reason for this is the use of oil-base, alkyd, or similar coatings on the original galvanized surface without a zinc-compatible primer. When such a condition exists, the best procedure is to remove all of the previous coating from the surface. This usually can best be done by chipping and scraping, since once the coating reaches this condition, it can be easily popped from the surface. Sandblasting or similar procedures are not recommended for removal of such a coating because of the damage to the zinc surface by the blasting process.

Where galvanizing remains uncoated for a considerable period of time and has begun to fail, the galvanizing usually fails by pinpoint rusting, with the rusting progressing over a period of time until it covers the entire surface. For practical repair, the galvanized surface should be coated prior to the time that the pinpoint rusting reaches the continuous stage. In this case, the repair procedure is to blast the surface lightly, removing all of the pinpoint rusting, and, if possible, touching up the previously corroded areas, using an organic zinc coating as a base coat. This renews the zinc in this area and allows the following organic coating adhesion to have a continuous corrosion-resistant base. The general procedure for making repairs to a galvanized surface is as follows.

1. Make certain that the zinc surface is free of all previous coating that has chipped and scaled. If corrosion has started in bare areas on uncoated galvanized surfaces, lightly blast or hand-sand the surface free of corrosion.

2. Touch up the previously coated areas with an organic zinc coating and allow it to cure as recommended in the instructions.

3. Select a primer for the surface that is compatible with the zinc and which will not react with it. Many such primers are available, usually based on inert resins such as vinyls and epoxies.

4. Brush the primer onto the surface to provide complete wetting and the best possible adhesion.

5. Following the primer, apply compatible topcoats of the type desired.

Cast Iron

Cast iron, being ferrous, has many of the properties of steel. However, since it has a large, crystal structure, it appears to be more porous than the conventional steel surface. Nevertheless, for repair purposes, the surface is quite similar to that of steel, with only one or two exceptions. Where there are breaks in the coating, the cast iron may be pitted, and the pits in cast iron are generally deeper than those in steel. It is necessary to thoroughly remove the corrosion product from the pits and to abrade the pits down to the solid cast iron surface. The pits in cast iron are often filled with graphite that may appear reasonably solid. This is called the *graphitic corrosion* of cast iron. If it exists, the material needs to be removed by blasting down to the solid cast iron surface.

The cast iron surface may also absorb moisture and other volatile materials into the surface to the point where, if a coating is applied over such contamination, it will fail again quite rapidly. In this case, it is recommended that the cast iron surface be heated with a blow torch or acetylene torch in the manner previously recommended for steel.

The general procedure for coating cast iron is as follows.

1. Remove all of the corrosion products from any breaks in the coating, feathering the coating, as recommended for steel.

2. Warm or heat the surface to the point where it is hot to the touch.

3. Once the surface is heated, the primer should be applied by brush to the warm surface. The primer should be applied in a full wet coat for maximum penetration to be obtained.

4. The selected topcoats can follow as soon as the primer has properly dried.

Aluminum

Coated aluminum is usually covered with a factory applied coating that has been applied under controlled conditions and usually baked to obtain maximum adhesion. Such coatings are usually rather thin. However, they are primarily an acrylic or vinyl acrylic which has excellent weathering resistance. For the most part, such coated aluminum surfaces are primarily recoated or repaired where damage has occurred or where the coating has weathered to the point where the general recoating of the surface is necessary. For the most part, there is little or no corrosion of the aluminum on such surfaces. Thus, maintenance and repair usually only involves the application of an additional coat or coats of a compatible topcoat over the previous coating. In this case, the required surface preparation primarily involves a thorough washing to remove any chalk, dirt, grease, or other impurity, followed by the recoating procedure.

Occasionally, coated aluminum surfaces are corroded. The corrosion product is usually a white aluminum oxide which may form in small pinpoints in and around the break in the coating. On aluminum surfaces such as those on the aluminum deck houses on ships, particularly where such structures are joined to a steel hull, aluminum corrosion can be quite extensive with undercutting and considerable pitting of the aluminum surface. Where aluminum is corroded, the surface should be abraded free of the corrosion product and the abrasion should be light. If blasted, the blasting abrasive should be very fine and the nozzle held a considerable distance away from the surface. Aluminum is a soft metal which both cuts and warps readily.

The general procedure for coating coated aluminum surfaces is as follows.

1. Thoroughly wash the surface of the coating with clean water to remove dirt and contamination. Oils and greases should be removed by a detergent-solvent-water combination prior to washing with clean water.

2. Lightly abrade any corroded areas to remove all of the white corrosion product. This can be done using a light sweep blast and fine sand. Hand sanding or a fine power disc may also be used on small areas.

3. Apply aluminum-compatible primer over any corroded areas of the surface.

4. Apply the proper compatible topcoats. These are usually either a vinyl or vinyl acrylic solvent-dry coating.

Other Metals

The coating of other metals, *e.g.,* copper, magnesium, stainless steel, and similar alloys, is usually an exception, so therefore the repair of coatings over these surfaces is likewise uncommon. Any repair procedure that may be required would be similar to that used for zinc or aluminum.

Wood

Coated wood surfaces, particularly around industrial plants, marine areas, and similar atmospheres, seem to be in constant need of repair. Wood is substantially different from the metals described earlier, as it is not a uniform or consistent surface. Coatings that are applied over wood fail primarily due to checking and cracking or chipping and blistering.

It should also be noted that coatings that are entirely satisfactory for metal surfaces may not be and generally are not practical for wood. Metal coatings initially may have good adhesion and appearance over wooden surfaces. However, because they are generally of high molecular weight, they do not penetrate the wood surface well. With the continual expansion and contraction of the wood, the coating tends to crack because of the different rate of expansion of the summer and winter grains.

The coatings that prove the most satisfactory, as previously discussed, penetrate the wood, cure within the wood pores, and therefore tend to have strong adhesion to the wood and are thus able to move with the wood surface to a much greater degree than coatings which merely lie on the surface. There are many water-based coatings on the market that are recommended for wooden surfaces, many of which work well over a previously coated surface that has no cracking, peeling, or scaling on the bare wood. However, these materials do not penetrate. Thus, it is recommended that where such materials are used, a penetrating primer should be applied to the bare wooden surface prior to the application of water-base topcoats.

With wood surfaces, it is always beneficial to apply relatively thin coats and to maintain and repair the coating before the surface becomes damaged by cracking, peeling, blistering, or scaling. The maintenance program should be scheduled so that surfaces are inspected at regular intervals, in order to renew the coating with an additional relatively thin coat prior to the exposure of any of the wood. This procedure is preferable to one that allows the wood to reach a cracking or scaling stage, when it is difficult to prepare the surface properly.

Wooden surfaces (Figures 15.11 and 15.12) are extremely difficult to repair and recoat because of the problem of properly preparing the surface. It is not possible to sandblast such surfaces without damaging the soft grain area of the wood. Chipping and scraping is only partially successful, and even heating the painted surface in order to scrape the worn paint from the wood is a long, difficult, and costly job. Figure 15.13 shows a chipped and scaled wooden surface. While the chipping and scaling was done as thoroughly as possible, it can be readily seen that the life of any coating applied over such a surface is in jeopardy.

Wooden surfaces are difficult to maintain. Where wood is exposed, the following general repair procedures are suggested.

1. Chip and scrape and hand- or power-sand the areas where the coating is failing and the wood is exposed. It is essential to make certain that all loose coating has been removed. On the other hand, the surface should not be oversanded, as it is easy to cut deeply into the wood in the areas where the coating is removed. This is particularly true when using a power sander.

2. The unbroken surface of any remaining coating should be lightly sanded and all dirt and chalk removed.

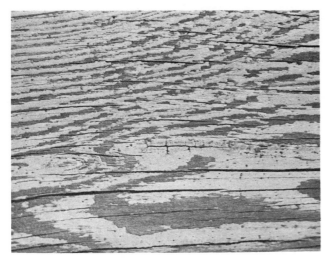

FIGURE 15.11 — Coating failure over a flat grain wooden surface.

FIGURE 15.12 — Coating failure over an edge grain wooden surface.

FIGURE 15.13 — Chipped and scraped wooden surface prior to recoating.

3. Apply a proper penetrating primer to the surface, working it well into the wooden surface by brush.

4. Once the primer has thoroughly dried, apply a compatible topcoat over the surface. It is not recommended that heavy coats be applied, but that thin coats be used, and that these be applied at closer intervals for proper maintenance.

Concrete

The repair of concrete surfaces requires one of the most difficult coating repair procedures to be followed. The problems are those inherent in the coating of concrete, as described in Chapter 11. In addition, however, the repair of concrete coatings usually involves a number of contaminants that may react with or penetrate into the concrete surface. The problem is complicated because of the vast number of areas where concrete is used in the coated condition. Examples of such areas are floors (which are undoubtedly the most difficult areas for coating repair), walls, ceilings, concrete pump bases, drainage ditches, sumps, etc.

That the coating and maintenance of concrete surfaces is difficult is evidenced by the many different and complicated methods of treating concrete floors, particularly where corrosion and wear is a problem. Treatments consist of procedures ranging from relatively thin coatings to heavy mastics and resinous cements. The ultimate in concrete floor protection is the use of acid-proof brick and cements. Quarry tile and brick floors are common in most areas where concrete is most affected by the service (*e.g.*, kitchens, food plants, areas of acid exposure, pump bases, etc.).

With all of the above variables, including the normal variables of a concrete surface, it is little wonder that repair of coatings on concrete surfaces is such a chore. The repair of coatings on overhead areas and concrete walls is much less complicated and less difficult than the repair of coatings on floors. In overhead areas and side walls, there is much less chance of damage to the coating and therefore a much greater possibility of an intact coated surface with only limited damaged areas.

Where the coating is intact, any surface contamination can be washed from the surface without penetration into the concrete itself. In fact, the primary reason for coating concrete walls and similar surfaces is to prevent splash and spillage of chemicals or other contaminants from penetrating into the concrete. This is particularly true in nuclear power and chemical plants where any penetration of radioactive material would cause substantial problems.

The primary cause of coating failure on walls is dampness from an outside source, *e.g.,* a roof or a drain, or because the concrete is below ground level. Where the concrete is damp from an outside source and the coating is porous, the dampness within the concrete will dissolve cementitious salts which will then be transferred to the surface as a white crystalline material. In so doing, the coating is pushed from the surface and damaged, usually to the point of requiring repair.

One of the first steps in the repair is to determine the source of moisture and to remove that source, if possible. If the source of moisture can be removed by waterproofing the opposite surface of the concrete, the damp concrete should then be allowed to dry prior to making the coating repair. In areas such as this, the efflorescence and the damaged coating must be thoroughly removed. This can be done by light sandblasting, power sanding, or the use of other power tools to remove any soft concrete down to the hard, dark gray concrete base.

If the area is damaged sufficiently so that coating alone will not be practical, a resinous concrete surfacer is recommended for the area. This is usually preceded by a highly penetrating concrete primer to obtain the maximum penetration into the concrete surface and bind the resinous surfacer strongly to the basic concrete. If the area can merely be recoated, a highly penetrating primer is also recommended to bind the coating to the surface as tightly as possible. With either of these procedures, concrete topcoats can then be applied.

Floors, as previously mentioned, are extremely difficult to repair. They are subject to water, chemicals, oils, and greases of all types. Not only are these materials spilled on the concrete surface, but the surfaces are also subject to wear by motorized equipment and foot traffic. The worn and damaged areas in the coating accumulate contamination which is usually worn into the surface. Contaminants can consist of pure water, which tends to etch the concrete surface; carbonated water, which will also etch concrete; other acids, such as acetic acid; acid salts; alkalies; various dry chemicals, such as those used in the fertilizer industry; and oils and greases. Vegetable and animal oils and greases are particularly difficult in the food industry where these materials are commonly encountered.

On walls and ceilings, where the coating is intact and tightly adhering to the surface, and on other areas that are not subject to damage, there is little problem of repair since the contaminating materials can be washed from the coated surface. It is only in the damaged areas where repair problems exist.

The first step in any repair procedure is to thoroughly inspect the area, and determine the type of coating damage and the extent of the damage to the substrate. Sources and types of contamination should be identified, if possible. Following the inspection to determine the condition of the coating and substrate, it may be well to test the substrate areas to determine suitable surface preparation methods. A four-step procedure has been proposed by an NACE Task Group T-6G-21 on Surface Preparation of Contaminated Concrete. The testing of substrate areas is outlined as follows.

Testing of Substrate Areas

The following four-step procedure can be used to determine a suitable preparation method:

Step 1. Treat a small area [0.09-0.9 m²ft² (1-10 square feet)] with a 1:1 solution of 36 percent concentration of commercial muriatic acid in water. If acid bubbles and leaves a clean surface, this treatment may be used for cleaning. If there is no reaction, try Step 2.

Step 2. Scrub a similar small area with a 20 percent concentration of lye solution, rinse and then treat as in Step 1. If the area bubbles and leaves a clean surface, this method may be used. If there is no reaction, try Step 3.

Safety Precaution: Workmen handling the lye solution, acid mixing, spreading and other operations must wear safety goggles, protective clothing, and rubber gloves and boots.

Step 3. Scrub a similar small area with a chlorinated solvent (trichlorethylene) to which has been added a strong compatible solvent. Do not let surface dry. As soon as thoroughly scrubbed, wash with clean water. This will remove both the solvent and the dissolved grease. Treat as in Step 1. If the area bubbles and leaves a clean surface, a high flash point solvent with low volatility [anything over 39°C (100°F) flash point] can be used for cleaning. If there is no reaction, try Step 4.

Safety Precaution: The high boiling solvents such as trichlorethylene and perchloroethylene should not be heated or evaporated with an open flame torch, but should be removed by the use of a high velocity circulating fan, or in the winter, with a salamander that has an outside fresh air inlet duct to prevent recirculation of the perchloroethylene fumes through the combustion chambers. Although perchloroethylene is not flammable, the fumes from the chlorinated solvents can decompose when heated to high temperature and emit toxic fumes of chlorides.

Perchloroethylene or trichlorethane to extract fats and oils from concrete is good from a fire safety standpoint. It is very important to remember when using perchloroethylene or a chlorinated solvent of this type that an abundance of air circulation (fresh air) is required. The threshold limit value of perchloroethylene is 100 ppm. Materials such as benzene or toluene should never be used to remove fats and oils because they are extremely volatile.

Step 4. Remove at least 1.6 mm (1/16 in.) of surface using a chipping hammer, chisel, or bush hammer. Treat as in Step 1. If acid bubbles, the entire surface may be prepared by mechanical means with acid etching. If there is no acid reaction, remove more concrete and repeat acid test to determine depth of contamination. Suggested methods of removal include: power tool grinding, scarifying or scabbling, dry abrasive airblasting, and airless, centrifugal wheel blasting.

As a general rule, monolithic organic corrosion resistant floor surfacings and tank linings used to protect heavy duty maintenance or immersion service areas require a stronger base than thin coatings. Weak concrete at the surface in contact with the topping or lining could result in failure of bond by the following:

1. Shrinkage stresses formed during the hardening and curing of the resin corrosion-proof overlay.

2. Excessive stresses, generated by the differences in thermal contraction of the concrete and corrosion-proof overlay.[7]

Concrete Preparation. The shrinkage stresses mentioned earlier are not at all uncommon, and epoxy toppings and even simple epoxy coatings will often develop stresses sufficient to pull up sizable pieces of the surface concrete. This emphasizes the fact that the concrete over which concrete toppings and coatings are applied must be solid, hard, and with as high a tensile strength as possible. In many cases, where concrete surfaces have been exposed to chemicals of a number of types found in industry, or where concrete has been subject to sewage conditions, the concrete can have very poor or marginal strength.

Where sewage corrosion is a problem, concrete may have to be removed to a depth of 1.5 or 2 in. to reach the solid gray concrete underneath the corrosion product. The same is true of a number of other chemicals, particularly sulfuric acid or sulfates. Where conditions such as this exist, the concrete is usually sufficiently rough so that tensile testing is not practical and the concrete must be removed by blasting or by using concrete scarifying tools. Blasting of such concrete surfaces is the most practical way of getting down to solid gray concrete. The color *gray* must be emphasized since sulfates develop a yellowish-white corrosion product, and, unless the concrete is removed below the color zone, the topping or cement used to build up the surface will not adhere properly. There is usually an obvious line of color demarcation between the corroded concrete and the solid gray concrete.

If the concrete has been saturated with oils and fats or petroleum products (and this is more common with floors than with other areas of concrete), the use of the Elcometer Adhesion Tester (Elcometer Adhesion Test Model 106) is an aid in indicating the type of surface preparation necessary and whether or not it is necessary to completely remove the surface from the concrete floor. The Elcometer test will usually identify improper or poor cleaning of the surface, or whether or not the surface is sufficiently impregnated so that cleaning is impossible. Under these conditions, the adhesive used in the test will separate cleanly from the surface and will not remove any concrete with it.

A practical method for determining the adhesion of a coating or the tensile strength of the concrete surface is shown in Appendix 15A. Any time that the floor is completely and deeply saturated with oil and grease, complete replacement of the top 1/4 to 1 in. of concrete may be less expensive than applying the floor surfacing or coating and having it blister or peel from the surface after a short period of use.

In places where the concrete has been severely eroded due to chemical attack or physical abuse, the use of organic floor toppings and surfacers is recommended as the best method of repair. Where concrete is seriously corroded and an inch or more must be replaced, concrete may be used to build up such areas. The floor toppings and surfacings can then be applied over this new surface. Generally, however, where a damaged area is 1/8 to 1/4 or even 1/2 in. in depth, the organic (epoxy) floor toppings provide excellent adhesion to the tight concrete and provide a good abrasion- and chemical-resistant surface.

Blasting. Abrasive blasting is a good method of preparing concrete surfaces. It will not only remove the old coating which may be loose on the surface, but it will also, if properly used, provide a good surface for the repair coating. However, there are some precautions to use in connection with abrasive blasting of concrete. Concrete is relatively soft, so that, if heavily blasted, the surface can be etched away for a considerable depth before the operator is aware of it. Thus, the blast nozzle must be kept at a distance from the concrete surface and a brush-off blast procedure used. It is also recommended that a fine abrasive be used to prevent too great an etch of the surface.

Floors which have a sidewalk finish (*i.e.,* where a dense cement sand topping is applied over the basic concrete floor surface) are much easier to prepare by blasting than where the concrete has been surfaced without the topping. In the latter case, once the smooth-troweled surface has been broken, a sand blast will etch into the underlying area quite rapidly, causing pits which then have to be filled with a surfacer rather than with the coating alone.

Water blasting with abrasive may also be used and is helpful in removing dust and other contamination. However, the same problem exists with water blasting as with abrasive blasting. Fine abrasive should be used and attention paid to the blasting process so that the concrete etch is not too deep. High-pressure water blasting without abrasive can also be used. This is particularly effective where the concrete has not been badly corroded or where there are considerable areas of loose coating to be removed. Water blast without abrasives will not, however, blast down through concrete corrosion to the solid gray concrete surface in the same manner as where an abrasive is used.

Acid Etching. Acid etching can be used in connection with repair of coatings on concrete. It will help remove a number of chemical contaminants and will also provide a mild etch to the surface where the coating is broken or removed. Excessive etching should not be used, as it may react to some depth, causing loose aggregate on the concrete surface. This would then have to be removed before a coating or a topping could be applied. The procedure for acid etching and the precautions for its use are outlined in Chapter 9 on Surface Preparation.

Power Tools. Power tools for cleaning the surface of concrete should be used with caution, unless the surface of the concrete is to be scarified and removed to some depth. Power wire brushing or power sanding can cut the surface of the concrete sufficiently so that usual concrete coatings will not provide a smooth surface. Heavily etched or cut surfaces would then require a troweled concrete topping of the epoxy type to provide a smooth surface for coating.

Repair of Coatings on Concrete. The coating of concrete surfaces will never be eliminated because of the various types of service to which the concrete surface is subjected, and the many variables inherent in the concrete surface. Because concrete is not a uniform material and because of its various uses, each repair situation is unique and must be carefully studied in order to reach a satisfactory and long-lasting solution. In spite of this, there are some general recommendations, as follows, that can be made for the repair of coatings on concrete.

1. Remove all of the damaged coating, or coating with loose adhesion, back to a point where the adhesion is consistent and satisfactory. Feather the edge of the remaining solid coating.

2. Remove all of the concrete corrosion product, such as the white carbonate deposit on the surface, the whitish or yellow corrosion product of sulfates, or any loose or projecting aggregate from the surface. If chemical

action has occurred on the concrete, remove all of the softened surface back to clean, gray concrete. This may require chipping, scraping, or sandblasting the surface. If the surface is saturated with oil or grease, use a strong solvent combined with an emulsifying agent. Remove all contamination. Wash and dry.

3. If the coating has failed because of water in the concrete, the surface should be thoroughly dried. This can be done by a blast from an air fan moving across the surface, or the concrete can actually be heated using a warm air source. At this point, it should be determined whether or not it is possible to eliminate the source of moisture or water within the concrete. If this can be done, it makes the repair of the coating much more permanent and satisfactory.

4. If the source of moisture cannot be removed, the surface of the concrete should be dried and a highly penetrating, highly alkali-resistant primer applied. There are several of these products on the market. The primer should be applied by brush and should be applied rather heavily for penetration. However, any excess should be removed from the surface so that no heavy areas or puddles are formed. Penetrating primers of the type described are generally based on liquid epoxy resins of a low viscosity. They are internally reacted and may be either amine- or polyamide-cured.

5. If the concrete is subject to severe abrasion or chemical use, even though it may not contain moisture, the above penetrating primer is also recommended. If vinyl or chlorinated rubber coatings are to be used, a very dilute primer of the proper generic type should be applied.

6. If the concrete has contained contaminants and the surface is clean and dry, use the penetrating epoxy primer to prime and seal the surface.

7. Should the concrete be deeply eroded or corroded, leaving a rough surface, the application of a heavy surfacer over the penetrating primer is recommended. The surfacer must be compatible with the primer in every way and generally will be an epoxy-type product.

8. Once the concrete surface has been primed and is smooth, compatible topcoats may be applied overall.

Plaster Repair. Plaster or stucco is generally quite similar to concrete, particularly where the plaster is a cement plaster. Where such surfaces have been damaged or corroded, a procedure similar to that recommended for concrete is suggested. If, however, these surfaces are only for light service and have been coated with water-base or emulsion-type products (*e.g.,* vinyl acetate or a vinyl acrylic), the following procedure is recommended.

1. Thoroughly wash the surface free of all dirt, grime, grease, or other contaminant. This can be done with clean water or a combination of an emulsion of solvent and water, followed by clean water.

2. Once the surface is cleaned and dried, reapply the water or emulsion-type coatings.

Cracking. One of the serious problems concerning concrete is its tendency to crack. These cracks may be very fine surface shrinkage cracks or substantial working cracks. The fine cracks, since they are not working cracks, are usually covered in any repair by the repair coating.

However, substantial working cracks of any type must be repaired prior to applying the repair coating. The general procedure recommended is to ''V'' out the crack using a chisel or similar tool. Once this has been done, the crack should then be filled with an epoxy or similar concrete surfacer and troweled smooth and level with the surface. A penetrating primer is highly recommended prior to the application of the surfacer in order to penetrate the crack as deeply as possible.

Repair of Failures

Earlier in the chapter, several types of failure were listed, along with ASTM specifications for their evaluation. However, because of the importance of the proper repair of such failures, more detailed information is included in the following sections, both as to the cause of failure and the method of repair. This information has also been outlined in a paper entitled *Practical Aspects of Coating Repair.*[8]

Blistering

Blistering of an original coating can cause substantial difficulties in coating repair. There are three types of blisters.

1. The area underneath the blisters is dry and the blister itself is raised by gas. Such blisters can be caused by trapped solvents or, where immersed, by the formation of hydrogen gas in areas underneath the coating.

2. The blisters are filled with liquid, and there is usually little or no corrosion underneath the blister. While the surface of the steel is wet and the blister filled with a watery liquid, the steel remains bright and uncorroded. This is because water will pass through a semipermeable membrane such as the coating to a much greater and faster degree than oxygen. Until oxygen passes through the coating and into the water area, no corrosion can occur. Oftentimes, the primer for the coating is also inhibited, and the inhibiting agent in the coating tends to prevent corrosion under the wet blister.

3. There is both liquid and corrosion underneath the blisters. Usually, such blisters have been in place for some time and oxygen has gradually diffused through the coating into the liquid, creating the possibility of local corrosion cells underneath the blister.

Regardless of the type of blister, the repair of these areas is difficult at best. If the blistering is widespread (Figure 15.14), as may be the case on the underwater hull area of ships, in ship ballast tanks, in water storage tanks, and in other similar areas, the repair of coatings is extremely difficult because of the numerous breaks in the coating extending over a large area.

Any repair on a blistered coating requires removal of the blisters. This can be done by scraping in order to remove the raised areas of the coating, as well as any coating around the blistered area that may be loose or have poor adhesion. Blisters usually are removed by sandblasting, sweeping the surface to remove the blisters, and then feathering the edge of the coating around the blistered opening as previously described. Star cracks between the blister areas due to the use of heavy sandblast particles may be new focal points for corrosion, even though the

FIGURE 15.14 — Fine blistering of epoxy coal tar coating after short exposure to potable water (ASTM Rating 6, Dense).

repair coating is applied overall.

It has often been observed that once a generally blistered surface, even with an ASTM rating of "few," is repaired and replaced in service, the blistering is rapidly renewed, usually starting at the edge of the old blisters and moving out onto the previous coating. This occurs even if the repair coating is tightly adherent to the steel in the area of the previous blisters. With the base coat having questionable adhesion, blistering will continue under the old film, even though the new coating may adhere tightly both to the steel and to the old coating. Thus, wherever there is general blistering of a coating, it is usually the best procedure to completely renew the coating (*i.e.,* remove all of the old coating from the surface and begin fresh with a new primer and complete new body and topcoats) instead of trying to repair it.

Isolated blistering areas, even though they may be reasonably extensive, can be repaired. In this case, it is necessary to remove all of the blistered area by blasting or other mechanical means back to where the coating has strong adhesion. The edge of the coating should be feathered mechanically. Following this, a primer compatible with both the steel and the previous coating can be applied, preferably by brush in order to work it into both the steel and the feathered edge area. Topcoats can then be applied as indicated by the coating service. Topcoats usually would be applied over the entire surface, unless the blistered areas were entirely local and covered only a small percentage of the coated surface.

Recoating

The recoating of previously coated undamaged sur-

faces can also create problems. In the case of a coating where the resinous materials are permanently soluble in coating solvents (*e.g.,* vinyls, chlorinated rubbers, and acrylics), the solvents in the added coat will penetrate into the previous coatings, swelling them slightly. Even though no actual coating disruption occurs and the application of the topcoat looks smooth and adhesive, if the coating is then immersed, or heated (such as by the sun), blistering may occur under the original coating. This is due to the softening of the original coating by the repair coat solvents, by the retention of these solvents in the old coating, or due to the moisture vapor being transferred more easily through the coating and down into the previous coats when it retains solvents.

The combination of the swelling of the original coating, the absorption of the solvents, and the absorption of moisture vapor can cause a release of adhesion resulting in blistering. This type of coating difficulty is not as frequent where highly cross-linked and insoluble coatings (*e.g.,* phenolics, epoxy phenolics, epoxy polyamides, and coal tar epoxies) are being used. However, even though these materials are generally insoluble in their previous solvents, some solvent penetration may occur with subsequent release of adhesion when immersion and absorption of water vapor also occur. The solution in any of these cases is to repair the coating as previously indicated, and then make certain that the repair coating is thoroughly dry or cured and that the undercoating is also thoroughly free of solvents prior to putting the coating in service. Use the maximum drying time possible whenever a repaired coating is to be immersed.

Pinpoint Rusting

Pinpoint rusting of coatings is a common type of failure. It occurs where there are pinholes in the coating, where there are thin areas in the coating, or where the coating has eroded to the point where pinpoint rusting occurs. Pinholes are usually defects in the original coating. If repair is started soon enough, pinpoint rusting can be limited to isolated areas and therefore successfully and easily repaired.

Pinpoint rusting of coatings is caused by the penetration of moisture and oxygen into pinholes or minute cracks or checks in the coating, so that a minute spot of rust is started on the steel beneath the coating (Figure 15.15). The corrosion product in this area tends to expand; the coating being pushed up into a pinpoint at the point of corrosion. The expansion of the coating at this point causes cracking, and the rust comes through to the surface. Each of the pinpoints of rust is a true corrosion cell, and if left unrepaired or unattended, the coating will tend to undercut, with consequent general rusting of the surface in these areas. Thus, the best repair procedure is to remove all of the coating in the pinpoint rust areas and out beyond the area to sound coating. This procedure is similar to that recommended for isolated blistering, with the coating being blasted free from the surface and the edge feathered out to where the coating is sound.

Repair by mechanically sanding the surface and removing the rust pinpoints from the surface is only a

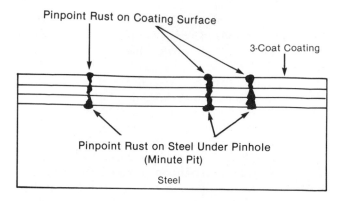

FIGURE 15.15 — Pinpoint rusting of a coating.

stopgap procedure. However, it can be reasonably successful in areas of light corrosion and where subsequent coats can be applied when the surface is warm and dry.

Under these conditions, it is necessary to apply a highly penetrating primer over the area, *i.e.,* one which will effectively wet the pinpoint rusted areas and penetrate down to the metal, followed by topcoats which have a low moisture vapor transfer rate and adequate thickness to thoroughly seal these areas away from both moisture and oxygen. This latter method should not be used where the coating is subject to severe service such as a marine environment, a chemical environment, or an environment where there is consistent high humidity or condensation.

Delamination

Another type of coating failure that poses problems from a repair standpoint is coating delamination between coats. A problem exists because the coating may look perfectly smooth and appear to have good adhesion; however, once the interface between the coats is penetrated by moisture or moisture vapor, the topcoat releases from the undercoat and the entire topcoat may be released in large areas. Intercoat blistering is one indication of this type of delamination.

Delamination most often occurs in coatings that are highly cross-linked and where the cure reaction is either from a catalyst or a reactive curing agent, or where the coating may react with moisture or oxygen from the air in order to become insoluble. It is less prevalent where the coatings are permanently soluble in their own solvents, *e.g.,* vinyls or chlorinated rubbers. In the case of the insoluble types (*e.g.,* epoxies, phenolics, coal tar epoxies, etc.), intercoat delamination can occur from the condensation of moisture on the surface, rain, or, in the case of coal tar epoxies (Figure 15.16), the coating having been subject to sunlight for more than short periods of time. The difficulty occurs where the coating which has delaminated comes off from the previous surface, leaving the original coating thin in that area and therefore subject to much more rapid corrosion than would have been the case with a full-thickness coating.

Repair consists of the removal of all of the delaminated coats back to a point where there is sound adhesion between all coats. If corrosion has started in the thin coating area, the best procedure is to remove all of

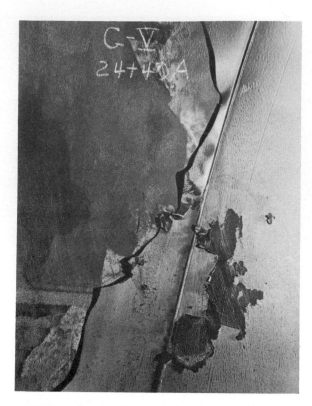

FIGURE 15.16 — Delamination of a coal tar epoxy coating.

the thin coating down to bright metal and carry the surface preparation out to where the full-thickness coating is tightly adherent to the steel. If an overall coat is to be applied, the surface of the epoxy or other insoluble coating must be roughened by brush blasting. Care to prevent any star cracking in the coating is essential for a proper repair job. Once the surface preparation is complete, the repair coating may be applied as previously described under blistering or underfilm corrosion.

Chalking

It has often been said that chalking is the most preferable type of coating failure since it results in the coating gradually decreasing in thickness to a point where the primer may show through or where pinpoint rusting may begin. Nevertheless, there are problems in the recoating of chalked surfaces, particularly when dealing with epoxies, coal tar epoxies, and similar insoluble coatings where chalking has been general yet substantial.

A chalked surface is one where the resins in the coating have been disintegrated by the actinic rays of the sun and other weather phenonema, releasing the pigment to remain on the surface as a chalk or soft powdery substance which, of course, must be removed. In the case of epoxies and other insoluble materials, the solid surface must also be abraded for proper adhesion of the repair coats. Removal of the chalk can be done by brush blasting the surface using extremely fine sand and holding the gun some distance from and at an oblique angle to the surface. This not only removes the chalk, but also will lightly abrade the surface for adhesion of the repair coat. Chalk can also be removed by high-pressure water, preferably accompanied by light abrasion or scrubbing.

Care in surface preparation is essential so as not to star crack the original coating and cause areas of incipient future failure. It is preferable, where a surface is simply chalked, to apply the first repair coat by brush. If this is not possible, as is ordinarily the case due to the magnitude of the surface area, the coating to be applied over the chalked area should be strongly penetrating and have a high wetting potential for the chalked surface. If this is not the case, any area where the chalk is not removed will be an area where the repair coat delaminates from the original coating, since it merely lies on top of the previous chalk without penetrating.

Epoxies and coal tar epoxies are undoubtedly the most difficult to recoat where the surface is chalked. Vinyls and chlorinated rubbers are somewhat less susceptible to delamination problems because of the penetrating character of the solvents used. Water-base coatings, being highly thixotropic, wet chalked surfaces poorly, and all chalk must be removed before recoating.

Repair of Coatings

Coal Tar Epoxies

It must be assumed that any coal tar epoxies that have been in use for a period of time will have an insoluble and impenetrable surface, and that any additional coating applied without careful surface preparation will tend to delaminate. This is particularly true where the coal tar epoxy is used for immersion. The surface of a coal tar epoxy must be roughened sufficiently to give some mechanical adhesion to the following repair coat. For best results, especially for immersion, it is suggested that a thoroughly compatible epoxy primer be used over the previous coal tar epoxy surface to provide a better bond for the coal tar epoxy repair materials. The epoxy primer is good insurance for a sound coal tar epoxy repair.

Inorganic Coatings

Repair procedures for inorganic coatings are different from those previously outlined for organic coatings. Where repairing inorganic coatings, only a single coat 50 to 100 micrometers or 2 to 4 mils in thickness is involved. The failure of inorganic coatings is primarily by pinpoint rust forming after a long period of life or use (Figure 15.17). The same type of failure is true of all zinc coatings, including galvanizing (Figure 15.18).

The failure continues slowly, over a long period of time, until the entire surface is covered with pinpoint rust. With inorganic zinc coatings, this type of failure occurs in thin areas. It has been observed a number of times that, with a single coat of inorganic zinc, the failure takes place in between the spray passes. In other words, where the spray is overlapped, the coating is slightly thicker. These areas will remain good, while the thinner areas between the overlaps will show pinpoint rusting.

During the aging of the inorganic zinc, there does not appear to be any loss in coating thickness. Usually, there is considerable zinc left in the coatings even though pinpoint rust is starting. Again, this is true for a pure zinc surface like galvanizing. Organic zinc coatings also tend

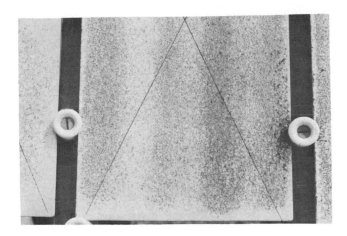

FIGURE 15.17 — Pinpoint rusting of an inorganic zinc coating.

FIGURE 15.18 — Rusting and failure of a corrugated galvanized surface.

to fail in a similar way when they are not topcoated. From the appearance of the failure pattern, the zinc reaction products which continue to form over many years tend to become more dense and seal the inorganic surface. This causes it to become inert and surrounds the zinc particles with the zinc oxide and carbonate reaction products to such an extent that they no longer tend to ionize and provide zinc ions in the area. The inert reaction products and the dense surface do not allow the repair coat of inorganic zinc to adhere satisfactorily to an old inorganic zinc or zinc surface. In order to obtain satisfactory adhesion, it is necessary to break up the surface of the original zinc coating, opening up and exposing the zinc which is present, in order to allow the repair coating to react with the zinc in the existing coating in the same way it reacts with the zinc particles in the new inorganic vehicle.

Sweep blasting is the preferred method of preparing the surface. Using this method, the rust stain is removed, any major failure area can be removed during the blasting process, and the sweep blasting will break up the surface of the existing inorganic zinc or galvanizing and allow the adhesion of the following coat. The sweep blasting also provides tooth for the repair coating, in addition to opening up the zinc for reaction with a new vehicle.

For small areas of failure, power sanding of the inorganic zinc surface can also be done. A medium sanding disc should be used that will cut the surface, but will not gouge or polish it. Once the surface of the inorganic zinc has been prepared, the repair coat of inorganic zinc, either water- or solvent-base, can be applied in a normal fashion.

Overcoated Inorganic Zinc

The repair of inorganic zinc or galvanized surfaces that have been overcoated poses another problem. Most of the failure is due to impact or abrasion, and such areas are usually relatively small in relation to the overall surface. Ship hulls, barges, or similar transportation equipment have the most extensive areas that might require this kind of repair.

The major problem is the two different types of surface over which the repair coating must be applied, *i.e.*, the inorganic base coat where the abrasion may have gone down to steel and the organic topcoat. The primary problem is the incompatibility of the inorganic coating with the organic topcoat. Satisfactory adhesion is not obtained when an inorganic zinc is applied over an organic surface.

The general surface preparation for repair of overcoated inorganic zinc surfaces is similar to that already described for the organic coatings, *i.e.*, abrading the corroded area (either by brush blasting or power sanding), feathering the edge of the organic coating, and lightly roughening the surface of the solid organic coating to obtain a good bond of the repair coating. In this case, however, in order to maintain the zinc undercoat, it is necessary to use a compatible organic zinc-rich coating which is applied to the corroded area, the feathered area, and to some extent out over the solid organic coating. The organic zinc-rich repair coating is required to obtain adequate adhesion to all surfaces—steel, inorganic zinc coating, and the organic topcoat. Once this is applied, the organic topcoats can be applied to complete the repair. While this type of repair may not be as effective as the original inorganic zinc base coat topcoated with the organic, it has been used in many instances, particularly in the marine field, with very good results.

There is one type of failure of overcoated zinc coatings that is as dramatic as it is catastrophic. It occurs in damp areas of high humidity and condensation and has been noted in chemical plants, paper mills, and particularly in marine areas. In this case, voluminous white zinc salts are formed under the topcoat, pushing them away from the surface with the appearance of blistering. In some cases, the white zinc salts will penetrate the topcoat and form on the surface. This type of failure occurs where a thin or very porous topcoat is applied over the zinc coating. Moisture penetrates to the zinc coating and accumulates there as a result of the dampness and high humidity. Zinc hydroxide and zinc oxide are formed (zinc oxychloride may be formed in a marine atmosphere), creating a strong physical pressure under the topcoat and forcing it off from the surface. The problem is that the topcoat allows the moisture to penetrate and accumulate in the zinc coating. Once the zinc salts are formed, they tend to form at an ever-increasing rate, thus making for rapid coating failure.

Repair of such a failure is difficult. All of the topcoat in the affected area must be removed, as well as all of the underlying zinc coating. If the failure is general, full coating renewal is required. If the failure is localized, these areas can be removed and repaired as previously indicated.

The solution to this problem is to make sure that the inorganic zinc coating is thoroughly cured and to apply topcoats which have a very low moisture vapor transfer rate. It is essential that the topcoats be applied in full thickness. Little failure of this type has ever been experienced where full-thickness, resistant topcoats have been applied.

Linings

The repair of lining materials is much more critical than the repair of exterior coatings. In order to be effective, the repair work must be the equivalent of the original lining material to both protect the substrate and minimize the contamination of the contained liquid. While thin coatings are also used as lining materials, such as linings for water tanks, the majority of lining materials are usually somewhat heavier than exterior coatings.

The original application of a lining material is usually more carefully done than that of an exterior coating, and failure of reparable linings is usually in relatively small, isolated areas. Failed areas may be due to physical damage or because of an isolated spot of poor adhesion causing blisters. The general failure of a lining in any container indicates that there is a basic problem with the coating material or its original application, and that the entire lining material should be removed. In this case, the repair actually consists of a completely new lining application. Only isolated spots can really be repaired.

The whole concept of repairing a lining is to duplicate the original lining work and to obtain equivalent resistance from the repaired material. The actual repair procedure should be to remove all of the corrosion products from the repair area, removing also any damaged lining out to where the original lining material has proper adhesion to the substrate. The edge surrounding the repair area should then be feathered to a thin film having an approximately 2-in. feathered edge, while thicker coatings may be feathered to a width of from 4 to 6 in.

All of the difficulties that have been previously described for the repair of coatings are inherent in any lining repair. Star cracking of the original coating must be prevented. If blasting is used to make the surface preparation repair, fine abrasive should be used and every care should be taken to protect the surrounding lining from any abrasion by whatever sand or grit may be used. Care must also be taken in the bottom of the container to prevent penetration of the lining by persons walking on any loose abrasive that might be on the surface.

In repairing tank linings, vacuum blast equipment provides a satisfactory method of surface preparation. The reclamation of the abrasive by the vacuum assures that damage to the surrounding lining is kept at a minimum. Power sanders and grinders may also be used to repair small breaks in the coating. Where this type of power equipment is used, care must be taken to prevent damage to the surrounding lining, and yet the corroded or damaged area must be ground sufficiently to remove all corrosion products. Power sanders or grinders may also be used to feather the edge of the coating around any breaks where repair is necessary.

In the repair of linings, the original number of coats should be applied during the repair procedure. In small areas, brushing of the coating may be practical. In more extensive areas, spraying is suggested, with the exception that the primer should be applied by brush, if possible, in order to make sure that the repaired surface is properly wetted by the coating. The repair material should be applied at the same thickness as the original coating, and it should be carried from the steel surface or substrate out over the feathered edge of the coating and on to a lightly abraded area beyond the feathered edge. It is usually not necessary or advisable to carry the topcoats out over the original coating if the remainder of the original coating is undamaged.

Once all coats have been applied to the proper thickness, curing of lining materials, particularly repaired areas, is extremely critical, and in almost all cases it is recommended that force drying or force curing be used. This essentially is the heating of the coating in order to thoroughly cross-link it to the same degree as the original coating, or, in the case of thermoplastic lining, to remove the solvents from the newly repaired area so that there is not more solvent remaining in the repair than in the original coating. This cannot be overemphasized, since the majority of failures after repair in a lining are due to poor curing or poor drying of the repair coating.

The equipment which may be used for force drying or curing internal linings has been outlined in NACE Standard RP-01-84 as follows:

2.3.1 Hot Air Heaters

2.3.1.1 Direct fired heaters using LPG or nature gas will produce 700 F (370 C) air temperature and metal temperatures up to 400 F (200 C).

2.3.1.2 Indirect fired heaters using LPG, natural gas, or other liquid fuels will produce approximately 110 F (60 C) air temperature rise and metal temperatures up to 175 F (80 C) when outside temperatures are 70 F (20 C) or above.

2.3.1.3 Heaters using electrical heated elements in the form of "heat guns" will produce air temperatures to 1000 F (540 C) and are most suitable for small repairs.

2.3.1.4 Do not use air heaters such as direct fired oil heaters that produce gases or particles which contaminate. Properly designed and adjusted natural gas or LPG heaters do not contaminate other than by condensation, discussed in Paragraph 2.3.1.5.

2.3.1.5 LPG or natural gas direct fired heaters may cause contamination in the form of condensation; however, a quick blush of condensation that evaporates is not necessarily damaging. The gases from direct fired gas heaters have a dew point of approximately 180 F (80 C), and a cold substrate may cause condensation. Any observation of red rust during or after a preheating period will require repeat of surface preparation.

2.3.2 Infrared source of heat lamps may be used for curing small areas. These heater units only provide enough heat to cure baked type coatings on small spots.

2.3.3 Ultraviolet source may be a sun lamp or suitable UV source that may be purchased for curing coatings which are formulated for ultraviolet cure.

2.3.4 Dehumidification equipment may be used by itself or in conjunction with heating equipment to keep the air in a space at a dew point of at least 5 F (3 C) below the substrate temperature to prevent condensation.[9]

Tank Cleaning

Any lining that is installed on the interior of a container will need some type of general cleaning prior to any repair procedures. Many contained materials will leave little contamination on the surface, as is the case with volatile petroleum products, solvents, and similar materials. On the other hand, a tank for the storage of vegetable oil will leave an oily, greasy surface. Food oil tanks leave a heavy oily deposit on the surface. Wine, beer, and similar food products often leave a precipitated deposit on the lining that is also quite difficult to remove.

Thus, the cleaning procedure for every type of contained product is somewhat different, and it also must be varied by the type of lining that is in the container. There are some linings which can be cleaned by the use of steam; others require ambient temperature cleaning in order not to damage the lining. In any event, the cleaning solution and the procedure for cleaning the tank must be such that it will not damage the lining or the substrate where it might be exposed. It must also have the capability of removing all of the contaminants from the surface of the coating. Special attention, of course, should be given to those areas where the coating has been broken, damaged, or has failed in order to assure that these areas are as clean as possible prior to attempting any repair procedure.

While repairing a lining, it must be remembered that the lining is a valuable part of the tank itself, and any additional damage to the lining can be extremely costly. The cleaning solution used in the tank should not have any effect on the substrate. For instance, it would be disastrous if a highly alkaline cleaning solution were used on the interior of an aluminum tank, even though the tank lining itself was fully resistant to that cleaning material. Neither should stainless steels be subject to chloride-bearing cleaning compounds. Detergents and similar materials may be used, as long as it is possible to thoroughly remove them from the surface and they do not have any effect on the substrate. Solvents may be used as long as they have no effect on the lining, and as long as the proper safety precautions are used to eliminate the possibility of physical harm to the workers from toxic exposure or fire.

It also must be remembered, as a general rule-of-thumb, that 1% solvent vapor in air is a maximum, from the lower explosive limit standpoint. While most solvents have a somewhat higher lower explosive limit, there are a few that are borderline. The air within the tank should be kept below the 1% limit under all conditions. Reference to Chapter 16 is suggested wherever solvents are involved. Any enclosed area is dangerous and all safety precautions should be used.

The general surface preparation requirements for various surfaces have been previously covered. However, it is important to mention that where linings are concerned and where immersion is a condition of service, the very best possible surface preparation must be used for all lining materials. There is no substitute for a perfectly clean surface over which to apply a lining material. In fact, it is essential if future failure is to be avoided. This applies whether the substrate is stainless steel, cast iron, aluminum, zinc, or concrete.

Lining Thickness

As outlined by NACE Standard RP-01-84, there are three types of lining materials in terms of thickness: (1) thin—those with less than 10 mils of thickness; (2) thick—those which range from 10 to 29 mils in thickness; and (3) thick film—those which are over 29 mils in thickness.[9] These categories reflect the linings that are commonly used on the interior of all types of containers. The thin linings (i.e., less than 10 mils) can be composed of almost any type of coating previously described. Inorganic zinc coatings with a thickness of 3 to 4 mils are commonly used as a lining for many solvent, lube oil, and petroleum storage tanks. Vinyl linings in this thickness range have commonly been used for water and brine tanks. Epoxy coatings in this range are used as linings for various mild chemical tanks where they are resistant to the contained product. Baked phenolic linings and baked epoxy phenolic linings are used for any number of food tanks including beer, wine, liquid sugar, etc.

Intermediate thickness lining materials (i.e., those from 10 to 29 mils) are often used to upgrade the thinner lining materials by providing additional thickness. These materials are usually the higher solids, reactive type of vehicles, such as epoxies, coal tar epoxies, some urethanes, and some thinner, rubber-type coatings. Many are used for lining water tanks and pipe, as well as for lining water process and waste water processing equipment. They may be used for the lining of tanks, tank cars, or ships for specific products.

The heavy thick film linings (i.e., those from 30 mils up) are the heavy-duty lining materials used in many areas of the chemical and mining industries, refining and chemical processing, marine and water reclamation plants, and piping systems.

Types of Linings

Sheet Linings. One of the lining materials that is not included in RP-01-84 on linings is the vinyl sheet lining material used to line concrete pipe structures in sewage treatment plants and outfall piping. It is also used as a lining for many chemical tanks, such as ferric chloride tanks, as well as chemical leaching tanks in the mining industry. These areas are subject to severe corrosion, and the lining may be applied to either steel or concrete structures.

Repair of the lining for any of these services is much the same. Small damaged areas may be repaired by using a patch of the original vinyl lining material over the break in the lining. This is attached to the previous lining with adhesive, followed by heat welding a weld strip over all edges of the patch. For larger areas, the damaged lining is removed, usually in a rectangular pattern. The repair sheet lining is then cut to fit the rectangular area, and the new material held in place with adhesive, or, under certain circumstances, stainless steel pins and washers have been used to mechanically hold the lining in place. These are driven into the surface using explosive charges. (This procedure is only applicable to concrete repair.)

Once the new lining is in place in the repair area, the joint between the new and the old lining is heat welded, using a welding strip over the joint between the two surfaces. The welding strip is usually 3/4 to 1 in. in width and is actually heat welded to the two surfaces. In this case, hot air guns are used to soften the two surfaces, i.e., the sheet lining and the weld strip, to a semiliquid state where they fuse together. Hot air welding and sheet lining repair are not simple tasks. Lining manufacturers instructions should be carefully followed if a satisfactory repair is to be made. Such repair, if properly done, can be considered permanent.

Rubber Linings. Rubber linings are used in tanks, pipe, and various production or process vessels. Like the sheet vinyl lining, they are used for difficult chemicals such as acid solutions or where abrasion and corrosion are combined. The rubber linings may be applied as a liquid or in sheets. The repair of both are similar, and, in general, is as follows.

1. Remove the lining around the damaged area back to an area of good adhesion. This may require grinding the rubber lining around the damaged area.

2. Prepare the substrate surface in the best manner possible (e.g., sandblasting, vacuum blasting, or power sanders). All edges of the rubber should be feathered or skived.

3. Apply a proper rubber bond coat or adhesive, and reapply the liquid rubber solution or apply a patch of uncured rubber.

4. Cure the patch in place by heating (in many cases steam is used) or by curing chemicals. The cure must be equivalent to the original lining.

Curing is the critical part of the process, and every care must be taken to cure the repair area so that the end result is equivalent to the original lining. Since there are many grades of rubber lining, and each has its own idiosyncrasies, manufacturer's instructions should be carefully followed.

Glass-Reinforced Linings. It is frequently necessary to repair glass-reinforced epoxy, polyester, or vinyl ester linings. The need usually arises from physical damage, although in the past there have been a number of adhesion problems resulting from continued polymerization, where these materials have actually pulled themselves away from either concrete or steel surfaces. These were often on concave or cylindrical surfaces where the shrinkage of the materials during cure pulled the lining away from the surface. This has also occurred on flat tank bottoms, particularly around the circumference of the tank.

In these cases, repair can usually be made by removing the damaged lining and reapplying the same or a similar material over the broken area, applying the same number of layers of glass matte and glass cloth and resin. In any case, the adhesion of the original material should be thoroughly tested to make certain that the repair is carried back to a solid coating with good adhesion to the substrate.

Glass Linings. There are many glass-lined tanks in service, as well as glass-lined pipe and equipment. Every care is ordinarily used to prevent damage to such linings, since they are difficult, if not impossible, to repair. Glass-lined equipment does chip, on occasion, and, if the contained material is corrosive, rapid damage can occur. Some repair is done with resinous linings if the exposure is satisfactory and the temperature sufficiently low. Epoxy phenolic lining materials are often used, heating the repair area to thoroughly cure the resin.

Most glass lining repair is made to small breaks in the lining. Some are sufficiently small so that gold or other inert metals have been used for repair. Large damaged areas usually require the vessel to be permanently taken out of service. The original manufacturer's recommendation should be followed for any glass lining repair.

Summary of Lining Repair

The repair of lining materials is a critical job, and every repair is a unique operation. Generalized procedures are ordinarily not practical. Nevertheless, in order to provide a quick overview, the following is a limited summary of lining repair procedures.

1. Clean the tank or lined surface free of all dirt or residue by the appropriate method.

2. Remove damaged or disintegrated coating back to an area of good adhesion to the substrate.

3. Prepare the metal substrate in the best possible manner—white sandblasting is recommended (NACE 1 or SSPC 5). Small repairs may be made with power grinders or sanders. Use particular care to protect the existing lining during surface preparation.

4. Prepare concrete substrates by removing all soft and damaged concrete back to solid gray concrete. Rebuild to original surface level with cement plaster or resinous surfacer (epoxy surfacer, if applicable).

5. Feather-edge existing coating 1 to 2 in. for thin linings, 2 to 3 in. for medium-thickness linings, and 4 to 6 in. for thick linings (over 30 mils). Lightly abrade original lining a short distance beyond the feathered area.

6. Apply appropriate primer or first coat to substrate and over feathered area. Extend a short distance out over the original lining.

7. Apply original number of intermediate or topcoats over the repair area. Use spray application for large areas, if practical; use brush for small areas.

8. An overall finish coat is usually not recommended or necessary. This is particularly true of epoxy, urethane, and phenolic linings, because of adhesion problems between coats.

9. Cure repaired area thoroughly. This step is vitally important. Use heat to force-dry or cure, if necessary, to obtain original coating resistance.

10. Specific suggestions for specific linings include:

a. Urethane linings usually require an epoxy primer for adequate adhesion to the substrate and original coating.

b. Where inorganic zinc is used as a repair base coat, remove the inorganic zinc from topcoat and feathered organic area by rubbing with screen wire or other mechanical means. This is essential for proper adhesion of the topcoats.

c. Extra care must be used with coal tar epoxy linings to roughen the original coating surface thoroughly to obtain adequate adhesion. Where possible, use a polyamide epoxy primer over the substrate and original coating for best adhesion.

d. High-baked lining materials (phenolic and epoxy phenolics) must be cured to the same degree as the original lining. Lack of full cure means certain failure for any repair.

e. Thick rubber-type linings, where damaged or blistered, must be cut away back to an area of good adhesion, the edges of the rubber skived, and the original rubber system replaced and cured. Use the manufacturer's recommended procedures.

f. Thermoplastic sheet linings (usually vinyl) are repaired by cutting sheet away from the substrate. The area is usually cut in a rectangle. A similar sized rectangle of lining is replaced in the repair area using an adhesive. The joint between the old and new lining is then heat welded using a weld strip. This makes a very effective repair. Use the manufacturer's recommended materials and procedures for making repairs of this type.

g. Glass linings are difficult to repair. Damage is usually in the form of small mechanical breaks. Glass cannot be replaced. Obtain the manufacturer's recommendation for repair.

h. Glass-reinforced epoxy or polyester linings can usually be repaired by using glass matte, glass cloth, and the original resin-catalyst combination.

These general recommendations provide a guide to the reconditioning of a coated surface. Coating or lining maintenance, however, includes much more than simply following rules for the repair of a surface. It is, as previously stated, the protective management of the surface, which is not limited to periodically touching up a few rust spots. As defined previously, it means management of the coating process from the beginning of the original application to its present and future states. It therefore includes such activities as material studies, specifications, inspection, record keeping, observation of the coating condition, and establishment of the limits of failure beyond which repair and recoating automatically take place.

Of all of these activities, the most important are writing proper specifications and following up the specifications with thorough inspection. Good specifications and inspection procedures can insure a satisfactory and long-lasting repair job, just as they can also assure a long life and effective service from the initial coating.

References

1. Department of Defense, Paints and Protective Coatings, Programmed Painting Section, Army-TM5-618, Navfac-MO-110, Air Force-AFM-85-3.
2. NACE, Industrial Maintenance Painting Program Technical Committee Report 6D160, Houston, TX.

3. Weaver, Paul E., Industrial Maintenance Painting, National Association of Corrosion Engineers, Houston, TX (1973).

4. ASTM Headquarters, 1916 Race St., Philadelphia, PA 19103.

5. Federation of Societies for Coatings Technology, Pictorial Standards of Coating Defects, Blistering, Philadelphia, PA.

6. Vind, H. P. and Drisko, R. W., Field Identification of Weathered Paints, Naval Civil Engineering Laboratory Technical Report # R-776, April (1972).

7. NACE, Proposed Technical Committee Report, Surface Preparation of Contained Concrete for Corrosion Control, NACE Task Group T-6G-21, 1982 Draft (subject to change), Houston, TX.

8. Munger, C. G., Practical Aspects of Coating Repair, Materials Performance, Vol. 19, No. 2, p. 46 (1980).

9. NACE Standard RP-01-84, Repair of Lining Systems, 1984, Houston, TX.

APPENDIX 15A[1]

Procedures for Adhesion Test[2]

Scope

This test method is intended to determine the surface strength of concrete and to evaluate or measure the bond of applied coating, and other corrosion protection materials, to concrete surfaces cleaned and prepared in accordance with the guidelines provided in this report.

Significance

The test can be used to evaluate the tensile bond strength of applied coatings, polymer lining (filled and unfilled), and flooring systems to concrete.

The surface integrity of concrete can be evaluated by this method. Concrete surfaces that have been exposed to chemical and other conditions may have poor or marginal tensile strength for application of corrosion protection materials. It is essential to measure the tensile strength of concrete surfaces before applying heavy duty floor toppings or linings. Many topping or lining products are considerably stronger than the concrete, and the adhesion of products to concrete also exceeds the tensile strength of the concrete. A tensile strength of 1.0 MPa (200 psi) for indoor exposure, or 1.4 MPa (250 psi) for exterior are the generally accepted minimum required values for heavy duty resin floor toppings or linings. At values lower than these, the concrete surface is too weak to accept a heavy duty floor topping.

Apparatus and Materials

Elcometer Adhesion Tester Model 106:
Scale No. 1 (0-35 kg/cm^2; 0.500 lbs/in^2)
Scale No. 2 (0-70 kg/cm^2; 1-1000 lbs/in^2)
Scale No. 3 (0-140 kg/cm^2; 0-2000 lbs/in^2)
Scale No. 4 (0-280 kg/cm^2; 0-4000 lbs/in^2)

Aluminum dollies (adhesive contact area 3.24 cm^2; 0.502 in^2)

Epoxy Adhesive
For testing the surface integrity and strength of concrete, fast set commercial adhesives can be found in hardware stores, but care should be taken to ensure that the strength of these adhesives is greater than that of the system being tested. For testing the bond of applied coatings and materials to concrete, it is best to use an adhesive based on the resin or polymer of the coating, topping, or lining being evaluated.

Hole saw (inner diameter 21 mm; 0.84 in)
A nominal 2.5 cm (1 in) OD hole saw with HSS Starret edge will produce this inner diameter.

Magnetic dolly clamps for vertical surface adhesion to steel substrates
Coarse sandpaper (grade 60)
Strong solvent (methylene chloride or equivalent)

Procedure

Drill 21 mm (0.84 in) diameter circular cut in the concrete and abrade the surface of aluminum dolly with the coarse sandpaper. Make sure the surface is clean and free of contaminants by wiping with a strong solvent. For greater bond strength, the dolly may be brush-blasted prior to adhesion.

On drilling large substrates, the use of a plywood or other device template has been found essential to controlling the hand-held drilling operation.

Bond the high tensile strength aluminum dolly to the test surface using a suitable adhesive. Any excess around the base of the dolly should be removed and the adhesive left the appropriate time to cure.

When the adhesive curing time has elapsed, the adhesion tester is placed in position by lowering the claw at the bottom of the unit so that it slides over the dolly. Make sure that the three swivel pads are positioned so as to transfer the force evenly; this will ensure that the claw will pull perpendicular to the substrate.

Tightening the handwheel or nut applies a pulling force to the dolly. At the point of break, the scale will rise a little, and the reading will be retained by the dragging indicator on the barrel. Estimate readings to the nearest half scale division.

Discussion

Three general types of data can be obtained from this test.

Tensile Strength/Bond Strength. An approximation of the mean tensile strength and/or bond strength can be evaluated from the readings taken from the barrel.

Mode of Failure Analysis. One of the most important diagnostic pieces of information in a bond test is the location of a failure in a composite system. Weak points may readily be defined by examining the fracture surfaces.

In most cases the coating will fully adhere to the dolly and the test can be claimed as 100% valid. This can be determined by examination of the dolly after completion of the test. In certain instances a portion of the coating will be removed and partial adhesion to dolly may occur. Allowances for this must be made when finally evaluating the results of test.

---

[1][SOURCE: NACE Proposed Technical Committee Report, T-6G-21, Surface Preparation of Contaminated Concrete for Corrosion Control, NACE, Houston, TX (1981 Draft, subject to change).]

[2]This is not a standard test method, but rather a practical field test derived from a review of current technology.

In instances where the concrete surface is weak, floor topping or lining failures are usually concrete failures, since 0.3 to 0.6 cm (1/8 to 1/4 in) layer of concrete adheres to the corrosion protection products when they separate from the concrete base.

Improper cleaning of concrete saturated with oils or fats or other contaminants is usually easy to identify, since the topping or bonding adhesive used in test usually separates cleanly, taking no cement with it.

The "go no-go" Test. This method of evaluation is of primary importance to field inspection work where minimum bond strengths have been written into the specifications of a system. If unusual application conditions are present during the installation of a product, its performance can be estimated from the mechanical efficiency displayed by bond testing.

Sources of Error

It is important to calibrate the adhesion tester from time to time to ensure that the spring constant for the Belleville[3] washers has not fatigued. This can be done by applying a known force to the tester and recording the difference in readings over the scale range.

It should be remembered that this is a method for estimating the average tensile strength and should not be confused with the actual tensile strength.

Sample preparation is important from the standpoint that the dolly must be parallel to the substrate surface to prevent shearing.

The available bonding surface on the dolly should be matched as closely as possible to the drilled hole. The hole saw recommended under the Apparatus and Materials Section will produce a surface matching error of 3%.

The choice of adhesive is important for consistency and significance in readings. If the adhesive is the weakest zone relative to the product, the product is not being evaluated. Care should be taken to choose adhesives with tensile strength well above that of the product under test.

In circumstances where finding a reliable adhesive becomes a problem, one might consider using the product itself as the adhesive, provided its bond to the dolly is sufficient.

Selecting the proper adhesion tester is also important. As listed in the Apparatus and Materials Section, there are four scales available in the Elcometer Model 106. The differences between these scales involve their capacities and accuracies; one would not choose a model with a capacity to 30 MPa (4000 psi) to evaluate a bond with a typical value of 0.7 MPa (100 psi).

When operating the 14 and 28 MPa (2000 and 4000 psi) capacity models, it is important to crank the wrench in a plane parallel to the plane of the substrate. Failure to do so, again, would result in the generation of a shearing force which would produce inconsistent readings.

If the coupon as pulled does not have concrete attached to the *whole* surface, the test is invalid and must be repeated.

---

[3]Trademark of Universal Products, Inc., 224 North Montello, Brockton, MA 02401.

16

Safe Application of Coatings and Linings

Safety requires a continuing level of consciousness which attempts to foresee and thus forestall problems before they occur. It is a process that attempts to overcome the commonly held theory that if it is possible for something to happen—it will. Safety involves taking the precaution necessary to ensure that the possible does not happen and should be a primary consideration when dealing with high-performance coatings, because of the many potential hazards related to both the material and its application.

Every coating assignment exposes maintenance personnel to conditions and situations that represent actual or potential danger to themselves and to others in the area. The frequent use of toxic and flammable materials, pressurized equipment, ladders, scaffolding, and rigging always presents potential hazards. Hazards may also be inherent in the environment, or they may be brought about by the ignorance or carelessness of the operator. Therefore it is extremely important to be aware of all potential hazards, since continuous and automatic precautionary measures will minimize problems and improve both the efficiency and morale of the coating crew.

These potential hazards that are a part of all coating operations make a continuing and enforced safety program absolutely essential. Adequate safety procedures will provide protection against the major types of hazards; namely, physical accidents, fire, explosion, and health problems. All personnel should be responsible for adhering to all established precautionary programs for their individual protection as well as that of others. Disregarding safety measures will significantly increase the potential dangers and the odds that an accident will occur.

Accidents can occur for a variety of reasons. The most common causes are outlined in the Department of Defense's *Manual of Paints and Protective Coatings,* as follows.

Accidents during painting operations are caused by unsafe working equipment, unsafe working conditions and careless personnel. Any or several of the following can cause accidents:

a. Lack of knowledge, experience and training in the use of painting materials and equipment.

b. Improper maintenance and storage of equipment.

c. Failure to pre-check equipment for mechanical and structural flaws.

d. Improper use of equipment.

e. Failure to consider environmental conditions and existing hazards in work areas before, during and after painting operations.

f. Personal carelessness.

Accidents most frequently involve commonly used equipment. The most common and serious accidents, by far, are falls either from a height or on the ground because of a loss of footing. Falling or moving objects are the next most serious hazard.[1]

Overconfidence resulting in personal carelessness is probably the greatest cause of all accidents. There are no good or practical safeguards against ignorance or carelessness. All personnel must be advised of all potential hazards and the precautions to take against them. "Short cuts" usually involve poor safety procedures and therefore must be avoided.

This chapter will emphasize the safety requirements inherent with the coating material rather than concentrating on the ordinary physical hazards involved with scaffolding, bosun's chairs, ladders, swinging stages, rolling towers, ropes, and similar equipment that is common to most construction jobs and to the coating process in general. There are literally hundreds of safety rules and regulations that are published and are requirements of various governmental agencies. These include city and county codes, state regulations, and, overall, the rules and regulations established by national government agencies such as the Occupational Safety and Health Administration (OSHA), the Environmental Protection Agency (EPA), the Department of Labor, and by the Toxic Substances Control Act (TSCA). Appendix 16A contains a number of references where additional information on safety rules and regulations may be obtained.

Introduction

One of the most comprehensive outlines for painter safety is NACE's *A Manual for Painter Safety* prepared by NACE Task Group T-6D-5 on Painter Safety. The entire manual is included here as Appendix 16B. By following this manual and the various governmental codes that may apply, the application of paint and coatings should be reasonably safe.

There are definite hazards associated with the application of coatings, particularly high-performance coatings. The greatest hazards exist while coating the interior of closed spaces such as tanks, processing equipment, or small rooms or chambers. These areas exist in almost all industries and can include the interior of water tanks, the interior of refined oil and crude tanks aboard ship, the interior of food vessels such as beer tanks, or the interior of tank cars or other transportation equipment. The existence of hazards involved in coating the interior of closed spaces is evidenced by the occurrence of numerous fatal accidents since the advent of high-performance coatings. However, the economic need for the product justifies the safety materials, methods, and procedures required to reduce the hazards involved in their use.

Many excellent examples of safety programs exist throughout industry. Two of the most outstanding are the programs established by the explosives industry and those employed in the development and use of nuclear energy. In both of these areas, safety precautions have been developed to drastically reduce the danger in working with these hazardous materials. The marine industry is also involved with many hazardous products in both the transportation and handling of such materials. The use of high-performance coatings also depends on the procedures and methods that are developed to assure their safe handling.

Since each coating job is unique, the hazards involved are highly variable. A coating applied to the exterior of structural steel, the outside of an offshore structure, or the exterior of a tank, certainly poses different hazards than one applied on the interior of a tank car, barge, or tank. As mentioned earlier, the most dangerous areas are those which are enclosed. Although in many large, enclosed areas, the hazards do not seem to be very real, a number of accidents have resulted, primarily from an attitude of nonchalance or carelessness. For example, a coating was being applied on the interior of a refined oil vessel which involved several men working in the tank. At lunch time when all the men left the tank, the ventilation was turned off, supposedly to save power. When they returned and entered the tank after lunch, the tank blew up, killing all of the workers. After the accident, a half-smoked cigar was found at the bottom of the ladder, indicating the type of carelessness that compromises safety.

Another example involved coating the interior of a floating roof tank, where the floating roof was held on legs or supports approximately six feet above the bottom.

After an explosion occurred, it was found that the workers had been using a string of unprotected light bulbs in the tank. Undoubtedly one of these dropped or was hit, causing the spark which set off the explosion. In both of these examples, the work was being done on overtime so that no superintendant was present. Unfortunately, overtime jobs where supervision is less adequate are often the jobs that incur safety problems. Unfortunately, once an incidence occurs, the obvious lack of understanding as to the importance of safety procedures is inconsequential.

Changes in the Coating Industry

Some major changes in the coating industry are resulting in changes in the composition of coatings, basically from those based on solvents to those which use no solvent at all or a reduced amount of solvent. These changes will have a marked effect on the necessary safety precautions. A look at some of these changes is given in *Applied Polymer Science* in a chapter on the chemistry and technology of solvents, as follows.

Government regulations regarding solvents and coatings will have a profound effect upon future products and practices. The regulations are concerned with the following topics:

A. Air pollution: federal activities under the Environmental Protection Agency (EPA) as well as state and local regulations.

B. Health: Occupational Safety and Health Administration (OSHA) as well as separate and overlapping laws from other agencies.

C. Safety: Department of Transportation (DOT) and other agencies concerned with flash point, labelling and related matters.

D. Energy conservation.

In virtually all cases, the regulations will tend to restrict solvent usage. As one consequence of restrictions, the introduction of new solvents will be inhibited in continuation of a trend already in existence. Indeed, the production of some existing solvents will be discontinued as has already happened in the case of some solvents where small volumes simply could not be sold profitably. Some producers of hydrocarbon products have eliminated many solvents from their lines.

Although normal growth of paint products approximates 3% per year, total solvent usage for paint will not grow commensurately. Trade sales coating for architectural use are already largely water-based and industrial coatings are now experiencing a strong impetus toward water-based types. Other types of coatings will grow but at a lesser rate than water-based; these include high solids coatings (at least 70-80% solids), powders and 100% convertible coatings where the reactive solvent becomes part of the solid final cured coating. It has been estimated that the chemical coatings market in 1972-73 was comprised of 64% conventional solvent coatings and 26% conforming (Rule 66) solvent coatings while in 1977-80 the percentages will be 18% and 34%, respectively. Nevertheless, switching from solvents is difficult, costly, and fraught with uncertainties. Replacing solvent paints by low-solvent or no solvent types may not be sound economically or environmentally in many cases. Because of long experience and some inherent advantages (wetting, adhesion, color-matching, durability) of solvent paints as a class, the need for less frequent or no repainting can often make up for greater initial cost or emission of solvents from solvent paints.

In product finishing, because of increased fuel costs, incineration of solvent fumes is becoming less popular, but solvent recovery is becoming of greater interest. Single-solvent systems or those with few components are more easily recovered than the usual multi-component solvent systems and will be favored where solvent recovery is concerned.

The product mix will be altered because of the impact of composition limitations of solvent blends under air pollution laws such as Rule 66. Aromatic hydrocarbons are quite vulnerable under air pollution laws; acceptable substitutes usually can be made from a mixture of about 20-50% oxygenated solvent and 80-50% of an aliphatic hydrocarbon. There is some pressure, only partially supported by technical data, to eliminate distinctions among solvents as to their oxidant or smog-producing capacity. Overall, most regulations have the two-fold effect of reducing total solvent usage and altering the product mix.

Since total solvent production will undoubtedly grow in the future because of increasing use of solvents as chemical intermediates, solvents will always be available for the conventional uses which make use of the solubilizing and viscosity-reducing characteristics.[2]

This last point, *i.e.,* the availability of solvents, is extremely important since many high-performance coatings cannot be modified materially and still retain their effective anticorrosion properties. As previously discussed, both water-based and high solids or 100% solids coatings present many difficulties in field application. Also, water-base materials, presently being made primarily from emulsions and dispersions, do not form the same continuous impervious type of film that is developed through solution techniques. Presently, emulsion-type coatings can only be converted to high-performance chemical-resistant products by heating to the point where the resinous vehicle fuses together into a solid film. It is the opinion of many persons in the high-performance chemical and corrosion-resistant coating field that it is going to require many years of development and research to eliminate solvent-base coatings from the high-performance field. Thus, from the standpoint of the corrosion engineer, solvents are still going to be involved in highly corrosion-resistant applications and, because of this, safety procedures for the use of solvent-base coatings will remain necessary.

Primary Hazards

There are essentially four primary hazards involved in the use of high-performance protective coatings. These are characterized as fire, explosion, reactivity, and health. Each one of these should be carefully considered.

Fire

All coatings which contain solvents are flammable. In many cases, the coating's binder resins are also flammable. This is particularly true when the coating is in the liquid stage. Any time that a solvent-based coating is liquid, whether in a container or applied to a surface, there is the possibility of fire. Thus, every safety precaution should be taken to prevent a flame or spark from contacting the paint while it is in the liquid stage. Coatings in the liquid stage generally are not explosive. This means that if the surface of the liquid coating in a container is on fire, it will not explode from the flame on the surface. This does not mean that a closed coating container that is subject to a fire will not blow its lid in a simulated explosion as a result of the vapor pressure of the solvents developed by the external heat. Such an exploding container materially adds to the fire by spreading the flammable material over a wider area.

None of this information should be interpreted to mean that fire is not a major hazard. For example, in the internal lining of a pipeline, where a workman was making repairs using a solvent-based material, the container caught fire, and in the excitement that followed was tipped over. The ensuing fire was sufficient to start the repaired coating area aflame, and before the workman could get out of the pipe, the entire pipe lining was aflame.

Enclosed areas are particularly dangerous where solvent-containing coatings systems are being applied. Therefore, every precaution must be taken to eliminate any possible source of flames or sparks which could ignite the coating.

Explosion

Explosion is differentiated from fire because it basically involves a different type of reaction. In the case of explosion, there is sufficient solvent evaporated into the air so that, if there is a source of spark or flame, the entire air-vapor volume will react at one time, creating an explosion. Explosions can occur without fire, although they are often combined, with the explosion igniting the liquid coating. Explosion is probably the greatest hazard from the standpoint of catastrophic loss of life and property, and there are a number of phenomena which are important in preventing explosion.

Lower Explosive Limit

The *lower explosive limit* of any solvent or combination of solvents can be defined as the percentage of solvent vapor in air which is the point at which an explosion will occur if the air and solvent mixture is ignited with a spark. The lower explosive limit of a solvent or solvent mixture is extremely important to the whole safety program in the application of coatings due to the fact that *if the lower explosive limit is never reached, no explosion can occur.* This means that every effort should be made to prevent the solvent-air mixture from reaching the lower explosive limit. There is also the *upper explosive limit,* above which the percentage of solvent vapor in air is sufficient to inhibit the explosive reaction. At this point, no explosion can occur, even though a spark may be available within the solvent vapor-air mixture. This is what prevents an automobile from starting when it is overchoked and the carburator is said to be "flooded." The upper explosive limit is much less important from a safety standpoint than the lower explosive limit. Information on the lower explosive limits of individual solvents has been published in various sources for many years.

Information on mixed solvent explosion limits has not been as well documented. Battelle Memorial Institute has done some work along these lines for private interest. In this research, they developed apparatus for accurately measuring the solvent volume in an air mixture and then determining the percentage of solvent vapor and air which would cause an explosion when set off by a spark. They checked both individual solvents and solvent mixtures used in high-performance coatings.

Table 16.1 shows the test explosive limit for individual solvents, and Table 16.2 shows the test limits for

TABLE 16.1 — Test Explosive Limit Data
for Individual Solvents

Solvent	Spark-Explosion Limit, Percent Vapor By Volume in Air	Pressure, mm Hg	Temp., Deg. F	Temp., Deg. C
Ethyl Alcohol	3.31	760	77	25
Isopropanol	2.59	760	77	25
n-Butyl Alcohol	No limit found	665	122	50
Tetrahydrofurfural Alcohol	No limit found	743	122	50
Acetone	2.88	760	77	25
Methyl Ethyl Ketone	2.15	760	77	25
Methyl Isobutyl Ketone	1.80	760	77	25
Cyclohexanone	1.49	760	122	50
n-Butyl Acetate	1.57	760	122	50
1-Nitropropane	2.32	760	122	50
Xylol	2.24	760	122	50
Aromatic Petroleum (toluol range)	1.19	760	122	50
Aromatic Petroleum (xylol range)	1.29	760	122	50
Petroleum, Naphthenic fraction	No limit found	626	122	50
Mineral Spirits	1.10	760	109	43

[SOURCE: Munger, C. G., Safe Application of Protective Coatings—Identifying the Hazards, Plant Engineering, Feb. 7 (1974).]

TABLE 16.2 — Test Explosive Limit Data
for Mixed Solvents

Solvent	Spark-Explosion Limit, Percent Vapor By Volume in Air	Pressure, mm Hg	Temp., Deg. F	Temp., Deg. C
Ketone Aromatic Thinner	1.13	760	82	28
Ketone Aromatic Cleaner	1.66	760	86	30
Wash Prime Solvents	1.98	760	86	30
Vinyl Tank Lining Solvents	1.40	760	86	30
Vinyl Exterior Coatings and Solvents	1.35	760	77	25
Asphalt Gilsonite Solvents	No limit found	760	122	50
Epoxy Ester Solvents	No limit found	760	122	50
Tank Lining Epoxy Solvents	1.48	760	95	35
Phenolic Tank Lining Solvents	2.49	760	88	31

[SOURCE: Munger, C. G., Safe Application of Protective Coatings—Identifying the Hazards, Plant Engineering, Feb.7 (1974).]

mixed solvents as they would be found in protective coatings. It is interesting to note in these tests that the lower explosive limit of the mixed solvents was less than the lower explosive limit of the individual solvents with the highest lower explosive limit (LEL), and the lower explosive limit of the mixture was above that of the solvent with the lowest LEL.

Flash Point

Flash point is another term which is commonly associated with solvent-type coatings. The *flash point* is commonly defined as the temperature of the solvent in degrees Fahrenheit at which it releases sufficient vapor to ignite in the presence of a flame. The higher the value, the safer the solvent (with respect to the flash point).

There are two ways of determining a flash point: (1) the closed-cup method, which evaporates the solvent in a closed vessel, measuring the temperature of the liquid solvent, and determining the temperature at which the solvent flashes in the presence of a flame; and (2) the open-cup method where the solvent is placed in an open cup, again with the temperature of the liquid solvent being determined, and the flash point, in this case, being the temperature at which the solvent in an open container flashes in the presence of a flame. The open-cup flash points are usually considerably higher than the closed-cup flash points. For this reason, the closed-cup procedure is used more often in determining this property.

Evaporation Rate

Evaporation rate, or the relative evaporation time of a solvent, is based on an arbitrary value of 1 for ethyl ether. In this case, the higher the evaporation time, the longer it will take for the solvent to evaporate and form solvent vapor in air. This is both good and bad. A solvent with a relatively slow evaporation rate will be slower in buildup to the lower explosive limit. On the other hand, by staying wet longer, the flammability is more of a hazard.

Evaporation rates are important from a coating safety standpoint because the solvents in a coating continue to evaporate over a period of time and not just during the application process. This means that once a coating is applied, the solvents are evaporating during the drying period in accordance with their respective evaporation rates. These can build up solvent concentration, particularly in a closed space, to the point where the lower explosive limit of the solvent is reached.

This was vividly demonstrated in one instance when a coating was being applied to the interior of a tank. When the coating was completed, the ventilation fans were removed. After an hour or so, a welder who was doing some work nearby needed a place to cut a piece of steel. He placed it across the manhole of the tank and proceeded to cut the steel with a torch over the manhole opening. The solvent in the bottom of the tank had built up to a lower explosive limit, and as the hot metal from the cutting torch dropped down to the lower portion of the tank, the solvent-air mixture exploded. While no one was hurt in this particular instance, there was a very surprised and frightened welder who will never again use a manhole as a convenient place to cut a piece of steel.

Solvent Vapor Density

Solvent vapor density is another safety factor to be seriously considered. Nearly all solvent vapors are heavier than air and therefore will concentrate in the lower portion of an enclosed space, which is what happened in the above example. The solvents in the tank had accumulated in the lower area in sufficient volume to create a zone of solvent vapor in air which was above the lower explosive limit in the bottom of the tank. This is often a cause for an explosion in a garage or home (*e.g.,* a water heater) where there is a source of escaping gasoline or natural gas. The concentration of the vapor of gasoline or natural gas residues spreads across a level floor to a point where the explosive limit of the air mixture is reasonably close to the flame. This is the point that an explosion occurs.

It must be recognized, then, that since solvent evaporation from a coating is governed by time, safety precautions must be taken in any closed area where solvent-based coatings are being applied. Even though there may not be sufficient solvent left in the coating to create an explosive condition within the total space of the enclosed area, for a considerable period following the coating application, the solvents (being heavier than air) accumulate in a low spot in the tank to the point of reaching a lower explosive limit in that area. This often occurs on the in-

terior of complex tanks, such as aboard ship, in complex closed areas in nuclear power plants, or in complex vessels where there may be one or more baffles in the container being lined.

Boiling Point

The boiling point of a solvent is relatively easy to determine. However, it has little to do with the actual hazards involved in working with the solvent. The *boiling point* of a solvent is the temperature at which the solvent rapidly changes from a liquid to a gas or vapor. Another way of determining this is to consider the temperature of condensation, where the solvent vapor condenses and forms into a liquid. This property, however, is not widely used in safety considerations.

Table 16.3 shows the properties of various solvents used in coatings, including the relative evaporation rate, flash point, explosive limits (both upper and lower), vapor density, and boiling point. These, of course, are all for individual solvents, and none of the figures would be necessarily representative of the solvent mixture in a corrosion-resistant coating. These figures do indicate that the boiling point has little to do with the lower explosive limit of a coating.

Some similar properties are given for several of the more common corrosion-resistant coatings in Table 16.4. This table shows that the percent of solvent vapor in air (lower explosive limit) for the various coatings (with the exception of oil-base coatings which might contain turpentine) is above a solvent vapor percentage of 1. This is important since it gives a relatively easy rule-of-thumb for the measure of safety using these various coatings. *If the percentage of solvent vapor in air is kept below 1% at all times, then no explosion can occur.* This concept is the key in the reduction of explosion possibilities to the barest minimum.

Calculation of Solvent Vapor Concentration

It has been calculated that a normal (*i.e.,* not high solids) vinyl coating, when sprayed into a closed space, would produce approximately 25 cubic feet of solvent vapor per gallon. Vinyl coatings have relatively low solids, so they have a relatively high solvent percentage in the vehicle. Therefore, it is a reasonably safe assumption to use 25 cubic feet of solvent vapor for every gallon of coating as a measure for the calculation of the lower explosive limit in any given enclosed space. For example, if a tank has a capacity of 37,500 gallons, it would have a volume of 5,000 cubic feet; 1% of this volume is 50 cubic feet. This would allow the spraying of 2 gallons of coating into the tank before the lower explosive limit is reached. However, this assumes that the solvent vapor is uniformally distributed throughout the tank and not concentrated in any one particular area. Keeping the percent of solvent vapor in air well below the 1% figure is one of the best safety precautions that can be taken.

Reactivity

Ordinarily, reactivity is not a major problem from a safety standpoint. However, it should be realized that

TABLE 16.3 — Properties of Flammable Liquids Used in Coatings and Lacquers

Solvent	Evaporation Relative Rate (Ethyl Ether 1)	Closed Cup Flash Point, Deg. F	Deg. C	Explosive Limits Percent by Volume in Air LEL	UEL	Vapor Density (Air = 1.00)	Boiling Point Deg. F	Deg. C
Acetone	4	0	32	2.15	13.0	2.0	134	56
n-Amyl Acetate	50	76	25	1.10	—	4.49	300	149
n-Amyl Alcohol	100	91	32	1.20	—	3.04	280	138
Benzene	8	12	−9	1.40	8.0	2.77	176	80
Sec. Butyl Acetate	—	66	16	1.70	—	4.00	234	110
n-Butyl Acetate	30	72	22	1.70	15.0	0.88	260	127
n-Butyl Alcohol	70	84	29	1.70	—	2.55	243	118
Carbon Tetra Chloride	8	—	—	—	—	—	—	—
Cellosolve	100	104	40	2.60	15.7	3.10	275	135
Cellosolve Acetate	32	124	50	1.71	—	4.72	313	155
Cyclohexanone	—	147	85	—	—	3.38	313	155
Di Acetone Alcohol	200	145	84	1.8	6.0	—	—	—
Ethyl Acetate	8	24	−6	2.18	11.5	3.04	171	77
Ethyl Alcohol	20	55	14	3.23	19.0	1.59	173	78
Fuel Oil No. 1	—	100-165	38-75	—	—	—	—	—
Furfural Alcohol	—	167	76	1.8	16.3	3.37	340	170
Gasoline	—	−40	−40	1.3	6.0	3-4	100-400	38-204
n-Hexane	—	−7	−21	1.25	6.9	2.97	156	70
Isopropyl Alcohol	25	65	16	2.5	5.2	2.07	181	82
Kerosene	800	100-165	38-75	—	—	—	—	—
Methyl Alcohol	10	54	13	6.00	36.5	1.11	147	65
Methyl Amyl Ketone	—	120	50	—	—	—	303	150
Methyl n-Butyl Ketone	—	—	—	1.22	8.0	3.45	262	128
Methyl Isobutyl Ketone	20	73	22	1.34	8.0	3.45	2.44	118
Methyl Ethyl Ketone	8	30	−1	1.81	11.5	2.41	176	80
Methyl n-Propyl Ketone	—	—	—	1.55	8.1	1.96	216	202
Mineral Spirits	150	100-110	38-43	1.10	6.0	—	300-400	149-204
Naphtha, Coal Tar	105	100-110	38-43	—	—	—	300-400	149-204
Naphtha, Safety Solvent	—	100-110	38-43	1.10	6.0	—	300-400	149-204
Naphtha, V.M. & P.	20	20-45-7- +7	—	1.20	6.0	—	212-320	100-160
n-Propyl Alcohol	—	59	15	2.50	—	2.07	207	96
Stoddard Solvent	—	100-110	38-43	1.10	6.0	—	300-400	149-204
Toluene	15	40	4	1.27	7.0	3.14	232	111
Turpentine	100	40	35	0.80	—	—	300	149
Water	100	—	—	—	—	—	212	100
o-Xylene	35	63	16	1.00	—	3.66	291	145

[SOURCE: Munger, C. G., Safe Application of Protective Coatings—Identifying the Hazards, Plant Engineering, Feb. 7 (1974).]

many of the new coatings, especially those with 100% solids, are internally reactive and can generate a substantial amount of heat if allowed to remain in a container in any significant volume after the catalyst is added. Epoxy coatings mixed with a fast curing agent, for instance, can develop sufficient heat so that the coating within the container will smoke and possibly catch fire. This is an internal reaction throughout the mixture, and the center of the mass of liquid is the hottest area since it has the most insulation. In fact, one painter was mixing coatings of this type, and, before he could get from the mixing area to where he wished to apply the coating, the epoxy coating was expanding out of the can, smoking, and, once the painter got away from the container, it actually caught fire. While it was thoroughly doused with water, the reaction continued unabated since the cooling capability of the

water on the outside of the container was not sufficient to cool down the internal mass of the coating.

Reactions such as this are primarily important when someone mixes the catalyst and the resin and then allows them to sit for some period of time (even overnight, if they are forgotten). Under these conditions, the exothermic reaction can build up to a flammable level. Fully polymerized coatings, such as vinyl resins, acrylics, chlorinated rubbers, and materials of this type do not have any internal reactivity so that they do not suffer from this type of hazard. Epoxies, polyurethanes, and similar reactive materials such as polyesters, however, can develop a substantial amount of heat, so that whenever they are being used, the mixed catalyst and resin should not be stored in any volume, even one gallon, without being used immediately.

TABLE 16.4 — Characteristics of Commonly Applied Coatings

Type of Coating	Material Resin	Solvent	Flammability Flash Point Open Cup Deg. F	Deg. C	% Solvent Vapor in Air, (L.E.L.)	M.A.C. Parts/Million Solvents	Toxicity Resin
Oil-Based Paint	Linseed Oil—Flammable	Aliphatic hydrocarbon may contain turpentine.	100 +	38	0.80 if contains turpentine, 1.1 if contains mineral spirits.	If turpentine, 100; If only mineral spirits, 500.	Nontoxic
Alkyds	Flammable	Aliphatic or aromatic petroleum	100	38	1.1	200-500	Nonirritating
Chlorinated Rubber	Chlorinated rubber and plasticizing resins, generally nonflammable.	Aromatics	40-60	4 - 16	1 - 1.3	100	Chlorinated rubber is nontoxic. Possible skin irritation due to aromatic solvent.
Vinyls	High molecular weight chlorinated hydrocarbons, self-extinguishing	Ketones and aromatics	30-100	– 1 - – 38	1.3 - 1.8	100	Nontoxic
Vinyl (water dispersed)	Vinyl acetate, will support combustion	Water	None	None	None	None	Nontoxic
Epoxies (solvent base)	Epoxy-amine Epoxy-polyamide, will support combustion	Ketones and aromatics	30-95	– 1 - – 36	1.3 - 1.8	100	Some cause dermatitis in sensitive individuals.
Epoxy-Coal Tar	Will support combustion	Ketone and aromatics	30-95	– 1 - – 36	1.3 - 1.8	100	Fumes very irritating. Skin irritation. Dermatitis.
Epoxy (water dispersed)	Epoxy, will support combustion	Water	None	None	None	None	Possible skin irritation.
Polyurethane	Flammable	Ketones aromatics	40-75	4 - 2.5	1 - 1.3	100	Contains isocyanates. Toxic fumes can cause irritation during application. Dermatitis possible.
Asphalt Gilsonite Cutback	Will support combustion	Aliphatic or aromatic petroleum	100 +	38	1.1	200-500	Nontoxic
Coal Tar Cutback	Flammable	Aromatics	40-100	4 - 38	1.1 - 1.27	200	Severe skin irritation.
Inorganic Zinc Silicate (water base)	Inorganic silicate, nonflammable	Water	None	None	None	None	Silicate mildly alkaline. Limited skin irritation.
Inorganic Zinc Silicate (solvent base)	Inorganic silicate, nonflammable	Ethyl Alcohol	55-60	15	3.2	1000	Silicate—possible mild skin irritation.

[SOURCE: Munger, C. G., Safe Application of Protective Coatings—Identifying the Hazards, Plant Engineering, Feb. 7 (1974).]

Health Hazards

Health hazards are last on the list of major difficulties involved in the use of high-performance coatings. This does not mean, however, that they are any less serious than the hazards described earlier. In fact, in many circumstances they may be more serious.

Any substance can be safely utilized industrially if the precautions and protective equipment that are necessary for the safe handling of the product by personnel are economically justified. Some hazards may be sufficiently dangerous because of the amount and character of the toxic material, so that even with maximum safeguards, they should be avoided. However, this is not the case with most protective coatings. Most are not so toxic that the proper equipment and protective clothing cannot provide full protection.

Physiological Action of Solvents

In order to provide some understanding of the psysiological action of solvents, the following are some excerpts from a book by A. K. Doolittle, *The Technology of Solvents and Plasticizers,* which examines the physical effects of solvents.

Foreign chemicals must reach susceptible cells in order to injure the body. To injure internally they must enter the body fluids, and their physical properties must allow transportation within the body. A compound in the digestive tract or in the lungs does no harm until it contacts susceptible cells. It may injure surface cells in the lining of these cavities by dissolving in the film of fluid upon them, and, if it is sufficiently irritating, it may kill these boundary cells. It may be absorbed into the circulating body fluids and contact susceptible cells at a distance. But, if it is not absorbed at all and if it is not sufficiently irritating to injure surface cells, it cannot harm the body, and eventually it will be eliminated from the cavity.

The most rapidly moving fluid circulation of the body is the blood stream, but it bathes directly only the cells lining its vessels. From its ultimate branches the capillary network, the fluid components reach every cell in the body.

...Chemicals soluble in fats are withdrawn from the blood fluid by tissues high in fat and can attain high concentrations and exert their action selectively upon particular organs such as the nervous system. Substances like ethyl alcohol and methyl alcohol will be found almost uniformly distributed throughout the body in the water of tissue and blood, but others, such as benzene, which are more fat-soluble, will be found in greater concentration in tissues high in fat.

When a material is absorbed from the digestive tract, much or all of it enters the blood stream and passes first to the liver... .Thus when a chemical is swallowed it reaches the liver promptly, if it is absorbed at all, where it may be so metabolized as to be harmless but where in the process it has an opportunity to exert serious injury.

Contrasted with this, when a material is absorbed from the lungs or through the skin it enters first the peripheral circulation and passes through the body before the liver has an opportunity to act upon it. It is for this reason that some chemicals exert qualitatively and quantitatively different actions when swallowed and when inhaled. It appears that during the eons in which our bodies evolved there was relatively little chance of injury from gases and vapors, and evolution accordingly did not lead to efficient defense against inhaled chemicals. Today (particularly with the application of protective coatings), we find ourselves with many opportunities for inhaling dangerous vapors, but, because defenses are not adequate to exclude them from our bodies, external controls must be provided.

The injury that a solvent can exert upon a cell is governed by the amount that is absorbed upon the cell wall or the concentration that can be built up within the cell. Either of these amounts depends upon the concentration of the chemical in the fluid surrounding the cell and hence upon the total amount that entered the body.

Absorption of vapors and gases through the lungs is an equilibrium process. The concentration of vapor in the inhaled air almost instantaneously attains equilibrium with that part of the blood which is in the tremendous surface area of the lung capillaries. The concentration in the inhaled air and the distribution coefficient of the vapor between blood and air. It is, of course, also a function of the rate of elimination from the body in urine and by other routes, as well as of the rate of oxidation or other form of metabolism within the body.

The maximum concentration that can be attained in the blood (and hence the maximum injury that can develop in a single exposure) is therefore governed only by the concentration in the air breathed. The blood concentration may never reach an injurious level when a man is breathing 500 p.p.m. of a vapor but may soon reach this level when he is breathing 1000 p.p.m.

The rate of attaining a particular blood level is dependent on the distribution coefficient of the vapor and the activity of the man as it affects his rate of respiration and circulation. Thus 30 minutes' inhalation of a high concentration may not result in a dangerous blood level, whereas 60 minutes may give too much, as may 30 minutes during violent activity with the resulting rapid respiration and circulation.

As soon as a man leaves contaminated air he starts to eliminate vapors from the blood to the exhaled air, and the same equilibrium is maintained as during absorption. His blood level drops, but usually more time is required to eliminate a given amount of vapor than was required to absorb it, because the concentration gradient is likely to be less during elimination, particularly when a fat-soluble vapor has reached fatty tissue. Thus, although a man may inhale a vapor for only 30 minutes, his blood may contain a significant concentration for an hour or more. It is not unusual to find in the morning that the blood still contains dissolved vapor from the day before and the peak blood concentration rises day by day until the end of the week because of slow elimination overnight.[3]

This has been experienced many times where solvents were readily noticeable on the breath and in the urine, even after an overnight rest from coating application. In these instances, the solvents were ketones and aromatics, both of which could be identified by their distinctive odor. This reaction is a common one where workers do not use proper protection. Doolittle continues:

The body mechanisms operate in such a way that the internal environment remains nearly constant. As soon as any foreign chemical or excess of a native chemical is present in the body, various processes tend to eliminate it. Elimination through the lungs is the most important route for a volatile chemical. Elimination through the kidneys into the urine is the next most important path. If a compound does not easily pass into the urine, it may be converted to an ionized acid which facilitates filtration, or it may be oxidized, reduced, or coupled with products such as glucuronic acid, sulfate, or hippuric acid. These chemical changes take place largely in the liver, and liver cells may be injured in the operation. The process is often known as detoxication, but it may result in the formation of a more toxic compound, as when benzene is oxidized to phenol or aniline to p-aminophenol. Elimination through perspiration is much the same as the kidney function, but less important in amount. Some materials leave the liver in the bile, entering the intestine where they may or may not be reabsorbed.

Skin Injury

Contact of chemicals with the skin may result in three kinds of injury. Some compounds cause primary irritation upon the skin of almost every person. This action may be only a mild reddening (erythema) as that resulting from xylene or from traces of organic acid in esters, due to dilation of the capillaries and increased permeability of their walls. It may be a severe erythema and formation of blisters from prolonged contact with turpentine. In more serious cases it may appear as a destruction of the skin (necrosis) leaving an open sore as from dimethyl sulfate, inorganic alkalies, or mineral acids.

Primary irritation is more severe when the solvent contacts the skin under clothing or gloves, because the covered skin is more moist, thus permitting easier penetration through the protective horny layer. Irritation from most compounds starts to subside shortly after contact ceases, and unless necrosis was produced no scar or permanent injury is to be expected, although an irregular distribution of skin pigment may persist for months. Prevention requires avoidance of contact by care, or impervious garments, and prompt removal of any chemical that may reach the skin in spite of precautions.

Compounds that are good fat solvents may injure the skin mechanically by removing the natural protective and plasticizing fats. These substances are also removed by prolonged contact with water or mild alkalies such as soap and phosphate cleaning solutions. Washing the hands with solvents after work is a frequent cause of skin defatting. The epidermis is left harsh and stiff, prone to crack, and a ready prey to bacterial and fungus infection. It will return to normal after infection is conquered if no further exposure is allowed. Prevention requires avoidance of frequent or prolonged contact with objectionable fluids, and replacement of fat removed by the application of hydrated lanolin or another suitable ointment.

Allergic dermatitis is seldom a problem in the handling of the commonly used industrial solvents because these materials do not sensitize many individuals. But when dermatitis does occur it is much more troublesome than the two other forms of skin injury because its manifestations may persist for weeks after contact with the causative agent ceases, and a victim may relapse later after a new encounter with a minute amount of the offending solvent or a closely related molecule. It appears that any compound that is soluble to even a very small extent in body fluids is capable of sensitizing some portion of the population. A resin from cashew nut oil is a frequent industrial offender. About three weeks after a person capable of being sensitized has the first contact with a particular chemical or after some ill-understood change following years of contact, an antibody concentration exists over the entire body. The next encounter after this state is established results in the development of a skin eruption, scaling areas, or blisters from the liberation of histamine in a type of anaphylactic shock. The reaction starts in a restricted area where the skin was exposed to the chemical, but it may eventually involve the entire body surface. Dermatitis is seldom a menace to life, but it may incapacitate a person for weeks. Impervious garments and care will reduce the number of workers who become sensitized, but after a severe attack the victim should not handle the offending chemical again. Barrier creams offer some protection to those who contact solvent infrequently but they must be spread thoroughly over the skin to be of value. When not formulated to match the particular solvent to be excluded they may actually intensify injury by dissolving the offending chemical and holding it in intimate contact with the skin.

Through a similar mechanism, sensitization reactions may also occur in the form of asthma and hay fever symptoms although this phenomenon has been rarely caused by solvents. These result from contact of the mucous membranes of the repiratory tract with chemicals in the form of dust, droplets, or vapor.[3]

Dermatitis

Dermatitis can arise in an even more difficult form from some of the high-performance coating resins such as the epoxies and some of their curing agents. The liquid epoxies are particularly hazardous in this regard, and many people have been seriously incapacitated because of allergic reactions and dermatitis from the epoxy resins. This type of dermatitis is apparently cumulative in that, once the person is sensitized, even limited contact with the resin will cause the dermatitis to break out again all over the body. In one instance a person became so sensitized that even if he was in the same room with the epoxy coating that caused the problem, he would again break out with a severe case of dermatitis.

The effect is very similar to poison oak, with leaking skin blisters and open skin sores developing. Oftentimes these areas are on the parts of the body which sweat most, causing even more discomfort. Not all persons are affected in this manner, and some can be in contact with the resins over a considerable period of time without any ill effect. Nevertheless, the problem is sufficiently serious so that every care should be taken to prevent contact of the epoxy resins with the skin.

The solvents in epoxy coatings can make the problem more severe because they dilute the epoxy resins and the curing agents and make them more penetrating from the standpoint of wetting the skin and the pores. While the solvents themselves might not contribute to the actual dermatitis problem, they do increase the danger from the epoxy resins several fold, because they thin the resins and make them more liquid and therefore more penetrating.

Inhalation of Solvents

Regarding inhalation injury, Doolittle states:

The volatility of most industrial solvents is not great enough to permit a concentration to be reached sufficient to reduce the oxygen content below that necessary to support life, even though the atmosphere were to become saturated with the solvent vapor. Accordingly, injury to a person from entering storage tanks and confined spaces saturated with solvent vapors is not usually caused by oxygen deprivation but is a true toxic process.

Injury from single exposures to the vapors of solvents is most likely to consist of partial or complete anesthesia, because most of them are narcotic agents. The earliest stages of narcosis are not infrequent. These consist of a sense of well-being (euphoria), impairment of judgement, and an increase in reaction time. It is this pleasant state which leads to addiction to some vapors, notably trichlorethylene, tetrachlorethylene, and ethylene. Vapors of the fluid hydrocarbons and the alcohols are too irritating when inhaled to allow frequent addiction to develop.[3]

Several cases are known where applicators have removed air masks and reacted as if they were in a happy state of drunkenness. In one instance, a worker was found sitting on the bottom of a tank, completely oblivious to where he was, singing at the top of his voice. If he had not been noticed by a fellow worker on the outside of the tank, he soon would have been in a much more serious state. In another situation, an applicator was finishing the coating of a tank and was standing so that he was half in and half out of the manhole. This reduced the ventilation to a minimum and he passed out hanging in the manhole. Again, a fellow worker saved him. This demonstrates that even a short exposure to strong solvent fumes can create serious physical problems.

Solvents of any sort in a closed area are a serious hazard. Doolittle continues:

The odor of almost any solvent is sufficiently strong so that the critical workman should feel warned against inhaling sufficient to cause anesthesia. However, complete anesthesia from entering storage tanks inadequately purged of vapor or from major spills of solvent is quite possible. If anesthesia persists for more than a brief interval, death may ensue, due to arrest of respiration or circulation. If an anesthetized person is removed to clean air promptly enough to regain consciousness, he suffers no permanent injury from most solvents, but some chlorinated hydrocarbons and ethylene oxide derivatives may cause serious kidney injury which will result in death after several days. Anesthesia from almost any solvent will cause some degree of lung irritation, and rest under medical care is advisable to guard against a serious respiratory infection.

Accumulation of solvent in the blood due to slow elimination overnight increases the tendency towards accidents. As blood concentrations increase day after day, the degree of narcosis rises until judgement and reaction time are impaired. However, this phenomenon is not what is meant by chronic injury. When enough vapor is inhaled each day to cause damage to liver cells, for instance, which is too great for natural processes to repair overnight, then each day increases the liver injury, and it is considered that chronic poisoning is taking place. Eventually the surplus of functioning tissue in the organ is eliminated and after months of daily inhalations the victim begins to perceive symptoms. At some earlier stage a clinical study might have revealed the injury, but unless careful periodic medical examinations are made the damage will be undetected until it has progress far. In most cases adequate function will be restored if the workman is shifted at the earliest symptoms to a job where he does not inhale the vapors. A notable exception is benzene, which has been known to kill by its injury to the bone marrow months after the last inhalation. The odor of most solvents cannot be depended upon to protect workmen from inhaling enough to cause chronic injury. Dangerous concentrations may build up in a working environment without causing discomfort, because of olfactory fatigue. The same phenomenon explains the lack of perception by a workman of the gradual failure of a chemical cartridge respirator or leakage in an air hood.

The commonest sites of chronic injury from solvents are kidney, liver, blood-forming organs, central nervous system, and peripheral nerves. The early symptoms of chronic injury are seldom sufficiently characteristic to lay the blame upon a particular solvent in a mixture. They are part digestive disturbances, a feeling of oppression in the chest, fatigue, insomnia, restlessness, nausea, shortness of breath, personality changes, anemia, and loss of weight. The diagnosis of the cause of such symptoms is difficult for the physician....In view of this lack of sensitivity in diagnosis it is desirable that every effort be made to control exposures so that no injury results.

When workmen first handle a new solvent there is likely to be a wave of complaints of weakness, nausea, insomnia, and the like. These should never be ignored, because they may have a physiological basis. On the other hand, many times the complaints have only a psycholgical basis. Usually after a few days or weeks no one feels the objectionable symptoms at all because the odor is now familiar. If processes and protective equipment had to be adjusted to the point where no one had any subjective complaint, vapor concentrations would have to be reduced to an extremely low level, far below the value at which the first physiological injury could develop. The above considerations show one of the difficulties encountered in trying to establish a fixed maximum allowable concentration for a particular material.

Injury by Skin Penetration

Skin penetration of solvents, and to a very minor extent of their vapors, occurs more frequently than is commonly supposed. Aniline and chlorohydrins are notorious offenders. Penetration through the skin leads to systemic injuries similar to those of inhalation. The factors that influence the rate of skin penetration of a particular compound are not well defined. It is wise that unnecessary skin contact with any solvent should be avoided and that clothing wet with a chemical should be removed at once and laundered before being reworn.

Injury by Swallowing

It is probably correct that no industrial solvent except water can safely be swallowed, and that there is no excuse for drinking any of them, because the desirable effects one expects from a beverage will not result, or will be accompanied by such unpleasant side effects that the experiment will be unhappy or fatal. Nevertheless, almost every solvent has been drunk either by accident or by intent. Fortunately, a frequent result is the protective reflex of vomiting. When a solvent is swallowed it will produce all the effects that it would if inhaled in high concentration. Even lung irritation is to be expected because much of the dose will leave the body as vapor in the expired air.

Eye Injury

The entry of most solvents into the eye in the form of drops will cause corneal necrosis. Fluid should be flushed out of the eye at once with copious amounts of water. No other flushing solution will serve so well, even when the solvent is completely insoluble in water. If the eye is apparently damaged, expert medical aid should be sought.

High concentrations of vapors may result in similar damage to the cornea, and lower concentrations can cause chronic inflammation of the conjunctiva. There have occurred a few cases of very spectacular injury in which the cornea was so damaged as to be completely opaque; yet the only known exposure was to a moderate concentration of a usually inoffensive solvent vapor. One of these cases was traced to a vinyl or similar unsaturated impurity present in small concentration in the solvent. Happily there was no permanent loss of vision in any of these victims.

The *acute toxicity* of a compound is the injurious effect that follows a single dose by mouth or on the skin or the inhalation of vapors for a single period up to one day. *Chronic toxicity* is the summation of small injuries acquired by many single doses or inhalation extending over a period of months or years.

Toxicity of Individual Solvents

Aliphatic hydrocarbons are so fat soluble or lipid soluble that they can attain a high concentration in the central nervous tissue. Their elimination from these areas is slow, primarily through the lungs. There is no significant metabolism of these materials within the body. The physiological effect is simply one of narcosis, which reaches its peak at about eight carbon atoms. The longer carbon chain molecules are less of a problem. In several of the hydrocarbons, the anesthetic dose is very close to the fatal dose, so that even though straight chain hydrocarbons like kerosene appear innocuous, they can still cause difficulties. Multiple bonds in the molecule increase the narcotic effect, and a single exposure can lead to inebriation, nausea, headache, unconsciousness, convulsions, and even death.

Aromatic hydrocarbons or alicyclic hydrocarbons are more strongly narcotic than aliphatic ones and somewhat more readily metabolized. Turpines like turpentine injure the skin, lungs, and kidneys more than any of the others, with skin sensitization the most likely result. Aromatic hydrocarbons concentrated in tissues high in fats or lipids are, however, more readily metabolized than the aliphatic hydrocarbons, and the water soluble metabolic products can be eliminated through the kidneys. Benzene, because of its oxidation products, can cause changes in bone marrow and a marked reduction in white blood cells. Aromatics such as toluene or xylene cause less of a problem, even though their narcotic effect is greater. All aromatic hydrocarbon vapors are lung irritants. The very high fat solubility of chlorinated hydrocarbons causes them to concentrate in the fatty tissues and the brain. Elimination of these is through the lungs, but metabolism in this case is even greater than with hydrocarbons so that the metabolic products also are eliminated by the kidneys.

Chlorinated hydrocarbons are much more narcotic than corresponding hydrocarbons. In industrial applications, it is their chronic toxicity that is most important. Foods that are rich in protein and carbohydrates help to reduce the chance of injury by chlorinated hydrocarbons. The lower aliphatic alcohols are rapidly distributed in all of the body fluids and tissues because of their water solubility. The same factor delays elimination through the lungs. As the molecular weight is increased, water solubility drops and lung elimination rises, as does the narcotic power.

Primary alcohols are oxidized in the body to aldehydes and acids. When acids are easily oxidized, the reaction products are CO_2 and water. Primary alcohols are oxidized to ketones, while tertiary alcohols are narcotic agents with very little chronic toxicity. Unsaturated alcohols of only a few carbon atoms are much more toxic than the corresponding saturated compounds, and skin penetration is also possible. Esters are even more powerful narcotic agents than alcohols, but they are easily saponified by the body enzymes, and their physiological effect is chiefly that of alcohol and acid. Few, if any, of the alcohols and carboxylic acids manifest chronic toxicity to any significant degree.

Ketones and aldehydes are largely devoid of chronic toxicity, although the introduction of the double bond between carbon atoms creates a tendency in that direction. Aldehydes are highly irritating to the skin, eyes, and respiratory tract, and the unsaturated representatives are outstanding in this respect. Ethers are little changed within the body, if at all. They react like narcotic agents, but their low water solubility leads to rapid elimination of the more volatile members through the lungs. Chronic toxicity is generally absent. Glycols and their derivatives, if sufficiently absorbed, injure the kidney to some extent and chronic poisoning is possible. All in all, chronic toxicity is expected of benzene and, to a limited extent, of the other aromatic hydrocarbons.

Chronic toxicity is also a characteristic of the saturated chlorinated hydrocarbons, many nitrogen compounds, glycol derivatives, and carbon disulfide. The other solvents, with a few exceptions, are harmful during industrial exposures absorbed during any one day. There can be some accumulation in the blood due to slow elimination of the solvent, but there is not the continuing day-after-day injury which is the criterion of chronic effect.

Table 16.5 lists a few of the more common solvents which may be encountered when high-performance coatings are used. The Threshold Limit Value (TLV) is given and is listed in two ways: (1) Threshold Limit Value—Time Weighted Average (TLV-TWA), and (2) Threshold Limit Value—Short Term Exposure Limit (TLV-STEL). The definition of these terms is given by the American Conference of Governmental Industrial Hygienists (A.C.G.I.H.).

(1) Threshold Limit Value—Time Weighted Average (TLV-TWA)—the time-weighted average concentration for a normal 8-hour workday and a 40-hour work week, to which nearly all workers may be repeatedly exposed, day after day, without adverse effect.

(2) Threshold Limit Value—Short Term Exposure Limit (TLV-STEL)—the maximal concentration to which workers can be exposed for a period up to 15 minutes continuously without suffering from (1) irritation, (2) chronic or irreversible tissue change, or (3) narcosis of sufficient degree to increase accident proneness, impair self-rescue, or materially reduce work efficiency, provided that no more than four excursions per day are permitted, with at least 60 minutes between exposure periods, and provided that the daily TLV-TWA also is not exceeded. The STEL should be considered a maximal allowable concentration, or ceiling, not to be exceeded at any time during the 15-minute excursion period.[4]

These limits should be used as guides in the control of health hazards and should not be used as fine lines be-

TABLE 16.5 — Threshold Limit Values of Common Solvents

| Common Solvent Names | Threshold Limit Value—Time Weighted Average | | Threshold Limit Value—Short-Term Exposure Limit | | Major Health Hazard |
	TLV ppm in Air	TWA mg/cu m Air	TLV ppm in Air	STEL mg/cu m Air	
Acetone	230	1780	1000	2375	Skin irritant, headaches.
n-Amyl Acetate	100	525	150	800	Narcotic effect, irritant.
Benzene (Benzol)	15 (skin)	30 (skin)	25	75	Toxicity by inhalation and skin absorption.
n-Butyl-Acetate	150	710	200	950	Irritation to eyes and respiratory tract, narcotic.
sec-Butyl Acetate	200	950	250	1190	Irritating to eyes and respiratory tract.
n-Butyl Alcohol	50	150	—	—	Irritating to eyes, nose, and throat; causes headache and dizziness.
Butyl Cellosolve (2-Butoxy Ethanol)	25	120	75	3600	Skin irritant
Cellosolve	200	—	—	—	—
Cellosolve Acetate (2-Ethoxyethyacetate)	100	540	—	—	—
Carbon Tetrachlorides	5 (skin)	30 (skin)	20	125	Narcotic, can cause organ damage.
Chloroform	10	50	50	225	Anesthetic; causes eye irritation.
Cyclohexanone	25	120	100	400	Eye and throat irritation, mild narcotic.
Diacetone Alcohol	50	240	75	360	Irritating to eyes and mucous membranes; mild narcotic.
Diisobutyl Ketone	25	150	—	—	Narcotic and anesthetic effect.
Ethyl Acetate	400	1400	—	—	Irritating to eyes and respiratory passages.
Ethyl Alcohol (Ethanol)	1000	1900	—	—	Irritant; narcotic.
Ethyl Ether	400	1200	—	—	Anesthetic

(Continued)

TABLE 16.5 — Threshold Limit Values of Common Solvents (Continued)

| Common Solvent Names | Threshold Limit Value—Time Weighted Average | | Threshold Limit Value—Short-Term Exposure Limit | | Major Health Hazard |
	TLV ppm in Air	TWA mg/cu m Air	TLV ppm in Air	STEL mg/cu m Air	
Ethylene Dichloride	10	40	15	60	Strong irritant
Ethylene Glycol Monobutyl Ether	50	240	—	—	Irritant
Ethylene Glycol Monoethyl Ether	200 (skin)	740 (skin)	—	—	Irritant; somewhat toxic.
Ethylene Glycol Monoethyl Ether Acetate	25 (skin)	120 (skin)	—	—	Irritant
Ethylene Glycol Monoethyl Ether	25 (skin)	80 (skin)	—	—	Can cause blood abnormality.
Ethylene Glycol Monomethyl Ether Acetate	25	121	—	—	Irritant
Furfuryl Alcohol	20	220	15	60	Somewhat toxic
Gasoline	300	950	500	1500	Irritant
Heptane	400	1600	500	2000	Irritating to respiratory tract; mild narcotic.
Hexahydrophenol (Cyclohexanol)	50	200	—	—	Irritant
Hexane	50	180	—	—	Irritant
Isobutyl Alcohol	50	150	75	225	Strong irritant
Isopropyl Alcohol	400	980	500	1225	Irritant; mild narcotic.
Kerosene	—	—	—	—	Irritant; headaches; moderate narcotic effect.
Methyl Acetate	200	610	250	760	Narcotic and irritant.
Methyl Alcohol (Methanol)	200	260	250	310	Strong narcotic and mild irritant.
Methyl n-Amyl Ketone	50	235	100	465	Irritant and moderate narcotic.
Methylene Chloride	100	360	500	1700	Dangerous to eyes, narcotic.
Methyl Ethyl Ketone (Butanone)	200	590	300	835	Irritant and narcotic.
Methyl Isobutyl Ketone (Hexone)	50	205	75	300	Irritating to eyes and mucous membrane; moderate narcotic.
Mineral Spirits	100	400	—	—	Irritant
Nitroethane	100	310	150	465	Irritant
Nitromethane	100	250	150	375	Irritant
1-Nitropropane	15	55	25	90	Moderately toxic
Perchloroethylene	50	335	—	—	Toxicity inhalation or swallowing.
n-Propyl Alcohol	200	500	250	625	Irritant
Propylene Oxide	20	50	—	—	Moderate irritant
Stoddard Solvent	100	525	200	1050	Mild irritant
Tetrahydrofuran (THF)	200	590	250	735	Irritating to eyes and mucous membrane; moderate narcotic.
Toluol (Toluene)	100	370	150	560	Moderate narcotic; can cause organ damage.
Tolene-2, 4-Diisocyanate (TDI)	0.005	0.04	0.02	0.15	Toxic
Trichloroethane 1-1-1	350	1900	—	—	Moderate narcotic and irritant.
Trichloroethylene	50	270	150	805	Narcotic and addictive.
Turpentine	100	560	—	—	Mild cause of allergy; toxic.
VM&P Naphtha	300	1350	400	1800	Causes intoxication.
Xylol (Xylene)	100	485	150	655	Moderate irritant

[SOURCE: Threshold Limit Values for Chemical Substances and Physical Agents in the Workroom Environment with Intended Changes for 1981, American Conferences of Governmental Industrial Hygienists, Lansing, MI.]

tween safe and dangerous concentrations. These limits were previously referred to as *Maximum Allowable Concentrations* or *MAC*.

Dusts

There are some dusty substances which may be involved in the application of high-performance coatings. These are materials such as quartz sand, flint grit, garnet, most beach or river sands, and a number of the synthetic abrasives such as copper slag. The ones containing silica are the most difficult from a health hazard standpoint. However, even copper slag has a TLV as it would be in the class of Nuisance Dusts (particulates). These are described by the A.C.G.I.H. in this way:

> Nuisance Particulates. In contrast to fibrogenic dusts which cause scar tissue to be formed in lungs when inhaled in excessive amounts, so-called "nuisance" dusts have a long history of little adverse effects on lungs and do not produce significant organic disease or toxic effect when exposures are kept under reasonable control. The nuisance dusts have also been called (biologically) "inert" dusts, but the latter term is inappropriate to the extent that there is no dust which does not evoke some cellular response in the lung when inhaled in sufficient amount. However, the lung-tissue reaction caused by inhalation of nuisance dusts has the following characteristics: (1) The architecture of the air spaces remains intact; (2) Collagen (scar tissue) is not formed to a significant extent; (3) The tissue reaction is potentially reversible.
>
> Excessive concentrations of nuisance dusts in the workroom air may seriously reduce visibility, may cause unpleasant deposits in the eyes, ears and nasal passages (Portland Cement dust), or cause injury to the skin or mucous membranes by chemical or mechanical action per se or by the rigorous skin cleansing procedures necessary for their removal.[4]

Table 16.6 provides a list of materials which, as dusts, are considered hazardous.

Pigmentation

In addition to the hazards involved in solvents and resins, there are also some hazards that result from the pigmentation of high-performance coatings. Many of the toxic pigments are used in prime coats as corrosion inhibitors. The vinyl wash primer coating systems contain red lead in the vinyl intermediate coat, which follows the wash-prime coat. In this case, it is a very effective inhibitor and well worth the extra safety precautions which should be used during its application. The wash primer also contains active chromates as inhibitors in addition to phosphoric acid, both of which can be a hazard if in contact with the skin or eyes.

Lead compounds, in addition to red lead, may be incorporated in many different binders, *e.g.*, alkyds, epoxies, vinyls, acrylics, etc. The types of lead compounds that may be found, in addition to red lead, are metallic lead, lead chromates, chrome yellow, chrome orange, chrome green, basic lead silica chromates, basic lead carbonate, basic lead silicate, blue lead, di-basic lead phosphite, and di- and tri-basic lead phospho-silicates. These are all good pigments. Unfortunately, all lead compounds are toxic to a greater or lesser extent.

Of all the "soluble" types of inhibitive pigments, the best known and probably the most effective are the chromates. These materials ionize in water to provide a source

TABLE 16.6 — Equivalent TLVs in mppcf and mg/m³ (Respirable Mass for Mineral Dusts)[1]

Substance	Count mppcf[4]	Threshold Limit Value Resp. Mass mg/m³	Total Mass[2] mg/m³
Silica (iO_2)			
Amorphous	20	(3)[3]	(6)
Cristobalite	1.5	0.05	0.15
Fused Silica	3	0.1	0.3
Quartz	3	0.1	0.3
Tridymite	1.5	0.05	0.15
Coal Dust	(12)	2	(4)
Diatomaceous Earth			
Natural	—	1.5	—
Graphite (Natural)	15	(2.5)	(5)
Mica	20	(3)	(6)
Mineral Wool Fiber	—	(5)	10
Nuisance Particulates	30	(5)	10
Perlite	30	(5)	(10)
Portland Cement	30	(5)	(10)
Soapstone	20	(3)	(6)
Tripoli	(3)	0.1	(0.3)

[1]Assuming that the mass median diameter is 1.5 μm and density is 2.5 g/cm³.

[2]Unless otherwise specified, respirable mass is presumed to equal approximately 50% of total mass.

[3]All values in parenthesis () represent newly calculated values based on equivalence of 6 mppcf = 1 mg/m³ respirable mass and respirable mass = 50% total mass.

[4]mppcf = millions of particles per cubic foot of air, based on impinger samples counted by light-field techniques.

[SOURCE: Threshold Limit Values for Chemical Substances and Physical Agents in the Workroom Environment with Intended Changes for 1981, American Conferences of Governmental Industrial Hygienists, Lansing, MI.]

of hexavalent chromium ions, the active inhibitor. Chromates are available in a wide range of solubilities, from completely soluble sodium chromate to relatively insoluble lead chromate. Among pigments, calcium chromate ($CaCrO_4$) is the most soluble, but when used alone, is too fugitive for long-term protection. The less soluble zinc-potassium complex (4 ZnO • K_2O • 4 CrO_3 • $3H_2O$, known in the U.S. as zinc yellow and in Europe as basic zinc chromate) is one of the more widely used pigmentary inhibitors in the coatings industry.

Still less soluble is strontium chromate ($SrCrO_4$). This pigment is soluble enough, however, to provide effective primer compositions, and its use in primers is rapidly increasing. It is used to a greater extent in high-performance coatings than in basic zinc chromates, because of its reduced solubility. Its lower solubility gives it better stability (compared with zinc yellow) in water-based coatings, which are highly sensitive to large amounts of hexavalent chromium ions and, consequently, are difficult to pigment. Unlike zinc yellow, strontium chromate is a non-complex pigment, free of hydratable material and, therefore, quite heat stable up to 1000 F.

Zinc tetroxy chromate (5ZnO • CrO_3 • 4H_2O, often called basic zinc chromate in the U.S.) is a lower solubility chromate with limited use in inhibitive primers, although it is the principal pigment of the WP-1 wash primer pretreatment [DOD-P-15328D (1978)]. Barium Chromate ($BaCrO_4$) and lead chromate ($PbCrO_4$) have even lower solubility and show little promise as corrosion inhibitors for coatings, although lead chromate is the prime ingredient in the pigments known as chrome yellows and chrome oranges which are used in finish coats because of their good color and opacity.[5]

Toxicity is an important consideration in the use of chromates. Hexavalent chromium has long been suspected of being a carcinogen, and is presently under investigation by several organizations around the world. Recent data from studies involving chromates indicate some connection with its use and the incidence of lung cancer.[5] Safety precautions should be used when applying or removing coatings which contain these pigments.

However, toxic pigments are felt to be less of a hazard than the danger of fire or explosion, since proper clothing and masks can effectively reduce any toxic hazard to a minimum. Also, pigments are incorporated into and coated by the coating resin and are reasonably well encapsulated. The danger from pigmentation is primarily due to physical contact of the coating on the skin, or inhalation of the resinous coated pigments due to dry spraying of the coating during application or during the removal of an existing coating. Hazards from pigmentation can usually be eliminated by the use of protective clothing and the proper respirator equipment.

Some of the tin compounds used in antifouling coatings are not nearly as easy to handle. Some of these materials are in liquid form, others are solids. Nevertheless, they have strong toxic characteristics, and persons spraying antifouling coatings can inhale sufficient fumes or dust to cause nausea, dizziness, and similar symptoms. Antifouling coatings may also contain such materials as Paris Green (copper acetoarsenite), mercury compounds, organic biocides, etc., as well as the common copper compounds (copper oxide, metallic copper). Again, protective clothing and the proper respirators are generally adequate to fully protect the applicator.

Other Chemical Compounds

There are also other chemicals which may be encountered during the coating of structures. Metal surface treatments for zinc and copper are commonly used. Usually these are acidic and may contain some organic liquid as well. Paint removers are often strongly alkaline and may contain strong and very penetrating organic materials, such as phenols. Proprietary metal cleaners of various types may be strongly acidic or alkaline. Brush cleaners often contain strong solvents, as well as other organic compounds such as phenols, and proprietary oil, grease, and sludge cleaners may contain aromatic solvents. All should be treated with caution, and proper protective clothing, masks, gloves, and goggles should be used when handling such compounds.

Urethane

Urethane coatings have, in addition to solvents and pigments which can cause safety hazards, isocyanate reaction products. These are toxic materials and add a special hazard to the use of urethane coatings.

Like several other substances used in coatings, isocyanates can irritate the skin, eyes, and respiratory tract of anyone coming into contact with the vapor or spray mist. Isocyanates may also cause an allergic sensitization or asthmatic response in some susceptible individuals. The distinction between irritation and sensitization-type (allergic type) reactions to isocyanate exposure is very important. Different measures should be taken to protect the individual for each case.

Irritation is a direct response resulting from the contact of isocyanates on the body surface, i.e., skin, mucous membranes of the nasal passages, throat and respiratory tract, eyes, etc. Irritation will take place at airborne levels of about 400 parts per billion in most individuals. Symptoms usually include watering of the eyes and a burning sensation in the nose and throat. The amount of irritation is dependent upon the dose, tissue exposed, and individual susceptibility, but it is generally independent of the individual's exposure history. These acute symptoms are generally reversible.

Sensitization, on the other hand, is a totally different type of response. Sensitization is believed to occur only in certain susceptible individuals as a result of numerous exposures and/or exposure at a high concentration. Exposures after this sensitization has taken place result in an allergic respiratory reaction similar to asthma. The response usually includes coughing, wheezing, shortness of breath, and tightness in the chest. The symptoms may be immediate or delayed. The airborne concentration of isocyanates necessary to cause a reaction in a sensitized individual is much lower than that which will cause an irritation response in most people, and may be as low as 5 parts per billion. Skin sensitization can also develop in some individuals as the result of direct contact with certain isocyanates. Once sensitized, these individuals, when exposed to isocyanate vapors at concentrations which have no effect whatsoever on most people, or when they come in direct contact with very small quantities of the liquid, may develop a rash, swelling of extremeties, redness, and scaling of the skin.

If an individual experiences an irritation response while handling urethane coatings, it should be determined whether or not contact with the isocyanate has caused the irritation. If isocyanate contact was the cause of irritation, a careful evaluation of the handling and/or application procedures should be made. If irritation persists in spite of proper protective measures, ventilation, and personal protection equipment, the individual should be removed from areas where isocyanates are processed or used.

Note: Persons having a sensitization response to isocyanates must not be exposed further to vapors or spray mist containing isocyanates.[6]

The isocyanate effect from the contact of liquid coating can be increased because of the solvents used in the coating. These may be ketones or aromatic solvents, both of which aid in the wetting and penetration of the skin. Proper masks and protective clothing should be used during the application of the urethane coatings.

Exterior Applications

There is no doubt that all of the previously described hazards exist during the application of protective coatings. However, if proper precautions are taken to eliminate the hazards, the actual risks involved can be quite low. The application of most protective coatings on the exterior of equipment, tanks, and structural steel generally provides rather low risks to the applicator. The reason is that there is free circulation of air surrounding the applicator, with a rapid dilution of any of the solvent or coating spray. With only a mild air movement, the solvent and any overspray from the resin coating are rapidly dispersed and moved away from the actual point of application. Unless the air movement is directly toward the applicator, he is in little danger of overexposure to the hazardous materials. There are, of course, areas in exterior applications where solvents might accumulate and not be dispersed as rapidly as indicated above. These areas should be considered the same as enclosed areas, and safety precautions taken in the same manner.

Enclosed Areas

As discussed previously, the primary areas of exposure to solvents and other coating ingredients are those which are enclosed *e.g.,* the interior to tanks, vessels, and ships, even the interior of rooms, pump enclosures, manholes, wet wells, or similar areas. When the coating is applied on the interior of such areas, solvents evaporate from the coating and build up a concentration of solvent vapor in the air. The most effective way of preventing a hazardous condition is to dilute the solvent concentration by mixing it with additional air, *i.e.,* proper ventilation. If the solvent vapor concentration in air cannot reach the level of 1% in the enclosed area, then no explosion can occur. Air dilution also applies to the maximum allowable concentration of solvent to which the applicators may be exposed. The greater the dilution of the solvent concentration in an enclosed area, the less the physical contact that is possible. Physical exposure, as well as fire and explosion hazards, are reduced by proper ventilation.

Ventilation

Proper ventilation is the key to safe application of coatings on the interior of closed areas. Its significance cannot be overemphasized from the standpoint of fire, explosion, or health. It is too often mistakenly assumed that proper ventilation is provided by placing a fan or blower in a manhole, doorway, window, or tank opening and blowing air into the area where the coating is being applied. Unfortunately, the dilution of the contaminated air occurs at the inlet, the pressure from the fan pushing air in and the internal pressure immediately pushing it out the same opening. The same is true with an exhaust fan under the same conditions. In this case, the exhaust fan pulls some of the contaminated air from the interior, but does it primarily at the opening. The suction pulls air from the outside around the fan and does not allow proper air dilution deeper in the interior of the closed area. Even if a short ventilation duct is used and air is blown in with a blower through the duct, it only changes the air in the area of the end of the duct, forcing the incoming air out through the opening where the blower is located.

These types of ventilation and the path of air circulation are shown in Figure 16.1. It can be seen that most of the air in the enclosed space is static under these condi-

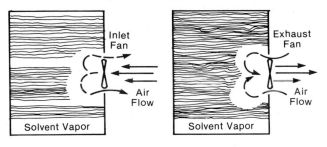

Not Recommended

Not Recommended

FIGURE 16.1 — **(a) Placing the blower in the tank opening is not recommended because air is circulated only at the opening; (b) Extending the short trunk from the blower may be unsatisfactory because air is not evenly circulated.** [SOURCE: Munger, C. G., *Safe Application of Protective Coatings—Recommended Practices*, Plant Engineering, March 21 (1974).]

tions and is not circulated or diluted. This allows the concentration of solvent and fumes to build up above the maximum allowable concentration or the lower explosive limit, both of which are hazardous conditions.

In order to provide proper ventilation to the interior of an enclosed area, air must be exhausted rather than blown into the area. Air blown into the area does not create the same degree of ventilation as does the exhaust of air from the same enclosure. When solvent vapors are heavier than air, as previously indicated, they tend to fall to the bottom of an enclosed space or tank and accumulate there. Thus, the exhaust system must draw the air from the lowest portion of the tank or enclosed space. This automatically draws air from the upper portion of the tank down across the bottom and into the duct. When this is done, clean air from the outside washes down from the upper portion of the tank, bringing the heavier-than-air solvents with it, and is immediately evacuated from the lowest portion of the enclosed area. This provides the most effective ventilation and the maximum removal of solvents.

The interior of a tank or container is often quite complicated, with a number of baffles, or, in the case of a tank aboard ship, the area is filled with reinforcing members, stiffeners, and similar structural members. These can divide the lower portion of the interior of the tank into isolated spots where the air circulation is not good unless the duct system is designed to bring the solvent fumes and air from each one of these isolated areas. For example, in some of the liquid natural gas vessels, there are approximately 600 individual compartments that are joined only by small manholes. The evacuation of the solvents in a complicated structure such as this is both necessary and extremely difficult. In cases such as this, holes actually had to be made in the hull of the ship to provide the proper ventilation required when the interior of these compartments was coated.

Again it must be stressed that suction blowers are the most satisfactory method of removing fumes from a closed area. Blowing air into a tank is not satisfactory since circulation of the air does not reach many portions of the tank. The suction fan or blower uses the force of the weight of the outside air to help remove the solvents from these low areas. Figure 16.2 shows some schematic diagrams of the various ways in which full removal of the solvents can be obtained.

Each individual tank is different. However, if the principle of pulling the air from the lowest portion of the tank is practiced, good air circulation and solvent removal should result. In addition, suction fans tend to lower the air pressure to some degree within the enclosed area,

making for more rapid solvent dispersion and solvent evaporation. Pushing air into a tank, however, has just the opposite reaction, partially increasing the air pressure and reducing the solvent dispersion.

During World War II, there were many underground concrete storage tanks lined with vinyl and other coatings to resist the petroleum products that were to be stored in these tanks. They were quite large, approximately 100 feet in diameter and 30 feet deep, and for the most part, since they were underground, had only one or two manholes in the tanks and these were in the top. The manholes were large, and 3-foot diameter suction trunks were dropped through the top manhole to the bottom of the tank with a large suction blower attached at the top. Even though the solvents for the vinyl coatings were extremely volatile and a large amount of coating was applied in any given period, air circulation in these tanks was sufficient so that there was no solvent problem in the interior of the tank. The 20,000-cubic-foot suction blower changed the air in the tanks every 10 minutes. Air circulation was sufficiently good so that the solvent-air mixture was kept far below the level of either a health or explosion hazard.

The coating of large ducts, tunnels, or pipe also presents a ventilation problem, even though it is an easy one to correct, provided the personnel working in the area understand how to cope with the ventilation. In this case, there should be two openings into the pipe or tunnel. An exhaust blower should be used at one of the openings and, if more air circulation is required, a pressure blower or intake fan should be placed at the other opening. The coating work should be started adjacent to the suction outlet, with the flow of air coming towards the worker, who then works toward the air inlet with his back towards the air flow. With this procedure, all of the solvent fumes and contaminated air are swept away from the point of application and the worker is constantly in a flow of fresh air. This type of operation is shown in Figure 16.3, which indicates the proper procedure for such work.

Buddy System

One mandatory safety procedure in the application of coatings in an enclosed area is the *buddy system*. Applica-

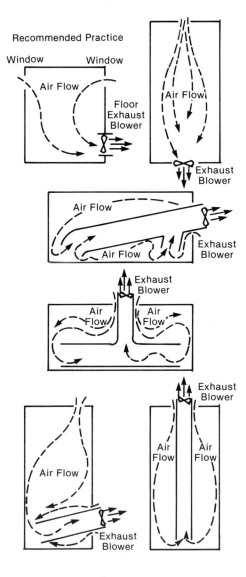

FIGURE 16.2 — Recommended ventilation procedure draws clean air from opening at top by exhausting air from the lower portion of the enclosed area. [SOURCE: Munger, C. G., Safe Application of Protective Coatings—Recommended Practices, Plant Engineering, March 21 (1974).]

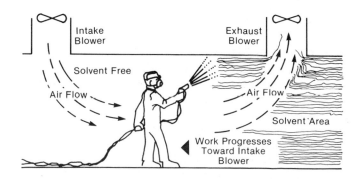

FIGURE 16.3 — A suction blower at one manhole and a pressure blower at a second manhole provide a solvent-free work area in pipes and ducts. [SOURCE: Munger, C. G., Safe Application of Protective Coatings—Recommended Practices, Plant Engineering, March 21 (1974).]

Corrosion Prevention by Protective Coatings

tion of coatings by a single individual in such places not only increases the hazards involved, but should the unforeseen take place (*e.g.*, solvents causing a worker to pass out), there is a much better chance of a warning and immediate remedial action by the second worker in the area. If the enclosed area is small, the second person should be watching through the manhole. If the area is large, more than one person usually will be working anyway. Two people working as a team should always be in reasonably close contact (by safety harness and rope if necessary) so one can provide immediate help to the other should any problem occur.

Solvent Retention

One of the items which must be kept in mind when applying solvent-containing coatings to the interior of a closed area is that the solvents are retained in a coating for a considerable period of time after it is applied to the surface. While it is true that a large amount of the solvent is evaporated in the first few minutes during and after application, solvents are retained in the coating film in sufficient volume so that the lower explosive limit can be reached, particularly in low areas of the enclosed space. Thus, it is necessary to continue the ventilation of the closed space for several hours following the last of the coating application on the interior. This is necessary to make sure that the solvent evaporating from the film does not build up to a lower explosive limit and create an explosion hazard (as was demonstrated by the welder mentioned earlier who used the manhole over a tank to cut a piece of steel).

Figure 16.4 shows the percent of retained solvent for different thicknesses of coating using different solvents. As this figure illustrates, higher boiling solvents are retained for a much longer period and in a greater concentration than the lower boiling ones. This same phenomenon is indicated in Figures 16.5 and 16.6, where two different coatings were applied and the solvent evaporation measured after each coat had been applied. Figure 16.5 shows the evaporation from the various coats of a 5-coat vinyl system, and indicates that most of the solvent evaporated within a matter of minutes, but that the solvent continued to evaporate over a period of several hours. As each coat was applied, it took a greater period of time to evaporate the solvent from the coating, since the solvent from each individual coat penetrated the previous one. With the greater thickness of coating after each coat, a longer drying period was required.

Figure 16.6 shows the same information for a 5-coat epoxy coating. In this instance, the epoxy was an amine-catalyzed coating. The drying of the solvents from the coating was somewhat slower than that for the vinyl, as indicated by the continuing slope of the solvent evaporation curve. This indicates that a considerable volume of solvent remained in the coating for a period of several hours, enough to create a hazard if the ventilation of the enclosed space is not continuous for several hours after the initial application of the coating. Even though the explosive limit in the tank probably would not be reached for the entire tank volume, there is sufficient solvent left in the coating for it to concentrate in the bottom of the

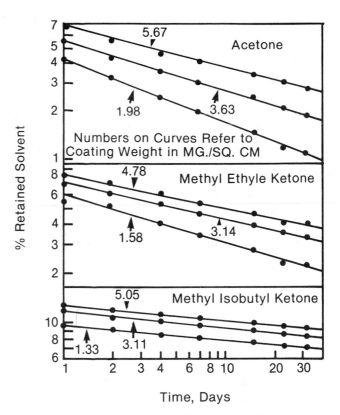

FIGURE 16.4 — Percentages of solvent retained by a vinyl film. (**SOURCE: Doolittle, Arthur K., Technology of Solvents and Plastisizers, John Wiley and Sons, New York, NY, 1954.**)

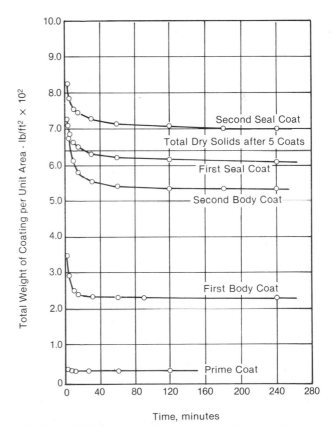

FIGURE 16.5 — Evaporation rate from a 5-coat vinyl tank lining. (**SOURCE: Munger, C. G., Coatings and Their Safe Application, Marine Section National Safety Congress, 1964.**)

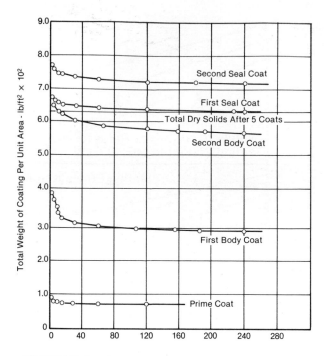

FIGURE 16.6 — Evaporation rate from a 5-coat epoxy tank lining. (SOURCE: Munger, C. G., Coatings and Their Safe Application, Marine Section National Safety Congress, 1964.)

enclosed space. If a spark of some type were to occur, a damaging explosion could take place with the solvent concentration in the lower areas of the structure. The primary safety precaution for enclosed spaces, then, is to continue the ventilation in a tank or enclosed area several hours after the application, not only to reduce the hazard from fire or explosion, but also to aid in the speed of the drying of the coating.

Summary

Taking all of the above information into consideration, the application of solvent-containing protective coatings can be reasonably free from fire, explosion, or health hazards if a few simple rules of safety are practiced. The rules are good for the application of any protective coating applied in an enclosed area or where solvents are an essential part of the coating.

1. Good ventilation is by far the most important safety rule. With adequate ventilation, there is little chance that the lower explosive limit of the solvent will be reached, and, if the lower explosive limit is not reached, no explosion can occur. Adequate ventilation also reduces health hazards. This is true whether the solvents are aliphatic hydrocarbons, or low toxicity or aromatic hydrocarbons with considerably more dangerous physiological properties. Good ventilation is the key to coating safety practice wherever coatings are applied in enclosed spaces.

2. When ventilating an enclosed space, always use suction blowers. Solvent vapors are always heavier than air, so that solvent concentration can build up in the lower areas of the enclosed space. Suction blowers should be operated in such a way that no solvent concentration can build up in pockets within the tank, and free air move-

ment is available from all isolated spots. Air should be sucked into the blower system from the lowest portion of the tank. In this way the solvents are swept down through the upper area and removed from the lowest area.

3. Ventilation should be continued in the coated area until the coating is sufficiently dry so that, with the ventilation removed, no area in the tank will build up to an explosive limit.

4. Explosion-proof equipment must be standard wherever coatings are applied on the interior of a closed space. Electrical equipment and lights must be explosion-proof. Tools and shoes must be spark-proof. Safety equipment must be readily available and considered an essential part of any safety program covering the application of coatings to enclosed areas.

5. Smoking, matches, fire, flames, torches, and welding must be prohibited within 50 feet of an area where coatings are being applied. This means 50 feet from where they are applied to the exterior of tankage and equipment, as well as 50 feet from the access openings to enclosed spaces.

6. Visitors and workers in enclosed areas must wear compressed air-type masks at all times to insure that everyone always has a supply of clean, uncontaminated air. This is a simple safety precaution to prevent inhalation of solvent fumes.

7. Where epoxies, coal tar epoxies, coal tar, phenolic coatings, or polyurethane coatings are being applied, protective clothing should be worn to eliminate contact with the liquid coating. In many cases, proper protective ointments and creams are good precautions wherever skin areas are exposed to such coatings.

8. The buddy system must be practiced wherever coatings are applied to enclosed areas.

These simple rules are easy to follow for any worker, foreman, or superintendent. If followed, they reduce to a minimum any hazards which are inherent in the coating process itself. Maximum safety can result for all workers in an area only if such rules are established for every coating application.

References

1. Department of Defense, Paints and Protective Coatings, Ch. 3, Section 2, Army 7M5:618.
2. Craver, Kenneth J. and Tess, Roy W., Applied Polymer Science, American Chemical Society, Washington, DC, p. 669, 1975.
3. Doolittle, Arthur K., Technology of Solvents and Plasticizers, John Wiley and Sons, New York, NY, 1954.
4. American Conferences of Governmental Industrial Hygenists, Threshold Limit Values for Chemical Substances and Physical Agents in the Workroom Environment with Intended Changes for 1981, Lansing, MI.
5. Federation of Societies for Coating Technology, Anticorrosive Barrier and Inhibitive Primers, Unit 27, Federation Series on Coating Technology, Philadelphia, PA.
6. Urethane Topcoats for Atmospheric Applications, NACE Publication 6H284, NACE, Houston, TX.

APPENDIX 16A
References—Safety and Environmental Control

1. "Air Pollution Engineering Manual," 2nd Edition, Chapter 4, Air Pollution Control Equipment for Particulate Matter, Danielson, J. A., Editor, U.S. Environmental Protection Agency, Office of Air and Water Programs, May 1973.
2. "Alkyd Coatings for Prevention of Atmospheric Corrosion," NACE Publication 6B165, National Association of Corrosion Engineers, Houston, Texas, 1965.
3. Brushwell, W., "Coatings Update. VII. Ecology, Pollution, and Toxicity," Am. Paint J., Vol. 65, Nos. 48, 49, 50, 1981.
4. Coatings and Linings for Immersion Service, NACE TPC Publication No. 2, National Association of Corrosion Engineers, Houston, Texas, 1972.
5. "Coating Effectively in Confined Areas," Marine Engineering/Log, Vol. 87, No. 4, 1982.
6. "Commodity Specification for Air, Pamphlet G-7-1" ANSI Z86.1-1973. Compressed Gas Association, Inc., New York, N.Y. 10036, 1973.
7. "Compressed Air for Human Respiration," Pamphlet G-7, Compressed Gas Association, Inc., New York, N.Y. 10036, 1973.
8. Dalager, N. A., et al., "Cancer Mortality Among Workers Exposed to Zinc Chromate Paints," J. Occup. Med., Vol. 22, No. 1, 1980.
9. "Industrial Health and Safety Criteria for Abrasive Blast Cleaning Operations," National Institute for Occupational Safety and Health, U.S. Department of Health, Education, and Welfare Report HSM 99-72-83. Arthur D. Little, Inc., Cambridge, Mass. 02140, June 26, 1973.
10. "Industrial Ventilation, A Manual of Recommended Practice," 12th Edition, American Conference of Governmental Industrial Hygienists, Committee on Industrial Ventilation, Lansing, Mich 48902.
11. Kronstein, M., "Toxic Releases from Applied Antifouling Paints," American Chemical Society, Division of ORPL, Papers 1981, Vol. 44, 1981.
12. Lankford, J. L., "Sand Blasting Safety Guide for Petroleum Storage Tanks," American Painting Contractor, Vol. 57, No. 8, 1980.
13. "Occupational Safety and Health Standards, Title 29, Chapter XVII," Occupational Safety and Health Administration, U.S. Department of Labor, 1972.
14. "Protection of the Environment, Title 40, Chapter 1, Part 122, Thermal Discharges," U.S. Environmental Protection Agency, 1973.
15. "Protection of the Environment, Title 40, Chapter 1, Part 126, Policies and Procedures for the National Pollutant Discharge Elimination System (NPDES)," U.S. Environmental Protection Agency, 1973.
16. "Recommended Safety Inspection Check List for Application of Interior Linings," NACE Publication 6F264, National Association of Corrosion Engineers, Houston, Texas, 1964.
17. "Respirators Approved by the Bureau of Mines, as of May 24, 1972," Information Circular 8559, Bureau of Mines, U.S. Department of Interior.
18. "Respiratory Protective Devices; Tests for Permissibility: Fees, Title 30, Chapter 1, Part 11, Bureau of Mines, U.S. Department of Interior, 1972.
19. "Safety and Health Regulations for Construction, Title 29, Chapter XVII, Part 1926," Occupational Safety and Health Administration, U.S. Department of Labor, 1972.
20. "Safety and Health Standards for Maritime Employment, Title 19, Chapter XVII, Part 1915, Safety and Health Regulations for Ship Repairing," Occupational Safety and Health Administration, U.S. Department of Labor, 1972.
21. "Safety and Health Standards for Maritime Employment, Title 29, Chapter XVII, Part 1916, Safety and Health Regulations for Ship Building," Occupational Safety and Health Administration, U.S. Department of Labor, 1972.
22. "Safety Requirements for Working in Tanks and Other Confined Spaces," ANSI Zi17.1-1974 (Proposed). American Petroleum Institute, Washington, D.C. 20006.
23. Solving Corrosion Problems in Air Pollution Control Equipment 1981, National Association of Corrosion Engineers, Houston, Texas, 1981.
24. "Straight and Modified Coatings for Atmospheric Service," NACE Publication 6B167, National Association of Corrosion Engineers, Houston, Texas, 1967.
25. "Vinyl Coatings for the Prevention of Atmospheric Corrosion," NACE Publication 6B163, National Association of Corrosion Engineers, Houston, Texas, 1963.
26. "Vinyl Coatings for Resistance to Atmospheric Corrosion," NACE Publication 6H177, National Association of Corrosion Engineers, Houston, Texas.
27. Weaver, P. A., Industrial Maintenance Painting, National Association of Corrosion Engineers, Houston, Texas, 1973.

APPENDIX 16B
A Manual for Painter Safety[1]

Table of Contents

Introduction

The Cost of Accidents

Disabling work injuries in all industries totaled 1.9 million in 1959. Of these, 13,800 were fatal, and 85,000 resulted in some form of permanent disability. The other resulted in financial loss and physical suffering to the employee and loss production to the employer.

The average cost per case resulting in fatality or permanent total disability was $13,862 in 1959. The average cost per case resulting in permanent partial disability was $1,911 in 1959. The average cost per case resulting in temporary total disability was $296 in 1959. Thus, the total dollar cost of accidents is more than $2.5 billion each year! The cost in suffering can never be measured, of course.

The Causes of Accidents

Records of the National Safety Council, the Insurance Rating Bureau, painting contractors, and safety engineers in several chemical plants and refineries show that as many as **70 percent of the most serious injuries to painters are caused by falls. Falling or moving objects** are the next most serious hazard, followed closely by **harmful substances. Back injuries** caused by lifting and handling equipment follow. Injuries resulting from **improper use of hand tools** are the least frequent.

Analysis of all types of injuries shows that several major factors are responsible:

1. **Unsafe Working Equipment.** Structural and mechanical failure because of broken, frayed, or chemically damaged rope and cables; towers that lose balance because of improper bracing, guying, or tying-in; stages of improper length for depth of side rails or stages that are not cabled: extension ladders and step ladders that are not rodded or that are light weight rather than heavy weight; outriggers and wood stages with knots and cracks and without straight grain required in stress areas.

2. **Unsafe Working Conditions.** Inadequate protection from falling objects; unguarded machinery: moving equipment; areas where heat is excessive or where chemical gases are prevalent; inadequate ventilation in enclosed spaces.

[1]SOURCE: A. Report of NACE Technical Unit Committee T-6D (C. W. Sisler, Monsanto Chemical Company, St. Louis, Missouri, chairman) on Application and Use of Coatings for Atmospheric Service. Prepared by Task Group T-6D-5 (Lowell S. Hartman-Walsh Painting Company, St. Louis, Missouri, chairman) on Painter Safety, 1963.

3. **Unsafe Personnel. Inexperienced personnel** who select improper scaffolding, attach stages and rigging to improper areas of buildings and structures, fail to provide protection against friction on ropes, improperly secure ladders and stages, tolerate surfaces slippery with wet paint and/or other products, over-reach, fail to look out for moving objects such as cranes, point spray or sandblasting guns at other workmen, and engage in horseplay and practical joking. **Personnel subject to stress, illness, and over-indulgence** that result in lack of concentration, carelessness, forgetting to comply with safety regulations, physical reaction, eye strain, taking a wrong step, and losing grip.

4. **Personality Traits.** Personality traits of the employee are important. In one industry, a study of accidents showed that 48 percent of the primary causes of accidents were **faulty attitude; impulsiveness; irresponsibility; failure to pay attention; nervousness; fear, worry, and depression.** If these six basic items that cause accidents could be eliminated, accidents could be reduced by 48 percent.

Steps to Improve Painter Safety

Study of the causes and frequency of accidents shows that a number of steps can be taken to improve painter safety. Among the more important of these steps, discussed more fully in later portions of this mannual, are:

1. A thorough medical examination prior to employment and at intervals thereafter.

2. Elimination of faulty attitudes.

3. A reasonable set of personnel safety rules and rules for safe working conditions.

4. Rules for the use of working equipment.

5. Instruction in the toxicity, health hazard, and flammability of many coatings materials, and in how best to protect the painter from harmful effects during exposure.

6. Instruction in the use of protective devices and protective clothing.

7. A continuing program of painter education.

I. Personnel Safety Practices

1. Medical Examinations and Medical Records

Reasonably good health and freedom from physical debilities that might tend to make an employee ''accident prone'' or susceptible to chronic illness should be made the first requisite of employment. Even the best safety precautions cannot help men who are physically or psychologically unsuited to do the work required.

A thorough medical examination that includes chest X-ray, back X-ray, electrocardiogram, check for presence of hernia, and blood and urine analyses should be made prior to employment. Permission to enter results of the medical examination and replies to questions pertinent to past medical history in a permanent employment record should be made a condition of employment.

After the employee has passed a rigorous initial medical examination, he should be ''checked out'' under various on-the-job working conditions. Some men naturally are unsuited for work in confined spaces or at high levels, regardless of physical condition.

These factors should be noted and made part of the employment record.

Follow-up medical examinations should be given at six-month or selected intervals, and results of these should be included in the employment record. Development of undersirable tendencies revealed by these medical examinations should be considered as an important factor in any decision to continue or to terminate employment.

Records of the employee's accidents should be kept up to date and should include accidents with former employers. Such a record will serve to determine whether or not the employee is ''accident prone'' and will help the employer decide whether or not the employee is suited to be an industrial painter under today's complex and demanding conditions.

Figures 1 and 2 illustrate typical medical history and medical examination forms that might be filled out and entered in the employee's record.

Figure 1

Typical Medical History Questionnaire

Please check each of the following items. If explanatory remarks are necessary, make them on an additional sheet and attach it to this form.

Have you ever experienced the following

	Yes	No	Date
Frequent headaches	___	___	___
Dizzy or fainting spells	___	___	___
Fits, epilepsy, convulsions	___	___	___
Loss of consciousness	___	___	___
Blindness, color blindness	___	___	___
Ear trouble, impaired hearing	___	___	___
Stiff joints	___	___	___
Deformities	___	___	___
Tuberculosis	___	___	___
Severe coughing, chest pains	___	___	___
Heart trouble	___	___	___
Shortness of breath	___	___	___
High blood pressure	___	___	___
Stroke or paralysis	___	___	___
Rupture or hernia	___	___	___
Hospitalization:			
For illness or injuries	___	___	___
For operation	___	___	___
Unusual weight gain or loss	___	___	___
Received:			
Workmen's compensation	___	___	___
Armed Forces disability	___	___	___
Compensation	___	___	___

Figure 2

Typical Items to be Recorded in Medical Examinations

Record results of the following examinations:

Eye Chart_____
Orthorater _____
Color Vision_____
Hearing _____
Urinalysis:
Sugar _____
Albumin _____

Reflexes _____
Electrocardiogram _____
Chest X-ray_____
Back X-ray_____
Blood Tests:
    VDRL_____
    Kahn _____

Examine and record under comments defects of the following:

Eyes_____
Tonsils _____
Skin_____
Abdomen _____
Arms _____
Legs_____

Ears_____
Throat_____
Lungs_____
Genitalia_____
Hands_____
Feet_____

Nose _____
Chest _____
Heart _____
Back _____
Fingers _____
Joints _____

Comments:

_____
_____
_____

2. Faulty Attitudes

How many times have you said about a man, "He's a good worker, but ...."?

What about the "but"? What about the fellow who shows irrational anger, extreme irritability, or constant grouchiness? What about the fellow who is always talking sex, who is over-inquisitive, or who lies just for fun of lying? What about the fellow who is a queer duck and whom you just can't figure out, and who has all kinds of peculiarities and mannerisms which, although they may be too small to resent, nevertheless keep people under a strain? What about those fellows who will do things on impulse and then be sorry--and then repeat the same stunts some other time?

These are the unstable, impulsive, and irresponsible persons who are constantly in need of warning or discipline. They not only get themselves into messes, but, in keeping others on edge, they may cause others to get hurt, too. They are too costly to keep on the payroll because they may cause permanent injury to innocent bystanders.

Take any one of your employees and make a personality inventory. How does he affect the others in the crew? Is he liked by the rest of the men? Does he mix with the others, or does he keep to himself? Is he a good influence in a group, or is there argument and friction wherever he goes? Does he take criticism and correction in the right spirit? If his feelings are hurt, does he get over it quickly or does he brood? Does he gossip and say malicious things about people? Is he considerate of the feelings of others? Does he talk too much? Ask yourself these and other questions, and then ask yourself, "Why?" People don't quarrel or gossip or brood or talk too much just for the sake of doing such things. There is a deeper

reason. They do it because it gives them an outlet for their feelings. They want to be happy, and people who quarrel and fight and gossip and brood are not happy. There is something behind these behaviors. They are only symptoms--signs that everyone of these people needs help.

If safety depends on the elimination of faulty attitudes, ask yourself what you have done to create these attitudes, and then get together with your employees. If safety depends on the elimination of fear, nervousness, worry and depression, find out what is behind these things and try to remedy the situation.

All of them are distractors of attention. **Nobody was ever hurt while paying strict attention to what he was doing.** Things which distract attention, whether brooding about something or arguing, make people unsafe.

3. A Safe Working Environment

Before painters are sent to work in any plant area, the working environment should be studied to detect hazards such as concentrations of noxious fumes, presence of splash or spillage of harmful liquids, conditions of excessive heat, or areas where oxygen content of the air may be dangerously low. Where these conditions exist, they must be corrected before the painter is permitted to enter the area, or the painter must be protected by materials or devices that will allow him to work without fear of injury or illness that might otherwise result from exposure. In most plants, the safety engineer knows where the hazardous areas are located and what measures are necessary to counteract the hazard. The project superintendent and the painter foreman should always consult with the safety engineer before starting each project, and his recommendations should be followed. Safety control measures stipulated by safety engineers should be developed in cooperation with the plant industrial hygiene engineer if available, with the plant chemist who knows the composition of compounds and groups of compounds being used, and with the plant industrial physician with his knowledge of toxicology and preventive medicine.

It is an excellent idea to prepare and use daily a safety check list of items that must be seen to before work commences. A safety check list might be a 3" × 5" card that provides blocks for checking the items shown in Figure 3. The project superintendent and the painter foreman should consult with the safety engineer to see that all pertinent items have been checked and that the painters have been informed of the relevancy of the items to his work during the day.

Figure 3

Typical Safety Check List

SAFETY CHECK LIST

☐ **Clothing**
☐ **Eye protection**
☐ **Respiratory protection**
☐ **Safety Belts**
☐ **Warning tags and signs**
☐ **Toxic materials**
☐ **Burns to skin**
☐ **Falling objects**
☐ **Electrical hazards**
☐ **Footing**
☐ **Moving objects, cranes, traffic, etc.**
☐ **Safety showers and eye baths**
☐ **Fire alarm station**
☐ **Fire extinguishers, fire blankets**
☐ **Nearest telephone**
☐ **Barricades**
☐ **Equipment grounded**
☐ **Sparkproof tools**

Corrosion Prevention by Protective Coatings

☐ **Safety or fire permits**
☐ **Flammability or flash point**

3.1 **Threshold Limit Values for Solvents, Fumes, Mists, and Dusts.** The following threshold limit values, taken from the most recent list published by the American Conference of Governmental Industrial Hygienists, have been modified to suit industrial safety experience. These values should be observed when work is done in confined areas.

TABLE I

Threshold Limit Values for Commonly-Used Solvents

Solvent	MAC[1]	Flash Point Deg F[2]
Acetone	1000	0
Benzene	25	12
n - Butanol	100	84
Butyl Cellosolve	50	141
Carbon Tetrachloride[3]	10	None
Cyclohexane	400	1
Ethylene Dichloride	100	56
Ethanol	1000	55
Gasoline	500	- 45
Methanol	200	52
Naptha, Coal Tar	200	100-110
Naptha, Petroleum	500	< 0
Perchloroethylene	200	None
Isopropanol	400	53
Stoddard Solvent	500	100-110
Toluene	200	40
Trichloethylene	100	None
Turpentine	100	95
Xylene	200	63

[1]Maximum allowance concentration of vapor in air (or hygienic standard) in parts per million by volume for breathing during continuing eight hour working day. These figures should be used as guides for evaluating exposures, and are subject to change when better information is available.

[2]TAG closed-cup tester: From "Flammable Liquids and Gases," National Fire Codes, Vol. 1.

[3]Because of cumulative health hazard, use of this cleaning solvent is not recommended.

TABLE II

Threshold Limit Values for Fumes, Mists, Dusts

Hazard	MAC[4]
Lead	0.2
Dust (no free Silica)	50.0
Silica:	
High, above 50% free SiO_2	5.0
Medium, 5-50% free SiO_2	20.0
Low, below 5% free SiO_2	50.0
Total Dusts below 5% free SiO_2	50.0
Chromic Acids and Chromates as CrO_3	0.1

[4]Maximum allowable concentration (or hygienic standards), milligrams per cubic meter of air for breathing during continuing eight hour working day.

3.2 **Ventilation**

3.2.1 **General Ventilation and Temperature Requirements.**

Safe working conditions require that outside air be provided to all workrooms at the minimum rate of 15 cubic feet per minute per person, or one-and-one-half (1-1/2) changes of air per hour, whichever is greater. (In most instances, leakage through walls, doors, and windows will produce at least one-and-one-half air changes per hour).

A minimum air temperature of 60 F is recommended in all workrooms where work of a strenuous nature is performed. A minimum air temperature of 65 F is recommended in all other workrooms unless prohibited by process requirements.

3.2.2 **Local Exhaust Ventilation.** Local exhaust ventilation systems should produce and maintain a movement of air toward the discharge opening sufficient to prevent escape of contaminant into the breathing zone beyond the limits shown in Tables I and II during working hours.

Velocity of air flow through branch and main ducts should be sufficient to transport the contaminant through the system without permitting it to settle.

Exhaust ventilation piping should be located so as to be accessible for inspection and maintenance.

Air flow equipment including hoods, pipes, fans, motors, and collectors must be adequately grounded.

Operations involving more than one contaminant must never be connected to the same exhaust system when combination of the contaminants can create a fire hazard, explosion hazard, or otherwise dangerous mixture.

It is imperative that the exhaust system take suction from some point at the bottom of the room or tank in which work is being done.

Processes or operations using or generating flammable dusts, gases, fumes, vapors, mists, fibers, or other impurities must be completely protected from all sources of ignition.

The capacity of an exhaust system shall be designed so that all hoods, booths, and enclosures connected to the system are fully open, except where the system is so interlocked that only a portion of it can be operated at a given time. In this case, the capacity shall be calculated on the basis that all hoods in the group requiring the greatest rate of exhaust are open.

Suitable air inlets must be sufficient to provide for replacement of exhausted air. Fresh air shall be furnished from a protected area, free of all contaminants. The intake pipe shall never be downwind from the compressor engines, truck, automobile, or other internal combustion engine. Exhaust systems handling dusts and discharging to the outer air shall be provided with suitable air-cleaning devices to remove air contaminants prior to discharge to the outer air, except under unusual conditions.

The discharge from any exhaust system shall be arranged so that no air contamination will enter any work space, window, door, or other opening in quantities sufficient to create a health hazard or a nuisance in surrounding areas.

Collected materials shall be removed from the exhaust system at frequent intervals.

4. Personnel Safety Rules

The project superintendent, foreman, and workmen should learn and follow all plant safety rules and regulations. Where such rules are absent or are not specific, the following requirements should be met:

1. Before painting work is begun in an operating area, the operating supervisor shall be asked to inform painters of all conditions which might cause injury. The supervisor shall also be asked to explain the area's safety rules.

2. Provide and use adequate protective clothing, safety shoes, helmets, gloves, safety glasses or goggles, respiratory protection, ventilation equipment, and skin protection as recommended by the safety engineer.

3. Obtain necessary fire permits, tank entering permits, elevated work permits, and other procedural safe-guards.

4. In confined spaces, check for explosion hazard with a gas tester recommended by the safety engineer. Since safe limits for breathing flammable vapors are far below the minimum explosive range, use proper ventilation to prevent vapors from accumulating or provide masks, respirators, etc.

5. When machinery, including agitated tanks, is to be painted, have fuses pulled and switches thrown and locked out. Attach "DO NOT OPERATE" warning tags.

6. Do not paint moving machinery.

7. Locate and know the use of department safety showers, fire alarm stations, fire extinguishers, stretchers, fire blankets, etc.

8. Keep aisles, fire and safety equipment, and exits free of obstructions.

9. Do not point spray guns or sandblast nozzles at anyone.

10. Dispose of all oily rags and waste at the end of each day. If disposal is not feasible, store oily rags and waste in metal containers with a tight fitting lid.

4.1 **Control of Surface Preparation Hazards.** Surface preparation requires the use of goggles, air helmets, neck scarves, and sometimes protective clothing, depending on the method of surface preparation employed.

Provide and use adequate eye protection in all instances. For hand cleaning, power tool cleaning, and water blasting methods of surface preparation, approved safety goggles will be adequate. For sand or grit blasting methods of surface preparation, full face protection is required. The most suitable equipment is safety goggles worn beneath a pressurized air helmet. Workmen in acid areas or areas in which corrosive liquids may be encountered are required to wear acid proof goggles rather than the conventional type.

Provide and use appropriate respiratory protection when sandblasting. Provide and use creams on exposed skin areas. Encourage careful washing of the hands and face before eating.

5. Rules for Safe Working Conditions

The project superintendent and the painter foreman, with advice by the safety engineer, should be alert to see that the following rules are obeyed:

1. Check tank or other enclosed space interiors for explosion hazard, toxic materials, and adequate oxygen content for breathing. Before tank entry, provide life lines and harness with an extra man stationed outside the tank. Ascertain if the tank opening is large enough to remove an injured man. Before entering a tank, check with operating personnel to ensure that all process lines entering the tank are blanked off.

2. Provide proper lighting in interior of tanks and other dark areas. Be sure that lighting does not create shadows or glare. All wiring must be so constructed that there is no danger of electric shock. In many instances electric head lamps such as are used in mines may be used.

3. When airline masks or hose masks are used, locate the intake air source or the blower in an area free from air contaminants. Check the output air for proper pressure and carbon monoxide content. Protect hose from damage and do not allow them to become tripping hazards.

4. Rope off or barricade work areas as required to protect passers-by.

5. Provide an adequate number of portable fire extinguishers.

6. Post warning signs which may include "WET PAINT," "NO SMOKING," and "MEN WORKING ABOVE."

7. Provide a safe storage cabinet for materials and equipment. Use ventilated metallic cabinets for overnight storage of paint brushes suspended in oil or solvents.

8. Protect machinery, equipment, and materials with fireproof tarpaulins.

9. Be sure that pressurized containers are fitted with perfectly functioning pressure gages and safety valves. Test safety valves daily before use.

10. Read the label and learn the properties of the materials that are being applied. If they are toxic, flammable, or harmful to any degree, find out what protective measures are necessary to prevent harm from exposure.

11. Follow the safety rules for working equipment that are given in following portions of this manual.

II. Safety Rules For Working Equipment

The safety of personnel depends to a large degree on the condition of equipment which must be used, and on the familiarity of personnel with safety procedures that must be followed when using the equipment. The following rules should be learned and enforced.

1. Rules for Safety With Ladders

1. See that all ladders conform to ASA A14.1 "Safety Code for Wood Ladders," and ASA A14.2 "Safety Code for Metal Ladders."

2. Inspect ladders frequently. Look for loose steps or rungs; loose screws, bolts, metal braces, and rods; split or broken side rails or rungs; and loose or bent hinge spreaders.

3. When ladders are found to be lacking in any safety device or to be defective in any manner, they shall not be used. Ladders found to be too defective for repairs must be destroyed.

4. Protect ladders with clear protective coatings so that cracks, splinters, or breaks will be readily visible.

5. Store ladders in a warm, dry place protected from the weather and from contact with the ground.

6. Equip all straight and extension ladders with safety shoes, unless special conditions prohibit their use.

7. Do not use portable ladders greatly than 60 feet in height. If greater heights are to be reached, provide intermediate landing platforms and use separate ladders.

8. Single ladders not constructed for use as sectional ladders shall not be spliced together to form a longer ladder.

9. Extension ladders must not be taken apart in order to use either section separately.

10. Extension ladders shall have a minimum of 15 percent overlapping of each section.

11. Ladders made by fastening cleats across a single rail shall not be used.

12. Do not use stepladders over 20 feet in height.

13. Never use a stepladder as a straight ladder.

14. Never use the platform of a stepladder as a top step on which to stand.

15. In placing portable ladders, be sure that the horizontal distance from the top support to the foot of the ladder is one-quarter of the working length of the ladder.

16. Straight and extension ladders must be tied at the top when in use. A helper must hold the ladder while it is being tied or untied at the top.

17. When it is necessary to place a ladder over a doorway, the doorway must be roped off and signs must be erected to warn users of the door of the ladder's presence.

18. Ladders must never be used as skids, braces, scaffold members, or for any other purpose than that for which they are intended.

19. DO NOT USE METAL LADDERS OR STAGES IN AREAS WHERE CONTACT WITH ELECTRIC POWER LINES IS POSSIBLE.

20. Use hand lines, not ladders, to raise or lower materials and tools.

21. Do not reach too far in any direction while working from a ladder. Frequent changes in a ladder's position will forestall accidents from this source.

22. Always face the ladder when ascending or descending.

2. Rules for Safety With Metal Scaffolding

1. Provide adequate sills or underpinnings on all scaffolds to be erected on filled or otherwise soft ground.

2. Compensate for unevenness of ground by using adjusting screws where blocking is not practical.

3. Be sure that scaffolds are plumb and level at all times.

4. Anchor running scaffolds to wall approximately every 28 feet of length and 18 feet of height. Use additional care when using pulley arms.

5. Do not force braces to fit. Adjust level of scaffolding until the proper fit can be made with ease.

6. Horizontal diagonal bracing must be used to prevent racking of all scaffolds at bottom and intermediate levels of 30 feet.

7. Handrailing must be provided on all scaffolds regardless of height.

8. **Use ladders** when climbing the scaffold rather than the cross braces.

9. Erect the scaffold in such a manner that a ladder is accessible and is lined up from top to bottom.

10. **Inspect all scaffolding parts before using.** Never use parts that are damaged or deteriorated. Remember that the strength of rusted materials is not known.

3. Rules for Safety With Wood Scaffolding

1. Lumber used in the construction of wood scaffolding shall be of good quality and reasonably straight grained. It shall be free from injurious ring shakes, checks, splits, cross grains, unsound knots, knots in groups, decay, and growth characteristics that materially decrease the strength of the lumber.

2. All nails used in scaffolding construction shall be driven full length. No nail smaller than 8d shall be used in scaffold construction, and a sufficient number of nails must be driven to support the design loading. Design of the scaffold must be such that no nail is subjected to a direct pull.

3. Follow the applicable rules given for METAL SCAFFOLDS, above.

4. Rules for Safety With Planking

1. All planking used on a scaffold shall be of sound quality lumber that is straight grained and free from knots. All planking shall have at least 24 inches overlap. Secure all planks to wood scaffolding.

2. Planking overhang shall be 3-5/8'' on stackup scaffolds. Wood cleats (1'' × 4'') shall be nailed across the top of planking at each frame and at outside end frames.

3. Planks used for platforms shall be of uniform thickness, laid close together. These planks must be overlapped at the bearers. An overlap of at least two feet must be allowed with the bearer in the center of the overlap. Planks should be fastened securely to the bearer at the opposite end to prevent tipping.

4. Platforms must be kept clear of slippery substances. They must be equipped also with handrails and ladders.

5. Planking must never be painted over; paint will conceal defects. Scaffolding boards shall be identified by painting each end. All scaffold boards shall be used exclusively for that purpose.

6. The nominal size of planking shall be determined from Table III. Values given are for planks with the wide face up, with loads concentrated in the center. Loads given in the Table are net. Allowance was made for weight of the planking. Loads in the Table may be increased 45 percent if select structural coast region Douglas Fir or merchantable structural square and solid Southern Pine are used.

TABLE III

Safe Center Loads for Scaffold Plank, in Pounds

Size of Plank

Span in Feet	2 × 10 Dressed to 1-5/8 × 9-1/2	2 × 12 Dressed to 1-5/8 × 11-1/2	3 × 10 Dressed to 2-5/8 × 9-1/2	3 × 12 Dressed to 2-5/8 × 11-1/2
6	256	309	667	807
8	192	232	500	605
10	153	186	400	484
12	128	155	333	404
14	110	133	286	346
16		116	250	303

5. Rules for Safety With Rolling Towers

1. Towers of a height greater than three times the minimum base dimension **must be guyed or tied-off when in use.**

2. **Apply all caster brakes** when the tower is not in motion.

3. DO NOT RIDE TOWERS!

4. Look where you are going when moving a rolling tower. **Do not attempt to move a rolling tower without sufficient help.**

5. Provide unit lock arms on all towers.

6. Do not extend adjusting screws more than 12 inches.

7. Do not use casters smaller than 6 inches in diameter, regardless of tower height.

8. The five-foot by seven-foot rolling tower now in general use is unstable at heights over 25 feet while in motion; stability must be maintained either with outriggers or with handling lines.

9. Use horizontal diagonal bracing at the bottom and at intermediate levels of 30 feet.

10. Guy or tie-off all fixed towers every 18 feet of elevation.

11. Do not use brackets on rolling towers.

12. Do not attempt to use scaffold as material hoist towers or for mounting derricks **without first determining the loads and stresses involved.**

13. **Inspect all tower parts before using.** Do not use parts that are damaged or deteriorated. Remember that the strength of rusted materials is not known.

6. Rules for Safety With Swing Stages

1. Be sure to read the manufacturer's instructions on the proper use and maintenance of the equipment.

2. Never overload equipment. Follow manufacturer's prescribed load capacities.

3. Always use a guard rail of sound lumber. Never substitute with rope.

4. Replace cables immediately that show 5 percent of the wires per lay broken or that show evidence of excessive wear or corrosion.

5. Cables must be replaced only by experienced personnel, and then with the utmost care.

6. Allow none but careful, experienced men to erect or operate the suspended stage.

7. When locating cables on the job site, CHECK NEARBY POWER LINES OR ELECTRIC SERVICE WIRING TO PREVENT ELECTROCUTION. When in doubt, consult the power service company for advice.

8. Wood stages shall be 20 inches minimum width, rated by Underwriters Laboratories for center load of 500 pounds with safety factor of 4. Wood stages over 24 feet in length shall be cabled.

9. Metal stages shall be not less than 20 inches wide, rated by Underwriters Laboratories for center load of 500 pounds with safety factor of 4. Stages over 24 feet in length shall be cabled, with no cracks in side rails.

10. Ropes and blocks of the stage shall be tested by suspending one foot off ground and loading at least four times the estimated work load.

11. Power stages shall have free fall safety devices with hand controls in case of power failures. Cables must be inspected regularly. Safety factor of cables must be four to one.

7. Rules for Safety With Ropes and Lines

1. See that ropes are of the proper size: Not less than 3/4'' manila for stages, 1/2'' manila for safety lines.

2. Discard ropes that have been exposed to acid or excessive heat. Twist strand open and check for dry rot, brittleness, and abrasions. Do not use frozen rope under any circumstances.

3. Inspect all wire ropes or cables frequently. Discard when 5 percent of the wires per lay are broken, when cable clamp damage is evident, or where there is evidence of corrosion.

4. Provide dry, empty drums for coiling.

8. Rules for Safety With Pressurized Equipment

The following safety rules apply to conventional and airless spray equipment, sandblasting equipment, and water blasting equipment that operate under pressure.

1. Use only pressure equipment that has been constructed as specified by the National Board Code **and** the ASME Code for Unfired Pressure Vessels. Make certain that this equipment carries a label certifying that such construction has been made. On equipment of this type, the manufacturer certifies that it has been built to rigid standards, that certain types and grades of steel have been used, that welding has been done by certified welders, that the machine has been tested to twice the expected working pressure, and that it has been inspected by state or insurance company inspectors.

If a coded vessel is not used, the job may be shut down in certain states that specify code requirements; the user is subject to an unrestricted judgment by a jury should an accident occur (if a coded vessel is used, liability usually is restricted to the amount set forth in the workmens' compensation laws of the state).

2. Conduct a hydrostatic test of pressure equipment at least once, preferably twice, each year.

3. Test safety relief valves used on pressure equipment daily.

4. Use remote control deadman valves with pressure equipment. With valves of this type, the pressure equipment will be shut off instantly when the operator releases his grip on the nozzle. Be sure that the deadman valves are pneumatic type that use the same air for activation that is used for blasting or spraying.

5. Be sure that the nozzle, tank, and pressure equipment are grounded as illustrated in Figures 4 and 5. Have an electrician check the grounding periodically with an ammeter to see that it is operative. (Ground electric or pneumatic power tool equipment used for cleaning operations as illustrated in Figure 5 also.)

9. Rules for Safety With Sandblasting Equipment

1. Use sandblast hose constructed with activated carbon to prevent electrical shock to the operator.

2. Be sure that sandblast hose is kept in dry areas when stored. When using the hose, be sure that it is running in as straight a line as possible, avoiding all 90 degree sharp bends. If a hose must be curved around an object, make a long radius curve; sharp curves will create rapid wear on the tube of the hose, causing a blow-out at the bend.

3. NEVER point a sandblast nozzle at any part of the human body—your own or another's.

4. Use a completely-contained safety helmet wherein a separate supply of air is fed to the operator to make certain that no dust can enter his respiratory tract. The helmet should be Bureau of Mines approved to make sure that danger of silicosis to the operator is minimized. It is important to note that air-fed helmets will not provide protection against fumes unless a separate air purifier or a special ring-free compressor is used to assure respirable air.

10. Rules for Safety With Waterblast Equipment

1. It is absolutely essential to have an automatic deadman safety control at the nozzle.

2. Water hose must be properly sized to suit the equipment and must be tested to withstand four times the maximum working pressure.

3. Fittings must be tested to withstand four times the maximum working pressure.

4. By-pass valves must be equipped with safety relief valves.

5. Nozzles must not be pointed at any person or any object that will break at working pressure, such as glass, scaffold supports, loose pipes, fittings, gages, lights, etc.

6. High pressure hose must be firmly secured at a distance not more than ten feet from the operator.

7. The operator must be protected with safety rubber gloves and a safety rubber suit that protects the head, face, neck, arms, legs, feet, and other parts of the body.

8. Face shields of heavy plastic shall be used in high pressure operations. Safety goggles may be worn when cleaning is done at low pressure.

11. Rules for Safety With Airless Spraying Equipment

Airless spraying equipment consists of a high ratio fluid pump equipped with an air regulator that may generate fluid pressures to 3000 psig, specially-designed fluid hose and fittings that will safely withstand such pressures, and an airless spray gun that sprays at a rate of 80 ounces per minute through small orifices from 0.007 inches to 0.036 inches in size. Material to be sprayed is finely atomized by low boiling solvents under pressure, seeking release through these very small orifices. It is obvious that a jet of material being released at such high pressures will pierce the human body if the operator reaches close proximity to the spray cap on the gun head; the material does not become atomized sufficiently to form a spray pattern until it has traveled from 12 to 14 inches. It is equally dangerous to pull the trigger on an airless gun when the cap has been removed. A large volume of material is then released at the same high pressure, and great bodily harm can result from contact with this stream. The following safety precautions should be observed and rigidly enforced:

1. Check all hose connections and fittings to be certain that they are tight and not leaking. Be sure that only components designed for high pressures are used.

2. Check the fluid hose to be sure that there are no weak or worn spots created by kinking, abrasion, contact with moving parts of machinery, friction from sharp edges and corners, deterioration due to exposure to chemicals, or ordinary wear.

3. When handling or carrying the gun but not actually spraying, hold the gun by the grip and remove the fingers from the trigger. This will prevent the gun from being activated if the operator should inadvertently tighten his hold when stumbling or slipping.

4. NEVER point an airless gun at any part of the human body—your own or another's.

5. Do not disconnect the gun from the fluid hose, or the hose from the pump until the pressure has been released from the hose. This is accomplished by first closing off the main line air pressure to the pump and then bleeding off the pressure on the fluid hose by triggering the gun before disconnecting it.

6. If spraying with flammable materials, it is necessary to have the object being sprayed as well as the spray gun grounded to prevent a spark from static electricity. Be sure that all other sources of ignition are eliminated as well.

III. Safety With Materials

1. Handling and Applying Coatings Materials

Coatings materials should be stored in safe, well ventilated areas where sparks, flames, and the direct rays of the sun can be avoided. Containers should be kept tightly sealed until ready for use. **Warning tags should be placed on toxic materials.**

Mixing presents many problems. Recommended safety rules for mixing operations include the following:

1. Use protective gloves.

2. Keep the face and head away from the mixing container.

3. Use protective face cream.

4. Avoid splash and spillage, and inhalation of vapors.

5. Use eye-protection goggles.

6. Mix all materials in well-ventilated areas away from sparks and flames.

7. Use low speed mechanical mixers whenever possible.

8. Clean up spillage immediately.

9. Check the temperature limitations of materials being mixed. Many materials are dangerous at high temperatures.

Protective devices and equipment required for application of coatings materials are determined by the type of coating as well as by the environment. **The coatings manufacturers should provide complete mixing and application instructions, including definite references to safety requirements.** Unless definite information regarding explosion and toxicity hazards inherent in the material are provided by the manufacturer, a written request for such data should be made **before starting the coatings application.** Records of previous applications using similar materials should be examined also.

Protective face cream is recommended for all spraying operations. Goggles should be worn wherever possible.

2. Health Hazards of Materials

A coatings material may be considered a health hazard when its properties are such that it can either directly or indirectly cause injury or incapacitation, either temporary or permanent, from exposure by contact, inhalation, or ingestion.

Degrees of health hazard are ranked according to the probable severity of injury or incapacitation, as follows:

1. Materials which on very short exposure could cause death or major residual injury even though prompt medical treatment were given. Types of these materials are:

a. Materials which can penetrate ordinary rubber protective clothing.

b. Materials which under normal conditions or under fire conditions give off gases which are extremely toxic or corrosive through inhalation or through contact with or absorption through the skin.

2. Materials which on short exposure could cause serious temporary or residual injury even though prompt medical treatment were given. Types of these materials are:

a. Materials giving off highly toxic combustion products.

b. Materials corrosive to living tissue or toxic by skin absorption.

3. Materials which on intense or continued exposure could cause temporary incapacitation or possible residual injury unless prompt medical attention is given. Types of these materials are:

a. Materials giving off toxic combustion products.

b. Materials giving off highly irritating combustion products.

c. Materials which under either normal conditions or fire conditions give off toxic vapors lacking warning properties.

4. Materials which on exposure can cause irritation but only minor residual injury even if no treatment is given. Types of these materials are:

a. Materials which under fire conditions give off irritating combustion products.

b. Materials which cause irritation to the skin without destruction of tissue.

5. Materials which on exposure to fire conditions offer no hazard beyond that of ordinary combustible material.

3. Flammability Hazards of Materials

A coatings material may be considered a flammability hazard when it will burn under normal conditions.

Degrees of hazard are ranked according to the susceptibility of materials to burning, as follows:

1. Materials which will rapidly or completely vaporize at atmospheric pressure and normal ambient temperature; which are readily dispersed in air; and which will burn. Types of these materials are:

a. Any liquid which is liquid under pressure and having a vapor pressure greater than 14.7 psia at 100 F.

b. Materials which may form explosive mixtures with air and which are readily dispersed in air, such as mists of flammable or combustible liquid droplets.

2. Materials that can be ignited under almost all ambient temperature conditions. These materials produce hazardous atmospheres with air under all ambient temperatures and are readily ignited. Types of these materials are:

a. Materials having a flash point of 100 F or below and having a vapor pressure not greater than 14.7 psia at 100 F.

b. Materials which ignite spontaneously when exposed to air.

3. Materials that must be moderately heated or exposed to relatively high ambient temperature before ignition can occur. Materials of this type are those having a flash point above 100 F but not greater than 200 F.

4. Materials that must be preheated before ignition can occur. These materials are those that will support combustion for five minutes or less at 1500 F.

5. Incombustible materials.

4. Toxicity of Materials

Virtually all solvent solution coatings are highly flammable in liquid form, and vapors released in the process of application are explosive in nature if concentrated in sufficient volume in closed or restricted areas. Even vapors from ordinary enamels and oil paints may be accumulated in such density as to result in explosive reaction if a source of ignition is present. Generally speaking, however, solvents used in solvent solution coatings are more volatile and dangerous than those used in conventional paints or coatings.

A wide variety of solvents are used in the formulation of many coatings now available. These include, but are not limited to, the ketone group which includes such solvents as acetone, methyl ethyl ketone (MEK), methyl isobutyl ketone (MIBK), ethyl amyl ketone (EAK), cyclohexane, and others; the hydrocarbons group of toluol, toluene, xylol, xylene, and others. In many instances, particularly in vinyl resin base coatings, the ketones act as the solvents which hold the resins in solution; the hydrocarbons are used primarily as diluents to control the viscosity or flow qualities of the coating. Most coatings have a combination of solvents from each group. The epoxy and synthetic rubber coatings use mostly solvents of the hydrocarbons group although ketones may be used in limited quantities.

All of these solvents are highly flammable and should be handled with the greatest of care.

Solvents, as well as other components of many modern solution coatings, present other hazards which must be guarded against at all times. Solvents of all groups are toxic to varying degrees and may cause serious effects to those working with them unless appropriate precautions are taken. Excessive breathing of concentrated solvent vapors may cause dizziness or nausea, excessive drying or irritation of the mucous membranes, and, in rare cases, allergic reactions on the skin.

The epoxides used in epoxy coatings and compounds are particularly irritating to the skin, and some persons are seriously affected by allergic reactions if proper hygiene is not practiced while these materials are in use. Common reactions include swelling around the eyes or lips, rashes of the skin, etc. Some epoxy coatings have polyamides as curing agents that react much like a mild acid on tender mucous membranes.

Mist and spray of lead based paints may be dangerous if these vapors are inhaled. These minute particles of lead may cause lead poisoning if exposure is not avoided. Good personal hygiene is necessary to control ingestion of lead from contamination by cigarettes or food.

The following basic safety precautions should govern the use of **ALL** coatings:

1. Always provide adequate ventilation.

2. Guard against fire, flames, and sparks, and **do not smoke while working.**

3. Avoid breathing of vapors or spray mist.

4. Use protective skin cream and other protective equipment.

5. Practice good personal hygiene.

5. Materials Requiring Special Handling

5.1 **Styrene Monomer (Rubber or Polymer Grade).** Styrene monomer is a volatile material. Adhere strictly to the manufacturer's specifications for mixing and application. Avoid contact with skin or eyes. Wash affected areas with soap and water. If irritation persists, see a physician.

5.2 **60 Percent Solution of Methyl Ethyl Ketone Peroxide in Dimethyl Pthalate Catalyst*.** MEK peroxide is an extremely strong oxidizing agent and must be handled in strict accordance with the manufacturer's specifications. Precautions include, but are not limited to the following:

1. Store in a cool place below 70 F and protect from direct sunlight at all times.

2. Keep away from heat, sparks, and open flames.

3. Do not add to hot materials. Explosive decomposition may occur.

4. Store in original container with special vented cap. Keep closed. Avoid shock and temperatures in excess of 70 F. Keep containers closed.

5. **Avoid contact.** If skin contact occurs, flush with water at once. If eye contact occurs, immediately flush with water for 15 minutes then obtain immediate medical care.

6. Do not mix more than four hours' supply at one time. Clean catalyst pot with acetone before mixing new batch. Release pressure on pot when not using catalyst.

5.3 **Acetone or MEK Solvents*.** Both acetone and MEK solvents are volatile and explosive liquids with low flash points. Handle these materials in strict accordance with manufacturer's specifications.

Avoid spilling acetone or MEK solvents on absorptive materials such as wood floors, ground in work areas, and asphalt paving to prevent fire and explosive hazards.

Remove **all** acetone or MEK solvents from equipment after cleaning.

*WARNING: Do not confuse MEK Peroxide Catalyst with MEK Solvent!!

5.4 **Vinyl Solvent Solution Coatings.** These coatings contain highly flammable solvents that may be toxic, irritate the eyes, nose, and throat, dry or irritate the skin, and that are explosive in concentration. Safety precautions for handling these materials include:

1. Store materials in well ventilated areas, away from heating equipment, open flames, and away from direct sunlight.

2. Keep containers tightly closed at all times.

3. Avoid sparks or open flames near open containers. **Do not smoke** while mixing or applying these materials.

4. Provide effective ventilation when handling or applying these coatings in closed or confined areas.

5. Use only vapor-proof and explosion-proof electrical equipment in confined areas.

6. Do not work in any area or closure when air is not clear and free of excessive solvent odor.

7. Use an approved type chemical respirator during spraying operations. (Use Binks No. 40-13, Mine Safety Appliance, or equal, organic type).

8. Exhaust from lowest level, where possible.

9. Extinguish all pilot lights on water heaters, furnaces, or other equipment, on all levels.

10. Wear only rubber soled shoes and clothing free of ferrous metal parts when working in confined areas.

11. Always use drop cloths under materials supply cans, spray

pumps and pots, and avoid dripping of cans or equipment in metallic or concrete closures.

5.5 Epoxy Coatings and Compounds. These materials contain flammable solvents that will dry or irritate the skin on contact. Solvent vapors may be toxic and may cause irritation to the eyes, nose, and throat. Material components may cause serious allergic reactions, and curing agents may be irritating to the skin and mucous membranes.

Safety precautions for handling these materials include the following:

1. Store materials in well ventilated rooms away from heating equipment and other fire hazards, and away from direct sunlight.

2. Keep containers tightly closed at all times.

3. Avoid sparks or flames near open containers.

4. **Do not smoke** while mixing or applying these materials.

5. Provide effective ventilation when using materials in confined areas.

6. Use only vapor-proof and explosion-proof electrical equipment in confined areas.

7. Do not work in any closure when air is not clear and relatively free of solvent odor.

8. Use an approved type chemical respirator when applying coatings by brush or roller in confined areas. (Use Binks No. 40-13, Mine Safety Appliance, or equal, organic type.)

9. Use a fresh air hood (Binks No. 40-54 or equal) when applying these coatings by spray in confined areas.

10. Avoid contact of coatings or compounds with the skin for prolonged periods. Wash face, hands, arms, and other exposed areas of the body at least every two hours, using soap and water. Trust commercial protective creams for short periods of time only. Wash clothing regularly. Personal hygiene is highly important when working with epoxies.

5.6 Neoprene-Hypalon (Synthetic Rubber) Coatings. These materials contain solvents that are flammable and that may dry or irritate the skin. Concentrated solvent vapors are explosive.

Safety precautions for handling these materials include the following:

1. Keep container closed at all times.

2. Store materials in well ventilated rooms away from heating equipment and other fire hazards, and away from direct sunlight.

3. Avoid sparks or flame near open containers.

4. **Do not smoke** while mixing or applying these materials.

5. Avoid excessive contact of coatings with the skin.

6. If coatings are used in other than outside applications, follow the safety precautions given for vinyl coatings, above.

IV. Personnel Protective Equipment

1. Respiratory Protective Equipment

When industrial work methods or processes create hazards of exposure to harmful vapors, gases, dusts, mists, or fumes, the work areas should be enclosed or well ventilated to eliminate or minimize the hazards. If enclosure, ventilation, or other engineering methods of control are not possible, however, respiratory protective equipment should be provided for workmen exposed to possible danger. Even when the area is enclosed or well ventilated, a supply of appropriate protectors should be readily available for emergency work or repairs under unusual conditions.

The American Standard Safety Code for Head, Eye and Respiratory Protection, ASA Z2.1-1959, suggests in broad terms where protection is needed. The code does not spell out in detail the specific hazardous conditions requiring respiratory protection; instead it provides that respiratory protection "shall be provided where a process presents the hazard of exposure to harmful vapors, gases, dusts, mists, or fumes." A complete listing of all situations requiring respiratory protection would not be possible, but every process which generates air contaminants should be investigated to determine the degree of possible hazard.

1.1 Approved Devices. At the present time, the United States Bureau of Mines approves respiratory protective devices in the following categories:

1. **Self-Contained Breathing Apparatus.** These apparatus include the compressed oxygen cylinder and the compressed air cylinder, demand type.

2. **Supplied Air Respirators.** These apparatus include hose masks with blower, Type A; special hose mask without blower, Type B; air-line respirator, Type C; abrasive-blasting helmet, Type CE; dispersoid (dust, fume, and mist) respirators; and non-emergency gas respirators (chemical cartridge respirators).

1.2 Classification of Hazards. Because the selection of respiratory protective devices is based on the type of hazards encountered, the American Standard Safety Code for Head, Eye, and Respiratory Protection, ASA Z2.1-1959, classifies hazards as follows:

1. Oxygen deficiency

2. Gaseous contaminants:

    a. Immediately dangerous to life

    b. Not immediately dangerous to life.

3. Particulate contaminants (dusts, fumes, smoke, mists, fogs).

4. Combination of gaseous and particulate contaminants:

    a. Immediately dangerous to life.

    b. Not immediately dangerous to life.

It is suggested that self-contained breathing apparatus be used where oxygen deficiency and certain types of gaseous contaminant hazards exist and that supplied air respirators be used for the remaining types. It is always important to consult the plant or area safety engineer as to which type of protective device is best for a given condition, however.

1.3 Safety Rules for the Use of Respiratory Protective Equipment, General

1. If the wearer is to enter a confined space containing a hazardous atmosphere, first connect a strong lifeline to his body. The lifeline will serve as a means of guiding him to the exit, as a means of

exchanging prearranged signals between him and the man at the fresh-air base, and as a means of aiding in rescue operations in case of an emergency or accident. The lifeline should be held by two attendants, one of whom is wearing a similar apparatus.

2. Where a lifeline is used, a signal code to enable the wearer to communicate with the lifeline tender, and vice versa, is necessary. The following code is suggested:

Tender to Wearer	Wearer to Tender
1 pull—Are you okay?	1 pull—I am okay.
2 pulls—Advance.	2 pulls—I am going ahead.
3 pulls—Back out.	3 pulls—Keep slack out of line.
4 pulls—Come out immediately.	4 pulls—Send help.

3. If upon entering a contaminated area the wearer is able to detect the contaminant by odor or taste, or by eye, nose, or throat irritation, he should return to the fresh air immediately and ascertain what is wrong with the equipment or the manner in which it is being worn.

4. Bear in mind the time limitations of the apparatus and allow an adequate margin of time for return to fresh air.

5. The mouthpiece and nose clip, or the facepiece, should not be removed until the wearer is certain that he is in respirable air.

1.4 Safety Rules for the Use of Hose Type Respirators

1.4.1 Hose Mask with Blower

1. Make certain that the hose mask is in good operating condition.

2. Set the blower in an assured source of respirable air.

3. Connect the proper length of hose (not over 150 feet) to the blower and to the facepiece, making sure that all gaskets are in place and that the connections are tight. Where more than one hose line is to be used, each should originate at the blower.

4. Operate the blower for a minute or two at a rapid rate to blow any dust out of the hose and to make sure that air is being delivered to the facepiece.

5. Adjust the body harness securely to the wearer. Connect a strong lifeline to the D-ring of the body harness.

6. Adjust the facepiece to the wearer so that it makes a gas-tight fit with his face.

7. Operate the blower, and adjust the flow of air to the wearer's satisfaction. The blower must be operated continuously during the use of the mask.

8. Check on the prearranged signals between the wearer and the blower operator.

9. Enter the contaminated area cautiously.

10. Be careful that the hose and the lifeline are not endangered by sharp edges or falling objects, and remember that the wearer must retrace his steps and leave by the same route by which he entered.

1.4.2 Air-Line Respirators and Abrasive-Blasting Respirators

1. Make certain that the air supply is respirable. Pay close at-

tention to the location of the intake to the air-supply device to make certain that the entering air is not contaminated from the exhaust of the compressor engine or from other sources. Provide a suitable filter to remove objectionable odors, oil, water mist, and rust particles from the air delivered to the air-supply line.

2. Make certain that the air-line respirator is in good operating condition.

3. Attach the proper length of air-supply hose to the source of compressed air and to the breathing tube.

4. Adjust the pressure of the air at the inlet to the air-supply hose so that it is within the proper pressure range.

5. Adjust the facepiece, helmet, or hood to the wearer according to the manufacturer's instructions.

6. When the rate of flow of air into the facepiece, helmet, or hood seems to be excessive, the wearer may decrease the flow of air by means of the air-regulating valve with which most air-line respirators are equipped. However, to prevent the contaminant in the surrounding air from reaching the wearer's breathing zone, the flow of air should not be decreased below 4 cubic feet per minute for facepieces, or below 6 cubic feet per minute for helmets or hoods.

7. Enter the contaminated area cautiously and leave by the same route.

8. In the case of abrasive-blasting respirators, make certain that the shatterproof eyepiece and the protective cover glass are in place. (Under no circumstances should regular window glass be used in place of the shatterproof eyepiece). Clean the inner and outer surfaces of the eyepiece. Make certain that the protective wire screen or perforated metal eyepiece is clean and in place.

1.5 **Instruction in the Use of Respirators.** For the safe use of any device, it is essential that the user be properly instructed in its selection, use, and maintenance. This is particularly important with respect to respirators. Competent personnel should give such instructions to the supervisors of all groups who may be required to wear respirators at their work. The supervisors, in turn, should instruct their men. No person should be allowed to wear a respirator of any type until he has received such instruction. Instruction should cover the following points:

1. An explanation of the need for using the respirator.

2. Its operating principle.

3. Steps to be taken to assure that it is in good operating condition.

4. Proper adjustment of the respirator to the wearer.

5. Proper use and maintenance of the respirator.

1.6 **Care of Respirators.** The life of the wearer may be dependent on the proper functioning and ready availability of respirators, so it is highly important that respirators be properly maintained and stored. Several rules can be fashioned for the adequate care and maintenance of this equipment:

1. All respirators shall be inspected at regular intervals to make sure that they are ready for use.

2. All rubber parts such as facepieces, mouthpieces, exhalation valves, breathing tubes, and headbands shall be inspected carefully for signs of deterioration such as hardening, checking, or tackiness. A check shall be made during this inspection to see that all gaskets

are present and that they are held tightly in place. Metal parts shall be checked for signs of corrosion, and plastic and glass parts shall be checked for breakage.

3. When it is necessary to replace worn or deteriorated parts, only those made specifically for the device shall be used, and repair work shall be accomplished by experienced personnel only.

4. Respiratory protective equipment shall be cleaned and disinfected after each use. The manufacturer shall be consulted for the cleaning and disinfecting method best suited to his products.

5. All types of respirators shall be stored in clean and dry compartments under conditions of moderate temperature.

2. Goggles, Safety Eyeglasses

Safety goggles and safety eyeglasses are familiar protective devices to those who work in modern industrial plants. In most chemical plants and refineries, their use is mandatory.

There are many excellent devices on the market today. The American Standard Safety Code for Head, Eye, and Respiratory Protection, ASA Z2.1-1959, is the Code authority for devices of this type. The main requirements are that the goggles and eyeglasses fit well, are not cumbersome, provide for adequate straight-ahead as well as peripheral vision, and be manufactured of tempered, unbreakable glass, or unbreakable plastic.

Glass or plastic shields should be readily replaceable. Spare goggles, eyeglasses, and shields should be kept readily available. It is mandatory that these devices be kept clean.

3. Protective Clothing

3.1 **Ordinary Work Clothing.** Tears and rips in work clothing are potential causes of accidents. Torn clothing can get caught in moving machinery or on a ladder or scaffold, and may lead to a nasty fall. Mend tears and rips.

Cuffs on trousers are dangerous, since they might also catch on projections and cause crippling falls.

Floppy pockets, dangling ties, and other similar items of clothing can catch in moving machinery, or may catch on projections and cause a fall.

Oil or chemical-saturated clothing (or clothing saturated with paint or other flammable material) easily can lead to dangerous or fatal burns. Clothing saturated with these materials can also cause painful skin irritations and burns. Oil saturated clothing should be stored loose, exposed to air.

Defective, poorly-designed, ill-fitting, or dirty clothing should not be worn to work. It is false economy to tolerate wornout or dirty clothing or to permit employees to wear to work old clothing not designed for safety under on-the-job conditions.

3.2 **Protective Clothing for High Temperature and Highly Toxic Atmospheres.**

Protective clothing capable of withstanding long operating temperatures of 500 F, and capable of providing protection from many highly toxic atmospheres has been produced. This type of clothing is a complete suit constructed of woven Teflon laminated to a film of FEP fluorocarbon with a layer of reflective metal vacuum deposited onto the FEP. The suit is complete with gloves, steel-toed boots, and helmet-type headpiece, all locked to the suit by sealing ring arrangements.

Details of this suit construction and use can be obtained from the Snyder Manufacturing Company, New Philadelphia, Ohio, or the E. I. du Pont Chemical Company.

3.3 **Hard Hats, Safety Shoes.** Most industrial plants require that workmen wear rigid safety helmets and steel-toed safety shoes as a matter of basic safety protection. The files of industrial safety engineers and insurance companies contain a sufficient number of case histories to prove to the most intransigent workman that his safety depends on these devices of protection. Project superintendents and painter foremen should not permit a man to begin work on a job until he agrees to wear these items.

Steel toed safety shoes should have rubber soles, with no nails or brads exposed, when the painter is to work in enclosed spaces where flammable vapors may be present.

V. Training In Safety

1. Education Is the Key

When working equipment is in perfect condition, hazards of a given area are compensated for, and knowledge of safe applications of coatings is complete, then there is a sound foundation for painter safety. Patient and persistent education and training of the individual painter are the building blocks with which to construct an enviable safety record.

The individual painter must be educated by indoctrination. This can be handled only by constant discipline and the example of trained foremen, and by close control by the project superintendent. Trainees should be taught basic safety principles first, and then be placed with an experienced man for development.

Nothing can be taken for granted. Even experienced men need refresher courses in elementary safety practices, such as the proper placement of ladders, securing of all ladders, staging, and scaffolding, etc.; since it is a human tendency to become careless as familiarity with work progresses, even the best painters need to be reminded from time to time to avoid over-reaching, and to avoid pointing nozzles at other workmen. Constant vigilance must be exercised.

Monthly safety meetings, particularly for foremen and superintendents, are imperative. These meetings should include open discussion of all near-miss accidents, and recommendations for improvements in all phases of coatings application. Foreman and key painter personnel should be given the benefit of a formal safety course when one is available.

1.1 **Material for Periodic Safety Meetings.** Much of the material and information in this manual is adaptable for presentation to a painter group during periodic safety meetings. Other information is available from NACE. Contact NACE Headquarters, P.O. Box 218340, Houston, TX, 492-0535, for further information.

It might be wise also, to invite safety engineers from industrial plants to meet with your men from time to time. These safety engineers can demonstrate safe and proper methods for lifting and handling materials and equipment, and can provide illustrations from their files of actual accidents and near-miss accidents that have been caused by carelessness, and by failure to wear hard hats, safety shoes, and safety goggles.

2. Responsibilities of Superintendents and Foremen

Painters in any crew depend upon superintendents and foremen to safeguard their lives. **No other responsibility transcends this in importance!!** With this responsibility in mind, it is suggested that superintendents and foremen be given the following instructions at the beginning of each project.

1. Maintain constant vigilance.

2. Be sure that each painter **understands and accepts** his personal responsibility for safety.

3. Be sure that you and your men are informed of all of the safety rules. **Never let a man be injured because of failure to learn the necessary precautions!**

4. Anticipate the risks that may arise from new equipment and materials. Secure expert advice in advance on such new hazards.

5. Encourage men to discuss the hazards of their work. **No job should proceed if a question of safety remains unanswered.** Be receptive to the ideas and suggestions of workmen; such suggestions may be the source of first-hand information and field knowledge that will prevent needless loss and suffering.

6. Insist with patience, persistence, and **determination** on men working safely. Enforce rules with disciplinary action if necessary.

7. **Set a good example.** Demonstrate safety in personal work and conduct.

VI. Appendix

1. Supplementary Data

1.1 **Recommended Supplementary Reading.** For those who are interested in pursuing the subject, the following treatise is recommended: "**The Human Side of Safety,**" Louis H. Bornoff, Bright Publishing Company, Boston 16, Massachusetts.

1.2 **Specific Acknowledgements.** Task Group T-6D-5 wishes to specifically acknowledge extraordinary assistance furnished by the following:

1. Use of safety manuals published by:

   a. E. I. du Pont de Nemours Chemical Company, Wilmington, Delaware.

   b. The Dow Chemical Company, Midland, Michigan.

   c. Union Carbide Chemicals Company, New York, New York.

2. Review of preliminary material by the following:

   a. The Safety Department of Humble Oil and Refining Company, Houston, Texas.

   b. The Safety Department of Monsanto Chemical Company, St. Louis, Missouri.

   c. The Medical Department of Continental Oil Company, Ponca City, Oklahoma.

17

Specifications

Irrespective of the type of coating job, a proper specification is the key to its success. A specification that describes the materials needed, the objective(s) of the coating, the key procedures required, the appearance of the finished coating, and the inspection requirements usually assures a good coating job. A poor specification, *i.e.,* one that is ambiguous, vague, lacks definition, or provides for a poor selection of materials, is an invitation to disagreements, possible lawsuits, and, above all, early coating failure. Even good specifications are sometimes disregarded, forgotten, unread, or made available only to the job estimator and not to the rest of the job crew. Such circumstances are also an invitation to early coating failure and costly, continual maintenance.

The task of writing any kind of specification is not easy. A *coating* specification is even more difficult because of the many variables involved. There are literally hundreds of materials to choose from, thousands of configurations of building materials, plus hundreds of specific types of exposures, unusual weather conditions, and interior and exterior surfaces. Added to these variables is a time frame within which the coating must be applied. Unfortunately, many specifications are reduced to little more than the following: "Painting. After fabrication and dispatch to the jobsite, clean steel and paint with one coat of good quality primer and one coat of good quality enamel." A statement such as this is not a specification, but a meaningless document which results in a protective system of the poorest quality, with a short life and continuing high maintanance costs for the life of the structure.

The procedure of writing an effective specification requires thorough planning for the structure, consideration of the needs and requirements of the coating, development of the specific procedures required for its application, consideration of environmental conditions during application, and development of a description of the finished product. All of this leads to obtaining the type of corrosion protection that is both necessary and desirable. A specification should be a practical document where the owner and designer effectively communicate to a coating contractor or coating superintendent what they want in a coating, when they want it, why it is needed, and what results they expect from the finished coating. The specification should not be unreasonable, but should take into consideration the problems that the contractor may encounter and must overcome in order to satisfy the specification.

A good specification should be a reliable guide for both the owner and the contractor or the coating crew and should be written to provide a clear understanding between them. A good specification should and will invite cooperation between the owner and the contractor; thus, it is more likely to result in a satisfactory job. On the other hand, some specifications are written with an adversary relationship in mind, which can only lead to the dissatisfaction of both parties and a poorer than ordinary chance for a satisfactory job. Dissatisfaction, arguments, hard feelings, and ultimately legal actions arise only from the use of poor specifications, not good ones.

In order to aid corrosion engineers in writing good, sound specifications, the National Association of Corrosion Engineers (NACE) has published *A Guide to the Preparation of Contracts and Specifications for the Application of Protective Coatings.*[1] This is detailed document and, as its title indicates, is a good guide to use in writing protective coating specifications. All parts of the document may not be applicable to any one specific job, and these parts may be eliminated or modified as desired. Even so, it should assure that the content of the specification is as close to what is desired as possible. It should result in minimizing the opportunity for poor materials and workmanship. Also refer to Appendix 17A which is an actual working specification that includes many of the requirements of a good specification.

On the other hand, comment on the useless specification is given in an article entitled *Specifications for Surface Protection, the Good, the Bad, and the Useless,* as follows.

> It is really only necessary to say that useless specifications contain nothing of the good or the bad, or the impractical, they are of no use at all. They are however being written every day without any knowledge of or regard for the job in hand. They are compiled from past information, irrelevant to the present situation and a history of coating failures.
>
> They are used and re-used without a second thought and are a waste of money and energy. They are generated by overnight experts who find it totally unnecessary to state the obvious. They result in costly maintenance problems, arbitrations, dissatisfied clients, disenchanted manufacturers and disgusted contractors. They should be abolished.[2]

A specification is, and should be considered, a legal document. It is a part of a contract between the owner and the contractor. While it is only a sheaf of paper, if the contract goes smoothly and a good application results, it can become extremely important in every detail if the job turns out to be a poor one and legal action results. Once legal action is taken, the specification, in all of its parts, becomes the center around which the entire case revolves. A well-written specification is well worth the resulting effort and cost, primarily because it usually insures a good job and good corrosion protection. It also becomes an invaluable factor should unforeseen circumstances result in legal proceedings.

Even the best of specifications, however, are not infallible in assuring the material, application, or protection intended. Any reliance on specifications must take into consideration their limitations and the dangers inherent in their use. The following statement, extracted from the *Steel Structures Painting Manual,* discusses some of these problems.

> Reliance on specifications to secure the desired protection may lead to a false sense of security. An owner's desires must be expressed in words which must then be interpreted by the contractor or applicator. Each party to the contract looks at the words through his own eyeglasses and what he sees is influenced by what he wants to see, or what is to his own advantage. Even with complete integrity, the unfortunate lack of exactness in communication, further complicated by differences in specialized meanings of words in different segments of the same industry, makes vigilance the price of security.
>
> Specifications do not guarantee that the protection or materials will be obtained at the lowest possible cost. They are designed to stipulate exactly and specifically what the purchaser desires. It is hoped that the specification will procure the most economical protection of surface, but this does not necessarily follow.
>
> . . .Also, specifications often invite the attention of dishonest operators who are searching for loopholes. A supplier or contractor is entitled to a reasonable profit; therefore, any unreasonably low bid should be suspect. An occasional below-cost bid may, for reasons unknown to the purchaser, be legitimate, but consistently letting contracts on the basis of extremely low bids openly invites inferior quality and indifferent performance. Dangers here, as in other areas, are minimized by dealing only with reputable concerns.
>
> . . .Unfortunately, there is often a divergence of viewpoints between the purchaser and the supplier or contractor. For the purchaser, a specification is minimal, establishing a floor under the minimum quality of work or product that will be acceptable. Too often the contractor looks on the same specification as maximal, a ceiling above which he goes only at a loss to himself. This latter point of view, though understandable, is negative and unproductive of repeat orders. The happy situation exists when the contractor or supplier accepts the specification as an absolute minimum, the purchaser recognizes the supplier's problems, and each cooperates with the other to secure the results the purchaser desires at a mutually satisfactory price. When this occurs, the owner gets a good job and the contractor makes a reasonable profit.
>
> . . .There is no fool-proof method of insuring compliance with specifications. Those who grant contracts to the lowest bidder on the basis of specifications should protect themselves against accidental or deliberate non-compliance by using reasonable inspection procedures and stipulating them in the contract.[3]

A specification is also a working document. It is important to understand this since, in many cases, specifications are more legalistic than practical. While the legal aspects of the specification are extremely important, the specification should be written in a way that the superintendent, the foreman, and the workers can understand and use it as a guide for their part of the operation. The information contained should be straightforward, simple, and written in a way that is easily understood by all those who may be required to use it.

For example, there is currently a strong tendency to use the metric system wherever temperatures, areas, coverages, and similar important information is used. However, in the United States, personnel such as foremen and applicators are not yet sufficiently acquainted with such terms to interpret them properly from the standpoint of their particular part in the job. In Europe, figures for temperatures and coverages *should* be metric terms since that is the system with which the people there are familiar. Where a specification is to be used in both areas, both terms should be included. A worker in the United States understands a mil of coating; however, 25 microns of thickness is generally a mystery. One foot is easy for an American worker to relate to, while 300 millimeters usually means little or nothing to him. Thus, the wording of the specification should be geared to the area where it will be used, and the terms and conditions should be readily understood by the people located in that area.

There usually are three, and perhaps four, entities involved in a specification. These include the owner, possibly an engineering and design company, the manufacturer of the product to be applied, and the contractor who will apply it. All of these people or companies should have a part in creating the specification. It is obvious that the owner is vitally interested in and ultimately responsible for the specification. The owner should provide the technical expertise needed for creating an effective specification, either from within his own organization or from the outside, seeking the best possible advice from various sources (*e.g.,* reliable engineering firms, coatings manufacturers, and/or contractors).

Since an engineering firm usually is involved exclusively in new construction, it is often given primary responsibility for developing the specification for the owner. If the engineering firm is knowledgeable in the area of coatings, a good specification should develop. On the other hand, coating specifications are often selected from previous contracts held by the engineering firm and thus may not be as applicable to the specific job involved as is necessary. Since the specifications are the owner's ultimate responsibility, the owner should carefully review any engineering firm specifications being used to make sure that they provide the information needed for his particular purpose.

The coating manufacturer also has some involvement in specification development. It is the manufacturer's responsibility to provide the best information available on the use of their materials for the job that is contemplated. The manufacturer represents the most knowledgeable source of both material and application information for their materials and thus should supply not only typical data about a coating (*e.g.,* percent of solids, drying time, etc.), but also practical information concerning the application of the specific material for the particular job. If the use of the coating being considered is questionable, the manufacturer should step forward, before the specification is written, and recommend a better material.

The coating contractor is also involved in the development of a specification. Since the contractor is generally an authority on the application of coatings, if he determines that there are impractical or useless aspects of the specification, he should communicate this to the owner and help develop a more workable specification which will provide the owner with the very best application possible.

The above entities should all be involved during the development of a specification so that both the best material and the best application procedures can result in the best possible job.

In many ways, there are two types of coating specifications, *i.e.,* those involved with new work and those involved with the maintenance of an existing coating. The specification for each may be quite different. In the case of a specification for new work, it may be drawn up by the overall architect or engineer for the job. In many cases, the specifications for new work may be much more detailed than those often used for maintenance work. Also, control of a specification for new construction is often with the engineering firm or the general contractor.

The specification for repair of an existing coating may be quite different, unless an addition to the structure is being made and repair of the existing coating is included in its specification. In this case, an engineering firm may also control and write the specifications. On the other hand, most specifications for maintenance coatings are developed in-house by the company engineering department or by the corrosion engineer involved in protecting the structure. This type of specification is often short and simple and thus more likely to result in trouble. The engineer drawing up the specification oftentimes has only a limited knowledge of coatings, and the corrosion engineer involved may be more familiar with cathodic protection than with the protection of structures by coatings. This leads to the short and simple specifications such as "apply in accordance with the manufacturer's recom-

mendation."

On the other hand, proper maintenance specifications may be more important than specifications for a new job, and thus may be more complicated. The reason for this, as brought out in Chapter 15 on the repair and maintenance of coatings, is that there are many more variables involved in coating repair than in a new application. In the case of a repair coating, failure has undoubtedly already occurred, so that the existing coating is contaminated enough to require remedial action. Cleaning of the coating, more difficult surface preparation, and the compatibility of the repair coating should be considered, as well as possible contamination from the atmosphere during the application process. Thus, the specifications for the repair of an existing coating on an existing structure should be much more detailed than is often necessary for the application of a new material. Even if the work is to be done by plant employees or a local contractor, a good specification for the guidance of these people and for proper inspection of the completed work is vitally important.

Parts of a Specification

In a specification for either new work or a repair coating, several different sections should be included. These are covered in some detail as follows.

General Information

In the section providing general information, there should be a good description of the work involved. This should include the location of the work, the type of structure involved, the scope of the work, the location of other structures in the area which may be affected by the application, and the degree of protection required for such structures.

General information should include a paragraph on the need for the coating. This is often left out of specifications. However, it is important in that it emphasizes to both the material supplier and the applicator the importance of the coating. For example, if the coating is to protect against acid fumes, this should be stated so as to alert both the coating manufacturer and the applicator that a superior coating is required, as well as a topnotch application.

The timing of the application is also important and should be outlined under general information. It may be even more important for a repair coating than it is for a new coating, since with a repair coating the application work must fit in with the operation of the structure during the repair period. This may mean that the application will have to be scheduled on weekends or on off-periods such as at night. This is important information, for both the manufacturer of the material and the applicator.

A statement regarding the storage of the material also should be included in the general information section. Unfortunately, the material is often brought out on a job, left in the open, and, especially during a long job, subject to considerable weather changes during that period of time. This can lead to many material difficulties such as jelling of the coating or its transformation to a heavy, thixotropic consistency. Such exposure also may increase the settlement of pigments in the coating which require substantial effort to bring back into suspension.

Exposure conditions may even allow water to get into the coating. In one instance, 55 gallon drums of coating were left in the open during a period of considerable rain. The drums were left upright, and because of temperature changes, water was drawn through the bung in the top of the drum, resulting in coating contamination. The pigments in the coating were precipitated because of the water and could not be properly redispersed. The manufacturer was accused of supplying poor material. Until the water was discovered, there was considerable rework of coated areas, which was quite costly. All of this could have been prevented by proper storage of the material in an area protected from the weather.

General information should also include some statement concerning weather protection of the coating work itself if there is a chance that this could be a factor during application. The specification should either include a statement that the work be done when the weather is satisfactory or provisions for the protection of work during inclement weather.

Safety is also a part of the specification that should be included under general information. This is an important part of the specification since accidents often result in lawsuits, and if proper safety practices have not been detailed in the specification, costly litigation may take place. If there are any special safety procedures required, they should be brought to the attention of all those involved through the specification.

An important part of the general information section is the arrangement of a pre-application meeting for all parties involved in the coating work. This includes the owner, the engineer, the corrosion engineer, the coating manufacturer, the applicator, and the inspector. Such a meeting can be extremely valuable in that it gives the owner an opportunity to communicate expectations to the contractor, and it gives the contractor an opportunity to discuss any foreseeable application problems. When these are discussed and settled at a pre-application meeting, a much smoother operation and better application result. Special safety procedures that may be required during the application of a repair coating can also be discussed and outlined at such a meeting.

Another important point for the general information section is the requirement that the superintendent or foreman of the application crew keep a diary of job progress. The same applies to inspectors, if they are involved. The diary should include (besides routine items such as weather) the areas coated, the coats applied during any period, and any exceptions to the specification. Such information might include moisture condensation on the coating shortly after application, welding in an area after the coating has been applied, or any unforeseen happening during the application. If there is a coating failure, such items can be extremely important in determining what may have caused the difficulty. If litigation results, such information can be extremely valuable to both the applicator and the owner.

The general information section of a specification

may include any number of other specific items that are important to the owner or the design engineer.

Materials

The materials section is always an important part of a coating specification. As stated before, materials represent the least expensive part of a coating application, with surface preparation and coating application usually amounting to several times the value of the materials. Thus, the percentage difference in the use of the very best materials versus poor materials is low in relation to the total coating cost. There is little economy in using poor materials, and they should never be considered where a serious corrosion problem exists.

Coatings are often procured or specified by such general terms as "a red lead paint," "a vinyl," "an epoxy," "a zinc chromate primer," or "a shop coat." These terms do not constitute a specification in any way, shape, or form. "Red lead" paint refers to any vehicle which contains red lead and is a red lead color. "An epoxy" coating may be a polyamide, an amine-cured material, a high solids material, or one with relatively low solids. Therefore, such terms should never be used in a proper specification. They are meaningless, wasteful, and only produce a continuing source of difficulty and dispute once the material is obtained.

One poor practice in material specifications is to specify a shop coat for shop-constructed structural steel or other items. "A shop coat" refers to a temporary material which provides color designation and little else until the structural components are put in place. If a shop primer or shop coat is to be applied, the type of material to be used should be specified so that following coats will be compatible and so that the shop coat itself will be sufficiently anticorrosive to provide a proper base for the topcoat materials.

Some coating manufacturers provide a low-cost shop coat primer which actually is the accumulation of the coatings remaining in various tanks at the end of a certain time period. These are mixed together and then pigmented to a solid color. This can result in either a good material or a poor one, depending on the composition of the various batches that are mixed together. More often than not, this type of material provides a very poor base for any following topcoats. Where a material such as this is applied to structural steel or other components in a shop, the specification should call for removal of the shop primer and repriming of the surface with a proper, compatible anticorrosive primer prior to the application of any topcoats.

Types of Specifications

There are a number of ways in which a material can be specified: (1) by trade name and number, (2) by trade name or equal, (3) by qualified product lists, (4) by material specification, or (5) by performance specification. Each of these methods has its advantages as well as its limitations.

Specification by Trade Name

Wherever the specification is controlled by a private

company, specification by trade name and number, using a reliable manufacturer, is often the best way to specify a material. This is particularly true where the purchasing company has had previous satisfactory experience with the material. Respecifying it under these conditions makes good sense. With a trade name product, a manufacturer usually develops a quality product to do a specific job, investing personnel and materials in research and testing. When the product is developed, an investment is made in equipment, quality control, advertising, and marketing. After the product is established, research in product improvement, cost reduction, new uses, and refinements in process application are continued. Charged against a single customer, these added costs would be prohibitive, but distributed among a large number of large-volume orders, the added services and assurances of quality may well make the proprietary product a bargain. The prospective user must gauge this for himself.

Also, when dealing with a reliable manufacturer and specifying by trade name, the manufacturer will stand behind the product and usually will give advice concerning its application to assure that it is applied in the best possible manner, and thereby protect the product's reputation. In this method of specification, the owner has full control of the material that is applied with some knowledge of the ultimate performance and life of the material. The limitation in this type of specification is that it does not allow competitive material bidding. On the other hand, with a critical corrosion area, competitive bidding often leads to the required acceptance of a poorer material, rather than the attainment of the particular material the owner needs and wants.

One method by which owners can be assured of receiving the material they want is to specify that the owner rather than the contractor will supply the material. In this case, the owner purchases the material and the contractor merely applies it so that there is no question about the quality of the material. Even though a material is specified by trade name, a contractor will often request to purchase another material (supposedly equal or better) for any number of reasons, and is often successful in making the switch. Owner-supplied material eliminates this problem.

Specification by Trade Name or Equal

The "or equal" to a proprietary trade name specification is supposed to allow competitive bidding. It implies that the owner has done considerable test work to determine that the materials on an "or equal" specification are actually equal. However, there is no practical method of evaluating or enforcing an "or equal" specification. The companies which wish to bid on this basis all claim that their materials are equal. However, without serious testing of the products, there is no basis for determining whether they are, in actual fact, equal.

Qualified Products List

If there has been serious testing of many materials by a company or a governmental agency, specification by qualified products list is a good method. It allows the specification of a number of materials and therefore allows competitive bidding, with all of the materials on the qualified products lists supposedly being equal. Many government agencies which use proprietary products have used the qualified products list very successfully. However, due to the length of time required to develop standard and complete testing, newer coatings and modifications are usually precluded.

Formula Specifications

In formula specifications, the coating is specified by its formula. Many federal specifications use the formula approach, as do some large companies. The Steel Structures Painting Council (SSPC), in their paint specifications, has developed many formula specifications materials which have been used since 1955. They are based on a continuing study of coatings that have performed well on actual steel structures. These were developed because many of the outstanding coatings that were being studied did not have available specifications or had specifications that were available for only limited distribution. The SSPC coating specifications were issued, therefore, to make it possible for anyone to specify these materials by formulation. In this case, the specifications are reviewed and revised periodically, and new and different types of coatings are added. The SSPC has developed the formulations and supplied this service to companies and various governmental agencies which wish to specify in this manner.

The coating from a formula specification usually must be formulated for each specific customer. As with all materials specified by formulation, the manufacturer supplies the coating exactly as it is formulated in the specification. Thus, no research or additional coating expertise is used, and the manufacturer has no responsibility other than to meet the formulation specification. Once it passes from the manufacturer to the customer, it is strictly the customer's responsibility, which can be a decided disadvantage.

There are a number of other serious limitations in specifications by formula. In many cases, the specified composition may represent a compromise since it may have been developed only through laboratory testing, without allowing for actual field experience. In one known case, a formula specification was written by a supposedly experienced individual, yet without any field testing. It was specified on a large job and the resulting failure was almost immediate. Since the material conformed to the formula specification, the only way to try to recover losses was for the purchasing agency to claim that the application had been improper, when it was actually the material that was at fault.

Also, in many cases, the specified composition coatings exclude the more modern, improved formulations or ones which have been more recently tested and proven to be better or at least cost reducing substitutions. In many ways, the formulation specification eliminates much of the research work done by coating companies to develop superior coating materials. However, the majority of the coatings used by governmental agencies are procured by formula specification. Many millions of gallons are procured in this manner by the federal government, and, where there is tight control, the materials obtained are satisfactory.

On the other hand, this method of specification can lead to the minimum quality available, and materials are often supplied which conform to the specification, but which perform very poorly when in service. These materials are too often supplied by the lowest bidder, regardless of his qualifications. If a formula material is specified and obtained from a reputable manufacturer, with close inspection of the material both during manufacture and upon receipt, it can be a good method of obtaining and specifying coating materials.

Performance Specifications

Obtaining materials by performance specification is also common. A performance specification does not designate the material by formula, but indicates the required material's performance. Any material which meets these performance requirements should be satisfactory. In many ways, it seems that performance requirements would be flexible and thus a good method of specification. On the other hand, many performance requirements have proven less than satisfactory. There is no way to determine whether two materials will perform equally and satisfactorily without putting them through actual performance tests.

However, because all coating jobs are different and each has many variables, only comparative testing of performance specification coatings under actual operating conditions will provide a positive answer as to which coating best meets the requirements. If performance tests are only laboratory tests, then they are limited to a few, short-time tests. If they extend into field testing, testing of the materials becomes very costly and time consuming. There are only a few companies and governmental agencies which have the testing background and the personnel and financial backing to satisfactorily determine whether several materials meet extended performance test requirements.

Of all of the above ways of specifying materials, the first three, *i.e.,* by trade name, by trade name or equal, and by qualified products list, provide the advantage of the manufacturer's backing of the material, since in each case the material is a trade-named item developed by the manufacturer through considerable research to provide the best service possible for the corrosive conditions where it is used. The backing of the manufacturer for their trade-named material is important and can actually be cost advantageous, particularly when the coating manufacturer or supplier backs up the product with good service and application assistance.

Areas of Specification

Surface Preparation

The specification for proper surface preparation is relatively easy due to the work that has been done by several organizations, such as the SSPC and the NACE, to develop standards for such specifications. Thus, when specifying one of the accepted standards, one can be relatively sure of obtaining the specified surface. NACE

and SSPC surface preparation specifications have become recognized worldwide, and most applicators and contractors are prepared to prepare surfaces in accordance with them.

In general, there are two items which should be covered in a surface preparation specification. One is the type of cleanliness required for the coating to be applied. The specification for cleanliness is covered by the NACE standards for blast preparation of steel and the SSPC specifications which cover not only blast preparation, but also preparation by lesser means. These standards or specifications are covered in Table 17.1, appearing in descending order according to the degree of surface cleanliness.

TABLE 17.1 — Specifications for Surface Preparation
(In Descending Order of Effectiveness)

1. White Blast	NACE 1	SSPC-SP-5-63
2. Near-White Blast	NACE 2	SSPC-SP-10-63
3. Commercial Blast	NACE 3	SSPC-SP-6-63
4. Acid Pickling		SSPC-SP-8-63
5. Brush Blast	NACE 4	SSPC-SP-7-63
6. Flame Clean and Power Sand		SSPC-SP-4-63
7. Power Tool Cleaning		SSPC-SP-3-63
8. Chip and Wire Brush		SSPC-SP-2-63
9. Solvent Clean		SSPC-SP-1-63

As outlined in previous chapters on coating application and on surface preparation, most of the high-performance coatings should be applied over blast-cleaned surfaces. Where any corrosion is a factor, surface preparation specifications less than acid pickling (which is number 4 on the list) generally would not be satisfactory. As previously stated, the cleaner the surface, the better the performance of the coating applied over that surface.

The second item which should be covered in a surface preparation specification is that of the surface profile. Much research has been done on this subject and it has been discussed throughout the industry and in technical literature on numerous occasions. Primarily, the surface profile must be practical and effective for the coating that will be applied over it. It is usually much more practical and effective in the long run to specify the type and size of the blasting media to be used in developing the proper profile, rather than specifying the surface profile as such. If surface profile alone is specified, field testing is required prior to the application of any coating. It is both possible and probable that the delay caused by surface profile tests could cause the freshly blasted surface to turn color and require reblasting.

Monitoring surfaces by profile alone is very difficult. It is the grit or the blasting media used that will primarily determine the profile obtained, and there is little that a sandblasting gun operator can do to change the profile. Naturally, the blasting pressure can be changed, as can the blast angle and other variables. On the other hand, the actual profile obtained depends on the blasting medium, all other conditions being equal.

There are many different blast media, as discussed in detail in Chapter 9 on surface preparation. These range from steel grit to steel shot, followed by flint abrasives, garnet, quartz sand, and synthetics or crushed slag. Materials such as steel grit and flint abrasives will create a rough profile. Materials such as silica sand or synthetic grits usually explode on impact and provide a scouring effect in addition to providing a profile. In most cases, for high-performance coatings, the exploding and scouring-type blast media are preferred compared with the cutting media for obtaining both a more uniform anchor pattern and one which is free of contamination. Nevertheless, all of the blast media have their respective uses.

In writing the specification for surface profile, the principle is to select the blast media which gives the best surface preparation for the coating that is to be applied. Where an automatic blast setup is used, steel grit is almost obligatory from a practical standpoint. In field applications, sand or synthetic blast media are required because of the waste factor. Therefore, the particular job and the physical circumstances of application usually determine the surface preparation specifications from a practical standpoint.

Table 17.2 indicates the comparative maximum heights of profile obtained by several abrasives. Again, it is suggested that the abrasive media and its size be specified in preference to specifying a 2- or 3-mil anchor pattern. It is difficult to measure such an anchor pattern in the field, so that specifying the abrasive media provides a more practical control of the anchor pattern.

The preferred abrasives for most high-performance coatings, and the ones which provide the most generally acceptable surface are:
1. 16-40 mesh silica sand or mineral grit
2. 20-40 mesh garnet
3. Crushed iron slag
4. G-50 iron grit

These materials provide a surface profile from 1.5 to 3.5 mils and are able to produce whatever degree of surface cleanliness is required.

Another item that is suggested for inclusion in a surface preparation specification is a requirement for the contractor to blast a section of the structure or work to be done and obtain approval for that surface before any coating is applied. This is an essential requirement since it allows all parties, *i.e.*, the owner, general contractor, painting contractor, inspector, and actual workers, to visualize the type of cleanliness and profile that is required by the owner and by the coating that is to be applied over the surface. This can eliminate many arguments and difficulties after the job has been started.

Application

Application is the area of a coating specification that should be the most detailed and therefore the most difficult to write. Oftentimes, however, this part of the specification simply states that the coating be applied in accordance with the manufacturer's recommendations. This is usually not sufficient since it only includes data put out in chart form by the manufacturer, an example of which is shown in Table 17.3. Such a chart provides a

TABLE 17.2 — Comparative Maximum Heights of Profile Obtained with Various Abrasives in Direct Pressure Blast Cleaning of Mild Steel Plates Using 80 psig Air and 5/16 in. Diameter Nozzle

Abrasive	Size[1]	Maximum Height of Profile In Mils	In Microns
Large River Sand	Through U.S. 12, on U.S. 50	2.8	70
Medium Ottawa Silica Sand	Through U.S. 18, on U.S. 40	2.5	62
Fine Ottawa Silica Sand	Through U.S. 30, on U.S. 80	2.0	50
Very Fine Ottawa Silica Sand	Through U.S. 50, 80% through U.S. 100	1.5	37
Black Beauty (Crushed Slag)	Estimated at minus 80 mesh	1.3	32
Crushed Iron Grit	G-50	3.3	82
Crushed Iron Grit	G-40	3.6	90
Crushed Iron Grit	G-25	4.0	100
Crushed Iron Grit	G-16	8.0	200
Chilled Iron Shot	S-230	3.0	75
Chilled Iron Shot	S-330	3.3	82
Chilled Iron Shot	S-390	3.6	90

[1]Sizes listed are U.S. Sieve Series Screen sizes or Society of Automotive Engineers grit or shot sizes.

[SOURCE: Steel Structures Painting Council, Mechanical Surface Preparation, Chapter 2, Steel Structures Painting Manual, Vol. 1, 1955.]

quantity of essential information; however, it cannot, because of its abbreviated form, cover many general conditions that are critical to the application of a coating. Manufacturers use this method of providing the essential information regarding a coating because of the many variables involved in application and the impossibility of covering all of these in any reasonably limited form.

In addition to this data chart, some manufacturers also supply more detailed application instructions. An example of this is shown in Table 17.4. In this case, detailed application instructions are provided which give the essential information required for the proper application of the coating. However, even these specifications may not sufficiently cover all of the variables involved in a sizeable or critical coating application. It must be remembered that every coating job is unique and there-

TABLE 17.3 — Example of a Coating Manufacturer's Data Chart

Physical Data	X Coating
Finish	Semigloss
Color	Aluminum
Applied over	Steel
Components	2
Volume solids (ASTM D 2697)	90%
Dry film thickness per coat	5 to 8 mils (125 to 200 μ)
Coats	1
Calculated coverage at	
1 mil (25 μ)	1444 sq ft/gal (35.4 sq m/ltr)
5 mils (125 μ)	288 sq ft/gal (7.1 sq m/ltr)
All for application losses and surface irregularities.	
Application	Airless or conventional spray
Pot life	8 hrs @ 70 F (21 C) 1½ hrs @ 90 F (32 C)
Drying time	
to touch	4 hrs @ 95 F (35 C) 12 hrs @ 70 F (21 C) 36 hrs @ 50 F (10 C)
dry through	10 hrs @ 95 F (35 C) 24 hrs @ 70 F (21 C) 72 hrs @ 50 F (10 C)
Mixing ratio (by volume)	1 part resin solution to 1 part curing solution
Temperature resistance	200 F (93 C)
Flash point (Setaflash)	
Resin (Component A)	112 F (44 C)
Cure (Component B)	116 F (46 C)
Thinner	Thinner X below 70 F Thinner XX above 70 F
Cleaner	Cleaner Y
Packaging	
2-gal unit	1 gal resin solution 1 gal curing solution
10-gal unit	5 gals resin solution 5 gals curing solution
Shipping weight (approximate)	
2-gal unit	Resin solution—12 lbs (5.5 kg.) Curing solution—8 lbs (3.6 kg.)
10-gal unit	Resin solution—60 lbs (27 kg.) Curing solution— 40 lbs (18 kg.)
Shelf life	6 months from shipment date for each component when stored indoors at 40 to 100 F (5 to 38 C)

[SOURCE: Ameron Corp., Monterey Park, CA.]

TABLE 17.4 — Coating Application Requirements

Surface Preparation

Maintenance - Steel
1. Round off all rough weld and sharp edges. Remove weld spatter.
2. Dry-abrasive blast according to Steel Structures Painting Council Specification SP-10-63 or SP-6-63. Use only steel grit (G-40 size), steel shot (S-230 size), graded flint, or silica sand (30-60 mesh). If reusing abrasives, clean them of contamination before reusing. Do not reuse sand or flint abrasives. Use air with minimum of 200 CFM (100 ltr/sec) per blast nozzle at minimum of 100 psi (7 kg/cm^2).
3. Apply coating as soon as possible to prevent blasted surfaces from rusting. Keep moisture, oil, grease, or other organic matter off surface, before coating. Spot reblast to remove any contamination. Solvent wiping is not satisfactory.

Maintenance - Concrete

Clean concrete surfaces, making them dry and free of previous coatings, disintegrated or chalky material, and surface glaze or laitance. Best cleaning is done by light sandblasting with 60-80 mesh sand and air at 50-60 psi (3.5-4.2 kg/cm^2). Note: Do not use form release agents, concrete curing compounds, or hardeners where a high-performance coating is to be applied.

Immersion - Steel
Prepare surface in same manner as for maintenance except blasting should be according to SSPC Specification SP-5-63 to "white metal."

Immersion - Concrete
Prepare surface in same manner as described above for concrete in maintenance service.

Equipment Required

For Airless Spray
Standard airless spray equipment such as Grace Hydra-Spray or others using a 28:1 pump ratio. Fluid tip with 0.015 in. or 0.017 in. (0.38-0.43 mm) orifice. Air supply: 80-100 psi (5.6-7.0 kg/cm^2) inbound.

For Conventional Spray
Pressure material pot with mechanical agitator.
Separate atomizing air and fluid pressure regulators.
Air supply: Compressor capable of supplying continuous volume of 20 CFM (10 ltr/sec) at minimum of 80 psi (5.6 kg/cm^2) to each nozzle.
Air hose for gun 5/16 in. (7.9 mm) I.D.
Material hose, 1/2 in. (1.25 mm) I.D.
Industrial spray gun—such as DeVilbiss MBC 510 with a 24D or 64D nozzle combination, 0.086 in. (2.1 mm) fluid tip, external atomization, 7 hole air cap.

Safety Equipment Required
(In Tanks or Confined Areas Only)

Explosion-proof lights and electrical equipment.
Fresh air mask, such as DeVilbiss P-MPH 527 and MPH 529, connected by 1/4 in. (6.5 mm). (Connect hose directly to air source.)
Exhaust fan capable of keeping solvent vapors below 20% of the explosive limit or 1/4% by volume of solvent vapor in air.

Table 17.4 (continued)

English

Tank Volume (Gallons)		Req'd Blower[1] (cu.ft./Min.)
500-	5,000	1,000
5,000-	20,000	2,000
20,000-	100,000	5,000
100,00-	250,000	10,000
500,000	—	15,000
1,000,000-	2,000,000	20,000

Metric

Tank Volume (kl)		Req'd Blower[1] (Ltr./Sec.)
4-	20	500
20-	74	1,000
75-	380	2,000
380-	1,000	5,000
2,000		7,500
4,000-	8,000	10,000

[1]All blowers to be suction type.

Application Procedure
1. Flush equipment clean with cleaner.
2. Thoroughly mix contents. Stir continuously throughout application.
3. Do not thin except for workability, and then with no more than 1 pint (0.5 ltr) of thinner per gallon of mixed coating.
4. Remove all dust from surfaces to be coated.
5. For conventional spray, regulate air pressure to 40 psi (2.8 kg/cm²) to spray pot and 70 psi (4.9 kg/cm²) to gun. NOTE: Required pressures may vary with hose length and temperature.
6. Make even, parallel passes. Overlap each pass 50%, and apply a heavy, wet coat. Double-lap spray all welds, corners, etc. Immediately "cross-spray" to achieve proper film thickness with no bare spots, pinholes, or holidays.
7. Allow first coat to dry at least 12 hours at 70 F (21.1 C) but not more than four days, before applying second coat. NOTE: Exposure to direct sunlight for more than 24 hours between coats will result in intercoat delamination.
8. Check for pinholes, bare areas, or holidays with nondestructive holiday detector. If coating is intended for use as a tank lining, check for pinholes, bare areas, and holidays with a nondestructive, field-calibrated sparking holiday detector, such as Tinker and Rasor high-voltage Model AP (approx. 2500 volts). Also check film thickness with nondestructive dry film thickness gauge, such as Elcometer or Mikrotest. If film is not up to required thickness, add additional material.
9. To repair coating, or to add thickness, clean area and remove all dust. Spray material on larger areas; brush may be used on smaller areas. NOTE: If coating has been applied more than four days, a "tooth" must be provided on the surface before recoating, brush blast with 60-100 mesh sand or roughen with coarse sandpaper. See note under #7 above.
10. Before placing coating into service, allow film to dry one week at 70 F (21.1 C) if intended for immersion or tank lining service (provide normal ventilation). NOTE: If intended as a tank lining for potable water tank, coating should be flushed, rinsed, or filled and dumped with fresh water before service to avoid taste pickup.
11. Clean all equipment with cleaner immediately after use.

[SOURCE: Ameron Corp., Monterey Park, CA.]

fore such application instructions pertain only to a specific coating. If several materials are specified, as in a qualified products list, this information would not necessarily be applicable to all the coatings listed.

Application Requirements

There are a number of application requirements which should be included, or at least considered, during the writing of a coating application specification.

Weather. Weather is an extremely important condition that must be taken into consideration during any coating application. The specification should include temperatures above and below which the coating should not be applied. It also should include humidity values above and below which the coating should not be applied. Wind is also a consideration, and it is usually preferable not to apply a coating under windy conditions. Much of the coating may be lost due to the wind creating dry spray which is blown away, or it may create a serious overspray condition on the finished coating or on adjacent structures.

Dew Point. The dew point and condensation also should be taken into consideration in the specifications. It is generally recommended that coatings not be applied where the temperature is within 3 C (5 F) of the dew point. Sometimes it is helpful to include a dew point chart in the specifications for ready reference. (Such a dew point chart is included in Chapter 10 on coating applications.) Coatings should not be applied on wet or damp surfaces, as adhesion will be affected. Surfaces subject to condensation or rain should not be coated until the surface has become thoroughly dry.

Method of Application. The method of application may be outlined in a specification. Many coatings are preferably applied by standard spray application; others may be preferably applied by airless spray; still others, as in the case of primers, may be better applied by brush in order to assure a thorough wetting of the surface. The structure and the surface to be coated may require a specific application method. For example, a badly pitted tank may be better coated by airless spray rather than by air spray since the material from an airless spray does not bounce or rebound out of small openings as it does from air spray. Also, a specific application of a coating may be such that only brushing or rolling can provide a satisfactory finished product. A specification for any coating job should include information on the method of application, even though the manufacturers of coatings supply considerable information in their application instructions.

Thickness. The number of coats to be applied is of vital importance since each specific coating may require more or fewer coats, depending on the severity of the exposure. The specification should not only cover the number of coats, but also should include the thickness of each coat to be applied and the total thickness of the completed coating. In this case, both maximum and minimum thicknesses should be specified, since it is almost physically impossible for an applicator to apply a specific number of mils of any coating.

These limits are important. It is obvious that a coating should not be too thin. On the other hand, excess thickness is a definite detriment to some coatings. Epoxy coatings applied with excess thickness can shrink due to internal stresses and therefore fail in a relatively short time. A good specification will include a maximum as well as a minimum thickness requirement.

Coverage. The coverage of the coating will determine the thickness of each coat and the thickness of the finished product. Most coating suppliers provide the coverage of a coating on a theoretical basis. This is a necessity for them, again because of the variability of the applications for which the coatings are used. For example, a windy condition will require more coating than a still condition, and a complex or heavily pitted surface will require considerably more coating than a plain surface. Therefore, the theoretical coverage must be modified from a practical standpoint.

Theoretical spreading rates are based on the volume solids of each coating and offer only a starting point from which to estimate practical coverage. It is always necessary to adjust the theoretical coverage rates since each job is characterized by conditions that will require more or less coating. The coverage on any application depends on the following factors.

1. The type of object being painted.
2. The accuracy of the square footage estimation.
3. The material needed to fill in the surface depressions caused by pitting and by blasting.
4. Excessive film thickness applied over the average thickness.
5. Material losses due to coating left in pots, hoses, brushes, rollers, etc.
6. Wastage because of overspray, dripping, and on some jobs, pilferage.
7. The spray equipment itself may cause a variation in the amount of coating required.
8. Proper lighting and staging of the work can influence the coating thickness.
9. The skill and experience of the applicators.

One rule-of-thumb on most plain or straightforward work is to reduce the theoretical coverage of the coating by 25%. This gives a practical coverage for many less complicated coating jobs. Because of all these variables, the coverage of the coating should be specified by the total dry thickness of the completed coating. This is a practical figure and can easily be determined by proper nondestructive thickness measuring equipment.

Equipment Cleanliness. Cleanliness of equipment is well worth including in a specification. Hoses can pose a particular problem, since they usually are not cleaned as thoroughly as they should be. On any large job, it is preferable to specify that new hoses be used. In this way, contamination of the coating to be applied is kept to a minimum since only one coating would be used in those particular hoses on that job.

If new hoses are not practical, used hoses may be satisfactory if they are thoroughly cleaned with strong solvents prior to being used with the new coating. If this is not done, the solvents in the newer coating often swell and release the previous coating materials from the interior of the hose, seriously contaminating the coating to be ap-

plied, as well as providing for innumerable work stoppages during the application. Material pots, guns, brushes, and all other equipment should be specified as being clean and in proper working order. There is nothing that delays a job longer than the use of poor and dirty equipment.

Drying Time. The drying time between coats and the final drying time of the coating are still other factors to be considered in the development of a specification. These are extremely important, and while they are keyed to a specific coating, the minimum that should be included in the specification is a statement that the drying of the coating between coats and the final drying time should be in accordance with the manufacturer's recommendations. On the other hand, application may be undertaken during specific weather and temperature conditions that may call for a shorter or longer drying period than recommended by the manufacturer.

The specific conditions of the job determine the drying times for the coating. Where heating of the coating is required, as is necessary for some tank linings, the specification should be explicit in requiring that the minimum temperature be above the curing temperature of the coating at all points in the tank or container. Exterior supports, braces, manholes, valve flanges, and other similar places are heat sinks and therefore require more attention and heat than the plain areas. Many coating failures have occurred in such areas because of incomplete curing. Failures occur much more often because of incomplete drying or curing than because of overcures. An even, minimum overall temperature is essential for satisfactory solvent removal and resin reactions.

Pot Life. Pot life of the material also should be considered in the specification. This is particularly true of the catalyzed systems where a definite pot life after mixing the two parts is critical. Temperature and weather conditions also affect the pot life of such mixed coatings, and many of the two-part coatings will react over a period of time to the point where either the application is difficult, or the adhesion of the material applied will not be satisfactory.

The pot life naturally varies according to the specific coating, and the recommendation of the manufacturer must be considered as well as the practical temperature and weather conditions at the time of coating. Any coating which has a critical pot life should have a reference to the pot life in the specification with a definite limit set on the time period in which the material is useable. Such a reference acts as a ''red flag'' to the applicator, indicating that difficulties can be encountered if the warning concerning the pot life is unheeded.

Thinning. Thinning is a common point of contention since many applicators tend to over-thin materials in order to make them easier to apply. It must be remembered that when a coating is thinned, the total solids in the coating are reduced so that additional costs may be necessary in order to develop the proper coating thickness. Over-thinning can also cause resin or pigment precipitation, and use of the improper thinner may coagulate or gel the coating, making it impossible to apply. If it does not do this, the drying time of the thinner may be such that it will adversely affect the adhesion or the durability of the coating.

Many coatings are now produced ready for application so that thinning is not necessary. If thinning is allowed, the specification should include a specific type and amount of thinner in strict accordance with the manufacturer's recommendations.

Manufacturer's Instructions. The manufacturer's application instructions should always be made a part of the coating specification. Any discrepancy between the requirements of the job, the job specifications, and the manufacturer's recommendations should be settled between the manufacturer and the owner prior to the time the specification is completed.

Coating Appearance. One item that is seldom included in a specification is a description of the final coating appearance. This can be done by specifying that the final appearance be in accordance with samples submitted by the coating manufacturer prior to the beginning of the job; by a written description of the coating detailing the color, texture, smoothness, and general appearance of the coating; or by specifying a sample application of the material by the contractor to be agreed upon by all parties prior to the beginning of the job. This last procedure is usually preferred since it provides a working sample which serves as a guide for both the applicator and his workers.

Specifications for Applicators. Another item that is seldom included in the specification, but which is well worth including is the requirement that a copy of the specification and the manufacturer's instructions be given to each foreman and each applicator for reference. Too often, the only reference available to the applicator is the information on the label of the can, which may amount to little or nothing. A copy of the specification in the hands of the foreman and applicator is also important from a legal standpoint, since this is evidence that those doing the work received a written description of the coating desired, the method of application, and the safety conditions required during the application. The fact that the coating applicator did or did not receive a copy of the coating specification has been the turning point in many legal proceedings where the coating application was in question.

If a complete specification is not practical, a synopsis of the specification which states in simple terms the requirements of the specification should be supplied. The short document should supply the information needed for the foreman and workers to do a proper job. It should be done in easy-to-ready type and in step form, *i.e.*, where one item follows another in logical sequence. If possible, it should be limited to one sheet so that it can be kept in a pocket for easy reference. These efforts are not often made, but can prove to be a great help in obtaining the best possible job.

Safety. A coating specification should include a section on the safe application of the required materials. It should specifically describe safe operating conditions, particularly where coatings are to be applied on the interior of tanks or structures. While this also needs to be geared to the manufacturer's application instructions, the owner and the general contractor are responsible for on-the-job safety.

Color Between Coats. Another item which is also good to include in any specification is the change of color between each coat applied. If the primer is red, the following coat can be black or gray, and, depending on the number of topcoats, these should be of a sufficient difference in color to contrast with the previous coat. However, the next to the last coat should be reasonably close to the color of the final topcoat so that an even color for the entire coating is obtained.

Differentiating between the colors of various coats not only makes it easier for the applicator to see where he has coated, but it also helps to eliminate holidays and other coating imperfections. It can also lead to better coverage per coat and a more even thickness than when coating with all the same color. The cost of changing colors is negligible. In fact, it should amount to the same or slightly less due to the benefit of increased visibility to the applicator, which is good insurance against poor application.

Inspection

Inspection should always be called for in the specifications. A clear, detailed specification and good inspection procedures are the best insurance for obtaining an effective coating application. On large jobs, an inspector or inspectors should be appointed to follow the job closely from the time of the pre-application meeting to full completion of the job for acceptance. On maintenance work, which may be less extensive, the corrosion engineer may take over the inspection function, making sure that the specification requirements are followed in order to produce the best possible application.

Inspection should be an in-process function. While it need not require full-time observation, the application of the coating should be followed closely. All difficulties and improper work should be called to the attention of the applicators or contractor as quickly as possible so that they can be remedied prior to the application of the following coat. Inspection is actually in-process quality control for coating application. The inspector should be required to know the key application properties of the coating, as well as be authorized to stop the work if the application is not satisfactory.

There are many points to be considered during the application, ranging from inspection of the material as it arrives on the job, to inspection of the mixing of the material to make certain that it is fully dispersed and that there is no settled material left in the bottoms of the containers. Each coat should be followed closely to make sure that all pinholes, holidays, and missed areas are eliminated. Varying the color of each coat makes inspection of the application much easier and aids in eliminating many of the coating defects.

The drying of each coat is a necessary point of inspection to make sure that the drying is adequate and also to make certain that condensation, moisture, rain, and atmospheric contamination are prevented so that satisfactory adhesion can be obtained between each coat. If the coating requires heat curing, this should be observed carefully to make certain that the coating surface is evenly and completely heated to the proper temperature to obtain a full cure of the coating.

Ventilation should be monitored where any interior coating is applied, since it is a key safety factor. Dehumidification is also required in most interior work. Again, this should be thoroughly monitored by the inspector to make certain that the equipment is operating properly and that there is adequate dehumidification to insure a condensation-free surface.

The inspector should also determine the thickness per coat and the final thickness of the coating in order to make certain that it meets the specification. This means the inspector should be familar with the determination of both wet film thickness and dry film thickness. Under certain conditions, the inspector should also be thoroughly familar with safety equipment such as an explosion meter in order to insure that the lower explosive limit is never reached in any portion of the enclosed area. Where the inspector finds coating imperfections, such areas should be marked for rework and should be followed closely.

Finally, as part of the inspector's job, a diary should be kept of the work progress, including all of the pertinent activities and conditions found during each day. This is important since it provides a check for follow-up on rework items and a basis of fact for any problems which may arise at a later data. Such diaries are always invaluable during any legal proceeding. Inspection details should be included in the specifications.

Cost of Specification

There is always some controversy over the additional cost of a job with a strong specification due to the fact that contractors may add to the contract price in order to cover possible additional work. Many contractors, particularly the poorer ones, prefer jobs that carry no specifications whatsoever since this allows them freedom to do whatever they want or think is best. On the other hand, most good contractors welcome a strong specification because, in many ways, it reduces their cost by specifically and clearly outlining requirements. With a strong specification there is usually much less rework, the cost of which is usually carried by the contractor.

While a strong specification may somewhat increase the bid price of a job over one which carries a more general specification, it will usually reduce the overall cost of the job due to the fewer requirements for rework and extra work and/or materials. Generally speaking, a strong specification is well worth any initial added expense as insurance that the job will be done properly and that many years of maintenance-free service can be expected from it.

References

1. A Guide to the Preparation of Contracts and Specifications for the Application of Protective Coatings, NACE Publication 6J162, National Association of Corrosion Engineers, Houston, TX.
2. Brett, Charles A., Specifications for Surface Preparation, the Good, the Bad, and the Useless, Corrosion and Coatings, South Africa, October (1979).
3. Steel Structures Painting Council, Discussion of Paint Specifications, Section 5, Vol. 2, Steel Structures Painting Manual, Pittsburgh, PA.

APPENDIX 17A
An Example of a Construction Specification:
Painting and Coating Works

Typical
Offshore Coating Work Specification

Section 1: General

1.1 Introduction. This Specification shall govern the design, materials, color-coding, surface preparation, application, and inspection requirements of all painting and coating of the jacket and appurtenances. It includes structural steel, caissons, and other appurtenances from −4.5 m with respect to ISLW to +21.9 m. It also includes markings on the piles, pile followers, and pile guides. The section of risers from below −4 m and above +8 m ISLW and the external surfaces of the buoyancy tanks and control chamber. In addition, monel sheathing for the section of risers from −4.5 m to +8.5 m ISLW is covered in Appendix C.

This painting and coating Specification gives details of metal surface preparation, the application of coatings, and the preferred types of coatings to be applied to the WORKS.

The details specified are minimum guidelines only and the CONTRACTOR should accordingly select an appropriately conforming system, and provide full details to the ENGINEER. All work will be subject to inspection by the ENGINEER who shall be given 48 hours notice prior to work commencement.

1.2 Exclusions. This Specification is not applicable to:

(a) Submerged structural steel and appurtenances below −4.5 m which are to be uncoated. Protection of this area is covered by Part 4 of Volume 4 of this Specification.
(b) Equipment or piping made of corrosion-resistant materials, such as stainless steels or nonferrous alloys.
(c) Flooding, venting, deballasting, grouting, and packer and hydraulic control lines.

1.3 Exceptions. In cases where the CONTRACTOR wishes to deviate from this Specification in any way, all details on materials, components, equipment, procedures for testing, and remedies shall be submitted for the ENGINEER'S approval. Proposals for changes shall be made within 3 months of award of CONTRACT, or within such time that should the proposals not be approved by the ENGINEER, the work, when executed fully in accordance with the Specification, will not affect the overall PLAN.

Section 2: Scope of Work

2.1 Labor and Equipment. The CONTRACTOR shall furnish all labor, materials, paints, and coating components required for properly preparing surfaces and coating them in accordance with this Specification.

2.2 Inspection Equipment and Personnel. The CONTRACTOR shall furnish experienced supervisors and all equipment necessary for accurately measuring wet and dry film thicknesses, and, where required in this Specification, for carrying out holiday detection at the specified voltage.
The CONTRACTOR shall be responsible for the proper calibration and functioning of all testing and inspection equipment at the time of inspection, and shall ensure that this equipment is being operated by skilled personnel.

2.3 Use of Subcontractors. Only SUBCONTRACTORS approved by the ENGINEER shall be employed.

2.4 Implementation of Specification. The CONTRACTOR shall submit detailed proposals for implementing this Specification with particular relation to type, make, and description of paint and materials to be used, manufacturer's specifications, application, and drying and overcoating times.
The proposed sequence and timing of blasting and painting activities shall be clearly shown in relation to the overall PLAN.
Full details are to be given of blast and paint facilities and resources, including dehydration, temperature and environmental control measures, access, handling, space, number of personnel, and the data which is proposed to be submitted to the ENGINEER on a weekly basis.

Section 3: Design

3.1 Codes and Standards. Unless otherwise required in this Specification, the latest editions of the following publications shall govern this work:

(a) *Person Protection Advice for the use of Marine Paints and Compositions* published by the Paintmakers Association of Great Britian Ltd.
(b) Swedish Standard SIS 055900, *Pictorial Surface Preparation Standards for Painting Steel Surfaces.*
(c) U.S. Specification of the Steel Structures Painting Council, SSPC: SP-1, SP-3, and SP-10.
(d) Australian Standards AS 1433 Color Card.

3.2 Systems. All painting and coating systems, as well as all suppliers and applicators of the painting and coating systems, must be acceptable to and approved by the ENGINEER.

Section 4: Suppliers

4.1 Approved Suppliers. With respect to the paint systems as specified in Appendix A of this Specification, suppliers shall be approved by the ENGINEER prior to any application.

4.2 Approved Systems. Only approved systems as detailed in Appendix A shall be used to form the basis of the CONTRACTOR'S proposal, as specified in Section 2.4 of this Specification.

Section 5: Application

5.1 Personnel. The CONTRACTOR shall submit a weekly list of personnel to the ENGINEER, specifying the number of foremen, skilled and semi-skilled laborers for blasting and painting, inspectors, and plant and equipment operators.

5.2 Plant and Equipment.

5.2.1 Equipment List. The CONTRACTOR shall keep an up-to-date record of any plant, equipment, tools, spraying and blasting guns, nozzles, brushes, etc., being used on the work, and shall submit a weekly list to the ENGINEER.

5.2.2 Blast Cleaning Equipment. The compressed air supply used for blasting shall be free of water and oil. Adequate separators and traps shall be provided and these shall be kept emptied of water and oil. Accumulations of oil and moisture shall be removed from the air receiver by regular purging.
The grit shall be discharged with a pressure of 700 kPa measured at the nozzle with a hyperdermic pressure gauge and shall not under any circumstances fall below 550 kPa.
Nozzles shall be discarded and replaced when wear reaches 50 %.
Where air-operated equipment is used, the operator's hood or head gear shall be ventilated by clean cool air served through a regulator filter to prevent blasting residues from being inhaled by the operator.

5.2.3 Paint Spraying Equipment. The spraying equipment to be used shall meet the recommendations and instructions set forth by the paint supplier for each specific paint system.
An adequate moisture trap shall be placed between the air supply and the pressure pot feed to the gun. The traps shall continuously bleed off any water or oil from the air supply.
Suitable working regulators and gauges shall be provided for air supply to the pressure gun.
Spraying units shall be grounded and precautions shall be taken to prevent build-up of static electricity.

5.2.4 Storage. All painting materials shall be delivered to the WORK SITE in the Supplier's original containers with labels and seals unbroken. They shall be kept in a locked, well-ventilated storage place assigned for this purpose. Storage areas shall be kept clean and neat.

5.2.5 Scaffolding, Staging, Accessibility. Fixed scaffolding or staging shall be used as required for surface preparation and painting and will be subject to approval by the ENGINEER. It shall be such that easy and sufficient access will be provided for correct painting and inspection of all surfaces. It shall be such that operators will be able to stand up with body and arms free of scaffolding or staging and the structure being worked on. The CONTRACTOR shall supply and maintain rigging and scaffolding equipment capable of enabling completion of the work in accordance with this Specification.

5.3 Protection.

5.3.1 Equipment and Structural Members. The CONTRACTOR shall be responsible for and shall protect all equipment and structures, and any other areas required by the ENGINEER to be protected from mechanical damage or damage from paint droppings or overspray to his entire satisfaction. Examples of areas to be protected are: valve spindles, glass, glass-faced pressure gauges, instruments, light fittings, machinery, cables or stainless steel pipework, gratings, control consoles, equipment identification plates, or any areas of the structure not being painted at that particular time.

5.3.2 Weather Protection. Where environmental circumstances and/or CONTRACTOR'S program and/or construction methods require a weather protection of all or part of the structure, the CONTRACTOR shall provide plant and equipment such as tents, heating and ventilating facilities, or scaffolding at no additional cost to the ENGINEER.

Section 6: Surface Preparation

6.1 Blast Cleaning.

6.1.1 General. Blast cleaning shall be in accordance with SIS 05 5900 standard Sa 2½ or SSPL-10 near-white blast cleaning at the time of painting.

Only dry abrasive blast cleaning techniques shall be employed. Abrasives shall be expendable copper slag or reusable iron and steel grit. The abrasives must be maintained free from dust, salts, and other impurities.

The type and size of abrasive for any particular job shall be selected to give a surface amplitude of 100 microns maximum peak height. The smallest size particle shall be an 80 mesh abrasive.

The surface amplitude shall be measured by direct microscopic assessment of replicas taken from the surface. At the beginning of the CONTRACT, blast-cleaned steel panels shall be prepared complying with this requirement to the ENGINEER'S satisfaction and subsequently used as a standard.

Where other methods of on-site determination of profile amplitude are used, the instruments concerned must be calibrated against such standard control panels.

6.1.2 Techniques and Restrictions. Blast cleaning shall not be conducted when the surfaces are less than 3 C above dew point or when the relative humidity of the air is greater than 80 %. The CONTRACTOR shall furnish a psychrometer to determine the relative humidity. Appendix C gives details of dew point determinations. Grit blasting shall not be done in areas close to painting operations and wet coated surfaces to prevent dust and grit contamination.

Grit blasting shall be permitted only during the daylight hours, except that rough grit blasting will normally be allowed during the night in the absence of, but with the approval of, the ENGINEER, providing that the surface shall be given a light blasting to the specified standard in daylight.

Maximum speed and most effective cleaning is obtained by systematic, even blasting. Work shall be blocked out in 300 mm squares and each square evenly blasted until complete.

All welded areas and appurtenances shall be given special attention for removal of welding flux in crevices. When electrodes are used with a basic coating or shielding, the welded area has to be freed from alkaline residues by rinsing with warm water prior to blasting. Welding splatter, slivers, laminations, and underlying mill scale not removed during fabrication and exposed before and during grit blasting operations shall be removed by the best mechanical means and the edges smoothed or rendered flush, as required by this Specification.

Where extensive rectification has been necessary on abrasive blast-cleaned surfaces, the dressed areas shall be reblasted to remove all rust and slag and to provide adequate paint key.

Grit blasting shall continue a minimum of 25 mm into adjoining coated surface. Any steel that is not primed and that is wet by rain or moisture shall be reblasted.

6.2 Power Tool Cleaning. Power tool cleaning shall be in accordance with SSPC-SP3, *Power Tool Cleaning* and to a visual standard in accordance with S.I.S. 05 59 00 to S.T.3.

Metal surfaces for which blasting is specified, but which because of their location cannot be so treated shall be 100 % power disc cleaned.

Where welds occur within areas which cannot accommodate a power disc, a Jason Hammer shall be applied.

Power tool cleaning shall continue a minimum of 25 mm into adjoining coated surfaces.

6.3 Solvent Cleaning. Solvent cleaning shall be in accordance with S.S.P.C. - S.P.I., *Solvent Cleaning.*

After blasting or discing and prior to priming, all surfaces shall be examined for traces or smudges of oil or grease. If any exist, they shall be removed by solvent washing and the area reblasted.

All metal surfaces to be coated which do not require blasting or power tool cleaning, and all galvanized metal shall be thoroughly cleaned with a high-pressure water or steam cleaning unit to remove all mud, oil, grease, or other foreign matter.

All bolt-holes shall be solvent cleaned prior to grit blasting. No acid washes or other cleaning solutions or solvents shall be used on metal surfaces after they are grit blasted. This includes inhibitive washes intended to prevent rusting.

Section 7: Coating

7.1 General. Any surfaces to be coated shall be rendered dust-free prior to the application of the prime coat. This shall be accomplished by blowing of the surface with clean dry air or by using an industrial vacuum cleaner.

All coating material shall be thoroughly stirred in a pressure pot with a power mixer for a time sufficient to thoroughly remix the pigments and vehicles.

Only thinners as specified by the manufacturer shall be used. Mixing and thinning directions as furnished by the paint manufacturer shall be followed.

If a coating material requires the addition of a catalyst, the pot life under application conditions shall be clearly stated on the label. The pot life must not be exceeded. When the pot life limit is reached, the spray pot must be emptied, the material discarded, the equipment cleaned, and new material catalyzed.

Paint which has livered, gelled, or otherwise deteriorated during storage shall not be used; however, thixotropic materials which may be stirred to obtain normal consistency will be accepted.

Coatings containing heavy or metallic pigments that have a tendency to settle must be kept in suspension in the pressure pot by a mechanical stirrer.

Paint application shall be carried out according to the paint manufacturer's recommendations. In case of conflict between this Specification and the manufacturer's recommendation, the conflict shall be resolved by a meeting of the ENGINEER and the manufacturer's representative prior to any application.

No coating shall be applied when the surfaces are less than 3 C above dew point, when the relative humidity of the air is greater than 80 %, when the air temperature is below 5 C, or when there is a likelihood of a change in weather conditions within 2 hours after application which would result in air temperatures below those specified or in deposition of moisture in the form of rain, snow, condensation, etc. upon the surface. Normally, paint is not to be applied to surfaces exceeding 50 C at the time of application. Dew points shall be determined in accordance with the table in Appendix C.

Blast-cleaned or power-disced surfaces shall be coated with the primer specified within four hours of blasting or such other time limits as may be specified by paint manufacturer, and prior to sundown of that day and also before any visible rusting occurs. A minimum of 5 cm around the edges of blasted areas shall be left uncoated unless adjoining a coated surface.

When primers are applied automatically after automatic blast cleaning, it is normal to apply 15-20 microns only, to facilitate subsequent welding and fabrication. In this case, an additional coat of epoxy primer shall be applied within 7 days to provide the required longer-term protection. When applying epoxy red oxide primers, care shall be taken to avoid damage to applied primer by adjacent blast cleaning. Primed surfaces must be cleaned properly before overcoating to ensure adequate adhesion.

Each coating shall be allowed to dry thoroughly at least for the specified time prior to application of a succeeding coat.

No coating shall be put on edges prepared for field welds or within 50 mm of these edges, except zinc primer applied at a maximum dry film thickness of 25 microns. Painting over blasted areas shall not be allowed within 50 mm of the unblasted areas.

For finish coats, color identification shall be as given in Appendix B.

7.2 Coating Application.

7.2.1 Spray Equipment. The spray equipment to be used shall meet the recommendations set forth by the paint manufacturers for applying each specific paint.

All equipment for spray application shall be inspected and approved by the ENGINEER before any application is made. Lines and pots must be clean before adding new material. An adequate moisture trap shall be placed between the air supply and the pressure pot feed to the gun. The traps shall continuously bleed off any water or oil from the air supply. Suitable working regulators and gauges shall be provided for air supply to the pressure pot, and the air supply to the pressure gun.

Spraying units shall be grounded and nonconductive hoses are to be used. The CONTRACTOR shall take such further precautions as may be required to avoid buildup of static electricity.

7.2.2 Spray Application. Lines and pots shall be thoroughly cleaned before addition of new materials.

The spray gun shall always be held at right angles to the surface and shall be held no closer than 450 mm nor more than 600 mm from the surface for the airless spray gun method, or no closer than 150 mm nor more than 250 mm from the surface for air spray equipment. Even, parallel passes shall be made with the spray gun. In application of material, each spray shall overlap the previous pass by 50 %. Large surfaces shall always receive passes in two directions at right angles to each other. Spray width adjustment on the gun shall be made and the readjustment of atomizing pressure at the regulators shall be made until the desired spray pattern is found.

Each coat is to be applied uniformly and completely over the entire surface. All runs and sags shall be brushed out immediately or the paint shall be removed and the surface resprayed.

Before spraying each coat, all areas such as corners, edges, welds, small brackets, bolts, nuts, and interstices shall be precoated by brush to ensure that these areas have at least the minimum specified film thickness.

Spraying of paints from a single boatswain's chair or spider stage will not be permitted.

A supply of tips with varying spray angles and washers, as recommended by the paint supplier for each specific steel configuration to be coated, shall be available.

7.2.3 Brush Application. Coating shall be brushed on to all areas which cannot be properly spray coated for any reason.

Surfaces not accessible to brushes shall be painted by other suitable means to ensure a uniform paint film of adequate thickness.

Brushes used in brush application shall be of a style and quality that will permit proper application of paint. Round or oval brushes generally are considered most suitable for rivets, bolts, irregular surfaces, and rough or pitted steel.

Wide flat brushes are suitable for large flat areas, but they should not have a width of over 120 mm. No extending handles shall be allowed on paint brushes.

The brushing shall be done such that a smooth coat, as nearly uniform in thickness as possible, is obtained. There should be no deep or detrimental brush marks. Paint shall be worked into all crevices and corners.

Runs or sags shall be brushed out.

In brushing any of the solvent-type coatings, care must be taken so that no lifting of former coats occurs.

During application of each coat, all areas such as corners, edges, welds, small brackets, bolts, nuts, and interstices shall receive additional coating material to ensure that these areas have at least the minimum specified film thickness and to ensure continuity of the coating.

7.3 Film Thickness. All coating dry film thickness limits must be adhered to strictly; these film thicknesses are to be checked with the calibrated film thicknesses gauges supplied by the CONTRACTOR. Where film thicknesses do not meet this Specification, additional material shall be applied. In order to achieve the specified dry thickness, frequent checks on wet film thicknesses shall be carried out by the CONTRACTOR.

The dry film magnetic type thickness gauge shall be calibrated by the CONTRACTOR using foils in the film thickness range being checked and over the type of surface being coated.

Calibration shall be carried out at least twice daily. The completed coating shall be free of defects such as runs, sags, pinholes, voids, bubbles, or other ''holidays''.

Section 8: Repair of Defects

8.1 General. Before application of any further coat of material, all damage to previous coats shall be repaired.

8.2 Inadequate Coating Thickness. Areas with inadequate coating thickness shall be thoroughly cleaned and, if necessary, abraded, and additional compatible coats applied until they meet this Specification. These additional coats shall blend in with the final coating on adjoining areas.

8.3 Contaminated Surfaces. Surfaces to be overcoated which become contaminated shall be cleaned by lightly brush blasting the surface free of all contamination prior to applying the following coats. After brush blasting any residual contaminants shall be removed by dry compressed air or wiped by hand with clean dry rags.

8.4 Coating Damage Not Exposing Steel Surface. Surfaces to be overcoated which become contaminated or damaged shall be cleaned by lightly brush blasting the surface free of all contamination prior to applying the following coats. After brush blasting, any residual contaminants shall be removed by dry compressed air or wiped by hand with clean, dry rags. The coating around the damaged area shall be chamfered, using an approved method, to ensure continuity of the patch coating. The full coating system shall then be reapplied strictly in accordance with this Specification.

8.5 Coating Damage Exposing Steel Surface. The damaged area shall be recleaned as originally specified for that item and the full coating system reapplied in accordance with manufacturer's recommendations. The recleaning shall carry over on to the secure surrounding coating for not less than 25 mm all around and the edges shall be chamfered by a method approved by the ENGINEER.

8.6 Epoxy Coatings. In case of:

(a) Repairing damage to fully cured epoxy coatings.
(b) Repairing damage to the paint over an epoxy coating.
(c) Application of the paint coat to an epoxy coating.
(d) Re-application of an epoxy coating.

The work shall only be carried out after the surface to be coated has been suitably abraded to afford an adequate key for the coating to be applied.

Section 9: Hot-Dipped Galvanizing

All grating, ladders, handrails, fence, stairways, walkways, cable trays, and other items as specified by the drawings (Volume IV) shall be hot-dip galvanized in accordance with British Standard 729: *Hot-dipped Galvanised Coatings on Iron and Steel Articles.* The minimum coating weight shall comply with the values laid down in Table 1 of British Standard, 729, but shall not be less than 610 g/m^2.

Small areas of galvanized coating damaged by cutting, welding, drilling, or any preparation during fabrication, erection, transportation, or installation, shall be well cleaned prior to repair in accordance with this Specification.

After being allowed to dry completely, the areas of damaged galvanizing shall be hand or mechanically wire brushed to clean metal finish. The areas so prepared shall then be renovated in accordance with the Appendix D of the British Standard 729 or painted with 0.1 mm of a zinc-rich primer given in Appendix A.

Section 10: Clean-Up

After painting and inspection have been completed, all plant, scaffolding, equipment, surplus materials, and waste resulting from painting work shall be collected and disposed of outside the working area. Over-runs, droppings, and smears shall be removed.

Section 11: Inspection

11.1 General. The ENGINEER shall have the right to inspect at all times any tools, instruments, materials, staging, or equipment used or to be used in the performance of the coating application. The CON-TRACTOR shall make all parts of the WORK accessible for these inspections.

The ENGINEER shall pinpoint a Coating Inspector who shall act for the ENGINEER and who will observe all application procedures during the time the work is in progress and who will approve the surface preparation prior to the application of any coating and approve the condition of each coat prior to the application of the following one.

11.2 Rejected Work and Equipment. The ENGINEER shall have the right to condemn any and all tools, instruments, materials, staging, equipment, or work which does not conform to the Specification. Condemned areas of coating applications shall be marked with a compatible paint of contrasting color.

Any condemned coating applications, defective preparatory work (*i.e.,* blast cleaning, staging) or any defective work not conforming to this Specification shall be rectified by the CONTRACTOR at no additional cost to the ENGINEER. Any condemned tools, instruments, materials, or equipment shall be replaced or rectified at no additional costs to the ENGINEER.

11.3 Approval. The CONTRACTOR shall notify the ENGINEER 48 hours before work or part of the work commences. Prior to final acceptance of part of or the complete work an inspection shall be made. The CONTRACTOR and the ENGINEER shall be present, and the CON-TRACTOR shall make an inspection report which shall be signed by both parties.

Section 12: Reporting

12.1 Status and Progress Reports. Together with equipment and man-power lists, the CONTRACTOR shall submit inspection reports giving details on weather conditions, air humidity, temperature, particulars of application (*e.g.,* blast cleaning, second coat, color, wet and dry film thicknesses, anomalies), and progress of work compared with the PLAN.

12.2 Completion Records. Before the WORKS are removed from the WORK SITE, a full report on outstanding and/or incomplete and/or remedial work shall be submitted to the ENGINEER for final approval, so that a percentage completion can be agreed.

APPENDIX A — Approved Coatings

The following is a list of approved coatings for the items required to be coated. All coatings shall be applied in accordance with the manufacturer's recommendations and Sections 5 through 8 of this Specification.

Atmospheric Zone

The following coating system shall be applied to all structural steel of jacket, caissons, appurtances, and oil and gas risers from +8 m to +21.9 m ISLW. It does not include flooding, venting, deballasting, grouting, packer and hydraulic line, buoyancy tanks, and pre-installed piles.

Coat	Type	Dimet	Napko/British Paints	Dulux	DFT (Micron)
1	Inorganic Zinc Silicate	Dimetcote 5	Napko 5-Z	Zinc Galv 6	75
2	High-Solids Epoxy (color different than 3rd coat)	Amercoat No. 383	Epoxycote PA HS 5800	Ferrodor EPX	100
3	High-Solids Epoxy (Grey)	Amercoat No. 383	Epoxycote PS HS 5800	Ferrodor EPX	125

Splash Zone

A heavy-duty coating shall be applied to all jacket structural members, caissons (including pipeline service caisson), and other steel appurtenances in the splash zone from -4.5 m to $+8.5$ m ISLW. It does not include risers, flooding, venting, deballasting, grouting, packer and hydraulic lines, buoyancy tanks, or pre-installed piles.

Coat	Type	Dimet	Napko/British Paints	DFT
1	Inorganic Zinc	Dimetcote 5	Napko 5-Z	75 Micron
2	Polymeric Epoxy or Vinyl Ester	Tideguard 171/Dimet Nukem 100 Polymeric Epoxy	Steel Shield VE	3 mm - 5 mm

Submerged Zone and the Inside Surfaces of the Caissons

Below -4 m ISLW the following coating system shall be applied to the risers and manifold piping. The monel sheathing on the risers shall be overlapped by 500 mm. This coating shall also be applied to the inside surfaces of the caissons over the full length.

Coat	Type	Napko/British Paints	Dimet	Dulux	DFT (Micron)
2 coats	Polyester	Flake Shield	—	—	400
		Or Alternative Type			
2 coats	Coal Tar Epoxy	Luxalar - 5	Amercoat 79	Duretar	400

Note: If the coal tar epoxy system is used for the inner surfaces of the caissons, the DFT should be 600 microns.

Risers in Splash Zone

The corrosion protection of the risers through the splash zone is achieved by the use of monel sheathing. This sheathing is to be applied from -4.5 m to $+8.5$ m ISLW.

Buoyancy Tanks

The following coating system shall be applied to the outside surfaces of the buoyancy tanks.

Coat	Type	Dimet	Napko/British Paints	Dulux	DFT (Micron)
1	Inorganic Zinc	Dimetcote 5	Napko 5-Z	Zinc Galv 6	75
Markings Only (White)	High-Solids Epoxy	Amercoat 383	Epoxycote PA	Durepon	N/A

Carried Piles

The carried piles shall be painted with the identification numbers and markings as shown on the Drawings. The following coatings are approved.

Coat	Type	Dimet	Napko/British Paints	Dulux
Markings (White)	High-Solids Epoxy	Amercoat 383	Epoxycote PA	Durepon Primer Surfacer

Surface Pretreatment—the piles shall be blasted or handcleaned in the regions where the markings are to be painted.

Pile Guides

The inside of the guide bell of each pile guide at −5 m ISLW shall be painted over their entire area and identification marks painted as shown on the Drawings. The following coatings are approved.

Coat	Type	Dimet	Napko/British Paints	Dulux	DFT (Micron)
1	Inorganic Zinc	Dimetcote 5	Napko 5-Z	Zinc Galv 6	75
2	HS Epoxy (White)	Amercoat 383	Epoxycote PA	Durepon	100

Internal Surface of Diesel and Drill Water Tanks

The following coating system shall be applied to the internal surfaces of storage tanks in jacket legs.

Coat	Type	Dimet	Napko/British Paints	Dulux	DFT (Microns)
1	Epoxy Primer	Amercoat 64	Thixopoxy	Ferrodor EPX	100
2	High-Solids Epoxy	Amercoat 66	Thixopoxy	Ferrodor EPX	100
3	High-Solids Epoxy	Amercoat 66	Thixopoxy	Ferrodor EPX	100

Note: Thixopoxy and Ferrodor-type coatings may be applied in two coats provided the total DFT equals 300 microns.

APPENDIX B—
Finish Color Schedule

Service	Color	AS1433 Code
Structural Steel		
Members above splash zone	Black	—
Members in splash zone	Black	—
Nonstructural Steel		
Gratings and Handrails	Galvanized	—
Piping		
Diesel Oil	Orange Brown	06D43
Buoyancy Tanks	—	—
Pile Guide Bells	White	—
Piles	—	—
Followers	—	—

APPENDIX C — Dew Point Determination

Dew Points (°C) at Various Relative Humidities

Air Temp.	30%	40%	50%	60%	70%	80%	90%	100%
−1	—	—	—	—	−6.5	−4	−2	−1
4	—	−6.5	−4	−2	0.5	1.5	3.5	4.5
10	−6.5	−3.5	0.5	2	3.5	5.5	8.5	10
15.5	0	2	4	8	10	11.5	14	15.5
21	3	6.5	10	13	15	18	19.5	21
26.5	7	12	15.5	19	21	23.5	25	26.5
32	13	16.5	20.5	24	25.5	28.5	30.5	32
38	18	22	25.5	29	31	33.5	36	38

NOTE:

It is essential to ensure that no condensation occurs on blasted steel or between coats during painting.

Air at a given temperature can only contain a certain (maximum) amount of water vapor. This proportion is lower at lower temperatures.

The dew point is the temperature of a given air-water vapor mixture at which condensation starts, since at that temperature its maximum water content (saturation) is reached.

In practice, a safety margin must be kept, whereby the substrate temperature is at least 3 C above dew point.

18

Inspection and Testing

The best coating available is of little value if it is not carefully and properly applied. Thus, coating inspections are specified to insure proper application. It has been stated that "A specification must be complied with if it is to serve a useful purpose...a good specification which is adhered to will produce optimum coating life and is economically advantageous."[1] This is because a good specification includes an inspection requirement to insure a quality coating application that will result in long coating life under even severely corrosive conditions.

Proper inspection is a primary key to effective coating performance and therefore should be a requirement of every good specification, as well as a periodic requirement over the lifetime of the coating. Effective inspection during the life of a coating can extend its useful life several times over by initiating coating maintenance and repair before major coating failure results.

Comparison of the results obtained when proper inspection has been carried out during coating application has shown that the life expectancy of the finished coating can be increased by a factor of two or three by the introduction of proper inspection procedures. It is shortsighted to develop detailed specifications and requirements for a coating application, yet fail to properly inspect the work to be sure that the specifications have been met. The primary purpose of inspection is to assure compliance with the specifications established for the job.

Unfortunately, inspection of the coating application is frequently overlooked by the owner of a structure, in spite of the fact that inspection is the area of the job where he is most likely to receive less than he specified. In fact, the application of protective coatings is often considered an indefinite and rather haphazard operation where quality control is more the exception than the rule.

Most other engineering operations, such as the installation of equipment, routinely require a thorough inspection of the installation process and end results. How-

Coating quality depends on level of inspection.

ever, installation of the material that will protect the equipment against the atmosphere, chemicals, and other destructive elements is often carelessly left to the discretion of the subcontractor who does the work. Instead, both owner and contractor should consider coating application as a manufacturing operation which creates a new product and therefore naturally requires thorough inspection.

Failure to provide inspection is also common in contracts for maintenance coating work, although it is no more excusable in this case than it is with new work. As stated by S. T. Thompson:

> The implementation of a good maintenance painting program depends upon four cornerstones: Good specifications, quality materials, qualified contractors, and effective inspection. Failure in any of these areas, like the proverbial weak link in the chain, will result in decreased performance and increased costs—often substantial.[2]

Actually, inspection of maintenance coatings may be

even more important than for new work because of the increased number of variables. In fact, it is these variables, all of which have a substantial effect on the quality of the protective coating, that dictate the need for competent inspection throughout the application process. The coating material, mixing of the coating (particularly where there are two or more parts included in the finished material), thinning, surface preparation, coating application, weather conditions during application, individual coat and total coating thicknesses, and surface treatment after coating application are all variables that must be carefully controlled if a high-performance coating is to perform properly. Without in-process inspection, conformance with application specifications is difficult to determine. The final coating appearance can be quite deceptive, giving no indication as to the probable effectiveness of the finished product.

In this age of critical labor-intensive operations, quality assurance is extremely important. The assurance of quality usually is dependent on in-process quality control, involving inspection of the basic material and its use or application to insure the long life and effectiveness of the product. There is no installation operation where this is more critical than in the application of high-performance coatings. It is a well-known adage in the coating industry that even an improperly selected coating or poor quality paint can provide good service if it is applied carefully and properly, while a high-performance coating can be completely ineffective due to a poor and careless application. This is the reason that inspection plays such an important role in a coating program.

As discussed in the previous chapter on specifications, the quality of a coating applied to an industrial plant, chemical plant, refinery, food plant, paper mill, offshore marine structure, or nuclear energy installation, is dependent on two important items. The first of these is the coating specification, which dictates the product to be used as well as its installation. The specification must be strong, stating the need and requirements for the final coating as well as providing a measure for the evaluation of the completed coating to assure that the desired results are obtained. A strong specification is indispensable and should include the requirements for in-process inspection.

The second important item (and the key step in insuring a successful coating job) is proper inspection. Inspection is necessary to insure that the intent of the specification, as well as its details, are carried out. The coating operation is usually the last phase in any construction or maintenance job. Thus, the pressure to complete the coating either during operation or prior to the start-up of a new operation is the cause of many coating failures. Without good specifications and thorough inspection, and under the above conditions where pressures to complete a job are extremely strong, serious errors and costly mistakes can be made during a coating installation.

Variables Involved in Quality Control

As previously discussed, there are so many variables involved in coating applications that it may seem surprising that any good applications are ever obtained. First of all, environmental conditions vary from hot, dry weather to cold, humid weather. Coupled with these outside variables are the variables inherent in the material itself, which can range from a fast-drying vinyl to a slow-curing polyamide epoxy. There are also variables associated with those doing the application work. For instance, one person may be proficient at applying one type of coating, while another may be much better at applying a second type of material. Inspection to provide in-process quality control is the only way in which variables such as these can be properly controlled.

Another possible variable is the degree of understanding regarding the specifications or the design engineer's coating requirements. In many cases, the general character and physical properties of a coating are not fully recognized by engineers, architects, contractors, and, more often than not, the actual owners. Many of these people regard a high-performance coating as no more than an average type of paint. While it is true that there may be no difference in the application equipment used, there is quite a difference in the handling and processing of the various coating materials.

A coating is a highly complex material. It is formulated from as many as twenty different materials mixed (or in some cases reacted) together, yet the end product is a very thin film of only a few microns or thousandths of an inch. As such, it must adhere to the surface; protect the surface from penetration or corrosion; and resist abrasion, high humidity, weather, water, chemicals, precipitated salts, or actual immersion in corrosive or sensitive solutions. In all cases, the coating is a thin film separating two reactive materials. Understanding the complex nature of coatings emphasizes the need for good specifications and inspection procedures.

Inadequate specifications also represent a variable, as discussed in the previous chapter on specifications. Many specifications are so inadequate as to be nearly useless in obtaining the desired coating. This is often caused by a lack of communication. However, it is more often caused by inept engineers using antiquated specifications or ones that are completely improper for the job at hand.

Selection of the lowest bid, irrespective of the quality of the bid, also presents a difficulty as far as quality control is concerned. Even though the bid may be a legitimate one, once problems begin to occur, there is the inevitable trend toward a reduction in quality to make up for the losses caused by various problems. Even small variations in the application process can make substantial differences in the end product. Efforts to reduce labor costs through quick applications inevitably lead to a lack of proper care during the application, which further deteriorates the quality that is necessary for the use of high-performance coatings.

Another difficulty from a quality standpoint is that too often the superintendents, managers, and even owners fail to understand the importance of corrosion control and the role that coatings play in preventing corrosion. Because painting is such a commonplace activity, supervisors and managers often do not realize that it is the greatest single factor in the cost of maintaining corrosion-free industrial structures. Thus, little attention may be paid to proper inspection of the final job. While this situation is not true in every case, it is common enough to be considered a serious difficulty with regard to coating quality.

The lack of proper inspection is another variable that can also cause serious problems. Even when a good specification calling for thorough inspection has been written, the personnel selected to provide the inspection often are not knowledgeable enough to interpret the specifications in either a practical or satisfactory manner. In addition to the fallacy that almost anyone can follow and inspect a coating job, another reason that paint inspectors are so often poorly qualified is that relatively few qualified guidelines applicable to the coating of industrial structures are available. This has been changing in recent years due to the extensive work on surface preparation and coating application by such organizations as the SSPC and NACE.

Types of Coating Inspectors

There are many types of coating inspectors. One is an inspector supplied by the coating manufacturer. This type can be very effective and helpful, particularly if they have had extensive field experience, are allowed to be on the job full-time, and are knowledgeable about the application of the supplier's products.

Another type is an inspector obtained from an application consultant. They are usually trained as inspectors and thus have had extensive experience on numerous coating application jobs. While they can be very effective inspectors, those obtained from general contract inspecting companies may be more experienced in setting up equipment than in actual coating inspection. Such individuals may be more interested in the details of the specifications than in their overall intent. In fact, some adhere to specification details so narrowly that a serious friction develops between the inspector and the contractor, while others are so lax, mainly due to little or no coating knowledge, that little benefit is obtained from their services.

A corrosion engineer assigned to a coating application often serves as a good inspector. This is because they are familiar with the company, the reason for the coating, the corrosion problems involved, and may even have taken part in writing the specifications.

A retired painter or coating foreman also can serve as a good inspector since he is familiar with application procedures and common applicator problems. However, they can sometimes be less than objective in their observations, depending on the attitudes and work habits developed over many years of field work.

Unfortunately, until recently there have been few inspectors who are thoroughly trained for the job. The good ones are self-made, as a result of an interest in coatings and long application experience. This situation may be changing due to the establishment of an NACE program for coating inspector certification. NACE's Coating Inspector Training and Certification Program, begun in 1983, provides extensive training in all aspects of coating inspection.

Another problem in obtaining a quality coating application lies with the applicators themselves. This is explained further in a chapter in an SSPC publication on inspection.

> The painting inspectors' problems will often be magnified by incompetence and lack of knowledge on the part of the painters themselves. Very rarely do painters have much detailed knowledge concerning the materials they use. This is particularly true in the case of the newer paint products. It cannot be assumed even that the painter will know what thinner is to be used in the particular type of paint he is applying. Not infrequently the man applying paint will be found to be not an experienced painter at all but a handyman who knows only a little about the many different jobs he does. The inspection difficulties inherent under such conditions need not be detailed. Personal prejudices of the painters must be overcome, since, if they feel their ''long experience'' is being ignored, the quality of the painting is apt to suffer.
>
> Despite the rather dark picture just presented, improvements in the quality of painting through the sensible employment of competent inspection procedures is distinctly attainable and should benefit all concerned except those few contractors who construe laxity in inspection as an opportunity to reduce cost.[3]

Good coating inspection is based on the concept of in-process quality control. This means that every step in the application of a coating, from surface preparation to final drying or curing, is monitored to assure its conformity with specifications. This in-process concept is important since coating inspection is too often done on a spot-check basis where the inspector may appear at the job site periodically to make a cursory examination of the situation. This type of inspection is not in-process quality control and does not lead to the type of application that is required for high-performance coatings. While periodic inspection is better than no inspection, continuous inspection of all steps involved in the application process is important for the best quality work. When each step is properly monitored, the quality of the end product is nearly always insured.

What Should a Qualified Inspector Know?

With this in mind, what should a qualified inspector know to provide the proper level of inspection for high-performance coatings?

Intent of the Specification

It is assumed in this case that the specification has been written. However, for an inspector to interpret the specification properly, it is necessary that he have a thorough understanding of the intent of the specification. As previously discussed, the specification should be a compilation of information provided by architects and engineers, company personnel (preferably a corrosion engineer), the applicator, and the top executives of the company. Each of these groups has a somewhat different perspective on what constitutes a proper coating job. It is the inspector's responsibility to interpret the specifications in such a way that will satisfy all of the parties concerned with the application.

The architects and engineers generally are interested in the protection of the structure and its appearance. Company personnel tend to be interested primarily in the operation of the structure and the costs involved in its maintenance. The corrosion engineer's interest lies in the

It depends on whose standards are followed as to whether a coating job is satisfactory.

resistance of the coating to the atmosphere. Company executives may be interested in the general appearance of the structure and how it affects the overall company image.

If the inspector understands the intent of the specification and then makes sure that each step is accomplished in accordance with that intent, the requirements of the specification are rather easily met. In so doing, the requirements of each individual involved also should be satisfied.

If dealing with an industrial structure in an acid atmosphere, the inspector should understand that the corrosion of any steel or concrete structure in such an atmosphere can be quite rapid and possibly catastrophic. The inspector should then interpret the specification in light of that knowledge. If the coating is for appearances only (which is unlikely where a high-performance coating is used), the principle of sidewalk inspection should be applied, which means that overall appearance must be good, irrespective of hidden areas. For instance, the appearance of the underside of an automobile generally does not matter. However, this certainly is not true where a corrosion problem exists, since any area left uncoated in that case could cause serious problems.

Substrate

The inspector should have a good knowledge of the substrate over which the coating is to be applied. The underlying surface (whether it is wood, aluminum, concrete, steel, galvanizing, or a specialty steel) is extremely important from the standpoint of coating durability. For instance, there are potential adhesion problems wherever a coating is applied over stainless steel. On the other hand, if there were an imperfection in a coating applied over stainless steel, the reaction at the imperfection would be considerably less and therefore less important than if the substrate were hot-rolled steel. Concrete surfaces, because of both their reactivity and their variable surface, present different problems than when coating a dense, smooth, clean steel surface. The coating that might easily be continuous over a steel substrate might not provide a continuous surface over a concrete substrate. The substrate can easily make the difference between a satisfactory coating application and a poor one, so that the in-

Surface Preparation

A good inspector should have a sound knowledge of surface preparation; not only for steel, but also for other surfaces such as galvanizing, concrete, aluminum, etc. The inspector should know the application characteristics of coatings over each type of surface, as well as the type of surface preparation that will provide the best base for a coating over the various substrates. While this subject of surface preparation has been thoroughly covered by both the SSPC and NACE, the standards and information presented by these groups usually is not sufficient unless augmented by some practical experience in the field with various types of surface preparation on various substrates For example, the field inspector may be required to rule on the difference between a white metal and a commercial blast according to the appearance of the surface. He must also know how the coating involved will react over such surfaces in order to provide practical evaluation of the surface preparation.

The inspector should be thoroughly familiar with the equipment used to obtain the various grades of surface preparation. This not only includes blasting equipment, but should include the equipment for hand preparation of surfaces as well. There is nothing quite so impressive as an inspector who is able to pick up the equipment and perform the proper type of surface preparation as a demonstration of how the work should be done. The inspector should also know the results of the surface preparation; *i.e.*, knowing the coating that is to be applied and the type of surface preparation available. He should be able to evaluate the effectiveness of the coating over such a surface and provide advice as to how to obtain the best possible coating application under given circumstances.

The Effect of Design

The coating inspector should be completely familiar with the design of the structure and with the effect that the design is going to have on the application and effectiveness of the coating, and on the degree of sandblasting or surface preparation possible. Unfortunately, structures are often designed without any consideration of coating application. They are designed with bolts and rivets, sharp corners, edges, hidden areas, crevices, and similar problems, which are all areas that can cause serious problems from a coating standpoint. The inspector should have a full knowledge of the application procedures that are required to obtain a full, thorough coverage of such problem areas. He should understand that coatings will tend to pull away from corners and edges, as well as from surface protrusions such as weld spatter. He should also realize that a coating must be applied from several different directions wherever bolts, rivets, or similar protrusions are encountered. This type of information will aid the inspector in properly evaluating the coating application and in obtaining the best possible job.

Coating Characteristics

The inspector should have a good knowledge of

various coating characteristics that can affect the application as well as the finished product. He should know, for instance, that vinyl coatings are generally relatively thin films and that additional care is required on rough areas or edges and corners in order to obtain full thickness. He should realize that epoxy coatings generally require a catalyst or an additive in order to cure properly, that such coatings have a limited pot life, and that, if that pot life is exceeded, adhesion characteristics are often damaged. The inspector should know the different application characteristics of a chlorinated rubber versus a vinyl and an epoxy. He also should know the general wetting characteristics of these coatings in order to evaluate their effectiveness over various types of surface preparation.

Even though the coating inspector may not be involved with the coating selection or specification, his knowledge of coating characteristics and basic properties influence the success or failure of the coating application. His knowledge usually comes from actual field experience with coatings. In fact, the coating inspector should have practical experience with high-performance coatings so as to provide the best evaluation of any given job.

The inspector should understand application characteristics of the coating.

Application

The coating inspector should have an excellent knowledge of the application procedures generally used for high-performance coatings. He should know the characteristics of air and airless spray, and should be familiar with the procedures used to apply coatings properly with such equipment. Such knowledge can help him in his evaluation of the job and in making timely suggestions to the contractor if application problems occur.

Application procedures frequently need to be changed during a job in order to obtain a proper coating. For example, if there is excessive overspray, a knowledgeable inspector should be able to suggest methods to correct the situation. If application is by air spray, perhaps the use of airless spray would overcome the problem without reducing the effectiveness of the coating or the labor involved from the contractor's standpoint. The inspector should also know that the mishandling of either air or airless spray can cause pinholing in various coatings. Thus, he also should know what to do about pinholes if

they develop. The inspector should realize that if pinholes are in the coating, spraying an additional coat over the surface will do little to overcome the problem. He should realize that working a coating into the pinholes by brushing is the only method which will obtain a satisfactory application.

Another thing the inspector should be aware of is that, if there is high humidity, some coatings will tend to blush or absorb water into the coating surface. This can result in a less than effective coating, and in many cases can change the coating adhesion characteristics of a following coat. This type of application expertise can only be gained through practical experience.

Application procedures may have to be changed during the job to obtain proper coating.

As part of his application knowledge, the inspector should thoroughly understand the application equipment, how it works, and its effect on various types of coatings. He should recognize the difficulties inherent with various types of application equipment. This is particularly important from a safety standpoint. For example, airless spray is under very high pressure and applicators must keep their hands away from the spray area of airless equipment. A number of accidents have occurred where the jet from an airless spray impacted a finger, hand, or arm with the jet, driving the coating into and through the skin, causing serious injury. A whipping air hose or a sandblast nozzle can cause serious injury to both the surface to be coated and the personnel in the area. Inspectors should understand this and be able to react quickly should a sandblast hose be dropped accidentally.

Cleanliness

Another item the inspector should be aware of is the cleanliness of the equipment. Lack of cleanliness has been the cause of more poor applications than almost any other problem. Equipment that has been used for oil paint, alkyds, or similar materials and is then switched over to high-performance coatings almost inevitably causes contamination of the coating and work stoppages. The same is true where organic coatings have been used, followed by inorganics. The inspector can easily save the contractor time and effort by making sure that the equipment is in proper order.

The Contractor

The coating inspector should also be able to recognize the contractor's problems, such as contending with poor structural design or learning how to apply a new type of coating. The experienced coating inspector may be able to help a contractor salvage a job that has gone sour and save money for both the contractor and the owner. In many ways, a good coating inspector can be the contractor's best friend. He can act as a consultant, teacher, helper, and, above all, an intermediary who promotes understanding between the contractor and the owner.[4]

The inspector, in all cases, should avoid developing an adversary relationship with the contractor. This happens all too often, with the contractor feeling that the inspector is interfering with his work, and the inspector feeling that the contractor is trying to speed the job up, use less material than is required, and generally exercise less than the required amount of care during application. This kind of a situation can best be avoided if the inspector makes a determined effort at the beginning of the job to develop the proper relationship with the contractor, making sure that his own activities on the job are helpful and practical. This does not mean that the inspector should allow the contractor to do things that are not proper. If the contractor feels that the inspector is being fair and that his objections or suggestions are sound, a good relationship can develop.

The Representative of the Owner

The inspector is often the only representative of the owner with whom the contractor will have any major contact. The inspector must have sufficient knowledge of the job, the specification, and the coating materials to act as the interpreter of the specification and as an intermediary between the contractor and the owner in order to assure that the owner receives the kind of job for which he contracted. The inspector must be sufficiently political to give in on items that are of little consequence and yet hold firm on those which are going to make a substantial difference in the effectiveness of the job. Communication is often a problem between the owner, the material supplier, and the contractor. The inspector can be the intermediary between all parties, providing the communication that is necessary to obtain the proper job. In this case, a pre-application conference can be the beginning of this type of open communication. As stated by Bayless in *Quality Control of Protective Coatings:*

> It is easy to draw up a list of the qualities required of the ideal painting inspector but again one must take a practical attitude. The job is not the most congenial, for instance including long hours and long periods away from home. The work like painting itself can be unpleasant and even with an element of danger. In practice the difference between a good and troublesome job is often a measure of the inspector's personality. A bad inspector is like a bad policeman and is either too petty and rigid or the reverse. Good quality control should result in the satisfaction of all parties that everybody has done the best they can, to the best of their abilities.[5]

One item that must be mentioned at this point is that during an application, particularly where the inspector is new to the contractor and to his personnel, the inspector will, as a rule, be constantly tested during the first part of the job by both the workers and the contractor to see whether or not he is knowledgeable and whether or not his suggestions are practical. If the inspector makes a favorable first impression, the job usually proceeds well, producing good results. If, on the other hand, he is not able to make a proper first impression it is more likely that he'll be baited and deprecated, and that almost anything he does will be side-tracked, overlooked, or carried out in a poor fashion. This situation can only result in a poor application and therefore poor overall results.

Areas of Coating Inspection

Material Inspection

The inspection of materials purchased for a coating application depends, to a great degree, on the method of purchasing. Several methods were outlined in the previous chapter on specifications. If a material is purchased according to the formula specification (which is usually the case where government purchasers are involved), considerable laboratory material testing is necessary to determine conformance with specifications. This kind of testing requires considerable laboratory equipment, personnel, and detailed procedures which cannot be carried out by the inspector in the field.

Trade name or equal specifications also require laboratory testing to determine whether or not the materials are equal. Since the field inspector cannot do this type of testing, it must be handled by the company or by a governmental agency.

Materials purchased on a qualified products list are usually pretested. This involves a long laboratory procedure to determine if the materials on the qualified products list meet a series of performance requirements for the purpose involved. Once the materials are on the qualified products list, no additional laboratory testing is necessary, unless it is desired by the agency purchasing the material. Again, these would be laboratory tests so that the field inspector would not be able to contribute in this area.

Materials purchased by trade name usually require no laboratory testing. Some purchasers may require infrared scans on different batches of materials to determine their equality. However, because these materials are backed by the manufacturer's name and reputation, there usually is little need for testing. A good discussion of materials inspection appears in Chapter 5 of the Steel Structures Painting Manual as follows.

> This section presumes that the paints are purchased on a specification basis, generally through competitive bidding. If paint has been purchased outright as the trade name or proprietary products of a single manufacturer, there is usually no point in testing it, since there is no agreement between paint supplier and buyer as to what the detailed characteristics of the paint must be. Similarly, there is little reason to test paint purchased on a 'similar or equal' basis because the determination of similarity and equality is for all practical purposes impossible within the time usually allotted for such testing work. The following discus-

sion presumes also that certificates of compliance with the specifications and certified test reports, originating from the paint manufacturer, are of limited value to the purchaser. When the buyer does not necessarily accept the lowest bid from any source but instead purchases the specification paint from a reputable concern in whom he has developed confidence, then such certifications may be acceptable.

The scope of testing and/or plant inspection of paint purchased on a specification basis must be determined by the buyer. His judgment in this matter will be influenced by: the volume of paint involved, the bid price per unit of supplied paint, whether competitive bidding was employed, whether there has been past experience with the particular supplier, the manufacturing facilities and reputation of the supplying organization, and lastly, the type of paint being furnished. The first of these considerations needs little explanation—the economic feasibility of testing several hundred gallons of paint is apparent when compared with the cost of testing forty or fifty gallons. When several thousand gallons of a paint item are involved, still more expense can be justifiably incurred in determining whether the material complies with the specifications. The need for considering the second point—unit bid price—in determining the scope of testing will be apparent to those who have received competitive bids on paint items. It is not unusual to receive quotations on a large quantity paint item ranging from the bare cost of materials to several times that figure. If the supply contract is awarded at a bid price which appears so low as to preclude reasonable profit, the buyer should take great precautions to insure that the material complies with the specifications. Although the private organizations can reject those bids which appear abnormally low, this highly desirable course of action is often nearly closed to municipal, state and federal government agencies. The effect of the buyer's past experience with a particular supplier on the scope of the test work to be performed requires no explanation.

Similarly, the supplier's manufacturing facilities and reputation may be important factors in determining whether or not the furnished materials should be tested thoroughly. The last consideration, *i.e.,* the type of paint being furnished, has a bearing on the thoroughness of the test work, as will be seen from the following examples. A relatively simple oil type paint such as that conforming to Federal Specification TT-P86c, Type I (Red lead-linseed oil paint), may be adequately tested merely by examination of the finished product. However, the buyer of a phenolic type red lead paint conforming to Type IV of the same specification could hardly be assured of full compliance with the specification without testing a separate sample of the phenolic varnish and the resin and oil used in making up the varnish, in addition to the paint. Indeed, some formulations for certain paints are so drawn up as to require inspection in the plant if the buyer is to be assured of obtaining the desired material.

In deciding how thorough the testing work to determine compliance with a paint specification should be, one additional group of factors should be given careful consideration: the contemplated exposure condition, the cost of making the surfaces to be painted accessible for repainting, the cost of surface preparation, and the cost of paint application. For example, a partially submerged miter gate in a navigation lock can be made accessible and the surfaces prepared for painting only at great expense. Clearly, the coating of such surfaces should make use only of those materials whose compliance with the specifications is beyond question. Conversely, the need for absolute assurance of paint quality is not so urgent for easily accessible steelwork subject to a rural atmospheric exposure.

As a general statement, it can perhaps be said that buyers who do not contemplate thorough testing of their specification paints should deal only with paint manufacturers with whom they have had satisfactory experience in the past and/or those who have an excellent reputation for integrity and ability. If he does not resort to comprehensive testing of specification paints, the purchaser should probably disregard the lower prices associated

with competitive bidding and buy from a technically competent manufacturer who believes a satisfied customer to be the most profitable one in the long run.

It is only fair to point out in closing this general discussion that not all of the trouble experienced in obtaining satisfactory specification paints is the fault of the paint manufacturer. Some of the specifications are so poorly written and so clouded with contradictions as to defy the interpretive ability of anyone except, and often including, the writer.[3]

Field Inspection of Materials

The inspection of materials on the job is only cursory at best. There are, however, a number of worthwhile steps that can be taken. The inspector should check the labels on the container to determine whether or not the material bears the designation and number that are detailed in the specification. It is important to check all containers, particularly on a large application job. This is because errors can be made in the shipping of materials, and the applicators could easily use these wrong materials, particularly if the material and the label on the container are the same color.

It is also worthwhile to check the date on the container which indicates the shelf life of the coating and the date beyond which the coating should not be used without further testing to see that its consistency and other properties are satisfactory. It is also important to read the label of the container to determine the level of safety hazard which may be involved in the use of the material.

Once the containers are open, the inspector should check for material problems, such as viscosity, livering, gelling, pigment separation, and very hard pigment settling, as well as for the solvent to be used if thinning is required. Most of these items can be checked purely by observation. If any abnormalities are found, these should be checked out with both the owner and the paint manufacturer to determine if the material is satisfactory. It is far better to run a check on the material before it is applied than to run one afterwards.

A material that is gelled, even though mechanically mixed, will probably have gel particles distributed through the coating which can create a weak point for failure. For example, solvent vinyl coatings that have been packed in an unlined steel container will often gel around the edges of the container. This is an irreversible gel, and if it is mixed into the coating and applied, it not only makes for a poor appearance, but it also creates an unsatisfactory coating from a resistance standpoint.

Some materials also settle very badly in the container, making it difficult to mix them back into proper suspension. Unfortunately, applicators often leave the badly settled materials in the bottom of the container and use only the liquid above. Of course, when this is done, the coating lacks solid content and the material applied is not in accordance with the formulation. The resulting application is a poor one from a resistance standpoint. Coating buckets have been found on the job with 1 to 3 inches of pigment settling which was never applied on the structure. Unless such settled materials are gelled or livered, they usually require some additional work to redisperse the pigments and therefore bring the coating back to a proper consistency.

Inspection of Surface Preparation

All of the information presented in Chapters 9 and 10 on surface preparation and coating application are applicable to the inspector at this point. As previously discussed, the performance of the best coating is usually no better than the surface preparation. If a high-performance coating is applied over a pitted surface with rust in the pits, little satisfaction is going to be obtained from the material. It is also true that once the coating has been applied to a surface, it is very difficult to determine the quality of the surface preparation. Coatings easily cover mistakes and improper surface conditions. It is only after the coating is in service that these imperfections become significant.

Unfortunately, in the case of surface preparation, the inspector is saddled with a considerable responsibility in determining whether or not the surface was properly prepared for coating. One of the best ways to deal with this situation is for the inspector, the owner, and the applicator to clean a section of the surface to the satisfaction of all parties. The inspector can then use this degree of surface preparation in determining whether the surface conforms to the specification.

In surveying surface preparation work, the inspector should check for a number of different items. One, he should check to see that the surface preparation conforms to the specifications. In the case of NACE 1 or SSPC 5 for white metal blasting, the inspection is relatively easy, since the surface must be of one color. On the other hand, near-white blasting or commercial blasting is much more difficult since closer evaluation is needed before a decision as to whether or not the blasting conforms to the specification can be made. Again, this is an area where a pre-application conference and demonstration help to assure a proper evaluation. Even so, inspection of old, heavily pitted surfaces, or those covered with old coating residues, involve many decisions that are difficult to make.

Pitted surfaces are particularly difficult because the surface is uneven and rust and contamination can remain in the pits, especially after the first blasting. This is more noticeable in areas where there are chlorides, because the chloride residue often is not readily visible until sufficient humidity is absorbed from the air and the pit begins to darken in color. If this is the case, the surface will have to be reblasted to obtain the desired cleanliness, and it will be up to the inspector to advise the owner that this is the case.

Contractors and sandblasters are often reluctant to go over the surface a second time because of the time and effort required. Some sandblasters complain that the surface is difficult to clean, that pitted steel cannot be cleaned properly, that heavy scale is almost impossible to remove, etc. Complaints such as these usually are not valid. If the sandblaster worked methodically across the surface, taking a foot-square area at a time and cleaning it thoroughly, these problems should not exist. Again, the inspector is the one responsible for advising the owner, and possibly the contractor, that the surface is either satisfactory or unsatisfactory.

In one case during the coating of a tanker, the sandblasting contractor claimed that oil and contamination were bleeding out of the surface because he would no sooner leave the surface than black areas would appear, particularly in the pits. There was no dehumidification on the job and condensation and absorption of moisture was very rapid. The problem was actually that the contractor was not thoroughly cleaning the surface, and the chlorides or chloride salts embedded in the surface were absorbing moisture and changing color very rapidly. This can be a difficult situation for an inspector to handle, and he must have the courage of his convictions and the backing of the owner in order to obtain proper results.

The inspector should check for surface impurities, both before and after blasting, since they are the cause of a considerable number of coating difficulties. It is often necessary to wash the surface with water or dilute phosphoric acid and to reblast in order to obtain the desired surface. The inspector must also be aware of the problem of humidity and condensation. The changes in temperature as the day passes can rapidly make a change in the possibility of moisture condensation on the surface. The surface temperature generally does not change as rapidly as the air temperature, and once the air temperature begins to drop, the humidity increases and the possibility of condensation becomes greater. Situations have been encountered where the surface has changed from dry to covered with condensation in a matter of fifteen minutes.

The temperature of the surface is also important to note. Hot surfaces are oftentimes difficult to coat under such conditions because the solvents in the coating boil off rapidly creating pinholing or poor adhesion. A warm surface is often advantageous during coating application. Cold temperatures cause the coating to dry and cure slowly. Temperatures that are too cold may inhibit the coating from curing, causing it to remain soft and sticky for considerable periods of time. On the other hand, water-based materials on cold surfaces may slide or refuse to evaporate properly, with the ingredients in the coating separating, leaving the coating open to early failure. This is true with water-base organic as well as inorganic coatings.

The inspector should also watch for damage to adjacent coatings during the surface preparation process. If a surface is being blasted, the adjacent coated surface can be badly damaged by the sandblaster waving his blast nozzle over into the coated area. While it may not be immediately obvious, these areas corrode rapidly as soon as the coating is put into service. Thus, ongoing inspection during the actual blasting process is needed to protect against such damage.

Inspection of other types of surface preparation may be even more difficult. For instance, if the process of acid pickling is thoroughly monitored and clean solutions are used, there is usually little difficulty. However, mill marks may have been painted on the steel, there may be pitting of the steel surface, or the surface may be spotted with oil and grease. Since all of these conditions create difficulties in the actual coating process, inspectors should be on the watch for them. Surfaces should be inspected and such items as mill marks, oil, paint, and similar impurities should be removed prior to the pickling process.

Pickling smut also poses a problem during the pickling process. The smut is a black deposit on the surface of the steel usually caused by poor pickling practice. It must be thoroughly washed off from the surface during the rinsing process and usually requires scrubbing with brushes for proper removal. Application of a coating over a smutty surface only leads to poor adhesion and coating difficulties. Under usual pickling conditions, the solutions are hot and the steel comes from the pickling process through a hot rinse bath. In this case, the steel is hot or warm as it comes out, the surface dries quickly, and it is an ideal time to apply the primer or first coat.

Surfaces prepared by power sanding and other power tools are also difficult to interpret. The inspector should realize that more contamination of the surface remains when using these types of preparation, and that a completely clean surface is impossible to attain by such means. Such surfaces usually are found during maintenance coating operations and cover only limited areas. Nevertheless, they should be as clean as possible, with all possible rust and contamination removed. The less contamination, the greater the chance for the coating in the repaired area to perform properly.

In any evaluation of surface preparation by blast methods, the inspector should pay particular attention to the type and size of the abrading media. As stated previously, surface preparation is most affected by the blast media, since this is the way in which the surface is cleaned and the surface profile obtained. The size and type of the blast media, therefore, primarily determines the conditions of the blasted surface. Surface cleanliness and profile can also be affected by the pressure of the blast nozzle, since the force which drives the blasting media against the surface determines the profile. A low blast pressure should provide for less profile than a higher blast pressure.

Most sandblast operators operate their equipment with a nozzle pressure of between 90 and 100 psi. However, the length of the blast hose and size of the compressor also can make a difference in the actual nozzle pressure. The inspector should realize this and make sure that the blasting air pressure at the nozzle is sufficient to properly clean the surface and obtain the necessary profile. The blasting air pressure should be determined at the nozzle and not at the gage on the compressor, due to the variables in the length of the line and restrictions such as moisture traps, etc. Air pressure at the blast nozzle can be determined using a hypodermic needle air pressure gage. The needle of the gage is inserted through the blast hose as close to the nozzle as possible, and the pressure reading should be taken with the nozzle in operation and the abrasive passing through the nozzle.

The inspector should also be aware of the cleanliness of the abrasive. Tator and Trimber describe a valuable method of evaluating the cleanliness of the abrasive in their paper entitled *Coating (Paint) Inspection Instrument Types, Uses and Calibrations*. The method is useful with not only metal abrasives, but synthetic and siliceous abrasives as well.

Although there is no inspection apparatus for determining the cleanliness of the abrasive used, a visual inspection must be made to assure that it is not damp or contaminated. When abrasive recycling systems are used, a simple test for the presence of oil or grease contamination should be made. Drop some of the abrasive (*e.g.*, teaspoonful) into a small vial of water (pill bottle size) and shake vigorously. Inspect the top of the water for a film of grease or oil which will be present if the abrasive is contaminated. Dirt and dust in the abrasive can be assessed in the same manner. Small abrasive "fines" will be held by surface tension at the meniscus and a dirty abrasive will color the water or cause turbidity. However, water soluble contaminants such as salt will not be detected using this test. If water soluble contaminants are present, a litmus paper test of the water in the vial will tell if they are acid or alkaline. If neutral, the only field method to detect their presence is to evaporate the water and look for salt crystals.[6]

The inspector should also understand that the air used for either blast cleaning, blow-down, or for the application of coatings should be free of contamination. There should be adequate moisture and oil traps in the lines to make certain that the air is both oil- and water-free. Even a small amount of oil and water is unsatisfactory. There have been incidences where there was so much water in the lines that once the air was turned on, water flowed from the air line for several minutes before any air was available. Even after the water was blown out, the air was not entirely satisfactory. There also have been instances where an oil-water emulsion was sprayed out of the air line, indicating a poorly operating compressor and creating a situation that was entirely unsatisfactory for either the surface preparation or the application of the coating.

It is relatively easy to determine the cleanliness of the air by simply blowing the air from the air line into a clean rag. If there is contamination in the line, it will readily show on the rag as either moisture or oil. If the air is clean, there should be little or no contamination on the rag surface. Should contamination be visible, the in-line moisture and oil traps should be investigated to make sure that they are operating properly.

There is a relatively easy test to determine whether or not a blast-cleaned surface is oil-free. This is described in the SSPC Painting Manual as follows.

An oil-free steel surface blasted with clean abrasive can be easily wetted with water. A representative area of surface thus cleaned when wiped with a clean, water-saturated cloth, should support a continuous film of water and there should be little or no tendency for the water film to draw up into globules or otherwise demonstrate the usual surface tension phenomena associated with water on oil metal surfaces. Such a surface is said to be free of "water break" if a continuous film of water is maintained. The water may also be applied by spraying or dipping. The ability, then, of a steel surface to support a break-free film of water is a positive indication of the absence of oil contaminants. It should be noted that the water-break test is quite sensitive and is suggested only for the specific case in which one wishes to determine if a freshly blasted surface (without further treatment) is free of oil contamination. Even a steel surface which has been thoroughly cleaned with solvents by the usual specified shop or field methods will generally not support a continuous film of water. In fact, the solvents themselves may leave sufficient hydrophobic residue to prevent the formation on the steel surface of a break-free water film.[3]

Such a test necessarily requires a reblast or repreparation in the area where water is applied.

Application Inspection

Unfortunately, the nature of the materials and operations involved in coating do not permit the owner who has contracted a coating job to make a few observations of the finished work and then determine whether or not he has received his full measure of coating performance. Once finished, an intrinsically poor coating job may appear, upon superficial examination, to be equal to a job of the highest quality. Moreover, the deficiencies of the former may not be apparent for an extended period of time, too late for the owner to negotiate for corrective action, beyond a guaranteed period. Observation of coating operations as they proceed is the only practical way in which the owner can assure himself that he is obtaining the coating job that was specified.[3]

Observation is the inspector's primary tool, and one of the inspector's primary points of observation is the coating applicator. The inspector should realize that the average applicator knows very little about the makeup of coatings, particularly the high-performance type of coatings that are used under highly corrosive conditions. Thus, he is much more likely to make mistakes in thinning and in the application of such coatings than in the past when paints were much simpler in character. The inspector should also realize that the personnel actually performing the work will probably never read the provisions of the specifications, so that the only information they have concerning the coating is what is printed on the can. Many applicators believe that they know all about coatings application, even though their information has been derived from the actual use of only one or two types. In many cases, even the contractor's supervisory personnel have not had access to the actual specifications. Thus, the inspector may have the delicate job of educating both the supervisors and the applicators with regard to the specific coating that is being applied.

One of the most obvious duties involved in inspecting a coating application is to check both the job specifications and the manufacturer's specifications, not only to make sure that he is fully aware of all details, but also that the two specifications agree. If there is a major discrepancy between the two, these should be reconciled at the pre-application conference between the owners, manufacturer, contractor, and inspector.

One of the first steps in the inspection of an application is the inspection of the surface over which the coating is to be applied. While surface preparation has previously been covered, the inspection of the surface should include assurances that the prepared surfaces are free of any dust, dirt, or sandblast abrasive and that the surface is ready for application of the coating. If the coating work is to be done on an exterior surface, the prepared surfaces should be coated with the first coat prior to nightfall in order to prevent turning of the surface or moisture condensation. Blast-cleaned surfaces should not be allowed to stand overnight without first being coated. If the surfaces to be coated are on the interior, there may be times when they can be allowed to stand overnight prior to coating. Even

in this case, however, it is not a good practice, and the inspector should thoroughly reinspect the surface the following morning prior to the application of any coating.

Inspection of the surface prior to application applies not only to bare steel, but also to previously primed or coated surfaces. It is important that each coat in the system be applied over a perfectly clean, dry surface, and that the previous coats be touched up for any damage prior to the application of the following coat. Care should be taken to make certain that the surface is free from precipitated salts or other impurities between coats.

Before the application of any coating, the inspector should assure the proper mixing and thinning of the coating prior to its use. Inspectors should make sure that the material in the containers is the proper one for the job, and that the solids in the bottom of the container are thoroughly mixed into a uniform, lump-free condition. If two-component materials are being used, each of the components may require proper mixing in order to thoroughly incorporate all of the materials into a uniform liquid. Following the mixing of each component, the mixing of the two components in the proper quantities should be observed. The mixture of the two materials should, when completed, be uniform in color, texture, etc. If thinning is called for in the specifications, the inspector should check to be sure that the proper thinner and the proper quantity of thinner are being used.

Applicators invariably tend to overthin paint, since this makes it easier to apply by either brush or spray. The inspector must realize that overthinning tends to produce thin films, thus requiring additional coats to make up for the difference in solid content. The average specification will call for no more than a pint of thinner per gallon of the original coating. This usually will adjust the viscosity as needed for varied weather conditions or use of specific application equipment.

Once the coating is ready to be applied, the inspector should know the coverage that is expected from the coating, and in this case he should know the practical as well as theoretical coverage. From this the inspector should be able to determine the wet film thickness to be applied and the dry film thickness of the final coating. With this information, the inspector can then check during the application for the thickness of the material as it goes onto the surface. With some experience, an inspector can actually determine quite accurately whether or not the material is being properly applied, even when standing some distance away from the surface. If he stands in such a position that he can see the gloss of the coating as it is being applied, the inspector can determine if it is being applied evenly, if the passes are being overlapped properly, if there are thin spots or holidays being left, or if the material is being applied as dry spray. All of this is part of the observation process during actual application. Inspectors should also make periodic checks of the total coating thickness with a wet film thickness gage. This is an easy instrument to use, and it should be used directly in the area where the coating is being applied.

The actions of the applicator, *i.e.,* how he handles the gun or the brush, the amount of coating being applied, whether or not he is flicking the gun at the end of a pass, and the gun distance from surface, are all important observations. The inspector can probably do more to assure a proper application at this point than at any other point in the operation. It is at this time that the inspector should observe all application conditions very closely.

The inspection should include checking each coat for imperfections such as holidays, pinholes, runs, blistering, and overspray. If at all possible, such imperfections should be taken care of immediately prior to the movement of the applicator away from a particular area. Before the coating is dry, runs can be easily brushed out, holidays can be recoated, pinholes can be brushed, and the gun can be adjusted for overspray conditions. If the imperfections are taken care of at this point, prior to the coating being dry, a solid, uniform coating is assured. If the coating has been allowed to dry and is then inspected, it is necessary to repair each of the above types of imperfections at that time, prior to the application of an additional coat. The coating should be repaired for damage as well as for imperfections, prior to the following coat.

Each coat should be checked for curing and drying conditions; the inspector should determine that the drying time is proper both between coats and for the final coat. He should make certain that there has been no condensation on the surface or fallout of contamination between coats. If such conditions have occurred, each coat should be cleaned prior to additional coats. Finished surfaces should be inspected. This inspection should cover overspray, pinholes, runs, holidays, and any area which appears to be rough or improperly applied. All such imperfections should be marked and repaired prior to acceptance of the coating.

During the application, the inspector should be aware of temperatures; not only the ambient temperature of the atmosphere, but also the temperature of the metal. The relative humidity should be checked to assure that condensation and moisture is not possible on the surface during the application. When the temperature is close to the dewpoint, even the evaporation of the solvent in the coating can reduce the surface temperature condition to the point where moisture can condense on the surface. When this happens, the coating may *blush* (which is the absorption of the condensed moisture into the coating), causing a poor film. The minimum air temperature usually permitted for the application of coatings is 40 F (5 C), and the temperature of the steel should never be much lower than that. Under certain conditions, and with certain coatings, the temperature limits can be lowered, but it is generally not good coating practice to coat when the metal temperature is below 35 F or 1 to 2 C.

Coating during below-freezing temperatures, although possible with some coatings, is always dangerous since any moisture that may have been on the surface turns to either frost or ice and will cause immediate delamination of the coating. High temperatures are also a problem. Generally, coatings should not be applied when metal temperatures are greater than 125 F (50 C). Some special coatings may be applied at higher temperatures, although these are usually the exception. The temperature at which it is uncomfortable to place a hand on the surface and usually not possible to hold it there for any period of time, is 125 F.

The temperature of the surface and the air is always a concern during an application. The contractor may want to continue coating even though the minimum or maximum temperature has been exceeded. At this point, the inspector must determine whether it is possible to exceed the limits or whether, because of conditions, the quality of the coating will suffer. Generally, it is best not to permit any broadening of the temperature range since coating problems only multiply at each end of the range.

If coating is done on the exterior surfaces, weather changes are always a problem, and may be harmful to the freshly applied coating. Rain, sharp decreases in temperature, increases in wind conditions, and increases in humidity should be carefully watched. If possible, the application should be halted prior to any damage to the coating, or, at least, time should be provided for the applied coating to dry before damage can occur due to such changes.

As previously discussed, condensation is a particularly difficult factor to not only observe, but also control. A surface can change within a matter of minutes from being dry to being so wet that you could write your name in the condensation. This is the reason that humidity should be checked periodically, making certain that the dewpoint is sufficiently removed from the ambient temperature that condensation cannot take place. Anytime that the steel temperature is lower than the ambient outside temperature, condensation problems are possible. This is particularly true where there are heavy steel masses, such as heavy plate or heavy cross sections of steel. The temperature of the steel, after being exposed overnight, takes a considerable period of time to increase to the ambient exterior temperature. In fact, in many cases it does not rise to the air temperature even over an entire day's period. This makes the surface thermometer an extremely handy tool for the inspector. Each day's work should be finished in considerable advance of the time when it is possible for condensation to occur on the surface in order to allow a reasonable drying period.

The inspector should have information concerning the drying time of the various coats being applied. Chlorinated rubber or vinyl coatings which dry by evaporation often can be recoated within a matter of minutes. This is sometimes necessary because of contamination fallout. Since they dry primarily by the evaporation of solvents alone, this rapid recoating is not a major problem. However, coatings (such as epoxies) which are internally cured usually require some additional time for curing prior to a following coat. This time should be known by the inspector so that he can judge whether or not the application is being conducted properly.

Also, with coatings such as epoxies, too long of a drying time between coats can also be a detriment and may cause intercoat delamination. This is especially true with coal tar epoxy coatings. Overnight cure is usually satisfactory for most coatings prior to a second coat. However, in the case of coal tar epoxies, if there is moisture condensation on the surface during the night, intercoat delamination may easily occur when the following coat is applied.

Under these conditions, the surface must be protected from moisture if the coating is to be allowed an overnight cure before application of a following coat.

In inspecting the coverage of the coating, the inspector should be guided more by the dry film thickness than the coverage per gallon, since it is the dry film thickness that is important from the standpoint of resistance. The coverage does not indicate whether or not the coating has been applied smoothly and evenly, and it cannot account for losses due to application conditions such as wind. The dry film thickness of the coating should be measured in numerous spots over the surface in order to determine the average thickness of the coating. Particular care should be taken in areas of possible thin application, as well as areas that may be overly thick since a thick film (particularly in the curing-type coatings) may become brittle and lose adhesion because of its thickness alone. Measuring the film thickness for each coat provides time prior to the second coat for any film thickness deficiency to be corrected.

Inspection Equipment

Knife

There are several items of inspection equipment that should be available to the coating inspector. A knife or a very sharp putty knife is essential in determining the adhesion of a coating (Figure 18.1). If there are any questions with regard to the adhesion, a knife cut should be made through the coating to the steel to determine the degree of adhesion. The knife should be held at an angle to the coated surface and a cut made which tapers down through the coating to the substrate. If the coating is satisfactory from an adhesion standpoint, the cut should come off as a ribbon and the edges of the cut should be smooth, with no evidence of interface between the coatings and substrate. If the coating shatters from the surface, the adhesion is poor.

The putty knife can be used in the same manner by taking a corner of the knife and making a linear gouge in the coating down to the substrate. With some practice, the

inspector can become very adept in determining the adhesion and flexibility characteristics of the coating with a knife. Of course, in all cases where such a cut is made in the coating, it should be repaired immediately.

The knife can also be used in the crosshatch or crosscut tape adhesion test described in ASTM-D-3359. This test consists of a series of crosscuts made in the coating. A 1-inch square is made on a side, with cuts approximately 1/8 inch apart made at right angles across the square. Masking tape is then rubbed onto the area and removed. The adhesion is judged by the number of squares that are delaminating from the surface.[7]

Magnifiers

Another item that is extremely valuable for inspection, of both prepared surfaces and the coating, is a 10X magnifying glass. This is a small glass that can be conveniently kept in a pocket for use at any time. An amazing amount of information can be gathered from the use of this small tool. Thus, it is recommended for use by every inspector to make the evaluation of contamination in pits, corroded areas, and pinholes or of other coating imperfections much easier (Figure 18.2).

There are also some small, hand-held microscopes that are valuable inspection tools. In fact, some will change magnification from 30 to 60X (Figure 18.3). One of the most practical scopes in the Panasonic Light Scope[(1)] which has a light source and 30X magnification (Figure 18.4). 30X is as high as is practical for hand-held field inspection instruments. Such microscopes are very effective for the fine study of surfaces, both before and after preparation, and for imperfections in finished coat-

---

[(1)]Tradename of Matsushita Electric Testing Co., Central Osaka, Japan.

FIGURE 18.2 — A geologist's 10X magnifying glass which is an essential inspection tool.

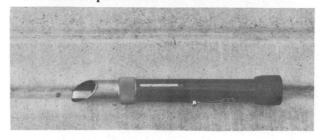

FIGURE 18.3 — A 30 to 60X field microscope.

FIGURE 18.1 — A knife or sharp putty knife is an essential inspection tool.

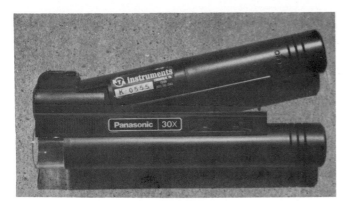

FIGURE 18.4 — A 30X magnification panasonic light scope.

ings. A great deal of information on coating failures can also be obtained with this instrument.

Other Equipment

There is also an illuminated magnifying glass which is much larger and requires flashlight batteries. They are usually five-power in magnification and cover a much larger area than the smaller glass. They are valuable in evaluating surfaces, and since they are self-illuminating, it is easier to inspect surfaces that may be in dark or dimly lit areas.

This equipment is also part of the Keane-Tator Profile Comparator.[2] This consists of a reference disc with five segments, each of which has a representative blast profile on the surface. The disc is laid on the prepared surface, and the profile of the blasted surface is compared with the reference disc profile. Using the illuminated magnifier, a good comparison of the two surfaces can be obtained, making it possible to determine by comparison the profile of the blasted surface (Figure 18.5).

There are actually three surface profile discs available, depending on the type of abrasive that is used on the job, i.e., sand, shot, or steel grit. Since the surface character and profile is quite different with each of these blasting media, the correct disc must be used in order to

properly determine the profile. Each of the discs is designated by letter and number. The number indicates the depth of the profile in mils, and the letter indicates the type of blasting media, e.g., "S" for sand, "GS" for steel or metallic grit, and "SH" for steel shot. A second number is also present, indicating the year that the master disc was formed, which generally is not that significant. For example, 1S70 indicates that the leaf of the disc was prepared using sand which created a one mil profile and that the master disc was made in 1970.

Another field instrument that is useful in determining the average profile depth is a depth micrometer. This consists of a needle or a pin which projects out from the flat base of the micrometer, and the distance that the needle or pin extends from the flat micrometer base is read on a dial in mils. Practically, the flat base is supposed to rest on the profile peaks and the pin extends into the valleys giving the profile depth reading. It is necessary to take a number of readings with this instrument in order to obtain an average profile. This kind of an instrument is also useful in determining the depth of pits and other depressions in a metallic surface (Figure 18.6).

Another method of determining the surface profile is through the use of a replica tape which consists of resinous film on a mylar tape (e.g., Press-O-Film Replica Tape).[3] The tape is pressed firmly onto the blast-cleaned surface, and the resins on the surface of the tape deform under the pressure used to apply it to the surface. The

---

[2]Tradename of KTA-Tator, Inc., Pittsburgh, PA.

[3]Tradename of Testex Corp., Wilmington, DE.

FIGURE 18.5 — The Keane-Tator surface comparator. (Photo courtesy of: KTA Tator, Inc., Pittsburgh, PA.)

FIGURE 18.6 — A surface profile micrometer or depth gage. (Photo courtesy of: KTA-Tator, Inc., Pittsburgh, PA.)

Inspection and Testing

resinous layer is penetrated by the profile peaks to the tough mylar backing so that when removed, it is an exact replica of the profile of the blasted surface, and it can then be measured with a light spring-loaded micrometer. The thickness of the mylar tape is subtracted from the micrometer reading, thus providing the average profile on the blasted surface (Figure 18.7).

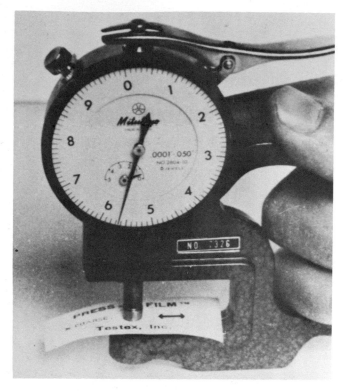

FIGURE 18.7 — Press-O-Film replica tape and spring loaded micrometer. (Photo courtesy of: KTA-Tator, Inc., Pittsburgh, PA.)

While none of these methods provide exact measurement, it is believed that the Keane-Tator Surface Profile Comparator is the most practical. There are other types of comparative discs available, and each provides a good comparison of the surface and the profile when used in the same manner. While it is up to the individual making the inspection to read the surface and the comparator, the profile on the comparator discs is sufficiently accurate so that a good correlation can be obtained with this method, even between different individuals. It is also portable for field use so that the actual surface can be measured on the job.

All methods of determining surface profiles have drawbacks. The depth micrometer depends on the nature of the surface, the number of readings taken, and the accuracy with which they are taken, and generally does not give the same degree of comparative accuracy as the comparator disc method. Replica tape depends on the pressure that is used to apply the tape, the amount of dust and dirt that may be on the surface, the ease and method by which it is removed from the surface, and the micrometer used to measure the thickness. It is important for the inspector to realize that there is no direct correlation between any of the above methods; however, as

previously stated, it is believed that the disc comparator provides the most convenient method and the best reading of the surface.

A viscosity cup is worthwhile in working with the liquid coating if it is necessary to determine the viscosity or consistency. There are two types: The Zahn Viscosity Cup and the Ford Cup. These are small cups with a known volume and with a sized orifice in the center. Each measures the time it takes the measured amount of coating to pass through the orifice. The faster the flow, the less the viscosity.

The viscosity cup is a good laboratory instrument and can be used to determine the viscosity of a coating fairly accurately. For a coating inspector on the job, it is also practical to check the coating viscosity between batches or where there may be a question of overthinning. It is strictly a comparison instrument when used outside the laboratory, and its use is only necessary where the consistency of the coating is questionable. The coating manufacturer should be consulted concerning the orifice size and the time involved for the coating to pass through the orifice, both unthinned and thinned (Figure 18.8).

Wet Film Thickness Gage

There are two types of wet film thickness gages. The first is the interchemical direct reading wet film gage. This is a circular gage which is rolled across the surface on the two outer rings. The center is an offset ring, calibrated

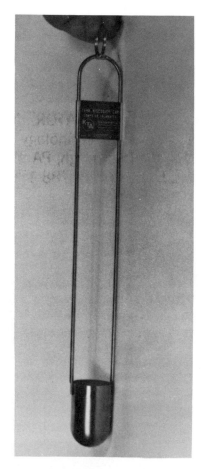

FIGURE 18.8 — Zahn cup for measuring coating viscosity. (Photo courtesy of: KTA-Tator, Inc., Pittsburgh, PA.)

in thousandths of an inch or mils so that when the gage is rolled on the surface, the point at which no liquid coating appears on the inner ring shows the wet film thickness of the coating. This is read on the exterior of the gage. This is a simple instrument which is very effective in determining the wet film thickness (Figure 18.9).

The second type of wet film thickness gage is an even simpler one. It is the notch-type gage that is made with two end points with progressively deeper notches or steps in between. Each step is a difference of 1 mil, so that when this is pressed into the coating, the wet film thickness is determined by the last step that is wet by the coating (Figure 18.10).

With either of these instruments, surface irregularities can distort the reading so that care should be taken during their use. Also, since they are an in-process check on the amount of coating being applied, they will not always yield an exact measurement. For example, if the coating has been allowed to dry for several minutes prior to the use of the gage, the wet thickness will measure less than when the coating was just applied. Neither can the wet film be exactly that calculated from the solid content of the coating due to evaporation and overspray between the spray gun and the surface. It is, however, a **reasonable measure of the film thickness** of the coating and provides a good evaluation of the uniformity of the material being applied to the surface.

The wet film thickness can be converted to the coverage of the coating in square feet per gallon by use of the chart in Figure 18.11. The wet film thickness, it must be remembered, is the actual amount of material being applied to the surface. Therefore, the spreading rate for coverage shown in the chart would be the practical and not the theoretical coverage of the coating. A wet film thickness gage has the advantage of determining a thin film application or an overly thick film application at the time it is being applied rather than after it is dry. Thus, the coating coverage can be corrected at the time of application and a much more even, uniform coating obtained.

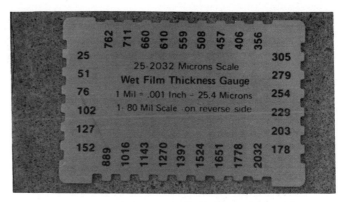

FIGURE 18.10 — A notch type wet film thickness gage.

Using the chart in Figure 18.11, it is possible to determine the coverage of the coating being applied, and from that, the approximate dry film thickness that will be obtained. For example, if the wet film thickness reads 8 mils, the coverage from the chart is 200 sq.ft./gallon. As was previously determined, the coverage of a 100% solids coating is 1600 mil sq.ft./gallon. If the coating in the example is 50% solids by volume, it would contain 800 mil sq.ft./gallon. The coating is being applied at 200 sq.ft./gallon, so that the dry film thickness should then be 4 mils. The formula is a simple one:

$$\text{Dry Film Thickness} = \frac{\dfrac{1600 \text{ mil sq.ft./gallon}}{\% \text{ solids by volume}}}{\text{actual coverage in sq.ft./gallon}}$$

$$(18.1)$$

$$\text{Example: } DFT = \frac{\dfrac{1600}{50\%}}{200} = \frac{800}{200} = 4 \text{ mils}$$

$$(18.2)$$

This provides the approximate dry film thickness that is actually being applied.

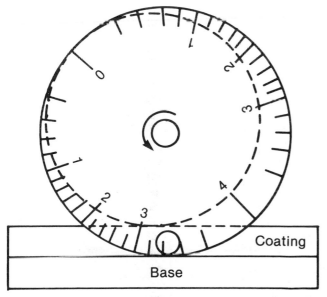

FIGURE 18.9 — An interchemical wet film thickness gage.

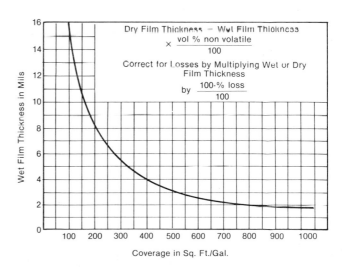

FIGURE 18.11 — The wet film thickness obtained at various coverages or spreading rates. (SOURCE: Good Painting Practice, Chapter 4, Vol. 1, Steel Structures Painting Manual, Steel Structures Painting Council, Pittsburgh, PA, p. 98, 1982.)

Tator and Trimber, however, have a different perspective.

Wet film thickness gages are only of value if one knows how heavy a wet film to apply. The wet film thickness/dry film thickness ratio is based on knowing the solids by volume of the specific material being measured. The old theory of doubling the desired dry film thickness to establish the wet film to be applied is only correct if the solids by volume of the coating material is 50%.

The solids by volume of the coating material is information that is readily available from the manufacturer and is commonly included in their product data sheets. The basic formula is DRY FILM THICKNESS = WET FILM THICKNESS × % SOLIDS BY VOLUME. A more workable variation of the formula showing the required wet film thickness for the desired dry film thickness is as follows:

$$\text{WET FILM THICKNESS} = \frac{\text{DESIRED DRY FILM THICKNESS}^{6}}{\%\ \text{SOLIDS BY VOLUME}}$$

[18.3]

As an example if the desired dry film thickness is four mils (0.004'') and the % solids by volume is 65% (0.65) the wet film thickness would be-wet film thickness 0.006 (6 mils) = 0.004 (4 mils)/0.65 (65%).

The above formulas are accurate provided the solids by volume of the material is not changed. The percentage will change, however, if any thinner is added to the coating. When thinner is added, the total volume of the material is increased without any corresponding increase in the amount of solids. Therefore, the thinned material will result in a lower percentage of solids by volume. Thus, when comparing thinned versus unthinned material in order to achieve a comparable dry film thickness, a heavier wet film application of the thinned material will be required. In order to evaluate what the new wet film thickness should be, the new percent solids by volume of the coating material after thinning must be determined. The following formula should be used to determine the required wet film thickness when the material is thinned:

$$\text{WET FILM THICKNESS} = \frac{\text{DESIRED DRY FILM THICKNESS}^{6}}{\dfrac{(\%\ \text{SOLIDS BY VOLUME})}{(100\% + \%\ \text{THINNER ADDED})}}$$

[18.4]

Again using the above example the desired dry film thickness is four mils (0.004). The coating is thinned by one quart per gallon. The wet film thickness then becomes:

$$\text{WET FILM THICKNESS} = \frac{\dfrac{0.004\ (4\ \text{mils})}{0.65\ (65\%)}}{100\% + 25\%} \quad \text{or} \quad \frac{0.004}{0.52}$$

$$\text{WET FILM THICKNESS} = 0.0077\ \text{mils}$$

or from a practical standpoint between 7.5 and 8 mils. [18.5]

Dry Film Thickness Gage

The measurement of the dry film thickness of a coating is probably one of the most important measurements that can be made. This measures the total dry thickness of the film as applied to the surface, and is therefore the working thickness of the coating. There are a number of methods of measuring dry film thickness. The most valuable are the nondestructive test instruments which are based on a magnetic principle. There are a number of quite accurate electrically operated thickness gages which are primarily for use in areas where power sources are available and where the instrument can be reasonably close to the coating being tested, such as on product finishes like washing machines, refrigerators, and other appliances. However, this type of gage is not practical for a field inspector working with steel tanks or similar structures.

For the field inspector, there are four principle types of magnetic instruments. While they are somewhat less accurate than the electric gages, they have the distinct advantage of portability and can be used on almost any steel structure with an accuracy sufficient enough to indicate whether or not the coating is applied within the limits of the specification.

The Elcometer

The Elcometer[4] is nondestructive and can be used to measure the thickness of any type of dry, nonmagnetic film on an iron or steel base (Figure 18.12). It has a two-point contact and utilizes a permanent magnet as the measuring force. Thickness is read on a calibrated scale, and it is a small instrument which is easily calibrated and easily carried.

The Mikrotest

The Mikrotest[5] is a single-point instrument using the magnetic pull-off principle (Figure 18.13). It is nondestructive and the thickness is read on a movable dial. It is lightweight, easily carried, and is one of the easiest of the thickness gages to use.

The PosiTector 2000

The PosiTector 2000[6] is another nondestructive,

[4]Tradename of Elcometer, Inc., Birmingham, MI.
[5]Tradename of ElectroPhysik, Cologne, W. Germany.
[6]Tradename of Defelsko Corp., Ogdensburg, NY.

FIGURE 18.12 — An Elcometer magnetic thickness gage. (Photo courtesy of: KTA-Tator, Inc., Pittsburgh, PA.)

FIGURE 18.13 — A Mikrotest III magnetic pull-off thickness gage. (Photo courtesy of: KTA-Tator, Inc., Pittsburgh, PA.)

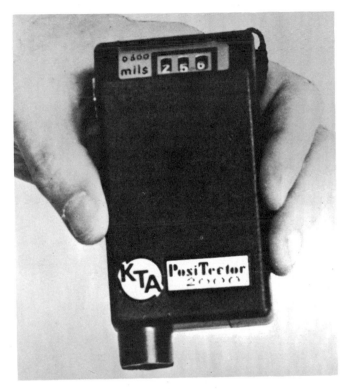

FIGURE 18.14 — A PosiTector 2000 nondestructive thickness gage. (Photo courtesy of: KTA-Tator, Inc., Pittsburgh, PA.)

magnetic, electronically-operated thickness gage (Figure 18.14). It is a more advanced type of instrument for measuring nonmagnetic coatings on iron substrates. It is small, easily hand-held with a single-point contact, and reads the thickness in mils on a digital scale. It is an accurate instrument and will give a continuous readout while being moved over a coated surface—something which other easily portable gages do not do. Where many thickness readings are necessary, this instrument is undoubtedly the quickest and easiest to use.

The Tinsley Gage

The Tinsley Gage is also a magnetic pull-off gage operated by a permanent magnet in a spring. It is about the size of a small pen or pencil and is a handy gage because of its size. However, it does not provide as ac-

curate a reading as the other types of gages described earlier.

There are numerous other proprietary-type, nondestructive thickness gages available, with essentially the same characteristics as those already mentioned. A more detailed description of the operation of the gages is given by Tator and Trimber in their paper,[6] and methods of calibration and operation are described in SSPC PA-2 *Measurement of Dry Paint Thickness and Magnetic Gages.*[8]

Of all of the gages presently available to the field inspector, the PosiTector 2000 is the easiest and most accurate, requiring only a point contact with the surface and giving an almost immediate digital readout. It reads in any position, and even reads as it is passed over the surface, so that variations in coating thickness can be read as it is moved. The Mikrotest-type thickness gage is very practical and provides consistent readings. The Mikrotest may be used in almost any position for a horizontal, vertical or overhead reading. Also, since the magnetic contact merely touches the surface, there is no damage to the coating, even though the coating may be somewhat soft. However, this can be a drawback of the Elcometer since it has two contact points, and if the instrument is held solidly against a soft coating, they may cause an indentation and therefore a false reading. The same problem may be true of the PosiTector.

These instruments generally should be calibrated for use with the thickness of coating that is to be measured. Calibration merely consists of placing a standard, with approximately the same thickness as the coating to be tested, on the bare metal surface to be coated and adjusting the instrument until it reads the correct thickness. The instrument should at least be checked to determine its variation from the standard.

It also should be considered that the magnetic gages will not read as accurately over a blasted surface as they will over a perfectly smooth steel surface. This is because the thickness of the coating varies between the peaks and valleys, and with the magnetic principle, the variation in the reading could be the depth of the profile. If the instrument is calibrated on a blast-cleaned surface, the test measurements will be more accurate and will generally read the thickness of the coating above the high points in the blast pattern. This, of course, is quite different from readings obtained on a perfectly smooth surface.

Steel surfaces that are badly pitted are also difficult to read with any of the magnetic equipment, due to the extent of variation in the steel surface. It should also be considered that red rust is nonmagnetic, so that if a coating is applied over a rusty surface, the thickness of the rust will increase the thickness of the reading. Care is necessary, then, in using these instruments to make sure that the coating being checked was applied over a clean, smooth surface. Edges, weld areas, and rough conditions interfere with an accurate reading with these instruments.

All gages based on permanent magnets will accumulate steel or other magnetic particles at the magnetic contact, causing inaccurate readings unless they are removed. These magnetic thickness gages are capable of fairly fast readings, particularly the PosiTector, so that any number of readings can be taken over a relatively short period of time. This is an advantage since it allows the inspector to check many spots and thus obtain a good measure of the average coating thickness that is being applied. The general accuracy of the magnetic thickness gages, when they are properly adjusted and in good working order, is between plus or minus 10% of the actual coating thickness. While there can be much larger variations due to external sources, the accuracy is much greater than the judgement of an applicator. A 10-mil film would have an accuracy of plus or minus 1 mil, which is entirely practical from an overall inspection standpoint.

There also are a number of destructive methods by which to determine the coating thickness. One of these, the Depth Gage Micrometer, was discussed previously. In this case, the coating can be scraped from the surface and the micrometer used to determine the thickness of the coating at that point. If it is possible to obtain chips or to peel the coating from the surface, a standard micrometer or a spring-loaded Micrometer can be used. Both of these provide an accurate measure of coating thickness.

The Tooke Gage[7] is a hand-held unit which uses a precision cutting tip to cut a groove in the coating (Figure 18.15). The groove is made with a known angle and is then viewed perpendicularly with a microscope. The microscope is calibrated, and the actual coating thickness can be determined from the scale in the eye piece. The advantages of this instrument are that it can be used for determining the thickness of multicoat systems, it can determine the thickness of the individual coats, and it can be used to read coating thickness on nonmagnetic surfaces. This allows the thickness of coatings on aluminum,

---

[7]Tradename of Georgia Inst. of Tech., Micrometrics Corp., Atlanta, GA.

copper, or concrete surfaces to be determined. The disadvantage of all destructive thickness testers is that the damaged area must be properly repaired after use.

Holiday Detectors

Where coatings are to be used in critical corrosion service, or as linings where sensitive or corrosive solutions will be in direct contact of the coating, inspection with a holiday detector is usually required. There are two principle types of holiday detectors: the low-voltage, wet sponge type and the high-voltage units. For relatively thin films, the low-voltage holiday detector is most useful. In this case, relatively low-voltage DC current is passed into a wet sponge which is passed over the coating surface. As the sponge contacts a pinhole or other coating imperfection, contact is made with the steel substrate and an audible alarm rings so that it is possible to locate the imperfection in the coating.

These are simple, nondestructive, easy-to-use testers that do not damage the coating even at the point of electrical contact. These units are designed for thin films, *e.g.*, a normal thickness of vinyl, epoxy, etc., where the wet sponge can supply enough water in the pinhole to make electrical contact. These are used wherever an imperfection-free lining or coating is required. They are self-contained, battery operated, and only require a ground contact with the base metal for them to operate. Figure 18.16 shows such a unit in operation, with the wet sponge on the coating surface and the ground lead attached to the flange of the tank.

FIGURE 18.15 — A Tooke coating thickness gage is a destructive gage which requires coating repair after use. (Photo courtesy of: KTA-Tator, Inc., Pittsburgh, PA.)

FIGURE 18.16 — A low voltage wet sponge holiday detector. (Photo courtesy of: Tinker & Rasor, San Gabriel, CA.)

Corrosion Prevention by Protective Coatings

The second type of holiday detector, the high-voltage unit, is available in a number of different forms, depending on the area of surface to be covered. One is a hand-held unit approximately the size of a flashlight, with a pointed probe that is passed over the coating surface. When the probe approaches a holiday or pinhole in the coating, the spark jumps from the point of the probe to the coating imperfection. This makes for a very accurate determination of the pinpoint imperfection. Some units require a 110 volt AC power source; others are self-contained, battery operated, and are effectively operated in remote areas. Such units can operate with voltages from 500 to 200,000 volts. Both the thickness and the dielectric strength of the coating impose limitations on the voltage used.

There are also larger units, using the same principal, with a wide rake-type probe that can cover several feet of surface at a time. In this case, the probe is made up of a section of conductive plastic or many fine wires, and, when an imperfection is found, a spark jumps at that point. These units may operate at somewhat higher voltages than the hand-held units (up to 35,000 volts) and are primarily used on thicker films such as vinyl sheet linings, tape wrapping, or coal tar (hot-applied) coatings. Where heavy coatings are applied to pipe, circular brushes which cover the entire circumference of the pipe can be used. The unit is rolled over the pipe exterior to pick up any possible coating defects. Many types of such units are available, and are extremely useful in the inspection of coatings, particularly those which have pore-free requirements. Figure 18.17 shows a circular unit for the automatic inspection of pipe coatings.

Some difficulties may be involved in the use of the high-voltage equipment. For example, when a coating contains conductive pigments, *e.g.*, graphite, aluminum, or copper, the high-voltage spark will pick out the conductive particles in the film and indicate an imperfection. If the high-voltage spark is left on the particle long enough, it will burn a hole in the coating at that point. Also,

because of the high voltage used, thin continuous areas of coating can be penetrated, causing a hole where none existed before. However, this also can be an advantage since such thin areas are potential failure areas.

Thermometers

A piece of inspection equipment that is useful under many circumstances is the surface thermometer. This thermometer is a dial-type unit that can be placed on the surface and that will read directly the surface temperature. This can be important in determining substrate or coating temperatures, particularly when the weather is unusually hot or cold and the temperature of the substrate is unknown. It can be very important to know when the substrate is approaching the temperature of condensation.

The thermometer is also a useful instrument in determining the high temperature of steel. This is particularly true where it is in the sun, such as on a roofdeck or a similar area subject to direct sun rays. This is important because excessive temperatures can cause blistering of many coatings during the application process. Figure 18.18 shows one type of surface thermometer that is easily portable for field work and can be placed on almost any surface for a quick reading.

Hygrometer

The combination of a thermometer and a hygrometer is a valuable inspection tool, particularly where there may be problems of dew or condensation. With the combined use of both instruments, it is possible to determine the probability of condensation on steel surfaces. This is done through use of a chart such as the one shown in Table 18.1 In this case, it is only necessary to know the metal surface temperature, the temperature of the surrounding air, and the relative humidity. If the relative humidity, as indicated by the hygrometer (Figure 18.19), is above a certain point, condensation is likely. For example, as indicated by the chart, if the air temperature is 70 F, the metal surface temperature 60 F, and the relative humidity above 56 F, the probability of moisture condensation on the steel is high.

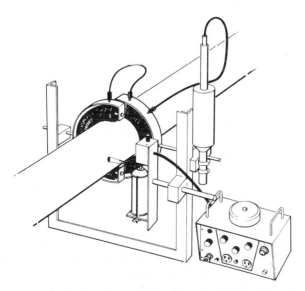

FIGURE 18.17 — Automatic holiday detector for pipe coatings using a circular high voltage probe. (Figure courtesy of: Tinker & Rasor, San Garbiel, CA.)

FIGURE 18.18 — A surface temperature thermometer. (Photo courtesy of: KTA-Tator, Inc., Pittsburgh, PA.)

**TABLE 18.1 — Percent Relative Humidity Above Which
Moisture Will Condense on Metal Surfaces Not Insulated**

Metal Surface Temp.	Surrounding Air Temperature F																
	40	45	50	55	60	65	70	75	80	85	90	95	100	105	110	115	120
35 F	60	33	11														
40		69	39	20	8												
45			69	45	27	14											
50				71	49	32	20	11									
55					73	53	38	26	17	9							
60						75	56	41	30	21	14	9					
65							78	59	45	34	25	18	13				
70								79	61	48	37	29	22	16	13		
75									80	64	50	40	32	25	20	15	
80										81	66	53	43	35	29	22	16
85											81	68	55	46	37	30	25
90												82	69	58	49	40	32
95													83	70	58	50	40
100														84	70	61	50
105															85	71	61
110																85	72
115																	86
120																	

% of Relative Humidity

[SOURCE: Ameron, Inc., Corrosion Control Principles and Methods, Section 7, Monterey Park, CA.]

FIGURE 18.19 — An hygrometer and temperature indicator.

Another useful piece of inspection equipment is called a Psychrometer or a Wet and Dry Bulb Thermometer. This is essentially two thermometers; one is covered with a wick or a sock that is saturated with water, the second is the dry bulb which gives the ambient air temperature. The most common type of Psychrometer is referred to as the sling psychrometer. In this case, the operating principle is to wet the wick on the wet bulb thermometer and then sling or revolve the thermometers through the air for a sufficient time to determine the cooling effect caused by the evaporation of the moisture on the wet bulb thermometer. After the two temperatures are determined, a psychrometric chart or table is used to determine the relative humidity and dewpoint temperatures of the air. While this instrument is commonly used to determine relative humidity, the combination of surface hygrometer and thermometer is much easier to use, particularly from the standpoint of the field inspector.

Adhesion Tester

The inspector is sometimes called upon to determine the adhesion of a coating on a numerical basis. The Elcometer adhesion tester can be used in the field and will provide a measure of the adhesion in pounds per square inch. It consists of a pulling unit with aluminum test dollies which are attached to the coating surface with a contact adhesive. After the adhesive has cured, the coating is cut down through to the substrate around the periphery of the dolly. The tensile unit is then put into place and attached to the dolly and the unit is tightened by hand. The test unit applies a direct tensile to the dolly, breaking it from the surface when the adhesion of the coating is broken. The point of the break is read on the scale of the instrument in pounds per square inch.

When this instrument is used, not only is the tensile adhesion value of importance, but the actual break from the surface is also important. If the break from the surface is clean, this indicates a possible adhesion problem. If the coating itself is broken, it means that the cohesive strength of the coating is less than the adhesion strength to the substrate. The break may also be in the adhesive itself. If this is the case, it indicates that the adhesive strength of the coating is greater than the strength of the adhesive.

Since this type of test is time consuming, a number of tests over a broad area is usually not practical. It is also a

FIGURE 18.20 — An Elcometer adhesion tester which provides numerical readings in pounds per square inch. (Photo courtesy of: KTA-Tator, Inc., Pittsburgh, PA.)

destructive test, so, as with all destructive tests, the area tested must be patched with the original coating in order to insure that full corrosion resistance is obtained. Figure 18.20 shows the pulling or tension unit by which the dollies are pulled from the surface by direct tension. It is the only field unit that is practical for on-the-job measurements where numerical adhesion ratings are desired or required.

Inspector's Diary

The diary of an inspector is an important document. Not only is it a record of the day-to-day progress of the work, but it is often the only one that is kept in a consistent manner. The diary is important as a record and as an aid to the memory of the inspector during meetings and conferences. It also constitutes documented evidence that is admissible in court proceedings should the coating fail prematurely. Detailed diaries should therefore be kept by the inspector on a daily basis, and should include such information as the weather, temperature, humidity, surface conditions, material used, drying or curing conditions, general appearance of each coat, any repairs that were necessary (including their extent and location), an evaluation of the quality of work being done, and the general appearance of the coating. Any particular difficulties during the application also should be recorded.

As evidenced by the above discussion, an inspector's job is not only important, but is often difficult as well. As stated by Tator and Trimber, "A few instruments, some knowledge, and good common sense are all that is required for good coatings inspection."[6] Sometimes the difference between a good coating job and one that is plagued with trouble can be attributed to the inspector's temperament and personality. A good inspector helps to insure good quality control of the coating application which should result in the satisfaction of the owner, general contractor, coating contractor, and all other parties involved.

References

1. Deacon, David H., The Importance of Painting Inspection, Construction Steelwork Metals and Materials, October (1970).
2. Thompson, S. P., Managing a Maintenance Painting Program to Reduce Costs, CORROSION/82, Preprint No. 146, National Association of Corrosion Engineers, Houston, TX, 1982.
3. Shanks and Rohwedder, Steel Structures Painting Manual, Vol. 1, Good Painting Practice, Chapter 5, Inspection, Steel Structures Council, Pittsburgh, PA.
4. Munger, C. G., Inspecting Plant Painting Projects, Plant Engineering, November 10 (1977).
5. Bayliss, D. A., Quality Control of Protective Coatings, Corrosion Control in Civil Engineering, I.C.E., London, England (1978).
6. Tator, Kenneth B. and Trimber, Kenneth A., Coatings (Paint) Inspection Instrument Types, Uses, and Calibrations, CORROSION/80, Preprint No. 254, National Association of Corrosion Engineers, Houston, TX, 1980.
7. ASTM, Standard D-3359, Philadelphia, PA.
8. SSPC, Measurement of Dry Paint Thickness with Magnetic Gages, Section 4, Steel Structures Painting Manual, Vol. 2, Pittsburgh, PA.

Additional information on equipment, manufacturers, and suppliers can be found in the annual *NACE Corrosion Engineering Buyer's Guide,* available from NACE, Houston, TX.

19

Typical Coating Uses

The uses of high-performance coatings have grown almost in an exponential fashion since they were first introduced in the early 1930s. Hundreds of useful applications have been developed and hundreds more have been tried. For example, inorganic zinc-coated washing machine tubs were once made with reasonably successful results. Inorganic zinc has also been used to coat the fan in a wind tunnel. A 5-coat vinyl lining was once tried on 55-gallon steel drums used to transport wine and on small metal battery boxes. Several thousand feet of small-diameter oil field piping was also once lined with a 5-coat vinyl system to prevent corrosion and paraffin buildup and actually remained in service for several years. Epoxy coatings have been used to coat toilet seats and vinyl coatings have been used on airline food trays. While most of these minor applications were successful, they did not represent the type of application for which the high-performance, corrosion-resistant coatings were designed.

The typical high performance coating is used for large corrosive industrial applications where no other type of material will provide the service or the life which is both required and expected. They are primarily applied on large, in-place structures and are not designed as product finishes (e.g., for appliances, automobiles, etc.). They are used worldwide in most of the industries where corrosion is a factor. Of necessity, the uses described in this chapter will be quite general, since the variety of applications in most industries is too extensive to cover the details of each one. Therefore, coating uses, as outlined in this chapter, will be typical of the industry being discussed.

The Chemical Industry

In this case, the chemical industry will be discussed in very general terms, since it is difficult to find an appropriate division between chemical, petrochemical, refining, or heavy chemical plants. This is because the processes they involve overlap and much of the general equipment used (e.g., piping, pipe racks, tanks, etc.) is common to all of them.

The chemical industry is extremely complicated and involves many different kinds of processes and plants. The extent of the corrosion encountered ranges from none to catastrophic, with no two plants having identical problems. This is due to the variety of processes that are used for producing the same product, as well as the variety of maintenance philosophies employed by the production companies.

For example, salt, or plain sodium chloride, constitutes the greatest tonnage of all chemicals and is produced in three different ways. It is produced from brine wells, where the salt water is actually pumped from underground structures. It is mined in large quantities as a solid in much the same manner as any underground mine is operated. It is also obtained through evaporation of seawater. Each one of these methods involves different corrosive processes; however, there is one corrosive chemical that is common throughout all three, i.e., saturated sodium chloride solutions which are recrystallized for purification.

While operations for the production of salt exist in isolated areas, salt itself is an intricate part of the entire chemical industry. For example, 45% of the salt produced is used to produce caustic soda and chlorine. In turn, caustic soda and chlorine are used throughout the entire chemical industry to produce both organic and inorganic products such as ethylene dichloride, calcium chloride, etc. Wherever salt is part of a chemical process, high-performance coatings are necessary. Inorganic zinc, top-coated with vinyls or epoxies, is the most common coating system used.

One factor that is common throughout the chemical industry is the environment. For the most part, chemical plants are open-work structures and are exposed to the weather and whatever fumes or contaminants that are in the area as byproducts from their own plant. Thus, the largest overall use of coatings is for the protection of these exterior structural steel units, including piping and tankage (Figure 19.1). The actual square footage involved in such steel surfaces cannot be calculated. It may run into several million square feet in any one chemical plant.

Corrosive conditions vary primarily according to the atmospheric conditions that are present. These range from coastal areas (where marine conditions prevail), to the dry plains, as well as to areas of extremely high humidity and high temperature. Chemical plants adjacent to the coast and in highly humid areas generally will have the most corrosive conditions, irrespective of what they produce. Therefore, all exterior surfaces must be protected against such conditions by coatings.

The actual fumes and contaminating conditions that make up the environment of a chemical plant have been, and are going to be within the next few years, greatly reduced. This is due to the environmental cleanup program dictated by federal, state, and local laws. Thus, un-

doubtedly, there will be less corrosion from atmospheric fumes than has existed in the past. However, to offset this decline, the rising cost of maintenance will be an economic factor in the use of better, more sophisticated and long-lasting coatings. It will no longer be practical for a chemical company to maintain a large staff of painters continually cleaning and coating structural steel throughout the plant.

A secondary economic factor is the rising cost of steel and steel installation, which make the cost of protecting steel structures a smaller percentage of the overall cost than ever before. The high-performance coatings, as compared with alkyds and other paints, are so similar in applied cost that the small difference is an insurance premium which the operating company cannot afford to do without.

Certain areas are common to all chemical plants where corrosion is a serious problem. One of these is the cooling tower. Almost all chemical plants, from refineries to the production of fertilizer, have extensive cooling tower facilities. Serious corrosion problems exist not only within the tower area, but also in adjacent areas which are subject to high humidity and spray, salt which accumulates in the cooling tower water, and condensation of moisture and fallout from atmospheric pollution. In fact, one well-known chemical plant uses the area adjacent to a cooling tower as a coating testing area, since it represents the most corrosive conditions in the plant.

Figure 19.2 shows a cooling tower riser that has been coated with inorganic zinc for several years. In this case, the single coat of zinc provided excellent protection. Most areas adjacent to the cooling towers (Figure 19.3), *e.g.*, piping, pipe racks, and similar structures, are coated with inorganic zinc and often topcoated with a high-build vinyl, since weather resistance in the cooling tower area must be the best.

Probably the most common structure in the chemical industry is the pipe rack with the pipe that it carries. They are usually made of structural shapes which are welded or bolted together, therefore creating a myriad of edges, corners, and other sharp protrusions. In addition, the pipelines may be placed close together and joined with flanges and bolts. Raised face flanges leave a space between the flanges so that from a coating standpoint, the pipe rack and pipe represent a substantial coating problem. The inorganic zinc coatings have vastly improved the protection provided these structures. In many cases, inorganic zinc has given adequate protection in a single coat and has provided many years of good corrosion protection (Figure 19.4).

On pipe racks that are in more difficult chemical areas, the addition of vinyl or epoxy topcoats has given excellent protection for many years without appreciable maintenance (Figure 19.5). It is, however, the inorganic base coat which has increased the life of coatings on pipe racks by several times that provided by ordinary organic coatings. The protection of edges, corners, and damaged areas and the elimination of undercutting and underfilm corrosion are the great advantages of the inorganic zinc base coat.

FIGURE 19.1 — Tanks and structures coated with inorganic zinc base coats and high-performance topcoats. (SOURCE: Ameron Corp., Monterey Park, CA.)

Corrosion Prevention by Protective Coatings

FIGURE 19.2 — Excellent coating protection of a cooling tower pipe riser by a single coat of inorganic zinc.

FIGURE 19.3 — Structures and piping adjacent to a cooling tower coated with one coat of post-cured inorganic zinc after seven years of service. The surface discoloration is due to cooling tower fallout.

FIGURE 19.4 — Typical pipe rack during installation coated with inorganic zinc without a topcoat.

FIGURE 19.5 — Refinery pipe and pipe rack coated with inorganic base coat and vinyl topcoats.

Another area common to all chemical plants is stacks. There usually are all types of stacks from different reactors throughout the plant; some may be hot, some may be cold, but usually they are both. Inorganic zinc has proven an excellent coating for stacks, particularly at the difficult operating temperatures of 200 to 500 or 600 F. Where the inorganic zinc is followed by a silicone aluminum, excellent, long-lasting appearance has been provided even at higher temperatures (Figure 19.6).

In at least one case, an inorganic zinc and an inorganic white topcoat were applied to both the interior and exterior of a series of reactor stacks. This combination, entirely inorganic, provided excellent protection to the stacks, both on the interior and the exterior, over a period of several years. This was in spite of a fire that burned one of two towers during the life of the coating. Even after being exposed to the fire, both the inorganic base and topcoat were in excellent condition (Figure 19.7).

Storage tanks pose a common problem throughout the chemical industry. The surfaces are all similar, *i.e.,* broad, flat, and relatively smooth. The atmospheric and weather conditions to which they are generally exposed are usually aggravated by chemical fallout from cooling towers, reactors, and adjacent industry. Therefore, any coating used must have excellent weather resistance as well as resistance to the local fallout conditions.

Inorganic zinc alone, preferably a post-cured or

Typical Uses of High-Performance Coatings

FIGURE 19.6 — Refinery stacks coated with inorganic zinc and silicone aluminum which is an excellent high-temperature system.

FIGURE 19.7 — Chemical reactor stacks coated with inorganic zinc base coat and white inorganic topcoat after several years of exposure and a stack fire which caused no coating failure.

FIGURE 19.8 — Floating roof refinery product tanks coated with one coat of post-cured inorganic zinc after ten years of service in a coastal marine atmosphere.

FIGURE 19.9 — The interior of a floating roof refinery product tank coated with one coat of post-cured inorganic zinc after ten years of service. The small rust streak in the center is due to the rubbing of steel shoes on the vapor seal.

high-zinc water-base product, has given excellent and long-lasting service on both field collection tanks (petroleum production) and refinery tanks where chemical fallout is not severe. Figure 19.8 shows refinery tankage coated with water-base inorganic zinc (one coat) after approximately 10 years in service. The interior of the floating roof tanks which stored distilled petroleum products were in a similar condition after the same period of service, as shown in Figure 19.9. Figure 19.10 shows a tank farm coated with inorganic zinc and white vinyl topcoats which is recommended for tank service to allow dirt wash-off.

Aliphatic polyurethane topcoats also provide excellent weather resistance and good service over inorganic zinc base coats. Properly pigmented polyamide epoxy topcoatings may also be used for such service, as well as alkyds or silicone alkyds, when a compatible organic intermediate coat is used and the chemical fallout in the area is not severe. Figure 19.11 shows bolted field petroleum storage tanks with such systems.

Water treatment facilities are another area common to nearly all chemical plants. The corrosion problems in this case vary from plant to plant, depending on the chemical fume conditions. However, the coatings involved generally would be similar to those used on any water treatment facility throughout the country. Coal tar epoxies, as well as polyamide epoxies, have given good service under water conditions. Where white or light colors are required, the polyamide epoxy is recommended. Properly formulated chlorinated rubber coatings have also given good service. Both the polyamide epoxy and the chlorinated rubber can be used on underwater con-

FIGURE 19.10 — A tank farm newly coated with inorganic zinc and white top coating so that any corrosion that may take place will be instantly evident. (SOURCE: Ameron Corp., Monterey Park, CA.)

FIGURE 19.11 — Bolted field storage tanks for crude petroleum coated with inorganic zinc and white epoxy polyamide topcoats which provides excellent protection to the bolted joints.

crete surfaces. The coal tar epoxies are most practical for submerged steel equipment.

All chemical plants also have waste disposal facilities. In this case, however, corrosion problems are unique to each chemical plant and therefore do not carry any universal coating recommendation. Each plant must be considered individually. The coal tar epoxies have proven very effective in many waste water facilities since they have good chemical as well as water resistance. Floating oil residue, especially aromatic hydrocarbon base, can cause many coating problems and thus should be carefully considered while selecting a coating for waste facilities. It is usually the minor waste products that cause both corrosion and coating problems. Figure 19.12 shows a small oil waste chamber in a refinery. All metal parts were coated with inorganic zinc because of solvent contamination in the waste water.

The chemical industry can essentially be divided into two major categories. The first is the heavy chemical category or the part of the chemical industry that deals primarily with inorganic material. This would include production of materials such as salts, sulfuric acid, soda ash, caustic, chlorine, etc. The second category deals with organic materials. This area includes all organic chemical process plants, petrochemical plants, and refineries. There are many other facets to the chemical industry as well, such as pulp and paper, mining, plating, soap manufacture, etc. In fact, nearly all manufacturing plants use chemical processes in one way or another.

Inorganic Chemical Plants

Salt

The heavy chemical plants are usually baesd on a particular process or basic material. An example of this is the production of salt. Salt is the largest of the basic chemical materials produced from a tonnage standpoint, and it is the basic raw material for many other chemical products. As mentioned previously, it is derived in three ways, *i.e.,* from wells, by mining, and from the evaporation of seawater.

The corrosion involved wherever salt is found is severe, as it easily pervades all processing equipment. The largest equipment involved is handling equipment, *e.g.,* conveyors, pumps, piping, and transportation equipment. Figure 19.13 shows a salt conveyor coated with an inorganic zinc coating. Such equipment accumulates salt dust and crystals in many inaccessible areas, so that without a thoroughly applied and fully resistant coating, corrosion would be severe.

Salt is usually handled in the form of rock salt which invariably contains a high quantity of moisture, so that wherever the salt accumulates, an area of serious corrosion results. Not only does corrosion take place rapidly, but when the surface is blasted for application of coatings, there is sufficient chloride remaining on the surface so that coating failure may be rapid as well.

FIGURE 19.12 — Concrete waste sump in a refinery where all metal surfaces are coated with inorganic zinc to resist solvent attack from floating sludge.

FIGURE 19.13 — A typical conveyor for handling crystallized salt. In this case, inorganic zinc is required for a base coat and may be used alone under certain conditions.

A thorough surface preparation is important for any coating that is to be applied. However, the one material which has helped to overcome most of the salt and chloride corrosion problems is inorganic zinc coatings. They have been used alone and in combination with topcoats to provide corrosion resistance to all types of conveyors and handling equipment, as well as an interior coating for salt barges. In the latter case, it must protect against abrasion from the crystaline salt as well as from corrosion.

One large salt operation is located at Gurerro-Negro in Baja, California and is an extensive seawater evaporation project. Most of the equipment involved has been coated with inorganic zinc and high-build vinyls, all applied over a sandblasted surface. The weather there is warm and humid and the above coating combination has given excellent corrosion protection over a period of several years. Not only is there extensive salt handling equipment at the plant, but there are several barges and additional handling and loading equipment on Cedros Island, which is a few miles at sea from the evaporation lagoon. This is the area where the very large salt-handling ships load and transport the salt to Japan. These ships are about the size of a medium-sized super tanker with complex conveyors for loading and unloading salt. These are

also protected with inorganic zinc and vinyl or epoxy topcoats. Many salt barges are coated on the interior with zinc and polyamine epoxy topcoats which provide a tough, abrasion-resistant as well as a corrosion-resistant coating.

Caustic and Chlorine

The production of caustic soda and chlorine must be considered together since they are primarily produced in the same process by the electrolysis of salt or salt solutions. They are often produced in captive plants, although the majority is produced in relatively large production units which are close to either a source of salt or a low-cost power supply.

The process is a relatively simple one of breaking down the salt molecule into a two-component part, chlorine and sodium, by electrolysis. In one process, molten sodium chloride is used and the electrolysis produces chlorine gas and sodium as a metal. In another process, mercury is used as the cathode. Chlorine gas goes off at the anode and the metallic sodium is dissolved in the mercury amalgam is reacted with pure water, and liquid sodium hydroxide is produced. When the sodium is dissolved out of the mercury, the mercury goes back into the process again. Approximately 45% of all of the salt produced in the U.S. is used for the manufacture of caustic soda and chlorine. Both products are produced worldwide by essentially the same process.

The chlorine, once produced, goes into a myriad of other industries and is used either as an intermediate or directly, as in the sanitation industry for the purification of water or sewage and in disinfecting swimming pools. More than 15% of the sodium hydroxide produced is used by the rayon industry. Thus, it can be seen that the production of caustic and chloride involves a basic process on which many industries throughout the world depend.

Corrosion in caustic soda chlorine plants can be quite severe. The concentrated caustic soda disintegrates any oil-containing coating, while the chlorine gas is a difficult material from a corrosion standpoint, readily forming ferric chloride. The corrosion problems are approximately the same, irrespective of the processes used to develop the two materials, and high-performance coatings are needed to provide adequate protection. All steel on the interior of such plants requires maximum protection, since even small quantities of chlorine will cause corrosion problems on any bare steel.

Floors, however, are the primary problem, due to the spillage and drip of chlorinated alkaline brine. Epoxy floor toppings have done a good job in protecting floor areas around and under the caustic/chlorine cells. Caustic/chlorine plants are often two-level structures, so that a high-performance floor topping is required to prevent seepage and corrosion from the second level. Coating systems used for structural steel should be based on inorganic zinc followed by vinyl coatings, which have performed very well due to their excellent resistance to the highly reactive chlorine gas.

One of the most serious corrosion problems, particularly with the flowing mercury cells, involves the spent sodium chloride solution which goes through the cells and from which some of the sodium chloride has been removed by the electrolysis process. As the sodium chloride solution flows across the mercury, some chlorine is dissolved in the salt solution with the formation of hypochlorite. This waste sodium chloride solution is an extremely corrosive product, and it may be recycled in the process by the addition of more sodium chloride, so that the solution circulates continually through the plant. The storage tanks containing this material require the best lining material available, usually a heavy inert sheet lining such as plasticized vinyl chloride or oxidation-resistant synthetic rubber.

Fertilizer

The production of fertilizer is another extensive segment of the industry. One of the primary processes involved is the formation of useable phosphates from the natural phosphate rock. Phosphoric acid is generally the first byproduct and is produced from phosphate rock in two ways: (1) the wet process, which uses sulfuric acid to dissolve the rock, and (2) the dry process, which produces the elemental phosphorous through the use of an electric furnace. Very pure phosphoric acid is produced by the latter method. Agricultural grade acid is produced by the wet process, which accounts for approximately 80% of all the phosphoric acid produced. Fifty percent of the sulfuric acid that is produced is used in the production of phosphoric acid. This combination of acids creates highly corrosive conditions, so that all areas of these plants require high-performance coatings.

The primary use for phosphoric acid is in fertilizer. This may be in the form of calcium phosphate (which is the super phosphate in fertilizer terms), ammonium phosphate (which is a crystalline chemical), or as crude phosphoric acid.

Even the warehouses storing phosphate fertilizers require good coatings. Super phosphate is made merely by the use of hot phosphoric acid on limestone or similar rock, which is then allowed to digest in large buildings. The super phosphate dust is corrosive and all of the structural steel, including the large conveyors, require good protection. A base coat of inorganic zinc followed by high-build vinyl or epoxy coatings has given good results.

All of the high-performance coatings have been used in connection with the production of phosphate fertilizers. There are many plants located in the phosphate sand country of Florida as well as large installations in Mexico where high-performance coatings have been used for 5 to 10 years.

Another fertilizer process is one which oxidizes the nitrogen in air to form ammonia and nitric acid. This is generally a high-temperature operation using steam, natural gas, and air. The majority of such plants are similar to any other chemical plant, with the structural steel surfaces subject to weather and local industrial conditions. There may be some nitric acid and nitrous oxide fumes present, which can aggravate corrosion.

Ammonium nitrate is one of the end products of such plants. This is a crystalline material with some tendency to absorb moisture and can break down to ammonia and nitric acid. Another end product is calcium nitrate, which is also hygroscopic. Dusts or other accumulations of these products can create corrosive conditions. High-performance coatings have proven entirely satisfactory for many areas in such plants.

Sulfur

Sulfur is produced in several ways: mining natural sulfur, oxidation of H_2S, and the Frasch process. Seventy-four percent of the native sulfur is produced by the Frasch process, which consists of drilling a well into a sulfur dome and injecting steam and compressed air into the structure. The steam melts the sulfur, with the compressed air blowing the molten sulfur to the surface through a pipe which is inside of the steam injection pipe. Sulfur arrives at the surface in a molten state and is piped into cooling areas where it solidifies. The combination of sulfur, compressed air, and steam results in a highly corrosive operation. However, the corrosion is primarily internal, and therefore does not involve exterior coatings.

Sulfur as an element is not necessarily corrosive; however, when oxidized it becomes extremely corrosive. Many of the sulfur operations are also located in high humidity and marine areas, so that most of the structures require high-performance coatings. Inorganic zinc, followed by either vinyls or epoxies, do very well in this instance. However, the process piping is usually at temperatures above the capability of organic coatings. One of the best known natural sulfur areas is the Freeport sulfur structure off Grand Isle, Louisiana. Post-cured inorganic zinc was used on the original structure and has provided many years of good protection in the marine exposure (Figure 19.14).

The other primary method of sulfur production is by the oxidation of hydrogen sulfide, primarily extracted from natural gas. In this case, the sulfur must be removed for the gas to be useable so that sulfur becomes a byproduct of natural gas production. There are a number of units throughout the United States, as well as in other areas of the world, where hydrogen sulfide is a part of the natural gas produced. Mexico has several large units in the new fields located in southeastern Mexico. Since hydrogen sulfide is a difficult and corrosive material, all of the equipment and piping involved in the production and separation of hydrogen sulfide from natural gas can be seriously corroded. Figure 19.15* shows the type of corrosion which is possible even on the exterior of unprotected equipment where hydrogen sulfide is involved.

The production of sulfuric acid is a process that varies from plant to plant. However, much of the acid is produced either from hydrogen sulfide extracted from natural gas, or from elemental sulfur which is burned to sulfur dioxide. Since either of these processes is corrosive, most plants are protected by high-performance coatings wherever temperatures allow their use. While much of the processing equipment is subjected to high temperatures, high-performance coatings protect most of the structural steel surrounding such equipment.

---

*See color insert.

Leaks and spills are always a problem in acid plants. The production of sulfuric acid is particularly difficult because of the highly oxidizing character of the concentrated acid. Where spills or leaks are possible, such as in the areas of pumps, acidproof brick floors are installed, using either furan or silicate cements as a mortar for the brick. These are used on pump bases, floors, trenches, and similar areas throughout the plant (Figure 19.16). Extreme corrosion is possible in many areas, and the best possible protection is required for both steel and concrete surfaces.

The production of hydrochloric acid creates extremely corrosive conditions, since chloride is in acid form. Much of the hydrochloric acid is produced as a byproduct from other chemical reactions and is therefore part of another chemical system. Sixty-five percent of the hydrochloric acid plants are captive, where the acid is used in the formation of other chemicals. About 30% of the hydrochloric acid is used in steel pickling.

Hydrochloric acid is a volatile acid, which means that it vaporizes easily. The vapor or fumes are extremely corrosive, forming iron chlorides wherever steel is ex-

posed to the fumes. Because of the volatility, coatings must be highly resistant to the acid and to the fumes, preventing their penetration to the steel substrate. This not only means that the coating must be resistant, but it must be applied in the best possible manner as well. Both heavy-bodied vinyl and epoxy coatings have provided good protection to structural steel in fume areas. Lesser coatings would not be suggested for this service.

There are several industries where hydrochloric acid corrosion is a serious problem. The production of titanium dioxide is one of these. Hydrochloric acid is used to extract the titanium from the ore, titanium tetrachloride being one of the intermediate products. This reacts very readily with any moisture, recreating hydrochloric acid, so that these processing plants require the best in high-performance coatings in order to provide the protection required for structural steel in the immediate process area, as well as wherever hydrochloric acid fumes may contact equipment or strutures. These are primarily fume conditions and are very similar to those existing in a hydrochloric acid plant.

The hydrochloric acid pickling of steel is another critical area from a coating standpoint. All areas in and around such a process require high-performance coatings for proper protection of the substrate. Heavy-bodied vinyl and epoxy coatings have provided good service in these areas when properly applied.

The production of silicones is also a potentially corrosive process, since silicone tetrachloride is one of the intermediate chemicals. This decomposes into silica and hydrochloric acid, creating conditions within the plant and surrounding area similar to those of a hydrochloric acid plant. Heavy-bodied vinyl coatings have provided good service in plants of this type.

The production of chlorine bleach is also a corrosive process which involves the reaction of sodium hydroxide with gaseous chlorine. The resulting sodium hypochlorite solution or calcium hypochlorite, if the chlorine is reacted with calcium hydroxide, is an extremely corrosive material in itself. The reason for its effectiveness as a bleaching agent is that it breaks down into oxygen and sodium chloride. The corrosive elements then are chloride, oxygen, and sodium chloride. The solution is unstable and tends to disintegrate into oxygen and sodium chloride with time and when in contact with many oxidizable materials.

Bleach plants are subject to heavy corrosion, both from fumes and wherever the liquid may be in contact with metal. High-performance coatings have been used in bleach plants for many years and, where they have been properly applied, have effectively protected the structural steel, tank exteriors, and equipment from corrosion. The coatings used have been primarily vinyls, due to the highly oxidizing conditions that prevail. Many of these plants are relatively small and much of the bleach is produced on a local basis. Plants which use bleach in their processes, such as paper plants, may produce their own as a captive operation.

All of the previously discussed plants are candidates for chemical-resistant coatings. The ones which are most

FIGURE 9.14 — A section of an offshore sulfur structure which acts as a production working platform as well as a base for steam, air, and sulfur piping.

FIGURE 19.16 — A typical pump installation handling acid solutions, showing the use of acid proof brick and dark furan mortar.

Corrosion Prevention by Protective Coatings

critical and have the most corrosion are those which involve halogen compounds. These are fluorine and fluorides, chlorine, hydrochloric acid, hypochlorites, chlorides, bromine and bromides, and iodine and iodides. Wherever these materials are found, either in inorganic or organic processes, high-performance coatings are required for the protection of steel and concrete surfaces.

Organic Chemical Plants

Organic chemical plants are usually larger than most of the inorganic chemical plants and the processes they involve are quite different (Figure 19.17). The corrosion conditions in organic chemical plants range from innocuous to extremely severe, depending on the products produced. Acetic acid and derivative products can be highly corrosive. Plants which use acetic anhydride and other anhydrides can be difficult from a coating standpoint, since acetic acid is both a strong organic acid as well as a good solvent for many other organic materials. Thus, many coatings are penetrated by the solvent action and the acid initiates corrosion as it contacts the substrate. Strongly cross-linked coatings are best for the above circumstances.

On the other hand, the production of other solvent materials (*e.g.,* alcohols) can be relatively noncorrosive. Once they are distilled and in their final form, many solvents are essentially noncorrosive and are handled much like gasoline or other petroleum products. In this case, the protective coating required is determined by the local atmospheric conditions, with humid, marine, and industrial environments requiring the best protection, while rural environments are satisfactorily maintained with less protection. An inorganic zinc base coat, followed by vinyl, acrylic, or polyurethane topcoats could provide years of corrosion protection under the most severe conditions. Any proposed coating system must be tested and proven under the specific set of chemical reactions required during production.

Refineries

Refineries are included in the organic category since they produce most of the feedstock for the other organic chemical plants. While a typical refinery unit (Figure 19.18) appears to be similar in structure to most other chemical plants, hot surfaces, some of which exceed the limits of organic protective coatings, are more characteristic of refineries.

Refineries begin with crude oil as it comes from the production fields and break it down in continuous operating plants (making use of high temperatures, high pressures, and catalysts) into many different organic, primarily liquid, products. These range from Bunker C oil to diesel oil, gasoline, jet fuel, solvents, waxes, and asphalt. Many of the processes are self-contained so that the only exposed product is the beginning liquid (crude oil) and the end products (gasoline, solvents). Thus, refinery equipment is primarily composed of tankage, pipe lines, pumps, reactors, distillation towers, and structural supports which require exterior coatings to protect the structure, piping, and equipment from the atmosphere of the particular refinery. Refineries that are

FIGURE 19.17 — A typical organic chemical plant, showing structural steel, reactors, stacks, piping tanks, and similar surfaces which require good protection.

FIGURE 19.18 — A typical oil refinery unit with distillation towers and stack coated with post-cured inorganic zinc.

located near the coast and those located in highly humid or industrial areas, usually require high-performance coatings, *i.e.,* inorganic zinc, vinyls, epoxies, or urethanes.

Areas of high temperature are a problem for organic coatings. However, inorganic zinc coatings have given excellent protection at temperatures of 600 F (315 C) or somewhat higher (Figure 19.19). Where combined with aluminum-pigmented silicone topcoats for 800 F (426 C) to 1000 F (538 C) service, such coating systems have given good results. Inorganic zinc base coats and in-

FIGURE 19.19 — A refinery stack coated with post-cured inorganic zinc, showing no corrosion after several years of exposure to humid marine conditions.

product materials. They also may contain a heavy layer of basic sediment, particularly in crude storage tanks. If hydrogen sulfide is present, scale from the tank top can fall to the bottom and create a severe pitting problem. The heavy-duty linings (*e.g.,* reinforced polyesters) are required not only because of the corrosion, but also due to the abrasion inherent in flow areas of tank bottoms, particularly when basic sediment and other deposits must be removed.

The production of petroleum, while it differs from the refining process, creates similar problems, particularly in the area of tankage. While tank bottom problems are similar, they may be even more severe due to the salt water that is mixed with the produced oil. The accumulation of basic sediment also can be greater, and much of the oil comes directly from the well at elevated temperatures. Where hydrogen sulfide is present, excessive corrosion of tank tops can occur, which in turn creates bottom corrosion as the iron sulfide drops to the bottom. Petroleum wax and gypsum deposits can also aggravate the corrosion in oil production tanks.

Nevertheless, the same high-performance coatings have provided good protection. Coal tar epoxy coatings have given many years of good service on production tanks in the hydrogen sulfide crude areas of Mexico (Figure 19.21). Figure 19.22 shows the vent on a similar storage tank in the same production area completely corroded in two years from hydrogen sulfide. The coal tar

organic topcoats have given very good service at elevated temperatures.

Inorganic topcoats are usually based on hydrolyzed ethyl silicate combined with nonmetallic but reactive pigments which create a hard silicate base product with excellent weather and temperature resistance.

Refinery storage tank interiors, since they usually hold products that are at ambient temperatures and atmospheric pressure, may require coating both to prevent corrosion of the metal and protect the product from contamination. Severe corrosion can take place in closed roof as well as floating roof tanks, depending on the nature of the environment (Figure 19.20). One of the first applications of inorganic zinc that was tried in the U.S. involved floating roof tank walls that were completely pitted with ¼-in. pits. While the first trial was not perfect, it prevented additional corrosion for several years.

Refinery products such as gasoline, lubricating oil, and most fuel oils and jet fuels, are noncorrosive in themselves. However, the combination of these materials with water, seawater, or marine or industrial atmospheres can be extremely corrosive. Thus, particularly where product contamination is also a factor, many refinery tanks are coated with inorganic zinc by itself; overcoated with organic coatings such as vinyls, epoxies, or urethanes; or coated using a good solvent-resistant organic system without the inorganic zinc base coat.

Tank bottoms usually pose a problem which requires use of heavy-duty materials such as glass-reinforced polyester and coal tar and coal tar epoxy. Tanks usually have a water layer in the bottom which contains dissolved

FIGURE 19.20 — The interior of a typical floating roof tank coated with post-cured inorganic zinc after storing gasoline for several years.

FIGURE 19.21 — A petroleum production tank coated on the interior with coal tar epoxy (two coats, 16 mils) and on the exterior with three coats of epoxy polyamide.

FIGURE 19.22 — A production tank vent completely corroded from hydrogen sulfide in two years.

FIGURE 19.23 — Metal gas separators at the production site exposed to a tropical atmosphere and high quantities of H₂S protected by an inorganic zinc base coat with epoxy topcoat.

FIGURE 19.24 — High-pressure gas holders at Tierra del Fuego protected with one coat of inorganic zinc.

epoxy on the interior of the tank gave excellent protection over the same period of time.

The production of natural gas usually accompanies petroleum production. The gas must be separated from oil and water and compressed for pipeline transportation. Hydrogen sulfide and other gas impurities must be removed and the gas recompressed for long-line transportation. All of these operations require better than average coatings, depending on the location of the production field and the hydrogen sulfide concentration in the natural gas.

The transportation of natural gas is most often handled by pipe, and high-quality coatings are required on the exterior to prevent corrosion and blowout due to the high-pressure gas. Hot-applied coal tar reinforced with fiber-glass wraps, vinyl tape with a butyl rubber adhesive, and epoxy powder coatings have been used. The greatest mileage of gas piping has been coated with glass-reinforced coal tar in multiple layers. Figures 19.23

through 19.27 illustrate coating uses in natural gas production and distribution.

The interior of gas transmission piping also must be lined with high-performance coatings. The problem in this case is one of friction from the flow of gas through the line. There is little corrosion on the pipe interior, unless the gas contains considerable H₂S, since the gas is "bone dry." Any water vapor is just that many more cubic feet of product to force through the pipe. The lining material is a specialized catalyzed epoxy applied in one coat, and it reduces the friction of the gas in the pipe to the degree that makes the lining both practical and economical.

Other Chemicals

In addition to petroleum and refinery products, there are at least 40 organic chemicals which are produced in excess of 1 billion lbs./yr. Ethylene is produced in by far the largest volume, and it reaches 20 billion lbs./yr. About 40th on the list is carbon tetrachloride at 1 billion lbs./yr. Other large-volume materials are propane, ben-

FIGURE 19.25 — Gas compression units for use on offshore platforms in the Gulf of Mexico protected with an inorganic zinc coating with epoxy topcoats.

FIGURE 19.27 — Gas transmission piping coated with hot-applied coal tar pitch and glass reinforcing. The white color is whitewash for heat protection for the coal tar. (SOURCE: Ameron Corp., Monterey Park, CA.)

FIGURE 19.26 — An expandable gas holder which usually is at a distribution point for natural gas protected by one coat of inorganic zinc, even in the water seal.

zene, ethylene dichloride, propylene, toluene, ethane, methanol, formaldehyde, vinyl chloride, butadiene, ethyl alcohol, and glycerine.

A characteristic of the organic chemical series, like the refinery products, is that many, in themselves, are not very corrosive, since many of them are produced and handled as gases in a completely closed atmosphere. Any fumes that exist come from leakage in valves or piping or

from stacks. The high-volume products that can be most corrosive are those organic compounds containing halogen atoms (e.g., fluorine, chlorine, bromine, and iodine). The end products may be ethylene dichloride or any of the other organic chlorides, including vinyl chloride. There is also ethylene dibromide and other ogranic bromides. Ethylene dibromide is the largest product in this group, since it is an additive to gasoline for the stabilization of tetraethyl lead. There are also a number of organic iodides. The production of all of these materials involves corrosive conditions.

The organic chemical plants are primarily open-work structures where the structures, processing units, piping, and storage tanks are all subject to the atmosphere and weather. Many of these plants are located in areas of high humidity and are subject to atmospheric fumes from other processes and other plants in the area. These organic chemical plants usually have extensive tank farms and miles of piping and pipe racks. All of these installations are costly to maintain since, in many instances, millions of square feet of surface need to be coated. High-performance coatings are the only practical answer to the preservation and protection of these structures. The most successful coating systems, and the ones which have been used extensively throughout the world, arc base coats of inorganic zinc followed by topcoats such vinyls, epoxies, urethanes, chlorinated rubbers, or other chemical-resistant coatings that may best meet the local corrosive conditions.

Tankage is a critical area involved in the production of organic chemicals. While there are many solid organics, there are also many liquid products, such as ketones, various aromatic materials, alcohols, aldehydes, esters, and chlorinated solvents. Many of the containers for these materials require coating for the protection of the organic liquid or to prevent the tanks from deterioration. There are many products used for such purposes, including vinyls, amine-cured epoxies, polyurethanes, modified epoxies (*e.g.*, phenolic epoxies, medium-temperature cured epoxies, and high-bake epoxies), phenolics, and inorganic zinc coatings. The coating depends on the organic product and its use.

Figure 19.28 shows the interior of an industrial alcohol tank being lined with a vinyl coating system. Where ethyl alcohol is to be used as a food or is not contaminated or denatured with strong solvent, vinyl chloride-acetate copolymers make good linings, since they are insoluble as well as noncontaminating to alcohols.

FIGURE 19.28 — The interior of an industrial alcohol tank being lined with a vinyl coating.

Pulp and Paper

Pulp and paper production is considered part of the chemical industry because many of the problem areas are similar, *e.g.*, structural steel, tank exteriors, weather exposure, quantities of water, chemical fumes, and fallout. It is a large industry represented by over 1000 individual plants, and consists of an inherently very corrosive process. To illustrate this, consider that there were about 103 to 104 million tons of paper produced in the western world in 1971. The cost attributed to corrosion throughout the plants producing this amount of paper amounted to 320 million dollars for the same year. In other words, over $3/ton of paper produced in 1971 was directly attributed to corrosion.

There are basically two types of areas to consider in a paper plant. Atmospheric areas are areas of the plant where coatings are subject to atmospheric corrosion, humidity, and some slop and spillage of water, usually resulting in wet and dry conditions. Immersion areas consist of tank and equipment interiors which are subject to continuous immersion in water, contaminated water, chemicals, etc.

Atmospheric

The entire paper plant, including all structural steel, tank exteriors, etc., is subject to atmospheric corrosion. A paper plant is subject to more severe corrosion than many other plants due to the chemicals involved in the manufacture of paper and the extremely large quantities of water that are required to wash the chemicals out of the pulp.

The chemicals involved are generally aggressive ones, particularly those used in the bleach area, *e.g.*, chlorine, sodium chlorate, chlorine dioxide, and sodium hypochlorite. Many paper plants actually manufacture these materials for their own use so that essentially there is an internal chemical plant producing the extremely corrosive chlorine-containing oxidizing chemicals, as well as using the chemicals in the pulp bleaching operation. The bleach manufacture and the pulp bleaching operation creates a fume problem, with chlorine, sodium chloride, and chlorine dioxide reacting with any exposed steel. Corrosion produced by chlorine extends into the pulp washing areas where extremely wet conditions exist. This compounds the severity of corrosion on equipment, piping, pumps, pump bases, and structural steel.

Because of the existence of chlorine, these areas require excellent surface preparation and well-applied high-performance coatings. For maximum protection, the coatings should be applied as the plant is built, using an inorganic zinc base coat, followed by heavy-duty coatings such as heavy-bodied vinyls, epoxies, urethanes, or the most resistant chlorinated rubbers. This area of a paper mill cannot tolerate poor surface preparation, poor application, or coatings which are less than the best.

The primary recovery area is another area where corrosion is a problem. It consists of large furnaces which burn the black liquor from the primary cooking operation in order to recover the basic chemicals involved, *i.e.*, sodium hydroxide, sodium sulfide, and sodium sulfate. Much of the difficulty in this area comes from the fly ash from the burning process. This type of fallout can occur throughout the paper plant, but is heaviest in the local recovery area. The fly ash consists of dry corrosive dusts, usually quite alkaline, which contain all of the previously mentioned chemicals, including carbon and various chlorides. Wherever these dusts accumulate, and then collect moisture from rain or high humidity, corrosion is severe.

Slaked lime also is used as part of the recovery process, which results in large quantities of calcium carbonate and calcium sulfate. These materials are recycled and are calcined in the recovery area, creating large quantities of calcium oxide dust which is highly alkaline. Wherever this accumulates and collects moisture, corrosive condition exist.

The third area which creates a serious atmospheric

problem is the blowdown area. This is where the pulp from the primary cooker is blown down into either a blowpit or blow tank, with the resulting steam spreading throughout that area of the plant. Steam carries many different types of organic solids, but also contains sodium hydroxide and sodium sulfide. Sodium hydroxide is extremely alkaline, while the sodium sulfate creates conditions of sulfide corrosion throughout the plant. Sulfides and iron react rapidly, followed by oxidation from moist air with regeneration of the sulfide as hydrogen sulfide. Thus, high-performance coatings are essential in these primary chemical areas. Inorganic zinc base coats, followed by vinyl, epoxy, or possibly urethane topcoats have been proven under these conditions.

In the actual paper-making area, water and humidity are the primary corrosive agents. The large amount of water and steam coming from the machines can create serious corrosion problems which cannot be tolerated because of the possible contamination of the paper. A drop of rusty condensation or a bit of scale from a rusty beam will cause the paper to be rejected. With the plants on 24-hour, 7-day operation, rejects and stoppages are costly. Therefore, tighly adherent water-resistant coatings (*e.g.,* vinyls and epoxies) are required.

The above conditions primarily exist in the kraft or sulfate-type paper mill. In the sulfite-type paper mill, the corrosive conditions are similar, but generally even worse due to the acidic type of digesting liquor they use. The sulfide cooking liquor is made in the paper mill by burning sulfur to form sulfur dioxide (SO_2). This is then reacted with calcium and magnesium carbonate, and calcium or magnesium bisulfite $[Ca(HSO_3)]_2$ is obtained. The fumes from the sulfite manufacture and from the blow pits create corrosive conditions wherever they are in contact with steel. The sulfite liquors are also corrosive to concrete.

Under almost any atmospheric condition, the use of an inorganic zinc coating underneath an organic coating will extend the life of the topcoat by several times. This is true even when the atmospheric corrosion is caused by an acidic or alkaline condition. This is because the inorganic zinc coating ties chemically to the steel surface so that there is no undercutting of the organic film, as there would be if it were applied directly to the steel surface. This is particularly true where sulfides are present, since they are particularly insidious from a corrosion standpoint. Inorganic base coats have more than proven their worth in paper mills.

One of the more recent problems encountered in paper mills involves environmental controls which have necessitated additional ducts, scrubbers, and both air and water treatment equipment. Many of these ducts and scrubbers are made from reinforced plastics, as is the pipe used for water, chemical, and pulp transfer. Reinforced plastics aid in overcoming the corrosion problem. On the other hand, all structural supports for the reinforced plastic equipment and all adjacent buildings require the best corrosion protection that coatings can provide.

A paper plant that is properly coated during its initial construction has taken a major step in reducing corrosion and maintenance. In many areas of a paper mill, particularly from a maintenance standpoint, sandblasting or an equivalent type of surface preparation may not be possible, since sand contamination in any of the machine areas or pulping processes could be extremely damaging. Therefore, much of the maintenance in a paper mill must be done by power sanding, power brushing, or the use of needle guns and similar means. Also, much of this work must be done during an annual overhaul period when the entire plant is down. Thus, equipment and time constraints, plus continually damp conditions make coating repair difficult in a paper mill.

During maintenance, where relatively small areas of bare metal are involved, it is recommended that these areas be primed using an organic zinc-rich coating over the bare steel areas and feathering it out over the old coating. While not as effective from an undercutting standpoint as inorganic zinc coatings, the zinc in an organic zinc-rich coating does provide a measure of cathodic protection. The proper topcoats can then be applied.

Where coatings must be applied under difficult maintenance conditions, there is an important step in the application process, following surface preparation, which can make the difference between success and failure. Once the surface has been thoroughly ground or prepared with a needle gun to remove all of the old coating and as much of the corrosion product as possible, it is recommended that, if possible, these surfaces be heated by the use of large propane torches to a temperature well above ambient. The temperature should not be so great as to damage the existing coating, but should be sufficiently high to drive all moisture out of the crevices, pits, corners, and similar areas. At the same time that the moisture is being driven out, any sulfides that may be present are also oxidized and the resulting hydrogen sulfide driven off.

Once the heating is carried out and the surface is dry and hot to the touch, the primer should be brushed on the surface. It should be thoroughly brushed into all pits, crevices, and similar areas to ensure full coverage. The heat from the substrate will lower the viscosity of the primer, so that with brushing, it will wet-out and penetrate much more thoroughly and quickly than on a cold application. Heat also has the effect of drying the primer rapidly so that topcoats can be applied rapidly as well. This, again, is insurance against intercoat contamination, which is a constant problem in paper mill applications.

While many coatings (epoxies, in particular) have given excellent service in paper mills, the fast-drying vinyl or chemical-resistant chlorinated rubber coatings have a decided advantage. This is because the continuous source of contamination throughout the plant from either calcium oxide, sodium hydroxide, sodium sulfide, sodium sulfate, or a number of different organic compounds combined with these materials, makes coating maintenance extremely difficult. With fast-drying coatings, second and third coats can be applied within even minutes following the first one. Since they usually dry by solvent evaporation, one coat may follow the first, even though the first is not thoroughly dry.

Immersion

Use of coatings in immersion areas in a paper mill should be approached with caution. These areas combine strong chemicals with relatively high temperatures, so that, for most of these areas, spray-applied, relatively thin coatings are not adequate.

Many types of linings have been tried under these conditions, and most of them have met with little success. Thus, wherever linings are required, the problem should be thoroughly studied, analyzing all corrosive conditions, temperatures, concentrations, and times involved before attempting any solution. Most paper plants use ducts and vessels of stainless steel where internal corrosion is a problem. Where even these are not practical, as in digesters, blowdown pits, and bleaching operations, the most successful and longest lasting answer is found in the use of membranes and brick or tile linings using furan or polyester-type cements. The primary materials involved are hard-burned red shale brick or tile, or carbon brick laid up with either furan or polyester cement. The furan may be used for most areas, except where oxidizing conditions are involved. Polyester cements are used where oxidizing bleaches are found. Such systems have been used for years in paper plants, and, where properly installed, have provided excellent resistance and service.

There are some areas where thin lining materials can be used. These are primarily for water, water treatment and water reclamation clarifiers, settling basins, aeration tanks, and other standard water processing equipment. Vinyls, epoxies, and coal tar epoxies have provided the required protection.

The Mining Industry

The mining industry generally has not used high-performance coatings in the past. This is because copper, zinc, nickel, and other metals have been roasted or smelted and converted to metals by hot molten reduction processes. These smelters were similar to those used for steel production, i.e., massive steel or cast iron units which generally received no coating whatsoever. The mining of more complex and lower grade ores has now become essential, and much of the dry processing of the ores is neither economical nor practical. This has resulted in the chemical processing of the lower grade ores, often by acid leaching. In the case of nickel, it is also leached by an ammoniacal ammonium carbonate process. The refining may then be by precipitation or electrolysis. Of course, the wet processing of ores is much more corrosive than the dry process, and therefore high-performance coatings are needed in the plants to protect them against serious corrosion.

Copper, zinc, and uranium generally use a sulfuric acid leach to extract the metal from the gangue or waste rock. This acid is relatively dilute, in the 5 to 15% range. In the case of copper, it is generally at ambient temperature. In the case of uranium, the temperature often is raised to 150 F (65 C) or slightly more during the processing. These acid solutions are extremely corrosive, and wherever they come in contact with either metal or concrete, rapid corrosion takes place. The solutions of copper

and nickel are even more corrosive, since even a neutral solution of these salts will displace iron wherever the solution comes in contact with it.

There are also a number of other methods of leaching these ores. Ferric chloride is one method used for copper, as well as nitric acid leaching. Ammoniacal ammonium carbonate is another leach that is used. The copper may be recovered from the pregnant solutions by electrowinning and solvent extraction. Irrespective of the leaching method used, the process involves corrosive solutions.

One quick test for copper in an ore is to dissolve it in almost any acid and then place a shiny nail in the liquid. If there is any copper in the solution, it will immediately precipitate out on the iron surface, creating a copper-red surface. In order for this to take place, an equivalent amount of iron goes into the solution. Therefore, any solution of copper or nickel can be quite corrosive to iron or steel.

This process is used in a practical way at most copper mines. The waste water from the mine is passed over a bed of scrap iron or steel, the copper precipitates, and the steel goes into solution. Thus, even small traces of copper can be removed. This is called *cement copper*, since it is a very fine paste. Many of the salts of copper, zinc, uranium, and nickel are acid salts. This means that when the copper, zinc, or other metal salts are put in solution, the solution is automatically on the acid side because these salts are a reaction of a relatively weak base, which would be the metal oxide, and a strong acid such as sulfuric.

Taking this into consideration, wherever there is a liquid processed or wherever liquid exists in the refining of these metals, corrosion is inherently present. All concrete and steel surfaces thus require high-performance coatings to protect them against rapid disintegration.

One of the largest and most critical areas to be protected in all of the extraction processes is the leaching tanks. The early copper leach tanks were lined with heavy asphalt linings. These were blocks of asphalt laid like tile or brick and fused together to form a lining. The bottom lining of the tanks was often several inches thick. Abrasion is one of the key factors here, because of both the dumping of the ore into the tanks, and, more importantly, the removal of the insoluble gangue after all of the metal has been extracted.

In some of the uranium extraction processes, large steel rubber-lined tanks have been used. More recently, some of the larger copper leaching facilities have been lined with plasticized vinyl sheet placed into the concrete structure as it was poured. The sheets of vinyl plastic are then heat welded to themselves to form a continuous membrane throughout the leaching tank. The physical attachment of the plastic sheet to the concrete through the use of tee extensions into the poured concrete structure is a decided advantage over coatings or sheet applied with adhesive alone. Good service has resulted from the use of linings of this type.

In addition to the leach tanks, numerous flotation units, precipitators, concentrators, washers, clarifiers, and similar equipment are used, many of which combine both concrete and steel construction. Each requires a heavy-duty lining or coating. Throughout the processing of the solutions, all coatings and linings must be applied with extreme care, since any pinholes or holidays will cause rapid failure of the metal or concrete. In general, multiple-coat, heavy, thick coatings such as high-build epoxies are required, except in areas where rubber or plastic sheet linings are necessary.

Many pipelines for metal solutions or leaching liquid are lined using vinyl plastisol. This is due to the continued abrasion throughout the process, as well as the need for a thick, heavy lining to fully protect the underlying surface. Thick, heavy coatings or linings provide the safety factor that is required. Wherever concrete tanks or structures are part of the process equipment, plastic sheet linings are a good recommendation, as they are continuous, pinhole-free, tough, and abrasion-resistant. Most importantly, they have the strength and extensibility to bridge the checks and small cracks that are always present in concrete.

In the case of liquid processed nickel, copper, or zinc, the final refining process is usually electrolytic. This is done by plating out the metal from the solution onto large electrodes, which may then be melted and cast into ingots or re-refined by the electrolytic process to obtain high-purity metal. The electrolytic refining area is highly corrosive inasmuch as acid solutions are again used and temperatures of the electrolytic solutions are elevated (120 to 160 F). For the most part, these processes take place in an enclosed building which creates fume problems. In this case, high-performance coatings are a must for protecting structural steel.

Sheet lead has been and is used as a lining material for electrolytic cells. This is extremely costly from a material standpoint, as well as from the standpoint of labor used in lead burning and placement of the lead in the cells. Nevertheless, as it was one of the secondary metals refined from the basic ores, it was readily available and resistant to the acid solutions. It had a serious drawback in that during the electrolytic process, part of the crude copper or zinc electrodes could fall and hit the bottom as they were gradually dissolved. If electrical contact between the electrode and the fallen piece was maintained, a hole was immediately burned through the lead. The cells in more recent installations have been lined with plasticized vinyl sheet. In this case, the burn-through does not occur since the heavy plastic liner acts as an insulating membrane.

Many auxiliary areas also exist, such as floors underneath equipment, tanks, and adjacent structural steel which also require coatings. Inasmuch as the floors are subject to leakage and spills, heavy-duty abrasion-resistant coatings must be used. Troweled epoxy floor toppings are recommended. On the structural steel, inorganic zinc and a vinyl or epoxy maintenance coating has proven to be satisfactory. In processing areas in cold climates where all processes are enclosed, humidity, steam, and fumes concentrate so that a high-performance coating system is required for all surfaces.

In many cases, mining operations are considered temporary, since the equipment, tankage, and structures will only be used for a limited period of time. However, where the ore bodies are large and years of production contemplated, the proper coatings, linings, and floor toppings are merely good insurance for a trouble-free operation for a number of years. Again, the cost of protection is a small part of the replacement value of costly equipment and structures.

Coal Mining

Coal mining is probably one of the largest segments of the mining industry. A study of the coal industry from a corrosion standpoint reveals three areas of concern:

1. The coal mine, which is the area where the coal is actually removed from the ground. It may be either an underground or a strip mine.

2. Coal preparation plants, which take the raw coal or the mine-run coal from the mine and prepare it for shipment and use.

3. Transportation of the mined coal by either railroad car or barge to the point of use.

An important conclusion in the study of coal mining from a corrosion standpoint was the fact that the corrosion problems found in the coal industry are not particularly sulfur oriented. The problems from sulfur in coal do not exist during the mining operation, but rather occur during the actual burning of the coal and oxidizing the sulfur in the coal to sulfur dioxide, which creates an acid condition that is a cause of both corrosion and air pollution.

The corrosion problem from the coal itself is due to the salt which is contained within the coal, primarily sodium chloride. The salt is extracted during the preparation of the coal, which is the operation where the coal is sized, washed, dried, and classified for use. The main corrosion problem, then, is found in the coal preparation plants.

The coal preparation process is primarily one of washing and classifying the coal into two grades: metallurgical and steam coal. Classifying the coal into sizes, and washing or cleaning the coal of debris such as rock, slate, sand, etc., all involve water, which is used throughout the entire preparation process. The water is collected and reused so that the salt or sodium chloride content builds up to a rather substantial point. In the dusty part of the preparation plant, the same wash water is used periodically to wash the accumulated coal dust from the structure, thus spreading the salt through all parts of the plant area. The lower areas of the structure are constantly wet due to the preparation process and the washing of the dust from the plant structure. The resulting corrosion is similar to marine corrosion, with tubercles of rust forming on the structural steel wherever moisture tends to accumulate. Coal dust and coal sludge also accumulate in all types of hidden areas. When these are wet, the coal aggravates the corrosion since it is cathodic to steel.

Coal preparation plants are usually large aboveground structures, although they are considered part of the actual mining operation. Most mines have their own coal preparation plant, although there are some cases where two or more mines may combine production and process it through one preparation plant. Coal preparation plants consist of a structure from six to ten stories high, with a long conveyor raising the coal from the mine or ground level up to the top of the preparation plant. The preparation plant is primarily a structural steel structure covered with aluminized or galvanized corrugated sheet. The principal equipment involved in the coal preparation plant are belt conveyors. Other equipment includes: numerous vibrating screens which grade the coal to size, heavy media separators (where the coal is separated from the rock and slate), coal crushers, magnetic separators, dicer tables (a specific gravity separation by vibration), froth or foam separators, dewatering screens, centrifugal dryers, cyclone water clarifiers, in-plant conveyors and chutes, silos for the finished coal, and loading equipment.

The flow of the coal through the preparation unit starts from the mine stock pile on a long conveyor to the top of the plant. As the coal enters the plants, it goes over primary screening units, which separate the coal into various sizes. The heavy media separator is primarily water. Coal passes through the plant primarily by gravity from one level to another. Since water is used even at the highest levels, the plant is damp or wet throughout the processing building. The amount of water and dampness increases as the coal goes down from the top to the bottom of the plant, and all corrosion within the plant is primarily caused by water. The water is reused, so that salts accumulate in the water, forming a highly conductive media which creates the extreme marine-like corrosion problem.

While in many small, temporary operations no protective coatings are used and corrosion continues with little or no maintenance, many of the larger units do protect against corrosion problems through use of high-performance coatings. The primary high-performance material used is inorganic zinc, as either a primary coating or as a base coat. Since the corrosion is primarily chloride oriented and there is little actual immersion, a properly applied inorganic zinc coating provides good protection to the structural steel used throughout the plant and on conveyor lines. However, there are areas where vinyl or epoxy topcoats are used over the inorganic zinc, both of which have proven to be satisfactory.

One operation that is part of some of the larger coal preparation plants is a drying process where the fine coal is dried for use by electrical utilities. When drying is required, waste coal from the plant is burned to create hot air, and the coal is passed counter-current to the hot air for drying. This can be a very corrosive process because, due to the appreciable sulfur content of the coal, the primary sulfur compound (iron pyrite) is oxidized by the hot air, creating sulfur dioxide or sulfuric acid. These gases must be scrubbed out of the hot air after the coal has passed through. Some of the pyrite remaining on or in the coal is also oxidized, so that any fallout of the coal dust is also extremely corrosive. Thus, the entire area around the coal drying sections of the preparation plant must be well protected by acid-resistant coatings. Fallout on either steel or concrete surfaces creates immediate corrosion, and these areas must be washed at frequent intervals to reduce the corrosion problem.

Except for the drying area in the preparation plant, there are no fumes involved in the preparation of coal.

Concrete floors, which are used throughout the coal preparation plants, are not affected by either the coal or the wash water. On the other hand, areas where the concrete is tied into any metal structure are severely corroded due to the accumulation of corrosive salts.

The remaining operation in coal mining is the transportation of coal to the point of use by coal trains, hopper cars, or barges. In this case, corrosion is also primarily chloride oriented, with the coals high in sodium chloride being much more corrosive than the ones with the low chloride content. Abrasion is a substantial problem in the transportation of coal, although use of coatings to prevent this problem has been limited up to this point. Some protective coatings are now being used in barges, particularly to protect hidden areas where coal dust and chloride salts accumulate.

The Steel Industry

The steel industry differs significantly from the chemical industry, in which entire plants are coated for corrosion protection. Unless a steel plant is located in a marine atmosphere or a severe industrial area, many of the structures are subject only to mild atmospheric corrosion. On the other hand, some extremely corrosive conditions do exist in fully integrated steel mills. The primary areas of concern are the pickling facilities, the coke plants, the byproduct plant of the coke operation where ammonium sulfate is collected, the smoke and fume scrubbers, the water treatment plant, and the acid regeneration area.

The pickling area, of which there may be several within one integrated plant, is probably the most corrosive area within the steel mill. This is because sulfuric acid is the primary product used to pickle steel. It generally exists in concentrations of 10 to 15%, and is kept as near the boiling point as possible by steam injection. There are two types of pickling, batch and continuous operation. First, and most important, is the continuous strip pickle lines where steel strip is continually passed through the acid bath prior to some other operation, e.g., cold rolling, galvanizing, tinplating, aluminizing, or some organic coating operations. The continuous strip pickle lines are for relatively light gauge metal and they pass through the bath at a rapid rate.

The pickling operation not only includes the sulfuric acid bath, but hot water washing, hot water rinse, and a quick drying operation prior to the strip going on to be galvanized or tin or organic coated. With steam injection, hot pickling solution, and the rapid movement of the steel through the pickling bath, there are plenty of fumes which can condense on any surface in the area.

The second type of pickling operation, the batch type, involves heavy plates which are pickled free of mill scale. Pipe, bars, wire, or even sheet are also pickled by the batch process. These operations primarily occur prior to galvanizing or some other coating process. The batch pickling operation is not only located in steel mills, but also may be found in galvanizing plants. This is the type of pickling operation used by steel fabricators for plates prior to making them into tanks and similar structures. All structural steel to be galvanized is passed through the batch-type pickling operation prior to hot dip galvanizing.

The primary corrosion problem in this process involves the pickle tank, which, in the case of a continuous process, may be several hundred feet in length. In the case of the batch process, the tanks are usually 50 feet or less in length. In each case, the protection for the interior of the tank against the hot acid is an acidproof brick lining system (Figure 19.29). One of the most important parts of this system is the membrane which prevents acid contact with the steel or concrete tank proper. This may be rubber or vinyl chloride sheet 1/16 to 1/8 in. thick. The acidproof cement and brick are primarily used for abrasion protection and heat insulation for the membrane itself.

FIGURE 19.29 — The inside of a batch-type pickling tank. In such tanks, the furan mortar between the brick may be more resistant to the hot acid than the brick itself.

In many cases, the exterior of the tank, as well as the surrounding floor area, must be covered with acidproof brick and cement to provide protection against hot acid spills. The steel areas adjacent to the pickling tanks (*i.e.,* the steel structure of the building) are subject to splash, spray, fumes, and moisture condensation. These areas require protection by high-performance coatings in order to prevent the steel structure from serious corrosion. They can be protected by heavy-duty vinyl, epoxy, or coal tar epoxy coatings.

The acidproof brick lining system has been used in this service for many years and is a standard of the industry. The acidproof cement is made with a furan resin with a backing between the membrane and the brick of sulfur and silica cement. Such a system is fully resistant to the hot dilute acid. Tanks where the brick has been eaten away after several years of use have been seen with the furan cement extending 1/2 to 1 in. beyond the brick.

The second area in the steel mill, and perhaps an even more important area from a coating standpoint, is the coke plant and its surrounding area. Coke is made in ovens by burning coal in an essentially reducing atmosphere which removes all of the volatile material from the coal and leaves the coke, or essentially pure carbon, as an end product. Coke ovens are made in such a way that the volatile materials from the burning of the coal are carried over into a byproduct plant. One of the byproducts is ammonia, which is treated with relatively dilute sulfuric acid to form ammonium sulfate. This material is used extensively as a fertilizer.

The primary corrosion problem around the coke plant is due to the quenching of the coke. Once the coking operation is complete, the hot coke is dumped from the furnace and must be quenched rapidly. This is done by the use of rather large quantities of water which, in many cases, come from the waste water from the rest of the steel mill. During the quenching process, large volumes of steam are formed, which carry any volume of salt that the water may contain to all adjacent steel areas surrounding and downwind from the quenching area. This can amount to a sizable area of a steel mill which needs to be coated. One large, multiple-story steel structure was completely coated with a vinyl primer and two heavy-body vinyl topcoats during the construction process. After several years of service, the coating was still providing excellent protection to the steel structure.

The coke oven byproduct plant is primarily for the processing of coal tar pitches and the extraction of various aromatic hydrocarbons from the coal gas. One is ammonium sulfate, which is the primary corrosive agent. The ammonium sulfate itself is a white crystalline material, much like sodium chloride. It is hygroscopic and is corrosive due to the hydrolysis in contact with water to form ammonia and sulfuric acid. Concrete and structural steel areas therefore require coatings to prevent disintegration. Vinyl coatings and heavy-duty epoxy coatings have been used successfully. Concrete floor areas can be protected with epoxy floor toppings.

The water treatment areas in the steel mill are an important part of the overall operation to prevent contamination of water sources. The water must be neutralized, the dissolved solids (*e.g.,* soluble iron compounds) precipitated, and the water clarified before it can be emptied into any other body of water. Much of the water is reclaimed and reused. While the primary effluent into the water treatment plant is acidic, the treating processes are similar to many water treatment plants. The same coatings would be used, *i.e.,* a coal tar epoxy for the underwater areas of tanks and equipment, and either heavy-duty vinyls or epoxies above the water. In the primary influent areas where strong acid solutions come into the plant, acid brick and cement-lined tanks and processing equipment should be considered.

Acid regeneration plants are also becoming a part of the steel industry. This is particularly true of the hydrochloric acid pickling facilities. The conditions in these plants are similar to the conditions where the acids are used for pickling. Similar coating and lining materials should be used.

Steel plants which manufacture stainless steel also have pickling facilities where nitric acid is used for both pickling and passivating the stainless to make it fully corrosion resistant. Nitric acid is both an oxidizing and a volatile acid and is, therefore, a very corrosive one. Coating systems for structural steel should be heavy-duty vinyls, while areas requiring acidproof brick would require polyester cements.

Corrosion Prevention by Protective Coatings

Because of environmental restrictions, more and more is being invested in smoke and fume abatement which requires large scrubbing facilities, many of which are subjected to extremely corrosive conditions. Where temperatures are sufficiently low, coatings are effectively used around and within the scrubbers.

The Power Industry

While each segment of the power industry has its own unique problems, they all have one particular concern in common, *i.e.,* the need to protect the structural steel, pipe racks, electrical distribution towers, and similar steel structures located throughout the plant. The need to coat these areas with high-performance coatings depends to a considerable degree on the atmospheric conditions and the geographical area in which the power plant is located.

Figure 19.30 shows the structural steel in a power station switch yard coated with a single coat of inorganic zinc. Many such stations as well as long load transmission towers have been coated in a similar way throughout Australia. In many cases, the inorganic zinc has outlasted galvanized towers and equipment.

Many power plants are located adjacent to the ocean in order to use seawater as cooling water, and are thus subject to marine atmospheric corrosion. This indicates the need for an inorganic primer, followed by the most weather-resistant topcoats available. These could be vinyl or vinyl acrylic coatings, and a typical system used might be an inorganic base coat, vinyl primer, and vinyl acrylic topcoat. Another good system would be inorganic zinc, epoxy primer, and polyurethane topcoats. The power industry generally requires excellent appearances in a coating, in addition to good corrosion resistance, for public relations reasons.

In most power plants, the need for coatings is on the exterior of equipment and structures, since most of the interiors (*e.g.,* control and generator buildings) are enclosed and therefore not subject to weathering or severe corrosion. Many of the operating parts of plants are open structures where piping and structural steel is subject to all of the atmospheric conditions that exist in the area. In this respect, power plants are similar to chemical plants and refineries, and often utilize the same coatings for corrosion protection.

Oil- or Coal-Powered Steam Plants

The largest areas in the oil-fired plant are the oil storage tanks. Depending on the type of oil used, and to a certain extent the geographical area where the plant is located, the storage tanks require coating on the interior as well as the exterior. On the exterior, an inorganic zinc primer base coat is recommended and has, in many cases, proven adequate without overcoating. However, it may be followed by the most weather-resistant topcoats. On the interior, a two- or three-coat epoxy system has given good service. In others, a single coat of inorganic zinc also has performed well.

In coal-powered plants, there are usually extensive conveyor systems in place of storage tanks. These are subject to both salt and acidic conditions due to coal handling. Most coal will contain some sulfur in addition to salt. When subject to moisture, rain, or high humidity over a period of several months, the sulfides oxidize to create corrosion problems wherever the coal dust accumulates and absorbs water. In addition to being slightly acidic, the coal itself is cathodic to steel and will therefore, when in direct contact with steel, set up an electrolytic cell, causing the steel to pit and corrode in that area.

Heat exchangers, cooling towers, and cooling piping are common to both oil- and coal-fired plants. Most of this piping and all of the cooling towers are above ground, with all of the inherent corrosion problems that go along with a cooling tower area. If the oil or the coal contains some sulfur, this increases the problems. If either of these types of plants are located near the ocean and are using salt water for cooling, the problems are multiplied, so that any piping involved would require proper coating to prevent corrosion due to the seawater. Also, if the piping is steel, it has to be properly lined to provide long-term protection. Properly applied coal tar epoxies, as well as concrete linings, have been used on the interior.

Many of the new plants have extensive pollution control systems, including large scrubbers and reactors to remove the sulfur oxides from the air. This equipment generally represents an area of high corrosion, and,

FIGURE 19.30 — Inorganic zinc-coated steel in an Australian power station switch yard after many years of trouble-free service from the use of the single-coat system.

depending on the temperature, requires high-performance coatings for corrosion-free maintenance.

All of these plants also have extensive water treatment facilities. This is necessary in order to supply deionized and deoxidized water for boiler makeup. Again, high-performance coatings are recommended for both the interior and exterior of this equipment, and the coatings used are the same as for any water treatment facility, *i.e.,* coal tar epoxy for immersion, and vinyl or polyamide epoxy for the exterior.

Stacks are also common to both of these types of plants. Where steel stacks are used, inorganic zincs have been used for exterior protection and have given excellent service over long periods. Figure 19.31 is an example of steel stacks in a power plant located in both a marine and industrial atmosphere. After 15 years of service, the post-cured inorganic zinc coating is still providing full corrosion protection.

Hydroelectric Plants

Hydroelectric plants are generally located near large dams or similar structures. The exposed steel areas on the dams and power houses may be exposed to high humidity and water, thus requiring high-performance coatings. Such areas include rotary drum gates, floating drum gates, tainter gates, and simple lift gates as well as trash racks. Such mechanisms are subjected to highly ox-

ygenated, rapidly flowing water, as well as to abrasion from floating debris and ice. Many such areas were coated with vinyl coatings (five to six coats) in the late 1930s and early 1940s. Some of the coatings are still intact and operating. Many other coating systems (*e.g.,* epoxies, coal tar epoxy, chlorinated rubber, and inorganic zinc) have been used since then, with much success.

Penstocks are a necessity for hydroelectric power. These have been coated on the interior as well as the exterior in many areas throughout the world. For example, Shasta Dam made some early tests where Bureau of Reclamation vinyl specifications VR-3 and VR-6 were tested against a large series of other coatings, including hot-applied coal tar. Both vinyl coatings gave excellent performance over this period of time compared with the other products. Hot-applied coal tar enamel and coal tar epoxy also showed up well.

Inorganic zinc coatings have been used on the exterior of many penstocks, particularly in Australia. One example is the Snowy Mountain Scheme which was put in service 20 years ago. There is a large penstock in Central Mexico which is coated on the interior with coal tar epoxy and on the exterior with inorganic zinc. Much of this penstock is in a rock tunnel so that the inorganic zinc is exposed to extremely humid, wet conditions most of the time. This unit has been in service for approximately 15 years.

There are two types of hydroelectric plants. One uses the Pelton Wheel, which is a very high-speed, high-pressure unit that can only be made of metal, except in the penstock. However, there also are many low-pressure penstocks which use a very large impeller and scroll case to turn the generator. These may be 30 feet or more in diameter. A large volume of water flows through the scroll case, but at relatively low pressure, and they usually are made up of heavy steel. Figure 19.32 shows the large baffles which direct the water into the impeller. The scroll case and the impeller were coated with Bureau of Reclamation VR6, and the photograph was taken after five years of service.

FIGURE 19.31 — Oil-fired power plant stacks in a marine and industrial atmosphere after 15 years of exposure.

FIGURE 19.32 — The scroll case of a large low-pressure high-volume water turbine developing hydroelectric power after five years of service from the B. of R. VR6 vinyl coating (six coats).

Geothermal Plants

Geothermal plants are relatively new, so that only a few exist around the world. Their corrosion problems are undoubtedly unique in that there is considerable waste steam which contains large amounts of salts and some hydrogen sulfide. Therefore, all of the structural steel, piping, any tanks or water processing equipment, and often some concrete areas require protection.

Only one of these plants has used high-performance coatings, and it is located just south of Mexicali in Baja, California. In this case, many of the hot areas of the plant were coated with inorganic zinc and an inorganic topcoat. This plant has been in service for from 15 to 18 years and the coating has performed in an excellent fashion. An addition to the plant was recently completed, and the inorganic coating system was used on it as well. Less critical areas were coated with epoxy polyamide coatings. It is anticipated that additional geothermal plants will be located around the world, wherever heat from the earth's crust can be tapped.

Figure 19.33 shows two barometric condensers coated with the two-coat inorganic system, after approximately eight years of service.

FIGURE 19.33 — A geothermal generating plant with barometric condensers with a two-coat inorganic system. Any corrosion would badly stain the white inorganic topcoat.

Nuclear Plants

Coatings for nuclear plants, except for the exterior structural steel areas, are quite different from those used in other power industry segments. The general intent is to apply coatings which will last 40 years, which in itself requires high-performance coatings. For the most part, the coatings are for standby protection, except in the areas where the coating is exposed directly to water, *e.g.,* on the interior of the torus, quenching mechanisms, and fuel storage areas. The primary reason for the majority of the coating is to prevent penetration of any radioactive material into the concrete structure or onto any bare steel area which would be difficult to decontaminate. The coating provides the decontaminating surface, and, except in the reactor areas where it is subject to radioactivity, it acts as a standby protection. It is therefore subject primarily to atmospheric conditions or wear from ordinary use or foot traffic.

An example is shown in Figure 19.34 where inorganic zinc and polyamide epoxy topcoats are applied to the exterior of the downcomers of the fast-quenching mechanism. Here, the coating is subject to some moisture, but otherwise is protecting against standby corrosion until an emergency occurs.

Because of the standby protection required of the coatings, there are tests that the coatings must pass before they can be used in areas where there is a possibility of radioactive contamination. Design *basic accident* and *loss of coolant accident* conditions are two pertinent ones. These tests simulate conditions as they might be if a power reactor went critical, with a failure of some of the mechanical components. These and other required tests are critical when it is realized that a relatively thin coating film must stand up to the accident conditions and still provide a surface which will decontaminate well enough to allow repair and early resumption of power production. The inorganic silicates, epoxy amine, epoxy polyamide, epoxy phenolics, specially cured epoxy coatings, or combinations of these products are now the coatings which provide the essential protection to the critical areas of nuclear power plants.

Since the beginning of the Manhattan Project, concrete has been the principal building material for all nuclear plants and facilities. These include the original plants for the development and extraction of plutonium, experimental reactors, heavy water plants, nuclear fuel plants, as well as the present day nuclear power plants.

The concrete areas in these plants are massive, with millions of square feet and hundreds of thousands of tons of concrete used not only as a structural material, but as a basic shield against high-density radiation. The structures themselves range from concrete tanks to house small experimental reactors, to concrete holding basins for hot uranium slugs from the original piles, storage areas for spent uranium fuel, massive underground chemical plants where all equipment is operated and maintained by remote control, and concrete containment areas around the primary nuclear power reactors.

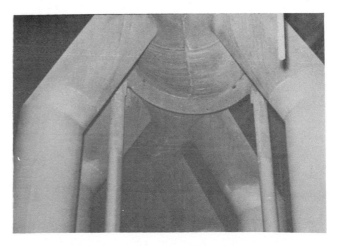

FIGURE 19.34 — Nuclear power plant downcomers coated with an inorganic zinc and polyamide epoxy system.

Unfortunately, concrete has its problems, the principal one being porosity. Most of the radioactive materials, whether in the form of a solution, a solid, a powder, or, in some cases, a gas, if they accidentally come in contact with a bare concrete surface, are quickly absorbed by it. Once absorbed, they cannot be readily removed except by destruction of the concrete surface. The improvement of the concrete surface is therefore essential, and this is where coatings enter the picture.

The magnitude of the problem can be better understood when it is realized that there are 2 to 5 times the area of concrete to be coated in a typical nuclear plant compared with steel areas. There are hundreds of thousands of square feet of complex surface to be coated in each plant. These may range from very critical to relatively noncritical areas. Nevertheless, the concrete surface must be coated to eliminate any porosity that would retain radioactive materials.

In order to provide a surface which would readily decontaminate if exposed to radioactive materials, special coatings were developed to eliminate the basic porosity of concrete and that would pass the tests for nuclear accidents. Fortunately, the epoxy resins have the necessary resistance to radiation so that epoxy surfacers were developed which would meet the other requirements. Epoxy surfacers and epoxy topcoats are now standard for concrete surfaces that could become contaminated.

Figure 19.35 shows the application of an epoxy surfacer by spray to a concrete surface in a nuclear power plant. In most areas, the surfacer was applied in this manner and then troweled for proper smoothness. Figure 19.36 illustrates the magnitude of the concrete surfacer in a nuclear power plant, as well as the application of the epoxy surfacer by trowel.

FIGURE 19.36 — An epoxy surfacer being applied by trowel to a large expanse of concrete on the interior of a nuclear power plant.

Steel surfaces, on the other hand, are smooth, dense, uniform, and present few of the coating application difficulties inherent in concrete surfaces. When sandblasted, mild steel presents an almost ideal surface over which to apply coatings. All of the principal types of coatings which have the required radiation resistance have been used. These include primarily vinyls, epoxies, and inorganic zinc. Single coats of inorganics have been applied to many areas of the containment vessel, torus, and downcorners. Other areas are coated with a combination of inorganic and epoxy coatings. Still others are coated with two or more coats of vinyl or epoxy. In the nuclear energy field, in order to permit proper decontamination, the same topcoating systems are required over both steel and concrete surfaces, which are expected to provide similar impervious surfaces and last for an equivalent period.

Figure 19.37* shows a steel reactor containment shell coated with inorganic zinc alone. High resistance to radiation and good corrosion resistance are required in this case. Figure 19.38 is a critical area, an air lock into the primary containment area. This area is coated with inorganic zinc and polyamide epoxies for the most effective decontamination.

The Food Industry

The food industry is a widespread, fragmented industry in which there are many small plants dispersed around the world in rural as well as in urban areas. There is one overriding corrosion problem inherent in the food processing industry: floors. They are a critical problem in every food plant for several reasons: (1) sanitation, (2) resistance to all types of food products, and (3) abrasion

FIGURE 19.35 — The spray application of an epoxy surfacer to a concrete surface in a nuclear power plant. The surfacer, when troweled, effectively fills the concrete surface to make a smooth, impervious base for the polyamide epoxy topcoats.

*See color insert.

FIGURE 19.38 — An air lock into a primary containment area coated with inorganic zinc and epoxy topcoats.

from foot traffic and mechanical handling equipment. There is no other single area in the food industry which is as important as a properly protected floor.

New floors are not difficult to lay or coat. Old floors, however, are among the most difficult of all areas to coat. This is because the concrete is dense and usually hard-troweled and smooth. Where the coating is worn or broken, oils, grease, fatty acids, proteins, and sugars penetrate the surface and are difficult, if not impossible, to remove. Floor maintenance coatings or surfaces must be capable of penetrating and adhering to such surfaces without blistering, peeling, or other failure.

There are several segments of the food industry, each with its own individual problems:

1. General food processing plants, which consist of meat packing plants, cheese and milk plants, fruit and vegetable packing plants, soft drink plants, aircraft service kitchens, chain restaurant kitchens, precooked and packaged food plants, candy companies, bakeries, etc. The galleys and kitchens aboard ships also are included in this segment.

2. Breweries.

3. Wineries, which are more highly developed in the United States than in any other area of the world.

4. Sugar mills.

5. Corn product plants.

General Food Processing

General food processing plants are located throughout the world in both rural and highly populated areas. The overriding problem throughout this segment of industry is sanitation. Equipment, floors, and wall areas (from the floor up to approximately 4 or 5 feet) are subject to constant washing, dampness, steam, and contact with disinfectants or bactericides such as sodium or calcium hypochlorite. In addition, these areas are subject to various food acids such as acetic, lactic, citric and phosphoric acid, as well as many kinds of sugars. Proteins, fatty acids, oils, and fats also may be present in these areas. Wherever there is an accumulation of these materials, bacteria, yeast (fermentation), and fungus are often found which then create acidic conditions.

Other areas are subject to highly alkaline cleaners and detergents. This includes the equipment area where processing vessels, bottle washers, pumps, and bottling and canning equipment are used or are washed with alkaline materials.

The most critical area in food processing plants is the floor. Not only is it subject to all of the previously mentioned materials, but because it is usually wet, it is a prime source of contamination by bacteria and fungus. Plain concrete floors are unsatisfactory, and coatings are not practical, primarily from an abrasion and thickness standpoint. Thus, the standard in the industry is acid-proof brick or cement or quarry-tiled floors grouted with chemical-resistant cements (*e.g.,* furan or epoxy grouting materials). Floor surfacers of the same resinous materials are often used in less critical areas with good success.

Figure 19.39* shows a quarry-tiled floor using furan grouting in a large meatpacking house. The quarry tile is impervious to grease, water, steam, protein, and microorganisms, as is the furan grout.

Breweries

Breweries are located throughout every country in the world, and, in many cases, represent the largest plant in their area. The problems in some areas of the plant are similar to those in the food processing industry; *i.e.,* conveyors, bottle washers, and bottling and canning equipment. Again, the greatest problem involves floors, since these are the areas where abrasion and constant damp or wet conditions exist. The same products can be used here as in the food processing industry.

A large area to consider in any brewery is the cold-rooms where the refrigerated holding tanks for finished beer are located. When the tanks are washed, the effluent often goes onto the floor and is directed out through the floor drains. These floors are therefore subject to beer, detergents, bactericides, and continuing moist-damp conditions. That is why one of the large breweries in Mexico, which adds several aging tanks at least every two years, has used only quarry-tiled floors with furan grout dating back to at least 15 years ago.

Since the temperature in beer cellars is 35 to 40 F, there is constant condensation on the exterior of the steel tanks and, in many cases, on the walls and ceilings. The exterior of the tanks often have been coated with vinyls or epoxies with good results over a number of years. This has been true not only in various areas of the United States, but in other countries as well. It must be emphasized that since breweries often serve as a local showplace, all parts are maintained for both sanitation and appearance. Therefore, semigloss or glossy coatings are preferred throughout the brewery. Alkyds are often used on noncritical walls and ceilings.

The interior of the brewery tanks that are not made of stainless steel may be glass lined or lined with high-bake phenolic coatings for maximum resistance to immersion and fermentation. The ones lined with phenolics require great care both in the application of the critical phenolic

*See color insert.

Typical Uses of High-Performance Coatings

and in the curing process. Several thin coats are usually applied, with a mild bake in between and a final high bake to cure all coats together. When properly done, many years of excellent service can be obtained from either the phenolic or glass lining.

Wineries

The segment of the industry that concerns wineries is quite different from either of the other two segments discussed because it essentially is a seasonal fermentation process. The plants are single-purpose, and, except for the bottling rooms, most areas are relatively dry with no corrosion. However, where wooden storage and fermentation tanks are used, there often is some leakage which can damage concrete floors. This makes floor toppings in these areas an advantage.

The primary concern in wineries is contamination of the wine, which makes tank linings extremely important. If any wine is exposed to iron, it changes to a bluish ink color and develops a definite metallic taste. Wine, therefore, usually is stored and fermented in wood, stainless steel, or specially lined steel containers.

The fermentation process is a difficult one from the standpoint of coatings, because it produces nascent carbon dioxide on the coating surface. Nascent carbon dioxide (i.e., single molecules of CO_2) is an extremely small molecule and can pass through most plastic materials in a manner similar to moisture vapor. Thus, the linings of fermentation tanks must be able to resist blistering and loss of adhesion due to the passage of carbon dioxide through the coating.

In the early days, multiple-coat vinyls were used in many wine tanks, primarily for finished wine after the fermentation process had been completed. However, vinyls were never very successful for service during the fermentation process. They were only successful for finished wine when the coating was heated for several days to eliminate the solvents. In fact, several of the champagne tanks which were lined at a California winery shortly after World War II were still in use and the vinyl was still providing full protection after 20 years of service.

However, a better type of lining is a tightly cross-linked thermoset coating. An epoxy urethane can provide excellent service as a fermentation tank lining for concrete tanks. In fact, there were several in California and a number in Mexico that were in service for over 10 years. Other highly cross-linked epoxy phenolics and high-bake phenolics have given equivalent service on steel. A strongly cross-linked epoxy amine coating should be satisfactory as a coating for concrete fermentation tanks.

The transportation of wine is an important part of the business. Most of the wine produced in California is shipped east. Thus, in order to be economical, it must be shipped in lined tank cars and bottled at the destination. Wine connoisseurs may shudder at the thought, but 10,000 or more cars are used in wine service in the U.S., with even more being used for transportation to Europe. The original cars were lined with multiple-coat vinyl coatings. In fact, vinyl coatings provided the opportunity that was needed for the wine industry of California to become a major business. Before the time of vinyls and tank cars, wine had to be shipped across the United States in bottles. Presently, epoxy phenolic and high-baked phenolic coatings are used in this type of service.

Sugar Mills

Sugar mills process sugar cane or sugar beets. They are crushed; the juice is pumped into tanks; and it is concentrated, crystallized, passed through centrifuges, recrystallized, dried, and bagged. The sugar syrup or sugar juice is relatively corrosive in that it ferments rapidly, forming acetic acid. Sugar solutions themselves are slightly acidic and therefore are quite damaging to concrete floors.

Epoxy concrete floor toppings in the crushing areas are an advantage. Vinyl or epoxy maintenance coatings have been used successfully on conveyors and other handling equipment, structural steel, and the exterior of equipment in various sugar plants throughout the world. Most plants now use stainless steel for concentrators, crystallizers, centrifuges, and most storage tanks.

Corn Products Plants

In many ways, corn products plants are more characteristic of the chemical industry than the food industry. There are numerous products produced from corn; however, the principal ones are corn oil, starch, and dextrose. They are usually produced in large, multiple-story plants, enclosed with a large amount of structural steel and concrete floors which must be protected.

The area which is most difficult from a corrosion standpoint in a corn products plant is the starch area. Once the starch is formed by a cooking and grinding process, it is then treated in several different ways to develop the required properties. Starches may be reacted with hydrochloric acid, sodium hydroxide, enzymes, sodium or calcium hypochlorites, peroxide, sulfur dioxide, or other chemicals. Some reaction equipment is rubber lined, and an oxidation-resistant coating is used for protection of both steel and concrete. The process using sulfur dioxide in some plants has caused concrete ceilings and columns to spall, due to the sulfur dioxide reacting with the concrete. Vinyl coatings have given good protection in the oxidizing and acid areas. Epoxy coatings and floor toppings have been used successfully in other plant areas.

Sewage Treatment

With the increasing emphasis on a healthy environment in the last several years, there has been a substantial increase in the number of sewage plants put into operation. Many of these are located in rural atmospheres where, with the exception of some possible hydrogen sulfide, there is little corrosion. Nevertheless, since these plants are permanent and therefore meant to operate with a minimum of maintenance, good long-lasting coatings should be used in all critical areas. Many of the larger plants in humid or coastal areas need the best protection possible on above-ground areas.

The influent in a sewage treatment plant is the primary source of corrosive conditions in the plant. The influent generally comes from a series of collection lines throughout a city or cities, which are concentrated into a central outfall which flows into the sewage treatment plant. The time involved in the flow of water from its entrance into the pipe from a plant or a home, may vary greatly. However, under most conditions transport through the pipeline has been sufficiently long so that the bacteria in the sewers have created a source of hydrogen sulfide (which is corrosive).

The conditions to look for which create the proper atmosphere for corrosion in a sewer depend on several factors. The generation of hydrogen sulfide is a bacteriological action. Thus, all conditions for bacterial growth must be present in order for sulfides to be produced. If any one of the required conditions is absent, then it is unlikely that gas will be produced, since the growth of the bacteria will be prevented or seriously retarded.

There are four primary factors in sewer gas production and the resulting corrosion of concrete and exposed steel.

1. The average temperature of the geographic location is extremely important. Generally, the areas of the country where the average temperature is 60 F or above are favorable for bacterial growth. The longer the high-temperature season, the more likely there will be hydrogen sulfide produced under the sewage conditions. Even in areas where there are several months of cold temperatures, bacterial action can be rapid enough during the summer months to create a serious sulfide problem. Thus, sewer corrosion has been found even in southern Canada.

2. The sulfate content of the city water is an important factor in the life of the anaerobic sulfate-reducing bacteria. In cities where the water supply is extremely pure and where sulfates are almost nonexistent, the probability of sulfide production is reduced. The normal sulfate content of city water of average hardness is sufficient for these bacteria to generate hydrogen sulfide. Additional sulfates also may enter the sewer by ground water infiltrating into the sewage system. The higher the sulfate content, the greater the quantity of sulfate bacteria that may be present and the greater the hydrogen sulfide produced.

3. The age of the sewage in any system is extremely important. A sewer which flows rapidly and where retention is short ordinarily will not maintain conditions that will produce hydrogen sulfide. If there are long outfalls or even long feeder lines into the outfalls, where the sewage becomes several hours old, then the formation of hydrogen sulfide is usually rapid. The age of the sewage depends to a great extent on the slope of the sewer. In areas where the flow is slow, the pipe may act as a septic tank, producing large amounts of sewer gases. This is an important factor to look for in determining the possible corrosion of concrete. The older the sewage, the greater the chance of corrosion.

4. High concentrations of organic matter lead to sewer gas production. Low percentages of organic matter ordinarily prevent the rapid growth of the anaerobic bacteria, as these organisms require some organic compounds as part of their reduction of sulfates or organic sulfur compounds to hydrogen sulfide. Food plants, fish canneries, meat packers, and plants of this type can materially increase the organic content of the sewage, as well as the hydrogen sulfide development.

Two important terms that may be encountered in any study of a sewage system are *BOD* and *oxidation-reduction (redox) potential*. These terms often are used as a measure of the "strength" of sewage. The BOD is the *biological oxygen demand* of the sewage and is a measure of the amount of oxygen absorbed by a given amount of sewage in the laboratory over a five-day period. Ordinarily, where the BOD is high, the above four conditions for bacterial growth are all optimum. The oxidation-reduction potential of sewage is again a measurement of the condition of the sewage. If the sewage is on the reducing side, this condition is favorable toward hydrogen sulfide production. If this potential is on the oxidizing side, then hydrogen sulfide will not be present.

An important point to remember is that hydrogen sulfide will not corrode concrete by itself. It is only when the hydrogen sulfide reacts with the bacteria to form sulfuric acid that there is any attack on the concrete. Any dissolved hydrogen sulfide that may be in the sewer water is inert to the concrete. Generally, liquid sewage is no more corrosive to concrete than plain water. It is only the area above the liquid which becomes corroded. Steel is not like concrete in that hydrogen sulfide can react directly with steel under almost any condition. Therefore, any steel that may be immersed in the sewage liquid must be effectively protected. This would include the bar screens, the scraper units in the grit chamber, and the scraper arms on the primary clarifier. Above the sewage, metal reacts with the hydrogen sulfide and sulfuric acid and is seriously corroded in the same way as concrete.

In modern sewage plants, there is a great effort to decrease obnoxious odors. Therefore, many of the chambers in the initial stages of the sewage treatment are closed chambers which create ideal conditions for the aerobic bacteria and, as a consequence, for severe corrosion of the concrete. Any concrete areas above the liquid level that are moist and enclosed can corrode at rates as rapid as 2 in./yr. Thus, most of these areas require the best in corrosion protection. Figure 19.40 shows a sewer chamber and the concrete corrosion that occurred even though the concrete was originally coated.

The majority of the surfaces that are in contact with the sewage are concrete. Concrete is difficult to coat because of its porous surface. In many ways, the concrete surfaces in a sewage treatment plant are similar to those in a nuclear energy plant. The surface must be completely impervious or the concrete underneath the coating will disintegrate. It must be remembered that the hydrogen sulfide is converted into sulfuric acid by the bacteria. The bacteria will find any imperfection in the coating and will create sulfuric acid in that area which will react with the concrete. The calcium sulfate that results has five times the volume of the original cement. Thus, in addition to the acid-cement reaction, there is a strong mechanical force involved which tends to break the coating away from the surface.

For instance, one coated surface showed no signs of any pinholes; however, after several months of service,

FIGURE 19.40 — A sewer chamber showing the original coating and heavy concrete corrosion.

the coating was covered with carbuncle-like areas where the acid had penetrated the coating and the sulfated concrete had pushed the coating out into a barnacle-like mound.

Although many coatings have been tried, only the best and most impervious ones with the highest adhesion have any chance of protecting these areas effectively. Even where operating areas (*i.e.*, the bar screen pit, grit chamber, primary clarifier, and even drum filters which are included in some plants) are left open, the area directly above the liquid level will remain damp and bacteria will feed on the hydrogen sulfide at that point to create a corrosive condition. These waterline areas are often badly corroded and must be effectively protected.

In many sewage treatment plants, these damp areas just above the liquid are protected by flexible vinyl plastic sheet which is designed to be cast into the concrete surface. All joints are heat welded. This is the most effective method of protecting concrete. The original application of this material in a sewage pump chamber in 1947 is still in service and effectively protecting the concrete in 1982. Millions of square feet of this material have since been applied to some of the largest sewage systems and treatment plants in existence.

Where such materials cannot be used or are impractical, the most effective way to prepare a concrete surface for use under sewage conditions is to use an aromatic amine-cured epoxy surfacer, followed by epoxy amine or coal tar epoxy amine topcoats. Using a system such as this to prepare the surface, the concrete itself need not be anything other than as left by the forms. If it is dirty, a light blast will clean it up. However, the heavy-bodied materials will fill the surface, after which they are smoothed by rolling or troweling, so that an impervious, heavy topcoating can be applied over the surfacer.

These types of materials should be used wherever there are wet areas in a sewage treatment plant, including the covered grit chambers or any other covered areas throughout the plant, *e.g.*, wet-walls, sedimentation basins, bar screen pits, etc. It must be noted that only amine-cured epoxy surfacers and seal coats are recommended. *Polyamide-cured epoxies are not effective under sewage*

conditions. The polyamide portion of the epoxy is consumed by various bacteria and soon turns to a slimy mush.

Heavily loaded, thick, amine-cured surfacers and topcoats are the only types of coatings that can be sufficiently well applied to concrete to seal it against the above-water corrosion in sewage treatment plants. Hundreds of other thinly applied materials from microcrystalline wax to chlorinated rubbers and vinyls have been tried, but generally have not proven satisfactory. Many are resistant to dilute sulfuric acid on a smooth surface, but fail quickly when applied to concrete and exposed to severe sewage treatment plant conditions.

Coal tar epoxy coatings (amine-cured) have been very effective topcoats over epoxy surfacers. Such coal tar epoxies have shown excellent resistance to hydrogen sulfide conditions. Steel, under these same conditions and particularly if it is immersed, can be effectively protected by a thorough blasting and the application of an amine-cured coal tar epoxy system. Steel areas such as bar screens and clarifier rake arms have been effectively protected by properly applied coal tar epoxy coatings (Figure 19.41).

FIGURE 19.41 — A typical clarifier mechanism coated with amine-cured coal tar epoxy for use in sewer service. (SOURCE: Ameron Corp., Monterey Park, CA.)

Corrosion Prevention by Protective Coatings

As with most plants, sewage treatment plants have a large above-ground area that needs coating. While there usually is only a minor amount of sulfide in the air, it is still necessary to provide for a minimum of maintenance. Thus, the maximum coating should be applied in the beginning. Whenever the sewage treatment plant is located in a highly humid area or an area along the coast, inorganic zinc is suggested for all above-water areas. This will provide the primary protection, and any organic coating (usually a vinyl or vinyl acrylic) applied in addition will provide the decoration and weather protection. If concrete areas above ground level (either poured concrete structures or concrete block) require protection, a number of coatings may be used. Except for the severe areas described previously, water-base vinyl acrylic coatings have given good service.

The Transportation Industry

The transportation of bulk chemicals may be one of the most difficult areas of all for protective coatings. While stainless steel and specialty steels and alloys are used extensively for transportation containers, mild steel is also used for the transportation of many commodities, due to cost constraints. Most means of transport (*e.g.,* trucks, tank cars, hopper cars, both dry and liquid barges, and ships) use linings primarily to protect the commodity being transported from any contamination. Bulk terminals, with their large number of storage tanks, should also be considered part of this overall industry, and involve the same types of linings.

The problem of container linings in the transportation industry is twofold. The principal problem is one of protecting the contained product from any contamination by iron or any other foreign material. For the most part, corrosion of the interior of the containers is secondary, except for specialty products which may be highly corrosive. Most of these are transported in special stainless steel or alloy cars or hoppers, with these cars being dedicated to that particular corrosive material.

There are two ways in which the transportation industry handles products. The first is by a dedicated container. This would be a tank car, a hopper car, a barge, or special compartments aboard ship which are dedicated to an individual product and are never used for anything other than that product. For example, there are hundreds of tank cars and barges which are lined specifically for the transportation of caustic soda. Very seldom are these cars or barges used for any other purpose. Similar transportation equipment may be dedicated to corn sugar, ammonia nitrate fertilizers, formaldehyde, molasses, and similar products. This type of equipment which is dedicated to one product represents a relatively easy type of service, since the coating that is used on such transportation equipment can be specialized for that particular service.

The second type of transport, however, is much more difficult. This is where the containers, whether they be tank trucks, tank cars, barges, or ships, are used for any number of different commodities. In this case, a coating with a broad resistance must be used in order to withstand the change in commodities as well as the cleaning required between commodities. For example, many different solvents may be transported in an inorganic zinc-lined tank. These include most of the petroleum products, acetone, other ketones, aromatic hydrocarbons, and even chlorinated hydrocarbons (which must be kept dry). Other products, such as molasses, glycols, corn oil, various vegetable oils, lards, and crude fats, are transported in epoxy-lined containers.

Epoxy coatings usually are specially developed for this purpose and may require force-curing at elevated temperatures in order to develop full resistance to the wide range of products. The interiors of tank cars, barges, and ship tanks are usually designed for easy coating application. There usually are no baffles in tank cars, very few in large tank barges, and the tanks aboard ship are large and completely smooth on the interior. The interior is prepared for coating, with all of the welds ground smooth and no edges or corners to create difficulties.

In addition, because the linings are specialized, the work is usually done by specialty lining contractors who have the necessary equipment for the application of coatings to tank cars and barges, or portable equipment which can be used for lining large ocean-going cargo carriers. Specialized blasting, heating, and dehumidification equipment is necessary for linings of this type. Application conditions are controlled so as to make the application of any coating used the best possible.

Many of the products used (*e.g.,* the inorganic zinc coatings for solvents and specialty epoxies for caustic soda or for a variety of chemical products) have been in service for as many as 10 years and have provided excellent service over that period of time.

The lining of large ocean-going vessels for the transportation of petroleum products (ranging from crudes to lubricating oil) also should be considered in this category. Refined petroleum products such as diesel oil and gasolines have been transported in lined ships since the middle 1950s, primarily using inorganic zinc coatings. There are many instances where ships have transported these products for over 20 years without corrosion or metal loss when coated with merely a single coat of the inorganic zinc. Figure 19.42 is an example of a refined oil (*i.e.,* gasoline, diesel, lube oil, etc.) tank aboard an ocean-going tanker after several years service in these commodities. Prior to high-performance coatings, the bulkhead steel in this type of service was corroded to one half its thickness in seven years. In this case, however, the tank lining was one coat of post-cured inorganic zinc coating applied as a retrofit lining after the steel was severely corroded.

Crude ships, because of their tendency toward deep pitting and the use of combustion gases for fire and explosion prevention, have required more sophisticated systems. For the most part, these have been base coats of inorganic zinc followed by an epoxy polyamide primer and coal tar epoxy topcoat system which has provided excellent service for a number of years in difficult crude service. In the use of refined oils, gasolines, jet fuels, and lubricating oils, where no contamination was allowable, systems such as an inorganic zinc coat and an epoxy polyamide primer and epoxy amine topcoats have provided excellent service.

FIGURE 19.42 — The interior of a refined oil tank in marine service lined with one coat of post-cured inorganic zinc showing no corrosion or rust stain after several years of service.

Marine Service

Marine applications undoubtedly account for the greatest use of high-performance coatings of any industry. The unusual problem in the area is salt. Figure 19.43* provides an overview of the uses of protective coatings in this segment of the industry. It not only includes ships, docks, and their facilities, but also all in-shore facilities within a reasonable distance of the seawater, since they are subject to increased corrosion problems due to the marine atmosphere. The one significant area not shown is that of offshore platforms and structures. However, the problems with offshore installations are similar to those for ocean-going vessels.

There are literally hundreds of different uses for high-performance coatings in the marine area. It has often been said that the inorganic zinc coatings led to a revolution in the protection of marine structures. These coatings have been the one great break-through in protection of steel in a marine atmosphere in the last century.

As an example of the type of protection offered by the inorganic zinc coatings, Figures 19.44, 19.45, and 19.46 show three distinct types of inorganic zinc coatings that have been tested at the International Nickel 80-foot Test Station at Kure Beach, North Carolina. The first is a sodium silicate-base material, showing panels exposed for over 25 years. The panels show no corrosion over this period of time, even along the edges and around the holes in the panels. Figure 19.45 shows a zinc phosphate-type coating that has been exposed to the same atmosphere for 16 years. Again, there has been no corrosion over this time period. Figure 19.46 shows a lithium silicate coating after 15 years, again in the same exposure. As can be seen, even the scribes have been well protected. While a panel with an equivalent exposure of ethyl silicate-based inorganic zincs is not shown, there is no reason to doubt that ethyl silicate-based materials, using equivalent zinc content and coating thickness, would perform in an equivalent manner.

*See color insert.

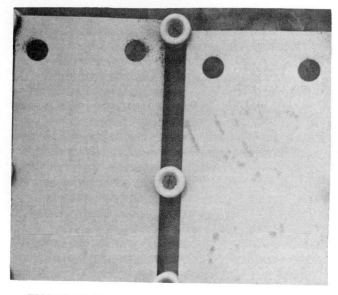

FIGURE 19.44 — A sodium silicate-base inorganic zinc coating after 25 years exposure on the 80-foot lot at Kure Beach, NC. Note the clean panel edges and the lack of corrosion around the holes.

FIGURE 19.45 — A series of zinc phosphate-base inorganic zinc panels after 16 years of exposure on the 80-foot lot at Kure Beach, NC. Note the lack of rust on the panels' edges.

FIGURE 19.46 — Panels coated with a lithium silicate-based inorganic zinc coating after 15 years of exposure to a marine atmosphere. Note the absence of corrosion on both edges and scribes.

Inorganic zinc coatings have become the standard base for marine exposure protection and, while overcoated with organic materials in many cases, still provide the basic protection needed for the steel surface. The zinc base coat also eliminates the problem of undercutting of organic coatings when they are applied directly to steel surfaces under similar conditions.

Ships of all types have been one of the major users of high-performance coatings. Figure 19.47* shows the deck of a large super tanker protected with one coat of postcured inorganic zinc coating after five years of service. Note even the lack of rust stain on this deck.

Figure 19.48* shows a closer view of a similar tanker deck. In this case, the coated area is close to the loading lines, which is one of the most corrosive areas on the decks of these ships. This is due to the abrasion caused by the large flanges connecting the loading and unloading hoses to the ship's piping. This coating is also a single coat of inorganic zinc, exposed for a period of five years. While in this case there is some rust stain on the coating, there is no active corrosion on any coated area.

Figure 19.49* is another example of a larger tanker deck, in this case protected with a base coat of inorganic zinc, followed by vinyl topcoats. Again, after five years of service there was no evidence of coating failure on the entire deck. (The glossy appearance of the deck is due to a rain storm that had just passed.) In this case, one of the reasons for the vinyl topcoats was to prevent contact of the inorganic zinc coating with highly acid soot from the boilers which precipitates on the deck during the boiler blowdown. Use of high-sulfur crudes for steam production creates the highly acidic soot which, if allowed to contact inorganic zinc, dissolves some of the zinc at the point of contact, creating a focal point for corrosion. The vinyl topcoats prevented this contact, and the coating system provided excellent protection to the deck in spite of the acid soot precipitation.

There are also examples of excellent service from high-performance coatings without the use of inorganic zinc. Such applications are becoming less frequent, however, due to the outstanding performance of the inorganic zincs. Nevertheless, Figure 19.50 shows a large Coast Guard buoy which was coated with vinyl primer and three coats of vinyl topcoat. The buoy was removed for fouling removal after several years of service, at which time this photograph was taken. There was no corrosion, blistering, or other coating failure on the buoy at the time of its removal, and it was expected to be replaced in service without appreciable coating repair, except to add a coat of antifoulant.

Figure 19.51 shows one of the largest offshore drilling platforms during construction. The coating used on all underwater portions of this semi-submersible drilling platform was a base coat of ethyl silicate-base inorganic zinc, followed by an epoxy polyamide primer and two or more coats of coal tar epoxy to provide the best possible corrosion protection.

Incidentally, corrosion tests for marine immersion

FIGURE 19.50 — A large Coast Guard buoy coated with a vinyl primer and vinyl topcoats. Note the lack of corrosion or blistering after several years of exposure. White spots on the surface are barnacle bases.

FIGURE 19.51 — An underwater area of a very large semi-submersible drilling rig coated with an ethyl silicate inorganic zinc base coat, epoxy polyamide primer, and two topcoats of coal tar epoxy.

that were conducted in Japan using coatings of different types from various areas of the world yielded similar results. Of all the coating systems tested, the one just described for use on the semi-submersible offshore drilling structure was the best of all systems tested with no corrosion, blistering, or other coating failure for the duration of the test.

Figure 19.52 shows the complex structure of an offshore jack-up drilling rig, all of which is protected by inorganic zinc and epoxy polyamide topcoats. The proper protection of such a complex structure could not be ob-

*See color insert.

Typical Uses of High-Performance Coatings

FIGURE 19.52 — A large jack-up drilling platform coated with inorganic zinc and epoxy polyamide topcoats. (SOURCE: Ameron Corp., Monterey Park, CA.)

FIGURE 19.53 — An offshore platform coated with inorganic zinc and epoxy polyamide topcoats, showing excellent protection of the welds after 14 years of service in the Gulf of Mexico.

FIGURE 19.54 — A saltwater desalinization plant along the Red Sea coated with inorganic zinc and various high-performance topcoats. (SOURCE: Ameron Corp., Monterey Park, CA.)

tained without the use of this type of high-performance coating system.

Figure 19.53 shows a close-up view of an offshore platform coated with inorganic zinc, followed by polyamide epoxy topcoats, after 14 years of service in the Gulf of Mexico. The yellow color was high-visibility yellow included in the epoxy topcoat.

Seawater desalination plants are areas where marine corrosion can be concentrated. Several large desalination plants have been installed along the Red Sea coast of Saudi Arabia, all of which must be protected by high-performance coatings. Figure 19.54 illustrates the complexity of such plants, which must withstand not only the marine atmosphere, but also the high humidity in that area of the world. Inorganic zinc base coats and various high-performance topcoats provide the basis for this protection.

All marine exposures are severe, whether they are full immersion, splash zone, or atmospheric. Marine structures of all types would be short-lived without the protection provided by the high-performance coating systems.

Bridges

In some cases, bridges can be considered in the marine category, since many are located in marine areas, along coastal sections, or over saltwater waterways. There are also thousands of bridges, particularly railroad bridges, which are subject to salt conditions even though they are located inland, away from the coast. Salt conditions in this case are due to refrigeration brines, which have caused serious corrosion in years past.

Bridges are usually difficult structures to protect because of the many angles, corners, braces, rivets, bolts, high-tension cables, and similar methods of construction. Bridges have always been a problem to maintain, and the best possible type of coating system generally has been employed to maintain them in a safe condition. Even in the days prior to high-performance coatings, bridges were coated with red lead, linseed oil, and graphite coatings, which were the best available at the time.

Inorganic zinc also has made an impact on the protection of bridges inasmuch as it provides long-term protection and reduces the constant maintenance which is otherwise necessary. Hundreds of bridges around the world have been coated with inorganic zinc and many coated with additional topcoats, ranging from alkyds to vinyls and epoxies. Many bridges have been in service for 10 or more years without appreciable maintenance when using the inorganic and high-performance topcoat system.

One of the most outstanding of these bridges is the Golden Gate Bridge across San Francisco Bay. This is an extremely difficult corrosion area, since it is in a marine atmosphere and there is constant fog in the area, although actual rainfall is relatively mild. Any salts that build up on the bridge, due to the prevailing wind across the entrance to the bay, are kept in a continually moist condition, creating a highly corrosive situation.

For at least 15 years, the Golden Gate Bridge has been coated with a base coat of inorganic zinc, providing the needed corrosion protection. The application consists of sandblasting the steel to a thoroughly clean condition and applying the inorganic zinc base coat, followed by two or more topcoats of the typical Golden Gate Bridge color. The bridge is an excellent example of the use of high-performance coating systems for maximum protection of steel structures. Figure 19.55 shows the Golden Gate Bridge under normal conditions, *i.e.,* moist sea air with a constant breeze, usually a mid-day sun, with the fog bank off the coast waiting to blow in in the early afternoon.

Figure 19.56 shows a bridge of an entirely different type. This is a bridge carrying a conveyor for loading iron ore aboard ship. The installation is located on the West Coast of Australia. The entire structure is coated with inorganic zinc without topcoats, and the exposure is typical of a marine coastal atmosphere.

Specialized Uses

There are literally hundreds of specialized uses for high-performance coatings throughout the world. One of these is the use of inorganic zinc as a base coat for all of the structural steel in the "Glass Cathedral" in Southern California, which is now considered a major showplace in the area. The inorganic zinc in this particular case was used in order to reduce maintenance on a very complex structure to an absolute minimum. The entire structure is coated with white topcoats for maximum light and visibility (Figure 19.57).

Another unique application of high-performance coatings is the use of epoxy primer and polyurethane topcoats on aircraft. In this case, the coatings are subject not only to abrasion from sand, rain, and hail, but also to deicing fluids, hydraulic oils, and other similar chemicals. Use of this coating system, not only on private aircraft (Figure 19.58), but in many commercial aircraft as well, provides the best appearance along with high performance and protection.

The use of high-performance coatings has been a major factor in reducing many different types of corrosion throughout the world. The applications for coatings range from exposures to rural atmospheres, industrial and marine conditions, splash, and fumes, to full immersion

FIGURE 19.55 — The Golden Gate Bridge across the mouth of San Francisco Bay coated with inorganic zinc base coats and characteristic orange topcoats. Coating maintenance has been reduced to a minimum since the use of the inorganic zinc base coat. (SOURCE: Ameron Corp., Monterey Park, CA.)

FIGURE 19.56 — A bridge structure carrying an iron ore conveyor on the West Coast of Australia coated with one coat of inorganic zinc and showing no sign of corrosion.

in many types of chemicals. This type of protection has only become available in the last 50 years, yet coatings have made many new plants and processes practical by allowing the use of low-cost materials through effective protection by coatings. Since steel and concrete are the most common building materials for industrial plants, ships, offshore structures, power plants, sewage and water works, etc., and with their costs escalating rapidly with inflation, the long-time protection of such structures is now a necessity. High-performance coatings, over the last half-century, have proven their effectiveness in providing this protection.

FIGURE 19.58 — A private jet aircraft coated with epoxy primer and aliphatic polyurethane topcoats. Maximum appearance and protection are obtained by using such a system. (SOURCE: Ameron Corp., Monterey Park, CA.)

FIGURE 19.57 — Structural steel for the "Glass Cathedral" in Southern California. Inorganic zinc was used as a base coat for all sections. (SOURCE: Ameron Corp., Monterey Park, CA.)

Color Insert Section

FIGURE 1.1 — A painting found in an Egyptian tomb reveals their use of synthetic blue pigment along with natural earth tones.

FIGURE 1.3 — In this centuries-old Algonquian pictograph, red pigment (finely ground iron oxide) is mixed with vegetable oil, blood, and egg white. (SOURCE: Reader's Digest Association, Inc., America's Fascinating Indian Heritage, Pleasantville, NY, p. 140, 1978.)

FIGURE 2.7 — The corrosion of a steel specimen submerged in a gelatin bath is represented by the blue anodes and pink cathode.

FIGURE 2.8 — Sandblasted steel in a gelatin bath shows blue anodes and pink cathodes. Scratches and areas of previous corrosion become strong anodes.

FIGURE 2.10 — An iron(blue)-copper(red) couple with an external connection submerged in a gelatin bath. Note the hydrogen bubbles forming on the cathode. These are also shown schematically in Figure 2.9.

FIGURE 2.13 — Zinc and copper coupled by an external connection in an indicator solution.

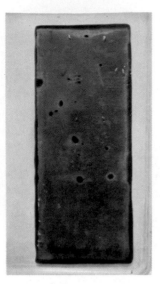

FIGURE 2.23 — Mill or blue scale steel submerged in gelatin indicator solution. Note strong blue anodes at breaks in the scale.

FIGURE 3.10 — The feather edge demonstrates excellent coating adhesion, with some of the steel substrate adhering to the coating curl.

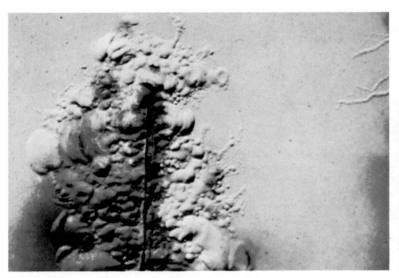

FIGURE 3.13 — Undercutting of an organic coating in a marine atmosphere with filiform corrosion starting at the right edge of the panel.

FIGURE 3.14 — A platform in service 14 years in the Gulf of Mexico coated with inorganic zinc and an epoxy polyamide topcoat.

FIGURE 6.61 — A white powdery deposit has penetrated the coating surface of some galvanized deck piping. The deposit formed beneath the coating has pushed the coating from the surface.

FIGURE 6.63 — View of boottopping and topside coating after 5 years service. Inorganic base coat-vinyl topcoat system. Note the corrosion-free surface.

FIGURE 7.3 — Perforation of the web of an H beam on a corroded offshore platform occurred after only a few years of service in the Gulf of Mexico.

FIGURE 7.4 — This coated, smooth pipe construction is free from excessive corrosion after 14 years of exposure due to its plane surface and the absence of corners, edges, and inaccessible areas.

FIGURE 7.38 — An aboard-ship open-work grating installation such as this should be heavily galvanized and protected by an organic coating to withstand marine exposure.

FIGURE 7.40 — Initiation of corrosion where condensation has occurred on the various coating danger points presented by this common type of pipe hanger.

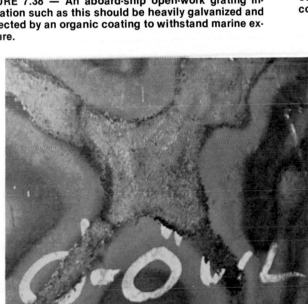

FIGURE 7.41 — Welding burn caused by welding on the exterior of a tank or opposite side of a bulkhead. The weld area is covered with iron oxide scale. The entire heat-affected zone surrounding the actual weld area must be reprepared before the coating can be reapplied.

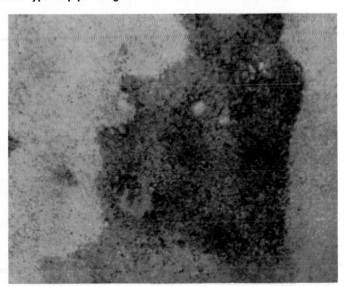

FIGURE 8.6 — Previously corroded steel was white sandblasted and allowed to remain uncoated in relatively high humidity for a short time.

FIGURE 8.9 — The two round blue anode areas represent contamination with hydrogen sulfide in a gelatin indicator bath. Blue indicates iron going into solution, while red indicates a cathodic reaction.

FIGURE 9.26 — This method of sandblasting ship hulls is not recommended due to the inefficiency created by the distance of the worker from the surface.

FIGURE 11.9 — The vinyl coating applied to the hull of this concrete ship has protected the surface from salt penetration for over 25 years.

FIGURE 11.10 — Corrosion of reinforcing in a concrete pipe subject to corrosive soil conditions.

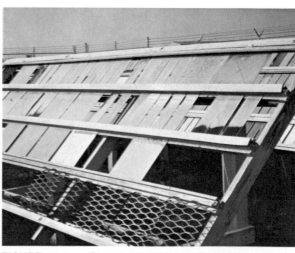

FIGURE 12.1 — Comparative test of five inorganic zinc coatings on expanded metal in a marine exposure. Best protection was provided by the center coating (no obvious corrosion), although originally all five were considered equivalent coatings.

FIGURE 12.3 — Test fence at the International Nickel Atmospheric Marine Test Station in Kure Beach, NC.

FIGURE 12.4 — Test panels, either fully immersed or subject to tidal variation, at the Battelle Memorial Institute's Tidal Marine Test Station in Daytona Beach, FL.

FIGURE 12.5 — Series of antifouling test panels comparing the effectiveness of various coatings.

FIGURE 13.6 — A zinc-iron couple in a phenolthalene indicator bath. The bright red iron panel has an excess of OH- irons and is being cathodically protected by the zinc.

FIGURE 13.8 — Calcareous deposits built up at holidays in the coating on the center pole of a domestic water tank. Deposits are due to impressed current cathodic protection.

FIGURE 13.29 — The boottopping of large tankers is an area where the combination of coatings and cathodic protection has proven advantageous.

FIGURE 14.2 — Red-pigmented coating specimens chalked and changed color after a short exposure to sunlight.

FIGURE 14.27 — Edge corrosion on a coated panel exposed to a marine atmosphere.

FIGURE 14.31 — Failure of a coating applied over bolt-heads.

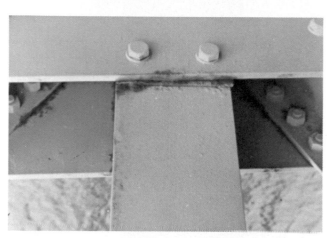

FIGURE 14.32 — Bolted structures can be effectively coated when proper care is taken during application.

FIGURE 14.44 — Undercutting of a coating due to corrosion of small surface abrasions.

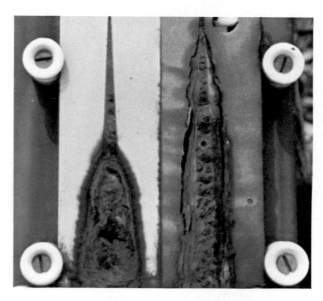

FIGURE 14.46 — Undercutting beginning at a break in the coating.

FIGURE 14.48 — Anchor-chain abrasion on a ship's hull.

FIGURE 14.49 — Erosion and abrasion of the coating on a ship's rudder.

FIGURE 15.6 — The edge of a repaired coating after blasting and recoating. The 1.8 mils area was blasted free of the original coating. Heavy impact areas can be noted above the 5.5 mils area where the original coating was chipped and star cracked by heavy blast media.

FIGURE 15.10 — Unsupported edge of a coating where the steel substrate had completely disintegrated. Repair of such a failure first requires repair of the metal substrate.

FIGURE 19.15 — Hydrogen sulfide corrosion of the exterior of a crude oil production tank. Corrosion of the interior is often even more severe.

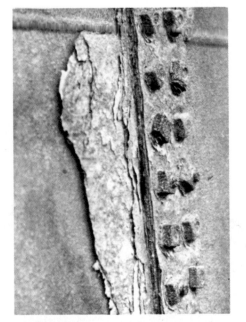

FIGURE 19.37 — A steel reactor containment shell (under construction) is coated in inorganic zinc without topcoats.

FIGURE 19.39 — Quarry tile floor subject to abrasion, chemical cleaners, grease, steam, proteins, etc. in a large meatpacking house. Floors such as this have given service for 10 to 20 years. (Photo courtesy of: Ameron, Inc., Monterey Park, CA.)

FIGURE 19.43 — Typical marine exposures such as ships, docks, loading facilities, bulk storage, and miscellaneous onshore equipment require better than ordinary protection against the salt atmsophere. (Photo courtesy of: Ameron, Inc., Monterey Park, CA.)

FIGURE 19.47 — Deck of a large supertanker with one coat of an inorganic zinc coating after five years of service.

FIGURE 19.48 — Coated ship deck near loading lines after five years of service shows rust stain from uncoated pipe, but no active corrosion of the deck itself.

FIGURE 19.49 — Ship deck with one coat of an inorganic zinc coating and vinyl topcoats after five years of service. Vinyl topcoat is used to protect against acid soot from boiler blowdown.

SUBJECT INDEX

Related NACE Publications

NACE Coatings and Linings Handbook

Contains the latest NACE technical committee reports and standards on coatings and linings for atmospheric and immersion services, plus three other NACE publications: Accelerated Tests for Marine Topside and Submerged Coatings, Coatings and Linings for Immersion Services, and Industrial Maintenance Painting. A complete reference source.

10 × 11½ in., loose leaf binder, 615 pages, figures, tables, and references (Item No. 52130)

Corrosion Control by Organic Coatings

Edited by Henry Leidheiser, Jr.

A compilation of classic papers from the 1980 International Conference on Corrosion Control by Organic Coatings. Forty-one papers ranging from highly theoretical to practical present information on a wide range of topics from investigation of cathodic detachment of coatings from surfaces to sample inspection plans.

8¾ × 11¼ in., hard cover, 300 pages, figures, tables, references, subject index (Iten No. 5115)

Coatings and Linings for Immersion Service (TPC-2)

Prepared by NACE Group Committee T-6 on Protective Coatings and Linings, this manual is a composite of factual and quantitative data on the performance of various materials commonly used as coatings and linings. Contains easy-to-read tables on the physical properties and chemical resistance of numerous coatings and linings in more than 230 chemical environments.

8¾ × 11¼ in., hard cover, 146 pages, tables (Item No. 52044)

NACE Surface Preparation Handbook

Practical handbook containing information on all aspects of proper surface preparation prior to coating application.

10 × 11½ in., soft cover, 42 pages, figures, tables, and references (Item No. 52131)

A Handbook of Protective Coatings for Military and Aerospcae Equipment

Edited by Sara Ketcham

Prepared by NACE Unit Committee T-9B on Systems for Materials. Contains valuable information and data for all professionals involved in the design, fabrication, and maintenance of military and aerospace equipment. Compiled by materials and corrosion specialists from the Air Force, Army, Navy, NASA, aircraft companies, and manufacturers.

8½ × 5½ in., soft cover, 109 pages, figures, tables, and references (Item No. 52270)

These and many other general and specific corrosion-related publications are available from the National Association of Corrosion Engineers, P.O. Box 218340, Houston, Texas 77218 (Phone: 713/492-0535)